AF880421

Springer Series in
MATERIALS SCIENCE 146

Springer Series in
MATERIALS SCIENCE

Editors: R. Hull C. Jagadish R.M. Osgood, Jr. J. Parisi Z. Wang

The Springer Series in Materials Science covers the complete spectrum of materials physics, including fundamental principles, physical properties, materials theory and design. Recognizing the increasing importance of materials science in future device technologies, the book titles in this series reflect the state-of-the-art in understanding and controlling the structure and properties of all important classes of materials.

Please view available titles in *Springer Series in Materials Science* on series homepage http://www.springer.com/series/856

Gennady G. Gladush
Igor Smurov

Physics of Laser Materials Processing

Theory and Experiment

With 514 Figures

Springer

Gennady G. Gladush
Troitsk Institute for Innovation and Fusion Research
Pushovikh str., Vladenie 12, 142190 Troitsk, Moscow region, Russia

and

Institute on Laser and Information Technologies RAS
1 Svyatoozerskaya St., 140700 Shatura, Moscow Region, Russia
E-mail: gladush@triniti.ru

Igor Smurov
Ecole Normale Supérieure de St-Etienne, DIPI Laboratory
Rue Jean Parot 58, 42023 Saint-Etienne, France
E-mail: smurov@enise.fr

Series Editors:

Professor Robert Hull
University of Virginia
Dept. of Materials Science and Engineering
Thornton Hall
Charlottesville, VA 22903-2442, USA

Professor Chennupati Jagadish
Australian National University
Research School of Physics and Engineering
J4-22, Carver Building
Canberra ACT 0200, Australia

Professor R. M. Osgood, Jr.
Microelectronics Science Laboratory
Department of Electrical Engineering
Columbia University
Seeley W. Mudd Building
New York, NY 10027, USA

Professor Jürgen Parisi
Universität Oldenburg, Fachbereich Physik
Abt. Energie- und Halbleiterforschung
Carl-von-Ossietzky-Straße 9–11
26129 Oldenburg, Germany

Dr. Zhiming Wang
University of Arkansas
Department of Physics
835 W. Dicknson St.
Fayetteville, AR 72701, USA

Springer Series in Materials Science ISSN 0933-033X
ISBN 978-3-642-19242-5 e-ISBN 978-3-642-19831-1
DOI 10.1007/978-3-642-19831-1
Springer Heidelberg Dordrecht London New York

Library of Congress Control Number: 2011934865

© Springer-Verlag Berlin Heidelberg 2011
This work is subject to copyright. All rights are reserved, whether the whole or part of the material is concerned, specifically the rights of translation, reprinting, reuse of illustrations, recitation, broadcasting, reproduction on microfilm or in any other way, and storage in data banks. Duplication of this publication or parts thereof is permitted only under the provisions of the German Copyright Law of September 9, 1965, in its current version, and permission for use must always be obtained from Springer. Violations are liable to prosecution under the German Copyright Law.
The use of general descriptive names, registered names, trademarks, etc. in this publication does not imply, even in the absence of a specific statement, that such names are exempt from the relevant protective laws and regulations and therefore free for general use.

Cover design: eStudio Calamar Steinen

Printed on acid-free paper

Springer is part of Springer Science+Business Media (www.springer.com)

Preface

We are glad to propose a new book *Laser Processing of Materials: Theory, Experiment* which combines theories and physical insights of industrial applications of lasers.

Despite the fact that interaction of laser beams with materials has been the subject of numerous books aimed to generalize the scientific backgrounds and techniques of laser processing some issues still remain unsolved. They build a long list of urgent tasks for communities of researchers and engineers around the globe. Many successful technological processes are known to have been developed empirically without a deep understanding of the triggering mechanisms and, moreover, lacking whatever quantitative theoretical or numeric description. Not all of the experimental regularities and phenomena are explained by theories. For example, in laser welding the origin of some weld defects is still open to question. The loss of efficiency at low welding speed in comparison with the electron-beam technique and generation of asperities on the weld surface at high speed is not currently understood. For the gas jet assisted cutting, there is a problem of the cut edge roughness. Selective laser sintering is known to be troubled by the quality losses, i.e. formation of droplets and at both slow and fast scanning speeds. This list can be continued to become a very long one.

Many resent books pay considerable attention to design and operation of laser systems, experimental setups and purely technological aspects. However, the presentation of the theories and simulations or any sort of mathematics is left rather fragmentary. For example, this can be very well seen in the excellent book by W. Steen, *Laser Material Processing*, Springer 2003. The fresh monograph by N.B. Dahotre, S.P. Harimkar, *Laser Fabrication and Machining of Materials*, Springer 2008, includes a lot of technical details of laser tools as auxiliaries to conventional methods of machining such as turning, grinding, drilling laser processing of equipment, manufacturing of three-dimensional objects with lasers, review of lasers in medicine and surgery. The introduction and review of theoretical models describing laser welding and cutting are brief and serve rather illustrative purposes. For a more detailed discussion of the mechanisms making the laser processing, one needs to see the book by E. Toyserkany, A. Khajepour, S. Corbin,

Laser Cladding, CRC press 2005. However, this book is devoted to the cladding techniques only. A lot of attention there is paid to the metallurgical processes which are studied in most cases empirically.

The book we present is intended to describe primarily the very mechanisms which make the basis of the laser processing techniques. A lot of attention is focused on the quantitative description. Here, using the similar vision angle, we aim to generalize the results including some new research directions. This should guide the reader through the diversity of parameters and help him determine the optimal processing modes.

The complexity of technological and physical processes is described using a unitary approach. We study a broad range of parameters which contribute to the nonresonant interaction of laser light with materials. These include pulse duration – from nanoseconds to the continuous wave mode, of the pulsed laser – from several units to hundreds, power density – from 10^4 to 10^{12} W/cm^2, beam power – from dozens of W to dozens of kW.

To make the book complete and self-consistent for every issue covered we are unwinding our description from elementary approaches to modern sophisticated models. The preference is granted to the models which have been verified in experiments. At the same time, we discuss important experimental results which have not yet obtained quantitative explanations as well as refer to theoretical models still waiting for their experimental verification.

Consequently, our book puts the stress on building the understanding of the cause-and-effect relations in physical processes employed in laser technologies. We expect it to become a useful supplement to the books published earlier. We hope it will help the engineers and researchers to improve the existing and develop new machining techniques. The book may serve as a textbook for under- and post-graduate students majoring in laser applications.

Acknowledgements

This book was made possible by the efforts of many people who offered their support and expertise. Without their encouragement, guidance, and sincere concern, many things would be different.

Our gratitude will go to all our colleagues and friends for fruitful discussions who have contributed their ideas, time, and energy, and especially to A. Gusarov, I. Yadroitsev, M. Doubenskaya, V. Mazhukin, W. Scharff, N. Rodionov, A. Krasyukov, S. Drobyazko, and Mrs. G. Antonova.

G. Gladush extends his heartily thanks to the staff of the Diagnostics and Engineering of Industrial Processes Laboratory (DIPI) of the National Engineering School of Saint-Etienne (ENISE) for the nice and friendly atmosphere during his staying with them, and especially to Mrs. I. Movchan for management and technical support.

G. Gladush also offers his profound thanks to M. Sapozhnikov, a valued specialist in English language for the precious time he gave up to go over the manuscript. To these acknowledgments must be added his deep gratitude to Mrs. E. Isaeva for the help in preparing figures.

G. Gladush takes this opportunity to express his profound gratitude to his beloved wife Galina and sons Maxim and Yuri. Without their support, cooperation, and preparation of the manuscript in LaTeX this book would not exist.

Finally, the authors thank all individuals who so graciously volunteered to dedicate their services, time, and support to our work.

Saint Etienne, Troitsk, Shatura
March 2011

Gennadiy Gladush
Igor Smurov

Contents

Acronyms

c_p	Specific heat
d	Size
E	Pulse energy
F	Focal distance
H_b	Boiling heat
H_m	Melting heat
L_b	Length
L_m	Enthalpy at boiling temperature
L	Enthalpy at melting temperature
l	Optical thickness
l_R	Rosseland's path
m	Atom mass
M	Mach number
Ma	Marangoni number
n	Refractive index
P	Power
Pr	Prandtl number
Pe	Peclet number
Q	Heat or off-duty ratio (in Chaps. 8 and 9)
q_t	Thermal radiation flux
q_{tr}	Transmitted laser intensity
R	Reflection coefficient
Re	Reynolds number
r_f	Radius of focal spot
S	Surface
T	Temperature
Tr	Transmission coefficient
T_s	Surface temperature
t	Time
V	Velocity

S	Surface
α	Absorption coefficient of surface
α_H	Absorption coefficient of H_{01} mode
α_E	Absorption coefficient of E mode
δ	Fresnel number
ε	Permittivity in Chap. 1 and fraction of pores in Chap. 2
ε''	Imaginary part of permittivity
η	Dynamic viscosity
θ	Angle
κ	Thermal conductivity
λ	Wavelength
μ_a	Absorption coefficient
μ_s	Scattering coefficient
ν	Kinematic viscosity
ξ	Surface impedance
Π	Hertz potential
ρ	Density of material or specific resistance in Chap. 1
σ	Stephan–Boltzmann constant in Chap. 9, surface tension in Chap. 2, and electric conductivity in Chap. 1
τ	Pulse duration
χ	Thermal diffusivity
ψ	Wave phase
ω	Wave frequency

Chapter 1
General Problems of Propagation of Laser Radiation in Gases and Plasma and Physical Processes on the Surface of Condensed Media

Abstract Laser processing of materials is usually performed by focused radiation, many operations requiring very tight focusing. The focal spot size cannot be made as small as one likes due to diffraction of a light beam. The laser beam focusing can be also restricted by nonlinear processes. Because laser radiation used for material processing has high average or peak powers, this radiation can change the optical properties of media in which it propagates. The so-called self-action effects such as radiation self-focusing or self-defocusing appear. In addition, light can be noticeably absorbed in a medium. Technical industrial conditions may require sometimes the transport of light over long distances in air or through condensed media (in the case of special technologies), and these effects can become significant. Plasma often produced during various technological processes can strongly affect the propagation of light, in particular, IR radiation at $\lambda = 10\,\mu\text{m}$. In this chapter, we consider this question only briefly. The propagation of radiation in the optical discharge plasma will be discussed in detail in Chap. 3. The efficiency of action of light incident on a sample depends on the absorption coefficient of the latter. For example, during laser welding and drilling light propagates in a narrow channel, where it is multiply reflected from walls. The properties of such propagation of light will be also considered in this chapter. Laser radiation can heat, melt, evaporate, and ionize the sample material. During heating in an oxidizing medium, the sample material can be oxidized. During melting of the sample, the motion of the melt can appear due to Marangoni effect and capillary instabilities can develop. If the radiation intensity is sufficient to initiate the evaporation of the target, specific evaporative instabilities can appear. We will briefly discuss all these phenomena in this chapter.

G.G. Gladush and I. Smurov, *Physics of Laser Materials Processing*,
Springer Series in Materials Science 146, DOI 10.1007/978-3-642-19831-1_1,
© Springer-Verlag Berlin Heidelberg 2011

1.1 Propagation and Focusing of Radiation in Vacuum, Gases and Plasma

1.1.1 Focusing of Light in Vacuum

As mentioned above, the tight focusing of light in vacuum is mainly prevented by diffraction. This phenomenon can be most simply described in the so-called quasi-wave approximation. The aim of this section is, in particular, the demonstration of the efficiency of this approximation. This is achieved by obtaining the known results on light focusing in vacuum. In this approximation, a wave remains quasi-monochromatic, and only its amplitude and phase change. Consider this approximation taking self-action into account [1]. The results will be used in the next sections.

Consider the propagation of a modulated wave in a medium with the permittivity depending on the wave intensity:

$$\varepsilon = \varepsilon_0 + \varepsilon_{\mathrm{nl}} \tag{1.1}$$

where ε_0 is the linear permittivity (the limiting value of ε for $(|E| \to 0)$) and $\varepsilon_{\mathrm{nl}}$ is the nonlinear addition. If $\varepsilon_{\mathrm{nl}}$ depends directly on the field strength, it can be represented as a series

$$\varepsilon_{\mathrm{nl}} = \varepsilon_2|E|^2 + \varepsilon_4|E|^4 + \ldots, \tag{1.2}$$

where the first term of the expansion corresponds to the polarization of the medium cubic in the field strength, and $\varepsilon_4|E|^4$ corresponds to fifth-order polarization over the field, etc. The value of $\varepsilon_{\mathrm{nl}}$ can depend on the field strength not directly but via a different quantity depending on the field strength. Then, the equation for the field should be supplemented by the equation describing the dependence of this quantity on the field. The nonlinear change in the permittivity produces self-action effects. The wave still remains quasi-monochromatic during self-action, the nonlinear effect being manifested in variations in the amplitude and phase modulation of the wave and changes in the shape of its angular distribution and frequency spectrum.

The stationary self-action of waves in a medium with permittivity (1.1) is described by the Helmholtz equation

$$\Delta E = \frac{\omega^2}{c^2}\varepsilon_0 E + \frac{\omega^2}{c^2}\varepsilon_{\mathrm{nl}} E \tag{1.3}$$

where ω is the frequency of light and E is the electric field strength vector of the wave.

In the case of the wave beams with a narrow angular distribution and the weak nonlinearity of the medium, the solution of the latter equation can be simplified by using the method of slowly varying amplitudes.

Lets us represent the wave field of a diffracting beam in the form of a quasi-plane wave

$$E = (e/2)E(\mu z, \sqrt{\mu}x, \sqrt{\mu}y)\exp[\mathrm{i}(\omega t - kz)] + c.c. \quad (1.4)$$

where e is the unit polarization vector; z is the beam propagation direction; x and y are the transverse coordinates; E is the slowly varying amplitude; μ is a small parameter characterizing the difference of the beam from a linear plane wave, which is manifested due to diffraction and the nonlinearity of the medium; and k is the wave-vector modulus. Expression (1.4) takes into account that variations in the complex amplitude across the wave beam occur faster than along the beam. By assuming that the nonlinearity of the medium has the same order of smallness, $\mu_1\varepsilon_{\mathrm{nl}} \sim \mu\varepsilon_0$, and substituting (1.4) into (1.3), we obtain in the first approximation in a small parameter the equation

$$2\mathrm{i}k\frac{\partial E}{\partial z} = \Delta_\perp E + k^2\frac{\varepsilon_{\mathrm{nl}}}{\varepsilon_0}E, \quad E = E_0(r,z)\exp[\mathrm{i}k\psi(r,z)] \quad (1.5)$$

where ψ is the phase.

Equation (1.5) corresponds to the quasi-optical approximation in the theory of nonlinear waves. It describes the stationary self-action of diffracting wave beams. The linear and nonlinear parts of the permittivity entering (1.5) are complex functions: $\varepsilon \approx \varepsilon' + \mathrm{i}\varepsilon''$.

If the linear and nonlinear absorption of the wave is small, ε and $\varepsilon_{\mathrm{nl}}$ are real quantities. Then, the term with $\varepsilon_{\mathrm{nl}}$ in parabolic equation (1.5) describes the dependence of the phase velocity on intensity.

Note that (1.5) in vacuum is transformed to a linear parabolic equation with the imaginary diffusion coefficient $D_k = -1/2\mathrm{i}k$. The diffusion equation has the general solution expressing the required function in terms of its boundary value. This solution shows that, if the beam has a Gaussian distribution on a focusing spherical mirror, this distribution is preserved during the propagation of the beam in vacuum:

$$E_0 = \frac{E(0)}{\sqrt{(1-z/F)^2 + D^2}}\exp\left\{-\frac{r^2}{a^2[(1-z/F)^2 + D^2]}\right\}, \quad (1.6)$$

where E_0 is the real amplitude of the field; F is the focal distance of the mirror; and $E(0)$ and a are the field amplitude and the beam width on the mirror for $z = 0$ and $D = 2z/ka^2$. The parameter $\delta = ka^2/2F$ (the Fresnel number) shows which of the two processes (diffraction or geometrical convergence) dominates. According to (1.6), the field amplitude on the beam axis first increases, achieves maximum $E(0)\sqrt{1+\delta^2}$ for $D = \delta/(1+\delta^2)$, and then monotonically decreases. The beam width first decreases due to focusing, achieves its minimum value (the beam radius in the focal plane $r_f = a/\sqrt{1+\delta^2}$) and then monotonically increases for $D > \delta/(1+\delta^2)$. If the Fresnel number is large, the minimal beam size is smaller than the initial amplitude by a factor of δ.

The known expressions obtained above show that the quasi-wave approximation well describes both the diffusion and refraction (see below) of electromagnetic radiation. The method of geometrical optics, which also describes the refraction of laser radiation, is restricted by the case of nonintersecting beams.

1.1.2 Propagation of Laser Radiation in Gases and Plasma

Due to the relatively high power of CO_2 lasers, the propagation of their radiation is often accompanied by nonlinear effects caused mainly by the formation of a "thermal lens" in the medium absorbing part of the laser beam energy.

The linear absorption coefficient for the $\lambda = 10.6\,\mu\text{m}$ CO_2 laser radiation in dry pure air at the atmospheric pressure at $T = 300\,\text{K}$ is determined by the interaction of radiation with carbon dioxide gas molecules and is quite small, $\mu = 0.072\,\text{km}^{-1}$ [2]. Gases in which laser radiation can propagate can differ significantly from pure dry air and can contain impurities of water, carbon dioxide, aerosols, etc. These impurities can enhance the absorption coefficient μ by many orders of magnitude. Here, we will consider for simplicity the propagation of laser radiation in pure dry air. The final expressions for other gases have the same form and can be easily obtained by the corresponding recalculation. The propagation of high-power CO_2 laser radiation has been studied in many papers. Here, we present the summary of the results [3, 4].

An initially focused laser beam (φ is the focusing angle) broadens due to thermal defocusing. Its minimal diameter is

$$d_1 = d_0 \exp(-\varphi^2/\varphi_0^2),\ \varphi_0^2 = \alpha_1 \mu P / 2\pi\kappa \tag{1.7}$$

where d_0 is the mirror diameter; κ is the heat conductivity of air; α_1 is the proportionality coefficient in the expression for the refractive index of air $N = 1 - \alpha_1 T$ and T is the air temperature.

The distance at which the laser beam is spread due to defocusing to the width equal to the initial beam diameter $d_0 = 2R$ can be estimated from (1.7) as

$$z^* \approx R\sqrt{2\pi\kappa/\alpha_1\mu P} \tag{1.8}$$

Let us make the estimate for air [$\kappa = 4 \times 10^{-4}\,\text{W/(cmK)}$, $\mu \approx 10^{-6}\,\text{cm}^{-1}$, $\alpha_1 \approx 10^{-6}\,\text{K}^{-1}$]. By assuming $R \approx 5\,\text{cm}$, we obtain $z^* \approx 2.5/\sqrt{P}$, where z^* is expressed in km and P in W. For example, a 10 kW parallel laser beam broadens twice at a distance of 25 m from a mirror.

The calculation of the thermal defocusing of a horizontally propagating beam may require the consideration of thermal convection, which results in the bending of the beam downwards towards a rising convective flow [see (1.10)] and in the decrease of the defocusing beam diameter due to convective cooling [5].

Let us assume that the transverse radiation intensity distribution on the mirror has a "hole" in the middle. Then, the air temperature distribution at the initial instant of time will be inhomogeneous, with a hole in the middle. Due to bending in heated regions, the inner beams will be focused on the axis, i.e. self-focusing appears. Later, the temperature of the heated gas will achieve its maximum at the beam center, and self-focusing will disappear. Self-focusing can exist during the time $\tau \approx R^2/4\chi$. This time in gas proves to be quite large. The thermal diffusivity of air is $\chi = 0.3\,\mathrm{cm^2/s}$; for $R \approx 5\,\mathrm{cm}$, we have $\tau \approx 25\,\mathrm{s}$. In the case of the parabolic intensity profile, the initially parallel beams intersect at one point

$$z^{**} = R\sqrt{c_p \rho / \alpha_1 \mu q_0 t} \tag{1.9}$$

(c_p and ρ are the specific heat and density of gas), which approaches the mirror with time. In air for $R = 5\,\mathrm{cm}$, we have $z^{**} = 15/\sqrt{tP}$, where z^{**} is expressed in km and P in W. For example, for $P = 10\,\mathrm{kW}$ and $t \approx 25\,\mathrm{s}$, the focus will be located at a distance of 30 m from the mirror.

Some technological operations require sometimes the blowing off of a sample by a gas flow. This can be related to the necessity of sample protecting or cooling.

A transverse (with respect to the beam axis) gas flow displaces the region of minimal density in the same direction and deflects the light beam. At the same time, gas cooling caused by this flow reduces thermal defocusing [2]. The beam bending and defocusing are also observed during processing of large surfaces by a scanning beam.

Calculations [3] show that in this case the focal point shifts upstream by the distance

$$d = 4\alpha_1 \mu P / \pi c_p \rho v \varphi^2. \tag{1.10}$$

For example, for $\varphi = 10^{-3}$ and $v = 1\,\mathrm{m/s}$ in air, we obtain from (1.10) $d = 4 \times 10^{-5} P$, where d is expressed in cm and P in W. For $P = 2\,\mathrm{kW}$, the displacement of the focus is ~1 mm, which exceeds the typical focal spot diameter of an industrial CO_2 laser. The calculation of radiation defocusing in a medium containing aerosols is complicated by the photorefractive effect for sol particles, which is quite complicated.

The bleaching of gas containing a suspension of liquid aerosols caused by the evaporation of drops during the propagation of a CO_2 laser beam was calculated in [6].

The main specific feature of the propagation of light in a medium with free charge carriers is related to another mechanism of the interaction light with medium. While the propagation of light in air is determined by a decrease in its density caused by heating, the interaction of light with plasma is determined by the motion of conduction electrons in the wave field. The plasma permittivity is well known [7]:

$$\varepsilon = 1 - \frac{\omega_p^2}{\omega(\omega + i\nu)} = \varepsilon' + i\varepsilon'',$$

$$\varepsilon' = 1 - \frac{\omega_p^2}{(\omega^2 + \nu^2)}, \quad \varepsilon'' = \frac{\omega_p^2 \nu}{\omega(\omega^2 + \nu^2)}, \tag{1.11}$$

where ν is the frequency of collisions of electrons with plasma atoms and ions and $\omega_p = 4\pi n_e e^2/m$ is the plasma frequency. The imaginary part ε'' of the permittivity is determined by the absorption coefficient μ of light in plasma:

$$\mu = k\varepsilon'' = \frac{\omega_p^2 \nu}{(\omega^2 + \nu^2)c}. \tag{1.12}$$

By substituting the real and imaginary parts of the permittivity from (1.11) and (1.12) into (1.5), we obtain the equation for the propagation of radiation in plasma:

$$-2ik\frac{\partial E}{\partial z} + \Delta_\perp E + k^2\Big(-\frac{n}{n_c} + i\frac{\mu}{k}\Big)E = 0 \tag{1.13}$$

where $n_c = (\omega^2 + \nu^2)m_e/4\pi e^2$ is the critical plasma density at which the real part of ε vanishes. For a CO_2 laser, $n_c \approx 10^{19}\,\mathrm{cm}^{-3}$.

The concentration n_e of free electrons in a plasma plume produced sometimes during laser processing of materials, is an equilibrium function of temperature. In the isobaric case (continuous optical discharge), the dependences of n_e and μ for different materials are presented in Figs. 3.3 and 3.4. One can see that, unlike air, the value of μ for plasma is quite large: $\mu \approx 1\,\mathrm{cm}^{-1}$. Therefore, it is necessary to take into account the attenuation of light due to absorption during propagation in plasma.

To describe the behavior of a laser beam in the optical discharge plasma, we will neglect first its attenuation and consider the propagation of beams assuming that the function ε' is specified. Considering only a plane case and using the geometrical optics approximation [4], we can obtain the equation for beam paths. For a beam focused at the point $z = F$, we have

$$x = (1 - z/F)x_0 + \frac{1}{4}z^2\frac{\partial}{\partial x}\varepsilon'\Big|_{x=x_0} \tag{1.14}$$

where x_0 is the coordinate of an elementary beam for $z = 0$ (on a lens or mirror). It is known that the equilibrium concentration of isobaric plasma is a rapidly increasing function of temperature. However, when the complete single ionization is achieved at $T = T^*$, the plasma density decreases due to the escape of particles from the heated region (see Fig. 3.3). If the plasma temperature $T_{\max}$ on the beam axis is lower than T^*, then $\partial\varepsilon'/\partial x > 0$ due to the decrease of temperature at the beam boundary. For $T_{\max} > T^*$, the function $\partial\varepsilon'/\partial x$ changes its sign at the point $T_{\max} = T^*$. One can see from (1.14) that the beams can be focused at one point only if the condition

$$\partial\varepsilon'/\partial x = \gamma x, \ \gamma = const$$

is fulfilled. By using (1.14), we can obtain the expression for the displacement ΔF of the focus in plasma:

$$\Delta F/F = \gamma\, F^2/4,\ |\Delta F| < F \tag{1.15}$$

It follows from (1.15) that, if $\gamma > 0$, the focus sifts along the beam due to defocusing. If $\gamma < 0$, the self-focusing of the beam takes place and the focus shifts towards the lens. The requirement of beam focusing at one point is difficult to fulfill because a parabolic dependence of ε' on the transverse coordinate is needed. The latter is unlikely. Indeed, we have

$$\frac{\partial \varepsilon'}{\partial x} = \frac{\partial \varepsilon'}{\partial T}\frac{\partial T}{\partial x}$$

In a continuous optical discharge, $\partial T/\partial x \sim x$ (see Sect. 3.3). As mentioned above, the quantity $\partial \varepsilon'/\partial T$ drastically depends on temperature. Therefore, in the general case the beams intersect at different points, the peripheral beams being deflected stronger because temperature at the plume periphery is lower and $\partial \varepsilon'/\partial T$ is higher than that at the beam axis. Thus, the beam shape should be distorted, i.e. aberration should be observed.

The displacement of the focus and the change in the focusing angle of light in plasma can be easily obtained from (1.11) and (1.15):

$$\Delta F/F = \frac{1}{4}\frac{n_{e\,\max}}{n_c \varphi^2},\ \varphi = R/F \tag{1.16}$$

$$\frac{\Delta\varphi}{\varphi} = -\frac{1}{4}\frac{n_{e\,\max}}{n_c \varphi^2}$$

It follows from (1.16) that the focus should draw aside from the lens with increasing power, while the focusing angle should decrease because the plasma temperature increases. The maximum density of the continuous optical discharge plasma in air at the atmospheric pressure is $n_{e\,\max} = 2 \times 10^{17}\ \mathrm{cm}^{-3}$ (see Sect. 3.1). Therefore, for $2\varphi \approx 0.1$, we have $\Delta\varphi/\varphi \sim |1|$. It follows from this that refraction can be considerable. Therefore, the theoretical problem of propagation of laser beams in plasma should be solved simultaneously for functions ε and E. This problem was solved in Chap. 3, where an optical discharge was investigated. In this chapter and in other chapters, the results of experiments on laser processing of materials are also described. Below, we will describe experiments on the refraction of a laser beam in the breakdown plasma in air. Although the experimental conditions strongly differ from the laser technology conditions, the results of measurements are clear and well illustrate the described phenomenon.

The refraction of a laser beam propagated through the optical breakdown plasma in the atmospheric air was measured in [8]. Experiments were performed by using a pulsed CO_2 laser. The pulse shape was typical for such lasers and consisted of a

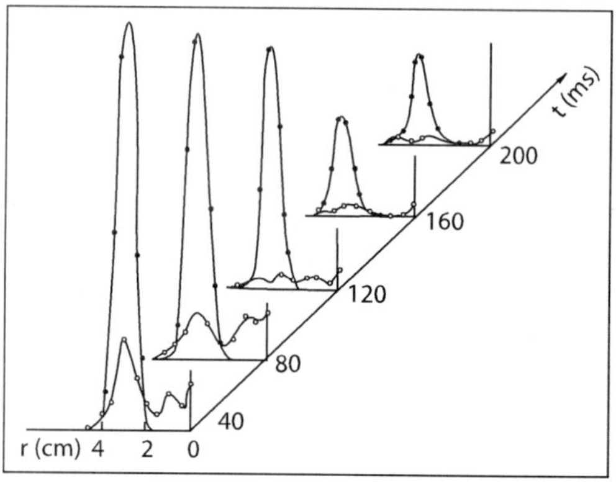

Fig. 1.1 Spatiotemporal laser radiation intensity distribution: (•) the initial distribution; (○) distribution after the breakdown

leading peak of duration $\tau_1 \sim 50\,\text{ns}$ and a tail of duration $\tau_1 \sim 2\,\mu\text{s}$. The radiation was focused by a metal mirror with the focal distance $F = 10\,\text{m}$. An unstable resonator generated light in the form of a ring with the external diameter 10 cm and the internal diameter 6.5 cm. The radiation intensity in the focal spot was $q = 5 \times 10^8\,\text{W/cm}^2$ for the total energy of the light pulse $E \approx 150\,\text{J}$. This radiation intensity in the dusty atmosphere was sufficient to produce the optical breakdown of air. The radiation intensity propagated through the optical discharge was measured with a germanium detector at a distance of 7 m behind the focal plane. Figure 1.1 shows the radial intensity distribution at different instants. One can see that the initial beam is strongly absorbed and is spread towards the ring center.

1.2 Absorption, Reflection, and Propagation of Radiation in Cavities in Condensed Media

In this section, we consider an important quantity for laser technologies – the coefficient of light reflection from materials. The main attention will be devoted to the reflection of light from metal surfaces.

1.2.1 Flat Surface

It is known that the reflection coefficient for light normally incident from vacuum on a flat surface is described by the Fresnel formula [9]

$$R = |(\sqrt{\varepsilon} - 1)/(\sqrt{\varepsilon} + 1)|^2 \tag{1.17}$$

Permittivity of metals, as for plasma, is determined by free electrons, i.e. it can be calculated from (1.11), where ν is the frequency of scattering of electrons by the

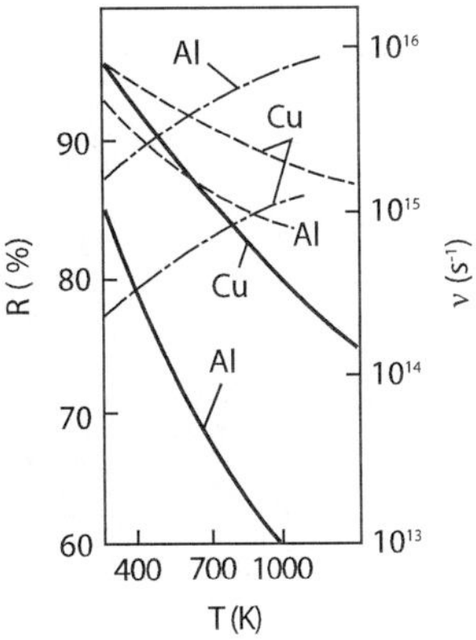

Fig. 1.2 Calculated temperature dependences of the reflectance R for laser radiation for wavelengths 1.06 μm (*solid line*) and 10.6 μm (*broken curves*) and the electron–phonon collision frequency ν (*broken point*)

thermal vibrations (phonons) of a lattice. The typical values for metals are $\omega_p \approx 10^{16}\,\mathrm{s}^{-1}$, $\nu \geq 10^{14}\,\mathrm{s}^{-1}$ (Fig. 1.2). While $\varepsilon'' \approx |\varepsilon' - 1| \approx 1$ for the optical discharge plasma, in metals for the CO_2 laser radiation, we have $\varepsilon', \varepsilon'' \approx 10^4$, i.e. $R \approx 1$. For most metals, especially at high temperatures, $\nu \gg \omega$ (Fig. 1.2), and therefore, as follows from (1.11), $\varepsilon'' \gg \varepsilon'$. In this approximation, we obtain from (1.17)

$$R = 1 - \sqrt{8/\varepsilon''}$$

By expressing ε'' in terms of the conduction $\sigma = n_e e^2/\nu m$ as $\varepsilon'' = 4\pi\sigma/\omega$, we obtain finally

$$R = 1 - \sqrt{2\omega/\pi\sigma} \tag{1.18}$$

Thus, the reflection coefficient R of a material is determined only by its conduction. In [10], the reflection coefficient R was calculated by (1.17) taking into account the temperature dependence for different laser wavelengths (see Fig. 1.2). In [11], the values of the absorption coefficient $\alpha = 1 - R$ are presented in a broader temperature range (Fig. 1.3). The temperature dependence of the conduction can be also estimated by using the known empirical relation for the specific resistance ρ

$$\rho = \rho_0(1 + \alpha_T T)$$

where ρ_0 is the specific resistance at $T = 0°\mathrm{C}$ and α_T is the temperature resistance coefficient. By substituting this expression into (1.18), we obtain the temperature dependence of α. For $T < \alpha_T^{-1}$, the increment of α will depend linearly on temperature (the Drude model). For example, $\alpha_T = 6 \times 10^{-3}\,\mathrm{C}^{-1}$ for iron, which means that α will increase linearly with temperature up to ~500 K. At higher temperatures of the metal, $\alpha \propto \sqrt{T}$, which is demonstrated in Fig. 1.3a.

It is know that the wavelength of light in a medium decreases by a factor of N compared to vacuum, where $N = \sqrt{|\varepsilon|}$ is the refractive index of the medium. In the case of $\nu \gg \omega$, the wavelength of light in a metal is $\delta = c/\omega\sqrt{\varepsilon} = \sqrt{c^2/4\pi\sigma\omega}$.

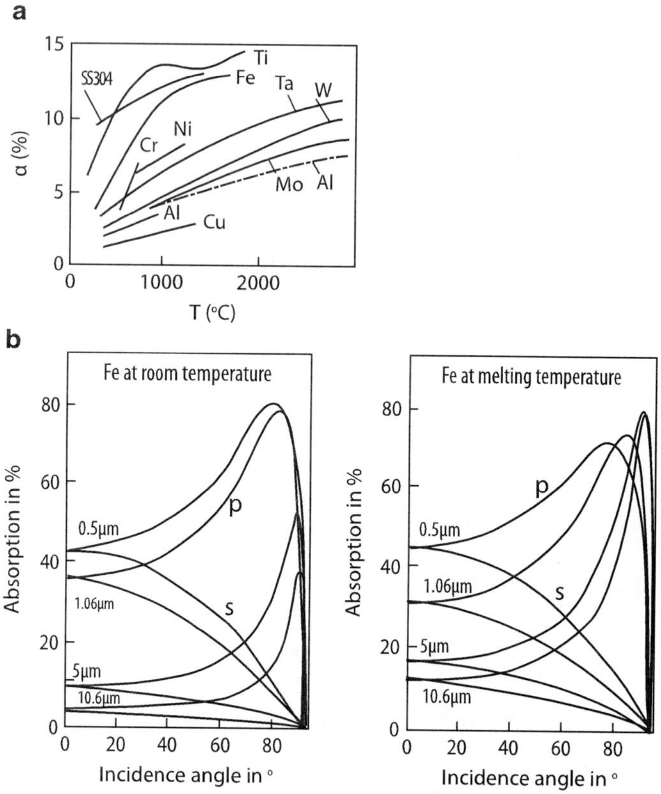

Fig. 1.3 Temperature dependences of the absorption coefficient of various materials at the CO_2 laser wavelength (**a**). Dependences of the absorption coefficient of iron on the angle of incidence, polarization, and wavelength at room temperature (*at the left*) and at the melting temperature (*at the right*) (**b**)

The skin depth[1] δ is very small because $\varepsilon \gg 1$. In this case, expressions for the reflection coefficients of light with different polarizations are rather simple [9]:

$$R_{\perp} = 1 - 4\xi' \cos\theta_0 \tag{1.19}$$

$$R_{\parallel} = |(\cos\theta_0 - \xi)/(\cos\theta_0 + \xi)|^2 \tag{1.20}$$

where $\xi = \xi' + i\xi'' = (1-i)\sqrt{\omega/8\pi\sigma}$ is the surface impedance of the metal; θ_0 is the angle of incidence of light; $R_{\perp}$ and $R_{\parallel}$ are the reflection coefficients for a wave with the electric field vector normal or parallel to the plane of incidence. In the case of normal incidence, $R_{\perp}$ is equal to $R_{\parallel}$ and coincides with (1.18). As the angle of

[1]The wavelength of light in a metal is at the same time the penetration (skin) depth of the field [9].

incidence is increased (i.e. on passing to grazing incidence), $R_{\perp}$ increases, while $R_{\parallel}$ decreases. For θ_0 not too close to $\pi/2$, we have

$$R_{\parallel} = 1 - 4\xi'/\cos\theta_0 \tag{1.21}$$

For $\theta_1 = \pi/2 - \theta_0 = |\xi|$, $R_{\parallel}$ has a minimum equal to $(\sqrt{2}-1)/(\sqrt{2}+1)$. In this case, $\alpha = 1 - R$ achieves the maximum value equal to 83%, and the grazing angle is $\theta_1 \sim 10^{-2}$. Figure 1.3b shows the dependences of the absorption coefficients of iron on the angle of incidence for light beams of different polarizations at the surface temperature equal to the room and melting temperature [12]. One can see that these dependences are similar in a broad wavelength range.

All the expressions for reflection coefficients are applied for pure, polished surfaces. For materials with unpurified and unpolished surfaces (supplied materials), the absorption coefficient depends on the surface state and can considerably exceed α for pure surfaces (Table 1.1) [13].

Note that upon heating of samples up to comparatively high temperatures, contaminants on the surface can be burned out and the roughness can disappear (upon melting). Therefore, if oxidation or other chemical processes do not occur at high temperatures, the absorption coefficient will be close to its value for a pure metal. The absorption coefficient can be increased artificially. It is especially important for the CO_2 laser radiation because $\alpha \leq 0.1$ at a wavelength of $\lambda = 10.6\,\mu m$. To increase the absorption coefficient, the surface of a sample is covered with special heat-resistant materials absorbing IR radiation, for example, zinc phosphate, which has the effective absorption coefficient $\alpha_e = 0.7$ at $T = 1{,}000°C$ [14]. This question will be considered in more detail in the discussion of laser hardening in Sect. 2.1.

The absorption coefficient can increase with increasing the surface temperature if heating is performed in an oxidizing atmosphere (for example, in air [15]). As an oxide film grows, the absorption coefficient periodically increases when the film thickness becomes multiple of half the wavelength of light. Thus, α_e oscillates in time (Fig. 1.4a,b), as it occurs on a copper target, where melting proceeds quite

Table 1.1 Absorption coefficient α at $\lambda = 10.6\,\mu m$, (%)

Material	Unpurified and unpolished surface	Polished surface
Stainless 316 steel	39	9
Stainless 416 steel	42	10
Aluminum	12	2
Copper	12	2
4130 steel	85	
Molybdenum	12	3
Tantalum	21	4
Silver	–	1
Gold	–	1
Rhodium	–	4

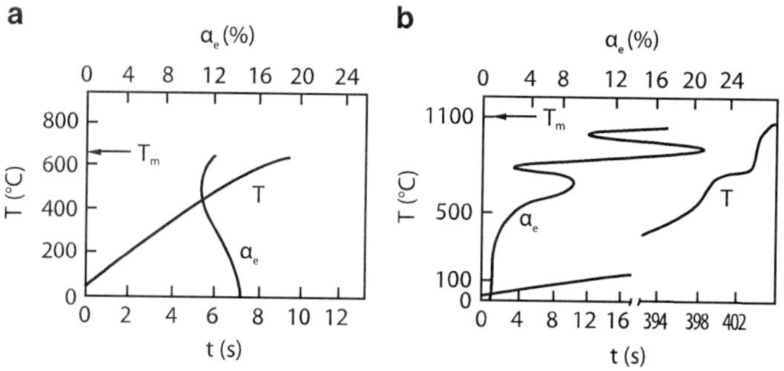

Fig. 1.4 Characteristic heating curves for thermally thin targets heated in air by the CO_2 laser radiation for $P = 30$ W and $q = 1.3$ kW/cm^2 and the corresponding effective absorption coefficients α_e: (**a**) duralumin ($m = 54$ mg); (**b**) copper ($m = 116$ mg)

actively. One can see from Fig. 1.4b that due to oxidation the absorption coefficient for a pure surface can be an order of magnitude higher. The situation is different when the aluminum surface is heated (Fig. 1.4a). In this case, the oxide film is very strong, and its thickness does not grow with time. The effect of laser-induced combustion of a metal surface is quite similar to the oxidation of a metal surface in the laser field. As many exothermal chemical reactions, the burning of a metal in a laser beam in an oxidizing atmosphere has a threshold and appears abruptly. Processes of laser thermal chemistry are described in detail in [15]. The oxidation of a metal surface plays a significant role in laser cutting of metals in the oxygen atmosphere (see Chap. 5).

1.2.2 Propagation of Laser Radiation in a Narrow Channel in a Metal

Some types of laser processing of materials (cutting, welding, drilling, see Chaps. 4, 5, and 8) require a deep penetration of the laser beam into material. In this case, the radiation intensity should be high enough. In this connection, it is often necessary to obtain the minimal size of a laser spot on the target surface. According to (1.6), the focal spot radius is $r_f \simeq \lambda/\varphi$, i.e. it is inversely proportional to the focusing angle φ. Therefore, it is necessary to use tightly focused radiation. But such radiation will be defocused behind the focal plane (usually coincident with the sample surface plane) already at a small depth $L \approx \lambda/\varphi^2$ and will fall on the side walls of the channel. If the absorption coefficient α is large (radiation is short-wavelength with parallel polarization), defocusing reduces a part of energy released at the channel bottom. If α is small, the greatest part of light will be reflected from the walls and fall to the bottom. Multiple reflections are especially important for normally polarized radiation for which the reflection coefficient increases with increasing the angle of incidence [(1.19), Fig. 1.3b]. During laser welding, cutting, and drilling, it is necessary to know the light intensity distribution in the channel because it determined the dynamics and efficiency of these processes.

The light distribution in the cannel can be relatively simply calculated in the geometrical optics approximation or waveguide approximation. An elementary light beam multiply reflected from the channel walls is either completely absorbed, if the channel is deep, or comes back, if the channel is shallow and melting is not through. If the beam is not spread due to diffraction to the width exceeding the channel width d, its behavior can be described within the framework of geometrical optics; otherwise, the beam propagation has the waveguide nature. We will assume for simplicity that radiation at the channel input is incoherent (for example, radiation from a multimode laser). In this case, it is convenient to describe radiation as a photon flux. Let us assume that the intensity distribution of light incident on a keyhole is bell-shaped. Due to limitedness of the beam, it has the angular spread $k_{\perp}/k \simeq \lambda/d$. Due to diffraction, the elementary beam is spread at the keyhole bottom up to the width $\lambda h/d$. To calculate the beam trajectory, this width should be smaller than the channel width d, as mentioned above. This gives the condition of the applicability of the geometrical optics approximation:

$$d^2/\lambda h \gg 1$$

Thus, the geometrical optics approximation is valid for large Fresnel numbers.

This inequality can be rewritten by introducing the channel aspect ratio h/d coinciding with the weld aspect ratio (see Chap. 4):

$$d/\lambda \gg h/d$$

In practice, $h/d \approx 5$–10, i.e. for $\lambda = 10.6\,\mu$m, the theory can be applied if $d \geq 0.1$ mm. The geometrical optics approximation is convenient in the case of shallow keyholes, when photons experience a small number of reflections and the reflected power is $P_r \sim P$ (in this case, the waveguide approximation is not valid).

In the geometrical optics approximation, the focused radiation can be represented by a set of N beams [16]. We assume that the angular distribution of beams at the focus is described by a Gaussian

$$f(\varphi) = \frac{N}{\pi\varphi_0^2}(2\pi)^{\nu}\exp(-\varphi^2/\varphi_0^2)$$

where $\varphi_0 < \pi$ is the focusing angle; $\nu = 0$ in the axially symmetric case and $\nu = 1$ in the plane case. Figure 9.45 presents the scheme of incident beams. The power of each beam at the input to the keyhole is P/N, where P is the laser power. The trajectory of each beam in the keyhole was numerically calculated by multiplying the power of each beam after the next reflection from the wall by the reflection coefficient calculated from (1.19) and (1.20) [16]. If the beam power became 10^4 times lower, the beam was excluded from consideration. The energy absorbed by a given element of the keyhole surface was calculated by summation of contributions from individual beams.

The dependence of the effective absorption coefficient $\alpha_e = (P - P_r)/P$ on the keyhole depth were numerically calculated by thus method. The absorption coefficient α_e increases with increasing the keyhole depth having a parabolic profile (Fig. 1.5). As expected, radiation with parallel polarization is absorbed more efficiently because the absorption coefficient α for this radiation is higher. The calculations were performed for steel with $4\xi' = 0.1$. It was assumed that a laser emits plane-polarized radiation with the electric vector directed either along a slit or across it. The focusing angle of light was $\varphi_0 = 30°$. In the case of an axially symmetric keyhole, calculations were performed for laser radiation with circular or radial polarization (Fig. 5.22). The effective absorption coefficient for laser radiation measured upon laser welding behaves similarly to the calculated α_e, increasing with increasing h/d and achieving unity for a large depth of the vapor-gas channel (Fig. 4.9).

Consider now the depth distribution of radiation intensity absorbed by keyhole walls. If the channel were cylindrical, the intensity of circularly polarized radiation

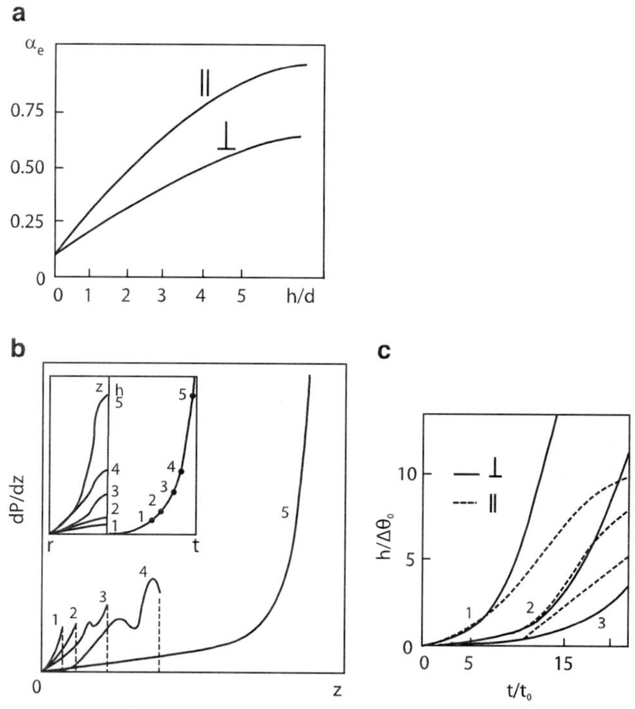

Fig. 1.5 Dependence of the effective absorption coefficient α_e of a steel target for the CO_2 laser radiation on the keyhole depth (the focusing angle is $\varphi = 30°$, $\alpha = 0.1$; ∥ and ⊥ are parallel and normal polarizations of light (**a**). The dependence of the radiation power dP/dz absorbed by the walls on the coordinate z inside each keyholes during drilling (**b**). The depthes of keyholes vs time at different polarization (**c**). Details in Chap. 8

absorbed by the walls would decrease with increasing depth. In the case of radial polarization, the situation is more complicated. At the same time, if absorption by walls were absent, the radiation intensity should increase with approaching the keyhole bottom due to the channel narrowing. This is illustrated by the elementary consideration of the paths of beams in a cone. Figure 1.5b demonstrated that both these tendencies are manifested in the calculation of the radiation intensity absorbed by the walls. This figure shows the keyhole shapes observed at subsequent instants of time during drilling of steel by radially polarized radiation from a CO_2 laser (see details in Chap. 8). The dependence of the radiation power $\mathrm{d}P/\mathrm{d}z$ absorbed by the walls on the coordinate z inside each keyhole is shown below. One can see that this power depends on the coordinate z and keyhole shape in a complicated way. Near the turning point of a group of beams propagating at an angle of $\varphi \sim \varphi_0$, a local maximum appears. As the keyhole depth further increases, the number of maxima increases and their amplitude increases. Finally, they merge to form one maximum located near the keyhole bottom (see curve 5 in Fig. 1.5b).

1.2.3 Waveguide Radiation Propagation Regime

The propagation of laser radiation in a deep narrow channel can be described as the propagation of radiation in a waveguide (with diameter slowly varying along its length and in time). It is known [9] that electromagnetic waves of two types (the E and H waves) can propagate in such a waveguide. The E wave will be excited when laser radiation is polarized along the beam radius (see Fig. 5.22). The attenuation coefficient of such modes in a waveguide with a circular cross section is [9]

$$\alpha_E = \alpha/2r, \tag{1.22}$$

where r is the waveguide radius. The H waves with the vector E parallel to the side wall of the waveguide are attenuated as the E waves (see Fig. 5.22). However, the absorption coefficient of the lowest H_{01} mode (without zeroes over the azimuthal angle) is very small:

$$\alpha_H = \frac{\alpha}{2r}(kr)^{-2},$$
$$kr = 2\pi r/\lambda \gg 1. \tag{1.23}$$

Knowing the absorption coefficient, we can write the equation describing the propagation of radiation in the waveguide regime:

$$\mathrm{d}P/\mathrm{d}z = -\alpha_{E,H} P, \tag{1.24}$$

where z is measured from the channel entrance towards its bottom. It is assumed in (1.24) that the bottom is located far from the channel entrance $h \gg 1/\alpha_{E,H}$ so that radiation reflected from the bottom can be neglected. By integrating (1.24), we obtain the dependence of the radiation intensity q_n absorbed by the walls on z for the E mode:

$$q_n = \frac{\alpha P_0}{4\pi r^2(z)} \exp\left[-\int_0^z \alpha \mathrm{d}z/2r(z)\right]. \tag{1.25}$$

Thus, $q_n(z)$ depends on α and the shape of the channel $r(z)$. Below, we will use (1.24) for calculating analytically laser drilling and an optical discharge in a channel (see Chaps. 3 and 8).

One or other type of modes can be excited in the channel by selecting the required polarization of a laser beam. As mentioned above, to excite the E wave, the electric field vector of the incident laser beam should be directed along the radius. For H wave the electric field vertor should be directed along the azimut. Note that the waves of both types are also excited in the case of plane-polarized incident beam, but their intensity ratio depends on the geometry and parameters of the problem. The radial or azimuthal polarization is obtained by inscribing corresponding grooves on mirrors of the CO_2 laser resonator (Fig. 1.6) [17]. In this way, it is possible to obtain the degree of polarization $\sim$50%: the nonpolarised TEM_{00} mode and radially polarized TEM_{01} mode (the TEM_{00} has no zeroes both over r and z, the TEM_{01} mode has no zeroes over r and has zeroes over φ). The laser beam power of $\sim$2 kW was typical for laser technology. Laser beams with such polarization are of interest not only for excitation of required waves in deep keyholes. In a number of problems, for example, in the description of gas-laser cutting such beams are used "directly" because due to a small number of reflections a keyhole weakly affects the laser beam structure. This is discussed in more detail in Chap. 5. It is clear that radially or axially polarized beams excite waves of the same polarization in a keyhole. But what will be the picture when the incident beam is plane-polarized? Such polarization is used in experiments most often. Before considering this situation, note that because this problem is quite complicated, a cylindrical keyhole is usually studied theoretically. The shapes of a vapor-gas keyhole in laser technologies strongly differ from cylindrical. In a better case, for example, during laser drilling a keyhole represents an axially symmetric deep hole resembling a cone. A keyhole

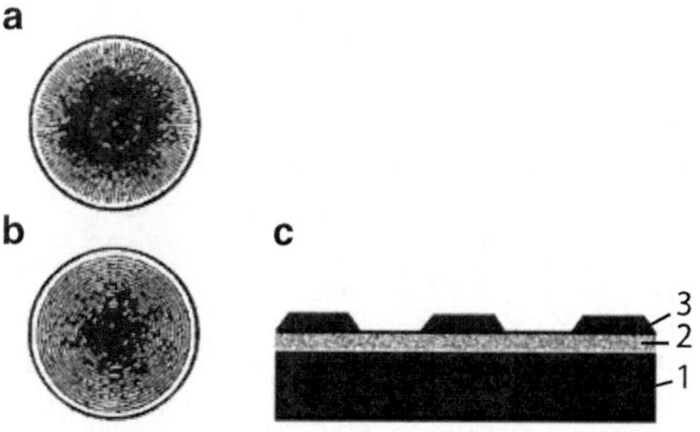

Fig. 1.6 Structure of grooves on optical elements for generating radially (**a**) and azimuthally (**b**) polarized radiation. The structure of an optical element (**c**): (**1**) silicon substrate, (**2**) Ti adhesion layer, (**3**) reflecting copper layer

in a welding pool has the three-dimensional geometry. It was shown experimentally [18] that the deviation from the cylindrical shape of the channel strongly affects the transmission of radiation through the channel. The problem is also complicated by the fact that the keyhole walls can be rough for a number of reasons. Thus, it is still prematurely to hope to describe adequately the propagation of radiation in real keyholes. Nevertheless, to study this effect qualitatively, it is reasonable to consider excitation of waves in a cylindrical keyhole by a plane-polarized beam.

1.2.4 Propagation of Plane-Polarized Radiation in a Cylindrical Keyhole

It is impossible to perform detailed measurements of absorption coefficients under real conditions of laser technological processes, and therefore we will consider only model experiments. Low-power radiation was incident coaxially on the input of cylindrical tubes of small diameter 0.5–1 mm (Fig. 1.7). The converging angle of a plane-polarized beam could be varied in a broad range with the help of a telescopic system. Figure 1.8 presents the dependences of the logarithm of radiation transmission on the ratio of the tube length to its internal diameter [19]. The straight lines fitting experimental points demonstrate the exponential decay of laser radiation in thin capillaries. The attenuation coefficient increases with increasing the converging angle φ of the laser beam.

We will describe the excitation of waves in a metal capillary by the method used for solving problems in waveguides [20]. The electric and magnetic field vectors are expressed in terms of scalar functions – the Hertz potentials [21]. This problem is cumbersome as a whole. To give only the general picture of the phenomenon, we consider only the formulation of the problem and results of its solution. The Hertz

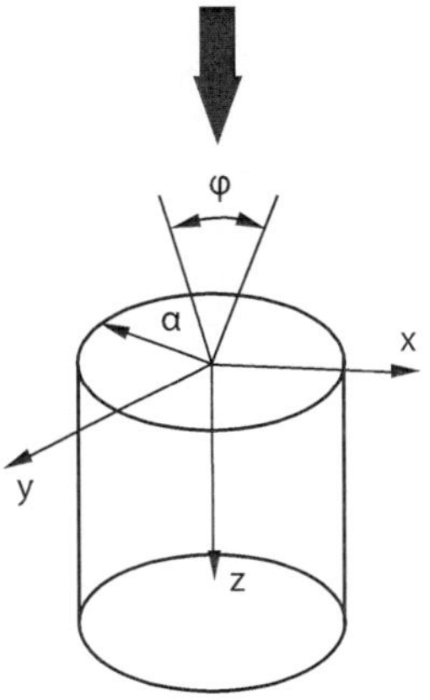

Fig. 1.7 Scheme of the propagation of radiation in a waveguide

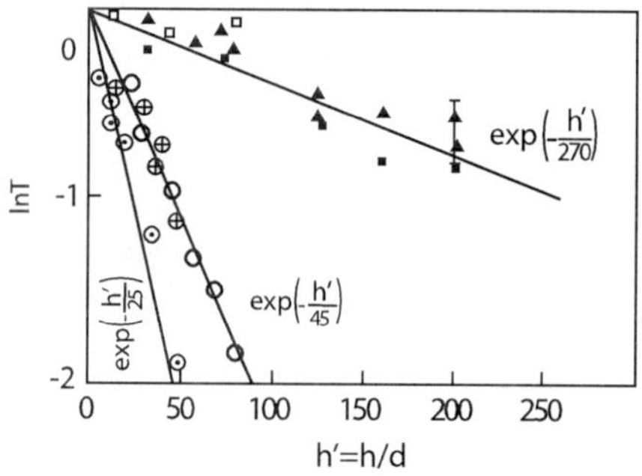

Fig. 1.8 Experimental dependencies of the logarithm of transmission of tubes on the reduced length: (*ring with cross*) Cr, $d = 1$ mm, $\theta = 0.05$; (*square*) Cr, $d = 1$ mm, $\theta = 0.003$; (*ring*) stainless steel, $d = 0.5$ mm, $\theta = 0.05$; (*dark ring*) German silver, $d = 0.78$ mm, $\theta = 0.05$; (*black square*) German silver, $d = 0.78$ mm, $\theta = 0.003$; (*triangle*) German silver, $d = 0.78$ mm, $\theta = 0.005$; (*ring with point*) German silver, $d = 0.78$ mm, $\theta = 0.12$

potentials satisfy the equations [21]:

$$\begin{aligned} &\mathrm{i}\frac{\partial \Pi}{\partial z} + \frac{1}{r}\frac{\partial}{\partial r}\left(r\frac{\partial \Pi}{\partial r}\right) - \frac{m^2}{r^2}\Pi = 0, \\ &\mathrm{i}\frac{\partial \tilde{\Pi}}{\partial z} + \frac{1}{r}\frac{\partial}{\partial r}\left(r\frac{\partial \tilde{\Pi}}{\partial r}\right) - \frac{m^2}{r^2}\tilde{\Pi} = 0, \end{aligned} \tag{1.26}$$

where $m = 0, 1, 2,$ is the azimuthal number. The cylindrical dimensionless variables are $r \to r/a$ and $z \to z/2\kappa a^2$. Due to the smallness of the impedance $|\xi| \ll 1$, the boundary conditions on the walls are selected by using the Leontovich conditions:

$$\begin{aligned} &\frac{\partial \Pi}{\partial z} = -\xi\kappa a\left(\frac{\partial \Pi}{\partial z} - \tilde{\Pi}\frac{m}{r}\right), \\ &\frac{\partial \tilde{\Pi}}{\partial z} = -\frac{\kappa a}{\xi}\left(\frac{\partial \tilde{\Pi}}{\partial z} - \Pi\frac{m}{r}\right). \end{aligned} \tag{1.27}$$

The absence of singularities Π and $\tilde{\Pi}$ on the capillary axis gives additional conditions

$$r\left(\frac{\partial \Pi}{\partial r}\right) \to 0, \quad r\left(\frac{\partial \tilde{\Pi}}{\partial r}\right) \to 0, \quad r \to 0$$

The Hertz potentials at the capillary input for $z = 0$ for a linearly polarized laser beam with the electric field distribution $E_0(r)$ and $E_z = 0$ will satisfy the relations

$$\begin{aligned}\frac{\partial \Pi}{\partial r} - \frac{\tilde{\Pi}}{r} &= E_0(r),\\ \frac{\partial \tilde{\Pi}}{\partial r} - \frac{\Pi}{2} &= -E_0(r).\end{aligned} \tag{1.28}$$

The solution of system (1.26)–(1.28) is represented in the form of an exponential depending on z and Bessel functions of order m [20]. The numbers n of a mode determining zeroes over coordinates r and z depend on m. The structure of modes depends on relation between n and $|\zeta|\kappa a$. When $n \gg |\zeta|\kappa a$, the E and H modes are excited, in the opposite case – superposition modes are excited. If $|\zeta|\kappa a \ll 1$, the entire mode composition is divided into the E and H modes. In the opposite limit, only highest modes are divided into the E and H modes. Each of the modes is characterized by its own attenuation coefficient. It usually increases with increasing n. Note here that $n = \kappa a \varphi$, i.e. the higher mode corresponds to the higher convergence angle of the incident laser beam. Thus, the absorption coefficient increases with increasing the convergence angle, which is observed in Fig. 1.8. For the same convergence angles, the attenuation coefficient of the E modes always exceeded that of the H modes. Note that at small converging angles for $\kappa a \zeta \gg 1$, mixed waves with the attenuation coefficient smaller than α_E are generated.

To perform quantitative comparison with calculations, a special experiment with the selected convergence angle was conducted. The mode composition and transmission of laser radiation were calculated numerically by (1.26), (1.28). The field at the input to a cylindrical tube was the result of diffraction of a linearly polarized wave from a circular aperture. The value of ζ was determined experimentally by measuring the absorption coefficient of the capillary material. The capillary transmission calculated numerically is presented in Fig. 1.9 for two converging angles [22]. In this case, $\zeta \kappa a \gg 1$, so that mixed waves with the attenuation coefficient that is considerably smaller than α_E correspond to small angles. According to (1.22), the attenuation length of the E mode under these experimental conditions is 2.5 cm, which is considerably smaller than attenuation lengths presented in Fig. 1.9. The agreement of calculations with the experiment confirms the validity of model (1.26). Note in conclusion that processes considered above are important

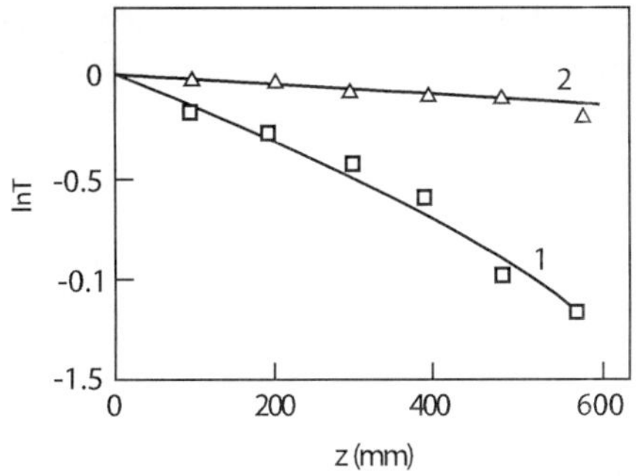

Fig. 1.9 Dependences of transmission on the length of a copper capillary of radius $a = 0.9$ mm for $\zeta = 0.02$–0.08. The capillary input is irradiated by a beam formed by a circular aperture. (1) $\varphi_1 = 0.019$, $\varphi_2 = 0.026$, $r_0 = 15$ mm, $\Delta r = 5$ mm; (2) $\varphi_1 = 0{,}003$, $\varphi_2 = 0$, $r_0 = 0$, $\Delta r = 2.5$ mm

for long-wavelength radiation, e.g., for practically important radiation of a CO_2 laser. For the wavelengths not exceeding 1 μm, a great part of radiation is absorbed already after the first reflection of beams from keyhole walls.

1.3 Physical Processes on the Surface of Condensed Media: The Interaction of Vapor with the Surrounding Gas

Consider processes on the surface of solids, which are important for laser processing of materials, at different radiation intensities. At low intensities, the surface will be only heated; at higher intensities, melting occurs[2] and as the radiation power density is further increased, evaporation begins. Surface heating processes are quite simple if the temperature dependence of the absorption coefficient is neglected. These questions will be discussed in detail in the consideration of laser hardening of materials in Chap. 2. Physical processes occurring during melting are more various.

1.3.1 Melting

One of the widespread processes is convection caused by gravitational forces in liquid heated from bottom (Benard convection). Because a sample is usually heated by a laser beam from above, such convection does not appear in this case. As was shown theoretically in [23], the Benard convection in a liquid layer can also appear (in the absence of the gravitational force) due to the temperature dependence of the surface tension. In this case, the sign of the temperature gradient is not important, i.e. convection appears upon heating from above as well [24]. Effects related to the liquid motion near the interface and caused by the dependence of the surface tension on temperature or impurity concentration are called Marangoni effects [25]. Thermocapillary convection in a thin liquid layer in the presence of the longitudinal temperature gradient was calculated in [26]. The motion of air bubbles in liquid caused by the Marangoni effect was observed in [27].

Based on the Navier–Stokes equations and the heat conduction equation (see Sect. 2.2), it was shown in [24] that, aside from usual gravitational-capillary waves $\omega_0 = (gk + \sigma k^3/\rho)^{1/2}$ [28], another type of undamped surface waves with the dispersion law of the sonic type $\omega = kc$, where $c^2 \sim \mathrm{d}T/\mathrm{d}z$ (thermocapillary waves) also exist in a liquid layer heated from above. The maximum increment γ_M is achieved in the resonance $\omega_0(k) = ck$, i.e. $k = k_1 = q|\partial\sigma/\partial T|/\kappa\sigma$, where σ is the surface tension coefficient and q is the radiation intensity absorbed by the surface. The value of γ_M is given by the expression

---

[2]Note that upon heating, a sample, remaining solid before melting, can experience a number of phase transitions; some such phase transitions during laser hardening are considered in Sect. 2.1.

$$\gamma_M \simeq \omega_0(k_1)/2$$

For $q \simeq 10^5\,\mathrm{W/cm}^2$, for steel $k_1 \simeq 10^2\,\mathrm{cm}^{-1}$ $\gamma_M \sim 10^3\,\mathrm{s}^{-1}$. Thus, the increment of this instability is quite large, and the instability can be observed during the melting of the sample by laser pulses of duration of a few milliseconds. It is possible that such instability was observed in experiments [29], where solidified waves are observed on the sample surface. In addition, such instability causes the mixing of liquid, which is important for laser alloying. The practical significance of this phenomenon is not clear enough because the nonlinear stage of the process has not been studied.

The Marangoni effect can be manifested when the limited size of the beam diameter is taken into account. In this case, the motion occurs at a large scale of the order of the light spot diameter on a target. This motion will be considered in more detail in Sect. 2.2.

1.3.2 Vaporization

Vaporization of materials irradiated by a focused laser beam takes places in many technological processes such as welding, drilling and cutting. Because the material vapor pressure rapidly increases with increasing the surface temperature, vaporization also appears very rapidly. Each material can be characterized by its vaporization threshold. If a sample is heated slowly enough, so that the heating time $t_h \gg r_f^2/4\chi$, the target surface temperature can be considered a function of the radiation intensity:

$$T_s = \alpha q r_f/\kappa$$

The vaporization threshold q_b can be found from the condition that the sample surface is heated up to the boiling temperature T_b:

$$\alpha q_b = \kappa T_b/r_f \tag{1.29}$$

Upon pulsed heating, the radiation intensity q_b producing boiling of the sample material at the pulse end can be found by replacing r_f by $2\sqrt{\chi\tau}$ in (1.29):

$$\alpha q_b \simeq \kappa T_b/2\sqrt{\chi\tau} \tag{1.30}$$

If $q \gg q_b$, the laser beam energy is completely spent to vaporize the sample material and the so-called vaporization wave is produced [30].

Due to the mass removal, the melt level under the laser beam lowers. The lowering velocity v_e (the vaporization wave velocity) can be easily found from the law of conservation of energy and mass:

$$\alpha q = L_b \rho v_e \tag{1.31}$$

where L_b is the enthalpy of the material unit mass at the boiling temperature (vaporization enthalpy). Generally speaking, this expression is valid for a broad beam, when the flowing of liquid in the lateral direction can be neglected (see Sect. 8.1).

1.3.3 Melting–Solidification Dynamics Taking Vaporization into Account

Formula (1.31) is a limiting expression for this case, which is valid within a long time after the onset of boiling. The boiling onset time depends on the radiation intensity and can be estimated from (1.30). To consider all the stages of heating, melting, vaporization, and passage to the stationary regime of the vaporization wave, it is necessary to solve the system of nonstationary heat conduction equations for a solid and melt:

$$\frac{\partial^2 T_l}{\partial x^2} = \frac{1}{\chi_l}\frac{\partial T_l}{\partial t}, \quad \frac{\partial T_s}{\partial x^2} = \frac{1}{\chi_s}\frac{\partial T_s}{\partial x^2} \tag{1.32}$$

$$q_0 = -\kappa_l \left.\frac{\partial T_l}{\partial x}\right|_{S_2} + \rho_l H_b \frac{\mathrm{d}S_2}{\mathrm{d}t} \tag{1.33}$$

$$\kappa_l \left.\frac{\partial T_l}{\partial x}\right|_{S_1} - \kappa_s \left.\frac{\partial T_s}{\partial x}\right|_{S_1} = -\rho_s H_m \frac{\mathrm{d}S_1}{\mathrm{d}t}, \quad T_l(x = S_1) = T_s(x = S_1) = T_m \tag{1.34}$$

$$\frac{\mathrm{d}S_2}{\mathrm{d}t} = V_* \exp\left(-\frac{mH_b}{k_B T_l(S_2)}\right), \quad V_* = p_0/2\rho_l\sqrt{2\pi\kappa T_a/m}\,\exp(H_b m/k_B T_b) \tag{1.35}$$

where T_l and T_s are the temperatures of the melt and solid, respectively; x is directed from the sample surface inside it; m is the material atom mass; subscripts l and s refer to the parameters of the liquid and solid, respectively; and S_2 and S_1 are coordinates of the melt–vapor and melt–solid surfaces, respectively. The lowering velocity of the melt surface S_2 (1.35) is obtained from the Hertz–Knudsen vaporization law, where p_0 is the atmospheric pressure, and T_b is the boiling temperature. Figure 1.10 presents the time dependences of the melting (S_1) and vaporization (S_2) fronts of a titanium sample obtained by solving the system of (1.32)–(1.35) [41]. One can see that the dynamics of these fronts is different. The melting front velocity decreases and tends to a stationary value. The vaporization front velocity monotonically increases and also tends to a stationary value. A stationary vaporization-melting wave can appear only when the velocities of phase boundaries are identical. Note here that the melting depths taking into account vaporization and neglecting it are close. This is explained by the fact that the temperature gradient decreasing on the growing melting layer front is compensated due to a decrease in the melt thickness due to vaporization.

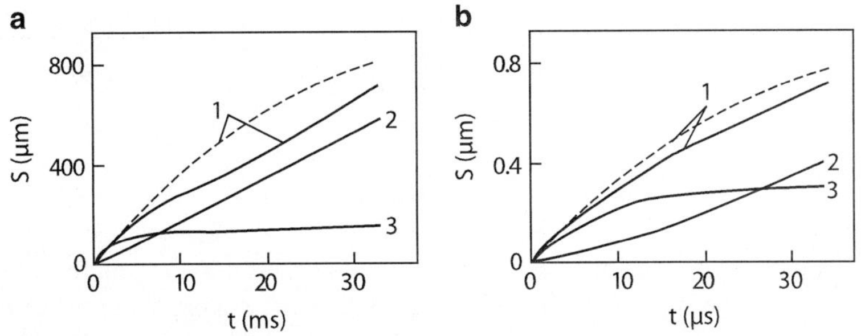

Fig. 1.10 Numerical calculations of thermal processes by laser irradiation $q = 10^5$ W/cm^2 (*left*), $q = 10^8$ W/cm^2 (*right*). Lines 1 are melting fronts, lines 2 are evaporation fronts, lines 3 are melt depthes. *Dash curves* are melting fronts without the evaporation

System (1.32)–(1.35) admits the self-modeling solution near the stationary state [58]:

$$v = q_0\{\kappa_s(T_m - T_0)/\chi_s + \kappa_l(T_a - T_m)/\chi_l + \rho(H_b + H_m)\}^{-1} \tag{1.36}$$

$$\Delta S = -\frac{\chi_l}{v}\ln\frac{\kappa_s(T_m - T_0)/\chi_s + \rho L_m}{\kappa_s(T_m - T_0)/\chi_s + \kappa_l(T_a - T_m)/\chi_l + \rho(H_b + H_m)} \tag{1.37}$$

$$T_a = -\frac{H_b m}{k_B}\ln^{-1}\frac{v}{v_*} \tag{1.38}$$

Comparison of the self-modeling solution with the results of numerical simulation shows that they coincide with an accuracy of 5%. For $T_0 \ll T_m$, expression (1.36) coincides with (1.31). The thickness ΔS of the melt layer is determined, as expected, by the thickness of χ_l/v of a thermal wave, and depends logarithmically on thermal parameters. Taking into account that $H_b m \gg k_B T_b$, one can see from (1.38) that the surface temperature T_a, as expected, is close to the boiling temperature T_b. As the laser radiation intensity is increased, the temperature T_a weakly (logarithmically) increases.

We considered above the simplest case of melting and vaporization of the surface of a thick metal sample. These processes can reveal interesting features when the interaction geometry is changed and a laser beam is focused in a small spot on a thin plate. In this case, the melt will be extruded on the edges of the melt pool under the action of vapor (see Chap. 8). Interesting properties can be observed when a thin metal plate is melted [58]. In this case, system (1.32)–(1.36) should be changed. Along with the energy loss due to vaporization, we introduce into (1.33) thermal losses described by the Stephan–Boltzmann law

$$q_\sigma = \varepsilon_1 \sigma T_a^4 \tag{1.39}$$

The same losses should be also introduced on the lower side of the plate:

$$-\kappa_s \frac{\partial T_2}{\partial x}\bigg|_{x=l} = \varepsilon_2 \sigma T_2^4 \tag{1.40}$$

Here, ε_1 and ε_2 are the radiation capacities of the upper and lower surfaces of the plate, respectively, and l is the plate thickness. Thermal radiation from the upper surface should be taken into account because the thermal story of the plate is considered after the laser switching off as well, when melting processes are changed by crystallization. Because radiation is focused into a small spot of radius r_f, temperature will change not only along the plate depth but also along its surface, i.e. along r. Therefore, equation (1.32) should be supplemented with a term describing heat transfer along the plate.

$$\kappa \frac{1}{r}\frac{\partial}{r} r \frac{\partial T}{\partial r}$$

In relation (1.34) at the melt–solid interface, derivatives and the front motion velocity normally to the curve $S_1(x, r)$ should be introduced. The conditions of the numerical experiment are chosen so that the plate thickness during irradiation decreases insignificantly due to vaporization. The most interesting melting-crystallization processes were observed, e.g., upon irradiation of a 0.27 mm-thick titanium plate. It was assumed that $\varepsilon = 0.25$, $r_f = 1$ mm, $q_0 = 1.5 \cdot 10^5$ W/cm^2, and $\tau = 7$ ms. Figure 1.11 presents the shapes of the melt pool after laser switching off. Because the melt surface temperature at the pool center by the pulse end exceeded the melting temperature by one and a half times, the melting boundary at the central region continued to move towards the plate bottom after laser switching off. At the same time, the peripheral region is cooled due to heat removal along the plate. The melt pool takes almost cylindrical shape. At later times, for $t > 30$ ms, surface thermal radiation losses begin to play a role. As a result, a liquid pool is produced inside the plate. The dynamics of crystallization fronts can determine the critical melting zone. Below, we will consider the two aspects of the vaporization process: the stability of a vaporization wave and the interaction of a vapor jet with the environment. The latter will be further used in the consideration of plasma phenomena near the surface.

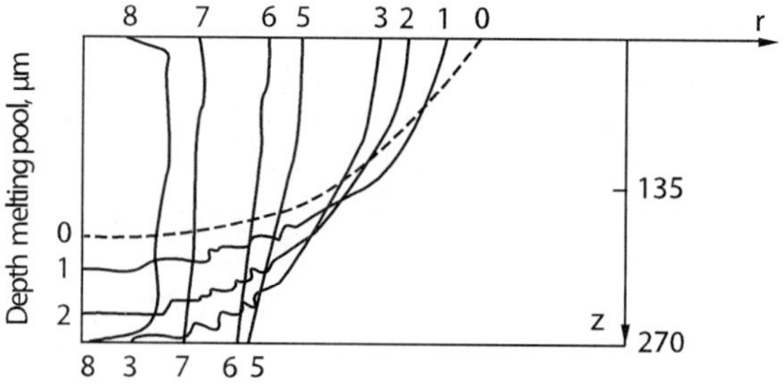

Fig. 1.11 Shape of a melt pool at different instants of time: (0) 7, (1) 8, (2) 10, (3) 12, (4) 14, (5) 20, (6) 25, (7) 35, (8) 40 ms

1.3.4 *Stationary Interaction of a Vapor Jet with the Surrounding Gas*

As was shown in the previous section, if $q \gg q_b$, the laser beam energy is completely spent to vaporize a material. The velocity of a vaporization wave propagating in the region of the material irradiated by the laser beam can be easily found from the mass expenditure $\dot{m} = \alpha q / L_b$ [see (1.31)]. Knowing the mass expenditure, it is easy to calculate the lowering rate v_e of the material level because the material density is known. As for the velocity of vapor, its determination is complicated because the density of the target vapor, depending on the pressure p_A of gas in which the target is located and the absorbed radiation intensity, is unknown. It is clear that for $q \geq q_b$, when $p_a \geq p_A$, vapor flies away from the target with a small subsonic velocity (p_a is the vapor pressure near the target). In this case, the vapor density near the target will be described by the expression $p_a \approx p_A$

$$\rho_a = p_A m / k_\mathrm{B} T_b \tag{1.41}$$

where m is the atomic mass of the target material. It is assumed in (1.41) that the gas temperature near the target surface is close to the boiling temperature T_b. Knowing ρ_a, we can determine from the law of conservation of mass and (1.31) the velocity v_1 near the target:

$$v_1 \approx \alpha q k_\mathrm{B} T_b / L_b p_A m \tag{1.42}$$

The values of ρ, v, and T change with distance from the target. We will discuss this question below, and now consider the behavior of the boundary values of gas-dynamic functions. According to (1.42), v_1 increases with increasing q and achieves the local sound speed $c_s(T_b)$ for $q = q_s$, where

$$\alpha q_s = L_b c_s p_A m / k_\mathrm{B} T_b \tag{1.43}$$

It is clear that as q is further increased, the gas velocity will no longer increase (for $T_s = T_b$) because the average velocity of atoms leaving the target surface cannot exceed the thermal velocity (see details in [30]). The law of conservation of mass flow for $q > q_s$ is satisfied due to the increase in the vapor density near the target:

$$\rho_a = \alpha q / L_b c_s \tag{1.44}$$

To find the surface temperature, it is necessary to recalculate the gas-dynamic values of v_1, ρ_a, and T_a to the corresponding values in the Knudsen layer on the target surface. For the supersonic flow, the recalculation expressions have the form [30]

$$T_a = 0.7 T_s \tag{1.45}$$

$$p_a = 0.2 p_s \tag{1.46}$$

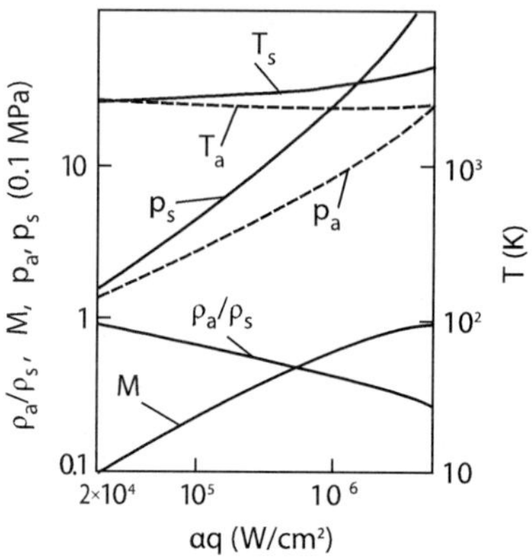

Fig. 1.12 Characteristics of the aluminum vaporization process in air

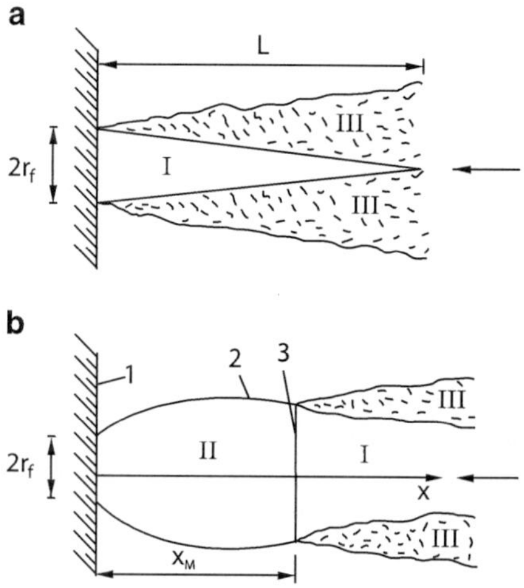

Fig. 1.13 Target vaporization process in air for subsonic (**a**) and supersonic (**b**) vapor flows; (1) target surface; (2) Barrel-shaped compression shock; (3) Mach disk; (I) Subsonic vapor flow region; (II) Supersonic vapor flow region; (III) Turbulent vapor flow region. The arrow shows the laser beam direction

where $p_s = p_0 \exp(-L_b/kT_s)$ is the saturated vapor pressure of the target material. In reality, the values of ρ, v and T_a change more smoothly with changing $q > q_b$. Figure 1.12 presents p_a, ρ_a, T_a, p_s, ρ_s, T_s and the Mach number M for an aluminum target [31]. One can see that for $\alpha q = 4\,\text{MW/cm}^2$, the flow near the target becomes supersonic. This value is well consistent with the value obtained for aluminum from expression (1.43). Thus, for $q < q_s$, the vapor flow is subsonic, while for $q > q_s$, it is supersonic.

Figure 1.13a shows the configuration of a subsonic flow [32, 33]. The estimate of the Reynolds number Re from the focal spot size gives the value 2,000–4,000, so that the subsonic jet flow is turbulent. It is known that the length L of a uniform central spot exceeds approximately by a factor of 9 the light spot radius r_f [34]. The thickness δ of the turbulent displacement region is determined by the turbulent kinematic viscosity ν_T: $\delta \approx \sqrt{\nu_T x/v}$. In the turbulent region, where turbulent layers are merged, the velocity and temperature decrease proportionally to x^{-1}.

The configuration of the supersonic flow is shown in Fig. 1.13b [32, 33]. The main characteristic of such a flow is the position of the shock-wave front (the Mach disc)

$$x_M/d \approx 0.6(p_s/p_A)^{1/2} \tag{1.47}$$

The pressure behind the shock-wave front increases and then slowly decreases down to p_A. The density in the supersonic jet considerably decreases:

$$\rho = \frac{\rho_a}{2}\sqrt{\frac{\gamma - 1}{\gamma + 1}\frac{d^2}{x^2}}, \quad \gamma = c_p/c_v \tag{1.48}$$

while the velocity almost does not change:

$$v^2 = v_1^2(\gamma + 1)/(\gamma - 1)$$

Of interest is the value of density behind the shock-wave front, where the flow is subsonic:

$$\rho_d = \rho(\gamma + 1)/(\gamma - 1) = mp_A/0.7\gamma kT_s \tag{1.49}$$

One can see from (1.49) that the density behind the shock-wave front weakly differs from the density near the target in the case of the subsonic flow.

Note that the properties described above are confirmed experimentally with good accuracy upon laser irradiation of a target in a gas medium [35]. During irradiation of graphite target, the Mach disc was observed. It was shown experimentally that $x_M \sim (q/p_A)^{1/2}$, which also follows from theoretical expressions (1.44), (1.46), and (1.47).

1.4 Vaporization Kinetics and Hydrodynamics

Above, by considering qualitatively the vaporization of the metal surface exposed to laser radiation, we observed the variation of the vapor velocity near the target from zero up to the sound speed. The analysis was performed for vaporization to the atmosphere, i.e. in a medium with counterpressure. In the case of subsonic velocities, the gas-dynamic parameters were determined by the metal surface temperature, while in the case of the sound speed, the temperature and pressure jump was introduced (1.45) and (1.46). It is clear that this jump appears not only for supersonic flows. It should increase smoothly from zero in the absence of the flow (i.e. in the case of the equilibrium) up to its maximum value determined by (1.45) and (1.46). Generally speaking, the jump itself appears due to the description of the problem in the hydrodynamic approximation (see, e.g., review [36]). In reality, the parameters of the problem drastically change at a distance of the order of the mean free path of atoms in the Knudsen layer. This problem is analogous, for example, to the concept of a thermal jump near the solid–gas interface in a strongly inhomogeneous temperature field (see, for example, [37]). The jump value in our case can be calculated similarly. We take into account that the velocity distribution function of vapor atoms flowing away from the metal surface is equal to the equilibrium Maxwell function at the metal surface temperature T_s:

$$f_M = n_s \left(\frac{m}{2\pi k T_s}\right)^{3/2} \exp\left(-\frac{mv^2}{2kT_s}\right), \ v_z > 0 \tag{1.50}$$

where $n_s = p_s/kT_s$ is the saturated vapor pressure. At a distance of the order of the mean free path from the surface, i.e. at the Knudsen layer boundary, the Maxwell velocity distribution function should be also established, but with the gas temperature T_a, density n_a, and hydrodynamic velocity u. The latter quantity is known only in the case of vaporization into vacuum or medium whose pressure is lower than the vapor pressure. Here, it is equal to the sound speed, i.e. the Mach number is $M = 1$. In other cases, the value of u is found by soling the gas-dynamic problem with boundary conditions n_a and T_a. Thus, the velocity distribution function at the Knudsen layer-gas interface has the form

$$f_r = n_a \left(\frac{m}{2\pi k T_a}\right)^{3/2} \exp\left\{\left[-(v_z - u)^2 - v_r^2\right] \Big/ \frac{2kT_a}{m}\right\} \tag{1.51}$$

It is reasonable to assume that a flow of particles incident on the metal surface is proportional to f_r [30, 38], i.e.

$$f_M = \beta_1 f_r, \ v_z < 0 \tag{1.52}$$

This assumption can be verified by solving a kinetic problem, which we will consider below. Then, it is necessary to satisfy the laws of conservation of particles, momentum and energy in the Knudsen layer. We obtain three equations for three unknowns n, T, and β. Figure 1.14 presents the results of numerical calculations of these equations depending on u or the Mach number. One can see that for $M = 1$, expressions (1.45) and (1.46) are obtained. This figure also presents the results obtained by other methods. Method 2 assumes the different velocity distribution function for atoms returning to the surface, which takes into account the influence of the parameters of a target on this function [39]. Model 3 assumes the dependence of this function only on the surface parameters T_s and n_s. All these functions are presented in Fig. 1.15. One can see that they most strongly differ from each other and numerical expression [40] in the region of small velocities. However, the moments of these functions, i.e. gas-dynamic parameters are close, which is well demonstrated in Fig. 1.14. Model 3 was used to study the dependence of the jumps of parameters in the Knudsen layer, when a part of atoms in vapor incident on the surface are reflected elastically or inelastically with the probability $1 - \alpha$ [48]. Here, α is the so-called accommodation coefficient. It is clear from general considerations that these atoms reduce the flow of atoms returning to the surface, thereby increasing jumps of n, T, and p on the Knudsen layer. This is especially

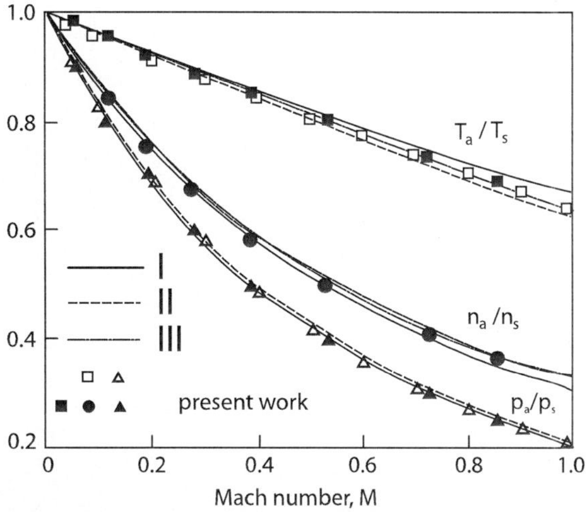

Fig. 1.14 Gas-dynamic boundary conditions for vaporization: the normalized temperature T_a, density n_a, and pressure p_a behind the Knudsen layer as functions if the Mach number M. Comparison of the numerical results of the presented model, the numerical results of Sone and Sugimoto [36], and analytic models I, II, and III (*curves*)

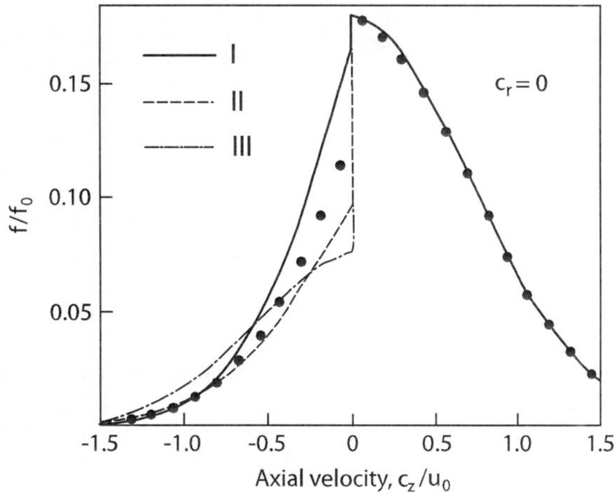

Fig. 1.15 Comparison of the numerical calculations of the velocity distribution function (*points*) and analytic models I, II, and III (*curves*) during vaporization for $p_a / p_s = 0.3$

pronounced for pressure Fig. 1.16. This figure also presents comparison with the case of diatomic vapor, which has a greater heat capacity than monatomic gas. One can see that the heat capacity weakly affects the pressure jump. A different situation is observed for the temperature jump Fig. 1.17. The jumps of the parameters are calculated based on the equality of flows. The expression for the equality of energy flows contains the heat capacities of vapors, and therefore the temperature jump will be smaller for vapor with the higher heat capacity. One can see from Fig. 1.16 that, as α decreases by half, pressures decreases approximately in the same way. Below, we will consider the inverse process of condensation and will show that the dependence of the jumps of parameters in the Knudsen layer on α is much stronger.

Thus, knowing the relation between the parameters of the gas and target at their interface, we can solve the problem of a sample heating in the case of developed vaporization. The propagation of heat in a solid in a moving coordinate

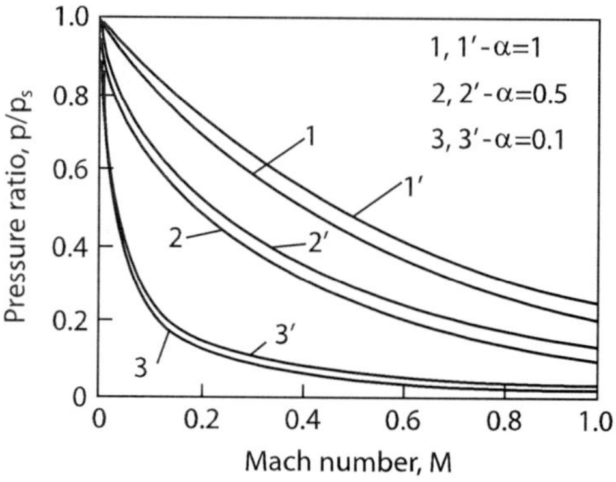

Fig. 1.16 Normalized pressure behind the Knudsen layer as a function of the Mach number for monoatomic (1, 2, 3) and polyatomic (1′, 2′, 3′) vapors

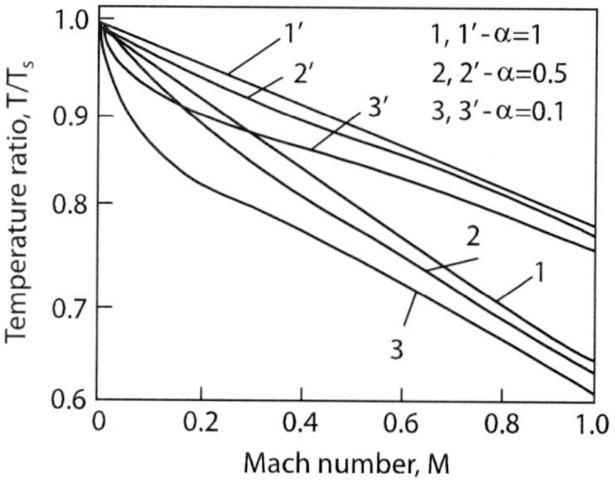

Fig. 1.17 Normalized temperature behind the Knudsen layer as a function of the Mach number for monoatomic (1, 2, 3) and polyatomic (1′, 2′, 3′) vapors

system coupled with the evaporating surface can be conveniently described for the enthalpy h

$$\frac{\partial h}{\partial t} + u_s \frac{\partial h}{\partial z} = \frac{\partial}{\partial z} \kappa \frac{\partial T}{\partial z} \tag{1.53}$$

where u_s is the velocity of the target-vapor interface, $u_s > 0$ if the target is evaporated and $u_s < 0$ if condensation proceeds. A change in the heat conductivity during melting is taken into account. The heat of melting can be taken into account by introducing enthalpy; if it is very small. then $h = c\rho T$. The continuity of the energy flow should be taken into account at the metal-gas interface. If the problem is considered within the framework of gas-dynamics, the relation

$$-\kappa \frac{\partial T_s}{\partial z} = q - E(n_a, T_a) \tag{1.54}$$

is used, where $E(n_a, T_a)$ is the heat flow carried away from the surface by vapor. Its value depends finally on n_s, T_s, and the Mach number M. In most of the problems of laser technology the absorbed power density q depends on time and coordinates. If vaporization occurs into vacuum or a medium with pressure that is considerably lower than the saturated vapor pressure at the target surface temperature, then

$M = 1$ and problem (1.53), (1.54) becomes closed and, correspondingly, solvable. However, here "underwater stones" can be encountered. It is necessary to be sure that the saturated vapor pressure at each moment exceeds the vapor pressure over the melt surface. As we saw above, this is a necessary requirement of the Anisimov–Knight model. For example, when the laser radiation intensity is drastically reduced, T_s will decrease so that this condition can be violated and condensation begins. Note that, if the decay time of the laser pulse intensity exceeds the unloading time due to the lateral expansion of vapor, we can assume that $M = 1$ when it is well known that the saturated vapor pressure exceeds the atmospheric pressure. This allows one to solve problem (1.53), (1.54) without the calculation of the dynamics of vapor. In this sense, two-dimensional problems can be simpler than one-dimensional ones, where the lateral unloading effect is absent.

It was pointed out above that relation (1.52) is generally speaking a priori. To substantiate it, a kinetic problem should be solved and compare the result with the gas-dynamic approach [42]. By solving the general problem of the action of a laser pulse on a target, it is necessary to find the energy flow leaving the surface at $z = 0$ [see (1.54)] by integrating the distribution function for $z = 0$:

$$E = \int v_z \left(H_b + \frac{mv^2}{2} \right) f \, d\mathbf{V} \tag{1.55}$$

where H_b is the heat of vaporization of target atoms. To find the distribution function on the metal surface, it is necessary to solve the kinetic equation. Such kinetic problems are often solved by using the collision term in the simplified BGK (Bhatnagar–Gross–Krook) form [43]:

$$\frac{\partial f}{\partial t} + V_z \frac{\partial f}{\partial z} = \nu_0 (f_e - f) \tag{1.56}$$

where ν_0 is the relaxation frequency of the distribution function to the equilibrium function (1.51). The relaxation frequency in this model is found by equating heat conductions in the BGK model and the Chapmen–Enskog model of solid spheres [44]:

$$\nu_0 = \frac{32}{15} n d^2 \sqrt{\frac{\pi \kappa T}{m}} \tag{1.57}$$

As the boundary condition for (1.56) at $z = 0$, we use the semi-Maxwell distribution (1.50). Now the system of equations (1.50), (1.51), (1.55)–(1.57) is closed. The problem was solved numerically with the initial conditions $T(0) = 298\,\text{K}$ and $f(0) = 0$. The calculation conditions were selected close to experimental conditions [45]. The laser pulse was bell-shaped, with the FWHM equal to 11 ns. Thermal parameters of a gold target are presented in Table 1.2. The results of calculations are presented in Fig. 1.18. One can see from Fig. 1.18a that the target surface temperature at the pulse trailing edge achieves the boiling point and then rapidly decreases down to the temperature at which vaporization ceases. Figure 1.18b

Table 1.2 Thermal parameters of gold

Heat capacity (solid)	28.8	J/(mol K)
Heat capacity (liquid)	31.3	J/(mol K)
Heat conduction (solid)	284	W/(m K)
Heat conduction (liquid)	155	W/(m K)
Melting temperature	1,336	K
Boiling temperature	3,150	K
Melting heat	12.6	kJ/mol
Boiling heat	331	kJ/mol
Atom diameter	0.236	nm

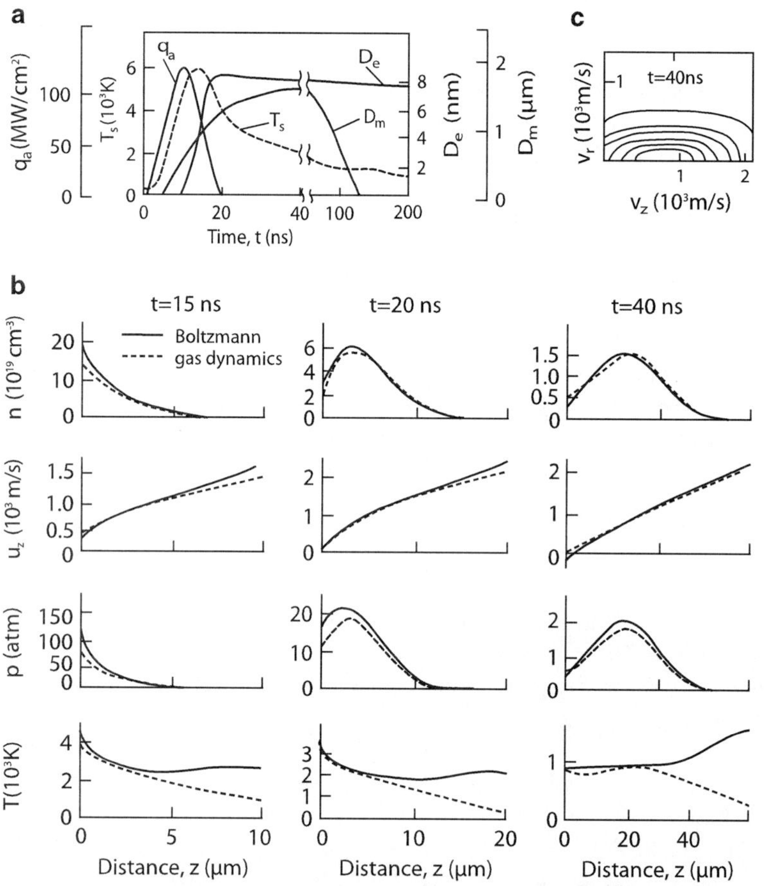

Fig. 1.18 The laser impulse form q_a, T_s is temperature of surface, D_e is evaporation depth, D_m is melt depth (**a**). The gas parameters vs z at different times (**b**). Velocity distribution function vs v_z and v_r (c).

presents a comparison of the results of kinetic calculations with calculation of usual gas-dynamic equations with the Anisimov–Knight boundary conditions at the solid–gas interface. The difference between the results of these calculations, as expected, are observed at distances from the surface no more than 1 μm. For $t = 15$ ns, the

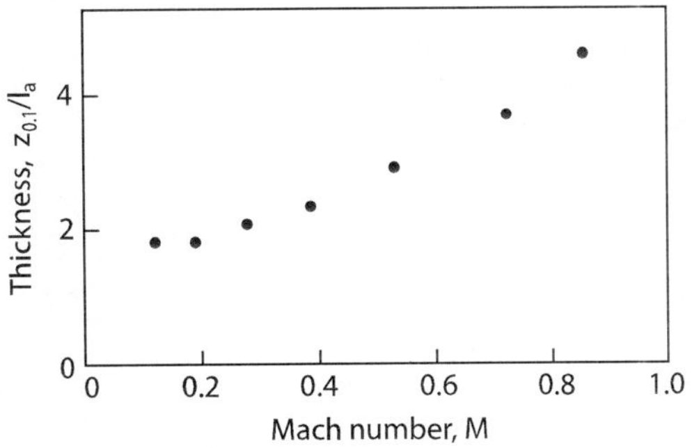

Fig. 1.19 Normalized thickness of the quasi-stationary Knudsen layer as a function of the Mach number during vaporization

mean free path l near the surface does not exceed 1 μm. Figure 1.19 presents the dependence of the Knudsen layer thickness on the Mach number. One can see that the Knudsen layer for the maximum value of M increases up to four mean free paths [40]. As expected, the strongest difference was observed between pressures, which also correspond to (1.46). At large distances $z > l$, where the hydrodynamic description can be applied, all parameters of a vapor cloud, calculated by different methods, coincide. In the cloud tail, at distances exceeding 5 μm, the mean free path becomes greater than the vapor cloud size. In this case, the kinetic and hydrodynamic approaches give substantially different results. As the vapor cloud expands in time, the region of coincidence of kinetic and hydrodynamic results also expands. It is interesting to observe the change in the velocity distribution of atoms with expanding the vapor cloud. This function is isotropic near the target (or at very low densities) and elongated in the expansion direction away from the target (Fig. 1.18c). These properties are important for PLD processes, which will be discussed in next chapters. Figure 1.18b demonstrates the limitation of the hydrodynamic model at $t = 40$ ns. One can see that the Anisimov–Knight condition is not fulfilled near the surface, and the density and pressure exceed their kinetic values. Condensation takes place, and the average velocity is negative. In this case, a comparison with experiment is also required because the kinetic equation is only a model one. Figure 1.20 compares the velocity and angular distributions of particles. Good agreement with experimental data is observed by selecting appropriately the laser pulse energy. The difference between distributions at large angles can be explained by the two-dimensional nature of vaporization in experiments due to a small focal spot.

1.4.1 Condensation

This process differs considerably from the vaporization process. In both processes, the relation between the saturated vapor and gas pressures is crucial. If $p_s/p_a > 1$, vaporization takes place, if $p_s/p_a < 1$, condensation occurs. However, as we saw, the gas temperature at the boundary during vaporization is a unique function of the

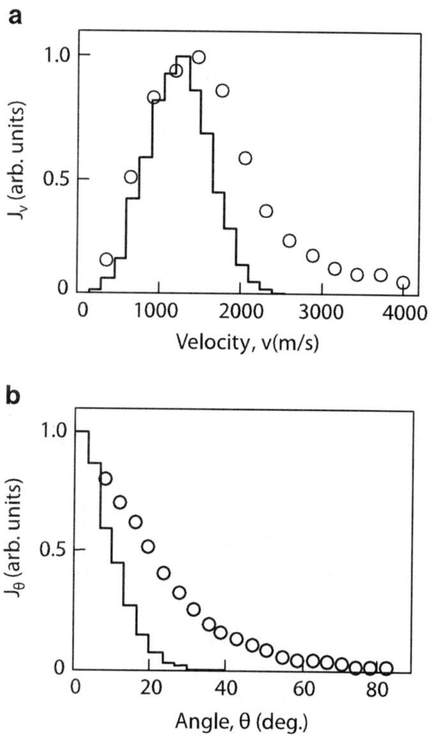

Fig. 1.20 Comparison of the calculated velocity $[J_v(v)]$ and angular $[J_\theta(\theta)]$ distributions of sputtered gold atoms (step lines) with experimental distributions (*points*) [45]: (**a**) velocity distributions averaged over the angular range from 5 to 29 degrees; (**b**) angular distribution averaged over the velocity range from 1,100 to 1,600 m/s. The calculated and experimental results are obtained for a 11 ns laser pulse. Other important parameters are following $E = 4.9$ J/cm^2 is the fluence of incident laser radiation; $E_\alpha = 1.5$ J/cm^2 is the absorbed laser fluence used in calculations; $t = 40$ ns is the time at which the distributions were calculated

Mach number. Condensation can occur at different relations between the gas and surface temperatures $T_s > T_a$ and $T_s < T_a$. This is demonstrated in Figs. 1.21 and 1.22, where $T_a/T_s = 0.2$ and $T_a/T_s = 4$. In both cases, the Mach number increases with increasing pressure, i.e. condensation is accelerated. However, the passage to the sonic flow occurs at different p_s/p_a. This is well demonstrated in Fig. 1.22, where the gas pressure at which $M = 1$ increases by a fact or of three with increasing wall temperature. However, these pressure variations are negligible compared to temperature variations. These figures present the results of analytic models based on the same method as vaporization models [38, 39]. The models differ in the choice of the form of velocity distribution functions for vapor atoms flying to the sample surface. The analytic expression for the dependence of p_a/p_s on T_a/T_s obtained in one of the models is too cumbersome and is not presented here.

Note in conclusion that the diffuse reflection of atoms from the target surface considerably affects the boundary conditions of condensation [46]. This process is characterized by the condensation coefficient α. For $\alpha = 1$, all the atoms are thermalised on the target surface. For $\alpha \neq 1$, an addition proportional to $(\alpha - 1)$ is introduced into distribution function (1.50). Calculations show that these reflected atoms considerably reduce the condensation rate of vapor, Fig. 1.23. Figure 1.24

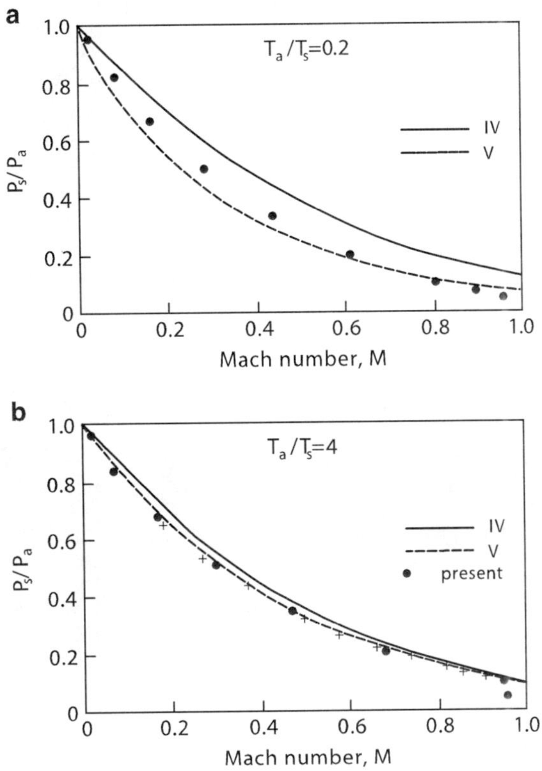

Fig. 1.21 Inverse pressure ratio p_a/p_s (**a**) as a function of the Mach number during condensation for $T_a/T_s = 0.2$ (**a**), $T_a/T_s = 4$ (**b**). *Circles* are numerical calculations; the *solid and dashed curves* correspond to models IV and V

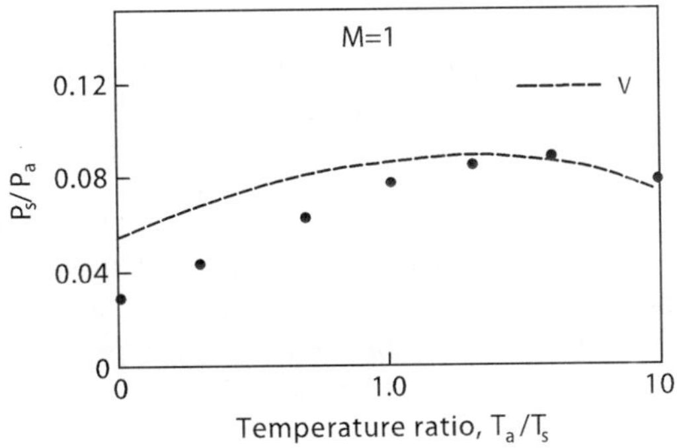

Fig. 1.22 Inverse pressure ratio p_a/p_s versus the temperature ratio T_a/T_s under sonic condensation (with $M = 1$): *points*, the numerical calculations; *broken curve*, Model V

[40] shows that the value of α determines the regions of temperature T_a/T_s where condensation can occur at the sound speed. Note that a small deviation of α from unity considerably affects the vapor condensation process. Thus, this circumstance should be also taken into account in the formulation of the pulsed ablation problem.

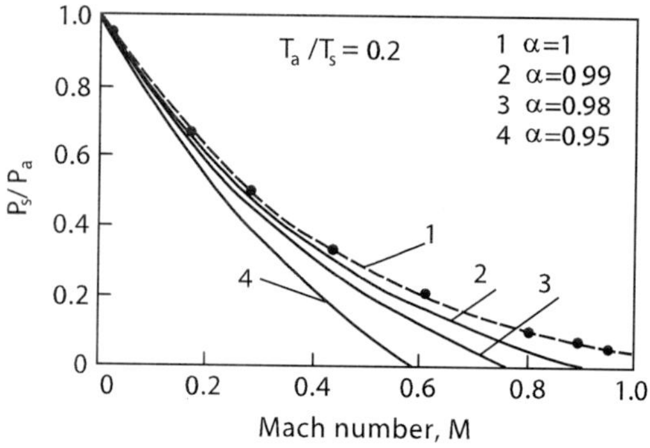

Fig. 1.23 Influence of the condensation coefficient on gas-dynamic conditions during condensation. The inverse vapor pressure ratio p_a/p_s as a function of the Mach number for $T_a/T_s = 0.2$ and different condensation coefficients. Circles are numerical values for $\alpha = 1$

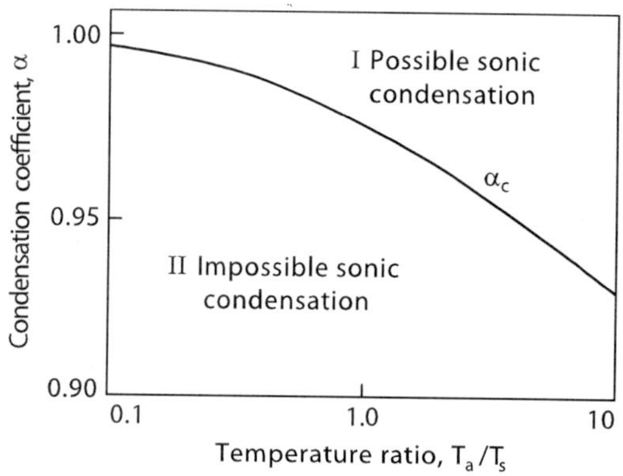

Fig. 1.24 The critical value of the condensation coefficient α_c as a function of T_a/T_s. Curve α_c divides two regions where sonic condensation is possible and impossible

1.5 Instability of the Laser-Induced Vaporization of Condensed Media

In the previous section, we considered effects taking place due to the nonequilibrium state of a condensed medium–vapor system. It is well known that in nonequilibrium media the instabilities often appear, resulting in the excitation of nonuniform oscillations with amplitude rapidly increasing in time. In this sense, the situation with the laser-induced vaporization of the condensed medium is very typical. Vaporization is an example of the first-order phase transition whose front is often unstable and has been long studied. Recall, for example, the formation of ice patterns during water freezing.

As for instability during vaporization, note that although this problem has been discussed in many papers (see, for example, [47] and references therein) the role of theoretically predicted instabilities in laser technology processes still remains unclear. This is explained by the fact that theories are restricted by the linear approximation when the oscillation amplitude is small. In addition, this problem is poorly studied experimentally.

For this reason, we will not consider in detail individual cases in the discussion of the theory of vaporization instabilities, as has been done before, and will restrict themselves to the general description and brief conclusions [48]. We assume for simplicity in the formulation of the problem that all the thermal parameters are independent of the liquid-solid target temperature and phase composition. The heat of melting is also neglected.

The unperturbed vaporization front in the laboratory coordinate system moves at the velocity v in the negative direction along the z axis in the condensed medium, which is described by the continuity, Euler, and heat conduction equations for an incompressible liquid with the constant thermal diffusivity χ, heat capacity c and density ρ_l. In a system coupled with the vaporization front, the equations for determining the velocity v, pressure p_l, and temperature T_l have the form

$$\begin{aligned}
&\frac{\partial \rho_l}{\partial t} + \rho_l \mathrm{div} v = 0, \\
&\frac{\partial v}{\partial t} + (v\nabla)v = -\frac{1}{\rho_l}\nabla p_l, \\
&\frac{\partial T_l}{\partial t} + (v\nabla)T_l = \chi\nabla^2 T_l + \frac{\alpha q}{c\rho_l}\mathrm{e}^{\alpha z}.
\end{aligned} \tag{1.58}$$

It is assumed here that the absorbed intensity q decreases according to the Bouguer law deep from the front surface $z = 0$ with the absorption coefficient α. Interference effects related to the diffraction of incident radiation from the inhomogeneities at the interface are not considered.

In the region $z > 0$, the evaporated material is located, which is treated as an ideal gas with the specific-heat ratio $\gamma = c_p/c_v$ and is described by a system of gas-dynamic equations of continuity, Euler and adiabaticity

$$\begin{aligned}
&\frac{\partial \rho}{\partial t} + \mathrm{div}\rho U = 0, \\
&\frac{\partial U}{\partial t} + (U\nabla)U = -\frac{1}{\rho}\nabla p, \\
&\left(\frac{\partial}{\partial t} + (U\nabla)\right) p\rho^{-\gamma} = 0.
\end{aligned} \tag{1.59}$$

where U, p, and ρ are the vapor velocity, pressure, and density, respectively. The quantity $p\rho^{\gamma}$ is the entropy of the ideal gas.

The region near the plane $z = 0$, as shown above, requires the kinetic description. Because the size of this region is of the order of the mean free path, for perturbations with a long enough wavelength it can be considered as a discontinuity where the laws of conservation of mass, momentum, and energy flows are formulated:

$$g_1 = \rho_l v_n = \rho_a U_n$$
$$g_2 = p_l + \rho_l v_n^2 = p_a + \rho_a U_n^2 + \sigma \nabla n \tag{1.60}$$
$$g_3 = g_1 \left(L_l(T_s) + \frac{v^2}{2} \right) - c\rho_l \chi \frac{\partial T_l}{\partial n} = g_1 \left(L(T_a) + \frac{U_1^2}{2} \right)$$

The condition of the equality of tangential components of the velocity has the form

$$v_\tau = U_\tau \tag{1.61}$$

In addition, two relations (vaporization boundary conditions)

$$\rho_a = \rho_a(T_s, M), \; p_a = p_a(T_s, M) \tag{1.62}$$

are satisfied, which take into account the nonequilibrium relaxation in the Knudsen layer (see Fig. 1.14), where $L_l(T_s)$ and $L(T_a)$ are the enthalpies of the condensed and gas phases depending on the surface temperature T_s and vapor temperature T_a, respectively.

The unperturbed stationary solution of system (1.58), (1.59) supplemented with boundary conditions for the heat conduction equation and $T_l(z = -\infty) = T_\infty$ has the form

$$v_x = 0, \quad v_z = v = const, \quad p_l = const,$$
$$T_l = T_\infty + \Delta T e^{k_0 z} + \frac{q}{c\rho_l \chi} \frac{1}{k_0 - \alpha} (e^{\alpha z} - e^{k_0 z}), \tag{1.63}$$
$$U_x = 0, \quad U_z = U = const, \quad p = const, \quad \rho = const,$$

where $\Delta T = T_s - T_\infty$ and $k_0 = v/\chi$. The surface temperature T_s is determined by the energy balance $q = \rho_l v[H_{\rm ne} + c\Delta T]$ in the condensed phase, in which the nonequilibrium heat of vaporization $H_{\rm ne} = H(T_s) + c_p(T - T_s) + U^2/2(1 - \rho^2/\rho_l^2)$ is introduced. Here, $H(T_s) = L(T_s) - L_l(T_s)$ is the equilibrium heat vaporization defined as the difference of enthalpies at the surface temperature T_s. The difference of $H_{\rm ne}$ from $H(T_s)$ is related to the consideration of the jump of thermodynamic parameters on the Knudsen layer and the motion of the condensed material and vapor. The term ρ^2/ρ_l^2 will be neglected below. The values of v, p_a, U, p and ρ are determined from the laws of conservation of mass and momentum flows and vaporization boundary conditions (1.45) and (1.46) for the specified Mach number $0 < M \leq 1$ on the external side the Knudsen layer.

Surface absorption is obtained from (1.63) by passing to the limit $\alpha \to \infty$. In this case, the right-hand side in the heat conduction equation in (1.58) is transformed to the singular delta function. Such a transformation of system (1.58) is equivalent to the passage to the homogeneous thermal diffusivity equation and to a new boundary condition, which contains, additionally to terms in (1.60), the radiation intensity q absorbed on the surface. These two different types of boundary conditions can be obtained from the solution for temperature in (1.63), which gives two different expressions for $\partial T_l/\partial z$ at $z = 0$ depending on the differentiation order and passage to the limit $\alpha \to \infty$.

The stability of system (1.58)–(1.63) can be studied by standard methods by linearizing equations (1.58) and (1.59) and boundary conditions (1.60)–(1.62). The dependence of the perturbed physical quantities on time and the transverse coordinate x is assumed, as usual, exponential: $\exp(i\omega_1 t + \omega_2 t - ikx)$. The modulation $\delta T_s = \delta T_e(0) + \xi \partial T_e/\partial z$ of the surface temperature takes into account its displacement ξ. The linear equations obtained for perturbed quantities can be solved due to a simple form of unperturbed stationary solution (1.63). The corresponding integration constants can be found from linearized boundary conditions (1.60)–(1.62) for $z = 0$. The condition of solvability of the obtained homogeneous system of equations gives the dispersion equation for the perturbation of the stationary vaporization front. All the calculations are omitted because they are too cumbersome. The algebraic equation for a quantity containing ω has the twentieth power [48]. This equation was analyzed numerically by using the thermal parameters for liquid silicon [49]:

$$
\begin{aligned}
&\rho_l = 2.52\ \mathrm{g/cm^3}, \quad \chi = 0.4\ \mathrm{cm^2/s}, \quad c = 1.05 \times 10^7\ \mathrm{erg/gK}, \\
&\sigma = 300\ \mathrm{erg/cm^2}, \quad L = 137 \times 10^9\ \mathrm{erg/g}, \quad m = 7.7 \times 10^{-23}\ \mathrm{g}, \\
&p_z = n_s k T_s = p_b \exp(\eta(1 - T_b/T_s)), \quad \eta = 12.7, \quad p_b = 10^6\ \mathrm{din/cm^2}, \\
&T_\infty = 300\ \mathrm{K}, T_b = 3514\ \mathrm{K}.
\end{aligned}
\tag{1.64}
$$

In the limiting case of $M = 1$, gas dynamics does not affect processes proceeding on the target, and instability is caused by temperature perturbations [50]. For surface absorption, this case is described by curve 1 in Fig. 1.25 for $T_s = 1.5 T_b$. In this case, the target vaporization rate is 74.2 m/s and the absorbed intensity is $q = 3.44\ \mathrm{MW/cm^2}$. The instability is developed at long wavelength up to a few microns, with the increment 0.5 MHz. As the radiation intensity is decreased, the Mach number decreases; for $q = 3.43$, 3.4, and 3.35 MW/cm^2, we have $M = 0.9$, 0.8, and 0.7, respectively. First the decrease in M causes the decrease in the increment and the narrowing of the instability region. But beginning from $M = 0.7$, when gas-dynamic perturbations become important [51], down to $M = 0.5$, the increment increases almost by an order of magnitude. Then it again decreases because the system tends to equilibrium for $M \to 0$, and instability should disappear Fig. 1.25b. For small Mach numbers and $\Delta T = 0$, the instability corresponds to the Darie–Landau instability [28] and a combustion wave [52]. In the case of bulk absorption of laser radiation, the increment and instability region increase compared to the case of surface absorption (see Fig. 1.26). Here, the nonmonotonic dependence of the increment on the Mach number is also observed, which is similar as a whole to the case of surface absorption (Fig. 1.25).

This instability can be clearly interpreted. It is explained by the fact that the temperature maximum for $\alpha^{-1} \neq 0$ is located inside a body under the irradiated

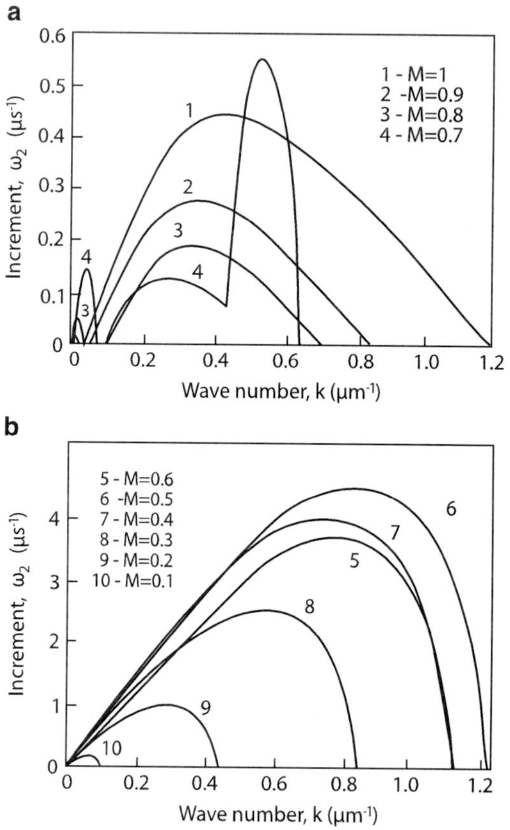

Fig. 1.25 Instability increment ω_2 as a function of the wave vector k for different Mach numbers (**a**), (**b**) (surface absorption, Si)

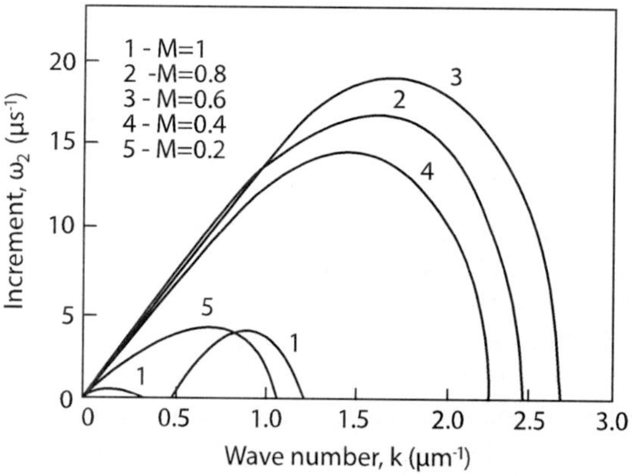

Fig. 1.26 Instability increment ω_2 as a function of the wave vector k. $M = 1$, $\alpha = 1\ \mu\text{m}^{-1}$, Si

surface. If the perturbation of the surface is a depression, the temperature in this site increases because this site proves to be located in a more heated zone. Due to the increase in temperature, the vaporization rate increases, the depression increases and this site is further heated. It is possible that such instability was observed in

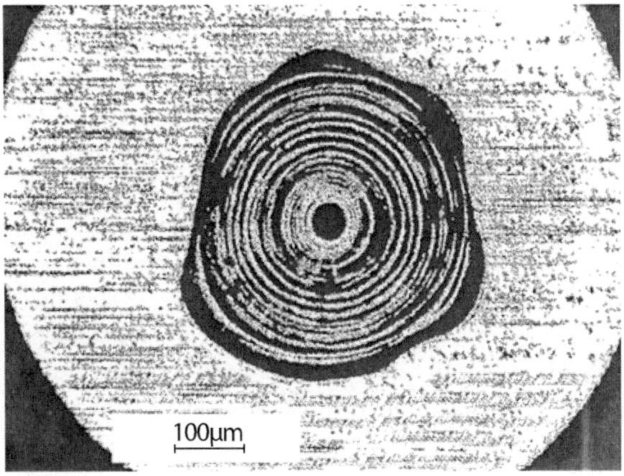

Fig. 1.27 Photograph of a spot produced on the back side of a 12 μm tick aluminum foil irradiated by $\tau = 10\,\mu s$, $E = 0.5\,J$ pulses from a CO_2 laser in vacuum

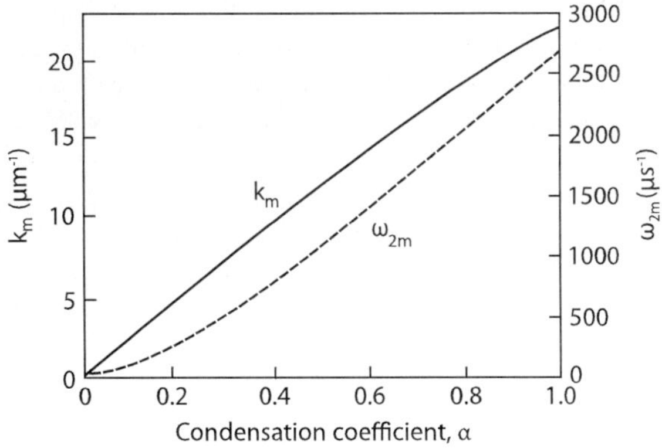

Fig. 1.28 Maximum value of ω_{2m} and corresponding k_m for water as functions of the condensation coefficient

experiments with a thin plate [53] (Fig. 1.27). The mechanism of this instability is similar to the instability of damage of metal wires heated by an electric current. In the latter case, the temperature maximum is also located inside metal [54].

By considering the kinetics of the Knudsen layer, we pointed out that the accommodation coefficient α considerably affects the value of the jump in gas-dynamic parameters. Numerical calculations show [48] that the instability of the vaporization front decreases with decreasing the accommodation coefficient (Fig. 1.28). In the numerical calculations presented above, all the three types of perturbations of heat, liquid, and vapor were considered together. The relative role of these perturbations was analyzed in detail in [47]. We illustrate this by two analytic expressions for instability increments. For the vapor velocity u_V smaller than the sound speed but close to it, we have

$$\gamma = k u_V (v_n/u_V)^{1/3}$$

Because $v_n/u_V = \rho_V/\rho_l \cong 10^{-5}$, $u_V \sim c_s = 10^5$ c/s, $\gamma \sim 2 \times 10^3 k$ we have $\gamma \approx 10^6\,\text{s}^{-1}$ for $k \cong 2\pi/r_f$, $r_f \approx 0.01$ cm. The thermal instability in the case of bulk absorption, mentioned above, is maximal for $k \simeq \alpha^{-1}$ [50]:

$$\gamma_T \sim H_b^2 q^2 / L^2 \rho_l \chi_l T_b^2$$

where H_b is the vaporization energy for an atom and χ_l is the thermal diffusivity of the melt. The absolute value of increments shows which laser irradiation regimes provide the stability of the phase transition front, which is important for many technological applications of laser ablation of solids. In particular, upon irradiation of strongly absorbing media, the front remains stable at times $T \leq 200$ ns, while upon bulk absorption, this time decreases down to $T \leq 50$ ns.

As mentioned in the beginning of this section, the experimental studies of the instability of the vaporization front are scarce. The situation considerably changes when samples are exposed to polarized radiation. Periodic structures are formed on the polished surface of strongly absorbing solids exposed to polarized laser radiation of certain intensities. These structures are caused by the interference of the incident wave with the waves excited due to the inhomogeneities of the boundary (see reviews [18, 47, 55]). A distinct relief with distances close to the laser wavelength was observed on the surfaces of metals, semiconductors, and dielectrics. However, although such structures can be easily produced, the possibility of their applications in laser technologies remains unclear. It was assumed that these structures can considerably increase the effective absorption coefficient of solids. Indeed, a strong increase in the absorption of optical radiation by a metal surface with a sinusoidal profile was experimentally measured in papers [56, 57]. However, these and other theoretical results were obtained for strictly periodic structures. The phases and amplitudes of real structures produced in experiments are correlated only within a restricted region [18]. This can probably explain the absence of a strong increase in the absorption of laser radiation.

In conclusion, we present the table of thermal constants for some metals. Here, H_m and H_b are the latent heats of phase transitions of melting and vaporization. We will use these data in the next chapters (Table 1.3).

Table 1.3 Thermophysical constants of metals

Material	T_m, C°	T_b, C°	ρ, g/cm^3	H_m, J/g	H_b, J/g	χ, cm^2/c	K, W/cm K	C_p, J/g K
Al	660	2,467	2.7	396	10,571	0.91	2.23	0.90
Be	1,227	2,970	1.85	1,092	–	0.42	1.47	1.89
Cr	1,875	2,665	7.19	403	6,564	0.20	0.67	0.46
Cu	1,083	2,596	8.96	214	4,813	1.14	3.95	0.39
Au	1,063	2,807	19.32	67	1,873	1.18	2.98	0.13
Fe	1,533	2,750	7.87	275	7,140	0.21	0.76	0.46
Mo	2,610	4,612	10.2	293	5,140	0.51	1.43	0.28
Ni	1,453	2,730	8.9	309	6,472	0.24	0.92	0.44
Pb	1,552	3,140	12.02	162	3,511	0.24	0.71	0.24
Pt	1,769	3,827	21.45	112	2,629	0.24	0.67	0.13
Si	1,410	2,355	2.33	1,814	10,647	0.53	0.84	0.68
Ag	961	2,212	10.49	105	2,335	1.71	4.2	0.24
Ta	2,996	5,425	16.6	155	4,200	0.23	0.55	0.14
Sn	232	2,270	5.76	60	2,394	0.38	0.63	0.23
W	3,410	5,660	19.3	184	4,830	0.62	1.68	0.14
Zn	420	906	7.13	101	1,787	0.41	1.13	0.39

References

1. M.B. Vinogradova, O.V. Rudenko, A.P. Sukhorukov, *Teoriya voln (Theory of Waves)* (Nauka, Moscow, 1979)
2. D.C. Smith, Proc. IEEE **65**, 1679 (1977)
3. A.A. Vedenov, *Fizika elektrorazryadnykh lazerov (Physics of Electric-discharge Lasers)* (Energoizdat, Moscow, 1982)
4. A.A. Vedenov, O.A. Markin, Zh. Eksper. Teor. Fiz. **76**, 1198 (1979)
5. B.P. Gerasimov, T.G. Elizarov, A.P. Sukhorukov, Zh. Tekh. Fiz. **53**, 1696 (1983)
6. V.E. Zuev, *Rasprostranenie lazernogo izlucheniya v atmosfere (Propagation of Laser Radiation in Atmosphere)* (Radio i Svyaz', Moscow, 1981)
7. E.M. Lifshitz, L.P. Pitaevskii, *Physical Kinetics* (Pergamon Press, Oxford, 1981; Nauka, Moscow, 1979)
8. M. Autric, J.P. Dufrense, Acad. Sci. B **288**, 237 (1979)
9. L.D. Landau, E.M. Lifshitz, *Electrodynamics of Continuous Media* (Pergamon Press, Oxford, 1960; Nauka, Moscow, 1970)
10. K. Ujihara, J. Appl. Phys. **43**, 2376 (1972)
11. Y. Arata, H. Maruo, I. Miyamoto, Tool. Eng. **25**, 24 (1981)
12. H. Hugel, F. Dausinger, W. Bloehs, B. Grunenwald, in *Laser Processing: Surface Treatment and Field Deposition*, ed. by J. Mazumder, O. Conde, R. Villar, W. Stean (Cluwer Academic Publishers, 1996)
13. I.P. Shkarofaky, RCA Rev. **36**, 336 (1975)
14. Y. Arata, H. Maruo, I. Miyamoto, Application of Lasers for Material Processing, IIW, Doc. IV-241-71
15. F.B. Bunkin, N.F. Kirichenko, B.S. Luk'yanchuk, Usp. Fiz. Nauk **138**, 45 (1982)
16. G.G. Gladush, E.B. Levchenko, A.A. Ezhov, et al., *All-Union Conference on Laser Applications in Mechanical Engineering Technology, Zvenigorod, 1982* (Nauka, Moscow, 1982), pp. 69–71
17. A.V. Nesterov, V.G. Niziev, V.P. Yakunin, J. Phys. D Appl. Phys. **32**, 2871 (1999)
18. R.V. Arutyunyan, V.Yu. Baranov, L.A. Bol'shov, D.D. Malyuta, A.Yu. Sebrant, *Vozdeistvie lazernogo izlucheniya na materialy (Interaction of Laser Radiation with Materials)* (Nauka, Moscow, 1989)
19. V.N. Anisimov, R.V. Arutyunyan, A.Yu. Sebrant, *Proceedings of the Interindustry Conference on the Interaction of Radiation, Plasma and Electron Fluxes with Matter* (Atominform Central Research Institute, Moscow, 1984), pp. 84–85
20. L.A. Bol'shov, K.A. Krivoruchko, V.P. Reshetin, Preprint ITMO no. 6 (ITMO Akad Nauk BSSR, Minsk, 1985)
21. A.A. Vainshtein, *Teoriya difraktsii i metod faktorizatsii (Theory of Diffraction and Factorization Method)* (Sov. Radio, Moscow, 1966)
22. V.N. Anisimov, L.A. Bol'shov, K.A. Krivorucko, Kvantovaya Elektron. **14**, 177 (1987)
23. J.R.A. Pearson, J. Fluid Mech.,vol.p. 489-500 (1958).
24. E.B. Levchenko, A.L. Chernyakov, Zh. Eksper. Teor. Fiz. **81**, 202 (1981)
25. L.E. Scriven, C.V. Sternling, Nature **187**, 186 (1960)
26. V.G. Levich, *Fiziko-khimicheskaya gidrodinamika (Physicochemical Hydrodynamics)* (Fizmatgiz, Moscow, 1959)
27. N.O. Young, J.S. Goldstain, M.J. Block, J. Fluid. Mech. **6**, 350 (1959)
28. L.D. Landau, E.M. Lifshits, *Mekhanika sploshnykh sred (Mechanics of Continuous Media)* (Gostekhizdat, Moscow, 1953)
29. N.N. Rykalin, A.A. Uglov, A.N. Kokora, Fiz. Khim. Obr. Metal. **6**, 14 (1972)
30. S.I. Anisimov, Ya.A. Imas, G.S. Romanov, et al., *Deistvie izlucheniya bol'shoi moshchnosti na metally (Interaction of High-Power Radiation with Metals)* (Nauka, Moscow, 1970)
31. G. Weyl, A. Pirri, R. Root, AIAA J. **19**, 460 (1981)
32. A.N. Barchukov, F.V. Bunkin, V.I. Konoiv, et al., Zh. Eksp. Teor. Fiz. **66**, 965 (1974)

33. A. Pirri, AIAA J. **15**, 83 (1977)
34. G.N. Abramovich, *Teoriya turbulentnykh strui (Theory of Turbulent Jets)* (Fizmatgiz, Moscow, 1960)
35. V.A. Batanov, F.B. Bunkin, A.M. Prokhorov, et al., Zh. Eksp. Teor. Fiz. **63**, 1240 (1972)
36. Y. Sone, H. Sugimoto, in *Strong Evaporation from a Plane Condensed Phase in Adiabatic Waves in Liquid-Vapor Systems*, ed. by G.E.A. Meier, P.A. Thompson (Springer, Berlin, 1990), pp. 293–304
37. I.N. Ivchenko, S.K. Loyalka, R.V. Tompson, Jr., *Analitical Methods for Problems of Molecular Transport* (Springer, 2007)
38. C.J. Knight, AIAA J. **17**, 519 (1979)
39. V.I. Mazhukin, P.A. Prudkovskii, A.A. Samokhin, Mathem. Model. **5**, 3 (1993)
40. A.V. Gusarov, I. Smurov, Phys. Fluids. **14**, 4242 (2002)
41. I. Smurov, A. Lashin. Proc. of NATO Fdvance Study Institute on Laser Appl. for Mechanical Indastry,, Erice, Tranpani, Italy, 4–16 Aprile 1992, pp.165–206
42. A.V. Gusarov, I. Smurov, Appl. Surf. Sci. **168**, 96 (2000)
43. P.L. Bhatnagar, E.P. Gros, M. Krook, Phys. Rev. **94**, 511 (1954)
44. J.H. Ferziger, H.G. Kaper, *Mathematical Theory of Transport Processes in Gases* (North-Holland Pub. Comp., Amsterdam, 1972)
45. J.W. Elan, D.N. Levy, J. Appl. Phys. **88**, 4352 (2000)
46. M.N. Kogan, N.K. Makashev, Fluid Dyn. **6**, 913 (1971)
47. A.A. Samokhin, Trudy. Inst. Obshch. Fiz. **13**, 3 (1988)
48. I.N. Kartashev, A.A. Samokhin, I. Smurov., J. Phys. D. **38**, 3703 (2005)
49. J.H. Yoo, S.H., Jeong, R. Greif, R.F. Ruso, J. Appl. Phys. **88**, 1638 (2000)
50. S.I. Anisimov, M.I. Tribel'skii, Ya.G. Epel'baum, Zh. Eksp. Teotr. Fiz. **78**, 1597 (1980)
51. E.B. Levchenko, A.L. Chernyakov, Prikl. Mekh. Tekh. Fiz. **6**, 144 (1982)
52. Ya.B. Zel'dovich, G.I. Barenblat, V.B. Librovich, et al., *Matematicheskaya teoriya goreniya i vzryva (Matematical Theory of Combustion and Explosion)* (Nauka, Moscow, 1980)
53. S.I. Anisimov, S.M. Gol'berg, O.L. Kulikov, et al., Pis'ma Zh. Tekh. Fiz. **9**, 226 (1983)
54. A.M. Iskol'dskii, V.K. Pinus, Ya.G. Epel'baum, Preprint of Institute of Atomic Energy, Siberian Branch, Acad. Sci. USSR, no. 59, Novosibirsk, 1977
55. S.A. Akhmatov, V.I. Emel'yanov, N.I. Koroteev, V.N. Seminogov, Usp. Foz. Nauk **147**, 675 (1985)
56. M.C. Hutley, V.M. Bird, Optica Acta **20**, 771 (1973)
57. M.C. Hutley, Optica Acta **20**, 607 (1973)
58. I. Smurov, A. Lashin, M. Poli, in *ICALERO (1992)*, Oct. 25–29, 1992, Orlando, Florida, USA

Chapter 2
Mechanisms of Laser Processing of Metal Surfaces

Abstract In Chap. 1, we considered the "elementary processes" providing the basis for all types of laser processing of materials. The laser processing of materials can be conventionally divided into two types: processing without considerable changes in the sample shape or by changing the sample shape. In the first case, the laser action causes the structural and phase variations in the surface layer of materials. This includes technologies of hardening, alloying, cladding, film deposition, polishing, and purification. The second type of laser processing includes welding, cutting, drilling, laser nanotechnologies for manufacturing parts, etc. As the laser radiation power absorbed in the surface layer is increased, structural and phase transitions occur successively in the layer. Structural transformations take place at a lower temperature, when the sample still remains solid, and their description is simpler. These structural transformations form the physical basis for sample surface hardening. Hydrodynamic models of laser-induced alloying of metal surface by taking into account of Marangoni flow are considered. Physical mechanisms of cladding and selective laser melting are described.

2.1 Thermal Model of Laser Hardening of the Steel Surface

The main advantages of laser hardening of materials compared to traditional methods are [1, 2]:

1. The selected sites of a sample surface can be locally thermally processed in region inaccessible for other processing methods; the configuration of the region being hardened can be three-dimensional.
2. Due to the high speed of laser thermal hardening, samples are deformed to a lesser degree than during hardening by other methods; in this respect, laser thermal hardening is similar to electron-beam hardening.

Unlike the electron-beam technology, laser thermal hardening does not require the evacuation of samples, and thermal processing can be performed in various (active

G.G. Gladush and I. Smurov, *Physics of Laser Materials Processing*,
Springer Series in Materials Science 146, DOI 10.1007/978-3-642-19831-1_2,

© Springer-Verlag Berlin Heidelberg 2011

and neutral) gases; in addition, samples do not emit X-rays (for $q < 10^8$ W/cm^2). Steel hardening requires the heating of a volume element up to temperature T_γ followed by its rapid cooling below temperature T_α. Here, T_γ is the phase transition temperature of iron from the α-structure to the γ-structure, and T_α is the temperature at which γ-iron is completely transformed to α-iron [3, 4]. When a sample is displaced, an element near its surface is heated up to a certain temperature for the irradiation time d_f/v (d_f is the focal spot size and v is the sample speed). After leaving the laser beam, each element is cooled due to heat transfer to the rest of the target. The elementary volumes of the sample located at different depths will be heated up to different temperatures and their thermal histories (the time dependence of temperature) will be different. It is the thermal history of the element that determines the hardening process proceeding in the sample material. The thermal history can be found by solving the problem of the temperature distribution in a sample heated by a surface source representing a focal spot.

2.1.1 Qualitative Consideration of the Stationary Thermal Model of Metal Hardening

The temperature distribution $T(x, y, z)$ in a metal occupying the half-space $z < 0$ and moving at the speed v along the x-axis is described by the heat conduction equation

$$c\rho v \frac{\partial T}{\partial x} = div \kappa \nabla T \tag{2.1}$$

with the boundary conditions

$$-\kappa \left. \frac{\partial T}{\partial z} \right|_{z=0} = \alpha(T) q(x, y), \qquad T(x = y = z = \infty) = T_0, \tag{2.2}$$

where T_0 is the sample temperature far from the focal spot; ρ, c, κ, α are the density, specific heat, heat conductivity, and absorption coefficient of the material, respectively (see Fig. 2.1). If the temperature dependencies of c,κ, and α are taken into account, the solution of this three-dimensional problem is quite cumbersome. If the coefficients c, ρ, κ, and α in the system of equations (2.1), (2.2) are temperature-independent, this system can be readily solved by the Green function method [5]. In the case of the normal (Gaussian) radiation intensity distribution in the focal spot

$$q = q_0 \exp(-r^2/r_f^2),$$

we obtain [5]

$$T(x, y, z, t) = T_0 + \frac{\alpha P}{2\pi\kappa\sqrt{4\pi\chi}}$$

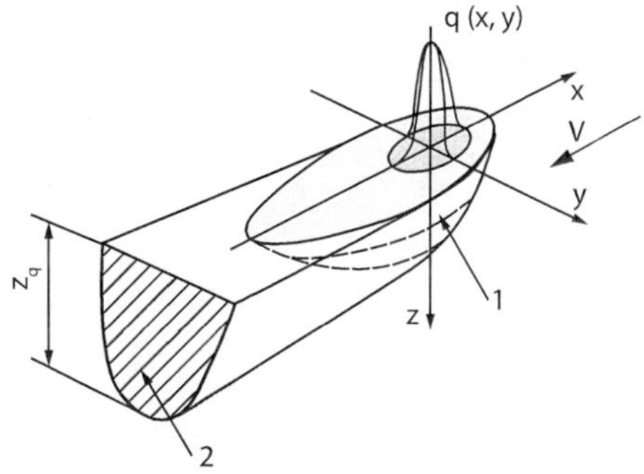

Fig. 2.1 Scheme of the laser hardening process: (**1**) isothermal surface $T = T_\gamma$; (**2**) hardening zone

$$\times \int_0^t \frac{dt}{\sqrt{t}\left(t + \frac{r_f^2}{4\chi}\right)} \exp\left[-\frac{(x+vt)^2 + y^2}{4\chi\left(t + \frac{r_f^2}{4\chi}\right)} - \frac{z^2}{4\chi t}\right], \tag{2.3}$$

where $P = \pi r_f^2 q$ is the total laser beam power; r_f is the focal beam radius; and χ is the thermal diffusivity. Integral (2.3) can be calculated numerically [6]. Before presenting the results of calculations, we obtain simplified analytic expressions for temperature in different limiting cases, which can be used to estimate the hardening depth. These expressions can be derived from (2.3). However, we will obtain them for clearness from qualitative considerations. A volume element is heated for the time that is approximately equal to the time $\tau = d/v$, $d = 2r_f$ during which a laser beam propagates over this volume. For this time, a heat conduction wave will penetrate to the metal by the depth $2\sqrt{\chi\tau}$. If this depth is much smaller than the beam radius, the problem can be treated as a plane one. In this case, the characteristic size of the temperature inhomogeneity is $2\sqrt{\chi\tau}$. The surface temperature T_s can be found from the equality of the energy fluxes absorbed and removed from the metal surface due to heat conduction

$$\alpha q \approx \kappa T_s / 2\sqrt{\chi\tau}\,. \tag{2.4}$$

Temperature decreases exponentially deep in the metal with the characteristic distance of $2\sqrt{\chi\tau}$. Taking this into account, we obtain finally from (2.4)

$$T(z) \approx \frac{\alpha q}{\kappa}\sqrt{\frac{4\chi d}{v}}\left(\exp - \frac{z}{2\sqrt{\chi d/v}}\right). \tag{2.5}$$

Here, the exact solution with the function erfc [5,7] is replaced by exp for simplicity. For small values of arguments used in the problem, the difference between these functions is insignificant [8].

After the propagation of the radiation source over an elementary volume of the metal surface, this volume is cooling. Because the cooling time is approximately equal to the propagation time of a cooling wave through the heated zone, it is close to the heating time d/v. For the typical laser hardening conditions ($d \approx 0.1$ cm, $v \approx 10$ cm/s or $d \approx 0.5$ cm, $v \approx 0.5$ m/s), we have $\tau = d/v \approx 10$ ms. Because heating is performed up to $T_s \approx 1{,}000°$C, the heating and cooling rates are quite high: $\mathrm{d}T/\mathrm{d}t \sim 10^5$ K/s, which is sufficient for hardening. The hardening depth h_h can be determined from (2.5) by assuming that to produce hardening, it is necessary to heat the volume element up to temperature T_γ:

$$h_h = 2\sqrt{\frac{\chi d}{v}} \ln \frac{\alpha q}{\kappa T_\gamma} \sqrt{\frac{4\chi d}{v}} \tag{2.6}$$

One can see from (2.6) that there exists the threshold intensity

$$q_h = \frac{\kappa T_\gamma}{\alpha} \sqrt{\frac{v}{4\chi d}}. \tag{2.7}$$

below which the hardening cannot be produced.

It follows from (2.6) that for $q \geq q_h$ the value of h_h linearly increases with q:

$$h_h = \sqrt{4\chi d/v}(q/q_h - 1) \tag{2.8}$$

For the specified intensity, the hardening threshold for speed is determined from (2.7), and the depth near the threshold is determined from (2.8):

$$h_h = \sqrt{4\chi d/v} - \kappa T_\gamma/\alpha q \tag{2.9}$$

At high powers ($q \geq 2q_h$), the hardening depth increases logarithmically with power. The purely thermal model (2.1) is restricted at high powers by the appearance of evaporation. The limiting admissible radiation intensity can be found from (2.4) for $T_s = T_b$ (T_b is the boiling temperature):

$$q_b = \kappa T_b \sqrt{v}/2\alpha\sqrt{\chi d} \tag{2.10}$$

The hardening depth at this intensity is determined by substituting (2.10) into (2.6):

$$h_{h\max} = 2\sqrt{\chi d/V} \ln T_b/T_\gamma \tag{2.11}$$

The above consideration concerned the case when the sample was moved at a sufficiently large speed, so that the heating depth was smaller than the light spot radius on the metal surface:

$$v > v_K = 16\chi/d \tag{2.12}$$

If hardening is performed at a lower speed $V < V_K$, the one-dimensional approach becomes inapplicable. Numerical calculations by using (2.1)–(2.2) showed that three-dimensional effects became noticeable already for $Pe = dv/\chi < 40$ [9]. The surface temperature, unlike (2.5), where $T \simeq \alpha P/d\ (\mathrm{Pe})^{0.5}$, depends on Pe weaker: $T \simeq \alpha P/d(\mathrm{Pe})^{0.4}$. It was also shown in [9] that the change in the temperature inside the metal can be approximated by a linear dependence $z/d \simeq 1$. As a result, the hardening depth can be defined differently than (2.8), and (2.9) [10]:

$$h_h = \frac{8\alpha(\chi)^{0.4}}{\pi\kappa T_\gamma}\frac{P}{(vd)^{0.4}} - d \tag{2.13}$$

As the hardening speed is further decreased, the temperature distribution far from the focal spot is determined by the action of a point thermal source [5] (see Sect. 4.1). In this case, the hardening depth is equal to the radius r_h of the hardened zone and is described by the expression [see (4.2)]

$$h_h = r_h = \alpha P/2\pi\kappa T_\gamma \tag{2.14}$$

At the high enough laser radiation intensity, the surface temperature achieves the boiling temperature. This restriction determines the maximum hardening radius and depth:

$$h_h = r_h = r_f T_b/T_\gamma. \tag{2.15}$$

By comparing (2.11) and (2.15), we see that the maximum depth of the hardened zone is achieved when $v < v_K$ and its value exceeds the light beam radius on the sample no more than by four times ($T_b/T_\gamma \approx 4$ for iron).

Technical conditions often require that the temperature of a hardened sample would not achieve the melting temperature. In this case, T_b in (2.10), (2.11), and (2.15) should be replaced by T_m.

2.1.2 Comparison with Experiments

The power dependence of the hardening depth at powers exceeding the threshold value by two to three times should be logarithmic according to (2.6) (Fig. 2.2) [11]. One can see from (2.6) that the slope of curves in Fig. 2.2 should not depend on α, which is also approximately observed in experiments. This means that a simple model neglecting the temperature dependencies of α,κ, and c well agrees with experiments. Simple expressions (2.7) and (2.8) can be easily compared with experiments. Note that this comparison is only qualitative since quantitative comparison is complicated because these expressions are approximate and thermal coefficients depend on temperature. One can see from (2.7) that the hardening threshold q_h decreases with decreasing the sample motion speed. It follows from Fig. 2.3, that approximately $q_h \sim v^{1/2}$, which corresponds to (2.7). The estimate of q_h from (2.7) ($\chi \cong 0.05\,\mathrm{cm}^2/\mathrm{s}$, $d = 0.3$ cm, $v = 1$ cm/s, $T_\gamma = 800°\mathrm{C}$) is consistent with the experimental value (curve 1 in Fig. 2.3), if we assume that the absorption coefficient is $\alpha \cong 0.3$. This value of α for the steel surface covered with graphite powder is quite real. One can see from (2.6) and (2.8) that the slope of the curves

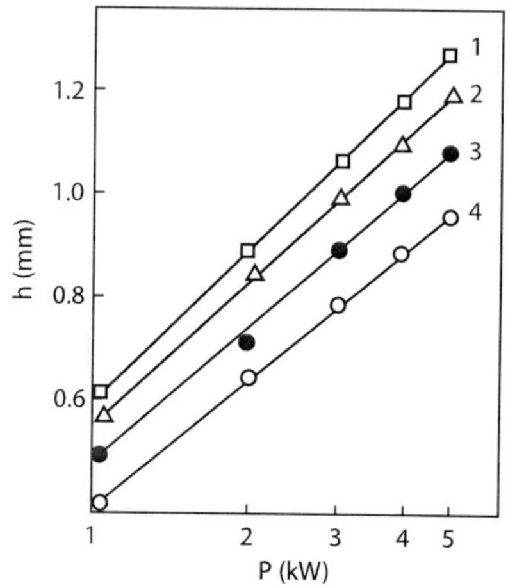

Fig. 2.2 Dependences of the hardening depth of B460-2 cast iron covered with manganese phosphate (**1**), black gouache (**2**), polygraphic dye (**3**), and without coating (**4**) on the CO_2 laser power for $v = 1$ cm/s

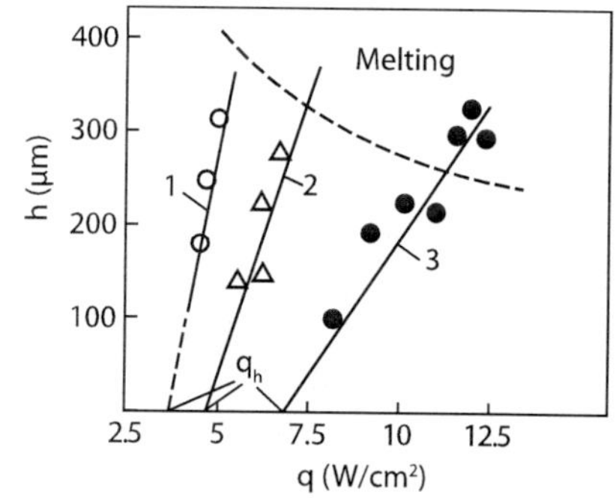

Fig. 2.3 Dependences of the hardening depth of AiSi4340 steel covered with colloidal graphite on the cw CO_2 laser radiation intensity for different hardening speeds (the laser spot diameter is 3 mm): $v = 1$ (**1**), 2.5 (**2**), and 3.3 cm/s (**3**)

with respect to the abscissa increases with decreasing the sample speed, which is also consistent with experiments (Fig. 2.3). Note that, according to (2.8), the value of h_h for $q = 0$ (although this value has no physical meaning because it is negative) is independent of α. The dependence of α on the speed V complicates a comparison with experiments. The ratio h_h for $q = 0$ for different v should be proportional to the square root of the ratio of velocities, which also agrees with experiments.

It follows from (2.7) and (2.10) (in the latter expression, the replacement $T_b \to T_m$ should be performed) that the radiation intensity at which the melting of a sample begins is approximately twice as large as the threshold value q_h, because $T_m/T_\gamma \sim 2$ for many steels.

Figure 2.3, which shows the melting boundary, demonstrates that this property is also observed experimentally.

Another parameter, which can be easily varied in experiments, is the hardening speed. One can see from (2.9) that the hardening depth linearly depends on $v^{-1/2}$, this dependence having a threshold. The angle between the straight line describing this dependence and the abscissa should increase with increasing the light beam diameter [see (2.9)]. By comparing curves 1, 3, and 4 in Fig. 2.4, we see that this tendency is also observed in experiments [12, 13]. As the radiation intensity

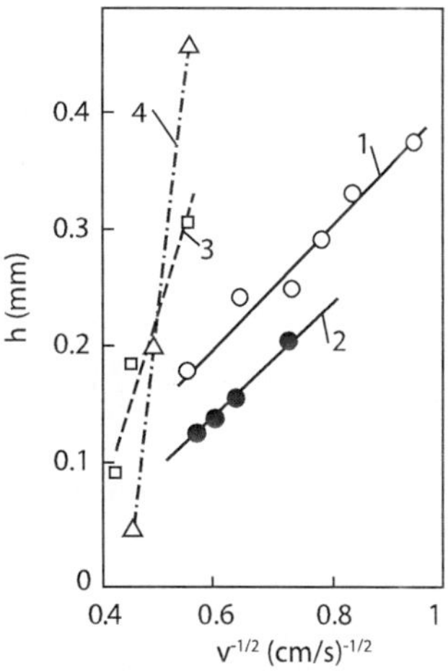

Fig. 2.4 Dependences of the hardening depth on $v^{-1/2}$: (**1**) 40X steel, $P = 1$ kW, CO_2 laser, $d = 3.3$ mm; (**2**) 40X steel, $P = 0.8$ kW, $d = 3.3$ mm; (**3**) 35 steel, $d = 5$ mm, $P = 5$ kW; (**1, 2, 3**) unstable resonator; (**4**) SK5 steel, $P = 1.3$ kW, $d = 7.2$ mm, multimode radiation regime (TEM00 + TEM11), the surface is covered with an absorbing $Mn3(PO_4){\cdot}2H_2O$ layer

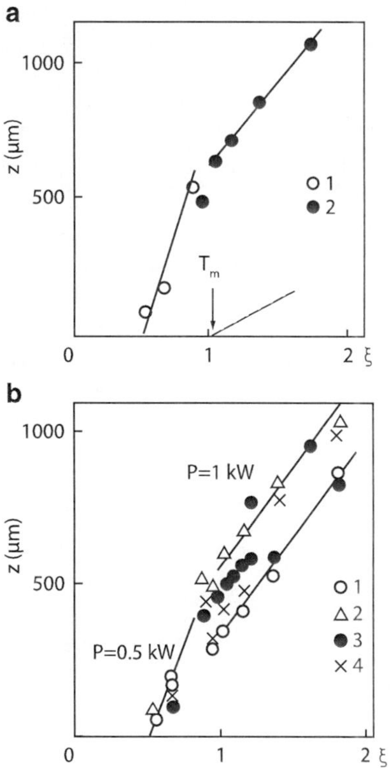

Fig. 2.5 Dependences of the hardened zone depth z for 45 steel on the parameter $\xi = P/(dV)^{0.4}$: (**a**) with yellow gouache coatings: $P = 0.5$ (**1**) and 1 kW (**2**); (**b**) with different coatings: (**1**) MTsS-510 coating; (**2**) yellow gouache; (**3**) black gouache; (**4**) SG-504 coating

is increased, the slope of the curves does not change; however, they should be displaced upward [see (2.9)]. This is also observed experimentally cf. curves 1 and 2 in Fig. 2.4). As for expression (2.12), it is shown in Fig. 2.5 that the hardening depth of 45 steel is approximately described by the dependence $P/(dV)^{0.4}$ [10]. The same dependence was found for the hardening of cast irons and other steels [15]. The close dependence $P/(dV)^{0.5}$ was derived by the semiempirical method in [16].

Thus, simple analytic expressions in the region of their applicability $h_h \ll d$ well agree with experiments. If the hardening conditions are so that $h_h \cong d$, the depth and width of the hardened zone, i.e. its shape should be found by numerical calculations from expression (2.3).

2.1.3 Numerical Calculations

Numerical calculations by (2.3) or calculations of a more general problem (2.1), (2.2) give the shape of the hardened zone. It is interesting to study the heating and cooling of elementary volumes located at different depths (thermal histories), which can be obtained by solving the stationary problem (2.3) by passing to a coordinate system moving with the sample:

$$T(x, y, z) \rightarrow T(x - vt, y, z).$$

One can see from Fig. 2.6 [13] that the maximum temperature in deeper layers is achieved later than on the surface (because heat later penetrates into lower layers). As for the characteristic heating and cooling times of different layers with the temperature above T_γ, they are close to d/v. Figure 2.6 shows that the cooling rate of upper layers is greater than that of lower layers; however, the cooling time of lower layers down to T_γ is smaller than that of upper layers. These differences in the heating kinetics determine finally the multilayer structure of the hardened zone. Numerical calculations showed that the maximum temperature at the specified depth and, hence, the hardened layer depth [17] comparatively weekly depend on the radiation intensity distribution in the light spot.

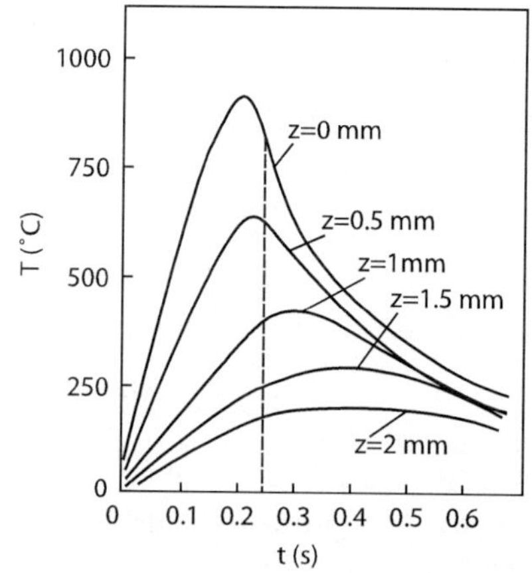

Fig. 2.6 Temperature story of different layers of a carbon steel sample for $d = 5$ mm, $P = 1$ kW, $v = 2$ cm/s, $\kappa = 0.24$W/(cm·K), $\chi = 0.05$ cm^2/s

Accurate calculations should take into account the temperature dependence of thermal coefficients and also the dependences of temperatures T_γ and T_α on the heating speed. Consider the first of these dependences. The absorption coefficient of light by a pure metal surface increases with temperature (see Sect. 1.2). One can see from Fig. 1.3 that α for some metals can increase by several times with increasing temperature. The heat capacity c and heat conduction of metals can strongly change with increasing temperature. Thus, e.g., the heat conductivity of carbon steel decreases from 0.57 to 0.27 W/(cm K) upon heating from 27 to 927°C [18]. The absorption coefficient increases approximately by three times in the same temperature interval (Fig. 1.3). These variations on the coefficients should be taken into account in calculations. The influence of the temperature dependence of thermal parameters on the properties of heating can be illustrated for molybdenum [19] (Fig. 2.7). The thermal parameters of this metal strongly depend on temperature (Fig. 2.8). The absorption coefficient α in Fig. 2.8 is inconsistent with data presented in Fig. 1.3; however, this is not important for the question under study. Curve 4 in Fig. 2.7 shows that the neglect of the temperature dependence of the coefficients can understate the temperature by several times. This error can be considerably reduced by using the average values of the coefficients in the solution of the linear problem. This circumstance and also the weak temperature dependence of the absorption coefficient of a surface covered with an absorbing material probably explain the agreement between linear thermal model (2.3) and experiments.

Note in conclusion another circumstance which should be taken into account in numerical calculations of the hardened zone shape. As mentioned above, laser hardening is characterized by high heating and cooling rates. While the latter factor is favorable for hardening, the former can adversely affect this process. It is known that upon heating a sample with the initial perlitic structure above the phase transition temperature T_γ, α-iron transforms to γ-iron, and carbon, which is contained in carbide inclusions, additionally dissolves in the iron [4]. The latter process proceeds slowly because it requires the diffusion of carbon atoms inside a grain from its surface. During cooling (proceeding at the rate close to the heating rate), the inverse process has not time to proceed (which results in hardening).

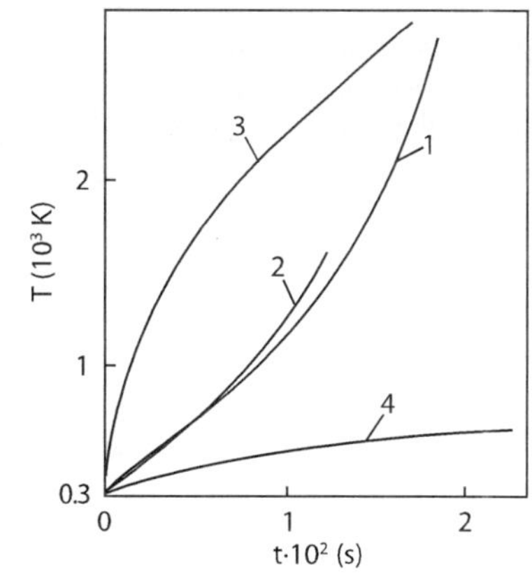

Fig. 2.7 Calculated time dependences of temperature at the center of a heated spot on a molybdenum sample for $q = 2 \cdot 10^5$ W/cm^2 and $r_f = 3$ mm: (**1**) calculation by (2.1) and (2.2) by using the approximate algorithm [19]; (**2**) numerical calculation by (2.1), (2.2); (3, 4) calculations by the linear model for average (**3**) and initial (**4**) values of thermal coefficients

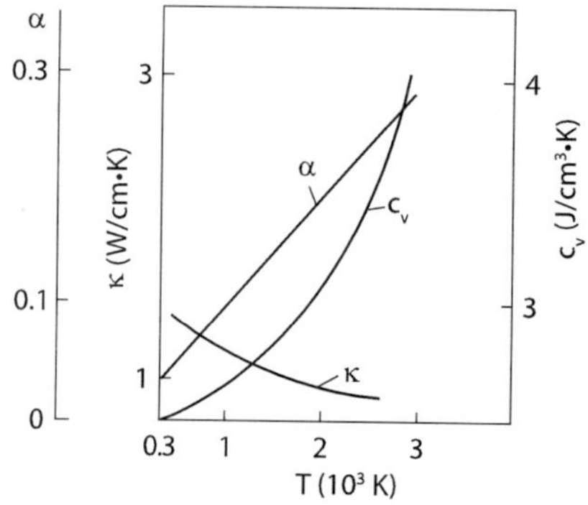

Fig. 2.8 Temperature dependences of thermal coefficients of molybdenum [19]

By considering the $\alpha \rightarrow \gamma$ transition, a rapid increase in the diffusion of carbon into iron with increasing temperature should be taken into account [20]. This increase is caused by the activation nature of the diffusion of atoms in a solid. The diffusion coefficient is $D \approx 10^{-6}\,\mathrm{cm}^2/\mathrm{s}$ at a temperature of 800°C. Carbon will diffuse into a grain of size $10\,\mu\mathrm{m}$ for the time $\sim$1 s. At the temperature close to the melting temperature, $D \simeq 10^{-4}\,\mathrm{cm}^2/\mathrm{s}$, and the diffusion time is reduced correspondingly to 10 ms. The minimal time of transition to the equilibrium austenite depends on many factors such as the initial structure of a material, its chemical composition and homogeneity, impurities, etc. Accurate calculations in this situation are complicated. Some authors [21] propose to take these kinetic effects into account approximately. The temperatures T_α and T_γ are measured for a certain material depending on its heating rate. Usually, the hardening depth in model calculations by (2.3) is set equal to the depth at which the sample temperature is $T_\gamma \approx 800$°C.

2.1.4 *Processing of Metal Surfaces by the Oscillating Beam of a CO_2 Laser*

It is often required industrially to harden the large areas of workpieces. This could be performed by defocusing laser radiation to obtain a large light spot. However, to maintain the required power density in this case, high-power lasers should be used. For this reason, a large hardened area is often obtained by hardening successively individual strips of width 3–5 mm with overlapped edges [22]. This method produces the inhomogeneous surface of the hardened region because tempering zones are formed in the overlapped region of the strips, which impairs the strength of the material. A wide hardened zone can be obtained by focusing radiation in a narrow strip (of width 2–4 mm and length a few centimeters) and moving a sample perpendicular to the large side of the strip. Lasers with output powers up to 5 kW are suitable for this purpose. However, it is difficult to obtain the uniform radiation intensity distribution in such a spot. Because of this, scanning devices are used for obtaining large hardened areas. These devices employ either an oscillating mirror or a rotating mirror pyramid. They are technological, reliable in service, and efficiently use radiation [23].

A scanning device with an oscillating plane mirror can provide the beam scan with any amplitude. However, because the spot motion on the surface is oscillatory, the energy release at cusps is enhanced, whereas in the middle of the strip the energy release is reduced due to the fast movement of the laser beam. Such a distribution of the heat release across the strip can lead to undesirable effects.

The second scheme with a rotating mirror pyramid gives a more uniform radiation intensity distribution across the strip and provides the required amplitude of beam oscillations (the strip width).

The most favorable operation regime of a scanning device is achieved at the high oscillation frequency of the laser beam, when the beam illuminates a specified surface element many times during the propagation by the sample of the distance equal to the diameter of the light spot of the target. Consider a specific case of hardening with the help of a rotating mirror pyramid [24]. The pyramid had 12 faces and was rotated at the speed of 3,000 revolutions per minute. A sample surface was covered with absorbing metal oxide coatings to increase the absorption coefficient up to 65% in a broad intensity range. Radiation was focused with a lens with a focal distance of 350 mm. Hardening was performed for St3, 75G, 40X steel and cast iron samples (Table 2.1). The value of hardness obtained by scanning radiation coincided for all the materials with the value obtained after single-pass scanning. Scanning extends the possibilities of choosing laser processing regimes. Thus, in the absence of scanning, the longitudinal and transverse sizes of a light spot are the same. Scanning allows one to change these sizes independently of each other, which simplifies the choice of the optimal hardening regime. For example, the hardening depth equal to 1 mm was obtained without melting for a strip width of 17 mm.

Because the melting and hardening temperatures of cast iron are close, it is difficult to perform laser hardening of this material without melting. Scanning provides stable hardening without melting, which is confirmed by good results obtained by using absorbing coatings and without them. By using coatings, a strip of width up to 30 mm and depth up to 0.24 mm was obtained. In the absence of coatings, the uniform hardening depth obtained in this regime was 0.1 mm. The hardened zone depth increases with increasing radiation power. Experiments showed that the tempering zone was absent in the longitudinal microsection. This suggests that hardening produced by a rapidly oscillating beam is identical to hardening produced by a stationary bean focused to a narrow strip.

Table 2.1 Parameters of laser hardening of various materials

					Microhardness, N/mm^2	
Material	Power, kW	Speed, cm/s	Width, mm	Depth, mm	Initial	Hardened
Steel	3.6	0.8	17	1.0	3,200	6,500
40X Steel	3.6	0.8	17	1.0	3,220	8,240
75G Steel	3.6	0.8	17	1.0	3,700	10,500

Scanning systems described above are useful for scientific studies, but they are too complicated for industrial applications. To produce a strip beam, it was proposed to use diffraction reflecting focusing devices [25]. Focal rectangular spots of lengths 15–25 mm and width 0.3–2 mm were obtained [26]. It is clear that in the case of the same radiation powers and areas of circular and rectangular focal spots, the hardening depth in the latter case will be smaller (Fig. 2.9) and the strip width and its homogeneity will be large enough (Fig. 2.10). The hardening structures of alloys and their properties are described in detail, for example, in [14, 27, 28].

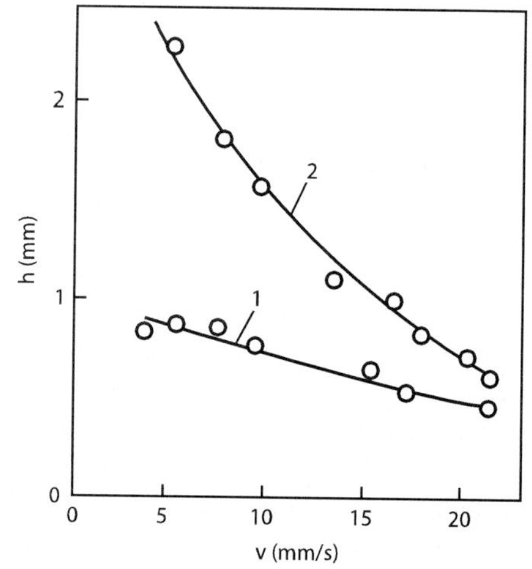

Fig. 2.9 Dependences of the laser hardening depth on the hardening speed at a constant irradiation intensity for stripe (**1**) and circular (**2**) laser beams

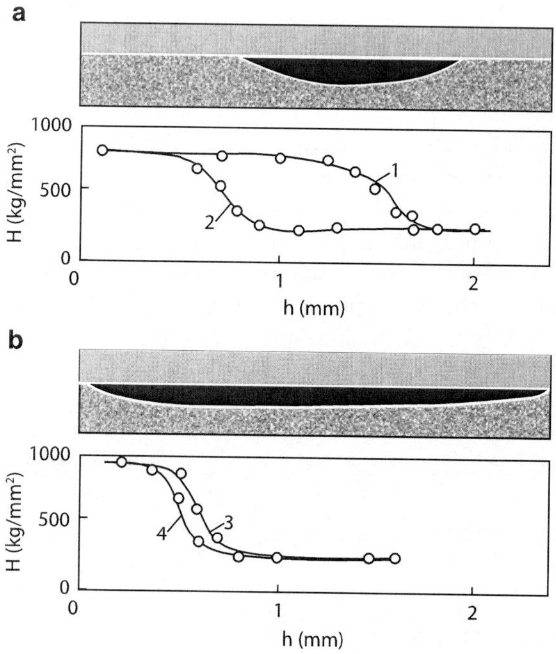

Fig. 2.10 Changes in the microhardness over the laser action depth upon irradiation by circular (**a**) and stripe (**b**) laser beams at the center of the irradiated zone (1, 3) and at a distance of 1.5 mm from its edge (2, 4)

2.2 Hydrodynamical Models of Laser-Induced Alloying of Metal Surfaces

Alloying and hardening are used to improve the mechanical and corrosion properties of the surface layer of materials. The fundamental difference of alloying from hardening is a change in the chemical composition of the surface layer of a sample. This is achieved by introducing alloying elements into the sample by melting the surface layer. Alloying materials can be applied on the surface of samples by electrolytic deposition or by the deposition of pastes containing metal powders and various binders. Sometimes an alloying metal powder is injected into the melting zone or is poured on the sample surface or is rolled in the form of a foil. If the melting boundary penetrates inside the material by a depth that is considerably smaller than the finite thickness of the applied coating, this process is called cladding. If the melting zone is thicker than a layer being alloyed, the process is called alloying [29]. The main requirement for alloying is mixing of an alloying material with the substrate melt. In this respect, the alloying process differs from surfacing where mixing is undesirable. Both cladding and alloying are performed in modern technologies by injecting powder into the laser action zone. It is obvious that the amount of injected powder in the case of cladding is considerably greater than upon alloying. Thus, alloying corresponds to the case when the mass of the introduced powder is considerably smaller than the mass of the remelted substrate material. The opposite case corresponds to cladding.

The nature of the mixing mechanism is the main problem of laser alloying. At present there exist three models of mixing of the alloying material with the sample material which explain the mixing process by the surface layer boiling, by capillary forces, or by molecular diffusion, respectively. Before analyzing these mechanisms, we recall that the melting depth during sample heating is described by the same expressions as for hardening in which the hardening temperature T_γ is replaced by the melting point of the material. The shape of a melt pool produced by a stationary laser beam can be easily calculated if the boundary between liquid and solid phases (melting isotherm) is located far away from a source. However, it follows from (2.15) that the maximum radius of the melting zone, for example, for steel is $r_m \sim r_f$, i.e. in this case the concept of a point thermal source cannot be used. We can find the upper bound of the melt volume. By using the analogy between the stationary heat conduction problem and the corresponding electrostatic problem, we can assume [30] that the volume of material melted by a laser beam of power P does not exceed the volume of a hemisphere of radius R [see also (2.14)]:

$$R = \alpha P(2\pi\kappa T_m)^{-1}.$$

Thus, as for hardening, the melt depth can be calculated until the laser power does not exceed the value P_b [see (2.10)] at which the surface temperature achieves the boiling point. When the laser power exceeds $P > P_b$, the liquid on the surface starts boiling and an indentation appears in the melt due to the vapor recoil pressure. The so-called keyhole melting takes place, which will be discussed in more details in Chap. 4. It is shown in this chapter that upon such melting a liquid metal in the melt pool is strongly mixed, the mixing rate considerably exceeding the sample travel speed. As a result, the liquid is completely mixed during the time d/v of laser irradiation. The estimates of the liquid motion speed in the melt pool are complicated because this motion can be caused by several reasons. The theoretical calculation of these speeds is strongly complicated because of the absence of the general solution of the problem. Alloying with vaporization is usually performed for a small keyhole aspect ratio (the melt depth h to width d ratio). While in the case of a large aspect ratio some theoretical models were proposed (see Chap. 4), the theoretical consideration for $h \sim d$ is more complicated. In this case, the required vertical flow can occur due to the action of pressure of vapors evaporating from the front wall and escaping outside (without touching the rear wall). This flow is similar to flowing around an obstacle slightly immersed into liquid with a free surface. The liquid flows around such an obstacle from sides and bottom.

2.2.1 Analysis of Experimental Data

Figure 2.11 [31] shows melt profiles in the case of keyhole melting (Fig. 2.11a) and when a vapor keyhole is probably shallow (Fig. 2.11b). These melts were obtained in chrome-plated (50-μm thick chrome layer) pure iron exposed to the CO_2 laser radiation of the same power. The keyhole shape of the melt zone was obtained when the laser spot diameter on the sample was 66 μm (Fig. 2.11a), while the triangle shape was obtained when the spot diameter was 250 μm (2.11b). Because the threshold boiling power was $P_b \sim r_f^{3/2}$ [see (2.10)], it is obvious that the developed evaporation regime was realized in the first case. One can see from Fig. 2.11d,e that

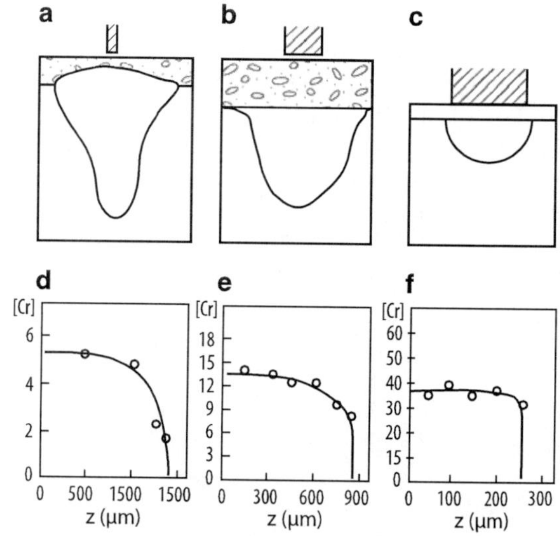

Fig. 2.11 Alloying zone shape for $P = 1.2$ kW, $v = 2.1$ cm/s and light spot sizes 66 μm (**a**), 250 μm (**b**), and 400 μm (**c**), and the relative chromium distribution in corresponding alloying zones (**d–f**)

in all cases chromium was quite homogeneously mixed over the entire depth of the melt. Figure 2.11c shows the cross section of the melt zone produced under the same conditions, but when the laser spot diameter was $d = 400\,\mu\text{m}$. One can see that in this case the keyhole melting does not occur: $P < P_b$ and the melt depth is smaller than its diameter. Despite the absence of evaporation, the mixing of chromium over the entire thickness of the melt zone is even more homogeneous than in previous cases (see Fig. 2.11f). The estimate of the chromium penetration depth due to diffusion gives the value an order of magnitude lower (the diffusion coefficient was assumed equal to 10^{-5} cm²/s [32]. It seems that mixing occurs under these conditions due to surface tension forces.

It follows from experiments that during the melting of metals by a laser beam, the liquid metal is mechanically mixed in the surface layer. The question arises of how this mixing affects the shape of the melt pool. It is known that mass transfer is accompanied by heat transfer. The influence of mixing on heat transfer can be taken into account in the simplest way by assuming that the heat conductivity increases in the melting zone. The increase in the heat conduction at $T > T_m$ in stationary problems changes the temperature profile only in the region $T > T_m$. The temperature distribution does not change in the zone where $T < T_m$. This follows from the fact that the dependence $\kappa(T)$ can be easily eliminated in the stationary thermal problem by introducing the thermal potential

$$\theta = \int_0^T \kappa(T)\mathrm{d}T.$$

By finding the distribution $\theta(r)$, which is independent of the form of $\kappa(T)$, we can reconstruct the dependence $T(r)$. It is obvious that the temperature distribution in the temperature region ($T < T_m$), where κ is independent of T, will coincide with the temperature distribution when κ is constant. The increase in κ at $T > T_m$ will only reduce temperature in this region. This can be also verified by solving the simplest problem of the temperature distribution from a point source. It is clear from general considerations that similar properties also take place in nonstationary problems. In particular, the thermal wave velocity is determined by the heat conduction of the unheated medium. However, the heat distribution profile behind the front ($T > T_m$) will be flatter than for $\kappa = const$. For example, the hardening process was described by expressions derived by neglecting the increase in the heat conductivity. We have seen that these expressions well agree with experimental data despites the fact that hardening was often accompanied by considerable melting. This occurs because the steel hardening temperature is lower than the melting point. Because of this, the possible increase in the heat conduction at $T > T_m$ did not affect the hardening depth. The increase in the heat conduction will only decrease the surface temperature of the melt pool, not changing its depth. The situation becomes somewhat more complicated when an alloying layer changes thermal properties of the material during heating and melting. Such a process occurs, for example, when alloying powders containing organic compounds are deposited. This method of powder deposition is not technological enough, but it well illustrates the properties of the process. The alloying of U10 steel was performed in [29] by depositing the 50% mixture of chromium powder with a binder burnt out at a temperature of 400°C (oxymethylcellulose). The thickness of a mixture layer was varied from 100 to 300 μm. Figure 2.12 shows the dependence of the alloying depth on the laser power for three different thicknesses d_n of the alloying paste layer ($v = 0.83$ cm/s, $d = 0.3$ cm). One can see that the alloying depth decreases with increasing the paste layer thickness. For $d_n \approx 100$–200 μm, chromium is mixed uniformly over the melt depth. Note that in this case for $P > 4$ kW the alloying depth increases with power stronger than for $P < 4$ kW. The estimate of the value of $\mathrm{d}h/\mathrm{d}\ln P$ from (2.6) gives the value ~0.3 cm; for $P \sim 4$ kW, the corresponding experimental value is ~0.04 cm. The discrepancy between these values is explained by the low heat conduction and burning out of organic from the paste. We see that this leads to a considerably decrease in the thermal diffusivity χ. At high powers ($P > 4$ kW), when chromium is probably rapidly melted, the value of $\mathrm{d}h/\mathrm{d}\ln P$ is 0.3 cm, which coincides with the theoretical estimate. At a constant power, the alloying depth increases linearly with increasing $v^{-1/2}$ [see (2.9)]. These experiments also reveal the power dependence of thermal coefficients. Figure 2.14 presents the dependences

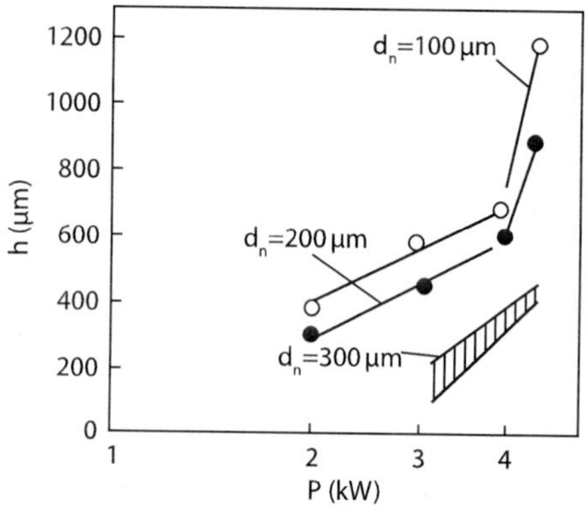

Fig. 2.12 Dependences of the chromium-alloying depth of C105W2 steel on the CO_2 laser power for different thicknesses of the initial alloying layer

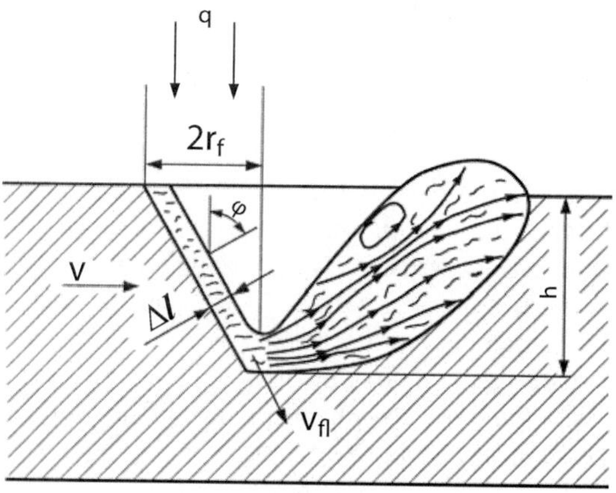

Fig. 2.13 Cheme of shallow keyhole with vapors evaporation from front wall

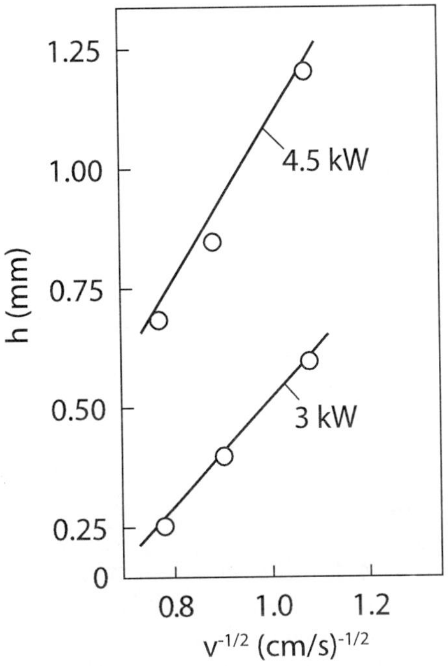

Fig. 2.14 Dependencies of the alloying depth of the alloying speed for different CO2 laser powers

of h on $v^{-1/2}$ for $P = 3$ and 4.5 kW [29]. The theoretical value of the slope of the curves equal to $\sqrt{4\chi d}$ is $0.25\,\text{cm}^{3/2} \cdot s^{-1/2}$ for parameters used in experiments. For $P = 3$ and 4.5 kW, the experimental slopes are 0.11 and $0.17\,\text{cm}^{3/2} \cdot s^{-1/2}$, respectively. Thus, these dependences also demonstrate the increase in χ with increasing the laser beam power (when the alloying material is deposited together with organic).

2.2.2 *Theoretical Consideration of Melt Motion During Alloying of Metals*

Motion of liquid during developed evaporation. As mentioned above, the motion of liquid with respect to a sample in the melt pool can be produced by the pressure of vapors evaporating from the front wall of a shallow vapor keyhole (Fig. 2.13). Consider the theoretical model of this phenomenon. A sample is being moved into a vertical laser beam. We assume that the thickness Δl of a liquid pool on the front wall of the keyhole is small compared to the laser beam diameter $2r_f$. This thickness is determined by the velocity of the approach flow of the target material. Estimates, similar to those performed is Sect. 8.1, show that for the specified travel speed of the sample V, the keyhole depth and liquid velocity V_l increase with the laser beam power. For the parallel polarization of laser radiation, we have

$$v_l = \sqrt{2c_s \alpha q / \rho L} \qquad (2.16)$$

where c_s is the sound speed in metal vapor. For the maximum admissible theoretical value of q determined by the condition $\tan\varphi = 1$ ($\alpha q_{\max} = 2V^4 r_l^2 L\rho/\chi^2 c_s$), we have $v_l/v = vr_f/\chi = Pe$. Because the theory is constructed by assuming that Peclet's number is large ($Pe \gg 1$), we have $v_l/v \gg 1$. We will show below that the local narrowing of the liquid flow enhances the diffusion of the alloying material.

Consider the case when $P < P_b$, i.e. evaporation is absent. In this case, it is most likely that the liquid motion is caused by the surface tension force.

2.2.3 *Analytic Consideration of Liquid Metal Motion Caused by Thermocapilary Forces*

As mentioned above, one of the main advantages of laser processing of materials is its selectivity. This property is related to a small size of the laser spot compared to the sample size. In this case, the temperature of the heated sample surface depends on the coordinate along the surface. The temperature is maximal at the center of the laser spot, decreasing to the spot edge. The surface tension depends on the surface temperature and usually decreases with temperature $-(\mathrm{d}\sigma/\mathrm{d}T) > 0$ [33]. As a result, a force appears on the surface, which is directed from the laser spot center to its edges. If the surface is melted, this force produced the motion of liquid (see Sect. 1.3), which is called the Marangoni flow. It is known that the surface tension depends not only on temperature but also on some surfactants on the melt surface [34]. Surfactants can often excite vortices directed from the melt pool periphery to its center [35], as in the case of $\partial\sigma/\partial T > 0$ [36–38]. It is likely that both thermocapillary and concentration-capillary effects can be manifested during laser alloying. It was pointed out in [39–42] that even low concentrations C of surfactants can strongly affect both $\sigma(T)$ and $\sigma(C)$. However, in many important cases, for example, for alloys of Fe with Mo, V, and Cr up to concentrations ~20%, the change in the surface tension does not exceed 10% [41, 42]. Therefore, in these cases $\sigma(T, C) \sim \sigma(T)$, and thermocapillary convection plays a key role.

The Marangoni flow has long been studied in connection with various phenomena not related to laser technologies. These studies are described in review [43]. Many properties of thermocapillary convection were adequately described theoretically. However, the specific features of the melt flow during laser alloying require additional investigations (see [43] and references therein). We begin our consideration with the simplest case on s thin liquid layer (see Fig. 2.15).

The force $\partial\sigma/\partial x$ acting on the unit surface of a sample is balanced by the viscous friction force $\eta\partial v_x/\partial z$:

$$\eta\partial v_x/\partial z = \partial\sigma/\partial x = (\partial\sigma/\partial T)(\partial T/\partial x) \tag{2.17}$$

Thus, the velocity rotor is specified on the liquid surface. When the Reynolds number is small ($Re \ll 1$), the change in the velocity vector is described by the Helmholtz equation [44]

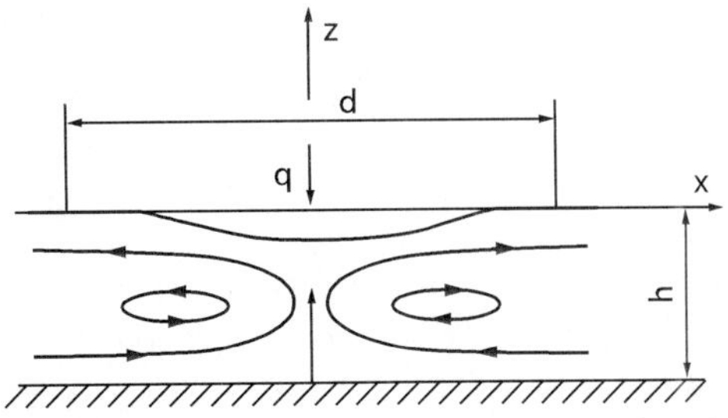

Fig. 2.15 Scheme of a liquid melt flow

$$\partial \mathbf{rotv}/\partial t = \nu \Delta \mathbf{rotv}, \tag{2.18}$$

where ν is the kinematic viscosity of the melt. This equation is analogous to (2.1) describing the heat propagation. The distance over which heat propagates for the time t is $z = 2\sqrt{\chi t}$.

Similarly, the characteristic distance of the action of the surface tension force is $\delta = 2\sqrt{\nu t}$, where t is the time measured from the beginning of the force action. By knowing δ, we find from (2.17) the velocity on the sample surface:

$$v_s = \frac{1}{\eta} \frac{\partial \sigma}{\partial T} \frac{\partial T}{\partial x} \sqrt{\nu t} \tag{2.19}$$

Apart from the vortex component of the velocity described by (2.18), there also exists the potential component of the velocity $v = \nabla \varphi$, where φ is the velocity potential. The potential φ satisfies the Laplace equation

$$\Delta \varphi = 0 \tag{2.20}$$

It follows from this equation that the characteristic variation of the velocity is determined by the geometry of the problem, in the given case, by the laser beam diameter d and the pool depth h:

$$v_x h \cong v_z d \tag{2.21}$$

If $h \gg d$, which is not typical for laser alloying, the only size of the problem is d and we have $v_x \sim v_z$. In this case, by using the continuity of the liquid mass flow, we obtain the relation

$$v_x d \cong v_s \delta. \tag{2.22}$$

between the velocities in the boundary layer and center of the flow.

By substituting (2.19) and the expression for δ into (2.22), we obtain [45]

$$v_x \cong v_z \cong \frac{1}{\rho} \frac{\partial \sigma}{\partial T} \frac{t}{d} \frac{\partial T}{\partial x}. \tag{2.23}$$

Not that the transient motion is independent of the liquid viscosity. In a simple case, when $h \ll d$, the boundary layers will merge within the time $t \cong h^2/\nu$

after the beginning of the source action, and the flow will transfer to the viscous thermocapillary regime [34]:

$$v_x \cong \frac{1}{\eta}\left|\frac{\partial \sigma}{\partial T}\right|\frac{\partial T}{\partial x}h. \tag{2.24}$$

Recall that all the results were obtained by neglecting the influence of nonlinear terms in the Navier–Stokes equation. Because the Prandtl number for metals is $Pr = \nu/\chi < 1$, the nonlinearity (convective heat transfer) can be neglected in the heat conduction equation as well. In this case, the temperature gradient in a shallow pool ($h \ll d$) can be readily estimated as

$$\partial T/\partial x \cong hq/\kappa d \tag{2.25}$$

The convective (nonlinear) terms in the Navier–Stokes equation can be neglected under the condition

$$Re = v_x h/\nu \ll d/h. \tag{2.26}$$

In the stationary case, this condition restricts the laser intensity. By substituting (2.24) and (2.25) into (2.26), we obtain

$$q \ll q_\nu = \kappa\rho\nu^2 d^2 h^{-4}|\partial\sigma/\partial T|^{-1} \tag{2.27}$$

The power density absorbed by a liquid metal (iron) estimated by this expression is $q_\nu \sim 10\,\text{kW/cm}^2$ for $d \sim h \sim 0.1$ cm. Laser power densities used upon alloying commonly greatly exceed this value, and therefore convective transfer is insignificant only if the expose time Δt is small. This time can be estimated from expression (2.26) by taking v_x from (2.23):

$$\Delta t \ll d^2\kappa\eta(h^2 q|\partial\sigma/\partial T|)^{-1} \tag{2.28}$$

In the case of small Reynolds numbers, the heat transfer is independent of the liquid convection. This simplifies the study of the problem also in a more complicated case when the form of the melt pool should be calculated as well [46, 47]. This situation is close to experimental laser alloying conditions [47]. The calculation of the melt pool form is quite cumbersome and necessitates the use of numerical methods. Below, we present for generality the total formulation of the problem for arbitrary Reynolds numbers.

2.2.4 Numerical Modelling of a Melt Flow During Alloying

We will simulate heat-and-mass transfer in a laser melt by using the system of equations for the velocity v of viscous incompressible liquid and the melt temperature T,

$$div\mathbf{V} = 0 \tag{2.29}$$

$$\frac{\partial \mathbf{V}}{\partial t} + (\mathbf{V} grad)\mathbf{V} = -\frac{1}{\rho_l} grad\, p + \nu \Delta \mathbf{V} \tag{2.30}$$

$$C_l \rho_l \frac{\partial T}{\partial t} + C_l \rho_l div(\mathbf{V}T) - div(\kappa_l gradT) = 0 \tag{2.31}$$

The temperature distribution in a solid phase is described the heat conduction equation

$$C_s \rho_s \frac{\partial T}{\partial t} - div(\kappa_s gradT) = 0 \tag{2.32}$$

The propagation velocity of the liquid–solid interface is determined by the difference of heat flows coming from the melt and outgoing inside the solid phase on the melting isotherm:

$$u H_m = \kappa_s \frac{\partial T}{\partial n}\bigg|_s - \kappa_l \frac{\partial T}{\partial n}\bigg|_l \tag{2.33}$$

Here, p is pressure in the liquid; C_l and C_s are the specific heats of liquid and solid phases; κ_l and κ_s the heat conductivities; ρ_l and ρ_s are densities; ν is the kinematic viscosity coefficient of the melt; H_m is the phase transition (melting) heat; u is the propagation velocity of the phase interface. The subscripts l and s refer to the liquid and solid phases, respectively.

It is assumed that the upper boundary is flat and the velocity satisfies the boundary conditions

$$v_z|_{z=0} = 0, \qquad \mu \frac{\partial v_r}{\partial z}\bigg|_{z=0} = \gamma \frac{\partial T}{\partial r} \tag{2.34}$$

Here, μ is the dynamic viscosity coefficient and $\gamma = \partial\sigma/\partial T$. The temperature at the sample boundary away from the zone of interaction of radiation with material was assumed constant and equal to T_0, while the heat flow distribution at the sample surface is specified as the function of time

$$\kappa(r) \frac{\partial T}{\partial z}\bigg|_{z=0} = \alpha(T) q(r,t), \tag{2.35}$$

where $\alpha(T)$ is the absorption coefficient for laser radiation and $q(r,t)$ is the laser radiation power distribution at the sample surface.

Thus, nonlinear terms in equation can be omitted in nonstationary case (2.28). The results of the numerical calculation of the system are presented in Fig. 2.16, which demonstrates that the pool size increases with time [47]. According to (2.23), the convection velocities v_r and v_z also increase (Fig. 2.17). Due to the increase in the sizes of the melt pool and vortex, liquid particles move along intricate paths, resulting in the mixing of the alloying additive with the matrix (see Fig. 2.18). Figure 2.19 compares the pictures of mixing additives introduced either in the flow from the gas phase (synthesis of Ti nitrides) or from a preliminarily deposited coating (molybdenum on Armco-iron, etc.). The main difference between

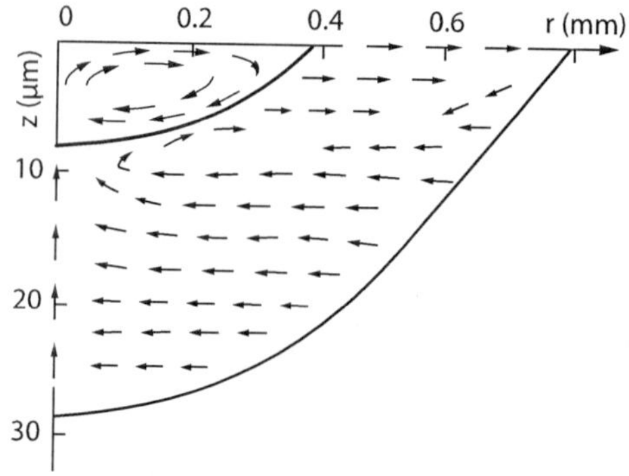

Fig. 2.16 Molten pool shape and field of directions for velocity vector in surface tension melt flow of titanium ($q = 5 \cdot 10^4$ W cm^{-2}, $r_f = 1$ mm) $t = 0.47$ ms and 1 ms

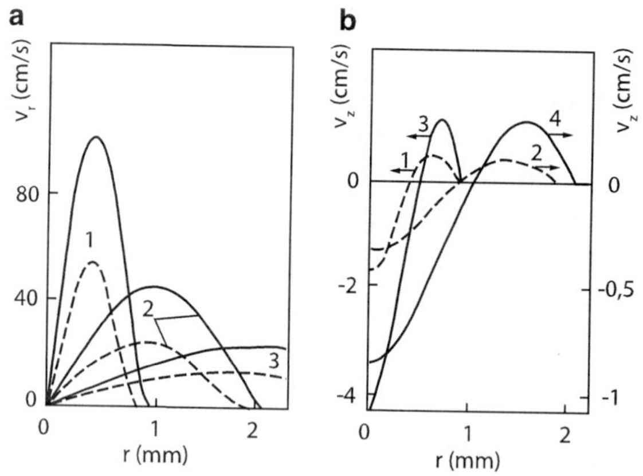

Fig. 2.17 (**a**) The Vr component (for $z = 0$) versus radius for $t = 1$ ms (*dashed curves*) and $t = 1.4$ ms (*solid line curves*). Material Ti, $q = 5 \cdot 10^4$ W/cm^2; curve 1 correspond to $r_f = 1$ mm; curve 2 to $r_f = 2$ mm; curve 3 to $r_f = 4$ mm. (**b**) The Vz component of melt velocity (on the depth $z = S(r, t)/3$) versus radius for $t = 1$ ms (*dashed curves*) and $t = 1.4$ ms (*solid line curves*). Material Ti, $q = 5 \cdot 10^4$ W cm^{-2}; curves 1 and 3 correspond to $r_f = 2$ mm; curves 2 and 4 correspond to $r_f = 1$ mm

these methods is that in the first case the amount of additives in the solution is continuously increased, while in the second one the alloying component is only redistributed. The alloying zone in these figures is conditionally divided into three regions. The concentration of the additive from 0.5 to 1 (max) is shown by most dark region, the second region corresponding to concentration from 0.1 to 0.5 is shown by less dark region, and the third region corresponding to concentration below 0.1 is shown by light region. One can see that in both cases the additive first collects near the edge of the melt pool. Due to the vortex motion of the melt, the alloying zone takes the characteristic spiral shape with alternating regions of

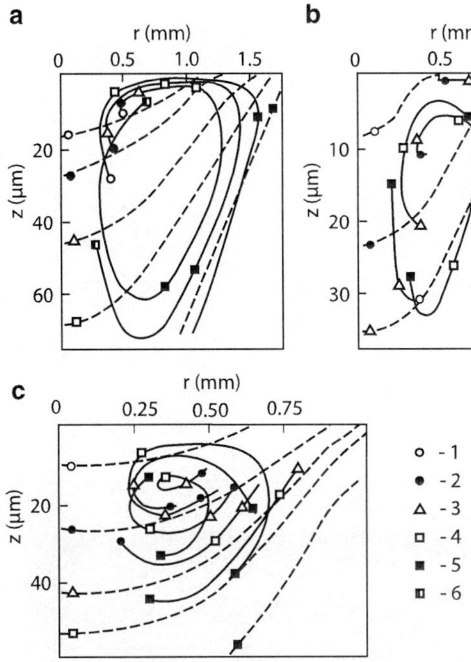

Fig. 2.18 Trajectories of particles of the melt: material – Ti: (**a**) $q = 5 \cdot 10^4\,\mathrm{W\,cm^{-2}}$, $r_f = 1.4\,\mathrm{mm}$; (**b**) $q = 5 \cdot 10^4\,\mathrm{W\,cm^{-2}}$, $r_f = 1\,\mathrm{mm}$; (**c**) $q = 5 \cdot 10^4\,\mathrm{W\,cm^{-2}}$, $r_f = 0.8\,\mathrm{mm}$. The time is indicated by the position of symbols 1–6 that correspond to the following moments: symbol (**1**) $t = 0.2\,\mathrm{ms}$, (**2**) $t = 0.4\,\mathrm{ms}$, (**3**) $t = 0.6\,\mathrm{ms}$, (**4**) $t = 1\,\mathrm{ms}$, (**5**) $t = 1.6\,\mathrm{ms}$, (**6**) $t = 2.3\,\mathrm{ms}$

different concentrations. In the case of alloying from gas, the alloying component is always present on the surface after some time, whereas during alloying from the coating the alloying film has the disruption region. Typically, the maxima of the alloying component are formed in both cases inside the melt and the impurity does not reach the melt pool bottom. The characteristic features of structures described above were observed in experiments [49,50]. Figure 2.19c demonstrates the layered structure of the alloying zone of iron by molybdenum, a thin layer of the latter being preliminarily deposited on the sample surface. The features of alloying from the gas phase will be described below. The qualitative agreement between calculations and experiments is demonstrated in Fig. 2.20, where the thermal surface tension coefficient was positive $\partial\sigma/\partial T > 0$ [47].

At large irradiation times, the velocity of vortices increases so that nonlinear terms in the system of equations (2.29)–(2.35) become important, i.e. convective flows change themselves and affect the temperature field. Let us analyze qualitatively these features by neglecting their influence on the form of the melt pool.

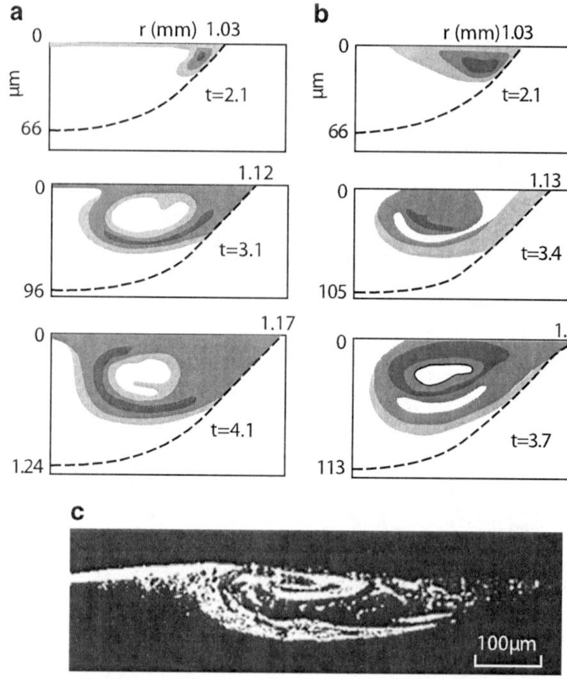

Fig. 2.19 (**a**) Concentration fields of the alloying element at indicated instants of time (ms). Absorbed constant flux of alloying material on melt free surface; base metal Ti; $q = 5 \cdot 10^4\,\mathrm{W\,cm^{-2}}$, $r_f = 1.4\,\mathrm{mm}$. (**b**) $q = 5 \cdot 10^4\,\mathrm{W\,cm^{-2}}$, $r_f = 1\,\mathrm{mm}$. Pulsed laser alloying of Fe from Mo coating. Cross-sectional photos left part (**c**)

2.2.5 Nonlinear Effects and the Instability of the Melt Surface Shape in the Marangoni Flow

It can be assumed that the flow structure will be preserved when nonlinear terms are taken into account (Fig. 2.21), however, quantitative properties will be different. In the case of rapid rotation, a liquid particle is subjected to alternating thermal action. A particle moving along the pool surface is heated, while a particle moving along the pool walls is cooled. The perturbation penetrates into liquid per revolution by the characteristic distance $\delta_T \cong \sqrt{\chi/\omega}$, where ω is the rotation frequency. When $d \sim h$, this distance can be represented in the more convenient form

$$\delta_T \cong d\sqrt{Pr}/\sqrt{Re} \tag{2.36}$$

It follows from (2.36) that the greater the rotation speed, the thinner the thermal boundary layer at the melt pool boundary (see Fig. 2.21). Thus, the temperature drop ΔT occurs over the characteristic distance δ_T. For the specified intensity, we have $\Delta T \cong q\delta_T/\kappa$. Thus, the temperature drop in a rapidly rotating liquid decreases

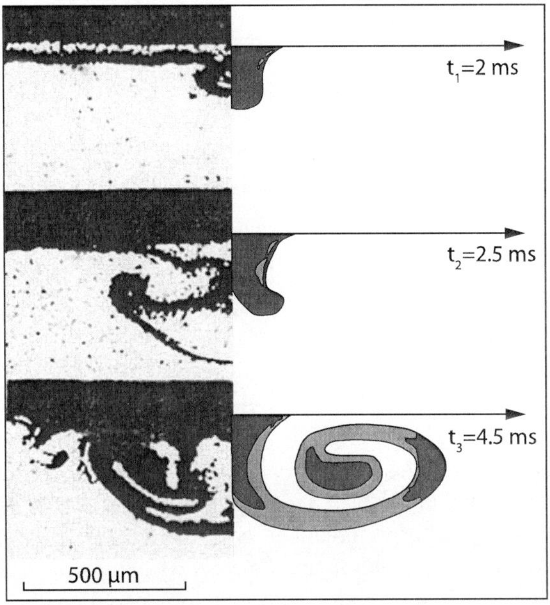

Fig. 2.20 The propagation of admixture in pulsed laser alloying from pre-deposited coating for positive value of surface tension coefficient. Numerical simulation: $q = 5 \cdot 10^4\,\mathrm{W\,cm^{-2}}$, $r_f = 1.4\,\mathrm{mm}$, the corresponding instants of time: 2, 2.5, 4.5 ms. Experiment: graphite coating 10 μm thickness on Ti, pulsed Nd:YAG laser

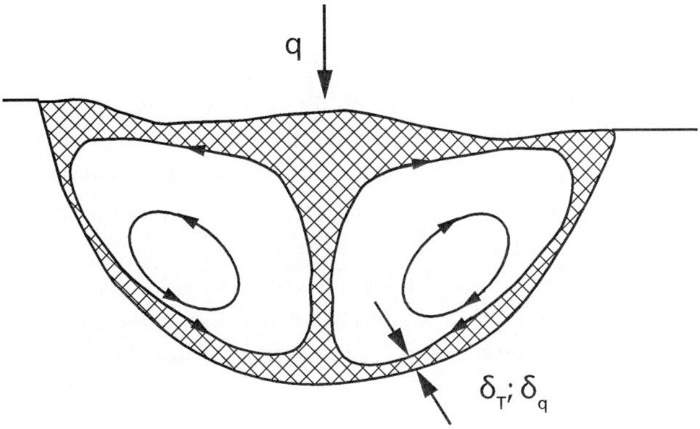

Fig. 2.21 Scheme of a liquid flow in a melt pool for the large Reynolds number

compared to a liquid at rest (or slowly rotating) by the factor $\sqrt{RePr} \gg 1$. In a particular case of a rotating liquid, similar to a rotating solid, this was rigorously proved in [45]. Note that the mathematically similar phenomenon of the ejection of a magnetic field from the flow region with closed current lines for the large magnetic Reynolds number was studied in the magnetic dynamo theory [48].

A similar picture takes place for the velocity distribution [45]. The velocity rotor in the pool volume is constant and changes only in a narrow region of viscous boundary layers:

$$\delta_\eta = d/\sqrt{Re}. \tag{2.37}$$

By equating the surface tension force to the friction force, as in the derivation of (2.36), and taking into account the characteristic variation scales of the velocity and temperature, we obtain

$$v \cong \left|\frac{\partial\sigma}{\partial T}\right|\frac{q}{d\eta}\delta_T\delta_\eta. \tag{2.38}$$

By substituting (2.36) and (2.37) into (2.38), we finally obtain

$$v^2 \cong \frac{q}{\kappa\rho\sqrt{Pr}}\left|\frac{\partial\sigma}{\partial T}\right| \tag{2.39}$$

Let us estimate the velocity of thermocapillary motion. By substituting $q = 5 \times 10^4$ W/cm^2, $\rho = 7$ g/cm^3, $|\partial\sigma/\partial T| = 5\times10^{-8}$ J/(cm^2 K), $\kappa = 0.23$ W/cm K, $\sqrt{Pr} = 0.2$, $d \cong 0.1$ cm into (2.39), we obtain $v \cong 2.6\,m/s$, $PrRe \cong vh/\chi \sim 26$, in good agreement with the assumption that $Re \gg 1$. By comparing (2.24), (2.25), and (2.39), we see that the liquid mixing rate increases first linearly with increasing laser radiation power and then as $\sqrt{q}$. The latter is explained by the decrease in the temperature gradient due to convection in the liquid.

The flow structure was compared with expression (2.39) by using a model medium where the appropriate measurements could be performed. An appropriate liquid for this purpose is paraffin in which the Marangoni effect can be stably observed. Paraffin was poured into a cell of size $10 \times 50 \times 100$ mm with transparent walls. The liquid velocity was measured with a laser anemometer, and the general structure – by the shadow method [51]. One can see from Fig. 2.22 that the liquid velocity near the melt surface in nonmonotonic. This corresponds to the vortex structure of the flow. The maximum of the velocity corresponds to the vortex center. As the laser power P is increased, the flow velocity increases and the vortex shifts from the beam center. This is also observed in experiments with another medium ($NaNO_3$ with aluminum powder) (Fig. 2.23) [52]. Figure 2.24 presents the flow velocity of the melt at different distances from the laser beam as functions of the absorbed power. The curves are constructed according to the dependence $v = AP^{1/2}$ (2.39) with the coefficient A depending on the distance of the observation point from

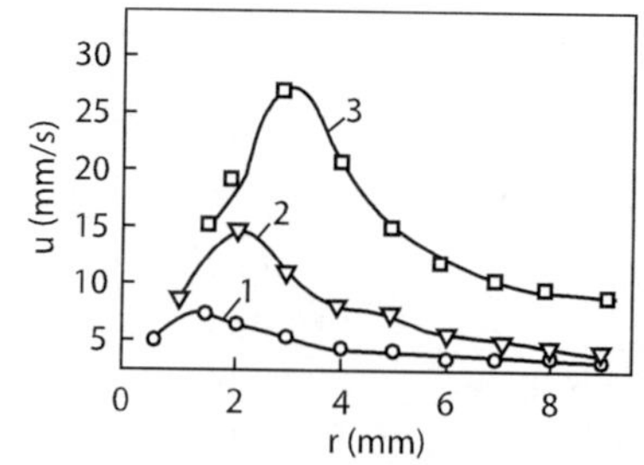

Fig. 2.22 Distributions of the liquid flow velocity in the surface layer for $P = 5.5$ W (**1**), $P = 8$ W (**2**), and $P = 17$ W (**3**)

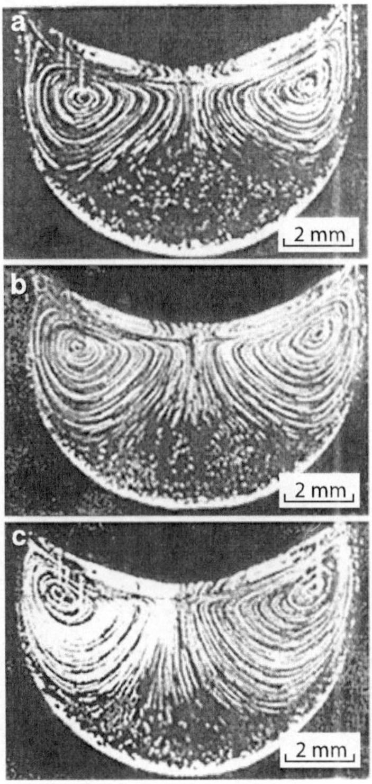

Fig. 2.23 Flow patterns induced by 25 W CO_2-laser beam at the following diameters: (**a**) $d = 5,9\,\mathrm{mm}$, (**b**) $d = 3,2\,\mathrm{mm}$, (**c**) $d = 1,5\,\mathrm{mm}$

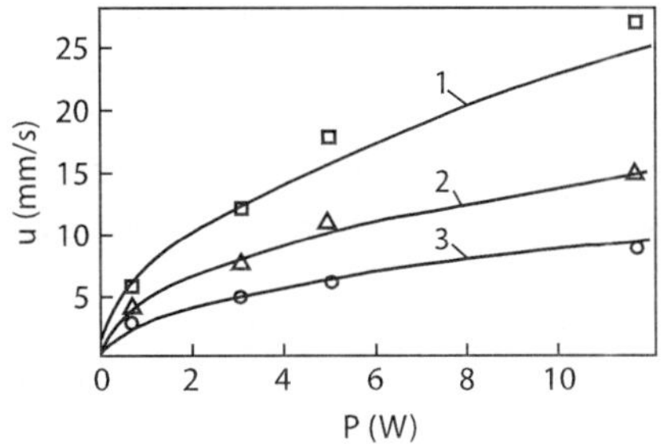

Fig. 2.24 Experimental (*points*) and calculated (2.39) (*curves*) dependences of the flow velocity on the absorbed power in the surface layer at distances of $r = 3\,\mathrm{mm}$ (**1**), $r = 5\,\mathrm{mm}$ (**2**), and $r = 8\,\mathrm{mm}$ (**3**) from the axis

beam center. The temperature field of the melt is shown in Fig. 2.25. A large region with the uniform temperature caused by the rapid rotation of the melt is observed. This is consistent with the theoretical model described above. Figure 2.26a,b shows a photograph of the paraffin melt pool at laser powers 11 and 20 W. The dark dots seen in paraffin are produced by graphite powder added to the sample visualize the image. The exposure time was 1/30 s. The focal spot diameter of the laser beam incident from top was $\sim$1 mm. The flow structure is similar to that considered above, two vortices being established on both sides of the beam. Directly under the beam, a stagnant zone exists. One can see that temperature in this zone is lower than temperature at the same depth away from the beam axis. This fact suggests that the heat transfer by the liquid flow dominates under these conditions.

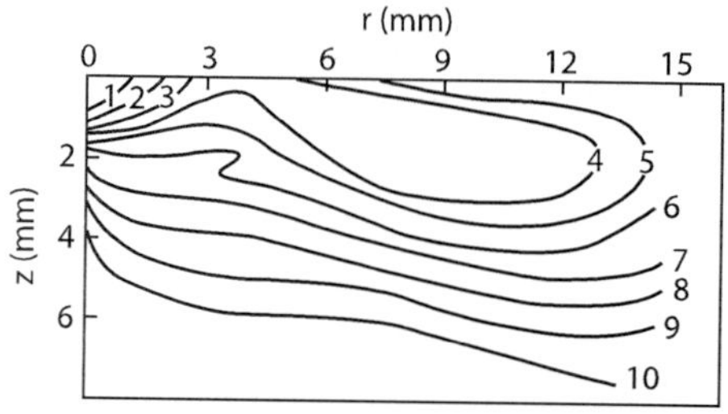

Fig. 2.25 Isotherms in a paraffin melt during a thermocapillary vortex flow induced by laser radiation directed along the z axis for $P = 15$ W, $T = 220°$ (**1**), 205 (**2**), 195 (**3**), 185 (**4**), 175 (**5**), 165 (**6**), 160 (**7**), 150 (**8**), 140 (**9**), and 130° (**10**)

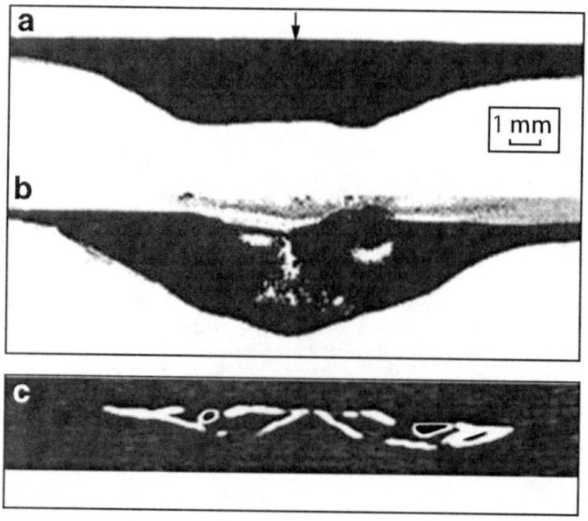

Fig. 2.26 Melt pool in paraffin for $P = 11$ W (**a**) and $P = 20$ W (**b**) (Photo by V.G. Niz'ev). $P > 17$ W (**c**)

As the laser beam power is further increased (above 17 W), new vortices appear against the background of a large-scale vortex in a shadow photograph (Fig. 2.26c). The appearance of additional vortices can qualitatively change the general picture of melting in the case of the Marangoni effect. As we will show below Fig. 2.34 [53], excitation of even one vortex affects the form of the metal melt pool. The influence of thermocapillary vortices on the melt pool form can be studied by using the same model material (paraffin).

We considered above the melt flows excited over the entire depth of the pool assuming for simplicity that the liquid surface is flat. However, the surface of the melt zone remains ribbed after solidification in experiments even in the case of shallow melting (alloying, welding of thin plates). Periodic flakes of length $\sim 10\,\mu$m extended along the weld middle are formed on the surface. To study this effect, the formulation of problem (2.29)–(2.34) should be changed. It is necessary to discard the assumption about the flat melt surface. The bent surface produces pressure fluctuations on the liquid surface. This pressure can be used as the boundary condition for the Navier–Stokes equation, which in turn should be transformed so that pressure would become the unknown function [54]. Thus, pressure on the free liquid surface is determined by the Stokes expression

$$P_s = \sigma \frac{h_{xx}(1 + h_y^2) - 2h_x h_y h_{xy} + h_{yy}(1 + h_x^2)}{2(1 + h_x^2 + h_y^2)^{3/2}}$$

where $h_x = \partial h/\partial x$, $h_{xx} = \partial^2 h/\partial x^2$, $h_y = \partial h/\partial y$, $h_{yy} = \partial^2 h/\partial y^2$, $h_{xy} = \partial^2 h/\partial x \partial y$. The heat transfer equation, unlike standard form (2.31), is written in the approximate form by making the replacements

$$\rho \frac{\partial}{\partial t} C_p T \to \rho \frac{\partial}{\partial t} C_p T + L\rho \frac{\partial f}{\partial t}, \quad \rho \mathbf{u} \nabla C_p T \to \rho \mathbf{u} \nabla C_p T + \rho \mathbf{u} L \nabla f \; ,$$

where f is the faction of the melt in a cell in the numerical solution of the system of equations: $0 \le f \le 1$. By omitting the details of the numerical method [55], we present the results. The conditions of numerical calculations are presented in Table 2.2. Note that we consider the case when $\partial\sigma/\partial T > 0$, i.e. the melt flow is directed from the pool boundaries to its centre. This explains the form of the longitudinal cross section of the melt pool in Fig. 2.27a. One can see from this figure

Table 2.2 Material properties

Focal spot diameter	0.001 50	m
Specific heat	940.0	$\mathrm{J\,kg^{-1}K^{-1}}$
Density	4098.360	$\mathrm{kgm^{-3}}$
κ	50.0	$\mathrm{W\,m^{-1}K^{-1}}$
Latent heat	20.00	K
Laser power	2500.0	W
Laser speed	0.050	$\mathrm{ms^{-1}}$
Viscosity	1×10^{-4}	$\mathrm{kgm^{-1}s^{-1}}$
Gravity	9.81	$\mathrm{ms^{-3}}$
$\partial\sigma/\partial T$	1×10^{-4}	$\mathrm{Nm^{-1}K^{-1}}$
Surface tension	1.0	$\mathrm{Nm^{-1}}$

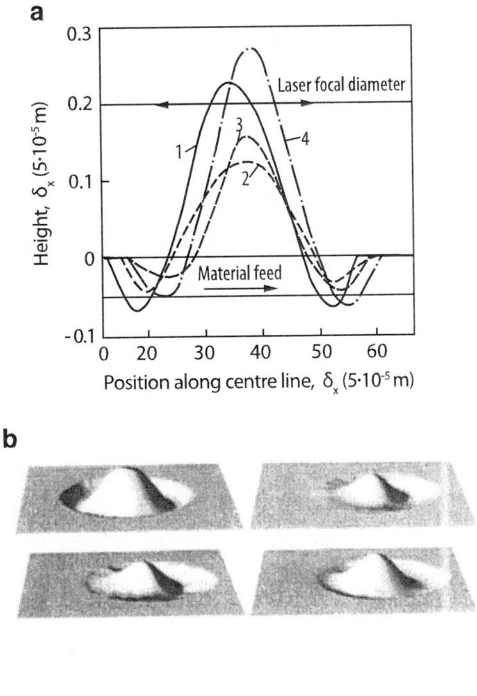

Fig. 2.27 (**a**) Transient surface deformation along the feed direction ordered according to elapsed time t. (**b**) Three-dimensional surface plot with amplified height. Time sequence: *top-left* $t = 0.124$ s, *top-right* $t = 0.12884$ s, *bottom-left* $t = 0.13135$ s, *bottom-right* $t = 0.1319$ s

that the sample melting oscillates with a period of 2–3 ms and amplitude ~30 μm. Figure 2.27b demonstrates the intricate picture of three-dimensional oscillations of the surface. The reason for such oscillations is unclear, although, as we have seen in Chap. 1, capillary waves can be unstable on the liquid surface heated from above.

2.2.6 *Development of the Multi-Vortex Structure of the Melt Flow*

As shown above, the rotation velocity of the liquid increased with increasing the laser radiation power. Under certain conditions, this leads to excitation of the second vortex in the metal melt, which in turn can qualitatively change the melt form in such a way that the ratio of the pool width to its depth can become smaller than unity, as in the case of deep-penetration laser welding of metals (see Chap. 4). Consider the conditions for the appearance of the multivortex structure.

The multivortex flow can be excited when the radiation intensity distribution in a laser spot is nonmonotonic with respect to the beam axis. The temperature distribution produced on the surface of a sample heated by such laser pulses can be also nonmonotonic. Numerical calculations show that this gives rise to vortices, their number being equal to the number of extrema of the laser radiation intensity profile (Fig. 2.28) [43, 56]. The multivortex structure of the flow also changes the alloying picture. Figure 2.29 presents the distributions of the impurity concentration at

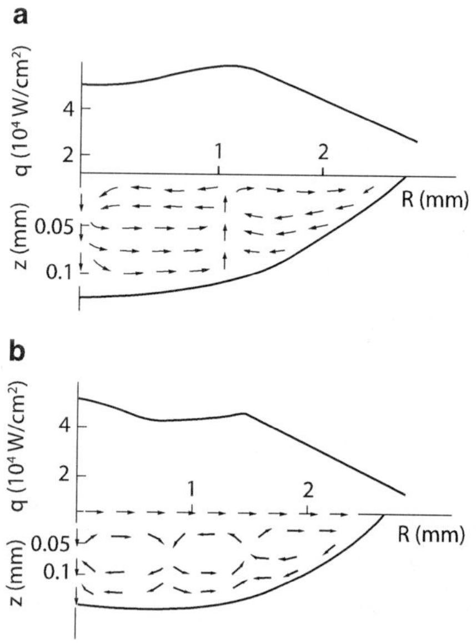

Fig. 2.28 Field of directions for velocity vector of surface tension melt flow of titanium at time moment (**a**) $\tau = 3.5$ ms, (**b**) $\tau = 4.3$ ms for the absorbed energy flux radial distribution, shown on upper curves

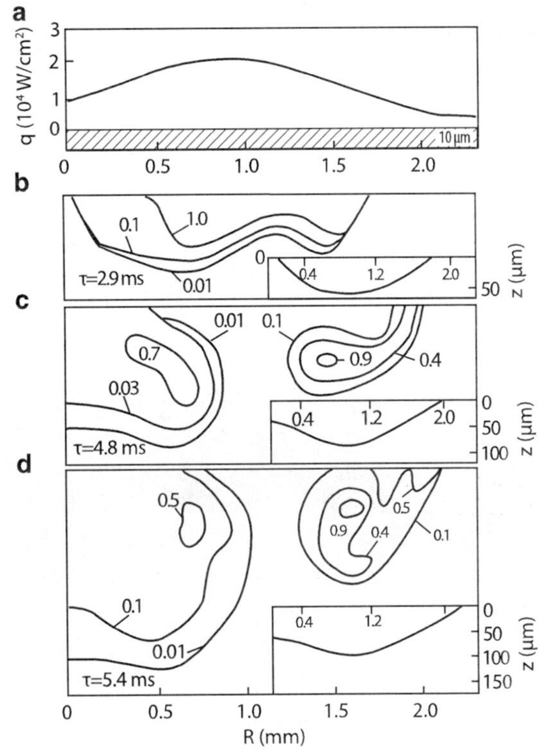

Fig. 2.29 Concentration fields of the alloying element at corresponding time moments (**b**) $\tau = 2,9$ ms, (**c**) $\tau = 4.8$ ms, (**d**) $\tau = 5.4$ ms. Base metal Ti; coating thick-ness 10 μm; continuous coating; a spatial distribution of energy density flow and of the coating initial position (**a**). The corresponding shapes of molten pool are shown in the frames in the reduced scale

different instants (similarly to Fig. 2.19). One can see that the alloying by a circular laser beam leads to the appearance of an additive at the focal spot centre as well, which was not observed for a Gaussian beam (see Fig. 2.19b). The theoretical study of the regimes of thermocapillary convection established in a cylindrical volume with the side surface heated increasing upwards [57] showed that, as ΔT increases and the Reynolds number achieves a certain value, the single-vortex regime transfers to the two-vortex regime, and when the Reynolds number further increases, the third vortex appears, etc. A similar change of thermocapillary flow regimes can also occur upon laser irradiation of the horizontal surface of metals. However, in this case it is impossible to perform a detailed theoretical analysis because the melt flow occurs in the region of a complex form and is interrelated with the melting process.

The conditions under which the passage to the two-vortex structure occurs can be determined by calculating numerically the system of equations (2.29)–(2.35) by varying the laser beam parameters. The absorption coefficients α of a solid phase (stainless steel) and a liquid phase at the laser wavelength were assumed equal to 0.08 and 0.16, respectively [58]. The results are presented in Table 2.3, where along with the parameters of a Gaussian laser beam, the parameters of the melt achieved at the instant of the creation of the second vortex are indicated. Here, $v_{\max}$ is the

Table 2.3 Parameters of radiation and a melt pool at the instant of the second vortex appearance

P_0, kW	r_f, mm	T_0, K	R/r$_0$	h/r$_0$	ΔT, K	V_{max}, cm/s	Ma$\times10^{-3}$	$Re^* \times 10^{-3}$
0.8	0.25	1,000	1.58	0.68	1,230	260	4.27	2.45
1.0	0.25	300	1.49	0.62	1,485	295	4.86	2.4
1.25	0.50	300	1.24	0.51	840	177	4.58	2.1
2.5	0.50	300	1.26	0.44	1,465	253	8.11	2.1
3.0	0.75	300	1.26	0.44	1,155	186	9.59	2.3
4.0	0.75	300	1.26	0.42	1,455	217	12.10	2.4
4.0	1.0	300	1.22	0.40	1,090	165	11.70	2.2
6.0	1.0	300	1.26	0.37	1,500	202	16.60	2.1
4.0	1.0	1,000	1.36	0.43	1,100	161	13.1	2.15
7.0	1.25	300	1.26	0.37	1,380	172	19.1	2.3

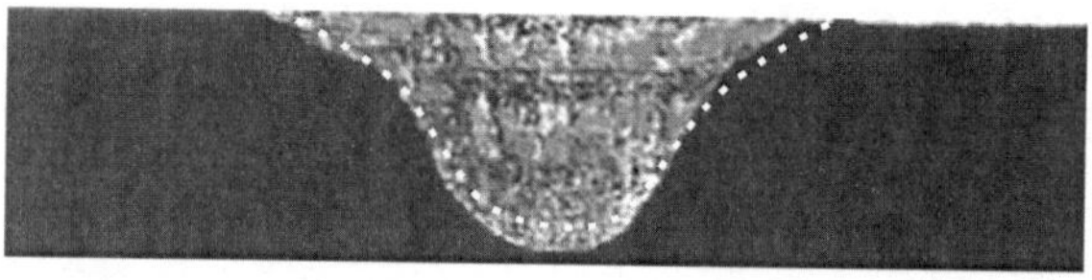

Fig. 2.30 Comparison of the experimental shape of a melt pool (shown by a *light shade*) with the calculated shape (*points*) for a 4 mm-tick SS304 stainless steel sample irradiated by a 6 kW, 4 s laser pulse

maximum velocity in the melt; $Ma = \gamma \Delta TR/\mu\chi$ is dimensionless parameter (the Marangoni number). Here, as in [59], the equivalent Reynolds number

$$Re^* = \frac{\partial\sigma}{\partial T}\frac{\nabla Th}{\mu\nu}\left(\frac{h}{R}\right)^3$$

is introduced, which takes into account the aspect ratio of the melt radius R to its depth h.

The Marangoni number was varied in calculation in a broad range from ~4 to 19. The most interesting is the fact that the appearance of the second vortex is characterized by the critical value of the equivalent Reynolds number, which is almost the same for all variants.

The measurement of the parameters of a melted metal pool during laser irradiation is a time-consuming problem. Because of this, the growth dynamics of a melt pool was measured by irradiating the surface of 4-mm thick stainless SS304 steel samples by rectangular laser pulses of the same intensity but different durations varied from 0.03 to 8 s [58].

Figure 2.30 shows the cross-sectional view of a 4-mm thick stainless SS304 steel sample after irradiation by a 4 s, 6 kW laser pulse ($d_f = 4$ mm). The dots show the phase interface calculated for these radiation parameters. The calculated and experimental dependences of the melt pool diameter on the laser pulse duration are compared in Fig. 2.31. One can see that in both cases the results are in good agreement, which confirms that the theoretical model adequately describes real

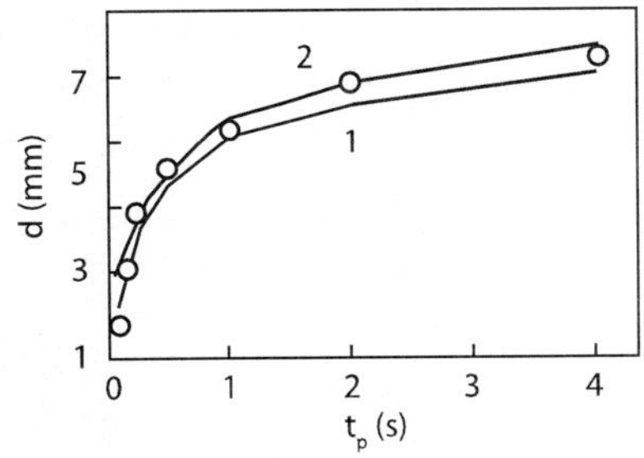

Fig. 2.31 Comparison of the experimental dependence of the melt pool diameter d on the 6-kW laser pulse duration t_p (**1**) with the calculated dependence (**2**) for a 4-mm thick SS304 stainless steel sample

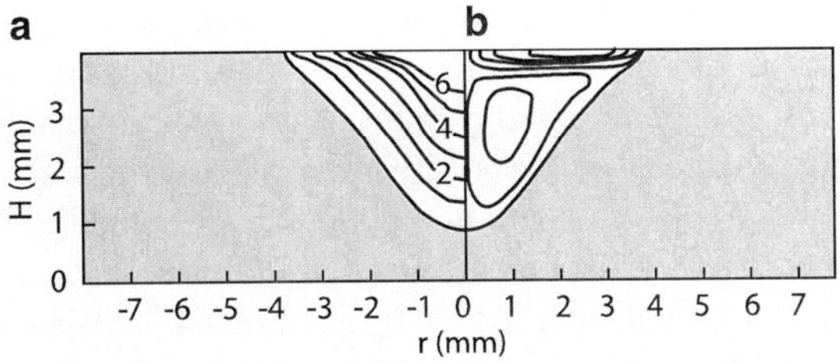

Fig. 2.32 Isotherm field (**a**) and stream lines (**b**) calculated for a 4-mm thick SS304 stainless steel sample at the end of a 6-kW, 4 s laser pulse for $T = 1{,}830$ K (**1**), 1,930 K (**2**), 2,030 K (**3**), 2,130 K (**4**), 2,230 K (**5**), and 2,330 K (**6**). The upper and lower vortices rotate clockwise and counter-clockwise, respectively

physical processes. The calculations show that the parameters of the melt pool produced during short irradiation of samples by laser pulses are insufficient for initiating flows with the multivortex structure, i.e. $Re^* < Re_c^*$. In this case, the melt flow has the usual single-vortex structure. In the surface layer the melt moves from the center to periphery, while near the bottom it moves in the opposite direction.

As the interaction time is increased, both the temperature gradient on the melt surface and the melt size increase. When the condition $Re^* > Re_c^*$ is achieved, the single-vortex flow becomes unstable and the two-vortex flow appears. Note that the melt in the surface layer still moves from the center to periphery. However, the vortex itself has a small thickness (of the order of a few tens of microns) and is pressed down to the free surface. In the rest of the region, the second vortex rotates in the opposite direction: the melt moves along the symmetry axis inside the pool, then it turns back near the bottom and flows along the phase interface from the center to periphery (Fig. 2.32). Such a flow structure more efficiently transfers heat from the surface inside the pool at the central zone of the melt, resulting in the formation of the characteristic "bell-shaped" pool (Fig. 2.32).

As pointed out in the beginning of this section, the liquid flow can be also initiated by surfactants or materials changing the sign of $\partial\sigma/\partial T$ [35–38]. Consider the influence of such materials on the metal melt flow.

2.2.7 *Influence of Surfactants on Heat-and-Mass Transfer During Laser Alloying*

Let us assume that a surfactant is insolvable in a sample melt. Then, its concentration C on the liquid surface is determined by the velocity of the liquid surface and diffusion:

$$\frac{\partial C}{\partial t} + \frac{1}{r}\frac{\partial}{\partial r}(rCv_r) - \frac{1}{r}\frac{\partial}{\partial r}\left(rD\frac{\partial C}{\partial r}\right) = 0 \qquad (2.40)$$

The concentration distribution gives the addition $(\partial\sigma/\partial C)(\partial C/\partial r)$ to the force on the liquid surface (2.34). At the melt boundary, we have $(\partial C/\partial r) = 0$. Taking the aforesaid into account, (2.40) together with the system of equations (2.29)–(2.35) compose a closed system of equations for solving the problem formulated. Although the equations describing the dynamics of the temperature and concentration of impurities are of the same type, the behavior of these quantities is different. This is explained by the fact that the temperature distribution is finally determined by an external source, namely, by a spatially limited laser beam, which is not the case for impurities. As a result, the initiation and formation of the melt flow is mainly determined by the Marangoni "force", while the concentration-capillary force only reacts to the kinetics of the latter. The relation between these forces can be conveniently characterized by the relation of dimensionless parameters [34]

$$\varepsilon_T = \frac{\partial\sigma}{\partial T}\frac{\nabla T}{\sigma_0}, \qquad \varepsilon_c = \frac{\partial\sigma}{\partial C}\frac{C_0}{\sigma_0}$$

where σ_0 is the surface tension coefficient for $C = C_0$ and C_0 is the equilibrium, i.e. initial impurity concentration. For liquid metals upon laser alloying, $\varepsilon_T \approx 1$ and ε_c can be changed in a broad range. Consider the type of the melt motion depending on the relation between these parameters.

As shown above, the radial velocity of the melt surface increases from zero at the focal spot center to maximum and vanishes at the boundary. As a result, the impurity concentration decreases in the region where $\partial v_r/\partial r > 0$ and increases at the periphery, where $\partial v_r/\partial r < 0$. The resulting impurity gradient reduces the liquid flow velocity where the impurity is carried away, i.e. at the pool periphery. As a result, the Marangoni flow continues on a part of the surface free of impurities, while at the periphery the forces are balanced and the flow ceases. It is clear that the interface r_c between these regions depends on the ratio $\varepsilon_c/\varepsilon_T$. Approximate estimates [53] give the simple expression

$$r_c/R = 1 - \sqrt{\varepsilon_c/\varepsilon_r} \qquad (2.41)$$

No special experimental studies of the influence of surfactants on the melt flow were performed. For definiteness, the system of equations (2.29)–(2.35), and (2.40) was calculated numerically under conditions close to experimental conditions [50]: Armco-iron covered with a molybdenum film was irradiated by 4 ms, 10 J pulses from a neodymium laser. The focal spot radius was 1.5 mm. It was assumed that

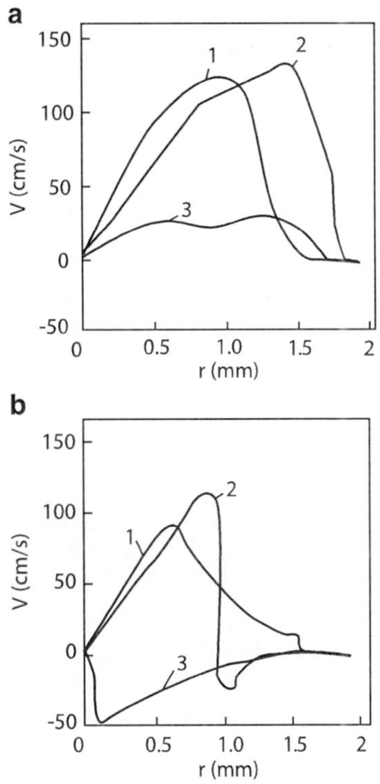

Fig. 2.33 Distribution of the melt velocity over a sample surface at instants $t/\tau = 0.5$ (**1**), 1.0 (**2**), and 1.5 (**3**) in the absence (**a**) and presence (**b**) of a surfactant

$|\partial\sigma/\partial T| \approx 5 \cdot 10^{-8}$ J/cm^{2}K, $\kappa = 0.7$ W/cm $\cdot$ K, $\sqrt{P_r} \approx 0.2$. Numerical calculations showed that, when the ratio $\varepsilon_c/\varepsilon_T$ was small, the stagnation zone was formed only on the pool boundary. For $\varepsilon_c/\varepsilon_T \approx 1$, according to (2.41), thermocapillary convection should be suppressed completely, which is confirmed by numerical calculations [53]. Of most interest is the intermediate case, when the interface r_c is located in the middle between the symmetry axis and boundary. According to (2.41), this should correspond to $\varepsilon_c/\varepsilon_T \approx 0.25$. Figure 2.33 [53] shows that for $\varepsilon_c/\varepsilon_T \approx 0.27$ and $t = 0.5\tau$, the concentration force did not manifest itself completely yet – the vortex motion exists in the entire pool. As the laser pulse ends, a zone is formed where the concentration force compensates the thermocapillary force. Convection in this region is weak, while in the region free of the impurity the convection is still strong. As a result, the melt flow at the interface of these zones abruptly turns downwards (Fig. 2.34) [53] and penetrates deeper that in the absence of impurities. After the switching off the laser pulse, the temperature gradient decreases, which leads to the melt flow to the pool center (Figs. 2.33, 2.34). This results in the equating of impurity concentrations and repeated appearance of the thermocapillary motion. These oscillations decay with time. Such an intricate character of the flow can facilitate a deeper penetration of the alloying material into the sample. Figure 2.35 [53] shows that the alloying additive for $\varepsilon_c = 0$ penetrates only by half the pool depth, whereas for $\varepsilon_c/\varepsilon_T \approx 0.27$, the additive reaches the pool bottom.

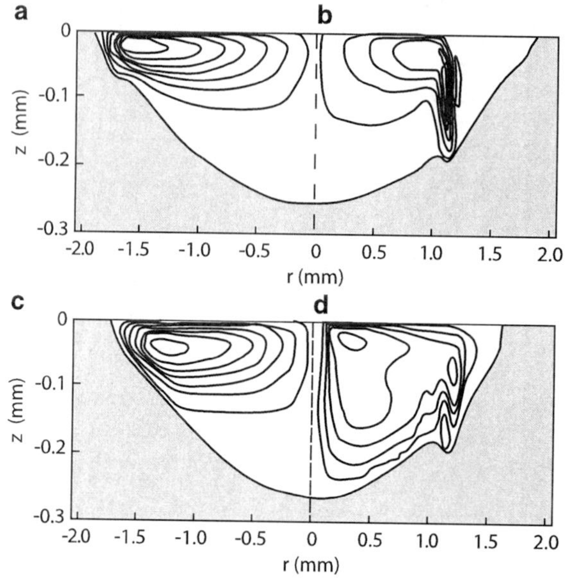

Fig. 2.34 Stream lines and the melt pool shape in the absence (**a, c**) and presence (**b, d**) of a surfactant at instants $t = \tau$ (**a, b**) and $t = 1.5\tau$ (**c, d**) (the *dashed straight line* is the symmetry axis)

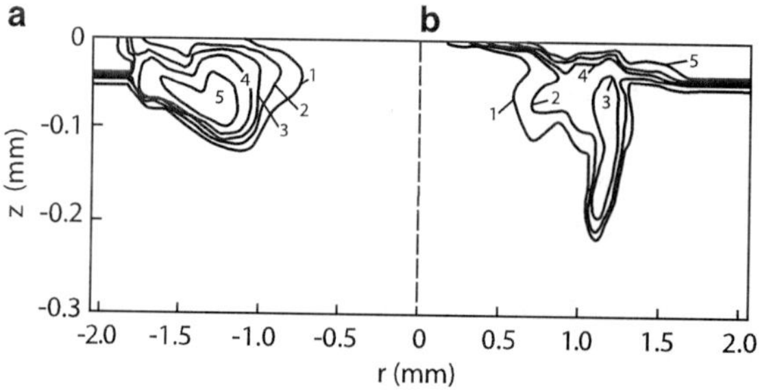

Fig. 2.35 Distributions of an alloying addition at the final instant in the absence (**a**) and presence (**b**) of a surfactant. The lines of the relative concentration of the alloying substance correspond to the concentrations 0.1 (**1**), 0.3 (**2**), 0.5 (**3**), 0.7 (**4**), and 0.9 (**5**) of the maximal concentration (the *dashed straight line* is the symmetry axis)

2.2.8 Mass-Transfer Kinetics During Gas-Phase Alloying

We considered above some questions concerning the gas-phase alloying of metal surfaces such as the mixing the alloying component due to Marangoni convection or the influence of surfactants on the melt flow. The properties of the mass transfer of allying elements through the gas–liquid surface remained unstudied. The mechanisms of the capture of gas particles and their interaction with the surface are

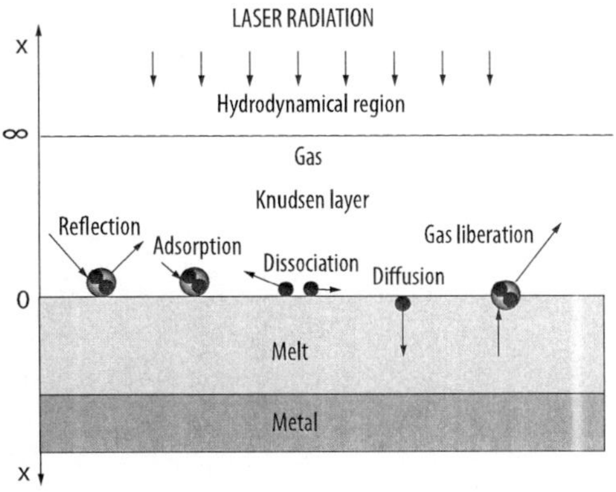

Fig. 2.36 Schematic diagram of the main molecular processes and structure of the flow at the gas–liquid interface

intricate and numerous, which complicates the construction of models predicting the properties of an alloyed layer [61]. In this connection, we will not consider here in detail the comprehensive material on the thermochemical processing of metals. Note that experiments were performed in two directions by using either millisecond Nd lasers [49, 62] or pulsed, cw, and repetitively pulsed microsecond CO_2 lasers ([61] and references therein). We will restrict our theoretical consideration to the simplest case of allying with the help of an Nd laser by neglecting chemical reactions and structural deformations in the substrate material and melt flows. Schematic processes proceeding on the metal surface are shown in Fig. 2.36. The model assumes that [63]:

- Only gas atoms can be dissolved in a metal.
- No gas ionization and dissociation occur in the gas atmosphere.
- Gas molecules dissociate only on the hot surface of a metal.
- The flow of gas atoms from a metal is determined by the surface concentration of the dissolved gas and the surface temperature.
- Atoms are transferred inside a metal due to diffusion.

The latter process is described by the diffusion equation

$$\frac{\partial C}{\partial t} = \frac{\partial}{\partial x}\left(D(T)\frac{\partial C}{\partial x}\right) \tag{2.42}$$

$$C(x,0) = C_0,\ C(-\infty) = C_0,\ -D\left.\frac{\partial C}{\partial x}\right|_{x=0} = j$$

The temperature distribution in a metal is determined by the heat conduction equation

$$\rho(C_p + L_m\delta(T - T_m))\frac{\partial T}{\partial t} = \frac{\partial}{\partial x}\left(\kappa(T)\frac{\partial T}{\partial x}\right) \tag{2.43}$$

$$T(x,0) = T_0, \quad T(-\infty,t) = T_0, \quad -\kappa\left.\frac{\partial T}{\partial x}\right|_{x=0} = q$$

The resulting flow j of particles from gas to the surface in the system of equations (2.43) is determined by the distribution function of atoms f:

$$j = -\int_{-\infty}^{\infty} mV_x f \, d\mathbf{V} \tag{2.44}$$

The velocity distribution function of molecules is found approximately, similarly to the methods used for studying the ablation of materials (see Chap. 1). The distribution function is represented in the bimodal form [64, 65]:

$$f(x,\mathbf{V}) = \alpha_s^{+}(x) f_s^{+}(\mathbf{V}) + \alpha_*^{-}(x) f_*^{-}(\mathbf{V}) + \alpha_\infty^{+}(x) f_\infty^{+}(\mathbf{V}) + \alpha_\infty^{-}(x) f_\infty^{-}(\mathbf{V}), \tag{2.45}$$

where $\alpha_i^{+-}(x)$ are unknown functions of the coordinate x, and $f_i^{+-}(\mathbf{V})$ are semi-Maxwell functions

$$f_i^{+-} = n_i \left(\frac{m}{2\pi\kappa T_i}\right)^{3/2} \exp\left[-\frac{m(v_x - u_i)^2 + v_y^2 + v_z^2}{2\kappa T_i}\right] \tag{2.46}$$

The subscript s in (2.45) denotes conditions near the metal surface, the subscript ∞ denotes conditions at the interface of the Knudsen layer and the hydrodynamic gas region, $u_s(0) = 0$, $u_* = (2\kappa T_s/\pi m)^{1/2}$, $n_* = n_s/2$, $T_* = (1 - 2/3\pi)T_s$, $\alpha_s^{+}(\infty) = 0$, $\alpha_*^{-}(\infty) = 0$, $\alpha_\infty^{+}(\infty) = 1$, $\alpha_\infty^{-}(\infty) = 1$, $\alpha_\infty^{+}(0) = 0$. In addition, it should be taken into account that gas molecules dissociate on the melt surface with the probability α

$$\alpha = \alpha_0 \exp(-T_d^*/T_s) \tag{2.47}$$

Here, T_d^* is the effective dissociation energy

$$T_d^* = (E_d - 2E_{\mathrm{gm}})/k_{\mathrm{B}},$$

where E_d is the dissociation energy of molecules in the gas phase; E_{gm} is the bond energy between an atom and metal, and k is the Boltzmann constant. Molecules escaping from the metal form the two groups of nondissociated molecules and molecules escaping from the liquid metal. The latter are characterized by the partial pressure p_s, which strongly depends on the surface temperature T_s. Here, the situation is similar to the evaporation of metals (see Chap. 1). As in the case of ablation, it is necessary to equate the flows of particles, momentum, and energy on the metal surface and the Knudsen layer boundary. The parameters n_∞, u_∞, and T_∞ at the interface can be related to the parameters P_g of the surrounding gas [65]:

$$T_g/T_\infty = (P_g/P_\infty)^{(\gamma-1)/\gamma}, \quad P_g/P_\infty = \left[1 + \frac{1}{2}(\gamma - 1)M_\infty\right]^{-2\gamma/(\gamma-1)} \tag{2.48}$$

where $M_\infty = \sqrt{2/\gamma} u_\infty/(2kT_\infty/m)^{1/2}, \gamma = C_p/C_V$. It is necessary to take into account in (2.42) that the diffusion coefficients of atoms in the condensed medium exponentially depend on temperature:

$$D = D_0 \exp(-T_0/T) \tag{2.49}$$

For example, for a nitrogen–iron pair, we have $D_0 = 4.1 \cdot 10^{-7}\,\mathrm{m^2/s}$, $T_0 = 7.6 \cdot 10^3\,\mathrm{K}$, $T_d^* = 2.3 \cdot 10^4\,\mathrm{K}$. The verification of model (2.42)–(2.49) for the simplest case of saturation of an iron melt with nitrogen at the equal gas and metal temperatures or during the inverse process, when the nitrogen concentration was initially higher than the equilibrium concentration, has demonstrated good agreement with experiments [66] (Fig. 2.37).

Consider now the application of this model for describing gas-phase alloying (a nitrogen–iron pair) produced by a millisecond pulse from a Nd laser ($q = 6.5 \cdot 10^4\,\mathrm{Wcm^2}$). Because the saturation time by gas is large (several seconds, see Fig. 2.37), the gas concentration will increase by the pulse end up to the value smaller than the equilibrium concentration corresponding to the finite temperature and gas pressure (see Fig. 2.38). For $T = T_m$, the equilibrium concentration of nitrogen in iron is $[\mathrm{N}]_c = 0.044\%$. After the switching off a laser, the situation can develops in two ways: the concentration can continue to increase or it will decrease due to nitrogen diffusion inside a sample. Calculations show that the latter process is more intense, and for this reason the surface density of nitrogen atoms begins to decrease after the end of irradiation (Fig. 2.38b). Then, after the decrease in the surface layer temperature, the diffusion coefficient abruptly decreases, according to (2.49), and the concentration profile "freezes". The thickness of a gas-saturated layer is $\sim 15\,\mu\mathrm{m}$. As mentioned above, the thickness of alloyed layers processed by millisecond pulses can achieve $\sim 100\,\mu\mathrm{m}$. The latter is caused by the hydrodynamic mechanism of impurity transfer (see Fig. 2.19a). The consideration of thermocapillary flows in model (2.42)–(2.49) is the problem of future investigations.

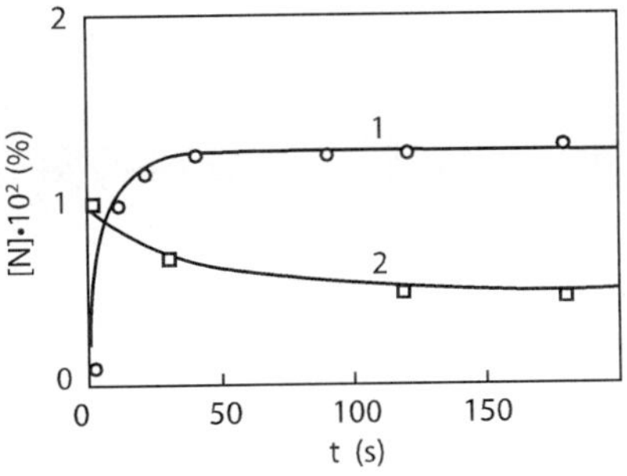

Fig. 2.37 Temporal variations of gas content in iron melt depending on the holding time in nitrogen atmosphere. Remelting regimes: $p_{N_2} = 0.01$ Mpa, $T = 1953$ k, $[N]_0 = 0.001\%$ (**1**) and $p_{N_2} = 0.008$ MPa, $T = 1873$ K, $[N]_0 = 0.01\%$ (**2**). *Points* – experimental data 7], *solid lines*– calculated results

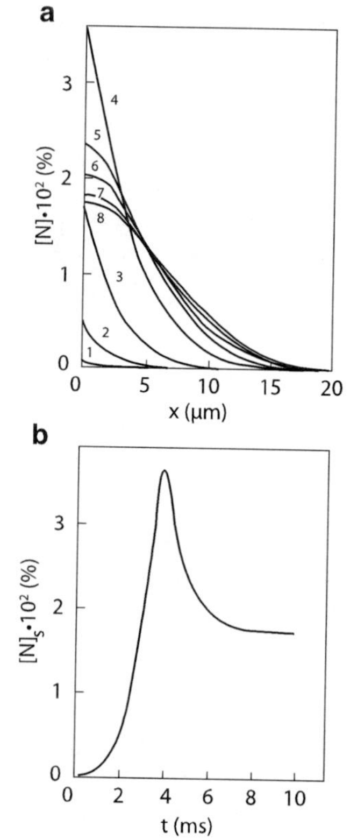

Fig. 2.38 Concentration profiles (**a**) and temporal behavior of subsurface content (**b**) of nitrogen in pulsed laser irradiation of iron. The numbers of the curves (**a**) denote time in ms. Alloying regime: $q = 6.5 \cdot 10^4$ W/cm, $\tau = 4$ ms, $p = 0.1$ MPa

The comprehensive consideration should also take into account plasma phenomena and chemical reactions between gas atoms and a metal, for example, the synthesis of titanium nitride, etc. [49, 67].

Alloying can be performed by using cw lasers. The question arises of how the vortex flow will develop under the action of a moving laser beam or a moving sample.

2.2.9 *Alloying of a Moving Sample Surface by Stationary Laser Radiation*

Before considering the mixing of additives caused by the Marangoni flow, we discuss the influence of this flow on the integrated parameters of a melt pool. As follows from estimated by (2.39) and numerical calculations (Fig. 2.17), the velocity of vortices can achieve large values $\sim$1 m/s. For typical travel speeds of samples of $\sim$0.1–1 cm/s, the vortex motion should considerably affect the parameters of the melt pool. The geometry of the problem is three-dimensional in real experiments and therefore its investigation is quite time-consuming. It is reasonable to use

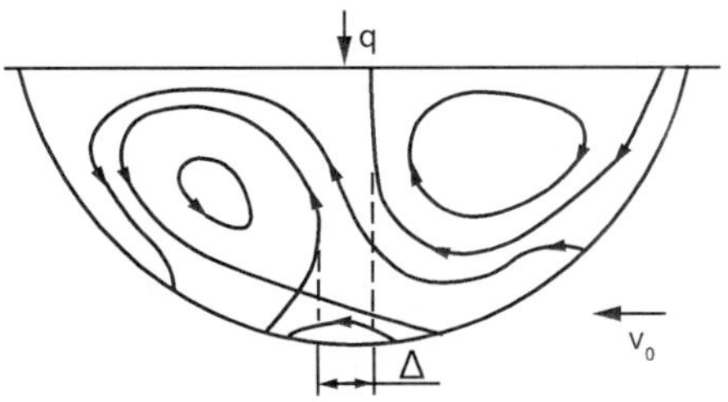

Fig. 2.39 Scheme of the liquid flow in a melt pool for a moving sample

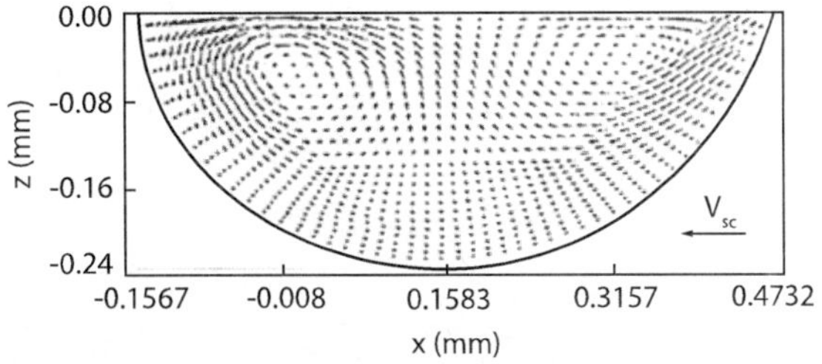

Fig. 2.40 Melt pool geometry and thermocapillary flow field for Al: $\sigma' = -0.35 \cdot 10^{-3}$ N/m K, $q = 3.3 \cdot 10^5$ W/cm^2, line focus, $d = 400\,\mu$m, $V_{sc} = 1.8$ m/min

the two-dimensional approximation by considering a rectangular focal spot. It is assumed that the spot size transverse to the motion greatly exceeds its longitudinal size (see [68] and references therein). The system of equations (2.29)–(2.35) was solved numerically by specifying the travel speed of a sample at the melting boundary. In addition, the form of the melt surface was considered by adding the equilibrium equation taking into account surface tension forces. Because the surface deformation is small $\approx$10% of the melt pool depth [68]), we will not consider this process. The picture of the resulting motion was qualitatively presented in paper [45] (Fig. 2.39). Figure 2.40 shows the velocity field in the melt pool of the surface of an aluminum sample moving at the speed $v_{sc} = 1.8$ m/s. The absorbed radiation intensity was 0.33 MW/cm^2 and $\partial\sigma/\partial T = -0.35 \times 10^{-3}$ N/mK. The velocity distribution coincides qualitatively with that in Fig. 2.39. The maximum speed was 0.65 m/s. Figure 2.41 demonstrates the influence of the Marangoni flow on the form and parameters of the melt pool. For $\partial\sigma/\partial T < 0$, the ratio of the pool length to its depth doubled compared to the case $\partial\sigma/\partial T = 0$. Calculations show that excitation of the Marangoni flow reduces the melt surface temperature by $\sim$30% [68].

It follows from calculations [68] that the depth of the origin of the path of an elementary particle of liquid at the melting front is equal to its depth at the freezing front. This means that the mixing of an additive cannot be caused by convection only, as was observed upon pulsed alloying. In the presence of vortices rotating with a high velocity, all the liquid entering the pool flows in a narrow region of width $\Delta \sim v_0 d/v$. The approach of current lines in the region between separatrices leads to an increase in the alloying depth l_V compared to that in the liquid at rest. Indeed, in a narrow region of width Δ, the layer thickness Δ' by which the additive will diffuse is $\sqrt{4Dd/v}$. As the flow is expanded up to the size $\sim d$, the diffusion

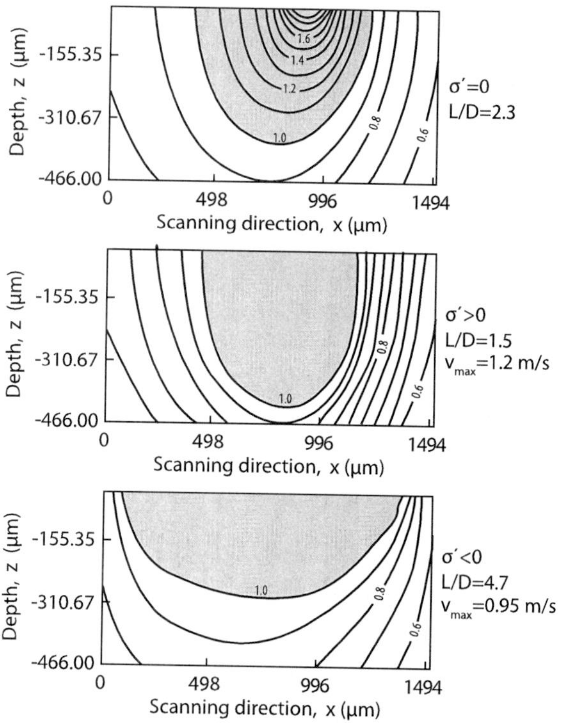

Fig. 2.41 Isotherms ($/T_m$), melt pool geometry for $S = 0$ (Rosenthal solution. $S = 396$ ($\sigma' > 0$) and $S = -264$ ($\sigma' < 0$), in the plane $y = 0$ (GGG50: Line focus, $d = 200\,\mu$m $b = 800\,\mu$m, $Pe = 3.1$, $Re = 25.5$. S is the surface tension number)

layer thickness in the rear part of the pool increases by a factor of v/v_0. This gives the relation

$$l_V = l_0\sqrt{v/v_0},$$

where $l_0 = \sqrt{4Dd/v_0}$ is the allying depth of the liquid at rest. Thus, we see that internal flows produced in the melt pool can considerably accelerate the mixing and dissolving of the allying additive in the surface layer of a metal.

The considerations presented above are qualitative. The influence of the Marangoni flow on the mixing of an allying element can be taken into account more rigorously [68]. For this purpose, it is necessary to add to the system of equations (2.29)–(2.35) the equations describing the diffusion of impurities, similarly to (2.40) and (2.42):

$$\frac{\partial C}{\partial t} + div(\mathbf{V}C) - D\triangle C = 0, \quad -D\frac{\partial C}{\partial z} = j \tag{2.50}$$

where the velocity is determined from (2.29)–(2.35) and j is the additive flow. The typical value of the diffusion coefficient of copper atoms in aluminum is very small 10^{-9} m^2/c, which complicates numerical calculations. The dependence of the alloying process on the diffusion coefficient was studied in paper [66] where model calculations were performed for the overestimated value $D = 5\times10^{-7}$ m^2/s. For this value of D, Peclet's number is $hv/D \approx 10^3$, which demonstrates the importance of

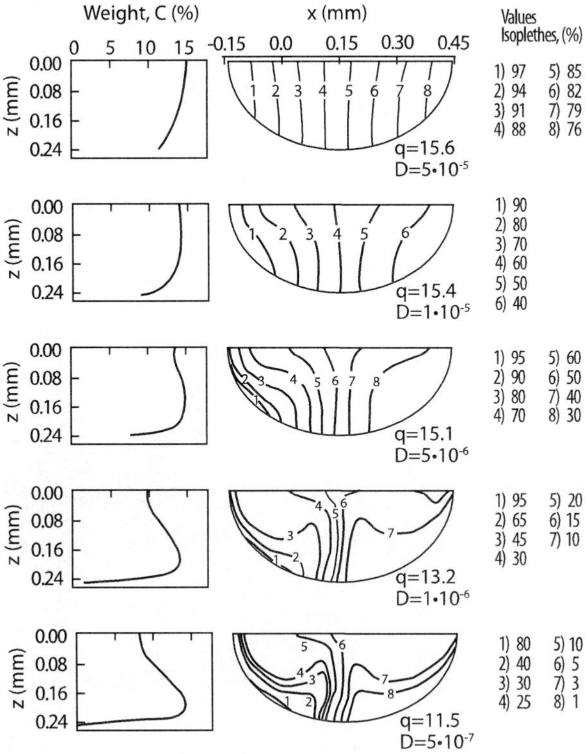

Fig. 2.42 Isopleths of Cu concentration distribution in the melt pool and resulting Cu distribution over the surface layer depth, variables see 2.40

convective transfer. Figure 2.42 presents the results of calculations beginning from $D = 5 \times 10^{-5}\text{m}^2/\text{s}$, when $Pe = 1$. One can see that the distribution of copper atoms is determined by diffusion. As the value of D decreases, the impurity distribution becomes more and more inhomogeneous, impurities are ejected from the region of vortices, and the size of this region and its temperature drastically change near walls in boundary layers. As a result, the copper distribution in the surface layer increases deep in the sample, as was observed upon pulsed alloying. The formation of a layer with the increased concentration of nickel in the depth of the melted surface of aluminum was observed experimentally [69].

The influence of the sample movement on the Marangoni flow can be considered rigorously in the case of a circular focal spot, which is more important in practice. We will use the fact that the travel speed of the sample ~1 cm/s is much smaller than the vortex motion velocity ~1 m/s. The influence of the travel speed of the sample on the temperature field and the pool form was taken into account in [70] by expanding three-dimensional equations in a small parameter (Figs. 2.43 and 2.44). These quantities change weakly due to a small travel speed of the sample. However,

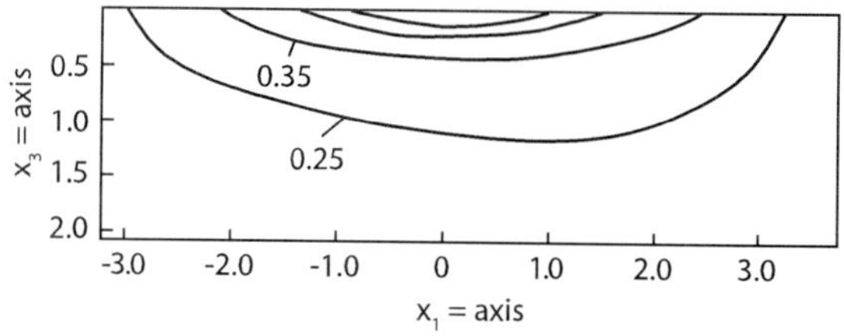

Fig. 2.43 Isotherms within the molten pool on the vertical plane at $X_2^* = 0.0$ for $Ma = 1500$, $Pr = 0.15$, $T_m^* = \kappa(T_m - T_\infty)/qr_0 = 0.25$

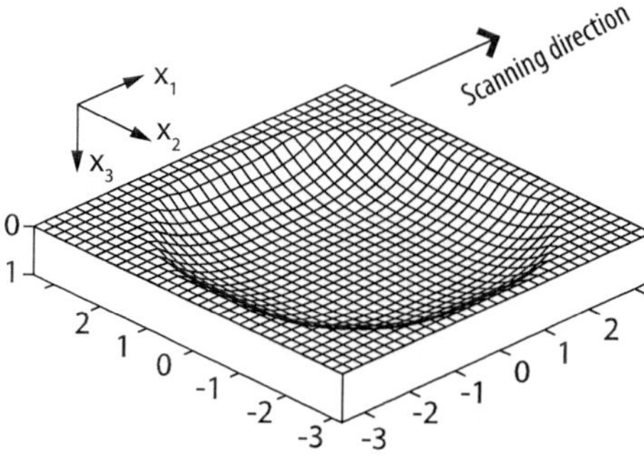

Fig. 2.44 Three-dimensional surface plot of the solid–liquid interface. $Ma = 1{,}500$, $Pr = 0.15$, $T_m^* = 0.25$

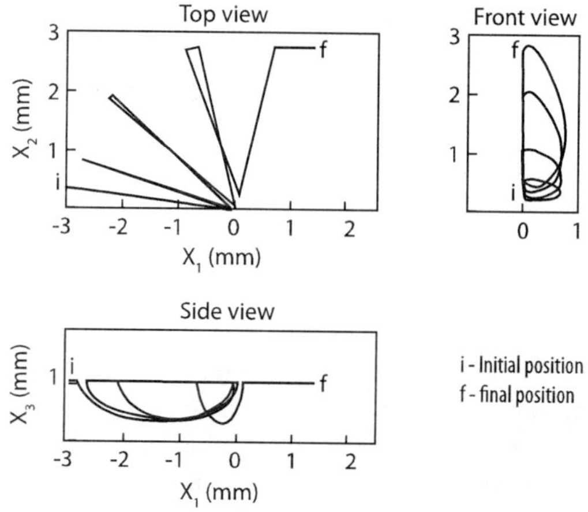

Fig. 2.45 Trajectory of a particle as it enters, recirculates, and ultimately freezes. The trajectory is plotted in *front, side, and top* view. $Ma = 1{,}500$, $Pr = 0.15$, $T_m^* = 0.25$

the sample movement considerably affects the trajectory of an individual liquid particle, as is well illustrated in Fig. 2.45. The involvement of a liquid particle into the cyclic motion can be caused either by the nonstationarity of the process, as is shown in Fig. 2.18, or by real diffusion or equivalent numerical diffusion. Figure 2.45 shows that a liquid particle circumscribes a trajectory passing over the entire volume of the melt pool, i.e. the additive can be well mixed during the illumination time d/v.

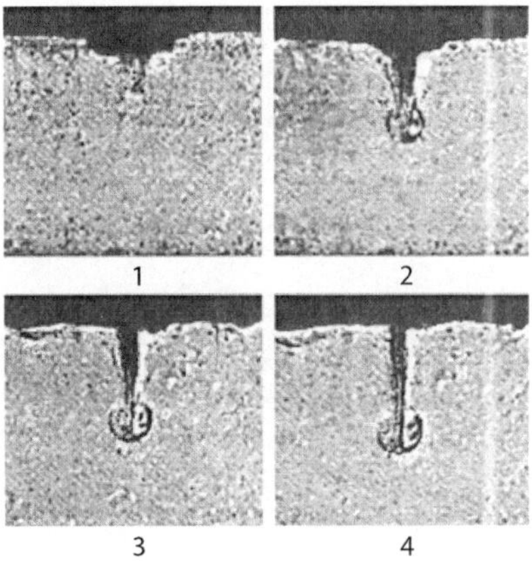

Fig. 2.46 Shadow photographs of excitation of flows in paraffin by a 40 μs CO_2 laser pulse $E = 4J$, $r_f = 2.7$ mm. The distance between the frames is 20 ms

2.2.10 Melt Flow Upon Pulsed and Repetitively Pulses Irradiation

We considered above the flows initiated by cw laser radiation or millisecond laser pulses, when the flow had time to become stationary. Consider now the situation upon irradiation of a surface by a short pulse of duration smaller than the settlement time of Marangoni vortices. Figure 2.46 [71] shows that within 50 ms after irradiation of melted paraffin by a 40 μs pulse, Marangoni vortices appear on the melt surface (the last frame). But even earlier, within 20 ms after the laser pulse, a narrow-directional flow is excited under the focal spot along the laser beam deep in the liquid. The velocity of this flow and penetration depth increase with laser intensity and independent on laser pulse time and focal size (Fig. 2.47a,b). The initial velocity of such a flooded jet can exceed the Marangoni flow velocity by a factor of six. Narrow-directional jets were also observed under the same conditions in other liquids such as water, alcohol, kerosene, stearin melt, and glycerol heated up to 160°C, where Marangoni flows were not always observed [71]. The Marangoni flow becomes more intense upon repetitively pulsed irradiation, when the surface is heated stronger. Figure 2.48 shows the picture of stationary flows in a finite paraffin pool for $f = 10$ Hz and increasing pulse energy [71]. One can see that, as the average power is increased, the Marangoni flow first appears and then flooded jets are observed. The mechanism of the appearance of flooded jets is unknown yet. It is found that their initiation threshold is close to the radiation intensity producing the boiling of the sample surface. It is possible that the liquid in the focal spot receives a momentum from the vapor recoil pressure and then moves deep in the pool by inertia. Numerical calculations show that a vortex specified on the surface moves inside the pool. A narrow-directional liquid jet is formed along the vortex axis (Fig. 2.49) [72].

Such flooded jets transfer heat inside the melt pool and can facilitate a deeper melting of a material, without the penetration of the laser beam inside the sample. This can explain the increase in the depth of pulsed steel welding by two laser

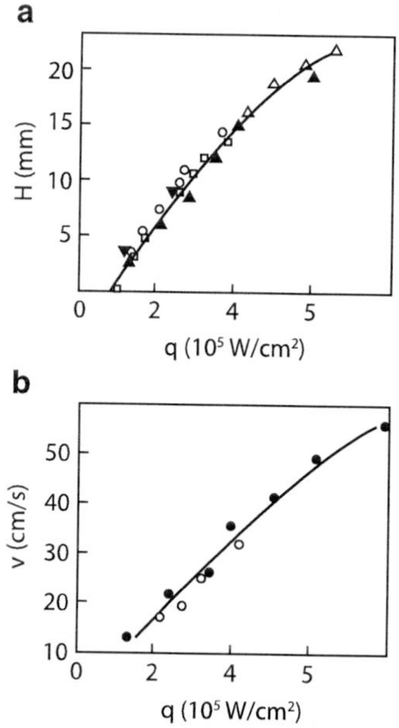

Fig. 2.47 Dependences of the penetration depth (**a**) and the initial velocity (**b**) of flooded jets on the laser radiation intensity at different magnitudes of τ, E, r_f

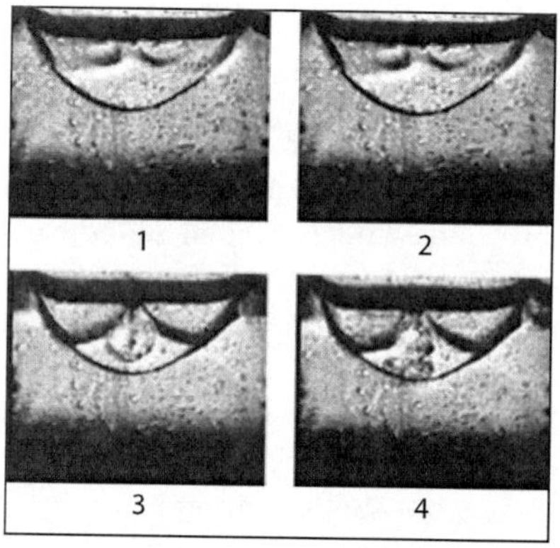

Fig. 2.48 Formation of flows in paraffin by repetitively pulsed radiation from a CO_2 laser with increasing the pulse energy

beams with energies differing by two orders of magnitude [71]. The excitation of such flows will undoubtedly also increase the alloying of surfaces.

Another unusual mechanism of excitation of closed flows upon melting of metals by laser pulses is the generation of a vortex flow simultaneously during melting and the action of the vapor recoil momentum. The vapor pressure acts normally to the melt surface and should not excite a shear flow. However, a more detailed study shows that the metal layers melted later absorb a smaller longitudinal momentum than layers located close to the surface. As a result, the upper layers are accelerated to a higher velocity and can turn inside the pool at its boundary. It seems that the possibility of such a mixing of the melt can be demonstrated only by numerical methods [61]. In this case, the vapor pressure on the melt surface should be

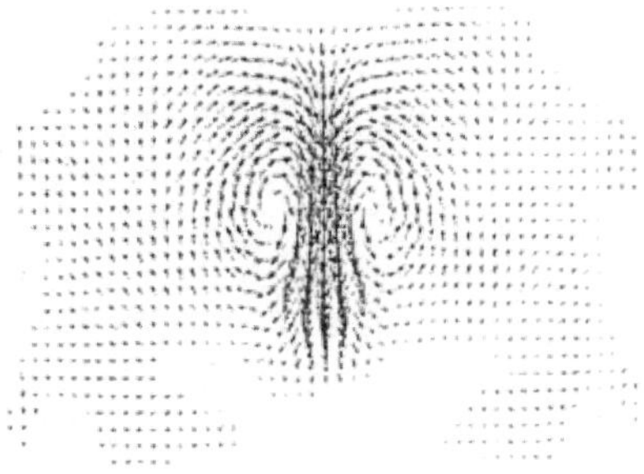

Fig. 2.49 Numerically calculated velocity field in a vortex excited on the liquid surface

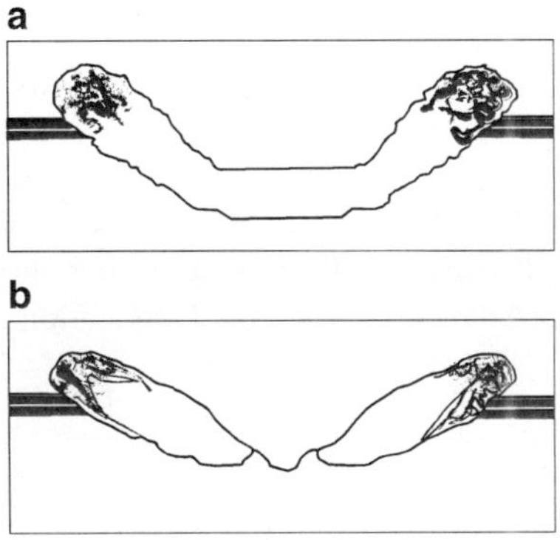

Fig. 2.50 Distributions of the alloying impurity over the melt pool cross section at instants 4 μs (**a**) and 7 μs (**b**) during melting of iron by a 5-μs CO_2 laser pulse. The characteristic laser spot diameter is 0.5 mm; the vertical scale is magnified

taken into account in the usual system of equations (2.29)–(2.35) (see Chap. 8). Figure 2.50 presents the forms of the melt and the distribution of the additive upon irradiation of iron of a 5 μs, 100 J/cm^2 pulse.

In shear flows, surface vortex waves can be excited [73]. It is possible that they transform at the linear stage to vortex structures having closed current lines. This can also result in the additional mixing of the metal melt.

2.2.11 *Thermocapillary Processes in the Dynamics of Gas Bubbles in a Melt Pool*

During laser processing of porous materials and welding and alloying of metals, gas pores can be produced, which impair the quality of these processes. The thermocapillary drift of a gas bubble has the same nature as the melt flow during alloying discussed above. The difference is that the bubble drifts to the high-temperature region, whereas the melt flow is directed from the hot region to the cold one (for $\partial\sigma/\partial T < 0$). Thus, the drift of the bubble near the melt pool bottom coincides with the melt flow, while their velocities near the surface are opposite to each other [36]. Therefore, the resulting motion of bubbles remains unclear.

The dynamics of laser degassing can be represented by three stages: the formation, floating up, and drift of a bubble at the surface for the time t^* and a passage through the melt surface for the time t_s. The time of the last stage should be much shorter than the time of the previous stage. Otherwise, the convective motion will carry the bubble at the pool periphery, where it will be either carried inside the melt or will form a pore during the solidification of the melt. The analysis of all the forces acting on the bubble showed [75] that in all cases the time t^* during which the bubble is found near the surface is approximately equal to the t_M during which it is carried out by the Marangoni flow.

The degassing rate of the melt pool can be analyzed in the case of a small volume concentration of bubbles $\alpha_0 \ll 1$:

$$\frac{\mathrm{d}N}{\mathrm{d}t} = \alpha_0 \frac{\mathrm{d}V}{\mathrm{d}t} - \frac{N}{\tau_b} \tag{2.51}$$

where $V(t)$ is the melt pool volume, $N(t)$ is the number of bubbles, and τ_b is the characteristic lifetime of the bubble in the melt. Taking into account that $\alpha = N/V$, we obtain from (2.51) the equation

$$\frac{\mathrm{d}\alpha}{\mathrm{d}t} = (\alpha_0 - \alpha)\frac{1}{V}\frac{\mathrm{d}V}{\mathrm{d}t} - \frac{\alpha}{\tau_b} \tag{2.52}$$

for the concentration of bubbles. Because the characteristic size of the melt is $h \propto \sqrt{t}$, we have $\frac{1}{V}\frac{\mathrm{d}V}{\mathrm{d}t} = \frac{1}{t}$. Taking this into account, we rewrite (2.52) in the form

$$\frac{\mathrm{d}\alpha}{\mathrm{d}t} = (\alpha_0 - \alpha)\frac{1}{t} - \frac{\alpha}{\tau_b} \tag{2.53}$$

If the time t_s of the bubble escape through the free melt surface is the longest, then $\tau_b = t_s$ and (2.53) can be easily integrated:

$$\frac{\alpha(t)}{\alpha_0} = [1 - \exp(-t/t_s)]t_s/t \tag{2.54}$$

In the opposite case, a bubble reaching the surface vanishes at once. Therefore, the total lifetime of the bubble is determined by its floating time:

$$t_{\mathrm{cs}} = h/v_q \tag{2.55}$$

where $h(t)/\alpha_0 \propto \sqrt{t}$ is the melt depth and $v_q = 1/3r_b q(\partial\sigma/\partial T)/\rho\nu\kappa$ [36, 76] is the drift velocity of the bubble. By substituting these expressions into (2.53) and integrating, we obtain [75]

$$\alpha(t)/\alpha_0 = \frac{\beta^2}{2t}\left[\exp\left(-\frac{2\sqrt{t}}{\beta}\right) + \frac{2\sqrt{t}}{\beta} - 1\right] \tag{2.56}$$

where $\beta = h_{\max}/v_q\sqrt{\tau}$ and $h_{\max}$ is the maximum depth of the melt pool by the end τ of the laser pulse.

Because the expression for degassing was obtained approximately, it is reasonable to perform a comparison with numerical calculations taking into account all the stages of this process. The simulation is based on the continual consideration of the carrier medium (2.29)–(2.31) and the discrete representation of the dispersion phase. The latter is described by the Newton equation for each bubble. Taking into account that the density of bubbles was small, the influence of their motion on the carrier medium was neglected. The radiation intensity distribution was assumed Gaussian with the radius r_f, the pulse duration was $\tau = 0.1$s, and the absorbed power was $P = 500$ W. The thermal parameters of steel were: $\rho = 7.8$ g/cm^3, $c = 0.56$ J/g·K, $\kappa = 0.4$ W/cm·K, $\nu = 5.5 \cdot 10^{-3}$ cm^2/s, $\partial\sigma/\partial T = 0.35$ dyne/cmK, $T_m = 1730$ K. The initial separation of gas-filled pores was assumed uniform over the volume, and the size of pores was varied from 0.1 to 10 μm. The characteristic escape time t_s of a bubble though the melt surface was also varied from 10^{-2} to 10^{-4}s.

The time dependences of the bubble concentration for $r_b = 10$ μm and $t_s = 10^{-2}$ and 10^{-3} s are compared in Fig. 2.51a [75]. The floating time (2.55) for such bubbles is $t_{cs} = 2.3 \cdot 10^{-4}$s. This time is shorter than t_s, and therefore $\alpha(t)$ can be approximated by expression (2.54). One can see from Fig. 2.51 that the analytic expression is consistent with numerical calculations. In the opposite case, which is realized for small bubbles of radius ~0.1 μm, the floating time t_{cs} is longer than the time $t_s = 10^{-3}$ s. Therefore, the concentration can be described by expression (2.56). Figure 2.51c [75] demonstrates good agreement between the analytic expression and corresponding numerical calculations. It follows from the above discussion that the removal of bubbles is caused by the thermocapillary force. Figure 2.52 presents the distributions of bubbles by the pulse end for $r_b = 1$ μm calculated by taking this force into account or neglecting it. In the absence of the thermocapillary force, the action of the buoyancy force proves to be insufficient and the Marangoni flow carries bubbles deep in the melt, preventing their escape from the pool. Numerical calculations showed that for $r_b < 0.1$ μm the degassing becomes inefficient even when the thermocapillary force is taken into account.

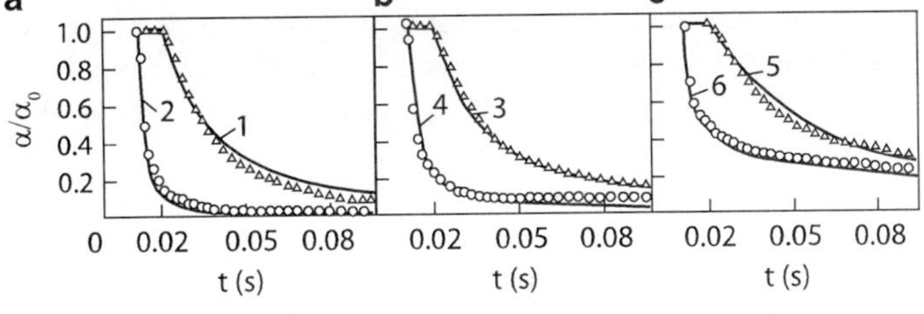

Fig. 2.51 Numerical (*points*) and analytic (*curves*) calculations of the time dependence of the volume concentration of a gas phase in a melt pool for bubbles of radii 10 μm (**a**), 1 μm (**b**), and 0.1 μm (**c**) for $t_s = 10^{-2}$ s (**1, 3, 5**) and 10^{-3} s (**2, 4, 6**)

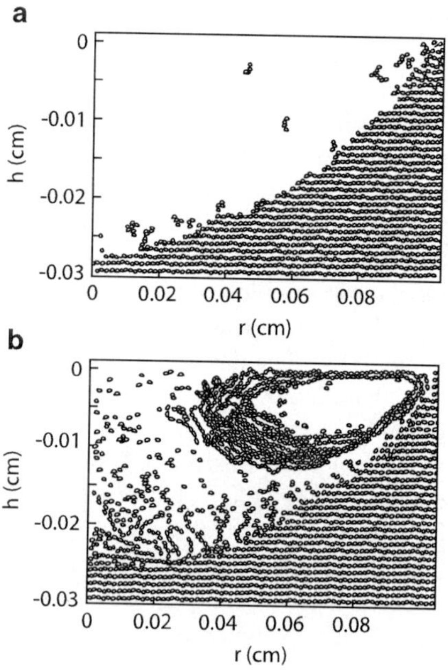

Fig. 2.52 Distribution of gas bubbles of radius 1 μm at the instant $t = 0.1$ s in a sample taking into account (**b**) and neglecting (**a**) the thermocapillary force

2.3 Physical Mechanisms of Cladding

As shown above, during alloying the sample surface, it is important to provide the uniform mixing of the surfacing alloy with the sample material. In technological processes, the problems can be encountered when an element surface should be covered by a different material and it is necessary that mixing or solution with the material of the sample would be minimal. A thin melt layer of the sample is required to provide good adhesion of the surfacing layer. The surfacing material is either preliminary deposited on the surface being processed and then melted or is supplied during laser irradiation. In the latter case, the surfacing material is supplied in the form of a powder jet into the interaction zone of the laser beam with the surface. The latter method is more efficient and widespread [77], and for this reason we will consider below only this method. The transport of powder is performed mainly by two methods. The powder carried by a gas jet is supplied either from the side of the laser beam (Fig. 2.53) [78] or coaxially with the beam (Fig. 2.54) [79]. Figure 2.55 [80] shows the form of a fused roller observed during surfacing. To achieve efficient cladding, it is desirable to provide several conditions. First, it is necessary to obtain the shallow and wide enough melting of the sample surface. Second, it is better to deposit the surfacing powder so that its particles would be at least partially melted. Otherwise the adhesion coefficient of the powder will be small.

Consider the simplest case of the motion and heating of powder particles in a gas jet flowing parallel to a laser beam (Fig. 2.54). Due to a low concentration of the powder, its heating in the laser beam will weakly affect the gas dynamics of the carrier jet. In this case, the problems of the dynamics of a powder particle and

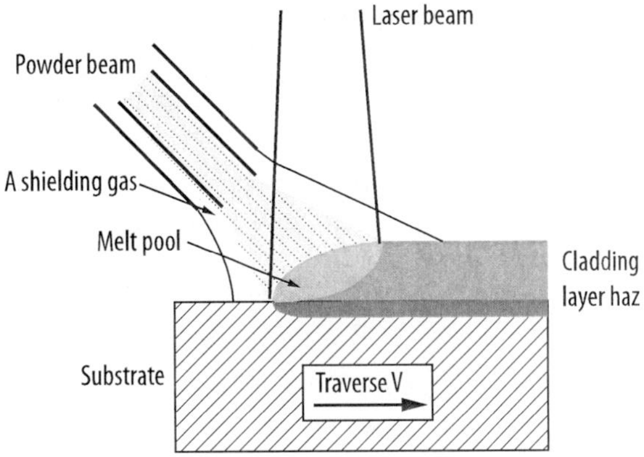

Fig. 2.53 Schematic of the one-step laser cladding process

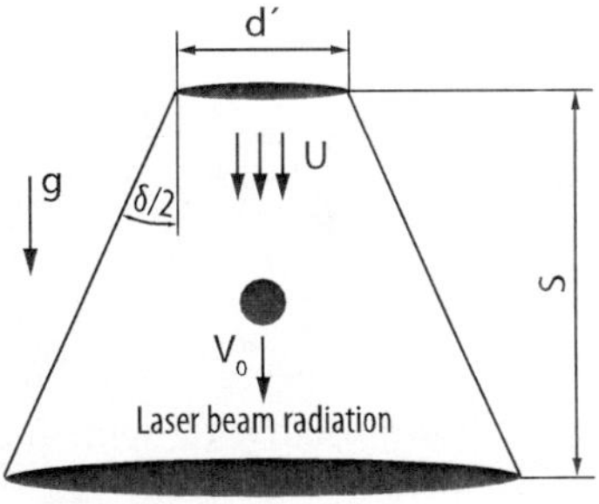

Fig. 2.54 Modeling of the gas-particle heating problem in coaxial laser cladding

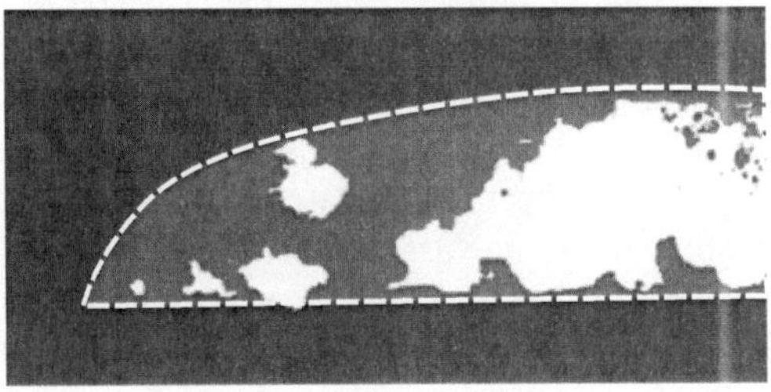

Fig. 2.55 The cladding layer, high-speed-video. White spots are patches of reflected light

its heating are separated. The motion of a particle in a turbulent jet flowing with velocity U is described by the equation [79]

$$\frac{dv}{dt} = \frac{3\mu ReC_D}{n\rho_p D_p^2}(U - v) + g \tag{2.57}$$

where v is the particle velocity; μ is the gas viscosity; ρ_p and D_p are the particle density and diameter; C_D is the friction coefficient; and g is the acceleration of gravity. The Reynolds number is

$$Re = \frac{\rho D_p |v - U|}{\mu}, \tag{2.58}$$

where ρ is the gas density. The expression for C_D for $Re < 800$ [79] has the form

$$C_D = \frac{24}{Re}(1 + 0.15Re^{0.687}) \tag{2.59}$$

The change in the particle temperature T_p is described by the balance equation taking into account heating in the laser beam and heat exchange with the gas flow [79]:

$$\frac{dT_p}{dt} = \frac{6Nu\kappa}{D_p^2\rho_p C_p}(T_g - T_p) + \frac{3\alpha q}{2\rho_p D_p C_p} \tag{2.60}$$

where κ is the heat conductivity of the gas; $Nu = hD_p/\kappa$ is the Nusselt number; h is the heat-transfer coefficient; and α is the absorption coefficient of the particle material. For a spherical particle [81],

$$Nu = 2 + 0.6Re^{0.5}Pr^{0.33}, \tag{2.61}$$

where Pr is the Prandtl number. Surfacing is often performed in a diverging laser beam:

$$q(x) = P/[r_f + xtg\delta/2]^2\pi \tag{2.62}$$

where r_f is the beam radius at the focus; $\delta/2$ is the half-angle of the beam focusing; and x is the distance from the focus.

By solving numerically (2.57)–(2.59), we find $v(t)$ and $x(t)$. Knowing $x(t)$, we can obtain the dependence of the temperature of particles on the coordinate x for different initial parameters of the problem. For example, for $\alpha = 0.25$, $P = 1$ kW, and the initial velocity $v_0 = 2$ m/s, the temperature of powder particles depends nonmonotonically on the distance from the nozzle edge (Fig. 2.56) [79]. This is explained by the rapid heating of particles near the focus followed by their cooling during downstream motion because the laser radiation intensity decreases with increasing x. As the flow velocity increases, the maximum temperature of particles decreases because each particle is irradiated by the laser for a shorter time. Figure 2.56 also presents the results of experimental measurements. One can see that these results are in good qualitative and quantitative agreement with calculations, which demonstrates the validity of the model.

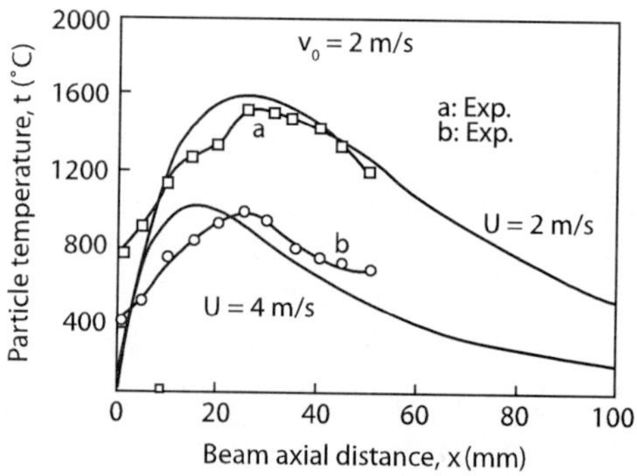

Fig. 2.56 The particle temperature profiles along the beam axis under various jet velocities with an initial particle velocity $v_0 = 2$ m/s

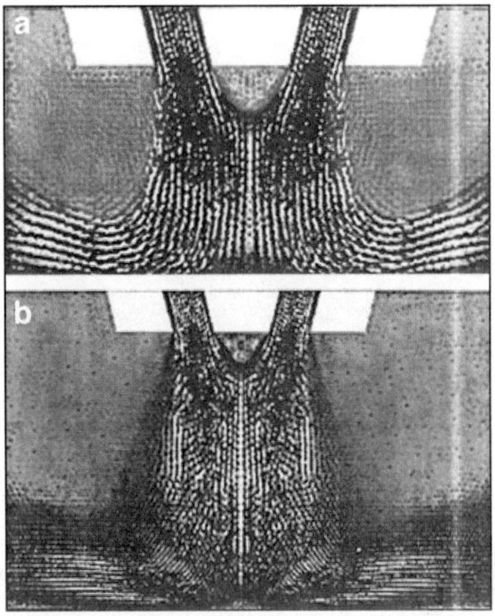

Fig. 2.57 Velocity fields of a gas jet for different distances between a nozzle cut and a sample surface: (**a**) small distance; (**b**) large distance

Surfacing should be performed so that the powder particles near a target would continue flying to the target surface although a gas jet is turned here by 90°. It is clear that this can be achieved by strongly accelerating particles, choosing particles of large sizes, and forming a powder jet converging to the focal spot center. The latter requirement is fulfilled by using special conic nozzles [83] (Fig. 2.57). The velocity field was calculated by solving the Euler equation in a cylindrical coordinate system. One can see from Fig. 2.57 that the velocity field considerably depends on the distance between the nozzle and the sample surface. Figure 2.58 shows the trajectories of iron powder particles calculated by (2.57)–(2.59) in the velocity field specified in Fig. 2.57. One can see that at small flow rates $v \sim 0.1$ m/s, light particles of radius 5 μm do not reach in fact the target (Fig. 2.58b). The flow of particles incident of the target can be formed by increasing their mass by a factor of 64. A narrow flow can be also formed by increasing significantly the rate of a carrier gas. When the rate is increased by a factor of 20, the flow on the target becomes no wider than the output hole of the nozzle (Fig. 2.59a). This figure also shows that there exists the optimal distance between the nozzle and target (cf. Figs. 2.59a and 2.59b).

The dynamics of powder particles depends on the type of a carrier gas. It is clear that, rate for rate, a heavier gas will accelerate powder particles to greater velocities (Fig. 2.60). This figure shows that the velocities of particles incident on the surface decrease away from the center, which is caused by the type of the velocity field of the carrier gas. The latter determines the different particle residence times in the laser radiation field and, finally, the particle heating temperature. In this case, the convective cooling by the carrier gas competes with heating. The peripheral regions of the powder jet a helium flow are cooled below the melting point due to the high heat conduction of the flow and do not adhere to a substrate (Fig. 2.61). In the example under study, this leads to the powder loss up to 80%.

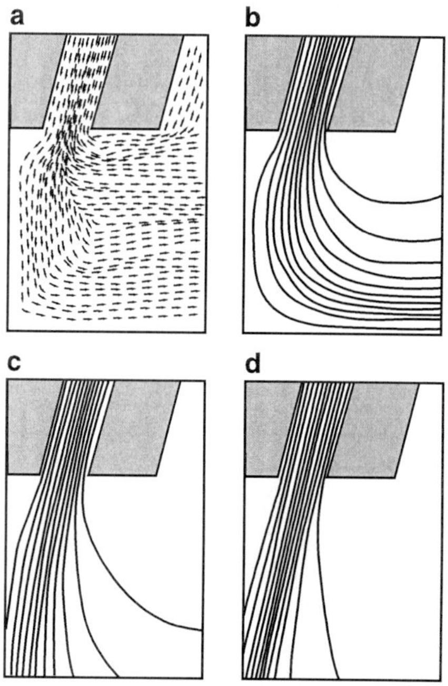

Fig. 2.58 Argon carrier gas flow field (**a**). Trajectories of powder iron particles of sizes 5 μm (**b**), 10 (**c**), 20 (**d**). Initial velocity at the head outlet is 0.1 m/s. The distance to the substrate is 3 mm

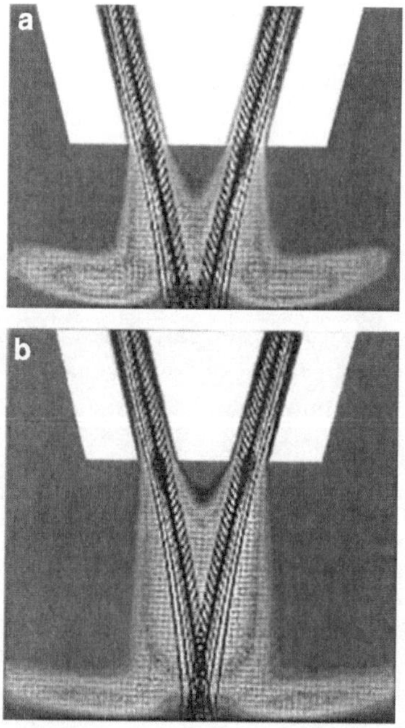

Fig. 2.59 Trajectories of powder particles above a sample surface: the intersection point of trajectories lies below (**a**) and above (**b**) the sample surface

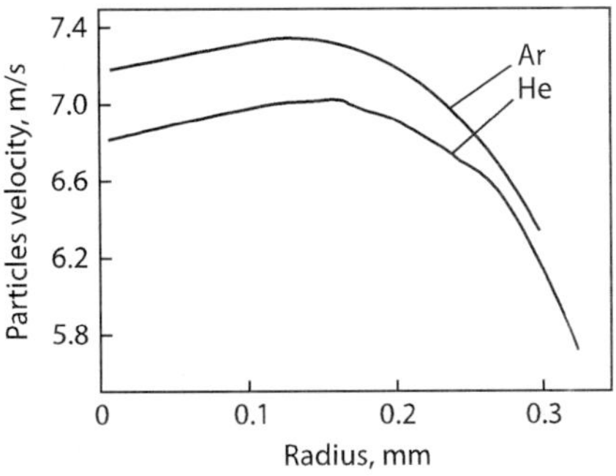

Fig. 2.60 Velocity distributions of particles along the gas jet radius

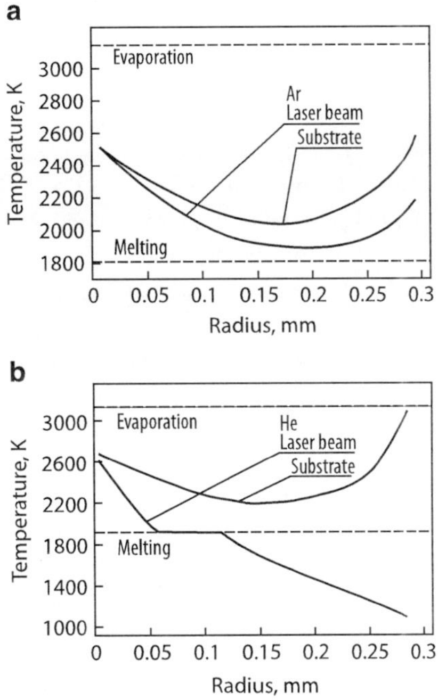

Fig. 2.61 Surface temperature profiles in the surfacing zone in the argon (**a**) and helium (**b**) jets. Line (Laser beam) is temperature of particles in laser beam

Consider the general cladding picture involving the heating of powder particles in a laser beam and its relation with surfacing parameters. When the sufficient amount of powder is supplied on a target, and the restricting factor is the necessity of powder heating up to the melting point, the cross section of a fused roller is determined by the energy balance:

$$S = \frac{\alpha_e P}{\rho v_t (H_m + c\rho \Delta T)} \tag{2.63}$$

where α_e is the effective absorption coefficient of the powder jet at the laser wavelength and v_t is the cladding rate. It is clear that α_e should depend on the concentration N_x of powder particles:

$$N_x = \frac{W}{\pi (r_c + x \tan\theta)^2 V_p M}, \tag{2.64}$$

where r_c is the nozzle hole radius, θ is the opening angle of the powder jet, V_p and M are the velocity and mass of powder particles. The absorption of laser radiation in the powder cloud is determined by the usual equation

$$\frac{dq}{dx} = -q\varepsilon N_x, \tag{2.65}$$

where ε is the optical factor describing the averaged absorption cross section for laser radiation.

In the simplest case, $\varepsilon \approx 2\pi r_p^2$. By integrating (2.64) and (2.65), we obtain the absorption coefficient α_e. By substituting it into (2.63), we obtain finally

$$S = \frac{P}{v_t \rho (H_m + c\rho \Delta T)} \left[1 - \exp\left(-\frac{3}{2\pi V_p r_p r_c \rho} \frac{W x_0}{r_c + x_0 \tan\theta} \right) \right],$$

where x_0 is the distance between the nozzle and substrate and V_p is the velocity of powder particles. Figure 2.62 [84] shows that the cross section of fused strips of satellite-6 powder increases with the powder consumption W and decreases with increasing the cladding rate. The efficiency η of the powder utilization is determined by the ratio of the cross section of the powder jet incident on the surface and the melt pool cross section [85]. It is clear that the latter quantity decreases with increasing the surfacing rate, which reduces the powder utilization efficiency (Fig. 2.63) [85].

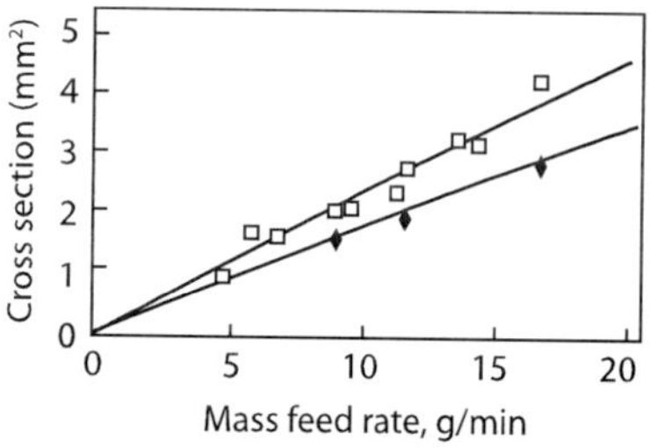

Fig. 2.62 Cross-section versus mass feed rate and scanning speed (*square*) – 150 mm/mn, (*black rhomb*) – 200 mm/mn). Comparison between calculations (*lines*) and measures. (P = 1,000 W, D_{bp} = 18 mm, powder side equals 104–150 μm)

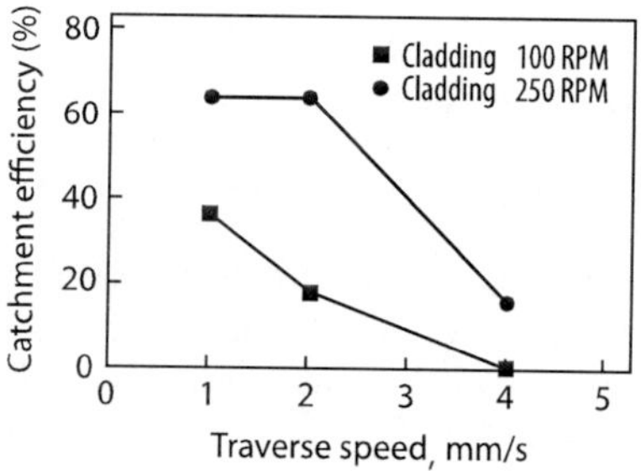

Fig. 2.63 Mass flow rate vs. catchment efficiency when cladding vertically downwards

Knowing the value of η and the melt width, we can find the area S of the cross section of the melted strip and its height h. The area S is found from the condition of conservation of the powder mass:

$$S\rho v = \eta W \tag{2.66}$$

where W is the powder consumption and v is the sample travel speed. It is assumed in (2.63) that all the particles entering the melting zone will participate in the formation of a fused strip. It is known from experiments that the shape of the cross section of the fused strip is close to the arc of a circle:

$$y(x) = \sqrt{R^2 - x^2} + h - R \tag{2.67}$$

where R is the circle radius and h is the maximum height of the strip. At the strip edge, $x = d/2$, and $y = 0$, which gives

$$h = R - \sqrt{R^2 - (d/2)^2} \tag{2.68}$$

The cross-sectional area of the strip S is found by integrating (2.67):

$$S = d(h - R) + \frac{d}{2}\sqrt{R^2 - \frac{d^2}{4}} + R^2 \arcsin\frac{d}{2R} \tag{2.69}$$

The values of d and η in [86] were determined experimentally, the melt width d being measured without the powder supply:

$$d = 2.56 - 0.05\,\mathrm{v}$$

(see Fig. 2.64a)

$$\eta = 0.34 - 0.016\,\mathrm{v}(W = 0.2\ \mathrm{g/s}) \tag{2.70}$$

The measurements were performed in the sample travel range from 4 to 14 mm/s by focusing the 2-kW radiation from a CO_2 laser to a 2.5-mm spot on a steel plate. Satellite-6 powder was supplied into the melting zone in an argon jet. Knowing η, we find S from (2.66). By excluding R in (2.68) and (2.69), we find h depending on the surfacing rate. Figure 2.64b [84] demonstrates good agreement between the model and experiment.

The dependence of the roller height of the sample travel speed (Fig. 2.64) is consistent with data presented in Fig. 2.65 [87], where the area of the strip fused per second increases with the laser power. Because the strip width weakly depends on the sample travel speed, we see from Fig. 2.65 that for any power $h \approx v^{-1}$. The latter is consistent with data presented in Fig. 2.64.

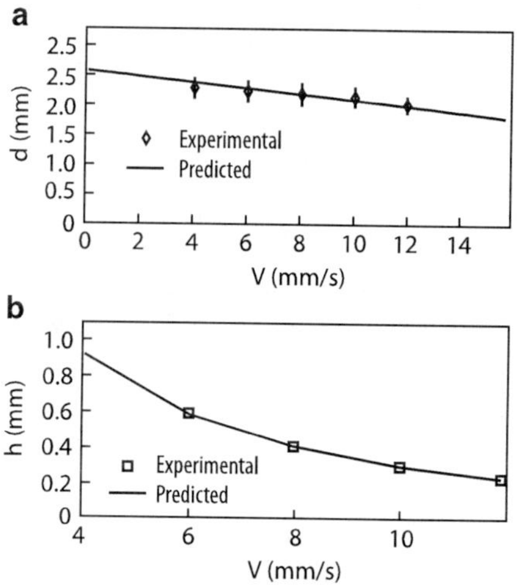

Fig. 2.64 Predicted and experiment values for d (**a**) and h (**b**) at a powder feed rate of 0.2 g/s

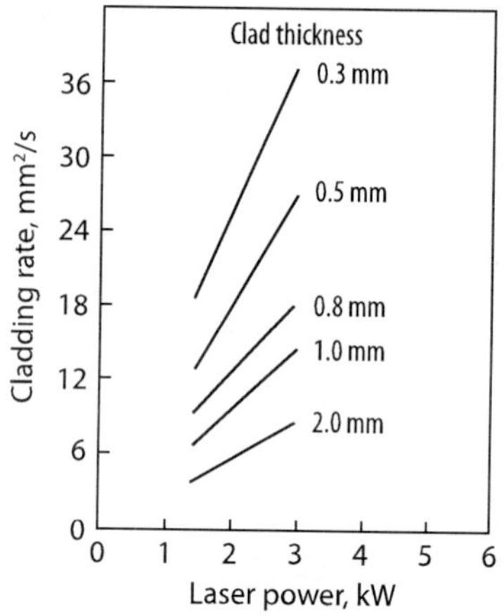

Fig. 2.65 Dependences of the cladding speed on the laser power for different heights of a cladding ridge

As mentioned above, laser cladding should be performed at the minimal depth of the substrate melt. It influence is characterized by the fusion value. It follows from Fig. 2.66 [78] that this quantity is mainly determined by the powder consumption and weakly depends on the surfacing rate and laser radiation intensity.

The properties of surfacing were described above qualitatively, by appealing to experiments. The detailed investigation of this process requires, as in the case of alloying, the use of numerical calculations. We saw that excitation of the Marangoni convection during alloying considerably changed the size and form of the melt pool (Fig. 2.41). This effect can be also important during cladding. It was shown in [80] that the size of the melting zone on the front of a fused roller almost doubled due to the Marangoni convection (Fig. 2.67).

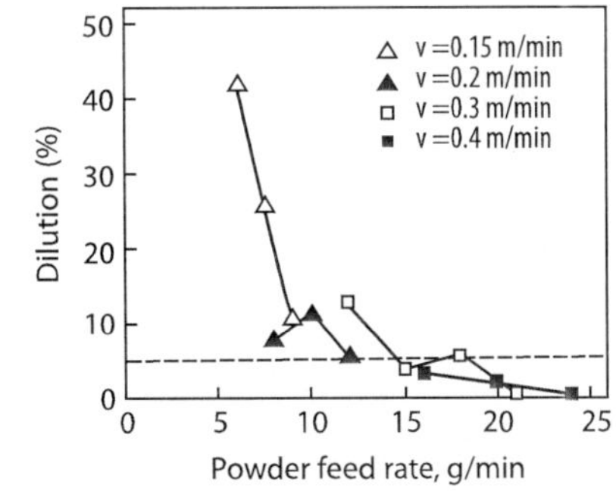

Fig. 2.66 Fusion depth as a function of the powder supply speed for different sample motion speeds

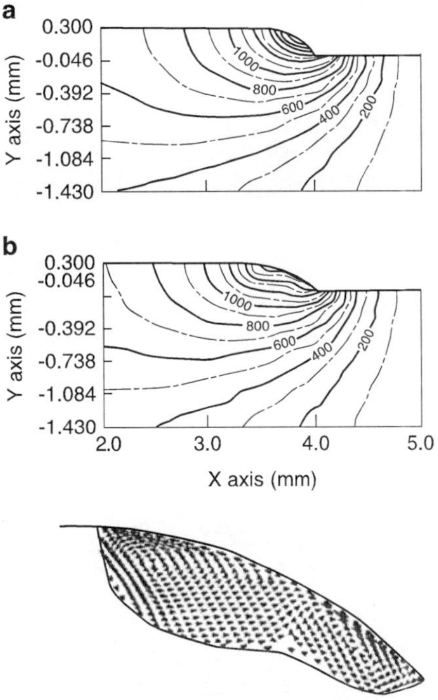

Fig. 2.67 Calculated temperature field, (satellite-6 on Ck 45, $V_{SC} = 250$ mm/min, $P_1 = 1.5$ kW/cm, $T_m = 1{,}300$ C, $L = 0.6$ mm) without (**a**) and with (**b**) Marangoni convection. Calculated velocity field ($v_{max} = 90$ cm/s)

Figure 2.68 [87] generalizes the properties of laser cladding. One can see that at a constant ratio of the laser power to d_f, the dilution decreases with increasing the powder consumption. However, the powder consumption W is limited because cracks appear at large consumptions. Thus, there exists a narrow region in the $P/d_f \div W$ plane where surfacing is efficient and has a high quality. If a cladding material should be deposited on a large area, strips are fused with some overlap ~30% (Fig. 2.69).

As shown above, the physical models of laser cladding assume the fulfillment of certain relations between various processes such as the heating and melting of powder particles, the heating and melting of the target, the adhesion of particles to the target, the fusion of particles with each other, the screening and scattering

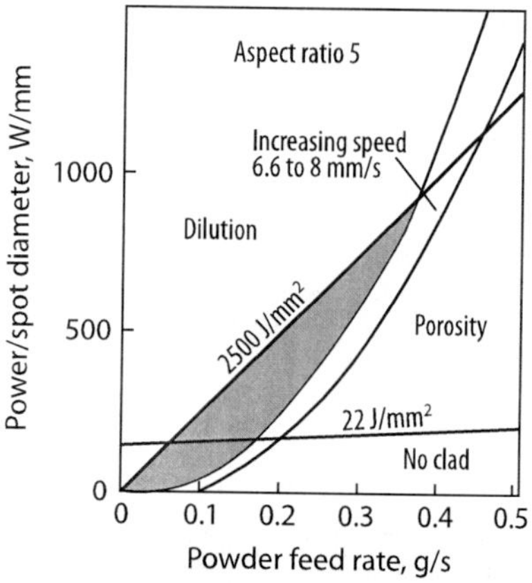

Fig. 2.68 The operating window for blown power laser cladding

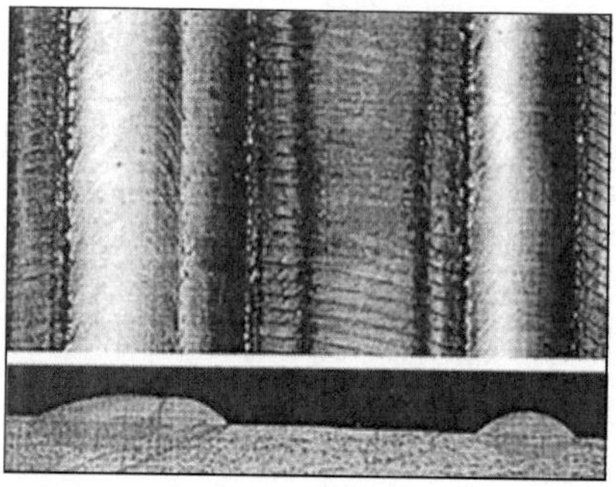

Fig. 2.69 Single and double cladding strips. *Top*: top view; *bottom*: cross section (photo by G. Antonova)

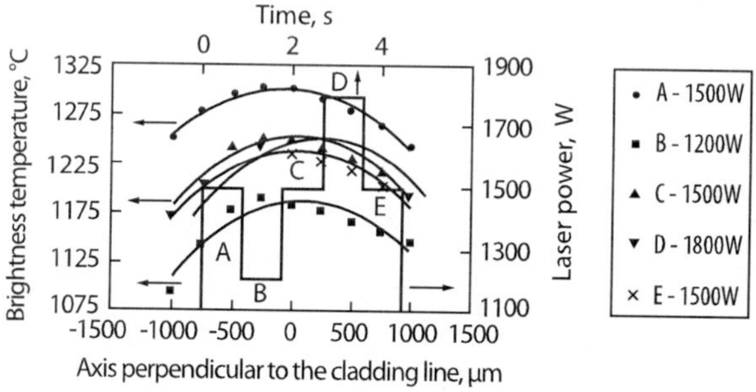

Fig. 2.70 Influence of laser power variation (five steps between 1,200 and 1,800 W) daring laser cladding of stellite on steel (6 mm head width) on surface temperature profile across the cladded head. Cladding parameters: $v = 3$ m/min, 50% of the maximum rotation speed of the powder distribution plateau (7.5 g/min)

of laser radiation. The measurements of internal parameters during cladding are complicated, as in most of the laser technologies. The most accessible contactless measurement method is the measurement of the temperature of a fused roller during surfacing. Figure 2.70 [88] shows the brightness temperature distribution in the hottest region of a broad fused strip measured when powder was supplied from the side. One can see that this temperature increases with increasing laser power. But its growth occurs somewhat slower than that of laser radiation, although $T \sim q$, according to the linear theory. The same was observed when the powder consumption was varied at a constant power (Fig. 2.71) [89]. All this suggests that the powder affects to some extent the laser beam, which was qualitatively taken into account in one of the models described above [see (2.63)–(2.65)]. The similar behavior was also observed during microsurfacing, when the strip width was ~0.3 mm (Fig. 2.72) [88]. During coaxial cladding, the temperature profile can have two maxima (Fig. 2.73) [88]. This suggests that the energy of powder samples heated in the laser beam plays an important role in the heat balance.

In the case of repetitively pulsed cladding with the coaxial injection of powder, a roller can solidify during an interval between pulses depending on the sample travel speed (Fig. 2.74) [88]. Each subsequent laser pulse starts surfacing at the beginning. As the sample travel speed is decreased, the melt has no time to solidify during the

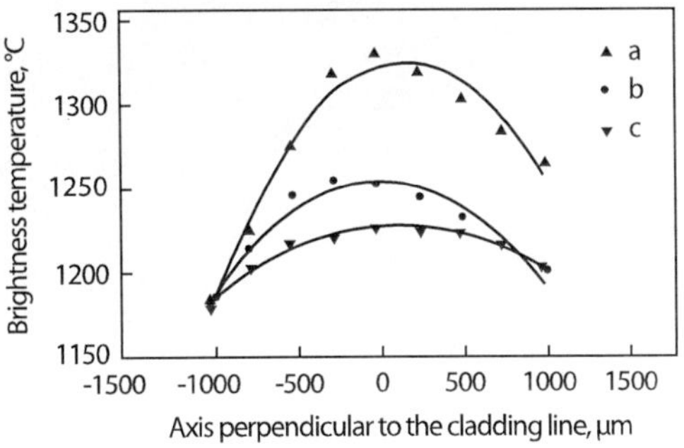

Fig. 2.71 Interface of powder feeding rate on surface temperature profile during laser cladding of stellite on steel (6 mm bead width). Cladding parameters: $P = 1{,}500$ W, $v = 3$ m/min, (**a**): 30% of the maximum rotation speed of the powder distribution plateau (4.5 g/min); (**b**): 50%; (**c**): 70%

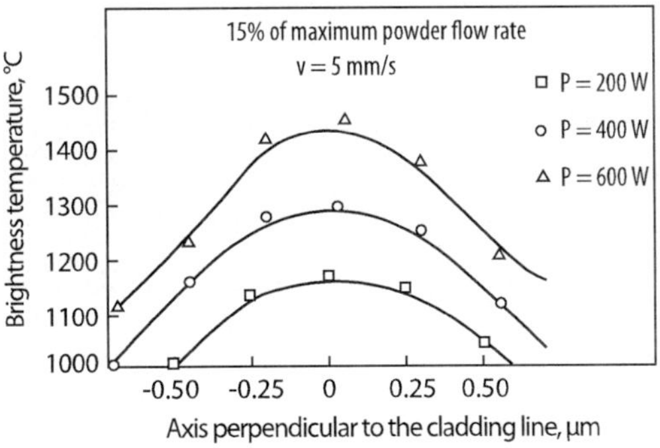

Fig. 2.72 Laser power influence on surface temperature during laser cladding of stellite on steel (300 μm bead width). Brightness temperature profiles recoded by the two-dimensional pyrometer ($\lambda = 0.86$ μm). Cladding parameters: $v = 0.3$ m/min. 15% of the maximum rotation sped of the powder distribution plateau (2.3 g/min)

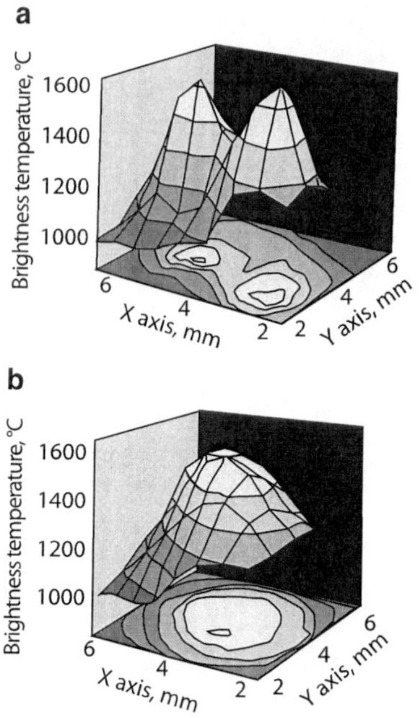

Fig. 2.73 Two-dimensional temperature fields in coaxial laser cladding of stellite + WC/Co (30%) on steel substrate; laser power 2 kW, velocity 1,000 mm/min: (**a**) non-optimised injection conditions; (**b**) optimized injection conditions

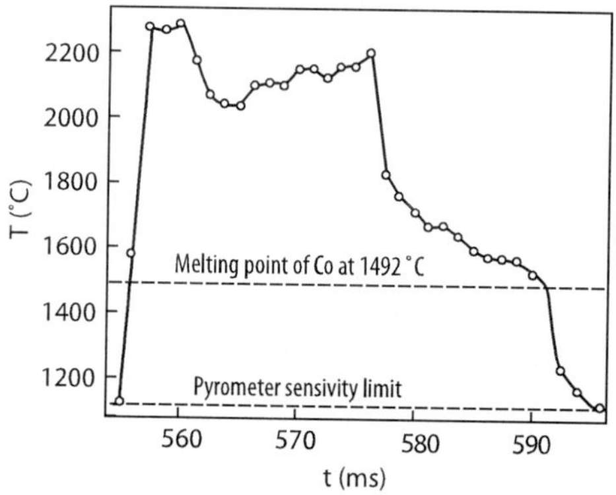

Fig. 2.74 Individual thermo-cycle of true temperature in pulsed-periodic laser cladding (20 ms pulse duration, rectangular pulse shape, 2 kW average power, velocity 500 mm/min) of CuAl and WC-Co (30% vol.) on steel

interval between pulses (Fig. 2.75) [88] and surfacing continues, resulting in an in crease in the roller height.

As mentioned above, there is no point in fusing strips of a high height because their quality decreases. If necessary, it is expedient to fuse one strip over another. This method can be used to construct thin and high walls consisting of different

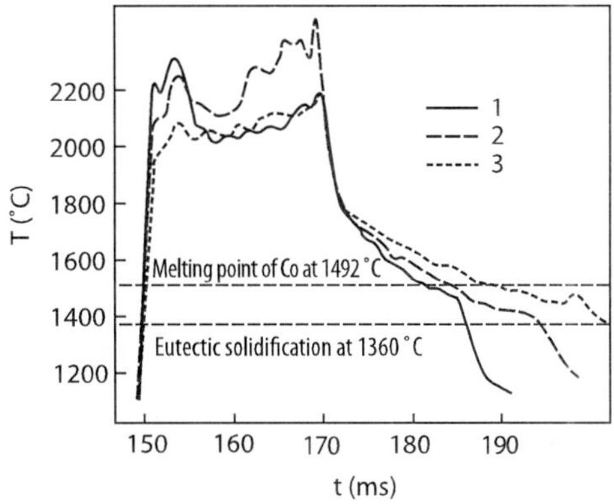

Fig. 2.75 Comparison of three individual thermo-cycles after the decrease of cladding velocity. Curve 1 is the thermocycle at high velocity. Curve 3 is the thermocycle at low velocity

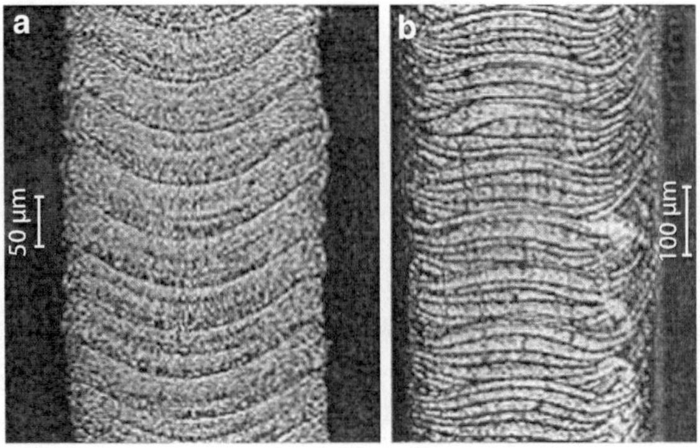

Fig. 2.76 The cross-section of the wall made by (**a**) CO_2 laser from stainless steel 316L powder, polished, etched; (**b**) pulsed Nd:YAG laser from stainless steel 316L powder, polished, etched

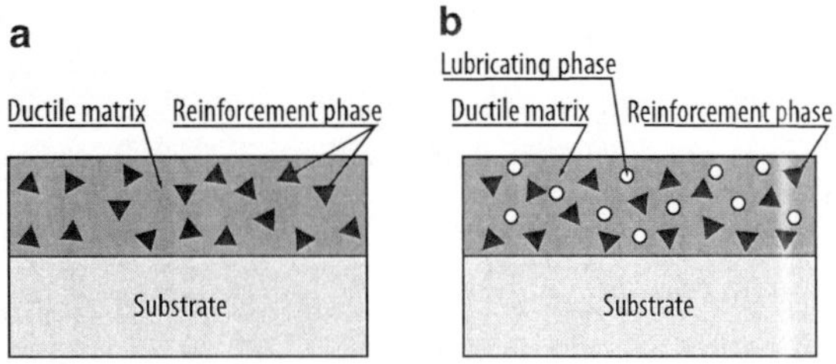

Fig. 2.77 (**a**) Coating from metal matrix composite with solid lubricant applied as ductile matrix. (**b**) Coating from traditional metal matrix composite with inclusion of solid lubricant

materials [89]. Microsurfacing can be performed by using both cw and periodically pulsed radiation (Fig. 2.76) [89]. In the latter case, the better quality of the fused walls can be obtained by selecting the proper irradiation regime.

The laser method can be used for cladding mixtures of metals and solid ceramics not dissolved in each other. For example, the mixtures of powders of 43OL stainless steel, WC/C_2 ceramics, and solid CuSm lubricant (Figs. 2.77, 2.78) and other mixtures [90] can be employed. Coatings made of such materials have low friction coefficients and are durable. The stellite/CuSn/WCCo coating has the friction coefficient ~0.12. The homogeneity and quality of surfacing produced by repetitively pulsed lasers are better than in the case of cw lasers [89].

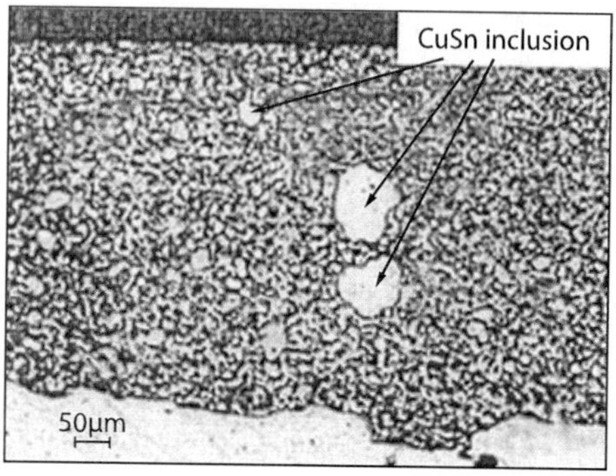

Fig. 2.78 The microstructure of metal matrix composite coating (SS439L/WC) with inclusions of solid lubricant (CuSn)

2.4 Mechanisms of Laser-Induced Surface Cleaning

The investigation of laser-induced cleaning of surfaces from various contaminations has been initiated in the 1990s. This technology can be used in industry, for the restoration of artworks, in medicine, and microelectronics. For example, the cleaning of surfaces from nanoparticles in the latter case becomes actual due to the constantly increasing integration degree of microinstruments [91]. The region of applications of laser-induced cleaning has expanded not only due to the development of reliable lasers with sufficient output powers but also due to the possibility of nondestructive cleaning of the internal parts of devices, cleaning from radioactive contaminations, etc. Laser deactivation can be used for processing not easily accessible places, this method does not require additional expendable materials, does not produce liquid radioactive waste, and is safe for maintenance personnel. The development of laser deactivation technologies considerably shortens the utilization process of nuclear energetics objects and provides the repeated industrial use of deactivated valuable materials [92, 93]. Laser methods are also proposed for cleaning the water surface from petroleum films and other contaminants in the environment [94, 95].

From the point of view of mechanisms, laser-induced cleaning can be divided into three classes: (a) Cleaning of surfaces from contaminating particles; (b) cleaning of surfaces from inhomogeneous films; and (c) cleaning of surfaces from films consisting of the sample material, which appeared during the operation of devices (oxide films and radioactive layers).

The cleaning of surfaces can be dry or wet. We will describe the mechanisms of laser cleaning by considering only macroscopic processes caused by the incoherent (i.e. thermal) interaction of laser radiation with matter. Physical processes proceeding during laser cleaning are described more comprehensively, for example, in [91]. These processes include the influence of microparticles on the structure of laser radiation, resonance effects, the damage of a substrate, excitation of acoustic and shock waves in a substrate, etc.

2.4.1 Cleaning of Surfaces from Microparticles

It is obvious that by exposing a surface to intense enough laser radiation, it is possible to initiate the evaporation of the surface, thereby cleaning it. However, it is often desirable to avoid the damage of the surface being cleaned. There exist mechanisms of surface cleaning at lower temperatures when the surface not only does not boil but even does not melt. The mechanism of such cleaning is obviously the thermal expansion of a solid:

$$\delta = \alpha_T \Delta T, \quad (2.71)$$

where δ is the relative increase in the size of the solid with increasing its temperature by ΔT and α_T is the linear thermal expansion coefficient. Expression (2.71) can be used when the heating of a body is uniform, i.e. in the case of thermally small particles. It is necessary that a particle should be heated over the entire depth during laser irradiation time τ

$$R < \sqrt{\chi\tau}, \quad (2.72)$$

where R is the particle size and χ is thermal diffusivity of the particle material. If a microparticle is transparent for laser radiation, but a substrate can be heated, the latter can transfer heat to the particle. The substrate region located under the microparticle can also expand (Fig. 2.79). In the case of dry laser cleaning, the heating of a particle can be determined from the thermal balance

$$Vc\rho\frac{\mathrm{d}T}{\mathrm{d}t} = \alpha qS/2 \quad (2.73)$$

where V and S are the volume and surface of the microparticle. The value of δ increases upon heating and the center of mass of the particle is displaced from the sample surface. As q decreases, the particle is cooled, the size S decreases, and the particle tends to escape from the sample surface. The appearing inertial force

$$F_0 = Rm\frac{\mathrm{d}^2\delta}{\mathrm{d}t^2} \quad (2.74)$$

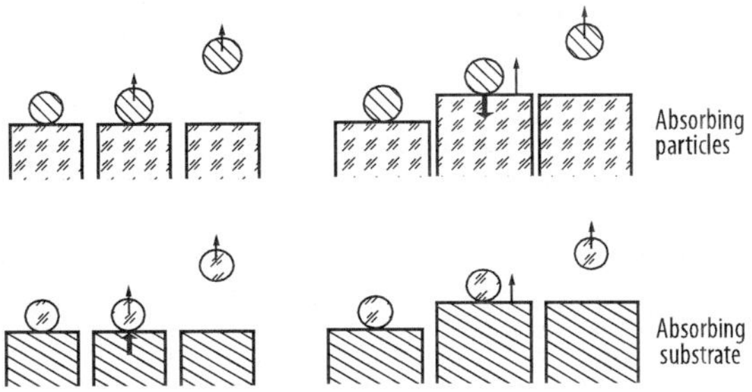

Fig. 2.79 Illustration of the physical mechanisms of dry laser cleaning

is opposed by the adhesive force F_a. The detachment of the particle occurs if $F_0 > F_a$. We obtain from (2.72)–(2.74) that

$$F_0 = \frac{\alpha}{2} \frac{S}{V} \frac{R\alpha_T m}{c\rho} \frac{\mathrm{d}q}{\mathrm{d}t} \tag{2.75}$$

For a triangle laser pulse of duration τ and the trailing edge duration τ_1, we have

$$\frac{\mathrm{d}q}{\mathrm{d}t} = \frac{2\varepsilon}{\tau\tau_1} \tag{2.76}$$

where ε is the pulse energy density. For example, by substituting (2.76) into (2.75), we obtain the detachment condition for a spherical particle:

$$\varepsilon > F_a \frac{\tau\tau_1 c}{2\pi\alpha\alpha_T R^3} \tag{2.77}$$

Condition (2.77) is necessary but insufficient for cleaning because the force F_0 disappears after irradiation, while the adhesive force remains. To prevent the returning of the particle back under the action of this force, the kinetic energy acquired upon expansion should exceed the work of the adhesive force. This additional condition leads to the increase in the threshold pulse energy density (2.76), but this increase in not large [96]. A drastic in the cleaning threshold with increasing the microparticle size was observed in experiments [98] (Fig. 2.80). It follows from expressions (2.72) and (2.77) that there exists the optimal size of particles being detached. The laser pulse energy should exceed ε corresponding to this size. It was shown experimentally [97] that after irradiation of a silicon plate by UV four 20-ns pulses from a KF laser producing the energy density on the plate equal to 300–500 mJ/cm^2, the 0.3–0.5 μm SiO, Al_2O_3 and other particles were removed without damaging the plate.

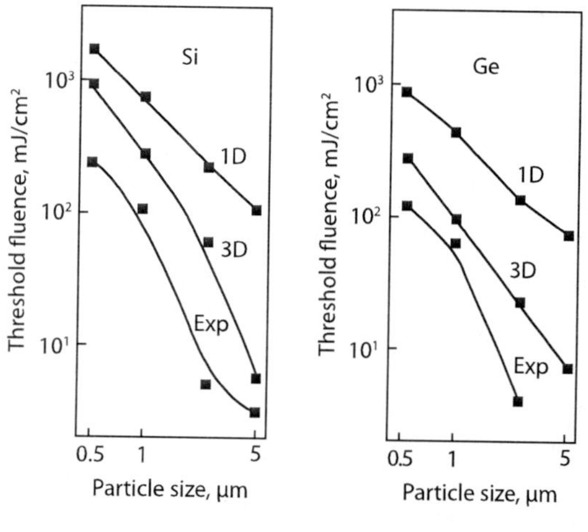

Fig. 2.80 Theoretical (for ID and 3D models) and experimental results of the threshold laser fluences for SiO_2 particles versus particle size for Si, Ge substrates. Excimer laser $\lambda = 248$ nm, pulse duration 23 ns

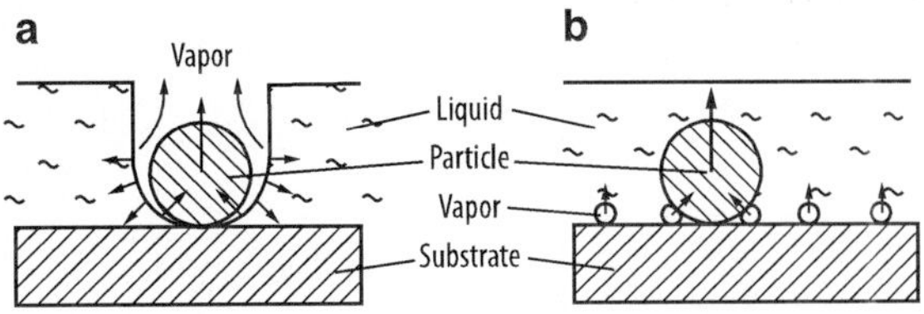

Fig. 2.81 Illustration of the physical mechanisms of steam laser cleaning: (**a**) – absorbing particle, (**b**) – absorbing substrate

During "wet" laser cleaning, the main mechanism is the heating of particles up to the boiling point of liquid. A particle is ejected by a vapor jet produced at the particle-liquid interface (Fig. 2.81a) [96]. In this case, the cleaning threshold coincides with the boiling threshold of liquid near the particle:

$$\varepsilon_b = \frac{\rho C R (T_b - T_\infty)}{\bar{\alpha}}$$

where $\bar{\alpha}$ is the effective absorption coefficient at the laser wavelength. This mechanism "operates" if the contact area of a particle with the sample surface is smaller than the maximum cross-sectional area of the particle. Vapor jets ejecting particles were observed experimentally [97]. In the case of transparent particles, transparent liquid, and absorbing substrates (Fig. 2.81b), the cleaning mechanism is more complicated and is determined by the boiling temperature of liquid, its surface tension, and the roughness of the substrate surface. It is necessary to select the irradiation regime so that the size of vapor bubbles would not exceed the thickness of the liquid heated up to the chosen temperature [96].

2.4.2 *Laser-Induced Solid Surface Cleaning from Films*

The laser cleaning of surfaces from films, as "dry" cleaning from microparticles, is often performed in the "pre-evaporation" regime. This provides a higher quality of a surface being cleaned. Such a regime is also based on the thermal expansion of a film or a substrate. This mechanism is realized during surface cleaning from metal films. Metals have the elasticity, which is also important for the realization of this mechanism. Thus, during the heating of the film region under the focal spot, the film expands. Because neighboring regions remain cold, the film bends under the action of the expanding force and is detached from the substrate (Fig. 2.82). We will estimate the film removal threshold and rate by using the simplified approach to the problem. According to Hooke's law, the compression stress σ is

$$\sigma_T = E \alpha_T T$$

where E is Young's modulus. The compression energy density per unit volume is

$$Q = \frac{\sigma_T^2}{2E} + \frac{E}{2}(\alpha_T T)^2 \tag{2.78}$$

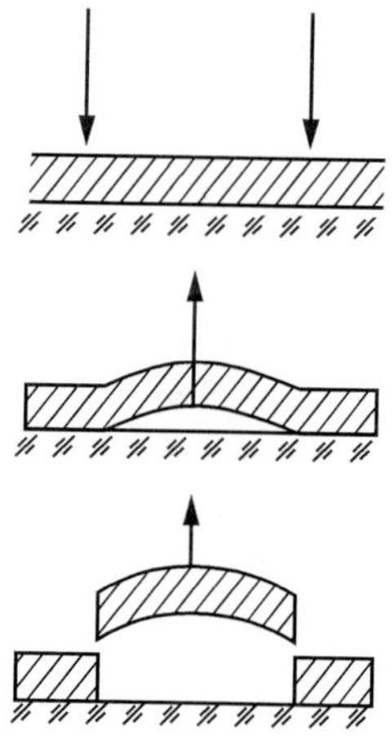

Fig. 2.82 Buckling mechanism of film removal

This energy can be transformed under certain conditions to the kinetic energy of the film fragments flying from the substrate. This mechanism can involve several different regimes. At moderate energy fluxes ε, when the displacement of the film by the pulse end instant τ is smaller than the film thickness, the velocity of the fragments will be proportional to the pulse duration. As ε is increased, the regime is developed when the film is displaced by its thickness for the time $t^* < \tau$. In this case, the maximum velocity of the fragments will be independent of the pulse duration τ because thermal expansion ceases after the film detachment.

Consider the simplest case when the laser radiation energy is spent completely to generate the thermoelastic energy:

$$Q = \frac{\rho v^2}{2} + Q_0 \tag{2.79}$$

where Q_0 is the adhesive energy per unit film volume. When adhesion is overcame due to evaporation at the film-substrate contact region, we have $Q_0 = 0$. In this case, by substituting (2.78) into (2.79), we obtain the maximum velocity of the film

$$V_1 = \sqrt{\frac{E}{\rho}\frac{\alpha_T \alpha \varepsilon}{hc\rho}} \tag{2.80}$$

where h is the film thickness. For $Q_0 \neq 0$, the velocity is somewhat smaller:

$$V_2 = \sqrt{\frac{E}{\rho}\frac{\alpha_T \alpha \varepsilon}{hc\rho}}\sqrt{1 - \frac{2Q_0 h^2 c^2 \rho^2}{E\alpha_T^2 \alpha^2 \varepsilon^2}} \tag{2.81}$$

One can see from (2.81) that the film detachment effect has an energy density threshold, which is determined by the condition $V_2 = 0$).

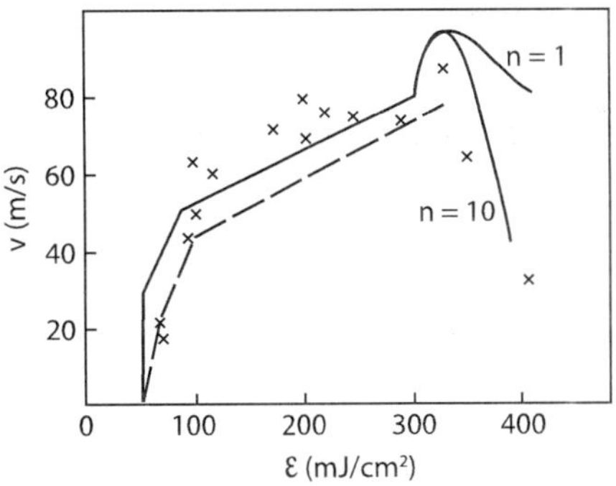

Fig. 2.83 The dependence of velocity of fragment of W film ($h_f = 10^{-7}$ m, $\tau = 10^{-8}$ s) on laser fluence; *dash line* – elimination of adhesion occurs due to compression energy, *solid line* – elimination of adhesion occurs due to other factors, x – experimental data, n is the number of "hot paints"

Expressions (2.78)–(2.81) were obtained for an elastic film, i.e. when $T < T_m$. As the laser radiation energy ε is further increased, the film can be heated above the melting temperature. At the beginning of this process, melting occurs in separate sites caused by the inhomogeneity of the absorbed laser radiation energy. In this case, the film can partially freely expand, resulting in a decrease in the compression energy density. This leads to the decrease in the velocity with increasing ε (Fig. 2.83) [99]. Figure 2.83 also shows the curves constructed by expressions (2.80) and (2.81). In the case of a transparent film and an absorbing substrate, the cleaning mechanism will be different. Under these conditions, apart from thermoelastic forces, the pressure of vapor produced by the evaporating substrate can be involved in the process. The latter mechanism operates in a "pure form" during liquid surface cleaning, which is discussed in the special section below.

Apart from thermomechanical shelling considered above, porous oxide films can be cleaned by increasing the pressure of gases in caverns and pores of the film. This is the so-called explosive mechanism [96], which is difficult to describe due to the uncertain structure of such films.

Finally, the evaporation mechanism of film cleaning always "operates", but it is characterized by a large energy consumption. Such cleaning can be efficiently performed by using a repetitively pulsed XeCl laser emitting at $\lambda \sim 0.3\,\mu$m. On the one hand, this radiation is better absorbed than radiation from Nd and CO_2 lasers, and on the other hand, radiation at this wavelength can be transported through optical fibers, while the shorter-wavelength radiation is already absorbed in fibers. The XeCl laser typically emits pulses of duration 10–100 ns. Pulses of such duration remove thin layers of a film being cleaned. This excludes the damage of the film surface. For a typical energy density of $\sim$1–3 J/cm^2, the laser radiation intensity in the laser spot achieves hundreds of megawatts per square centimeter. These intensities obviously exceed the evaporation threshold (1.30). The film is evaporated and a plasma is produced in vapor, and then this erosion plume flies away from the target. We will discuss these processes in detail in next sections. Note here that within a long time after the pulse end ($\sim$1 μs), the reverse motion of the evaporated material to the target is observed. This process depends on the pressure and type of the surrounding gas, which is demonstrated in Fig. 2.84 [100]. As the gas pressure increases, the film material is partially again deposited in the laser spot region [101].

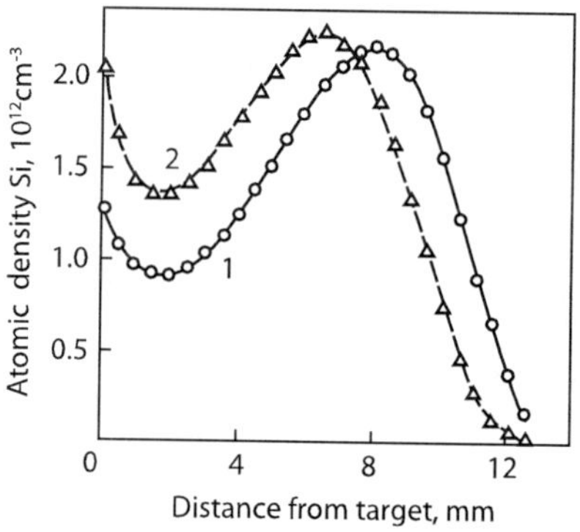

Fig. 2.84 Dependences of the density of Si atoms on z at the instant $t = 2.4\,\mu$s obtained in two-dimensional calculations by the Navier–Stokes equations for the energy density 2.4 J/cm^2 and argon pressures 400 (**1**) and 700 mTorr (**2**)

We considered above the removal of contaminants from solid surfaces. In practice, it is necessary sometimes to clean the surface of liquids, e.g., water. Most often water should be purified from petroleum products. Laser radiation can be also used to clean water from thin films ($\sim$100 μm) [94, 95].

2.4.3 Physical Model of Water Surface Cleaning from Thin Films of Petroleum Products

It is convenient to investigate the mechanism of removal of petroleum films from the water surface by using model experiments because quantitative measurements during the cleaning of water from petroleum involve difficulties. Model experiments should be performed with a substance whose removal can be easily controlled and whose properties are close to those of petroleum. The optical and thermal properties of paraffin at a wavelength of 10.6 μm at temperatures above 60° weakly differ from these properties of petroleum [102].

Because the absorption coefficient of paraffin at the CO_2 laser wavelength is considerably lower than that of water (the absorption lengths of the CO_2 laser radiation in water and petroleum products are $\sim$20 and $\sim$100–200 μm, respectively [103]), radiation is weakly absorbed in paraffin during cleaning from thin films ($<$100 μm) and mainly heats and evaporates the water surface. Water vapor detaches and removes a paraffin film. The production of water vapor from the unit surface can be found from the heat balance

$$L_b \frac{\partial m}{\partial t} = q_0 \exp(-h/l) \tag{2.82}$$

where L_b is the vaporization enthalpy, h is the film thickness, and l is laser-beam absorption length in the film. The dynamics of the paraffin film is described by the Newton equation

$$\rho h \frac{\mathrm{d}^2 y}{\mathrm{d}t^2} = p - p_0 \tag{2.83}$$

where y is the vapor layer thickness and p and p_0 are the vapor pressure and external pressure, respectively. The vapor pressure can be found from the Clapeyron–Mendeleev equation

$$py = \frac{m}{\mu} RT \tag{2.84}$$

From (2.83) and (2.84), we obtain the expression

$$\rho hyy'' = \frac{m}{\mu} RT - p_0 y \tag{2.85}$$

One can see from (2.85) that at the beginning of the process, when $y \approx 0$, we can neglect the external pressure p_0. By differentiating (2.85) and substituting dm/dt from (2.82), we obtain the nonlinear equation for $y(t)$. It has a particular solution satisfying the initial condition $y(0) = 0$.

$$y = \frac{2}{\sqrt{3}} \left[\frac{RT}{\rho\mu} \frac{q_0}{L_b h} \exp(-h/l) \right]^{1/2} t^{3/2} \tag{2.86}$$

The vapor layer thickness y increases with time and the role of the external pressure increases. When the external pressure becomes equal to the water vapor pressure, the right-hand side in (2.85) vanishes and the film will move at a constant velocity. This velocity can be determined by equating the right-hand side of (2.85) to zero. By differentiating it and using (2.82), we obtain

$$V = \frac{RTq_0}{\mu L_b p_0} \exp(-h/l) \tag{2.87}$$

The time after which the film velocity becomes constant and equal to (2.87) is determined by differentiating (2.86) taking (2.87) into account:

$$t_1 = \frac{2}{3^{3/2}} \frac{RT\rho h q_0 \exp(-h/l)}{\mu L_b p_0^2} \tag{2.88}$$

For $q_0 = 10^4\,\mathrm{W/cm^2}$, $h = 100\,\mu\mathrm{m}$, and $t_1 = 100\,\mu\mathrm{s}$, the film movement velocity will be of order of 100 m/s, according to (2.87). After the end of the laser pulse, if $\tau < t_1$, the film movement will be slowed down. However, the decomposition of the film into drops can substantially change the model described above.

The acceleration of the film by water vapor resembles the well-studied effect of a thermonuclear target compression by laser radiation [104] in which the target walls are also accelerated by metal vapor. It is known that this process proves to be unstable due to the development of the Rayleigh–Taylor instability. In the case of high accelerations a, the increment of this instability has a simple form [103]

$$\gamma \approx \sqrt{ak}, \tag{2.89}$$

where k is the wave vector of the instability. If the instability development time is shorter than t_1, the film decomposes into drops before its velocity achieves the maximum value (2.87), the vapor pressure becomes equal to the atmospheric pressure, and acceleration ceases. One can see from (2.88) and (2.89) that the time t_1 increases with increasing q_0, while γ^{-1} decreases with increasing q_0. When q_0 exceeds a certain value q_0^*, the inequality $\gamma^{-1} < \tau_1$ takes place and the film dynamics will be determined by its decomposition into drops. If $q_0 < q_0^*$, the instability appears after switching off the pulse, when the film begins to slow down.

All the expressions presented above have the qualitative nature. For example, it is assumed that the laser beam energy is completely spent for evaporation, i.e. it is assumed that the developed evaporation regime was established at once. In addition, it is assumed that the laser beam is strictly rectangular. The film acceleration a depends on time [see (2.86)]; therefore, the development time of the instability should be determined by averaging. All these require the refinement of the equations and the use of numerical methods. Instead of (2.82), the balance equation for the vapor mass and energy should be introduced [103]:

$$\frac{\mathrm{d}m}{\mathrm{d}t} = \frac{1}{L_b}\kappa \left.\frac{\mathrm{d}T}{\mathrm{d}z}\right|_{z=0} \tag{2.90}$$

$$c\rho\frac{\partial T}{\partial t} = q(t)\mu_b \exp(-h/l)\exp(-z\mu_b) + \kappa\Delta T$$

where κ is the heat conduction of water and μ_b is the absorption coefficient of water at the laser wavelength. These equations also describe the boiling threshold. Figures 2.85 and 2.86 show the results of numerical solution of the system of equations (2.83), (2.84), (2.90) for the velocity and lift of a 30-μm thick film. The laser pulse shape was nearly trapezoidal with half-widths of 30 and 20 ms. One can see that the cleaning threshold under these conditions is 3.1 J/cm^2. The average development time of the instability is ~11 ms, the velocity of the film and produced drops being 0.2–1 m/s. This determines a small lift of the drops (1–5 cm).

When microsecond $\tau = 22\,\mu$s pulses are used, the intensity increases by the three order of magnitude. The cleaning threshold decreased by 3–5 times compared to the previous case (Fig. 2.87). Numerical calculations showed that the average destruction time of the film in the energy density interval from 1 to 7 J/cm^2 was

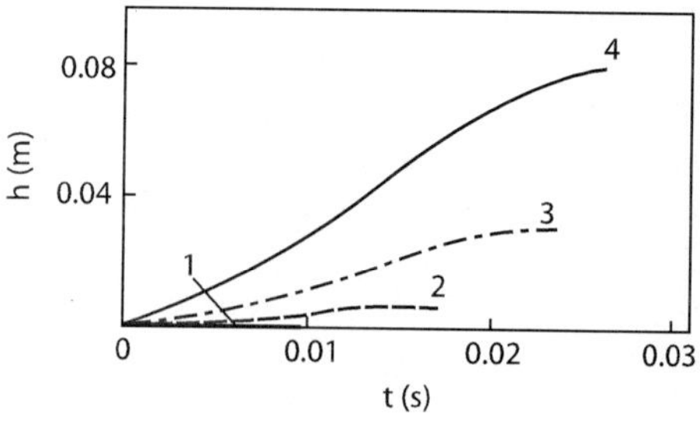

Fig. 2.85 Rise dynamics of a 30-μm thick oil film as a function of the 30-ms pulse energy density: $\varepsilon = 3.1$ J/cm^2 (**1**), 5.0 (**2**), 10 (**3**), 20 (**4**)

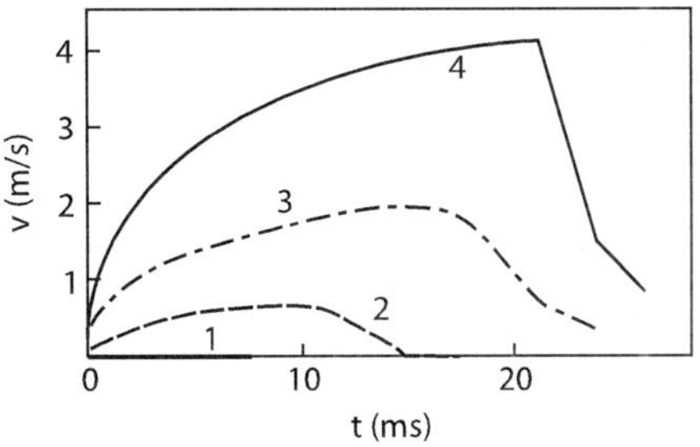

Fig. 2.86 Rise speed dynamics of a 30-μm thick film as a function of the 30-ms pulse energy: $\varepsilon = 3.1\,\text{J/cm}^2$ (**1**), 5.0 (**2**), 10 (**3**), 20 (**4**)

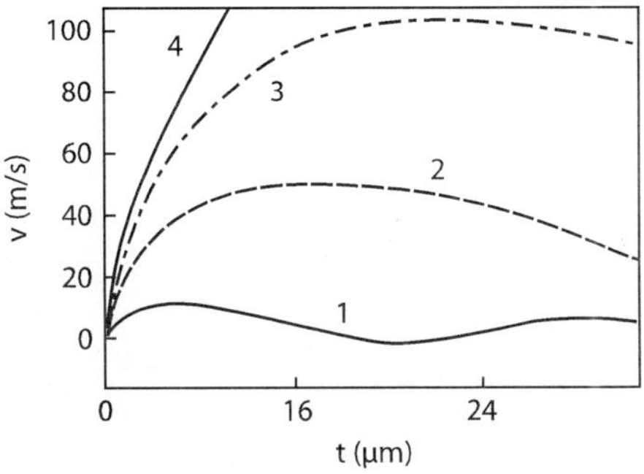

Fig. 2.87 Dependence of the 30-μm thick film speed on the 22-μs pulse energy density: $\varepsilon = 1.0\,\text{J/cm}^2$ (**1**), 2.0 (**2**), 4.0 (**3**), 7.0 (**4**)

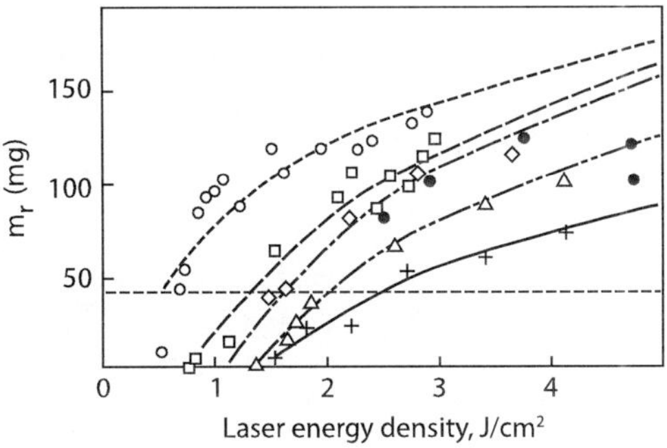

Fig. 2.88 Dependencies of the removed paraffin mass m_r on the energy density in a laser spot for different laser pulse durations τ and $S_n = 0.1\,\text{cm}^2$, number of pulses $N = 50$ and $h = 87\,\mu\text{m}$: $\tau = 42\,\mu s$ (*ring*), 50 (*square*), 75 (*rhomb*), 150 (*triangle*), 160 (*black ring*) and 270 (*cross*)

0.5–5 μs. This leads to considerably smaller lift velocities of the film compared to the case of the absence of the instability (5–50 m/s, see Fig. 2.87). Figure 2.88 shows that the removed paraffin mass is inversely proportional to the pulse duration. The cleaning threshold (energy density) decreases with decreasing the pulse duration. All this corresponds to analytic and numerical calculations.

Note that the curves in Fig. 2.88 are not saturated with increasing the energy density. The removed mass can exceed the film mass in the focal spot by 2–3 times (Fig. 2.89). The theoretical analysis shows [103] that the detachment of a paraffin film produced by short laser pulses occurs not only in the focal spot where water starts to boil. Water vapor, detaching the film, propagates along the water surface outside the focal spot, the detachment velocity being quite high. For example, for a pressure drop of 1 atm, a focal spot of 0.5 cm, a film lift of ~0.1 mm, and a film

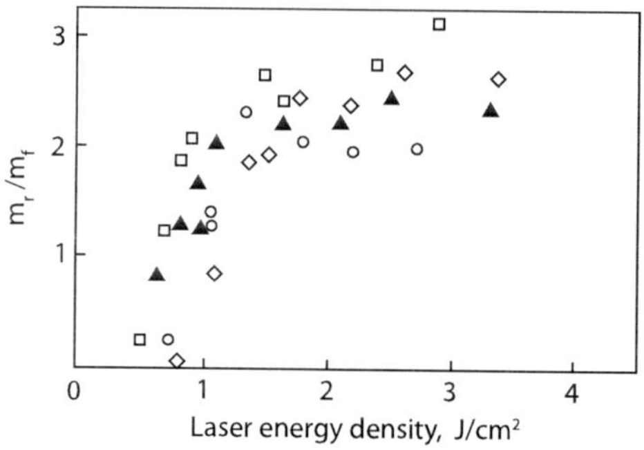

Fig. 2.89 Dependences of the ratio of the removed paraffin mass m_r to the paraffin mass in a laser spot m_f on the energy density in the spot for different thicknesses of a paraffin film h and $S_n = 0.1\,\text{cm}^2$, number of pulses $N = 50$ and $\tau = 42\,\mu\text{m}$: $h = 22\,\mu\text{m}$ (*ring*), 44 (*black triangle*), 65 (*rhomb*) and 87 (*square*)

thickness of $h = 50\,\mu\text{m}$, the film detachment velocity achieves 250 m/s. It is clear that due to such a high velocity, the mass of the removed film can considerably exceed its mass in the focal spot.

2.4.4 *Laser-Induced Metal Surface Cleaning from Radionuclides*

The cleaning methods described above are based on the removal of condensed substances such as solid or liquid films and solid particles. In particular, it is necessary to clean surfaces from radionuclides. Radionuclides mainly contaminate oxide films formed on the surface of the elements of constructions. They can be removed by laser cleaning methods described above. Laser cleaning is used, for example, for the deactivation of energy facilities taken out of service [92, 93, 100, 105]. As mentioned at the beginning of this section, the use of laser technologies provides a considerably smaller volume of radioactive waste to be buried. This method is highly efficient and considerably reduces requirements for the radiation safety of works.

The mechanisms of laser deactivation can be studied by simulating deactivation processes. For example, the cleaning of the radioactive ^{60}Co isotope of the most widespread impurity in oxide films can be simulated by introducing and removing the nonradiactive cobalt isotope [106]. The distribution of cobalt over the cross section of a sample was investigated by the method of X-ray microprobe spectral analysis. This analysis showed that cobalt atoms penetrated by a depth of $\sim 10\,\mu\text{m}$ into a $100\,\mu\text{m}$ thick iron hydroxide film (Fig. 2.90a). The surface was processed by rapidly scanning the beam of a cw CO_2 laser, similarly to hardening by using a scanner (see the beginning of this chapter). The rate of rapid scanning along the coordinate perpendicular to the sample travel axis was ~ 2 m/s and the strip width was ~ 30 mm. The sample travel speed was $V \sim 1$ cm/s. The laser beam was focused to a spot of diameter ~ 1 mm. Thus, each point of the processed surface was illuminated ~ 7 times. The power density was $q = 4 \cdot 10^5\,\text{W/cm}^2$. It was found that the energy density required for the complete removal of the film was

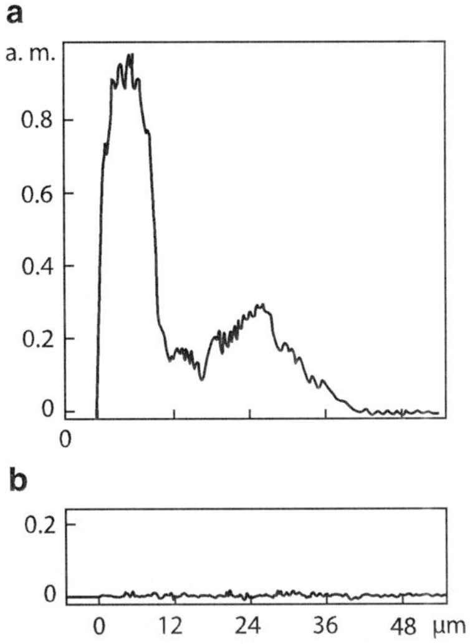

Fig. 2.90 Co concentration distributions over the sample depth before (**a**) and after (**b**) laser-induced cleaning

100 J/cm^2 (Fig. 2.90b). A comparison with the evaporation enthalpy of iron oxide ($\sim$500 J/cm^2) shows that the film is removed not due to evaporation.

Sometimes radioactive substances penetrate (or are formed) inside a material. If the penetration depth is small, laser cleaning can be used in this case as well. A typical example is cleaning from the radioactive hydrogen isotope (tritium). The problem of cleaning the elements of constructions from tritium is quite urgent for the development of thermonuclear energetics [107, 108], in particular, for the International Thermonuclear Experimental Reactor (ITER) [107]. For certain operation regimes of a tokamak-reactor, the penetration depth of tritium is relatively small, of the order of 100 μm. In this case, one laser pulse can be sufficient for heating a layer of such thickness and initiating the diffusion flow of tritium from a sample. The diffusion outside is caused by a drastic increase in the diffusion coefficient of tritium with temperature. This process is reverse with respect to the gas-phase alloying of a metal, which was considered above in this chapter.

In the tokamak-reactor, tritium is most intensity deposited in secondary graphite deposition layers. According to estimates, the thickness of such films formed in the ITER operating for a week should achieve 60–600 μm [107]. The penetration of tritium directly into the construction material (stainless steel) is less efficient, but nevertheless, the problem of tritium removal is quite urgent in this case as well. Laser cleaning from tritium is performed by nanosecond pulses or cw radiation from CO_2 and Nd lasers. In the case of pulsed cleaning, UV radiation is used. This is related, in particular, to the fact that tritium can be also removed by using other UV sources [108]. Indeed, the energy of the ArF laser photon equal to 6.3 eV exceeds the bond energy of hydrogen and carbon atoms. A competing mechanism can be in this case the ablation of the material surface. Figure 2.91 shows the thickness of a graphite material removed by a 25-ns pulse from an ArF laser as a

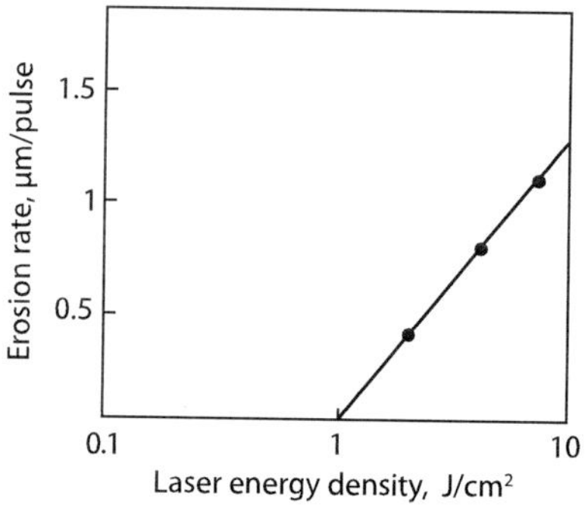

Fig. 2.91 The ablative erosion rate of JT-60 codeposits

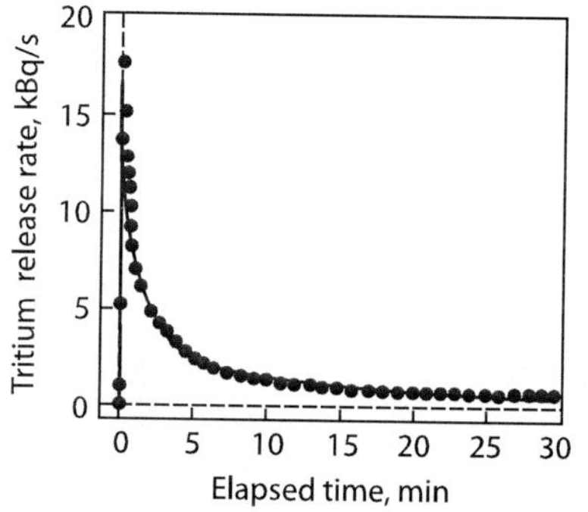

Fig. 2.92 Tritium release rate during the UV laser irradiation on the stainless steel shutter

function of the energy density ε. The ablation threshold equal to 1 J/cm^2 is observed. However, the removal of tritium is also observed for the energy density lower by an order of magnitude ($\varepsilon = 0.1$ J/cm^2). Figure 2.92 shows the removal rate of tritium after irradiation by an ArF laser for 20 s with a pulse repetition rate of 5 Hz. Experiments were performed by measuring the radioactivity of the Ar flow into which the sample was placed. Numerical estimates by expression (2.4) show that in this case the temperature of the graphite surface should achieve 918°C ($\kappa = 1$ W/cmK, $C_p = 0.8$ J/gK, $\rho = 2$ g/cm^3). This temperature is sufficient to initiate the removal of tritium due to a drastic increase in the diffusion coefficient. It is known that the heating of a secondary graphite deposition layer up to 350 °C completely removes tritium [109]. However, because the average power of this laser is low ($\sim$1 W), the laser pulses do not affect each other, the heating depth is small, and the removal rate of tritium is also small.

If the removal mechanism is thermal, it is appropriate to use repetitively pulses or cw laser radiation with a high average power. Solid-state lasers are preferable in this case because their radiation can be transported through optical fibers. Laser cleaning with the help of cw Nd lasers is performed by using scanners, as in the case of CO_2 lasers described above. Figure 2.93 illustrates the behavior of temperature at a certain site of the surface measured with a pyrometer with a spatial resolution of 0.7 mm. The lower limit of temperature measurement was 500°C. The laser

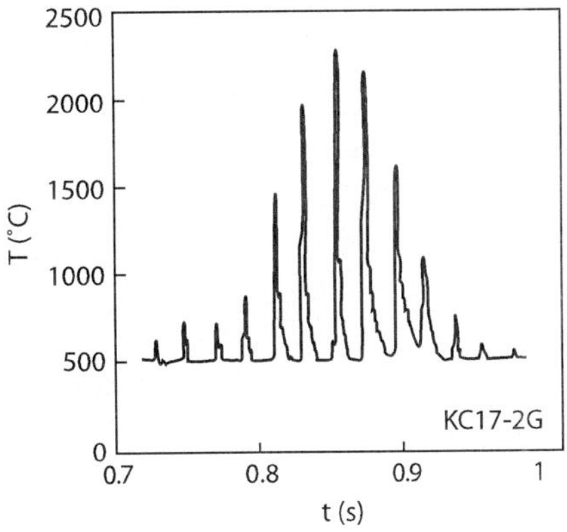

Fig. 2.93 Temperature time history as the laser rasters over the viewed by the pyrometer successive times. The pyrometer operating range is 500–2,300 C

power was 300 W and the power density at the focal spot was 8 kW/cm^2. Optimal irradiation regimes can be selected by varying the scan speed, the sample travel speed, and the focal spot size. For $q = 8$ and 3 kW/cm^2, the amount of tritium removed from the secondary graphite deposition layer achieved 46% and 84%, respectively.

The mechanism of tritium removal by a scanned laser beam was studied for a material with a structure more definite compared to that of the deposited graphite. Information on the tritium distribution inside a stainless steel sample was obtained by the method of layer-by-layer material removal by using a magnetic microscope [110]. The CO_2 laser scanning parameters are close to these of the Nd laser used for removing tritium described above. The speed of scanning along the transverse coordinate was 2 m/s (the scan time was ~15 ms) and along the processing direction – ~1 cm/s. The focal spot size was ~0.7 – 1 mm and the power density at the focal spot was ~40 kW/cm^2. Measurements show that the initial tritium distribution considerably changes after irradiation. The tritium concentration near the surface at a depth of ~10 μm decreased by an order of magnitude and at a depth of ~80 μm by 2-3 times. The concentration dip behind the boundary layer was filled after irradiation (Fig. 2.94). According to estimates, the surface temperature was close to the melting temperature but did not exceed it. A layer of depth ~100 μm was heated during one pulse. The diffusion coefficient of tritium in stainless steel strongly depends on temperature as

$$D = D_0 \exp(-E_d/T),$$

where $D_0 = 6\times10^{-3}$ cm^2/s and $E_d = 0.53$ eV. The tritium diffusion length during a pulse even at the maximum temperature 1,400°C is ~5 μm. Taking into account that each site of the surface is illuminated ~7 times, the total diffusion length obtained by summation is still smaller than the thermal length equal to 100 μm. It seems that diffusion over grain boundaries can play a certain role.

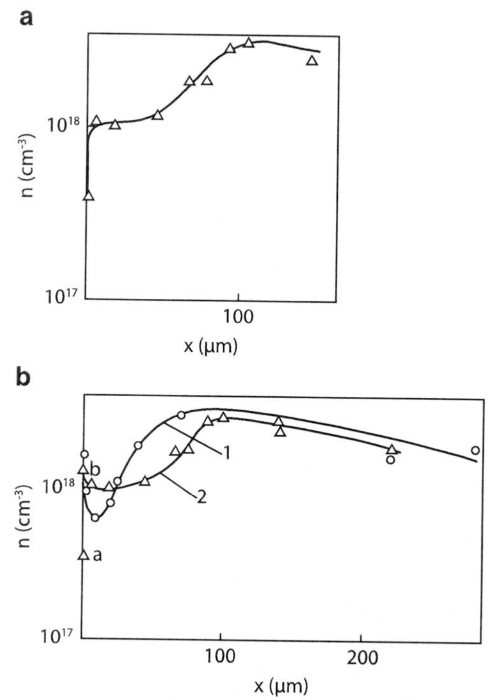

Fig. 2.94 Tritium concentration distributions over the sample depth obtained with a magnetic microscope before (**b**) (**1**) and after (**a, b**) (**2**) laser irradiation

2.5 Modelling of Selective Laser Melting

Unlike laser cladding, where powder is supplied on a sample surface simultaneously with laser irradiation, in the case of selective laser sintering, powder first is deposited in the form of a thin layer on a substrate. A normally incident laser beam moves along the powder surface by sintering powder particles, thereby producing a narrow metal strip of width close to the focal spot diameter. Then, the next powder layer is deposited and the second strip is constructed over the previous strip. By controlling the beam displacement with a computer, it is possible to create metal elements of any intricate shape, which are difficult or impossible at all to create by other methods. Such sintering can be performed most easily when powders with different melting temperatures are mixed. In this case, a low-melting powder is used as a binding material to hold together a high-melting powder [111–113].

2.5.1 Structures

Somewhat different situation is observed during laser sintering of a single–component powder. Processes with the complete melting of powder are often unstable, which results in the appearance of regular waists in the melt strip [114]. It

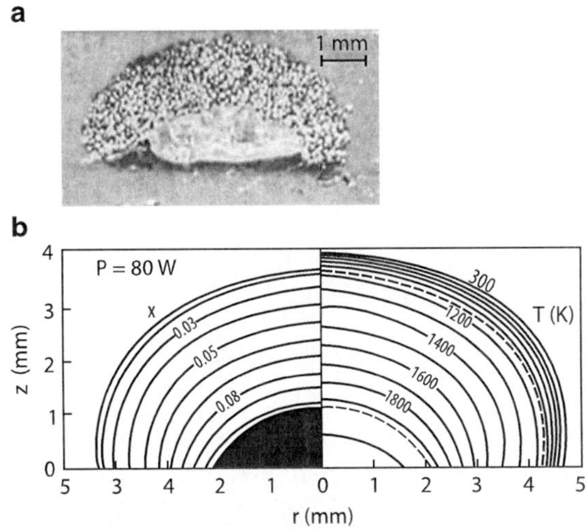

Fig. 2.95 Cross-section of laser-sintered Ti powder of spherical particles 63–200 μm in diameter (**a**). Laser power $P = 80$ W. (**b**) – optical micrographs of the relative neck size x (*on the left*) and the temperature T (*on the right*) for the monodisperse powder of spherical particles 160 μm in diameter and the spot diameter $d = 5$ mm. Isotherms corresponding to the – phase transformation, $T = 1{,}156$ K, and to the melting point, $T_m = 1{,}944$ K, are shown by broken lines. The remelted domain is filled by the black colour on the left half of the diagrams

is clear that sintering can occur under certain condition during the melting of only the surface of microparticles followed by the connection of unmelted cores by the melt [115]. Between these two limiting cases, other, more complex structures can also exist. This depends on the size and shape of particles, the powder material and the laser power. Sintering structures were studied in special experiments. Powders with particles of different sizes and shapes were poured by a 10-mm thick layer on an aluminum substrate. A 10–100 W laser beam was focused on the layer surface to a spot of diameter from 2.7 to 5.3 mm. The irradiation time was 10 s. The analysis of experimental results showed that the powder was melted at the focal spot center, and the melting zone represented a spherical segment. As a whole, the sintered sample also resembles a spherical segment of large radius whose particles are connected by short necks (Fig. 2.95). The latter fact can strongly restrict the adequacy of calculations of such systems.

2.5.2 *Heat Conduction of Powders in Vacuum*

To calculate sintering processes in powders, it is necessary to know the heat conductivity of such complex media [116]. The heat transfer in powders is performed by the gas filling pores, due to heat transfer (κ_r) through these pores, and heat transfer

(κ_c) through contacts between particles. The radiation heat conduction caused by the isotropic scattering of photons was calculated by Zel'dovich and Raiser [117]:

$$\kappa_r = \frac{16}{3} l \sigma_B T^3, \tag{2.91}$$

where l is the photon mean free path and σ_B is the Stefan–Boltzmann constant. In our case, l is the pore size, which is of the order of the powder particle diameter $2R$.

The contact conduction is determined by the linear size of a contact [118] and the heat conduction κ_0 of a powder material:

$$\kappa = \Lambda \kappa_0 b / R \tag{2.92}$$

where the numerical coefficient Λ depends on the packing type of powder particles. For body-centered, cubic, and diamond-type packing, this coefficient is $\Lambda = 1.7$; 1; and 0.43, respectively. The average relative density of powder also depends on the packing type and is $\bar{\rho} = 0.68$, 0.52, and 0.34, respectively. The relative size $x = b/R$ of the contact is described by the Kuczynski equation [119]

$$\frac{dx}{dt} = \frac{1}{t_0 x^4}; \quad t_0 = \frac{N_0 k_B R^3}{8\sigma D_s} \tag{2.93}$$

where N_0 is the solid material density; k_B is the Boltzmann constant, σ is the surface tension coefficient, and D_s is the volume diffusion coefficient. Knowing the heat conductivity, we can calculate the heat distribution, the temperature field, and the melting region size:

$$\frac{\partial L}{\partial t} = \frac{\partial}{\partial z}\left(\kappa \frac{\partial T}{\partial z}\right) + \frac{1}{r}\frac{\partial}{\partial r}\left(\kappa r \frac{\partial T}{\partial r}\right), \tag{2.94}$$

where L is the enthalpy equal to the product of the relative density $\bar{\rho}$ of the medium and the enthalpy L_0 of the solid material. The heat conduction equation is supplemented, as usual, by the boundary conditions

$$-\kappa \left.\frac{\partial T}{\partial z}\right|_{z=0} = \alpha q(r) - \varepsilon \sigma_B T_s^4, \tag{2.95}$$

where T_s is the temperature of a powder layer surface, σ_B is the Stefan–Boltzmann constant, and ε is the emission ability of the medium. The formulation of the problem in [120] assumes that the absorption length of laser radiation is considerably smaller than all dimensions of the problem. In this sense, the formulation of the problem is similar to that in thermal problems of the interaction of laser radiation with metals (see the beginning of this chapter). Numerical calculations were performed with parameters presented in Fig. 2.96a and Table 2.4. It is taken

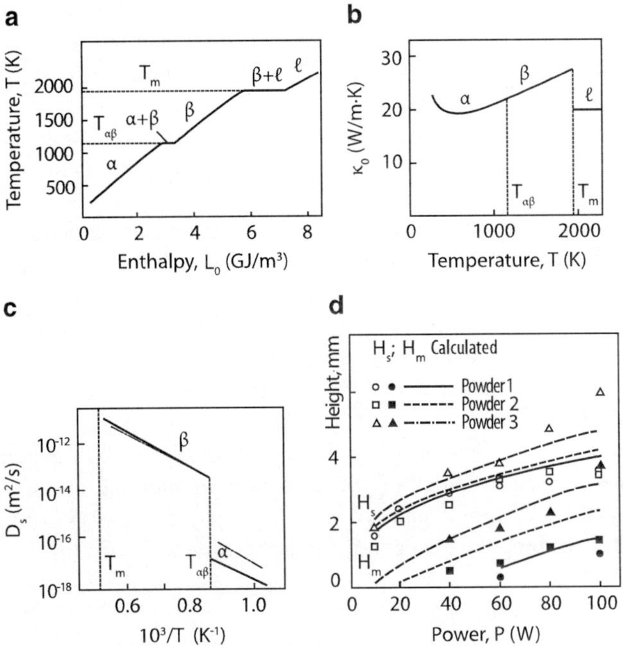

Fig. 2.96 Temperature-dependent properties of titanium: (**a**), Temperature T versus enthalpy L_0 [139] (*full curve*); (**b**), thermal conductivity κ_0 versus temperature [140] (*full curve*); (**c**) self-diffusion coefficient D_s versus reciprocal of temperature [18], *full curve*; [141], chain curve. Phase state is marked above each segment of the curves: α, $\alpha + \beta$, β, $\beta + l$, l. The $\alpha - \beta$, transition temperature, $T_{\alpha\beta} = 1{,}156$ K and the melting point $T_m = 1{,}944$ K are marked by broken lines. (**d**) Characteristic dimensions of the cakes obtained by 10 s laser irradiation of the titanium powder bed versus the laser power P: heights of the sintered, H_S, and remelted, H_m, domains

Table 2.4 Parameters of the Ti powders used in the experiment and the corresponding parameters accepted in the modelling

Parameters		Powder 1	Powder 2	Powder 3
Experiment	Particle shape	Spherical	Spherical	Irregular
	Particle diameter (μm)	63–200	200–315	160–315
	Bulk density, ρ (g/cm^2)	3.0	2.3	1.4
Modelling	Particle shape	Spherical	Spherical	Spherical
	Particle radius, R (μm)	80	130	120
	Relative density, p	0.680 (BCC)	0.524 (SC)	0.340 (Diamond)
	Normalized contact conductivity, Λ	1.732 (BCC)	1.0 (SC)	0.433 (Diamond)

into account that the enthalpy L_0, the heat conductivity of solid titanium, and the volume diffusion coefficient depend on α and β modifications of the solid titanium structure. The calculations were carried out for the three packing types of powder. A comparison with experiments was performed for media with identical relative densities because the real packing of powder particles in experiments was unknown. Figure 2.95b shows the temperature field and the spatial distribution of the relative contact size x.

Melting zones are observed. The boundary of the sintering zone is close to the temperature front of the $\alpha \rightarrow \beta$ phase transition. The diffusion coefficient D_s and x drastically decrease at lower temperatures. For $x < 0.01$, the contact heat transfer is replaced by thermal radiation transfer. Because κ_r drastically decreases with decreasing T, this leads to large temperature gradients on a heat wave front.

Figure 2.96d presents the heights of melting and sintering zones as functions of the laser radiation power, the size and shape of powder particles. One can see that the results of calculations are qualitatively consistent with experiments. The measurements and calculations of the diameters of melting and sintering zones are also in qualitative agreement. All this demonstrates the validity of the model proposed for describing the heat conduction of powders in vacuum.

2.5.3 *Calculation of Thermal and Optical Constants of Initial Materials*

As pointed out above, the heat conductivity of powder sintered in a gas medium will be determined by the heat conduction of the gas filling pores. Therefore, it weakly depends on the powder material and is determined by the size and shape of particles and the fraction of pores in the powder. A simple consideration of this problem was proposed in [121]. The heat conductivity of stainless steel powder consisting of particles of size 10–50 μm in the temperature interval of the order of the melting temperature is $\kappa_p = 0.3$ W/m·K. The heat conductivity κ of the powder-melt system is determined by the expression [122]

$$\kappa = \kappa_p + (\kappa_d - \kappa_p)\phi$$

where κ_d is the heat conductivity of the corresponding solid and ϕ is equal to zero for $T \geq T_m$ and unity for $T < T_m$.

The heat conductivity of mixtures of different powders was determined by comparing calculated temperatures with thermocouple measurements in a powder layer. Laser irradiation of the powder was adjusted to provide the temperature interval required for selective laser sintering experiments [123]. The heat conductivity of the powder mixture was estimated from the relation [124]

$$\kappa = \theta\kappa_a + \kappa_m(1 - \theta)$$

Table 2.5 Thermochemical and optical parameters of powder compositions

Powder	Size (μm)	Thermal conductivity of material (W/cm K)	Thermal conductivity of powder (10^{-3}W/cm K)	μ_s' (mm^{-1})	μ_a (mm^{-1})	μ_s'/μ_t'	R (%)
Ni	50–160	0.6–0.7	5–11	–	–	0.80	30
Polyamide	60–80	7×10^{-4}	0.7–1.8	–	–	0.99	92
10:1 mix	–	–	1.3	11.7	1.5	0.80	30

where θ is the powder porosity, $\theta = 1 - \rho_p/\rho$, ρ_p is the bulk powder density, ρ is the solid material density, κ_a is the heat conductivity of air, and

$$\kappa_m = (V_1\kappa_1 + V_2\kappa_2)(V_1 + V_2)^{-1}.$$

Here, $V_1\kappa_1$ and $V_2\kappa_2$ are the volume fractions and heat conductivities of pure powders, respectively. Table 2.5 presents the thermal parameters of components and mixtures of metal powders (nickel, brass) and a polymer (polyamide). One can see that the heat conduction of the powder is considerably lower than that of the initial material. The addition of a weakly heat-conducting polymer further reduces the heat conductivity of the powder. Due to a great difference between the melting temperatures, the selective laser sintering of these powders occurs in the liquid phase. Bimetallic powder compositions are prepared by soldering.

2.5.4 Volume and Surface Absorption Coefficients

We described above the model for calculating the heat conductivity of powder. The geometry of the region for calculations was selected according to this aim: the size of the sintering region was several mm, which allowed us to assume that a laser beam is absorbed on the surface, i.e. the absorption length of laser radiation is fractions of mm (see below). The size of a sintering strip in a real technological process does not exceed 1 mm and the absorption of laser radiation will be volumetric. Therefore, it is necessary to calculate the volume absorption coefficient. The propagation of laser radiation in powder is accompanied by its scattering and absorption. The optical properties of scattering media are described by the scattering coefficient μ_s and absorption coefficient μ_a, which determine the probabilities of these processes per unit path length. Inverse quantities are called the absorption and scattering lengths. The loss of photons from the incident beam due to absorption and scattering is described by the coefficient $\beta = \mu_a + \mu_s$. Experiments [125] showed that this coefficient, as the heat conductivity, depends on the size and morphology of particles. The relevant problems are solved by the three methods:

- Geometrical optics, which requires the knowledge of the powder structure and the geometry of particles.
- The Monte-Carlo method, which can be used for studying inhomogeneous (over surface) cases. The values of μ_s and μ_a required for this method are found by comparing calculations with experiments.
- The simplest model of identical spheres reflecting specularly laser radiation with the reflection coefficient ρ.

The coefficient β in this simplest case of spherical particles of diameter D homogeneously spatially distributed with the density n is described by the expression [126, 127]

$$\beta = \frac{3}{2}\frac{1-\varepsilon}{\varepsilon}\frac{1}{D},$$

here ε is the fraction of pores in powder and $\varepsilon = 1 - \pi D^3 n/6$. Knowing the volume scattering coefficient, it is useful to introduce the optical thickness $\lambda = \beta L$, where L is the powder layer thickness. For example, for $D = 20\,\mu\text{m}$ and $L = 50\,\mu\text{m}$, the optical thickness is $\lambda = 3.75$ for $\varepsilon = 1/2$ and $\lambda = 1.875$ for $\varepsilon = 2/3$. Photons leaving the beam are scattered and absorbed by powder particles and a substrate. For spherical powder particles and isotropic scattering, the problem of radiation distribution in powder is solved analytically [126, 127]. The radiation intensity in powder is the sum of directed radiation and radiation reflected from the substrate and scattered radiation:

$$\begin{aligned} q = {} & \frac{\rho\sqrt{1-\rho}}{(4\rho-3)D}((1-\rho^2)\mathrm{e}^{-\beta L}[(1-\sqrt{1-\rho})\mathrm{e}^{-2\sqrt{1-\rho}\xi} + (1+\sqrt{1-\rho})\mathrm{e}^{2\sqrt{1-\rho}\xi}] \\ & - (3+\rho\mathrm{e}^{-2\beta L})\{[1+\sqrt{1-\rho}-\rho(1-\sqrt{1-\rho})]\mathrm{e}^{2\sqrt{1-\rho}(\beta L-\xi)} \\ & + [1-\sqrt{1-\rho}-\rho(1+\sqrt{1-\rho})]\mathrm{e}^{2\sqrt{1-\rho}(\xi-\beta L)}\}) \\ & - \frac{3(1-\rho)(\mathrm{e}^{-\xi}-\rho\mathrm{e}^{\xi-2\beta L})}{4\rho-3} \end{aligned}$$

where L is the thickness of the powder layer uniformly distributed on the substrate. For stainless steel powder and radiation of a Nd:YAG laser, $\rho = 0.7$.

The heat release u provided by a volume radiation source required for our calculations is described by the expression

$$u = -\beta Q_0 \frac{\mathrm{d}q}{\mathrm{d}\xi},$$

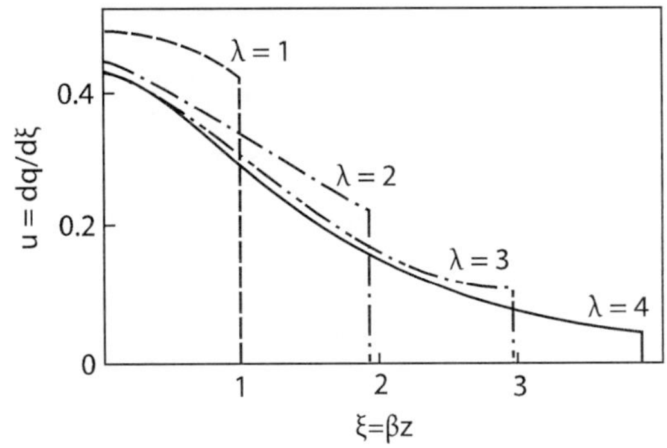

Fig. 2.97 Volumetric heat source u in the powder layer versus dimensionless depth ξ for various values of optical depth λ at the reflectivity of dense material $\rho = 0.7$

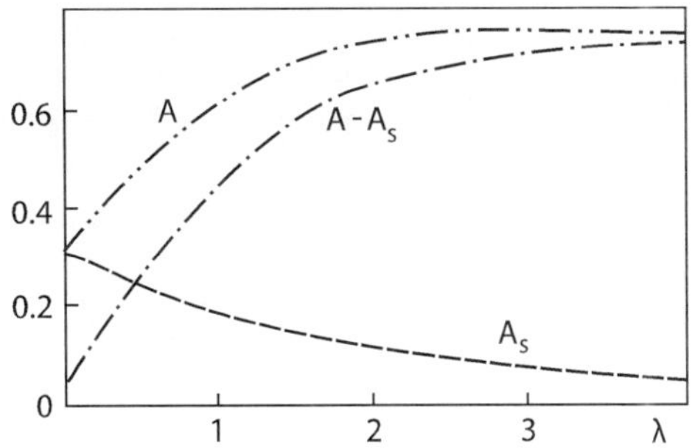

Fig. 2.98 Total absorptivity of the system powder-substrate A and fractions of the incident laser radiation absorbed by the surface of the substrate A_s and in the powder $A - A_s$, versus optical depth of the powder layer λ at the reflectivity of dense material $\rho = 0.7$

where $q = Q/Q_0$, $\xi = \beta z$, $Q_0 = Q_m \left(1 - \frac{r}{r_f}\right)^2 \left(1 + \frac{r}{r_f}\right)^2$, $0 < r < r_f$, r_f is the laser beam radius, Q_0 is the laser radiation distribution on the powder surface, and Q_m is determined by the total laser power P:

$$P = \frac{\pi}{3} r_f^2 Q_m.$$

Because the expression for u is too cumbersome, we consider only Figs. 2.97 and 2.98. These figures show the radiation intensity distributions over the powder layer depth for different optical lengths for the reflectance of the material $\rho = 0.7$. It is clear that in the case of small β^{-1}, the intensity q and volume heat release u decrease exponentially deep in the powder layer. As the optical thickness increases, the absorption coefficient of the substrate decreases, while the total absorption o laser radiation increases (Fig. 2.98). These properties are generalized in Fig. 2.99 for different absorption coefficients of the powder material and substrate [135]. Figure 2.100 demonstrates a comparison with experiments for optically thick powder layers and different laser wavelengths [135].

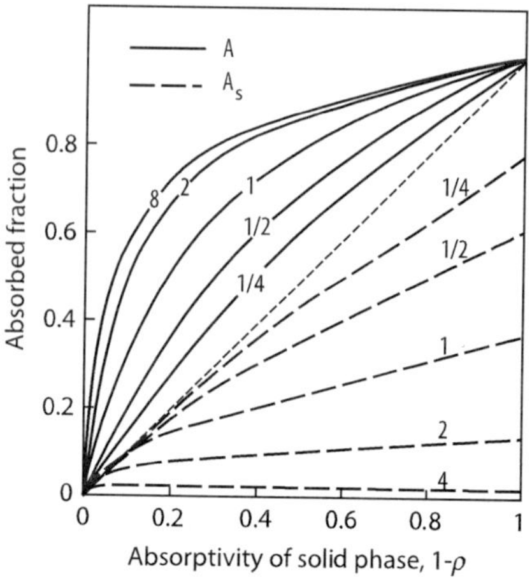

Fig. 2.99 Fractions of the incident laser radiation absorbed by the system powder layer-substrate of the same material A and by the substrate A_s versus the absorptivity of the solid phase $1 - \rho$ at various optical thicknesses of the layer (marked near the curves). The dotted line $A = 1 - \rho$ corresponds to the limiting case of a not-covered substrate

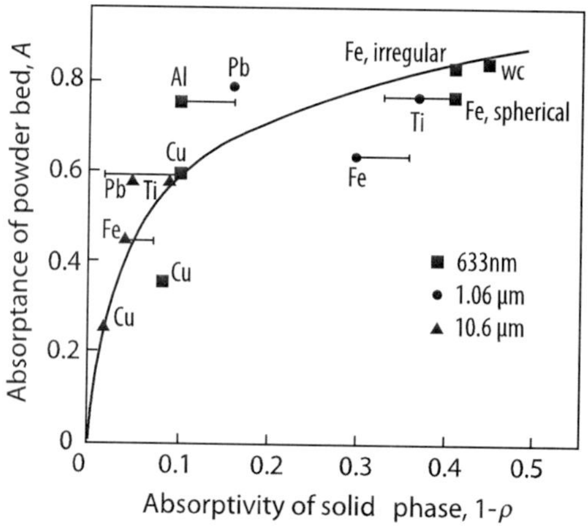

Fig. 2.100 Absorptance of an optically thick powder layer A versus the absorptivity of the solid phase $1 - \rho$. Comparison of the theoretical estimation (line) with experimental points for micron sized powders at the different wavelengthes

2.5.5 *Powder Mixtures*

Consider a more complicated case of powder mixtures. For the specified angular distribution of radiation, a medium is also characterized by the mean scattering angle cosine $\bar{q}$, and the quantity $\mu_s' = \mu_s(1 - \bar{q})$ is called the transport scattering coefficient. The mean free path is $\Lambda = (\mu_a + \mu_s')^{-1}$. The scattering probability $\mu_s'/(\mu_s' + \mu_a) \equiv \mu_s'/mu_t'$ at the mean free path is called the transport albedo.

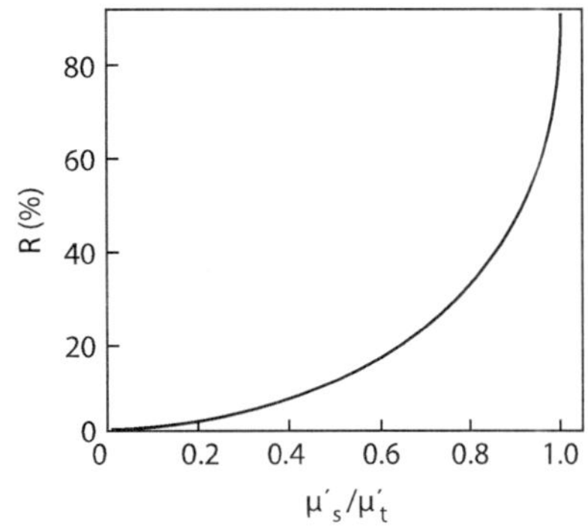

Fig. 2.101 Dependence of the reflectance R of a semi-infinite layer from the transport albedo μ_s'/μ_t'

Strong scattering of radiation in powders considerably complicates the measurement of the volume absorption coefficient or, in other words, the in-depth distribution of the absorbed energy. Radiation can be partially backscattered, making contribution to the integrated reflection coefficient. In addition, radiation can be scattered perpendicular to the laser beam direction. Because the laser beam is focused upon selective laser sintering, such a broadening of a volume source can significantly reduce the sintering efficiency. The latter circumstance can restrict the use of traditional models used in the calculations of light propagation through scattering media. Taking into account the above discussion, the most convenient is the approach based on the Monte-Carlo method. If the values of μ_s' and μ_a are known, this is sufficient for calculations of radiation distributions in different geometries. The values of μ_s' and μ_a were determined by comparing the results of calculations with the measurements of reflection and transmission of radiation by a powder layer of the specified thickness [128]. For example, by measuring the reflection coefficient of an infinite layer, the transport albedo can be calculated (Fig. 2.101). Figure 2.102 shows the distributions of the absorbed energy density over depth and radius for different transport albedo. One can see that radiation penetrates deeper into a powder mixture when scattering dominates.

The albedo of a purely metal powder can be calculated in the following way. The ratio of the absorption and scattering coefficients can be treated as the probability of photon absorption in a scattering event. This is the coefficient of photon absorption by a metal surface. This is confirmed by the data presented in Table 2.5.

Knowing the transport albedo, we can determine the light penetration depth (corresponding to the radiation intensity attenuation down to 0.1 of the initial value) by universal curves. For example, for $\mu_s'/(\mu_a + \mu_s') = 0.99$, light penetrates by a depth of the order of seven mean free paths. However, the mean free paths should be calculated.

In the model of identical specularly reflecting spheres ($\rho = 0.7$), the scattering of laser radiation along the surface also can be taken into account [135]. Figure 2.103 shows that, as the laser beam diameter is decreased, the energy absorbed by powder on the symmetry axis decreases. Thus, for $\beta r = 1.2$, this decrease for $\beta z = 2$ will be 30% compared to the homogeneous case.

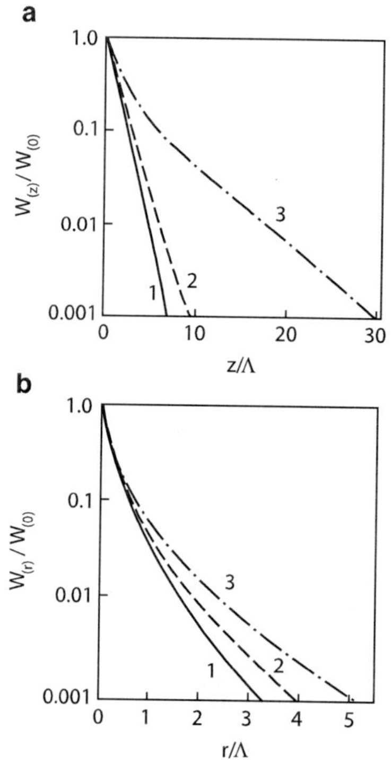

Fig. 2.102 (**a**) Normalized distributions of the absorbed energy density $W(z)/W(0)$ over the depth for the transport albedo $\mu_s'/\mu_t' = 0.7$ (**1**), 0.92 (**2**), and 0.99 (**3**); (**b**) Radial distributions of the absorbed energy density $W(r)/W(0)$ over the radius for the transport albedo $\mu_s'/\mu_t' = 0.7$ (**1**), 0.92 (**2**), and 0.99 (**3**)

2.5.6 *Thermal Model of Selective Laser Sintering*

Consider again the case of powder consisting of identical spheres made of the same material. Knowing the volume heat release, we can calculate the heat distribution in a powder layer moving at the velocity v along the x axis:

$$\frac{\partial L}{\partial t} - v\frac{\partial L}{\partial x} = \frac{\partial}{\partial x}\left(\kappa\frac{\partial T}{\partial x}\right) + \frac{\partial}{\partial y}\left(\kappa\frac{\partial T}{\partial y}\right) + \frac{\partial}{\partial z}\left(\kappa\frac{\partial T}{\partial z}\right) + u, \tag{2.96}$$

where only one phase transition (melting) is taken into account in enthalpy.

$$T = \begin{cases} L/C_s, & L \le C_s T_m \\ T_m, & C_s T_m < L < C_s T_m + H_m \\ T_m + (L - C_s T_m - H_m)/C_l, & L \ge C_s T_m + H_m. \end{cases}$$

For steel, $T_m = 1{,}700$ K, the melting heat is $H_m = 2.18$ GJ/m^3, $C_s = 4.25$ MJ/m^3·K, and $C_l = 5.95$ MJ/m^3·K.

Unlike similar equation (2.94), here the propagation of heat along all the three coordinates is taken into account. It is assumed that a volume heat release source depends only on the coordinate z (Fig. 2.97). Its transverse size is determined by the

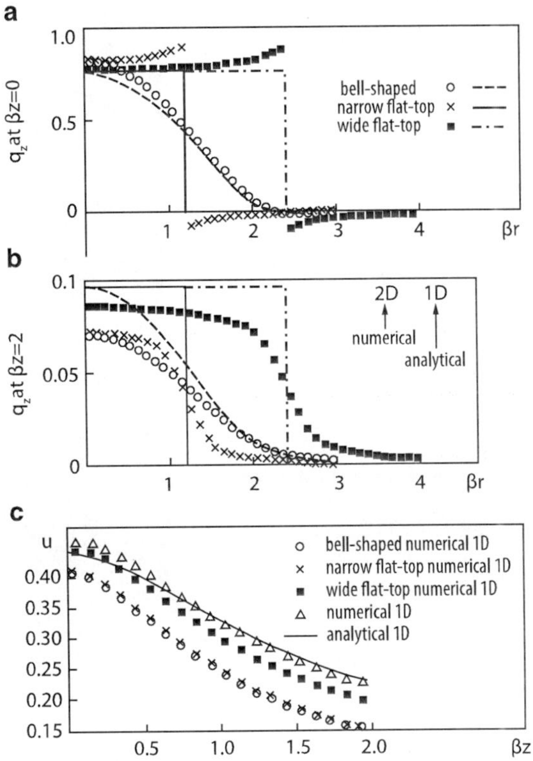

Fig. 2.103 Comparison of radiation transfer in the powder layer with optical thickness 2 and albedo $\omega = \rho_1 = \rho = 0.7$ for bell-shaped (radius at half-maximum $\beta r = 1.2$), narrow flat-top (radius $\beta r = 1.2$) and wide flat-top (radius $\beta r = 2.4$) laser beams: the normalized axial energy flux q_z at the surface of the powder (**a**) and at the surface of the substrate (**b**) and the normalized deposited energy u at the axis (**c**). Numerical 2D and 1D calculations (points) and the analytical 1D approach (lines)

focal spot size if the beam diameter exceeds the powder layer thickness. The melting of upper layers within the focal spot in hydrodynamic model (2.96) may screen the lower levels from a direct heating. In addition, as follows from calculations, the upper layers are overheated and start boiling. The reasonable results of calculations are obtained by neglecting the formation of such a continuous melt layer. The discontinuities can be caused either by a finite merging time of melted powder particles or air jets escaping from pores during powder heating. The importance of the latter process during processing even solids was pointed out in paper [136]. Because the powder structure in the experiment with which a comparison was made was unknown, thermal calculations were performed for two optical thicknesses $\lambda = 2$ and 3. Figure 2.104d presents the temperature fields of the powder layer surface calculated for increasing travel velocities of the laser beam. As usual, the surface temperature decreases with increasing the beam travel velocity and the width of the melting zone also decreases. However, the length of the melting zone increases, and sharp temperature gradients are observed near the melting isotherm. All this is caused by a low heat conductivity of the unmelted powder. The usual properties are also observed during the propagation of heat inside: the cross sections of the melting zones of the powder and substrate at the strip end decrease with increasing

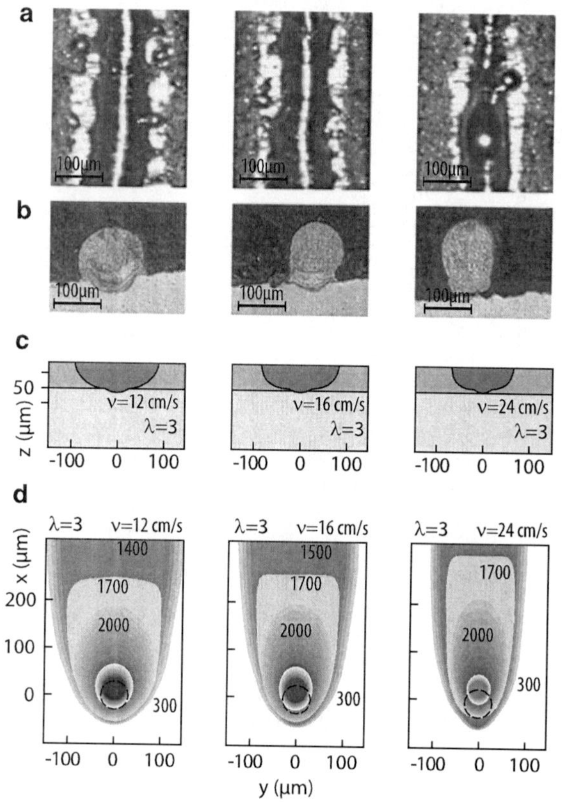

Fig. 2.104 Comparison between experiments (**a and b**) and calculations (**c and d**) at various scanning velocities (marked on the top of each column): (**a**) top view of remelted tracks; (**b**) transverse micrographs; (**c**) calculated phase distribution in transverse cross-section with light substrate, grey powder, and dark remelted zone; (**d**) surface temperature distribution (K). Experiments at 45 W laser power. Calculations at 30 W laser power. 60 μm FWHM. and the optical thickness $\lambda = 3$. The thickness of the powder layer is 50 μm for all the experiments and calculations. Broken-line circles on diagrams (**d**) are projections of the laser beam

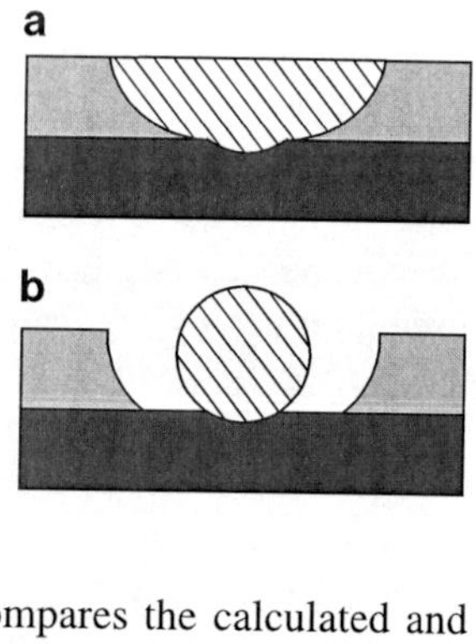

Fig. 2.105 Qualitative evolution of phase distribution in the cross-section of a track with substrate (*black*), powder (*grey*), and melt (*broken*): (**a**) just melted; (**b**) after minimizing the melt surface

the beam travel velocity (Fig. 2.104c). Figure 2.106 compares the calculated and measured widths of melting zones of the substrate and powder. The last value in the experiment was the width of the region free of powder. It is assumed that the melted powder forms under the action of surface tension forces a cylindrical melt strip (Fig. 2.105). One can see from Fig. 2.106 that calculations and experiment are in good agreement. Thus, although the laser beam diameter is close to the powder layer thickness, the scattering of radiation along the surface was insignificant.

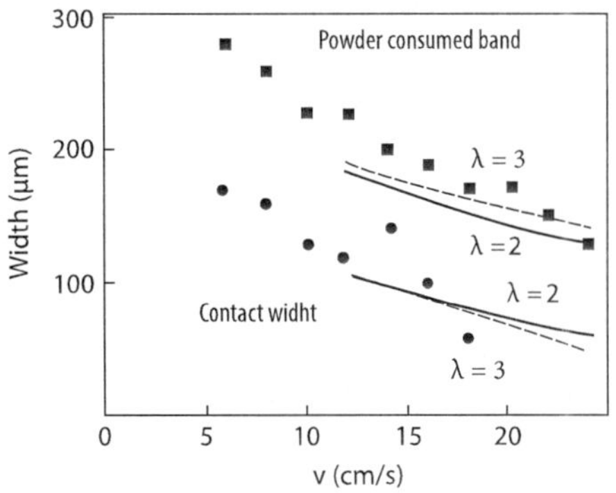

Fig. 2.106 Widths of the powder-consumed band and the contact of the remelted material with the substrate vs. the scanning speed: experiment (*points*) and calculations (*lines*) at two values of the optical thickness, $\lambda = 2$ (*solid lines*) and $\lambda = 3$ (*broken lines*)

2.5.7 Instability of Selective Laser Melting

Experiments show that a melt strip resembles a "liquid" cylinder (Fig. 2.104b). In this case, the Rayleigh instability can be expected [129, 130]. One can see from Fig. 2.104a that at large beam travel velocities, the strip decomposed into drops. It is known that this instability is excited when the cylinder perimeter becomes smaller than its length. As mentioned above, this is observed with increasing the beam travel velocity. Under the conditions of selective laser melting, it is necessary to take into account the additional factor that can affect the process stability. This is fusion with a substrate. However, this stabilizing factor weakens with increasing the beam travel velocity because the melting zone under these conditions also decreases (Fig. 2.106).

It is interesting to consider approximately the quantitative aspect of this stabilization mechanism of the Rayleigh instability. It is assumed for simplicity that the displacement of a liquid cylinder in the melt zone is zero and the size of the melt pool of a substrate does not change. As the cylinder thickness is decreased, its center should descend. In the region where the cylinder size is increased, on the contrary, the cylinder center should ascend. Thus, the displacements become not axially symmetric, as in the case of the Rayleigh instability for a free liquid cylinder. As a parameter characterizing the connection of the cylinder with substrate, we can take, for example, the angle Φ between a vector directed from the cylinder center to the point of intersection of the cylinder with the substrate and vertical [137]. For $\Phi = 180°$, the cylinder touches the substrate. For $\Phi = \pi/2$, the liquid cylinder is immersed by half into the substrate. Such a cylinder will be stable if a perturbation does not increase its area. By omitting details of calculations, we present the results in Fig. 2.107. One can see that the deeper the cylinder "immersed" into the substrate, the more it stable. Foe $\Phi = \pi/2$, the instability does not develop at any relative elongations λ/D of the cylinder. On the contrary, for $\Phi = \pi$, the cylinder only touches the substrate surface. This, nevertheless, makes perturbations nonaxially symmetric, such perturbations being more stable compared to axially symmetric perturbations, although this difference is small. Despite the simplicity of the model,

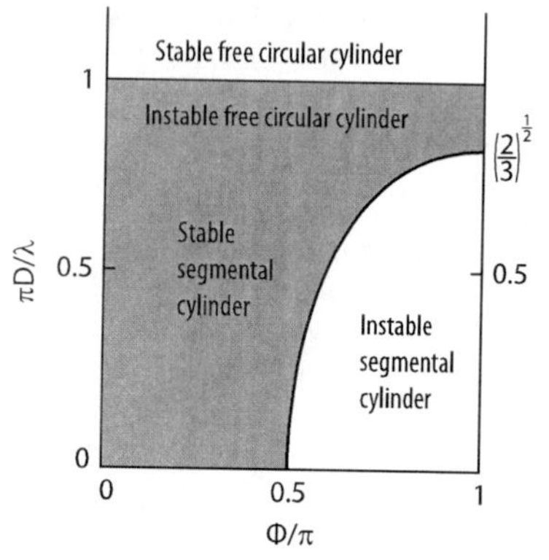

Fig. 2.107 Physical model describing of stability region

it describes the tendencies of the development of the instability of a powder melt track during changing the processing conditions. The instability is enhanced with increasing the powder layer thickness because in this case the width of the zone of melting with the substrate decreases. The same is true when the stainless steel powder is replaced by copper and nickel powders with higher heat conductions.

The mechanism of decomposition of a melt strip into drops can be also more complicated. Thus, upon selective laser melting of steel powder with impurity oxygen, the instability enhanced compared to the case of pure steel [131]. However, it is known that the oxygen impurity in steel reduces the surface tension [132], which mainly initiates the Rayleigh instability. On the other hand, it is known that the oxygen impurity affects the temperature dependence of the surface tension, in particular, the sign of $\partial\sigma/\partial T$ can be even changed [20]. As shown above in this chapter, the change of the sign of this derivative leads to the change in the Marangoni flow direction to the opposite and to a narrowing of the melting zone. The strip narrows down, which facilitates excitation of the Rayleigh instability.

2.5.8 Thermal Hydrodynamic Model of Selective Laser Sintering

The model considered above assumes that the action of capillary forces can be very strong and plays the main role in the formation of a melt strip. To consider this problem in more detail, it is necessary to take into account the flow of melted powder [133]. The consideration of the hydrodynamics of the melt is similar to the description of the Marangoni flow excited by a moving laser beam (2.29)–(2.35). This system of equations should be supplemented by a term taking into account the influence of contraction appearing due to the porosity of powders ($\theta < 1$). Because the contraction is caused by the gravitational force, this velocity is directed vertically. The velocity value is determined by the displacement S of the melt-solid surface:

$$V_{zs} = \theta\left(\frac{\partial S}{\partial t} - v_b\frac{\partial S}{\partial x}\right), \quad z < S.$$

The pressure at the melt surface $S = S_0$ is equal to the atmospheric pressure. The pressure at the rest part of the liquid–solid interface is determined by the Laplace equation:

$$p_l = p_A - \frac{2\sigma}{r_{\text{eff}}}, \quad z = S.$$

The heat conductivity of powder is described by the empirical expression depending on the porosity and heat conductivities of the powder (κ_p) and gas (κ_g) where sintering occurs [134]:

$$\kappa_{\text{eff}} = \kappa_g(1 - \alpha_0)\frac{\theta f_0 + (\kappa_p/\kappa_g)(1 - \theta f_0)}{1 - \theta(1 - f_0) + (\kappa_p/\kappa_g)\theta(1 - f_0)}$$
$$+ \alpha_0 \frac{2(\kappa_p/\kappa_g)^2(1 - \theta) + (1 + 2\theta)(\kappa_p/\kappa_g)}{(2 + \theta)(\kappa_p/\kappa_g) + 1 - \theta},$$

$$f_0 = 0.8 + 0.1\theta,$$

$$\log \alpha_0 = \begin{cases} -4.898\theta \quad \text{for} \quad 0 \le \theta \le 0.0827, \\ -0.405 - 3.154(\theta - 0.0827) \quad \text{for} \quad 0.0827 \le \theta \le 0.298, \\ -1.084 - 6.778(\theta - 0.298) \quad \text{for} \quad 0.298 \le \theta \le 0.580. \end{cases}$$

Unlike the previous model, the sintering model considered here assumes that laser radiation does not penetrate inside powder and is absorbed by the powder surface. Such a situation was justified in studying the heat conduction of powders in vacuum considered at the beginning of this section. In this case, the thickness of powder layers was very large. The model considered here allows one to study the influence

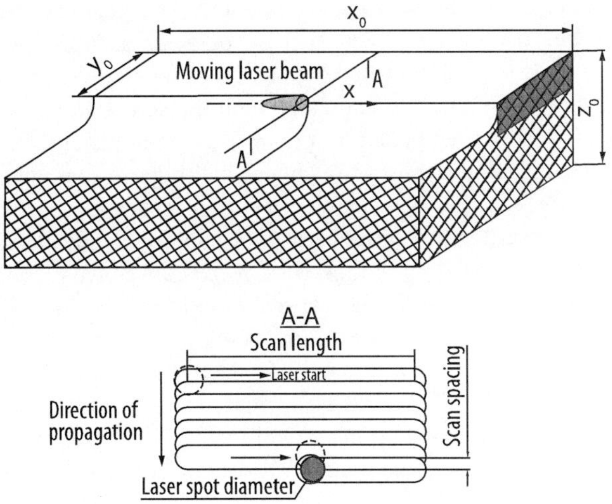

Fig. 2.108 Optimization of laser intensity and scanning velocity on the sintering process ($\theta_i = -1, 0$, $\varepsilon = 0.4$, $N = 1$). (*upper*) Longitudinal view at $y = 0$ and (*bottom*) transverse sectional view at $x = 0$

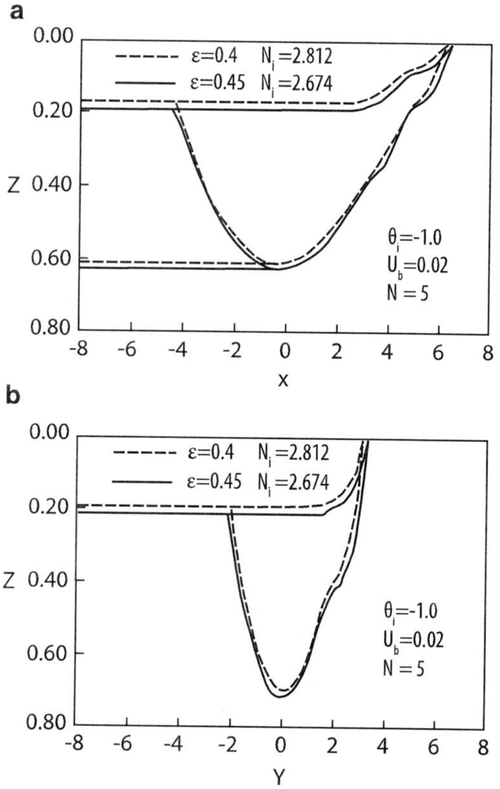

Fig. 2.109 Cross sections of melting zones along the laser beam direction (**a**) and perpendicular to it (**b**). Effect of porosity on the sintering process. $U_b = V_b r_f / \chi$, $Ni = \alpha P / \pi r_f \kappa_l (T_m - T_i)$; N is number of the existing sintered metal layers. $\theta_i = (T_i - T_m)/(T_m - T_i))$, T_i is initial temperature

of the contraction due to the porosity of powders. The geometry of the problem is complicated because not one melt strip is obtained, as in the previous model, but strips are obtained by the repeated deposition of overlapped strips (Fig. 2.108). Figure 2.109 shows the cross sections of melting zones along the laser beam direction ($y = 0$) and perpendicular to it ($x = 0$). The liquid surface in both cross sections is not horizontal, which is caused by the melting and subsequent contraction of powder. One can see that powder with a high porosity is contracted stronger, which is obvious. The depth of a melting pool increases with increasing powder porosity. The latter is caused by the increase in the convective heat transfer inside the sample due to the increase in the vertical component of the melting velocity. This velocity increases with increasing porosity and the material contraction.

Note in conclusion that the investigation of selective laser sintering and melting is very labor-consuming. The SLS models considered above study only some aspects of this complicated technological process. For example, the first (heat conduction) model [122] takes into account the volume absorption coefficient, while the second model neglects it. On the other hand, the second model [133] takes into account the melt flow, which is neglected in the first model. Both models considered above neglect a number of more complicated processes such as diffraction effects during the propagation of laser radiation in powder pores, side scattering of the laser beam, etc. The field of studies and applications of SLS is very broad. This is related first of all to a variety of material used in this technology. Reactive powder compositions and laser processing of powders in active gases (oxygen and others) require the theoretical consideration of complicated physicochemical processes. These are undoubtedly the problems of investigations in the future [138].

References

1. J.F. Ready, *Lasers in Modern Industry* (Society of Manufacturing Engineers, Dearborn, Michigan, 1979; Mir, Moscow, 1981)
2. G.A. Abil'siitov, V.S. Golubev, Vestnik Akad. Nauk SSSR, **11, 37**, 44 (1982)
3. E. Goodremon, Special Steels, vol. 1. (Gos. nauchn-tekh. izd. chern. tsvetn. metallurg., Moscow, 1959)
4. Ya.S. Umanskii, Yu.A. Skakov, *Fizika Metallov (Physics of Metals)* (Atomizdat, Moscow, 1978)
5. H.S. Carslaw, J.C. Jaeger, *Conduction of Heat in Solids*, 2nd edn. (Clarendon Press, Oxford, 1959; Nauka, Moscow, 1964)
6. Y. Arata, H. Mario, Application of Lasers for Material Processing, IIW, Doc. IV-241-71, 18 p
7. N.N. Rykalin, A.A. Uglov, A.N. Kkra, in *Handbook, Laser and electron beam processing of material*, ed. by M. Mashinostroenie (1986), 496 p
8. E. Jahnke, F. Emde, F. Losch, *Tables of Higher Functions*, 6th ed. (McGraw-Hill, New York, 1960; Nauka, Moscow, 1968)
9. S. Kon, D. Sun, Y.P. Le. Met. Trans. A. **14A**, 643–653 (1983)
10. V.S. Kraposhin, I.F. Kraposhina, Fiz. Khim. Obrabot. Mater. **6**, 19 (1989)
11. V.M. Andriyakhin, N.T. Chekanova, Poverkhnost' **1**, 129 (1983)
12. V.V. Korneev, F.K. Kosyrev, S.F. Moryashchev, Fiz. Khim. Obrabot. Mater. **3**, 3 (1980)
13. H. Kawasumi, Techocrat **11**, 11 (1978)
14. V.S. Kovalenko (ed.) *Uprochnenie detalei luchom lazera (Strengthening of Details by a Laser Beam)* (Tekhnika, Kiev, Ukraine, 1981)
15. V.S. Kraposhin, Fiz. Khim. Obrabot. Mater. **6**, 88 (1988)
16. W.M. Steen, C. Courtney, Met. Technol. **6**, 456 (1979)
17. V.V. Korneev, A.N. Yavokhin, Fiz. Khim. Obrabot. Mater. **2**, 7 (1980)
18. I.K. Kikoin (ed.), Tablitsy fizicheskikh velichin. *Spravochnik (Handbook of Physical Quantities)* (Atomizdat, Moscow, 1976)
19. N.N. Rykalin, A.A. Uglov, Dokl. Akad. Nauk SSSR **267**, 377 (1982)
20. C. Kittel, *Elementary Solid State Physics* (Wiley, New York, 1962; Nauka, Moscow, 1965)
21. Y. Miwa, J. Jap Instrum. Met. *39, 3*, 255 (1975)
22. A.G. Grigor'yants, I.N. Shiganov, V.V. Ivanov, et al., Svaroch. Proizvod. **9**, 29 (1982)
23. L.Z. Kriksunov, *Spravochnik po osnovam infrakrasnoi tekhniki (Handbook of Fundamentals of IR Techiques)* (Sovetskoe radio, Moscow, 1978)
24. G.F. Antonova, S.F. Moryashchev, A.A. Startsev, Svaroch. Proizvod. **6**, 1 (1983)
25. M.A. Golub, S.V. Kariev, A.M. Prokhorov, I.N. Sisakyan, V.A. Soifer, Pis'ma Zh. Tekh. Fiz. **7**, 618 (1981)
26. P.E. Dubovskii, I.B. Kovsh, M.S. Strikalova, I.N. Sisakyan, Kvantovaya Elektron. **21**, 1183 (1994)
27. A.G. Grigor'yants, A.N. Safonov, *Osnovy lazernogo termouprocheniya splavov (Fundamentals of Laser Thermal Strengthening of Alloys)* (Mashinostroenie, Moscow, 1988)
28. A.G. Grigor'yants, A.N. Safonov, V.V. Shibaev, Svaroch. Proizvod. **12**, 11 (1983)
29. V.M. Andriyakhin, N.B. Edneral, Kh.A. Mazorra, et al., Poverknost **10**, 134 (1982)

30. G. Polia, G. Sege, *Izoperimetricheskie neravenstva v matematiheskoi fizike (Isoperimetric Inequalities in Mathematical Physics)* (Fizmatgiz, Moscow, 1962)
31. P.A. Molian, Scripta Metallurgia. **16**, 65 (1982)
32. Ya.I. Frenkel, *Kineteicheskaya teoriya zhidkosti (Kinetic Theory of Liquid)* (Nauka, Leningrad, 1975)
33. L.E. Scriven, C.V. Sternling, Nature **187**, 186 (1969)
34. V.G. Levich, *Fiziko-mekhanicheskaya gidrodinamika (Physico-mechanical Hydrodynamics)* (Fizmatgiz, Moscow, 1959)
35. V.S. Maiorov, M.P. Matrosov, Kvantovaya Elektron. **16**, 806 (1989)
36. M.E. Tompson, J. Szekely, Heat Moss Tranfer **32**, 1007 (1989)
37. H.G. Krans. Welding Res. Suppl. **7**, 269 (1989)
38. C.R. Heiple, J.R. Roper, R.T. Stagner, R.J. Aden. Weld. J. **62**, 72 (1983)
39. C.R. Heiple, J.R. Roper, Weld. J. **61**, 97 (1982)
40. C.R. Heiple, P. Burgarrdt, Weld. J. **64**, 159 (1985)
41. G.S. Ershov, Yu.B. Bychkov, *Properties of Metal Melts and their Interaction in Steelmaking Processes* (Metallurgia, Moscow, 1983)
42. T. Zacharia, S.A. David, J.M. Vitek, T. DebRoy, in *Recent Trends in Welding Science and Technology TWR'89, Proceedings of the 2nd International Conference on Trends in Welding Research*, Gatlinburg, Tennessee, 14–18 May 1989, ed. by S.A. David, J.M. Vitek (ASM International, Tennessee, 1989), pp. 25–30
43. I. Smurov, L. Covelli, K.I. Taguirov, L. Aksenov, J. Appl. Phys. **71**, 3147 (1992)
44. L.D. Landau, E.M. Lifshits, *The Classical Theory of Fields*, 4th English edn. (Pergamon Press, Oxford, 1975; Nauka, Moscow, 1967)
45. G.G. Gladush, L.S. Krasitskaya, E.B. Levchenko, A.A. Chernyakov, Kvantovaya Elektron. **9**, 660 (1982)
46. C. Chan, J. Mazumber, M.M. Chen. Metall. Trans. A. **15A**, 2175 (1984)
47. A.A. Uglov, I.Yu. Smurov, K.I. Taguirov, A.G. Guskov. Int. J. Heat Mass Transfer. **35**, 783 (1992)
48. S.I. Vainshtein, Ya.B. Zel'dovich, A.A. Ruzmaikin, *Turbulentnoe dynamo v astrofizike (Turbulent Dynamo in Astrophysics)* (Nauka, Moscow, 1980)
49. N.N. Rykalin, A.A. Uglov, Fiz. Khim. Obrabot. Mater. **4**, 3 (1985)
50. N.N. Rykalin, A.A. Uglov, V.A. Grebennikov, M.B. Ignat'ev, Dokl. Akad. Nauk SSSR **272**, 1114 (1983)
51. G.G. Gladush, S.V. Drobyazko, V.V. Likhanskii, A.I. Loboiko, Yu.N. Senatorov, Kvantovaya Elektron **25**, 439 (1998)
52. C. Limmaneevichitr, S. Kon, Welding Res. Suppl. **126**, (2000)
53. G.G. Gladush, V.V. Likhanskii, A.I. Loboiko, Kvantovaya Elektron. **24**, 274 (1997)
54. S.V. Patankar, *Numerical Heat Transfer and Fluid Flow. Series in Computational Methods in Mechanics and Thermal Sciences* (Hemisphere, New York, 1979)
55. M.S. Gross, I. Black, W.H. Muller, J. Phys. D. Appl. Phys. **36**, 929 (2003)
56. I. Smurov, N. Ignat'ev, Proceeding of *Laser Processing Surface Treatment and Film Deposition*. ed. by I. Mazumber, O. Conde, R. Villar, W. Steen, NATO ASI Series E: Applied Sciences, vol. 307
57. Yu.V. Sanochkin, Teplofiz. Vysok. Temp. **27**, 1029 (1989)
58. G.F. Antonova, G.G. Gladush, F.K. Kosyrev, A.G. Krasyukov, V.V. Likhanskii, A.I. Loboiko, Kvantovaya Elektron. **25**, 5 (1998)
59. L.G. Napolitano, C Golia, A. Viviani. *Proc. 5th European Symp. Material Sciences in Microgravity condition*, ESA-SP-222, 1984
60. A.A. Uglov, I.Yu. Smurov, M.B. Ignat'ev, et al., Teplofiz. Vysok. Temper. **28**, 530 (1990)
61. R.V. Arutyunyan, V.Yu. Baranov, L.F. Bol'shov, D.D. Malyuta, A.Yu. Sebrant, *Vozdeistvie lazernogo izlucheniya na matrialy (Interaction of Laser Radiation with Materials)* (Nauka, Moscow, 1989)
62. A.L. Galliev, L.L. Krapivin, L.I. Mirkin, Dokl. Akad. Nauk SSSR **251**, 336 (1980)

63. A.G. Gnedovets, O.M. Portnov, I. Smurov, G. Flamant. *Proc. of the Laser Materials Processing Conf. ICALEO'94 LIA*, Vol. 79, Orlando, FL, USA
64. T. Ytrehus, Phys. Fluids **26**, 939 (1983)
65. T. Bergstorm, T. Ytrehus. Phys. Fluids **27**, 583 (1984)
66. V.V. Averin, A.V. Revyakin, V.I. Fedorchenko, L.N. Kozina, *Nitrogen in Metals*, Metallurgiya, Moscow, 1976
67. M. Ignat'ev, E. Kovalev, I. Melekhin, I. Smurov, S. Sturlese. Wear **166**, 233 (1993)
68. N. Pirch, E.W. Krentz, B. Oliver, X. He, in *Laser Processing: Surface Treatment and Film Deposition*, ed. by J. Mazumder, O. Conde, R. Villar, W. Steen (Kluwer Academic Publishers, 1996)
69. T.W. Krentz, N. Pirch, M. Rozsnocki, *Laser Treatment of Materials*, ed. by Morddike, ECLAT Proc. DGM, 269, 1992
70. C.L. Chan, J. Mazumder, M.M. Chen, J. Appl. Phys. **64**, 6166 (1988)
71. L.I. Antonova, A.F. Glova, S.V. Drobyazko, Yu.M. Senatorov, Kvantovaya Elektron. **32**, 1029 (2002)
72. G.G. Gladush, S.V. Drobyazko, N.B. Rodionov, Yu.M. Senatorov, L.I. Antonova, IQEC/LAT Moscow, June 22–27, 2002
73. S.I. Anisimov, M.I. Tribel'skii, Ya.G. Epel'baum, Zh. Eksp. Teor. Fiz. **78**, 1597 (1980)
36. A.A. Uglov, I.Yu. Smurov, A.G. Gus'kov, Kvantoavaya Elektron. **18**, 1081 (1991)
75. V.V. Likhanskii, A.I. Loboiko, Kvantoavaya Elektron. **30**, 827 (2000)
76. A.S. Povitskii, L.Ya. Lyubin, *Osnovy dinamiki i teplo-masso-obmena zhidkostei i gazov pri nevesomosti (Fundamentals of the Dynamics and Heat-and-Mass Exchange of Liquids and Gases in the Case of Zero-Gravity)* (Mashinostrienie, Moscow, 1972)
77. E. Toyserkani, A. Khajepour, S. Corbin, *Laser Cladding* (CRC Press, 2005)
78. B. Grunenwald, W. Hennig, St. Novotny, F. Dausinger, H. Hugel, *Laser Processing: Surface Treatment and Film Deposition*, ed. by J. Mazumber et al., NATO ASI Series, vol. 307, Sesimbra, Portugal, July 3–16, 1994
79. J. Lin, Opt. Laser Techn. **31**, 565 (1999)
80. B. Ollier, R. Pirch, E.W. Krentz, H. Schluter, *ECLAT1992*, ed. by B.L. Mordike (Gottingen, Germany. 1992)
81. R. Clift, J.R. Grace, M.E. Weber, *Bubbles, Drops and Particles* (Academic Press, UK, 1978)
82. X. Chen, E. Pfender, *Plasma Chem. Plasma Proces.* **2**, 293 (1982)
83. A.V. Gusarov, I. Smurov, Appl. Surf. Sci. **253**, 8316 (2007)
84. F. Lemoine, D.F. Grevey, A.B. Vannes, ICALEO (1993)/203
85. S. Carty, I. Owen, W.M. Steen, B. Bastow, J.T. Spencer, *Laser Processing: Surface Treatment and Film Deposition*, ed. by J. Mazumber et al., NATO ASI Series, vol. 307, Sesimbra, Portugal, July 3–16, 1994
86. R. Colaco, L. Costa, R. Guerra, R. Vilfr, ibid.
87. W. Steen, *Laser Material Processing*, 3rd edn. (Springer, London, 2003)
88. M. Doubenskaia, Ph. Bertrand, I. Smurov, Thin Solid Films **453–454**, 477 (2003)
89. A. Yakovlev, Ph. E. Trunova, D. Grevey, M. Pilloz, I. Smurov, Technology **190**, 15 (2005)
90. A. Yakovlev, Ph. Bertrand, I. Smurov, Thin Solid Films **453–454**, 133 (2003)
91. B. Lukyanchuk (ed.) *Laser Cleaning* (World Scientific Publishing, Singapore, 2002)
92. K.A. Nikishina, V.M. Stekhin, V.I. Gorynin, *Sbornik trudov Mezhdunarodnoi konferentsii po fizike radiatsionnykh yavleniii i radiatsionnomu materialovedeniyu (Proceedings of the International Conference on the Physics of Radiative Phenomena and Radiative Materials Technology)* (Alushta, Crimea, Ukraine, 2002)
93. A.G. Krasyukov, V.G. Naumov, V.E. Cherkovets, A.B. Senyavin, Byulleten' "Voprosy utilizatsii atomnykh podvondnykh lodok (Bull. Problem Utilization Atomic Submarines) **2**, 24 (2004)
94. G.A. Askar'yan, E.K. Karlov, R.P. Petrov, V.B. Studenov, Pis'ma Zh. Eksp. Teor. Fiz. **18**, 665 (1973)
95. A.I. Kutsenko, V.A. Gurashvili, A.G. Krasyukov, et al., Laser 96, International Conference, Portland, Oregon, USA, 1996

96. V.P. Veiko, E.A. Shakhno, in *Physical Mechanisms of Laser Cleaning, "Laser Cleaning"*, ed. by B. Lukyanchuk (World Scientific Publishing, Singapore, 2002)
97. W. Zarka, in *The road to steam laser cleaning, "Laser Cleaning"*, ed. by B. Lukyanchuk (World Scientific Publishing, Singapore, 2002)
98. B.S. Lukyanchuk, M. Mosbacher, Y.W. Zheng, et al., *Optical resonance and near-field effects in dry laser cleaning, "Laser Cleaning"*, ed. by B. Lukyanchuk (World Scientific Publishing, Singapore, 2002)
99. Z. Toth, B. Hopp, Z. Kantor, F. Iqnacz, T. Szorenyi, Z. Bor. J. Appl. Phys. A. **60**, 431 (1995)
100. M.L. Santis, F. Delaport, B. Martin, O. Uteza, Kvantovaya Elektron. **30**, 495 (2000)
101. I.A. Movtchan, W. Marine, R.W. Dreyfus, H.C. Le, M. Sentis, M. Autric. Appl. Surf. Sci. **96–98**, 251 (1996)
102. L.I. Antonova, B.V. Dobrov, S.V. Drobyazko, A.G. Krasyukov, V.V. Likhanskii, Yu.M. Senatorov, N.G. Turkin, Prepring RINITI0049, TSNIIATOMINFORM, 1998
103. L.I. Antonova, B.V. Dobrov, S.V. Drobyazko, A.G. Krasyukov, V.V. Likhanskii, Yu.M. Senatorov, Kvantovaya Elektron. **32**, 165 (2002)
104. I.G. Lebo, V.F. Tishkin, *Issledovanie gidrodinamicheskoi neustoichivosti v zadachakh termoyadernogo sinteza (Study of Hydrodynamic Instability in Problems of Thermonuclear Fusion)* (Fizmatgiz, Moscow, 2006)
105. M. Harada, R. Walter, S. Seitfert, *Chemical Oxygen-Iodine Laser for Decontamination and Decomissioning of Nuclear Facilities. Laser 99*, December 13–16, 1999, Quebec, Canada
106. V.Yu. Dubrovskii, F.K. Kosyrev, V.G. Naumov, L.V. Shachkin, *Proc. of the International Conference on Lasers 2000*, Albuquerque, USA, 2000
107. W.M. Shu, Y. Kawakubo, S. O/hira, et al., Fusion Sci. Technol. **41**, 690 (2002)
108. C.H. Skinner, C.F.Gentile, G. Guttadora, et al., Fusion Sci. Technol. **41**, 716 (2002)
109. J.W. Davis, A.A. Haas. J. Nucl. Mater. **266–269**, 478 (1999)
110. V.E. Cherkovets, E.A. Azizov, F.K. Kosyrev, V.G. Numov, A.I. Markin, N.I. Syromyatnikov, Vopr. Atom. Nauki Tekh., Ser. Meterialoved. Nov. Materl. **2**(67), 356 (2006)
111. D.L. Bourell, H.L. Marcus, J.-W. Barlow, J. Beaman, Selective laser sintering of metals and ceramics. Int. J. Powder Met. **28**, 369 (1992)
112. M. Agarwala, D.L. Bourell, H.L. Marcus, J.-W. Barlow, Direct selective laser sintering of metals. Rapid Prototype J. **1**, 26 (1995)
113. J.-P. Kruth, P. Peeters, T. Smolderen, J. Bonse, T. Laoui, L. Froyen, Comparison between CO_2 and Nd:YAG lasers for use with selective laser sintering of steel-copper powders. Int. J. CAD/CAM Comput. Graphics **13**(4-6), 95 (1998)
114. N.K. Tolochko, S.E. Mozzharov, N.V. Sobolenko, Yu. V. Khlopkov, I.A. Yadroitsev, V.B. Mikhailov, Main relationship governing laser sintering of loose single-component metallic powders. J. Adv. Mat. **2**, 151 (1995)
115. H. Yasuda, I. Ohnaka, H. Kaziura, Y. Nishiwaki, Fabrication of metallic porous media by semisolid processing using laser radiation. Mater. Transact. **42**, 309 (2001)
116. A.V. Luikov, *Spravochnik po teplomassperenosy (Handbook on Heat-and-Mass Transfer)* (Energia, Moscow, 1971)
117. Y.B. Zeldovich, Y.P. Raiser, *Physics of Shock Waves and High-temperature Hydrodynamic Phenomena*, vol. 2 (Academic Press, New York, 1967)
118. V.V. Skorohod, S.M. Solonin, *Fiziko-metallurgicheskie osnovy spekaniya poroshkov (Physico-metallurgic Fundamentals of Powder Sintering* (Metallurgiya, Moscow, 1984)
119. G.C. Kuczynski, Self-diffusion in sintering of metallic particles. J. Metals **1**, 169 (1949)
120. A.V. Gusarov, T. Laoui, L. Froyen, V.I. Titov, N.K. Tolochko, Numetrical simulation of laser solid state sintering of loose titanium powder. in *Proc. 10th European Conference on Rapid Prototyping and Manufacturing* (Paris, France, June 7-8, 2001)
121. M. Rombouts, L. Froyen, A.V. Gusarov, E.N. Bentefour, C. Glorieux, J. Appl. Phys. **97**, 024905 (2005)
122. A.V. Gusarov, I.A. Yadroitsev, Ph. Bertrand, I. Smurov, Appl. Surf. Sci. **254**(4), 980, (2007)
123. A.M. Ivanova, S.P. Kotova, N.L. Kupriyanov. A.L. Petrov, E.Yu. Tarasova, I.V. Shishkovskii, Kvantovaya Elektron. **25**, 433 (1998)

124. V.V. Levandovskii, et al., in *Vozdeistvie kontsentrirovannykh potokov energii na meterialy (Interaction of Concentrated Energy Fluxes with Materials)* (Nauka, Moscow, 1985)
125. M. Rombouts, L. Froyen, A.V. Gusarov, E.N. Bentefour, C. Glorieux, J. Appl. Phys. **98**, 013533 (2005)
126. A.V. Gusarov, J.-P. Kruth, Int. J. Heat Mass Transfer **48**, 3423 (2005)
127. A.V. Gusarov, E.N. Bentefour, M. Rombouts, L. Froyen, C. Glorieux, J.-P. Kruth, J. Appl. Phys. **99**, 113528 (2006)
128. A.M. Ivanova, A.A. Kovalenko, S.P. Kotova, N.L. Kupriyanov, V.M. Petropavlovskii, in *Trudy XXIV shkoly-simoziyuma po kegerentnoi optike i golografii* (Proceedings of XXIV School-Symposium on Coherent Optics and Holography, Yaroslavl', 1997)
129. Lord Rayleigh, Phil. Magazine **34**, 145 (1892)
130. B.V. Schueuren, J.P. Kruth, in *Laser Assisted Net Shape Engineering*, Proceeding of the LANE 1994
131. H.J. Nin, I.T.H. Chang, Scripta Mater. **41**, 1229 (1999)
132. B.J. Keene, Int. Mater. Rev. **33**, 1 (1988)
133. B. Xiao, Yu. Zhang, J. Phys. D Appl. Phys. **40**, 6725 (2007)
134. M. Kaviany, *Principles of Heat Transfer in Porous Media*, 2nd edn. (Springer, New York, 1995)
135. A.V. Gusarov, I. Smurov, *IMSE-2008*, Santander, Spain, 7–10 July, 2008
136. Yu. Chyivel, M. Petrushina, I. Smurov, Appl. Surf. Sci. **254**, 816 (2007)
137. A.V. Gusarov, I. Smurov, *Proc. 15 Int. WLT-Conf. on Laser in Manufacturing 2009*, Munich, June, 2009
138. I.V. Shishkovskii, *Laser fusion of functional gradient mesosttructures and volume wares*, Moscow, Fismatlit, 2009
139. Thermodynamic properties of substances, L.V. Gurvich, I.V. Veitz, V.A. Megvedev, et. al., Nauka, (Moscow, 1978).
140. C.Y. Ho, R.W. Powel, P.E. Liley, J. Phys. Chem., 1, 279, 1972.
141. CRC Handbook of chemistry and physics, D.R. Lide, CRC, (New York, 1985).

Chapter 3
Plasma Phenomena in Laser Processing of Materials

Abstract The appearance of plasma during optical breakdown near a target can strongly change the interaction of radiation with the target, by enhancing or weakening this interaction depending on the processing conditions. For short laser pulses ($\tau \sim 1\,\mu$s), when a plasma cloud has no time to depart for a target during the laser pulse the interaction is enhanced. In this chapter, we consider the conditions for producing a plasma plume (optical breakdown near a surface) and also questions related to the physics of a developed optical discharge (nonstationary and stationary) near the sample surface. In addition, we discuss the absorption and refraction of laser radiation in the plasma plume, continuous optical discharge in a longitudinal and transverse gas flows, and optical discharge in a keyhole.

3.1 Thermal Properties of the Plasma of Noble and Molecular Gases and Metal Vapors

Because laser radiation represents an ac electric field having, as a rule, a high strength (e.g., a 10-kW laser beam focused into a spot of diameter 1 mm produces the electric field strength $E = 50\,\text{kV/cm}$), it can produce, as a dc electric field, a breakdown in gases. An optical breakdown in homogeneous gases was first observed in the late 1960s with the advent of high-power pulsed lasers [1]. The radiation intensity q_{br} required for breakdown (the breakdown threshold) in air at the atmospheric pressure is $10^9\,\text{W/cm}^2$ for radiation from a CO_2 laser ($10^{11}\,\text{W/cm}^2$ for radiation from a neodymium laser). The laser radiation intensity used for material machining is considerably lower, not exceeding, as a rule, $10^7\,\text{W/cm}^2$. However, even lower-power laser beams interacting with targets can produce an optical breakdown, and a plasma with temperature of tens thousands of kelvin appears which strongly affects the interaction of radiation with materials. For pulses of duration $\tau = 0.1$–$1\,\mu$s, we have $q \approx 10^7\,\text{W/cm}^2$ [2] the value of q_{br} decreases with increasing pulse duration and can be fractions of megawatt per

G.G. Gladush and I. Smurov, *Physics of Laser Materials Processing*,
Springer Series in Materials Science 146, DOI 10.1007/978-3-642-19831-1_3,
© Springer-Verlag Berlin Heidelberg 2011

square centimeter for continuous radiation [27]. Note that the radiation intensity at which the vaporization of metals begins (the developed vaporization threshold) is also ~1 MW/cm^2. This means that laser processing involving the damage or deep melting of materials is often accompanied by plasma effects [3,4]. As a pressure is decreased, the breakdown threshold increases, whereas the vaporization threshold decreases, and therefore the developed vaporization regime can be observed without plasma formation [5]. As a pressure is increased above the atmospheric pressure, the breakdown threshold becomes lower than the vaporization threshold, and plasma is formed without damage or even without melting of materials [6] (a similar phenomenon is also observed for refractory materials at the atmospheric pressure [5]).

Because radiation of a CO_2 laser is weakly absorbed by a clean surface of metals, the energy transfer to a sample can increase due to a stronger absorption of laser radiation by plasma, which is accompanied by the heat transfer to the sample due to the heat conduction or radiation from plasma. Because the maximum of the spectral density of thermal radiation from a plasma plume falls usually in the visible or UV region, this radiation is absorbed by the sample stronger than the IR radiation of a CO_2 laser (see details in Sect. 5.2).

The main effect observed in the cw regime or for long pulses is the attenuation of the intensity of radiation focused on the sample surface due to absorption and defocusing in plasma. The negative role of a plasma plume was pointed out already in the first welding experiments with a CO_2 laser [4]. It was necessary either to blow it away or to perform welding in the helium atmosphere, where the formation of a plasma plume is hindered.

The thermal and optical properties of gases and vapors in which discharges are burning play an important role in the discussion of plasma phenomena. Therefore, we describe briefly the properties of gas media which are most often used in laser processing of materials. Thermal coefficients used in our numerical calculations are presented in Figs. 3.1–3.3, [7,8,79,80]. The coefficients of the electron and atomic heat conduction are described by the expressions

$$\kappa_e = \frac{5}{6} k n_e \frac{\bar{v}_e}{\sigma_{ea} N + \sigma_c n_e}, \quad \kappa_a = \frac{5}{6} k \frac{\bar{v}_a (n_e + N)}{\sigma_{aa} N + \sigma_c n_e} \tag{3.1}$$

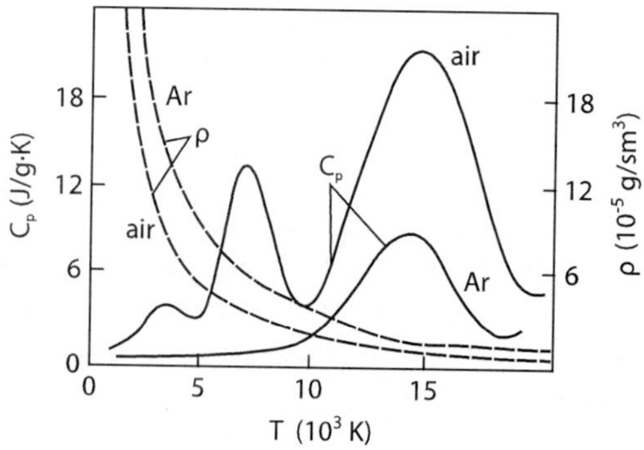

Fig. 3.1 Temperature dependences of the density ρ and specific heart C_p of the equilibrium argon and air plasma [1,79,80]

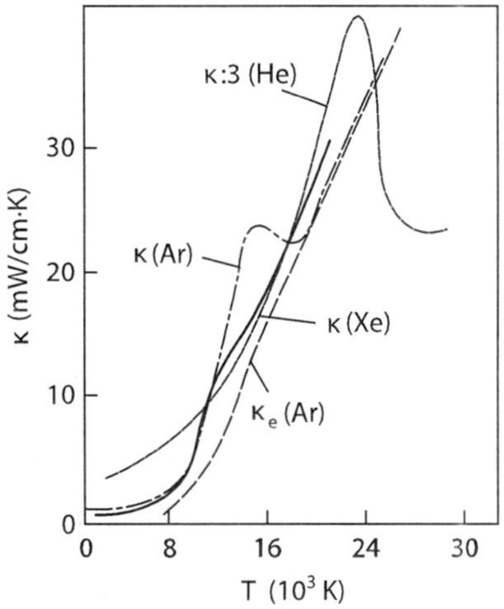

Fig. 3.2 Temperature dependences of the heat conductivity of the equilibrium helium, argon, and xenon plasma [7, 8]

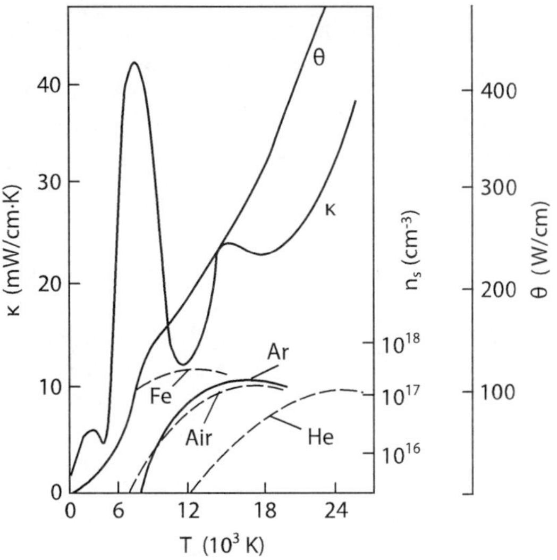

Fig. 3.3 Temperature dependences of the heat conductivity κ, the thermal potential θ of air, and the electron density n_s of some materials

where n_e and N are the electron and atomic densities, respectively; $\sigma_c = 10^{13}/T^2$ is the Coulomb scattering cross section (in cm^2); T is temperature; $\bar{v}_e$ and $\bar{v}_a$ are the thermal velocities of electrons and atoms; σ_{ea} and σ_{aa} are cross sections for scattering of electrons and atoms by atoms. At high electron temperatures, heat transfer due to ionization should be also taken into account. In the case of molecular gases, heat transfer due to the dissociation of molecules should be also considered. The peaks caused by dissociation are observed in Figs. 3.1 and 3.3. The concentration of charged particles determines not only the heat capacity and heat conduction of plasma but also the absorption coefficient μ for the CO_2 laser radiation in plasma, which is caused by the acceleration of free electrons in the ac electric field of a light wave and their scattering by atoms and ions. This process is in fact the Joule heating in the high-frequency field of the light wave [1]. The absorption coefficient is described by the expression

$$\mu = \frac{\nu}{c}\frac{\omega_p^2}{\omega^2}, \tag{3.2}$$

where ν is the frequency of collisions of electrons with atoms and ions; c is the speed of light; ω_p is the plasma frequency; and ω is the laser radiation frequency (see Sect. 1.2). At high temperatures, when Coulomb collisions become important, expression (3.2) in the isobaric case can be rewritten in a more convenient form [1]:

$$\mu = \frac{10.4 \times p^2 x_e g}{(T/10^4)^{7/2}}; \quad g = 0.55 \ln\left(\frac{2.7 \times 10^{-4}}{T p_e^{1/3}}\right) \tag{3.3}$$

where μ is expressed in cm^{-1}, T in kelvins; $x_e = p_e/p$; p_e and p are the electron and total pressure, respectively (10^5 Pa). The ionization degree x_e is determined by the Saha

$$\frac{x_e^2}{1 - 2x_e} = 6.7 \times 10^3 \frac{g_+}{g_a}\frac{(T/10^4)^{5/2}}{p}\exp(-I/kT) \tag{3.4}$$

where g_+ and g_a are the statistical weights of ions and atoms, respectively. By using (3.3) and (3.4), we can calculate absorption coefficients for the CO_2 laser radiation at 10.6 μm in different media (Fig. 3.4). At the gas temperature $T \gg 1$ eV, the multiple

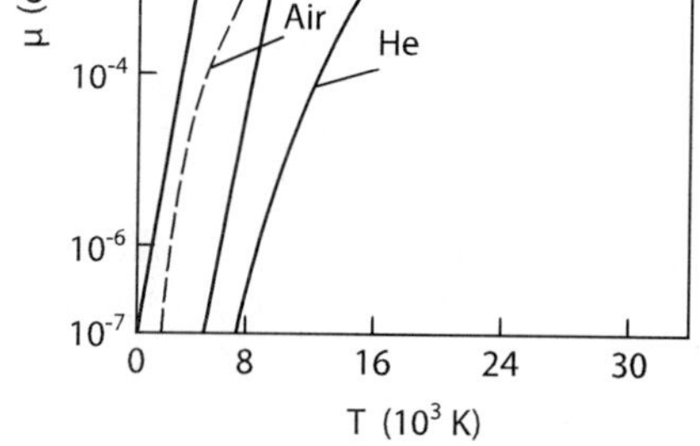

Fig. 3.4 Dependence of the absorption coefficient for radiation from a CO_2 laser on the plasma temperature

ionization of atoms becomes important. In this case, μ increases with increasing the ion charge ($\sim T$ [1]):

$$\mu_{\omega} = \frac{3.1 \times 10^{-31} Z^3 N^2 g}{T^{3/2} (h\omega)^2}; \quad g = 0.55 \ln \left(\frac{2.4 \times 10^3 T}{Z^{4/3} N^{1/3}} \right) \tag{3.5}$$

where N is the ion density. When Z approaches its limiting value, equal to the nuclear charge, with increasing the gas temperature, the value of μ begins to decrease. For example, in air this occurs for $T \approx 25$ eV [1]. The photon path in this case ($N \sim 10^{19}$ cm^{-3}) is about 2 μm.

3.1.1 Plasma Emission

The escape of emission from plasma is mainly determined by the absorption coefficient. In the case of a high absorption coefficient, when the photon path is smaller than the plasma size, emission is considered closed. When the photon path is larger than the plasma size, the plasma is called optically thin, and in this case losses can be very simply calculated. The loss of emission from plasma in a continuous spectrum is described by the Kramers–Unsold formula [65] for the isobaric case:

$$\Phi = \frac{7.6 p^2 x_e^2}{(T/10^4)^{3/2}} \left(1 + \frac{h\nu_g}{kT} \right) \tag{3.6}$$

where Φ is expressed in kW/cm^3; p is pressure (10^5 Pa); and $h\nu_g$ is the binding energy for the first excited level. The recombination of electrons to the ground level is neglected in (3.6). The loss of emission from plasma, taking into account recombination to the ground level, are described by the different expression

$$\Phi = \frac{280 p^2 x_e^2}{(T/10^4)^{5/2}} \left(1 + 0.027 \frac{T}{10^4} \right) \tag{3.7}$$

where T is expressed in kelvins.

A comparison of (3.6) and (3.7) shows that for $T \sim 10^4$ K the emission loss, taking into account photorecombination to the lower level, exceeds by several times photorecombination losses to excited levels. Photons created due to the capture of electrons by the ground level have the energy $h\nu > I$, where I is the ionization potential. They are readily absorbed by unexcited atoms. In molecular gases, where strong molecular absorption bands exist, emission with the photon energy smaller than I can be also closed. For example, only photons with $h\nu < 6$ eV can escape from the region of a continuous optical discharge (COD) of size $h \sim 1$ cm at the atmospheric pressure.

Real volume energy losses from plasma can differ from the values calculated from (3.6) and (3.7) mainly due to emission in spectral lines. Although the main part of linear emission is absorbed again, its contribution to emission losses can be

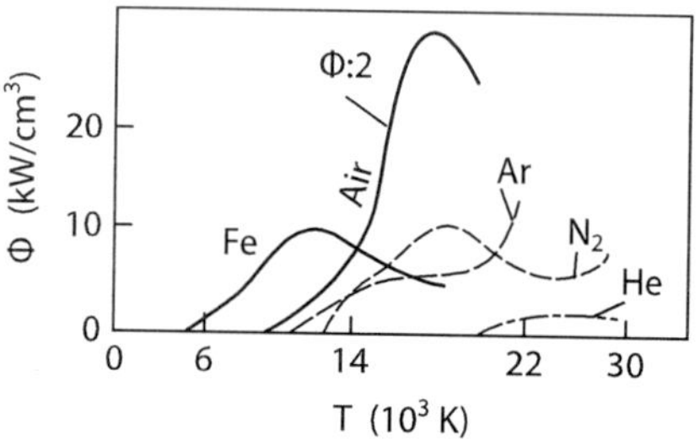

Fig. 3.5 Volume energy losses due to radiation from plasma in different gases: air [1], N_2 [11], Ar [44]; He, Fe [calculated by (3.6)]

considerable. Figure 3.5 presents the integrated dependences of emission losses for different gases. If emission is closed in some spectral regions and freely escapes from plasma in other spectral regions, to calculate emission losses, it is necessary, strictly speaking, to solve the complete radiation transfer equation [26]. This method was used to calculate the development dynamics of a pulsed optical discharge [12] and COD [36, 39, 42, 52, 53]. The radiation transfer is sometimes described in the Schwarzschild approximation ("forward–backward" approximation) [26]. This approximation was used, e.g., for the two-dimensional calculation of the development of a plasma plume near a target surface [13] (see Sect. 7.3).

3.2 Mechanisms of the cw Laser-Induced Breakdown of Gases Near Solid Surfaces

3.2.1 *Stationary Breakdown of Gases in the Absence of a Target*

A target placed in a laser beam focus is heated and can be evaporated. In a gas surrounding the target, the target material vapor appears, the gas is heated and its density changes. In addition, the presence of the reflecting target enhances the light intensity. All these factors can in principle reduce the breakdown threshold of the gas near the target surface. Here, we will consider briefly the mechanism of the gas breakdown without a target and analyze the possibility of its application to describe the gas breakdown near a target. In addition, the breakdown of gases away from a target is also of independent interest for laser technologies because materials can be processed by high-peak-power lasers, which can produce breakdown in the absence of a target.

As mentioned in Sect. 3.1, the interaction of radiation from a CO_2 laser with gases is similar to the interaction of microwave radiation with gases. It is known that the gas breakdown in a microwave field is similar to that in a dc electric field [1]. The difference is that the plasma conduction in the microwave field is ω^2/ν^2 times lower than that in a dc electric field [1] or the microwave field strength E is ω/ν times lower than the dc electric field strength $\bar{E}$ [see [9]]: $\bar{E} = \nu E/\omega$.

The rigorous substantiation of this statement (under the condition $h\omega \ll \varepsilon_e$, ε_e is energy of atom excitation) follows from the Boltzmann equation for the time-averaged electron distribution function in an ac electric field [1]. For the convenience of calculations, we will replace the ac electric field E by the effective dc electric field $\bar{E}$ and use the results of calculations and measurements of the required quantities (ionization and excitation frequencies, etc.) performed for a dc electric field $\bar{E}$ [15, 65]. The electron distribution function in the low-density plasma considerably differs from the Maxwell distribution. This function will be close to the Maxwell function when the electron–electron collision frequency ν_{ee} considerably exceeds the collision frequency ν_ε of an electron with atoms in which the electron loses (or acquires) energy. Because the frequency ν_{ee} is proportional to the density of charged particles in the plasma, while the frequency ν_ε is proportional to the atomic density, it is clear that $\nu_{ee} \ll \nu_\varepsilon$ when the ionization degree is small enough. An electron acquiring energy in the electric field imparts it to atoms in elastic and inelastic collisions. Because the cross section for inelastic collisions has the electron-energy threshold, inelastic collisions occur very rarely in weak fields where an electron acquires insufficient energy before a collision. As the electric field strength is increased, the electron energy increases, resulting in the increase in the inelastic collision frequency. It is clear that the mean collision frequency is determined by the electron distribution function. The latter depends on the reduced electric field strength $\bar{E}/N$. The electron distribution function and inelastic collision frequency are usually calculated numerically. Figure 3.6 shows that, as $\bar{E}/N$ increases, the processes of electron energy losses change each other. We are interested first of all in electron energy losses caused by ionization. A breakdown is produced when the electron generation rate exceeds electron losses caused by diffusion [1]:

$$\nu_i(\bar{E}/N) > \nu_d \tag{3.8}$$

where $\nu_d = D_e/d^2$; D_e is the diffusion coefficient of electrons; and d is the characteristic size of the problem. For the optical breakdown, d is the focal spot diameter. The approximate numerical expression $\nu_d \approx 6 \times 10^{14} I^*/\nu d^2$ is given in [1], where I^* is the first excitation potential of an atom (in eV). The elastic collision frequency $\nu = \nu_1 N$, where $\nu_1 = \langle v_e \sigma \rangle$, is equal to 8.5×10^{-8} for He, 2×10^{-7} for Ar, and 4×10^{-7} cm^3/s for Xe. At the atmospheric pressure and $d = 0.05$ mm, we have $\nu_d \approx 2 \times 10^7$ s^{-1} for Ar. It seems that the electric field

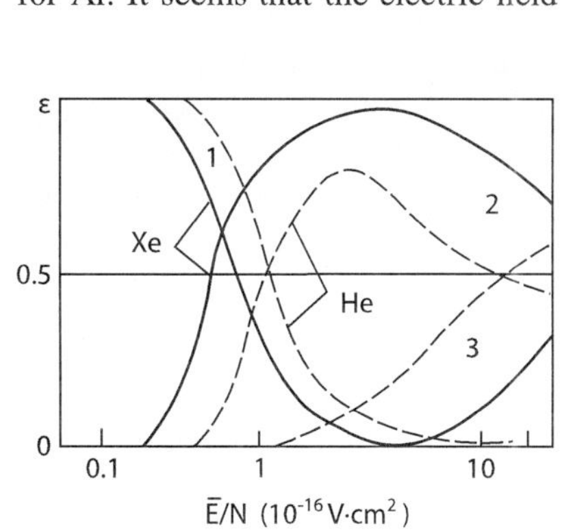

Fig. 3.6 Relative energy input ε to different degrees of freedom of He and Xe atoms: (**1**) translational (elastic) loss; (**2**) electronic excitation; (**3**) ionization

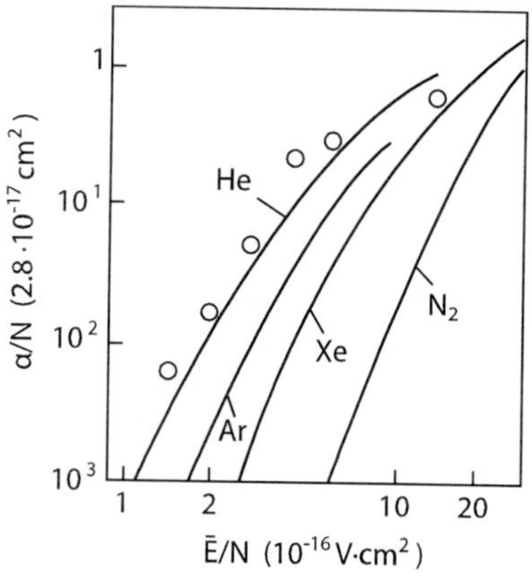

Fig. 3.7 Dependences of the ionization coefficient from the ground state from $\bar{E}/N$ for different gases

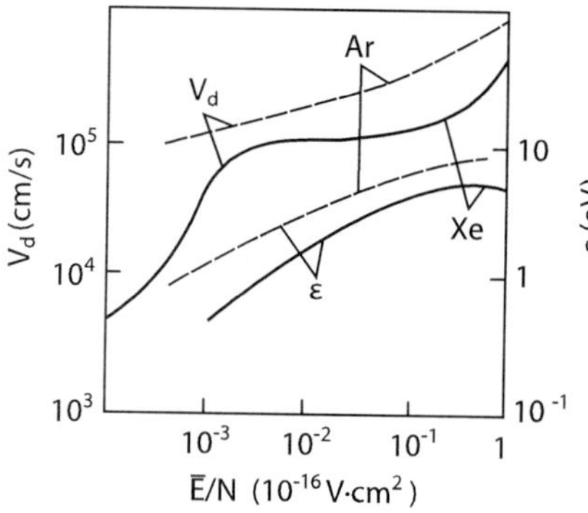

Fig. 3.8 Dependences of the drift velocity and average energy $\varepsilon = D/\mu$ on $\bar{E}/N$

strength at which the ionization frequency is equal to this value of ν_d can be found from Fig. 3.6. However, we will show below that this cannot be done because the obtained value of $\bar{E}/N$ proves to be small. The ionization frequency at low values of $\bar{E}/N$ can be calculated from the ratio α/N, which is measured experimentally or calculated numerically [15, 65]. The ionization frequency is relate to α/N by a simple expression $\nu_i = \nu_d \alpha$, where ν_d is the drift velocity. Figure 3.7 presents the dependences of α/N for helium, argon, and xenon [15], and Fig. 3.8 [15] shows the dependences of the drift velocity and the mean electron velocity on $\bar{E}/N$. By extrapolating the data in Fig. 3.8 to the region $\bar{E}/N = (1 \div 2) \times 10^{-16}$ Vcm^2, we can determine the value of $\bar{E}/N = 5 \times 10^{-16}$ Vcm^2 at which $\nu_i = \nu_d = 2 \times 10^7$ s^{-1}. One can see from Fig. 3.6 that in this case the electron energy is mainly spent to excite the metastable levels of atoms. Knowing this value $\bar{E}/N$, we can calculate the electric field amplitude of laser radiation during

$$E = (\bar{E}/N)\omega/\nu_1 \tag{3.9}$$

Table 3.1 The breakdown threshold in xenon

p, MPa	0.01	0.1	0.5
$\tilde{E}$/N, $V{\cdot}cm^2$	10^{-15}	5×10^{-16}	3×10^{-16}
E, V/cm	3.4×10^5	1.7×10^5	1.1×10^5
E_{exp}, V/cm	–	3×10^5	2×10^5

For argon, $E = 1.2 \times 10^5$ V/cm. By using the expressions for ν_i and ν_d, we can write the breakdown condition in the form

$$\frac{\alpha}{N}\left(\frac{\bar{E}}{N}\right) = \frac{10^{16} I^*}{(dN)^2(\bar{E}/N)} \tag{3.10}$$

where the expression $\mu = e/m_e\nu_1 N$ for the electron mobility is used and I^* is expressed in eV. By using the data in Fig. 3.7 and this approximate expression, we can estimate the breakdown threshold for xenon, see Table 3.1.

The breakdown threshold E_{exp} in xenon irradiated by a CO_2 laser was measured for $\tau \approx 1\,\mu$s pulses and $r_f \approx 0.04$ mm [1]. The difference between calculated and experimental values can be caused by the uncertainty in the rate constants of processes and in the focal spot size. Thus, as the gas pressure was increased by 50 times (from 0.01 to 0.5 MPa), the breakdown threshold with respect to the electric field strength decreased only by three times.

It is interesting to compare breakdown curves for gases with substantially different characteristic values of ν_1 and α/p. Let us take, e.g., helium, argon and nitrogen. The ionization curves for these gases are presented in Fig. 3.7. At a low pressure $p \sim 0.01$ MPa and a focal spot size of ≤ 0.1 mm, when the right-hand side of (3.10) is equal to $(0.3 - 3) \times 10^{17}$ cm^2, the difference between breakdown field strengths in different gases is small (Fig. 3.7). According to (3.9), this difference under these conditions is caused by different values of ν_l. Because ν_l for N_2 and Ar are close, while $\nu_l(He) \approx 0.2\nu_1(Ar)$, breakdown thresholds in N_2 and Ar will be also close, while the breakdown threshold in He will be higher. As the gas pressure is increased, when the right-hand side of (3.10) decreases, the threshold values of $\bar{E}/p$ for these gases will be already substantially different (Fig. 3.7). In the pressure range from 1 to 10 MPa, the threshold breakdown field in nitrogen will be higher than that in argon, and the difference between breakdown intensities in helium and argon will decrease. These trends are manifested in experiments with a ruby laser (Fig. 3.9), although the situation here can be more complicated. This is discussed in more detail in [1]. At a very high pressure $p \approx 100$ MPa, when $\nu > \omega$, the breakdown threshold is determined only by equation (3.10) [in this case, expression (3.9) is no longer valid, and the optical field is in fact static, $E = \bar{E}$] and data in Fig. 3.7. Figure 3.9 shows that experimental and theoretical threshold field strengths in different gases are qualitatively consistent in this pressure region as well.

As for the breakdown of gases by radiation from a CO_2 laser, in this case the difference between the values of frequency ν_l [see (3.9)] for different gases is manifested weaker. The dependence of E on the gas type, as in the case of a ruby

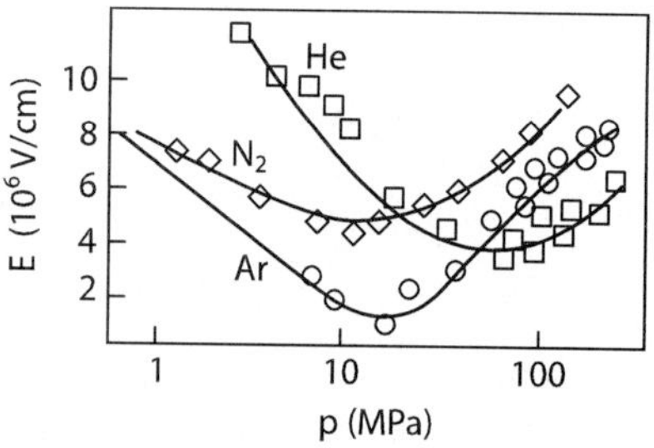

Fig. 3.9 Dependences of the breakdown field strength for 50 ns pulses from a ruby laser on the gas pressure in He, N_2, and Ar

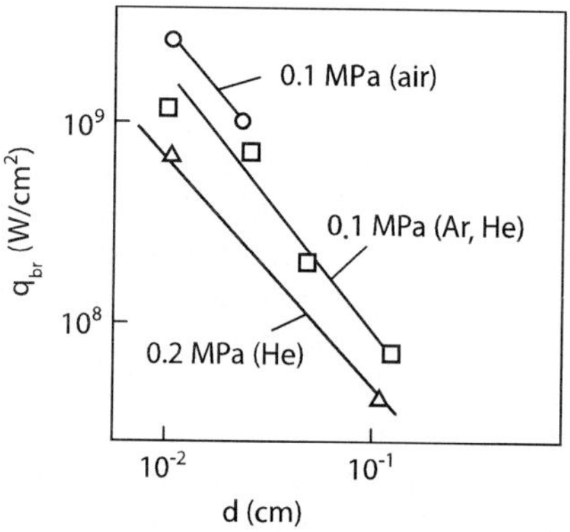

Fig. 3.10 Dependences of the breakdown intensity in air, Ar, and He for $\tau = 200$ ns pulses from a CO_2 laser on the focal spot size

laser, for $\nu > \omega$ is determined by expression (3.10). Figure 3.10 [1] presents the dependence of the breakdown intensity on the focal spot diameter: $q_{\rm br} \sim d^{-1}$ or $E_{\rm br} \sim d^{-1/2}$. One can see from Fig. 3.7 that $\alpha \sim (E/p)^3$ in the range $\bar{E}/N \sim (1 \div 10) \times 10^{-16}$ Vcm^2 studied, so that we obtain that $E \sim d^{-1/2}$, in accordance with the experiment. Note that the breakdown threshold can depend on the dust content of gas under study. Dust strongly reduces the breakdown threshold [1, 16], which weakly depends on the gas type in this case. Figure 3.7 shows that the dc electric field strength at which breakdown occurs in molecular gases is higher than that for atomic gases: $\bar{E}/p$ for nitrogen is twice that for xenon. However, even at such a high strength, only a small part of energy is spent for ionization. This is explained by the fact that a great part of electronic energy in molecular gases is spent for excitation of vibrational degrees of freedom. As a result, the values of $\bar{E}/p$ at which ionization losses are considerable are several times higher in molecular gases compared to atomic gases. Finally, note also another circumstance that can affect theoretical results. The matter is that expression (3.9) for recalculating a dc electric field to an ac field is strictly speaking valid in the approximation of a constant collision frequency, which is independent of the electron energy (and $\bar{E}/p$). The dependence of ν_l (or mobility μ) on E/p exists for many gases (see, e.g., Fig. 3.8). Therefore, ν_l in (3.9) depends on $\bar{E}/p$). Fig. 3.11 shows the ionization constants of argon calculated for dc and ac electric fields [17]. The curves were plotted for the reduced ac electric field strength multiplied by ν/ω ($\nu = \nu_1 N$, $\nu_l = 2 \times 10^7$ s^{-1}cm^3). It follows from

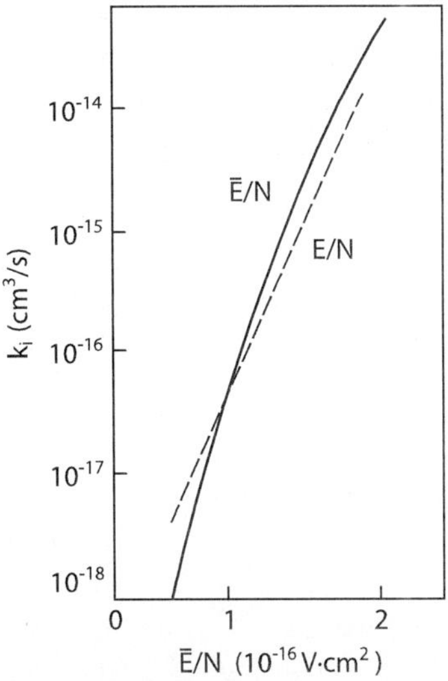

Fig. 3.11 Dependences of the ionization constant $k_i = \langle \sigma_i v_e \rangle$ on $\bar{E}/N$ (d.c. field) and E/N (a.c. field)

the data in Fig. 3.11 that for $\bar{E}/N < 10^{-16}$ Vcm^2, the effective collision frequency is higher than $\nu = \nu_l N$, while for $\bar{E}/N > 10^{-16}$ Vcm^2, the effective frequency is lower than this value. The dependence of ν_l on E/N in this example proves to be weak. However, to calculate accurately the breakdown field strength for other gases and other ranges of values of E/N, it is necessary to calculate $\nu_l(E/N)$ numerically. For approximate estimates, expressions (3.9) and (3.10) can be used by assuming ν_l constant.

3.2.2 Nonequilibrium Mechanism of Optical Breakdown in Gases Near a Target

The study of experimental results on the optical breakdown of gases near a target surface shows that the theory of breakdown in a gas volume considered above proves to be invalid. The heating of the target leads to a decrease in the gas density in a light spot. For example, heating up to 3,000 K reduces the gas density by an order of magnitude, which corresponds to a pressure of 0.01 MPa. If the target well reflects laser radiation ($R \approx 0.9$ for tungsten at $\lambda = 10.6\,\mu$m), the radiation intensity in the focal spot will be twice its initial value. As shown in the previous section, the decrease in the gas pressure by an order of magnitude compared to the atmospheric pressure reduces the breakdown intensity by four times. Taking the reflection of light into account, the heating of the target up to 3,000 K would double the breakdown threshold compared to the breakdown threshold away from the target. However, experiments [5] showed that the breakdown threshold considerably decreased when a target was placed to the laser beam focus. Thus, during the irradiation of a tungsten target in the argon atmosphere, the breakdown occurred at $q_{\text{br}} \approx 10$ MW/cm^2 when he target surface was heated up to 3,000 K. The breakdown threshold in argon without a target at the atmospheric pressure is ~300 MW/cm^2 [1].

The main factor reducing the optical breakdown threshold near a target is the presence of a seed thermal plasma and easily ionized target vapor. Both these factors can be manifested together because the concentration of the seed plasma depends itself on the concentration of target vapor atoms. Consider in more detail the factors determining the seed plasma concentration near a target heated by focused laser radiation. We assume for simplicity that the target temperature T_M in the light spot is specified. The metal vapor density increases with this temperature according to the expression for the equilibrium vapor density

$$N_m = N_0 \exp(-H_b/kT_M) \tag{3.11}$$

where N_0 is a constant depending on the material type and H_b is the vaporization heart of the target material. We consider first the case of high temperatures T_M, when the plasma concentration is high enough, so that the plasma is quasi-neutral (the calculation of a non-quasi-neutral plasma is much more complicated). To provide the plasma quasi-neutrality, the Debye radius should be smaller than the focal beam diameter d, which gives the condition for the density value

$$n > kT_e/4\pi e^2 d^2 \tag{3.12}$$

As we saw by considering the gas breakdown near a target, the electron temperature (or their mean energy) is a few electron volts. It follows from (3.12) that for $d \approx 0.3$ mm the quasi-neutrality appears for $n \approx 10^{10}$ cm^{-3}. The density of the quasi-neutral plasma in the absence of the light field can be determined from usual boundary conditions [18]:

$$j_c - 1/4 n_0 \bar{v}_i \exp(-\Delta\varphi/kT_M) = -D_a \nabla n|_{x=0} \tag{3.13}$$

$$j = j_R \exp(-\Delta\varphi/kT_M) - n_0 \bar{v}_e/2 \tag{3.14}$$

where $j_c = N_m \bar{v}_i \exp[-(I - \varphi)/kT_M]$ is the ion Saha–Langmuir current; j_R is the thermal emission current; I is the ionization potential of an atom; φ is the work function of a metal; n_0 is the plasma density at the potential jump boundary $\Delta\varphi$; $\bar{v}_i$ and $\bar{v}_e$ are the thermal velocities of ions and electrons, respectively; and j is current in the system. Because j is much smaller than $n_0\bar{v}_e/2$, we can determine $\Delta\varphi$ from (3.14) and substitute into (3.13) to

$$j_c - \frac{1}{8} n_0^2 \bar{v}_i \bar{v}e/j_R = -D_a \nabla n \tag{3.15}$$

Equation (3.15) means that a part of ions emitted by a metal returns back to it, while another part diffuse from the focal region. If the density n_0 (and metal temperature) is small, the second term in the left-hand side of (3.15) can be neglected, i.e. the entire flux produced on the surface is carried away by diffusion. The dependence of

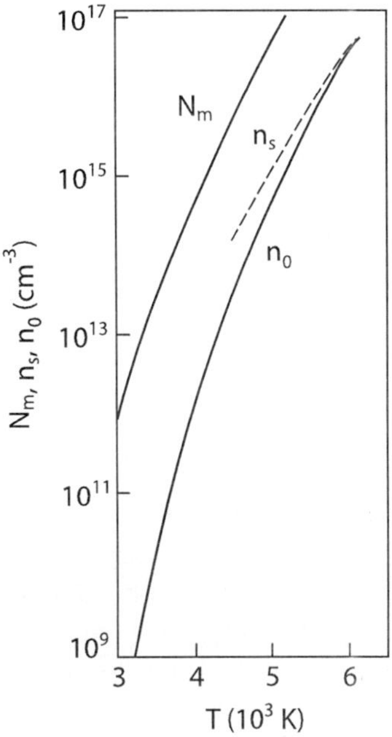

Fig. 3.12 Dependences of the density N_m of tungsten vapor, thermal plasma density n_0, and equilibrium plasma density n_s on the surface temperature (the laser radiation focal spot diameter is $d = 0.3$ mm, the surrounding Ar gas is at the atmospheric pressure, $I_W = 7.9$ eV, $\varphi = 4.54$ eV, and $\sigma \sim 3 \cdot 10^{-16}$ cm^2)

n_0 on T_M can be determined approximately from (3.15) as

$$n_0 \approx 3N_M d\sigma N \exp[-(I - \varphi)/kT_M] \qquad (3.16)$$

As T_M is increased, when the plasma density increases according to (3.16), so that the second term in the left-hand side of (3.15) becomes greater than $D_a n_0/d$, all ions will return back to the metal. The equilibrium will be established, i.e. the plasma concentration will be determined by the Saha formula. One can see from the data in Fig. 3.12 calculated by (3.16) that, if the plasma heating by laser radiation is neglected, the plasma concentration becomes equilibrium at the rather high surface temperature $T_M \approx 5{,}000$ K. In this case, the plasma density is $\sim 10^{15}$ cm^{-3}.

According to Fig. 3.12, $n_0 \approx 10^{10}$ cm^{-3} at $T \approx 3{,}500$ K, and according to (3.12) the plasma is quasi-neutral. The spatial density distribution is described by the ambipolar diffusion equation. The plasma produced on the red hot surface diffuses and disappears away from the target or on the cold part of the target surface adjacent to the focal spot, i.e. the problem of plasma diffusion is a two-dimensional one. To understand the nature of breakdown initiation, it is sufficient to study a simplified model, which, however, preserves the basic properties of the complete problem. We will assume that the plasma are located between two planes, the hot ($x = 0$) and cold ($x = d$). In this case, the ambipolar diffusion equation has a simple form

$$D_a \frac{\mathrm{d}^2 n}{\mathrm{d}x^2} = \nu_i(E) n \qquad (3.17)$$

The boundary condition in the hot plane for $x = 0$ is given by expression (3.15), while in the cold plane for $x = d$ it has the form $n = 0$. For $T_M < 5{,}000$ K, as mentioned above, the second term in (3.15) can be neglected. Then, by solving (3.17) and finding constants from boundary conditions, we obtain

$$n(x) = \frac{j_c}{\sqrt{D_a \nu_i}} \left[\frac{\sin kd}{\cos kd} \cos kx - \sin kx \right] \tag{3.18}$$

where $k = \sqrt{\nu_i / D_a}$. The plasma concentration $n_{\max}$ achieves the maximum value on the target surface (on the hot plane):

$$n_{\max} = j_c (D_a \nu_i)^{-1/2} \tan kd \tag{3.19}$$

One can see from (3.19) that $n_{\max} \approx j_c d / D_a$ for small values of ν_i, i.e. $n_{\max}$ is determined by thermal ionization. As the light intensity is increased, $n_{\max}$ increases and tends to infinity at $kd = \pi/2$, resulting in the breakdown (see curve 1 in Fig. 3.13). By substituting the expression for k into this formula, we see that it coincides with (3.8), where D_e should be replaced by D_a. Because $D_e/D_a \approx 100$ and the density of atoms at $T = 3{,}000$ K decreased by an order of magnitude compared to that at the atmospheric pressure, according to (3.8), the threshold breakdown field coincides in this case with the breakdown field away from the target. Because of the reflection of radiation by the target, the breakdown intensity will be half the gas breakdown intensity in the absence of the target. The experimental breakdown intensity at $T_M = 3{,}500$ K is approximately an order of magnitude lower that he theoretical value [5]. Note that the breakdown intensity in

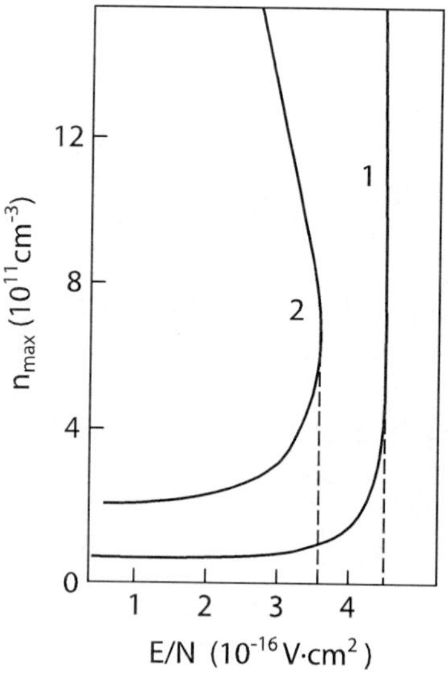

Fig. 3.13 Dependences of the maximal plasma density $n_{\max}$ near a target on the electric field strength at different target temperatures (initial concentration), Ar(W): (**1**) $T_M = 3.5 \cdot 10^3$ K, (**2**) $T_M = 4.1 \cdot 10^3$ K

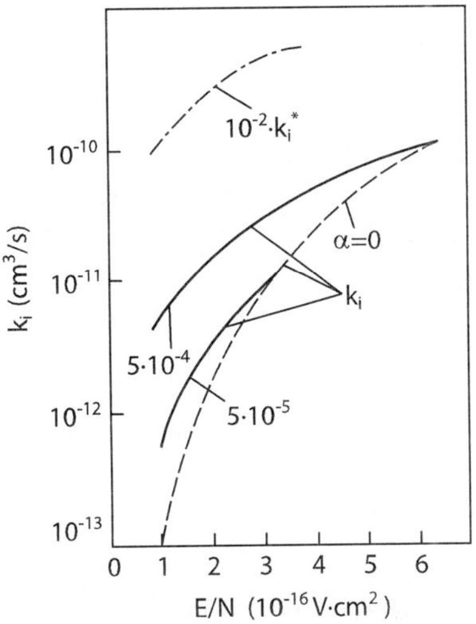

Fig. 3.14 Ionization constants from the ground (k_i) and excited k_i^* states for vapor–gas mixtures for different values of $\alpha = N_m/N$, where N_m is the W vapor density and N is the Ar density (I. Kochetov)

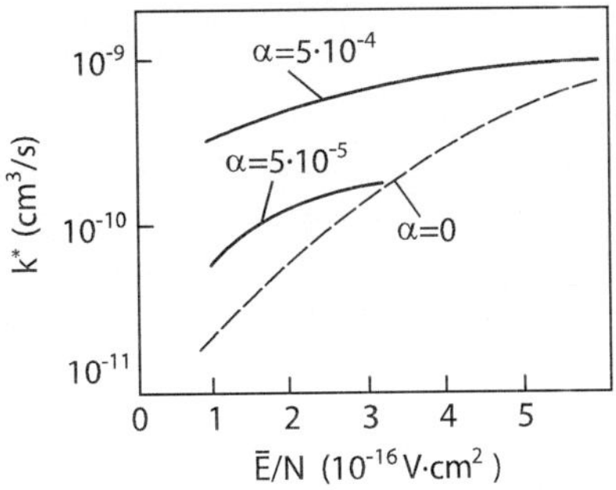

Fig. 3.15 Dependence of the excitation constant of metastable levels on $\bar{E}/N$

the model under study is independent of the target surface temperature and material, although the concentration of the seed plasma depends on these quantities. The vapor density at a temperature of ~3,500 K is very low, as is seen from Fig. 3.12 ($N_m/N \approx 10^{-5}$). In this case, the thermal plasma density is determined by vapor, but the laser radiation field ionizes mainly gas atoms. The latter circumstance is confirmed by the results of calculation presented in Fig. 3.14. A similar dependence also takes place for the total excitation constant of metastable levels of the gas–vapor mixture (Fig. 3.15). Thus, excitation and ionization at $T \approx 3{,}500$ K are mainly determined by gas. It seems that the difference between theoretical and experimental values of q_{br} is explained by the role of step processes, i.e. the ionization of excited argon atoms. The step ionization is described by two equations

$$D^* \frac{d^2 N^*}{dx^2} = k^*(E)Nn - k_q n N^* \tag{3.20}$$

$$D_a \frac{d^2 n}{dx^2} = k_i N n + k_i^* N^* n \tag{3.21}$$

where D^* and k_i* are the diffusion and ionization coefficients of excited atoms, respectively; k^* is the excitation constant; N^* is the density of excited atoms; and k_q is the quenching constant. If the density of excited atoms is lower than its equilibrium value $N_p^* = Nk^*/k_q = N\exp(-I_1/kT_e)$, their approximate concentration can be found from (3.20):

$$N^* \approx k^* N n d^2 / D^* \tag{3.22}$$

where I_1 is the excitation potential of a metastable level. By substituting (3.22) into (3.21) and taking into account that direct ionization can be neglected under the condition

$$n > n_1 = (D^*/d^2) k_i / k^* k_i^* \tag{3.23}$$

we obtain the equation

$$-\frac{d^2 n}{dx^2} = k_i^* k^* N_a d^2 n^2 / D_a D^* = A n^2 \tag{3.24}$$

The study of the solution of this equation showed that the dependencies of A and $\bar{E}/N$ on the maximum plasma concentration $n_{\max}$ are nonmonotonic (see curve 2 in Fig. 3.13). The field corresponding to the maximum values of A is the breakdown threshold. Calculations give the equation

$$k_i(E) = \frac{D_a}{d^2} \frac{n_1}{n_0} \tag{3.25}$$

for determining the breakdown threshold, where n_0 is described by expression (3.16). One can see from (3.23) and (3.25) that the threshold breakdown decreases with increasing temperature because n_0 increases with temperature. For example, for $T_M \approx 3{,}700$ K and $n_0 \approx 2 \times 10^{11}$ cm^{-3}, the value of $\bar{E}/N$, determined from (3.25), is approximately 10^{-16} Vcm^2. This corresponds to the breakdown intensity ~10 MW/cm^2 (Fig. 3.16). As the surface temperature is further increased, the seed plasma density n_0 and $\bar{N}$ increase [see (3.22)]. For $n_0 = n_2 = (D^*/k^*)\exp(-I_1/kT_e)N^*$, N^* achieves the equilibrium value N_e^*. The ionization rate in (3.21) will now depend linearly on the electron density [see (3.17)]. Therefore, the breakdown condition in this case will be similar to (3.8):

$$k_i^*(\bar{E}) N_e^* = D_a / d^2 \tag{3.26}$$

For $T_M > 4{,}000$ K, estimates by (3.26) give the value $\bar{E}/N \approx 0.35 \times 10^{-16}$ Vcm^2, which corresponds to $q_{\rm br} \approx 3$ MW/cm^2. In the spherically symmetric case, system (3.20)–(3.21) was solved numerically with boundary condition (3.15). The geometry of the model is shown in Fig. 3.23. The calculation was performed for

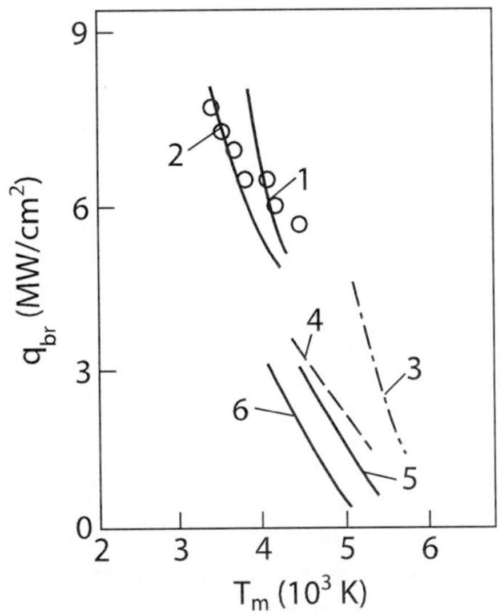

Fig. 3.16 Dependencies of the optical breakdown intensity in gases on the target temperature (φ is the focusing angle) (**1**) calculation by (3.25) for Ar(W); (**2**) numerical calculation by (3.20), (3.21) for Ar(W); (**3**) numerical calculation by (3.27), (3.28) for $\varphi = 0.3$ Ar(W); (**4**) calculation by (3.38) for He(W); (**5, 6**) numerical calculation by (3.27), (3.28) for He(W) and Ar(W); dots are experimental values for Ar(W) [87]

$2r_f = 0.2$ mm and the focusing angle $\varphi \approx 1/7$. The dependences of E and N on the coordinate r were taken into account. The results of the calculation are presented in Fig. 3.16.

It is clear that at high surface temperatures $T_M > 4{,}000$ K and high seed plasma densities, electron–electron collisions become important. The losses of excited and charged particles should be also taken into account, i.e. it is necessary to calculate the level population kinetics of atoms. We will, however, omit this inconvenient temperature range and consider the case of high surface temperatures and high seed plasma densities, when the plasma is in the thermodynamic equilibrium state. As will be shown below, this limiting case is realized at target temperatures 5,000–6,000 K.

In conclusion of this section, we point out the characteristic properties of the breakdown threshold, which follow from basic expression (3.25). The breakdown intensity rapidly decreases with increasing target temperature because n_0 strongly depends on temperature. The value of n_0 at a fixed temperature depends on the target material. Thus, if the breakdown threshold is measured in the same gas but for different targets, the temperatures at which the thresholds are identical should be, according to (3.25) and (3.16), proportional to the vaporization heat $I + \lambda - \varphi$ of the target ion. This property was observed experimentally [5]. Because constants k^* and k_i^* strongly depend on $\bar{E}/p$, the dependence of the breakdown intensity $q_{\rm br}$ on the focal spot size should be weak; the same is true for the pressure dependence of q_{br} at the constant target temperature. However, it follows from experiments that $q_{\mathrm{br}} \sim p^{-1}$ [87].

The model of breakdown of gases near a target taking into account the presence of the seed thermal plasma and ionization of the target vapor well describes as a whole the basic experimental properties.

3.2.3 *Thermal Model of Optical Breakdown in Gases Near a Target*

At high surface temperatures $T_M > 5{,}000$ K, the thermal plasma density is so high that the production of particles due to ionization is balanced not by diffusion but recombination, i.e. the ionization equilibrium takes place. In this case, the breakdown can be described based on the thermal balance equation rather than the balance equation for charged particles. The thermal model assumes that due to heat conduction a hot target heats surrounding vapor and gas. Thermal ionization is accompanied by the heating and ionization of the vapor and gas in the laser beam [19]. The interaction of the laser beam with the target is described by a system of two thermal balance equations for electrons and a heavy component (ions and neutral atoms):

$$-\frac{1}{r^2}\frac{\mathrm{d}}{\mathrm{d}r}r^2\kappa_e\frac{\mathrm{d}T_e}{\mathrm{d}r} = q\frac{R^2}{r^2}\mu - \frac{3m_e}{M}n_s k(T_e - T)\bar{v}_e\left(N\sigma_e + \frac{M}{M_M}n_s\sigma_c\right) \tag{3.27}$$

$$-\frac{1}{r^2}\frac{\mathrm{d}}{\mathrm{d}r}r^2\kappa\frac{\mathrm{d}T}{\mathrm{d}r} = \frac{3m_e}{M}n_s k(T_e - T)\bar{v}_e\left(N\sigma_e + \frac{M}{M_M}n_s\sigma_c\right) \tag{3.28}$$

where $\kappa_e(T_e)$ and $\kappa(T)$ are the heat conductivities described for an atomic gas by expressions (3.1); T_e and T are the electron and gas (ions and neutral particles) temperatures, respectively; n_s is the equilibrium concentration determined by the Saha formula (3.4), in which $\exp(-I/kT)$ is replaced by $\exp(-I/kT_e)$ [18]; $\mu(T_e)$ is the absorption coefficient for laser radiation [see (3.2)–(3.3)]; M and M_M are masses of the gas atom and ionized atom of the target. The system of equations (3.27) and (3.28) is written for the spherically symmetric case simulating the real picture of the optical gas breakdown near the target. We assume that the breakdown occurs near the surface of a metal sphere of radius R. The density of the metal vapor, which determines the plasma density, is assumed equilibrium, as usual, and its change due to diffusion is neglected. The system of equation (3.27) and (3.28) is supplemented by the boundary conditions

$$T_e(R) = T(R) = T_M; \quad T_e(\infty) = T(\infty) = T_\infty \tag{3.29}$$

where T_∞ is the gas temperature away from the target. Equations (3.27) and (3.28) neglect diffusion and radiation losses due to their smallness.

The absence of the discontinuity between the electron and surface temperatures [see (3.29)] follows from the consideration of the balance of heat flows for electrons, similarly to the balance of particles (3.13) [18]. For the potential jump $\Delta\varphi$ slowing down electrons escaping from a metal, we have

$$(T_e - T_M)T_M^{-1} = \sqrt{m_e/M_M} \ll 1 \tag{3.30}$$

The system of equations (3.27), (3.29) describes the transfer of energy from the laser beam to electron gas accompanied by its removal outside the boundaries $r_{1,2}(r_1 = R,\ r_2 = \infty)$ due to the electron heat conductivity and transfer to the heavy component (gas, vapor), where it is also removed outside the boundaries, but already due to the gas heat conductivity. It is clear that, if electrons slowly impart their energy to atoms and ions, the value of T_e will be considerably greater than T. The maximum difference $T_e - T$ can be estimated by equating in (3.27) the heat release and losses caused by elastic collisions:

$$T_e = T + \frac{4\pi e^2}{cm\omega^2 k}\frac{M}{3m}\frac{R^2}{r^2}q = T + T_1 \tag{3.31}$$

In argon for $q \approx 1$ MW/cm^2 and $r = R$, the value of T_e exceeds T by 6,000 K; in xenon, this difference is even greater. Because the electron heat conductivity is proportional to the plasma concentration, the atomic heat conductivity at low temperatures will considerably exceed the electron heat conductivity (they become equal under the condition $n_s/N > (m/M)^{1/2}$. Thus, the electron energy loss at low temperatures is mainly determined by elastic collisions followed by the heat removal due to the gas heat conductivity. Because $T_e = T$ on the target surface, the electron heat conductivity becomes important near a metal.

In the case of a molecular gas, the energy loss in collision of an electron with molecules will be more significant. To estimate this loss, the term describing elastic losses in (3.27) and (3.28) is multiplied by the coefficient $\delta \simeq$10–20. The difference between the electron and molecular temperatures in a molecular gas is small [65].

3.2.4 *Theoretical Model*

It is convenient to study the general properties of the solution of the obtained system by using a simplified model of the phenomenon. We will assume that the plasma temperature is low enough to neglect the influence of the electron heat conductivity. As in the previous section, we consider a plane case, when one of the plates is heated up to the temperature T_M, while another, located at a distance of R from the first plate, has the zero temperature. We assume that the gas heat conduction is independent of temperature. By introducing the dimensionless variables $\theta = IT_M(T/T_M - 1)/(T_M + T_1)^2$ and $\xi = r/R$ and expanding the exponent in the expression for the ionization degree neat T_1 (see (3.4) and [20]), we obtain from (3.28)–(3.31)

$$d^2\theta/d\xi^2 = -D \exp\theta \tag{3.32}$$

$$\theta(1) = 0, \quad \theta(\infty) = -IT_M(T_M + T_1)^2 \tag{3.33}$$

where

$$D = \frac{q_{br}R^2}{\kappa}\frac{1}{2}\frac{2-\alpha}{(T_M + T_1)^2}\mu(T_M + T_1) \tag{3.34}$$

The obtained system (3.32) and (3.33) describes a problem of asymmetric ignition. The Zel'dovich ignition condition [21] has the form

$$D > \frac{1}{32}\frac{I^2T_M^2}{(T_M + T_1)^4} \tag{3.35}$$

Equation for q_{br}, we find asymptotics for $q_{br} \to 0$ and $q_{br} \to \infty$. For small q_{br} and large T_M the critical intensity q_b at which the plasma temperature drastically increases, resulting in the optical breakdown, can be found from (3.34) to (3.35). By solving the transcendent we have

$$q_{br} = B\exp\left(\frac{I + L_b}{2kT_M}\right) \tag{3.36}$$

where

$$B = \frac{2-\alpha}{64\pi}\left(\frac{T_M}{T_M + T_1}\right)^2 \frac{I\kappa}{R^2}\frac{cm\omega^2}{e^2\bar{v}_e\sigma_e}\left(\frac{h^2}{2\pi m}\right)^{3/4}\frac{(kT_M)^{1/4}}{p\sqrt{N_0}} \tag{3.37}$$

In the opposite case of large q_{br} and small T_M, we obtain

$$q_{br} = \left[\frac{I}{2k}\left(\ln\frac{q_{br}}{B} - \frac{L_b}{2kT_M}\right)^{-1} - T_M\right]\frac{cm\omega^2k}{4\pi e^2}\frac{3m}{M} \tag{3.38}$$

One can see from (3.36) to (3.38) that q_{br} drastically decreases with increasing temperature T_M. The surface temperature T_M of a bulky target at rest is related to the intensity by the expression (see Sect. 1.3)

$$T_M = \alpha qR/\kappa_m \tag{3.39}$$

where α is the absorption coefficient of the target surface target for laser radiation and κ_m is the heat conductivity of the metal. By substituting (3.39) into (3.38), we obtain the final solution

$$q_{br} = \left[\frac{I}{2k}\left(\frac{\alpha R}{\kappa_m} + \frac{4\pi e^2}{cm\omega^2k}\frac{M}{3m}\right)^{-1} + L_b\kappa_m/2k\alpha R\right]\ln^{-1} q_0/B \tag{3.40}$$

For $q > q_{\text{br}}$, temperatures T and T_e increase jumpwise, the plasma density rapidly grows, and the atomic heat conductance (as we will call for brevity the atomic an ionic heat conductivities) becomes smaller than the electron heat conductivity (see Fig. 3.2), which determines mainly the value of T_e. The role of elastic collisions becomes insignificant. Knowing T_e, we can find from (3.28) the temperature distribution for heavy particles. At the electron temperature of $(1.5–2)\times 10^4$ K, the surrounding gas will be completely ionized, and $\Delta T = T_e - T$ can be easily estimated from (3.28) as

$$\frac{\Delta T}{T_e} \approx \frac{1}{12}\sqrt{\frac{M}{m}}\frac{T_e^2}{R^2}\sigma_e^2 n_s^2 \tag{3.41}$$

For argon at the atmospheric pressure and $R = 1$ mm and $T_e \approx 2 \times 10^4$ K, we obtain $\Delta T/T \approx 0.001$. This means that both temperature T and T_e drastically increase after breakdown and the continuous optical discharge plasma becomes isothermal. However, as one can see from (3.41), as the intensity further increases (and, therefore, T_e also increases), the electron temperature can again become different from T.

The optical breakdown model considered here uses a substantial assumption that the plasma is in equilibrium. Let us estimate the applicability of this assumption in the absence of laser radiation. Expression (3.4) for the ionization degree can be used for $T_e \neq T$ if the ionization frequency ν_i exceeds the diffusion loss frequency in the plasma ν_d. For the plasma temperature and density typical for the pre-breakdown state near the surface of refractory metals, ionization and recombination are stepwise. This is fulfilled for $\nu_1 > \nu_i$, i.e. when the excitation frequency ν_1 of the first level of a metal exceeds the ionization frequency. Thus, the conditions

$$\nu_1 > \nu_i > \nu_d \tag{3.42}$$

should be fulfilled. The frequencies are described by expressions

$$\nu_1 = N_m\sigma_1 v_e \exp(-I_1/kT_e), \quad \nu_i = \beta n_s^2, \quad \nu_d = D_a/R^2$$

where σ_1 is the excitation cross section for the first level; β is the step recombination coefficient; for tungsten $I_1 = 2$ eV and $\sigma_1 \sim 10^{-16}$ cm^2 and β depends on the individual properties of the target material.

The values of β for metals of interest for us are not available in the literature. We can make the estimate by using the general theoretical expression [22]:

$$\beta = 5e^{10}m^{-1/2}(kT_e)^{-9/2} \tag{3.43}$$

Let us estimate the applicability of the model for tungsten in the argon atmosphere for $T_m = 5{,}000$ K. It follows from the data in Fig. 3.12 that $N = 1.5 \times 10^{18}$ cm^3, $n_s \approx 10^{15}$ cm^{-3}, and $N_m = 2 \cdot 10^{17}$ cm^{-3}. For these values of parameters, the characteristic frequencies are $\nu_I \approx 3\times10^5\,\mathrm{s}^{-1}$, $\nu_i \approx 3{\cdot}3\times10^4\,\mathrm{s}^{-1}$, $\nu_d \approx 10^3\,\mathrm{s}^{-1}$, and $n_s/N \approx 10^{-3}$, i.e. condition (3.42) is fulfilled.

For lower values of T_m, the equilibrium state of the plasma is violated because of the drastic temperature dependence of the ionization and recombination rates. In this case, the consideration of the plasma diffusion [23] and the deviation from the equilibrium distribution function, etc., becomes important. The dependence of q_{br} on the target temperature was studied in [24] within the framework of the thermal model taking into account plasma diffusion for $T_m < 5{,}000$ K.

3.2.5 Numerical Calculation of the Thermal Model

We presented above the optical breakdown theory in the case of plane geometry. In addition, the neglect of the electron heat conductivity overstated the electron temperature near a metal. To study the thermal model of the breakdown under conditions closer to physical reality, it is necessary to calculate numerically the complete system of equations (3.27)–(3.29) of the model. The problem was solved by the establishment method [20], taking into account the time derivative, which allows one to study the breakdown development dynamics and to obtain the high-temperature solution describing the developed optical discharge (Fig. 3.17). Figure 3.17 presents the results of calculations for argon at $q = 0.5$ MW/cm^2. When the temperature of the tungsten surface was maintained at ~5,000 K, the intensity was lower than q_{br}. One can see from the distributions of T_e and T (curves 2 and 3) that in this case the electron temperature at a distance of 1 mm from the surface is higher than the temperature of atoms by 2,000 K. The maximum of T_e is located at a distance of ~0.5 mm from the surface. The difference of the electron temperature from T in xenon in a similar situation, as expected, is considerably greater. When the surface temperature is increased up to the values at which q_{br} becomes smaller

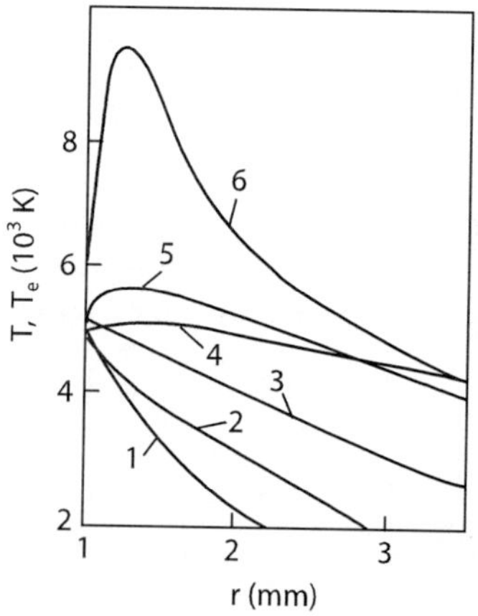

Fig. 3.17 Temperature distributions for electrons and argon atoms for $q = 0.5$ MW/cm^2 (tungsten target); (**1**) initial distributions of T_e and T; (**2, 3**) stationary distributions of T_e and T for $T_M = 5{,}000$ K and $q < q_{\mathrm{br}}$; (4–6) distributions of T_e for $T_M = 5{,}200$ K and $q > q_{\mathrm{br}}$ at instants $t = 0.1$, 0.2, and 0.3 ms

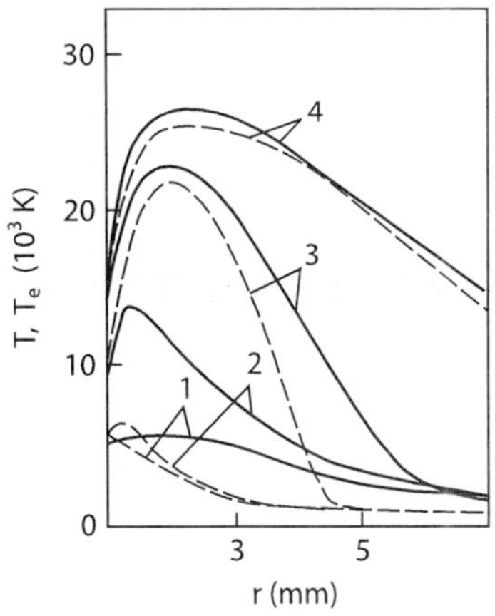

Fig. 3.18 Temperature distributions of electrons (*solid curves*) and atoms (*dashed curves*) in argon during the breakdown development ($q = 0.5$ MW/cm^2); **(1–3)** successive time instants $t_1 < t_2 < t_3$; **(4)** stationary state

than q, the electron temperature rapidly increases (see Fig. 3.17). The plasma density grows even faster. The electron-ion collision frequency increases. This leads to a drastic increase in the atomic temperature, which approaches the electron temperature (Fig. 3.18). The difference between these temperatures in the stationary state is small. In the case shown in Fig. 3.18, plasma transmits to the surface ~80% of the incident radiation. Figure 3.16 presents the dependencies $q_{\text{br}}(T_m)$ calculated for argon and helium near a tungsten target. One can see that q_{rb} drastically increases with decreasing T_m. Calculations showed that, for the same surface temperature, q_{br} decreases with increasing the gas molecule mass (He, Ar, Xe), in accordance with dependence (3.38) and experiment [5]. The calculation by (3.38) for helium (see the dotted curve in Fig. 3.16 is in good quantitative agreement with the numerical calculation. Thus, the theoretical model qualitatively adequately describes the basic properties of the solutions of the complete system of equations of the thermal breakdown.

Let us discuss the dependence of q_{br} on the focal spot radius R. In the isothermal plasma of molecular gases, as follows from (3.36), the optical breakdown depends only on the laser power. If qR^2 is smaller than a critical value, the breakdown will not occur. The situation changes in atomic gases. As follows from (3.38), q_{br} decreases with increasing R much weaker than in (3.36). This is also confirmed by exact calculations. It was shown in experiments [5] that the breakdown intensity in molecular gases decreases with increasing the focal beam radius as $\sim R^{-2}$, whereas in atomic gases the value of q_{br} is independent of R.

Figure 3.17 shows that the size of a region with the high electron temperature is 1–2 mm. This means that, if a laser beam propagates within an axially symmetric cone, the "side thermal losses" from plasma can become considerable. To study approximately the influence of these losses, it is necessary add the terms $\kappa_e T_e/(\varphi r)^2$, and $-\kappa T/(\varphi r)^2$ into the right-hand sides of (3.27) and (3.28), respectively (here, φ is the beam focusing angle, see Fig. 3.23). Figure 3.16 presents the dependence $q_{\text{br}}(T_M)$ calculated for the case $\varphi = 0.3$ (a tungsten target in argon).

The exact dependence for an axially symmetric focused beam is probably located between curves 3 and 6.

Figure 3.16 presents the dependencies of the breakdown intensity on the target intensity [5]. One can see that theoretical dependencies are in qualitative agreement with experiments as a whole. The estimates of the breakdown intensity in xenon neat a tungsten target [25] also give values from ~1 to 10 MW/cm^2.

3.2.6 *Optical Breakdown of Chemically Active Gases Near a Target*

Many molecular gases are chemically active at high temperatures. First of all this concerns air because oxygen forms oxides with metal vapors. Nitrogen forms nitrides with metal vapors. An important role of chemical reactions in the optical breakdown near a target was pointed out, e.g., in [5, 28]. The binding of atoms in an oxide is not sufficient for increasing the plasma concentration because many oxides have ionization potentials exceeding these of metal atoms. However, due to the oxidation reaction the total number of metal atoms (bound in the oxide and free) can considerably increase at a certain temperature. This can lead to an increase in the plasma concentration and decrease in the optical breakdown threshold. A more considerable increase in the plasma concentration can occur in the absence of chemical equilibrium in the system.

The investigation of the nonequilibrium case taking into account the ionization kinetics is complicated because the values of corresponding ionization constants are unknown. The nonequilibrium mechanism of chemiionization can be qualitatively discussed by the example of the nitrogen association reaction with the well-known constants [26]. Nitrogen atoms at $T \approx 4{,}000$–$5{,}000$ K can form either molecules (constant k_0 with the participation of three bodies) or molecular ions (constant k_e with the participation of two bodies). The density of molecules n_{Me} in the equilibrium considerably exceeds the density n_{ee} of molecular ions. At the initial moment only nitrogen atoms are present (whose density is n.) After "switching on", reactions will proceed over these competing channels. Although $\partial n_M/\partial t > \partial n_e/\partial t$ at the atmospheric pressure, the total transition time τ_M to the stationary state in the neutral association reaction proves to be greater than this time τ_e in the associative ionization reaction: $n_{\mathrm{Me}}/k_0 n^3 > n_{\mathrm{ee}}/k_e n^2$ because $n_{\mathrm{Me}}^p \gg n_{\mathrm{ee}}^p$. As a result, the ion concentration $n_e = n\sqrt{k_e/\beta}$ for $\tau_M > t > \tau_e$ can considerably exceed the equilibrium value n_{ee} (β is the dissociative recombination coefficient). The calculations show that $n_e/n_{\mathrm{ee}} > 10$ at $T = 5{,}000$ K (Fig. 3.19). This effect can take place for a mixture of metal and gas vapors during optical breakdown near a target. Figure 3.20 shows the decrease in the breakdown threshold (on the ordinate the breakdown power is plotted because it is independent of a light spot radius) with increasing the oxygen concentration in nitrogen. This demonstrates a strong influence of oxidation processes on the breakdown power.

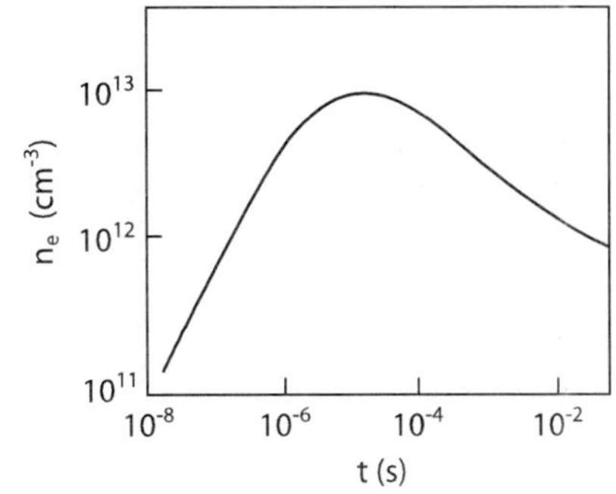

Fig. 3.19 Time dependence of the electron concentration during the association of nitrogen atoms (atmospheric pressure, $T = 5{,}000$ K)

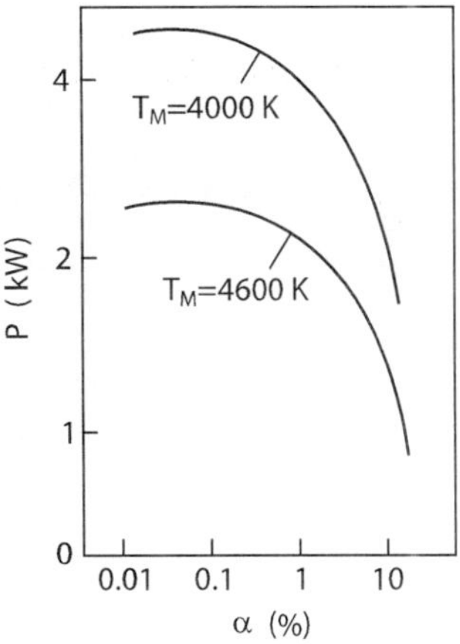

Fig. 3.20 Dependencies of the breakdown threshold in gases near a tantalum target for a CO_2 laser on the oxygen concentration in nitrogen $\alpha = N_{o_2}/N_{N_2}$ [87]

3.2.7 Optical Breakdown During Laser Welding

We considered above the situation when the surface temperature exposed to laser radiation was rather high (>3,500 K) and the vaporization of the material was weak. Such a situation is realized upon irradiation of refractory materials. During the processing of steels, aluminum alloys and other non-refractory materials, developed vaporization regimes can be realized without optical breakdown as well. However, vapors can be heated up to temperatures exceeding the boiling temperature of the material. A vapor cloud or jet is mixed with the surrounding gas (Fig. 3.21a, top). Such a hot gas–vapor erosion plume is formed during welding with the help of low-power radiation. At the beginning of welding, when a laser beam has not yet penetrated into a sample, an intense erosion plume is formed perpendicular to the target surface. When the laser radiation intensity exceeds a certain value, the optical breakdown in the vapor jet is observed. In [29], a beam from a 5-kW quasi-cw CO_2 laser was focused on a sample surface at an angle of ~45°. When the breakdown of the erosion plume occurred, the second bright plasma plume was formed along the beam in the surrounding gas (Fig. 3.21a, middle). As the

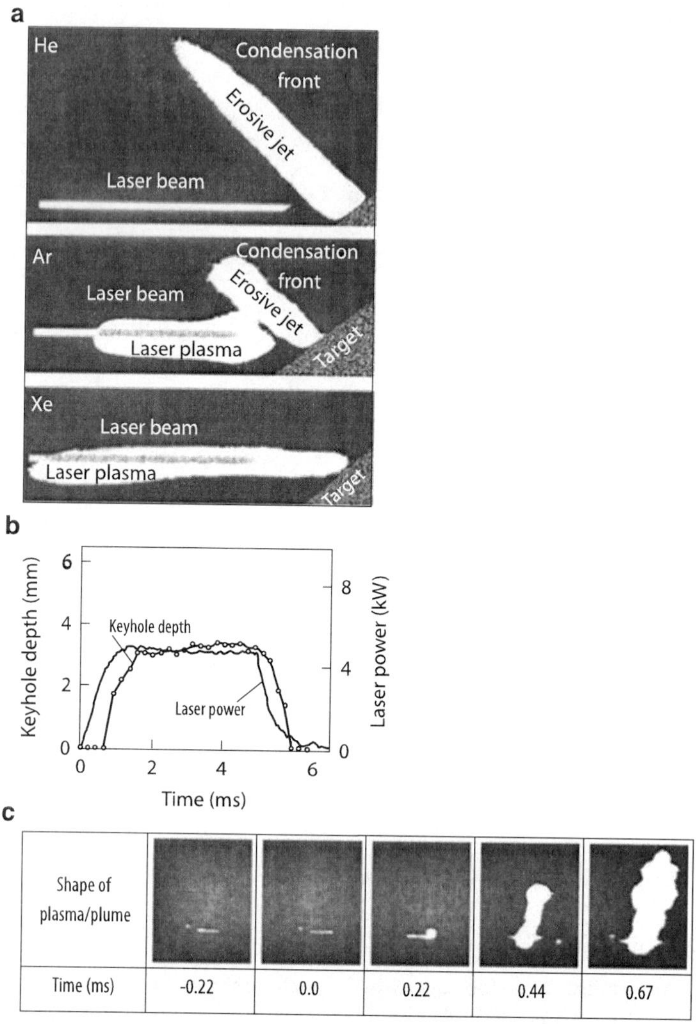

Fig. 3.21 (**a**) Plasma plume structures generated near a target in He, Ar, and Xe by an inclined laser beam; (**b**) Variations in the welding depth and laser beam shape (a rectangular laser beam was used in experiments); (**c**) Plasma plume generation during laser welding of a plate

laser power was further increased, the plasma plume propagated along the laser beam and "burned" in the surrounding gas (Fig. 3.21a, bottom). A similar picture is observed at a constant radiation power but for different gases having different breakdown thresholds. This is demonstrated in Fig. 3.21a [84]. In the case of the normal incidence of a laser beam on a sample surface, the picture is similar as a whole. An optical discharge is also produced due to breakdown in the surrounding gas. During welding by periodic millisecond laser pulses with a small of-duty ratio

(∼2), each pulse produces a keyhole anew (see Chap. 9). The question arises of what is produced earlier, a keyhole or a plasma plume? It follows from experiments [78] that a plasma plume appears earlier than a gas–vapor keyhole. Figure 3.21b,c present the laser pulse shape, the time dependence of the keyhole depth, and the generation of a plasma plume during the welding of steel S304 by 4-kW pulses from a Nd laser. The focal spot diameter was ∼1.1 mm (see details in Chap. 9). In the case described, a plasma plume weakly affected welding. In the case of continuous welding, the plume departed from a sample after breakdown, and a new plume is formed on the sample surface (Fig. 3.22a, type 3). Finally, when the laser beam power is high enough, the optical discharge is burning stably in the surrounding gas (Fig. 3.22a, type 4 [30]). Depending on conditions and the surrounding gas, such a plume can move away from the sample by a distance up to 10 cm [31]. The breakdown of the erosion plume during welding of metals occurs when the CO_2 laser power achieves 7–8 kW [4]. A comprehensive summary of experimental

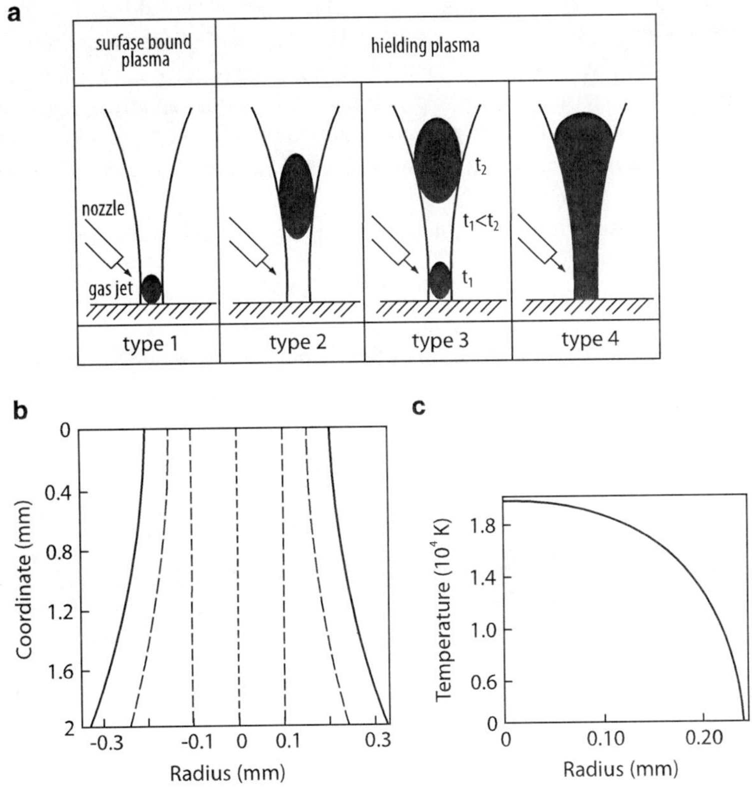

Fig. 3.22 (**a**) Different shapes of plasma plumes during laser welding of metals. The laser power increases from left to right [30]; (**b**) Thermal defocusing of a laser beam in the plasma of a gas–vapor keyhole during welding; (**c**) Plasma temperature distribution in a gas–vapor keyhole during laser welding

studies of plasma phenomena observed upon deep penetration of a laser beam in metals is presented in [32]. The erosion plume has smaller dimensions and the lower plasma temperature and density. The parameters of this plume depend on the target material. Thus, in the case of a steel sample irradiated by a 2.5 kW cw CO_2 laser ($d_f = 0.3$ mm) measurements by the spectral lines of iron give the temperature ~5,000 K. The plume temperature above an aluminum sample is higher (~6,000 K). Note that in the latter case the aluminum vapor density is two orders of magnitude lower than the density corresponding to the atmospheric pressure. Thus, the plasma density in this plume is low ($\sim 5 \times 10^{15}$ cm^{-3}). The measurements were performed at a distance of 5 mm from the sample surface [41]. The low vapor density is probably explained by the intense mixing of vapor with air. A considerable decrease in the plume temperature and density with distance from the sample surface was also observed in other papers. Thus, the plume temperature directly above an iron target in helium was 6,000 K, while at the point higher by 1 mm, it was 5,200 K [33].

Although the parameters of an erosion plume are mainly determined by the target vapor, the surrounding atmosphere can also affect somewhat these parameters. Thus, the plasma density at the keyhole input in an aluminum alloy welded by a CO_2 laser in the helium atmosphere was $n_e = 3.2 \times 10^{16}$ cm^3 and in the argon atmosphere $n_e = 7 \times 10^{16}$ cm^3. Because such a plume is burning in the surrounding and partially in a mixture with external gases, the properties of the latter can affect, e.g., the heat balance. The heat conduction of helium is higher than that of argon. In addition, the absorption coefficient μ in argon is also higher because the frequency of collisions of electrons with argon atoms is twice that with helium atoms (see Sect. 3.1). Therefore, the plume temperature in argon is higher that in helium, resulting in the higher plasma density in argon. It was shown in experiments [34] that the parameters of the erosion plume generally weakly depended on the laser radiation wavelength (CO_2 and CO laser were used), although it is known that $\mu \sim \lambda^2$. This is probably explained by the fact that the CO laser radiation intensity was twice that of the CO_2 laser [34].

The construction of the breakdown model for the erosion plume is complicated by the problem of description of a vapor–gas jet in a keyhole during welding.

3.3 The Numerical Model of an Erosion Plume During Welding

Welding is usually performed in a medium of protective gases, which are supplied to a welding zone with the help of gas nozzles of different shapes (see Chap. 4). Protective gas jets collide with a jet of the sample material vapor flying away from a gas–vapor keyhole. As a result, turbulent mixing occurs and plasma processes are developed in a mixture of the metal vapor and protective gas. Because of the complexity of these processes, we describe them qualitatively based on the model approach. Consider the dependence of the mixture temperature on the content of target vapor in the mixture. The heat conduction equation for a isothermal equilibrium plasma, obtained by adding (3.27) and (3.28), was solved in the cylindrical geometry [86]. The radius of a cylinder was twice the laser beam radius. The focal spot radius was 0.2 mm and the cylinder radius was 0.4 mm (Fig. 3.22b). The temperature on the cylinder walls was set equal to 6,000 K. The laser beam attenuation due to absorption was neglected in the calculation of the temperature field. The total gas pressure was assumed equal to the atmospheric pressure. The CO_2 laser power was 2 kW. The numerical calculations of the heat conduction equation showed that if the partial pressure of iron vapor is lower than 0.22 atm, the maximum temperature of a plasma cloud only slightly differs from 6,000 K. As the vapor pressure was increased above 0.22 atm, the situation changed considerably and temperature approximately doubled. The thermal breakdown in the erosion plume takes place. As the vapor pressure was further increased above 0.24 atm, the second temperature jump up to 18,000 K occurs, resulting in the ionization of the protective argon gas (Fig. 3.22c). Thus, a breakdown in the surrounding gas occurs. The picture resembles the thermal breakdown of gases near a refractory target considered above, when the breakdown threshold q_{br} also decreases with increasing the target temperature and, hence, with increasing the metal vapor pressure (Fig. 3.17), and for $q > q_{\text{br}}$ the gas temperature near a solid surface drastically increases (Fig. 3.18). Here, it is appropriate to present the results of calculations of the laser beam refraction after breakdown in the erosion plume. Calculations performed in the ray-optics approximation under considered conditions showed that even in the erosion plume, despite its small size, the beam radius increases due to refraction by factor of $\sim$1.5 (Fig. 3.22b). This, of course, affects welding parameters. The influence of a protective gas on the welding efficiency will be described in detail in Chap. 4.

3.4 Optical Discharge Burning Near a Sample Surface

Consider the simplest case when the role of a target in the development of a plasma plume after breakdown is minimal. After the gas breakdown by laser radiation focused on a target, a plasma cloud at temperature $\sim$20,000 K is produced near the target surface for the time 1 ms (Fig. 3.18). This plasma cloud will move away from the target with time. Then, it will either stop at some distance from the target and will stabilize or will continue to propagate further away. In the first case, the so-called continuous optical discharge (COD) is formed, while in the second one, a light combustion wave (LCW) is produced [1]. Which of these possibilities is realized - depends on the experimental conditions. For example, if the focusing angle is large enough, the COD will be established after the breakdown, and when this angle is small, the LCW will be produced after the breakdown. The combustion wave either decomposes with time or transforms to the COD near optical elements such as lenses and mirrors. Below, we will discuss these cases of optical discharges in more detail. Let us begin with the COD.

3.4.1 Theoretical Models of a Continuous Optical Discharge

A freely burning COD was studied after the advent of kilowatt cw CO_2 lasers. A spherically symmetric model of the optical discharge is considered in [1]. Although the discharge is burning in reality with a light cone, some of its properties observed in experiments are well described by the spherical model. The COD burning within a light cone was studied numerically in the one-dimensional approximation [35]. The presence of a target near the laser beam focus leads to the appearance of reflected light (if the target well reflects light) and to additional cooling. If the target is evaporated in this case, this also can affect the COD. If the COD is established at a great distance from the target (more than its thickness) and the target well absorbs radiation but is not evaporated, such a discharge will not differ from the freely burning COD. We consider theoretical models of a discharge burning near a target in gas and compare the expressions obtained for estimating discharge parameters with two-dimensional calculations.

As shown in Sect. 3.2, the optical breakdown plasma has a low temperature (see Fig. 3.18). By adding (3.27) and (3.28) at $T_e = T$, we obtain the thermal balance equation for this plasma:

$$\frac{1}{r^2}\frac{\mathrm{d}}{\mathrm{d}r}r^2\kappa\frac{\mathrm{d}T}{\mathrm{d}r} = \frac{\mu P}{\pi(r\varphi)^2}\exp\left(-\int_r^\infty \mu dr\right) - \gamma\mu T - \frac{\bar{\kappa}T}{(r\varphi)^2} \tag{3.44}$$

where φ is the beam focusing angle and $\bar{\kappa} = T_K^{-1}\int_0^{T_K}\kappa dT$, $T_K \approx 20{,}000\,\mathrm{K}$. Expression (3.44) takes into account the absorption of laser radiation in the optical breakdown plasma. The latter term in the right-hand side of (3.44), as in the consideration of the breakdown, describes approximately side thermal losses due to heat conduction (Fig. 3.23). Radiation losses are expressed for convenience in terms of the absorption coefficient, this relation following from (3.3) to (3.6), (3.7). The numerical factor γ depends on the expression [(3.6) or (3.7)] used in calculations. For example, expression (3.6) is more convenient for air, and in this case the value of γ was set equal to $\sim$1 W/cm$^2\cdot$ K. The light spot radius on a target in the model is $r_f = R\varphi$ (see Fig. 3.23).

Equation (3.44) should be supplemented with boundary conditions for temperature. Because the maximum temperature of the optical discharge is rather high ($T \sim 20{,}000$ K), the target and gas temperature at infinity can be set equal to zero:

$$T(R) = T(\infty) = 0 \tag{3.45}$$

Model (3.44) neglects the heat carrying out due to the natural convection of non-uniformly heated gas. The Peclet number $Pe = u_\kappa l_\parallel/\chi_0$, characterizing the ratio of the convective and conductive heat flows, is smaller than unity under conditions under study. Because of the gas flow acceleration in the COD, convection velocity should be increased by a factor of $\sqrt{\rho_0/2\rho_h}$ [36], where ρ_0 and ρ_h the density of the cold and hot air, respectively. Numerical estimates give $u_\kappa = 6$ m/s, which is close

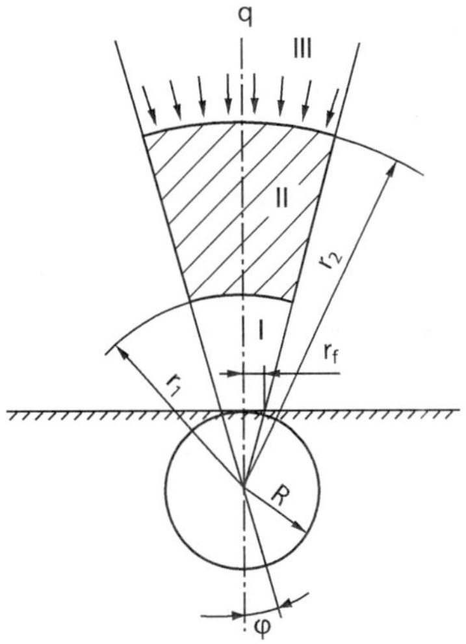

Fig. 3.23 Schematic presentation of an optical discharge burning near a target

to numerical calculations of the convection velocity in the COD for $P = 5\,\mathrm{kW}$ (Fig. 3.24). The experimental studies of a COD in argon by using a $P = 1.5\,\mathrm{kW}$ CO_2 laser showed that velocities in the range 0.8–32 cm/s, which are typical for natural convection, weakly affect the discharge parameters [37]. It was also shown in experiments [38] that the convection velocity was small (~20 cm/s). The thermal diffusivity of air at the COD temperature ~20,000 K is $\chi \approx 10\,\mathrm{cm^2/s}$ (Figs. 3.1 and 3.3). The COD length taken from Fig. 3.25 is $l < 1$ cm for $P < 6\,\mathrm{kW}$. Under these conditions, $Pe < 1$ even for maximal velocities. For high powers and large dimensions (length and width) of the caustic, natural convection becomes more important in the thermal balance of the COD [39]. The COD model, taking the thermal convection into account, will be discussed below, and now we return to expression (3.44). Because the dependence $\mu(T)$ is close to the step dependence (Fig. 3.4), we will assume that $\mu = 0$ for $T < T^*$ and $\mu = const > 0$ for $T > T^*$ [1]. The boundary temperature and the value of depend on the gas type and will be determined below. The radiation intensity is now expressed by a simple relation

$$q = P \exp[-\mu(r_2 - r)]/\pi\varphi^2 r^2$$

where r_2 and r_1 are the coordinates of the leading and trailing edges of the discharge, which are determined by the relations $T(r_1) = T(r_2) = T^*$, where $T > T^*$ between r_1 and r_2 (Fig. 3.23). Because the dependence $\kappa(T)$ is nonmonotonic, it is convenient to pass from T to another variable [1]:

$$\theta = (\theta^*)^{-1} \int \kappa(T)\mathrm{d}T, \quad \theta^* = \int_0^{T^*} \kappa(T)\mathrm{d}T$$

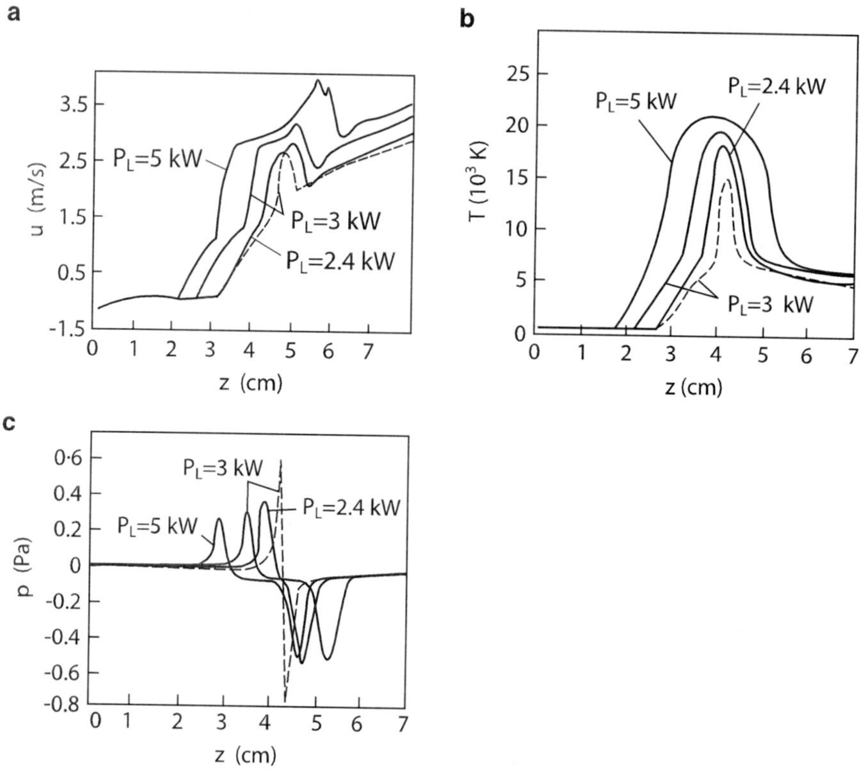

Fig. 3.24 Velocity (**a**), temperature (**b**), and pressure (**c**) distributions on the COD axis as functions of the laser beam power (dashed curves correspond to the unstable regime)

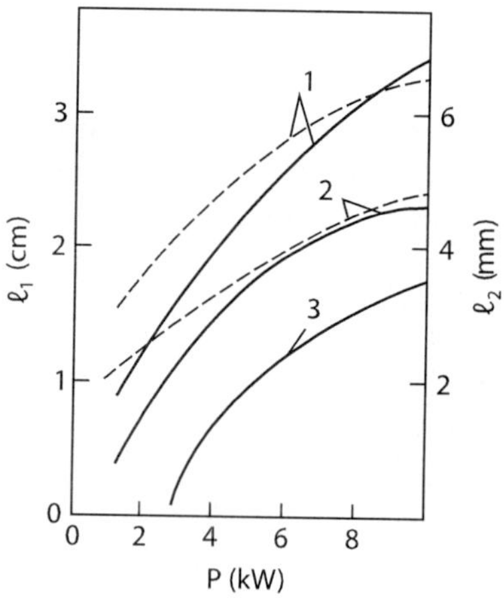

Fig. 3.25 Power dependences of the longitudinal l_l (*solid curves*) and maximal transverse l_2 (*dotted curves*) discharge thicknesses calculated by the isotherm 10^4 K: (**1, 2**) numerical calculations for iron and air; (**3**) calculation for air by (3.51)

By introducing also the variable $x = \mu r$, we rewrite (3.44) in the dimensionless form

$$-\frac{1}{x^{\nu}}\frac{d}{dx}x^{\nu}\frac{d\theta}{dx} = \frac{\bar{P}}{x^2}e^{x-x_2} - \xi\theta - \frac{\theta}{(x\varphi)^2} \tag{3.46}$$

where $\bar{P} = \mu P/\pi\theta^*\varphi^2$; $\xi = \gamma/\mu\kappa$; $\nu = 0$; 2. By introducing the notation $x_i = \mu r_i$, where $i = 0, 1, 2$, we rewrite (3.45) in the form

$$\theta(x_0) = \theta(\infty) = 0 \tag{3.47}$$

Solutions obtained in all three regions $r < r_l, r_l < r < r_2$, and $r > r_2$ should satisfy the "sewing" conditions at the boundaries of functions and their derivatives:

$$\theta(x_1) = \theta(x_2) = 1 \tag{3.48}$$

$$\left.\frac{d\theta}{dx}\right|_{x=x_1+0} = \left.\frac{d\theta}{dx}\right|_{x=x_1-0}; \quad \left.\frac{d\theta}{dx}\right|_{x=x_2-0} = \left.\frac{d\theta}{dx}\right|_{x=x_2+0} \tag{3.49}$$

Because the solution of (3.44) is cumbersome (and it is difficult to obtain expressions for discharge parameters), it is reasonable to consider first a model in which radiation losses are neglected (these losses will be taken approximately into account later).

3.4.2 The Heat-Conduction COD Model

Thus, we assume that $\xi = 0$ and $\nu = 2$. By solving (3.46) in three regions and using conditions (3.47)–(3.49), we can obtain the system of equations for determining the boundaries x_1 and x_2 of the optical discharge. The analysis of this system, which is too cumbersome to be presented here, shows that two values of x_1 and x_2 and, hence, two values of δ correspond to one value of the laser beam power.

As in the spherically symmetric case [1], there exists the critical power $\bar{P}_{\min}$ below which solutions are absent, i.e. the discharge is quenched. The correctness of this mechanism of discharge quenching at low powers is confirmed by two-dimensional numerical calculations based on a more complete model of a COD in a focused laser beam, which takes into account the transfer of thermal radiation and air convection in the gravitational field [41]. This radiative-gas-dynamic model will be described in the section devoted to an optical discharge in a gas flow, while here we consider only its result (Fig. 3.26). Figure 3.26 illustrates the existence of two stationary discharge burning regimes with the high and low plasma temperatures at the same laser radiation power. Figure 3.24 shows the profiles of COD parameters corresponding to the stable and unstable branches in Fig. 3.26. Thus, the inclusion of additional processes to the COD model changes the situation neither qualitatively nor quantitatively (cf. Figs. 3.26 and 3.27). Let us continue the analytic study of the COD.

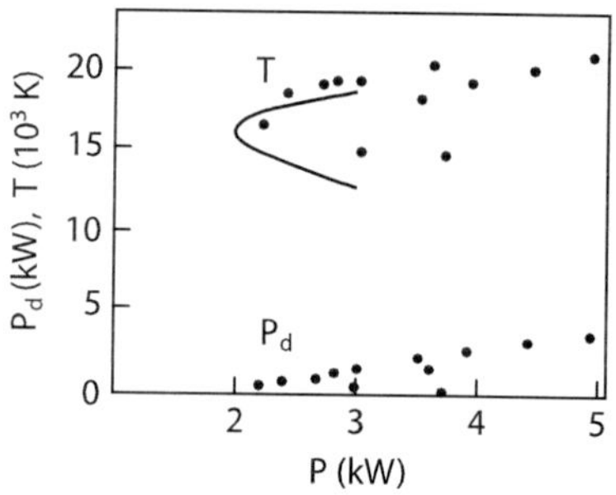

Fig. 3.26 Dependences of the maximum temperature and dissipate power P_d of the COD on the laser power (solid curves are calculated based on the spherical model [1])

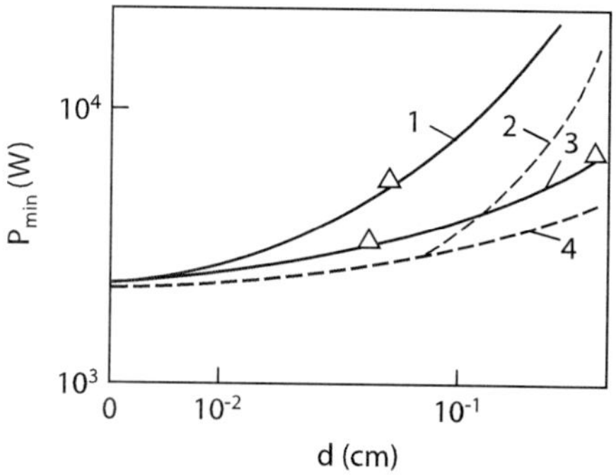

Fig. 3.27 Dependences of the lower limit P_{min} on the laser beam geometry in air; (**1**, **2**) experiment [27] and calculation by (3.72) for $v \sim 20$ m/s; (**3**, **4**) experiment [27] and calculation by (3.52) for $v \sim 0$

Thus, the first solution describes the discharge which approaches the target with increasing radiation power, the discharge thickness $\delta = x_2 - x_1$ being decreased (curve 1 in (Fig. 3.28)). The numerical calculation of nonstationary equation (3.44) shows that this solution is unstable and is not realized. The second solution corresponds to the discharge moving away from the target with increasing radiation power. The discharge thickness increases in this case, but remains restricted by the value [40, 42]

$$\delta_{max} \approx 3\varphi/\mu \qquad (3.50)$$

In this respect, the COD within a light cone considerably differs from a spherical discharge [1] whose width increases linearly with increasing power. Here, we will not present the expression for the lower threshold P_{min} for maintaining COD burning near the target because the more general expression for P_{min} will be obtained taking radiation losses into account. By using this model, we can study the influence of radiation reflected from the target on the COD. Because for $P \approx P_{min}$, the optical thickness of the discharge is small, P_{min} decreases by a factor of $(2 - \alpha)$. As for the maximum COD thickness, due to the consideration of reflection, δ_{max} tends to a constant for $\alpha = 0$ with decreasing the focusing angle φ:

$$\delta_{max} = 4/\mu(2 - \varphi), \quad \varphi < 1.$$

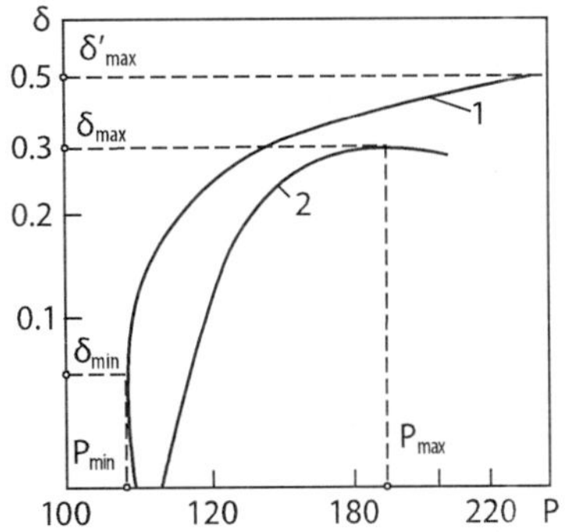

Fig. 3.28 Dependences of the plume thickness on the laser power for $\varphi = 0.133$ and $x_0 = 0.1$; (**1**) nonradiative model; (**2**) model (3.51)

3.4.3 COD Model Taking into Account the Heat Conduction and Emission of Plasma

Usually, the longitudinal size of a discharge is larger than its transverse size because the beam focusing angle is $\varphi < 1$. Therefore, thermal losses from the COD along the beam are considerable only near its boundaries x_1 and x_2 [43]. Away from discharge boundaries inside the discharge, the equilibrium exists: the energy absorbed by the plasma is removed due to heat conduction in side directions. The role of emission especially increases with increasing the COD volume. According to the above discussion, we will seek the solution of (3.46) in region II in the form ($\nu = 0$)

$$\theta = c_1 \mathrm{e}^{-k_1(x-x_1)} + c_2 \mathrm{e}^{-k_2(x_2-x)} + \frac{\varphi^2 \bar{P}}{1+\xi\varphi^2}\mathrm{e}^{x-x_2}$$

This expression takes into account that thermal losses along the beam are most considerable along the discharge boundary. Therefore, the first two terms of the solution are written similarly to the boundary-layer approximation in hydrodynamics. By neglecting the contribution of $\mathrm{d}^2\theta/\mathrm{d}x^2$ inside the discharge, we obtain from (3.46) the third term of the solution.

From conditions (3.47)–(3.49), we find c_1, c_2, k_1, k_2, and the system of equations for determining the boundary coordinates x_1 and x_2. The expression for x_2 can be obtained in the explicit form

$$x_2 = \left[\frac{1}{\xi}\left(\sqrt{\bar{P} + \frac{\alpha^2}{4\varphi^2}} - \frac{\alpha}{2\varphi}\right)^2 - \frac{1}{\xi\varphi^2}\right]^{1/2}, \quad \alpha = 1 + \frac{\varphi}{2} \qquad (3.51)$$

The lower power threshold can be found from the condition $\delta = 0$:

$$\bar{P}_{\min} = \alpha^{\varphi} x_0^2 \xi + \varphi^{-2} + \alpha\sqrt{\varphi^{-2}(\xi x_0^2 \alpha^{\varphi} + \varphi^{-2})} \qquad (3.52)$$

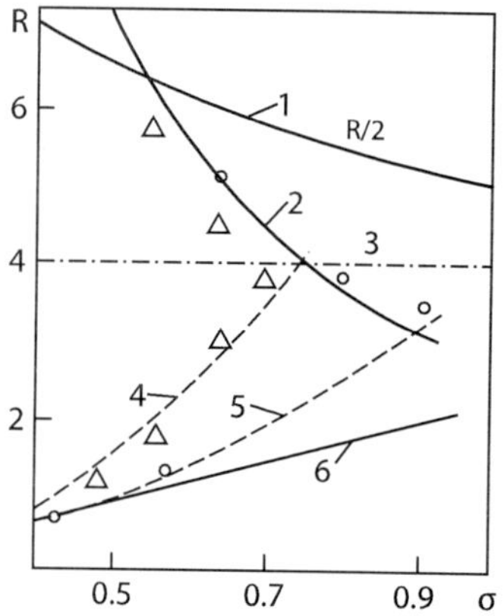

Fig. 3.29 Limits of the support of stable COD combustion: calculation of $\bar{R}_{max}$ for $\alpha = 0.1$ (**1**) and $\alpha = 1$ (**2**); (**3**) estimate of $\bar{R}_{max}$ by model [81]; (**4**) $\bar{R}_{min}$ calculated by (3.52) for argon; (**5**) calculation of $\bar{R}_{min}$ for nitrogen; (**6**) calculation of $\bar{R}_{min}$ by (3.55)

In the COD model taking into account only heat conduction losses and admitting the exact solution, we have $\delta_{min} > 0$ (see Fig. 3.28). Because the absolute value of the thickness δ remains in this case quite small, the difference between the values of P_{min} for these two models is small. An important specific feature of the dependence of x_1 and x_2 on $\bar{P}$ is the presence of a region at large enough $\bar{P}$ where the function $\mathrm{d}\delta/\mathrm{d}\bar{P}$ is negative, which corresponds to the unstable solution in this region. This gives the upper limit $\bar{P}_{max}$ of the stationary burning as the power at which the condition $\mathrm{d}\delta/\mathrm{d}\bar{P} = 0$ is fulfilled.

The dependence of $\bar{P}$ on the beam and medium parameters is quite complicated in the general case. However, in the case of small φ, this dependence can be reduced to some universal dependence $\bar{R}_{max}$, where $\bar{R}_{max} = \bar{P}_{max}/\xi$ and $\sigma = (\xi\varphi^2)^{-1}$.

Figure 3.29 shows the solution obtained numerically (see curve 2). In the region $\sigma > 0.5$ important in practice, this function is well approximated by the dependence $\approx 3{+}0.8\sigma^{-2.32}$. As mentioned above, the decrease in φ or ξ leads to the approaching the thresholds P_{min} and P_{max}. Analysis shows that the value of φ_{min} and ξ_{min} can be approximately found from the condition $\sigma \approx 1$

$$\varphi_{min} \cong (\mu\bar{\kappa}/\gamma)^{1/2}; \quad \xi_{min} \cong \varphi^{-2}; \quad \mu_{max} \cong \varphi^2\gamma/\bar{\kappa} \tag{3.53}$$

The presence of such restrictions on the discharge burning conditions was pointed out in experiments [27, 44, 65]. The leading front position weakly depends on the absorption coefficient α of a target. However, the additional heating of the COD by reflected light considerably affects the trailing edge position. In this case, the power range of stable burning can increase almost by three times. The dependence of the upper threshold on σ for $\alpha = 0.1$ is presented in Fig. 3.29.

Finally, we estimate the transparency coefficient of the discharge, which is equal to the ratio of the radiation power propagated through the plume to the laser power. This coefficient in model (3.46) is described by a simple expression $\eta = \exp(-\delta)$. For high pressures close to the values at which the discharge is still burning, when $\mu = \mu_{max}$ [see (3.53)], we have

$$\eta \approx 2/\varphi^2\bar{P} \tag{3.54}$$

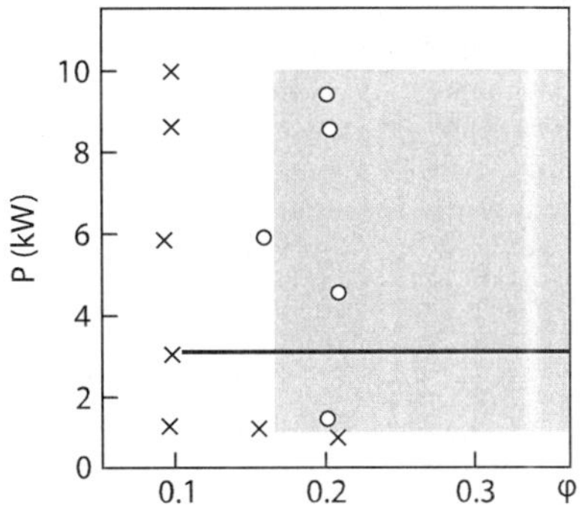

Fig. 3.30 Stable combustion region in air at the normal pressure (*grey color*); (∘) stationary COD combustion; (x) quenching of the discharge; *solid line* is dependence $P_{\min}(\varphi)$ calculated by (3.52)

In the general case, the calculation of the transparency coefficient in model (3.46) is reduced to a rather simple numerical calculation. To compare experimental data with analytic dependences obtained above, we determine the numerical values of quantities entering these dependences. Note that the choice of the values of T^*, κ, and μ is arbitrary to some degree. We assume below that at the atmospheric pressure $\mu = 0.63\,\text{cm}^{-1}$ for air and $\mu = 0.4\,\text{cm}^{-1}$ for Ar: $T^* = 1.5 \cdot 10^4$ K, $\theta^* = 54$ W/cm, and $\gamma/\bar{\kappa} = 110\,\text{cm}^{-1}$ for argon. By substituting these values into (3.53), we obtain $\varphi_{\min} \approx 0.1$ for air. This value is consistent with experiment [83], where it was found that $\varphi_{\min} \approx 1/8$. The dependence $P_{\min}(\varphi)$ calculated by (3.52) is indicated by solid line in Fig. 3.30. The COD maintenance threshold in air obtained in [27] is equal to 2 kW. Although the beam geometry was not reported in [27], the average values of φ and r_f used in [27] demonstrate the qualitative agreement between experimental data and calculations by (3.52). If $\varphi \ll 1$ and $(\varphi x_0)^2 \xi \ll 1$, this expression gives a simple relation

$$\bar{P}_{\min} \approx 2\varphi^{-2} \tag{3.55}$$

This estimate was obtained in [1] for the spherically symmetric model of the discharge and is consistent with experimental data at low pressures (~0.1 MPa), when energy is mainly dissipated near the lower threshold due to heat conduction. Expression (3.52) also takes into account radiation losses, and therefore the agreement with experiments can be expected in a broader range of values of φ and ξ. To verify this, we plot in Fig. 3.29 the values of $P_{\min}$ [44] recalculated for argon (they are indicated by the index Δ). We assume that $\varphi = 0.35$. One can see that at a high pressure, when radiation losses become considerable, experimental points are located above curve 6 calculated by (3.55). The lower threshold calculated by (3.52) for $r_f = 0.18$ mm corresponds to curve 4. As expected, the consideration of radiation improved agreement with experiments. Experimental dependencies obtained in air [44] substantially differ from those obtained in other gases, and therefore it is difficult to expect the agreement between theoretical calculations and experiments in this case. We compared the results of calculations with recalculated data [44] obtained in nitrogen (they are indicated by the index ∘). One can see that dependencies $\bar{P}_{\min}$ and $\bar{P}_{\max}$ are consistent with experiments.

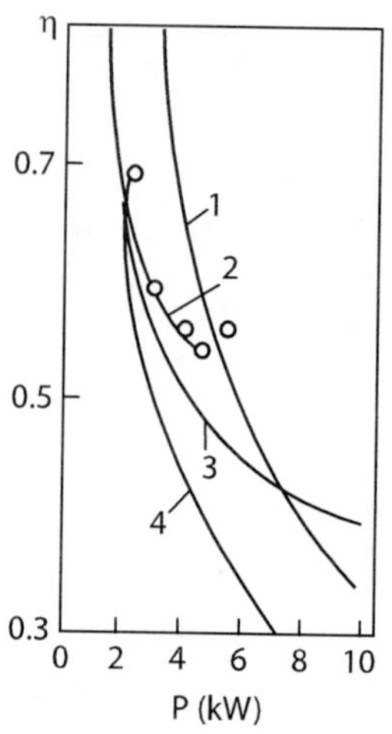

Fig. 3.31 Dependences of the COD transparency on the laser power: (**1**) calculation fore air by (3.51); (**2**) experiment [27]; (**3, 4**) numerical calculation for air and argon

Figure 3.27 presents a comparison of $P_{\min}$ measured in [27] with calculations by (3.52). Figure 3.25 presents the results of the numerical experiment (which will be described below) and the calculation of the COD thickness by (3.51) for $\delta \approx x_2$. Taking into account that the size is estimated analytically by the isotherm at 1.5×10^4 K and numerically by the isotherm at 10^4 K, we can conclude that the obtained results are consistent. The estimate $l_1 \sim \sqrt{P}$ obtained in [10], assuming that the discharge fills the light cone directly behind the cross section where the radiation intensity is equal to the threshold value, is valid only at high powers, as follows from (3.51). Note that the expression for $P_{\max}$ obtained in [10] is based on this assumption. This dependence in the coordinates $\bar{R}$ and σ shown by the straight line $\bar{R} = 4$ in Fig. 3.29 can be used only for rough estimates of the upper limit. Curve 1 in Fig. 3.31 describes the dependence of the transparency coefficient of the COD on the radiation power. This dependence was calculated by using the COD thickness δ presented in Fig. 3.25 (curve 3) [[$\eta = \exp(-\delta)$]]. Data [27] presented here well agree with calculations for $P \approx P_{\min}$. However, as the radiation power is increased, the theoretical value of transparency decreases, while the experimental value tends to a constant. This question remains open due to the absence of other measurements of η.

3.4.4 *Numerical Calculations of Optical Discharge Parameters*

We studied the COD in previous sections by using models roughly taking into account side thermal losses, the solution method being approximate itself. In addition, because only stationary states were considered in the theoretical study, the instability of the obtained solutions remained unstudied.

To investigate the COD more completely, it is necessary to solve numerically the nonstationary thermal balance equation in plasma [43]. Because a light beam is usually axially symmetric, the COD combustion is considered in a cylinder with the specified surface temperature T_M. Radiation is focused by a lens through one face of the cylinder on another face of the cylinder – a target. The irradiation conditions provide the absence of target vaporization. We are interested in a region where plasma is in the nearly stationary state. Therefore, the plasma displacement velocity is lower than the sound speed. For this reason, as in previous sections, the plasma pressure was assumed uniform and time-independent. By using these assumptions, we can write the thermal balance equation

$$\rho c_p \frac{\partial T}{\partial t} - div(\kappa \Delta T) = \mu q - \Phi(T) \tag{3.56}$$

where ρ is the gas density and c_p is the specific heat (see Fig. 3.1). The term $\Phi(T)$ describes radiation losses from plasma. This term was calculated for air [1] $\Phi(T)$ by (3.6), as in Sect. 3.1. The radiation loss in the argon plasma was calculated by (3.6) taking into account corrections for linear radiation (see Fig. 3.5).

The laser radiation field propagating in the volume under study was described in the parabolic approximation [see (1.5)]:

$$2\mathrm{i}k\frac{\partial E}{\partial z} + \frac{1}{r}\frac{\partial}{\partial r}r\frac{\partial E}{\partial r} + k^2\left(-\frac{n_e}{n_c} + \frac{\mathrm{i}}{k}\mu\right)E = 0 \tag{3.57}$$

Here, E is the complex amplitude of the electric field strength of laser radiation:

$$\tilde{E}(r,z,t) = \frac{1}{2}E(r,z)\exp(\mathrm{i}kz - \mathrm{i}\omega t) + k.s.$$

Recall that the first and second terms in parentheses (3.57) describe the refraction and absorption, respectively, of the laser beam in the discharge plasma. The free electron concentration n_e was assumed equilibrium at the given temperature, as usual for the COD, and was found from the Saha equation. The quantity $n_c = \omega^2 m_e/4\pi e^2$ is the critical plasma density at which the plasma permittivity vanishes ($n_c \approx 10^{19}$ cm^{-3} for CO_2 laser). The radiation intensity is related to E by the expression $q = cE^2/8\pi$. The volume under study in the plane $z = 0$ is restricted by a collecting lens with the focal distance F through which a laser beam propagates with the initial Gaussian field distribution

$$E(0,r) = E_0 \exp[-(r/r_a)^2 - \mathrm{i}kr^2/2F]$$

where r_a is the incident beam aperture. (The characteristic beam radius r_0 is related to r_a by the expression $r_0 = r_a/\sqrt{2}$). The field on the side boundaries of the specified volume was assumed zero.

The initial state of the COD was specified by the Gaussian temperature distribution $T = T_i \exp\{-[r^2+(z-z_i)^2]/r_i^2\}$, where T_i changes in the range $(1.5–2\times10^4)$K. Such an artificially specified plasma bunch begins to move toward the laser beam. Depending on the radiation power P, the initial plasma position with respect to the focus, and the values of T_i and r_i, the combustion regime can be quite different. When the radiation power is lower than a threshold value $P_{\min}$, the discharge is rapidly quenched even in the focal region. For $P > P_{\min}$, a LCW is formed whose velocity decreases during its movement toward the laser beam. Below, we present the results of calculations performed for the size of the calculated region and laser beam parameters corresponding to the slowing down of the combustion wave and its stoppage outside the zone of influence of the region boundaries, so that the obtained stationary position would correspond to the freely burning COD.

The numerical study of the described model was performed for two media (air and argon at the normal pressure) in the range 1–10 kW and for the focusing angle from 0.2 to 0.05. We found that the stationary combustion of the discharge can occur only the laser beam is focused tightly enough. If φ is smaller than some limiting value $\varphi_{\min}$, the plasma decomposes after some time at any power level. The hatched region in Fig. 3.30 corresponds to the laser beam parameters at which the COD can be maintained in air. Note that the calculations were performed both for air and argon for $P < 10$ kW. In this case, the upper power limit $P_{\max}$ of the discharge maintenance was not achieved even for angles close to $\varphi_{\min}$. The properties of the COD were studied in more detail for $\varphi = 0.2$ by using the geometrical parameters $F = 3.3$ cm, $r_0 = 0.64$ cm, the calculation region length $z_{\max} = 4$ cm, and the beam radius at the focus $r_f = 0.17$ mm. The COD was initiated by plasma located in the focus of a lens ($z_i = z_f = 3.3$ cm, $r_i = 4$ mm). Figure 3.32 presents the results of calculations. Curve q describes the radiation intensity in the COD plasma. For comparison, the radiation intensity distribution

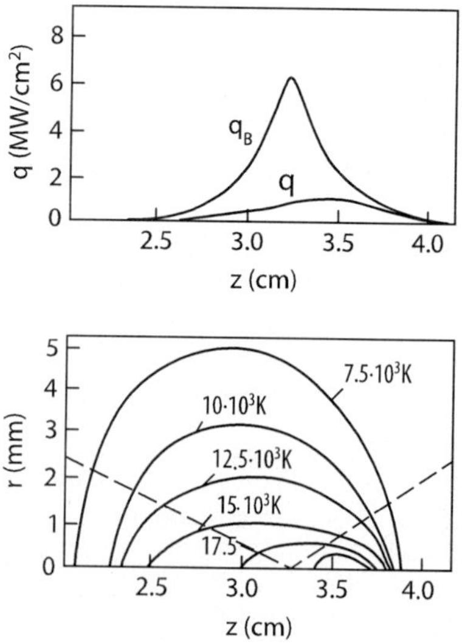

Fig. 3.32 Light intensity distribution along the laser beam (*top*) and the temperature field (*bottom*) I a discharge burning in air for $P = 5$ kW and $T_{\max} = 22.5 \cdot 10^3$ K, $T_{\min} = 6 \cdot 10$ K (the *dotted line* indicates geometrical boundary of the laser beam, beam is directed from left to right)

$q_B(r, z)$ in the absence of interaction between the medium and radiation is also given. One can see that the COD plasma considerably affects radiation parameters. Due to absorption and refocusing, the radiation intensity in the focus decreases from 7 to 1 MW/cm^2. Note that the focus position itself is displaced along the beam by 0.4 cm. At the same point the maximum plasma temperature is also achieved. The stationary position of the temperature maximum depends on P, and when the radiation power is high enough, this maximum is already located between the lens and its focus. The important parameters of the discharge are its transparency for laser radiation and the refraction broadening of the light spot on the target. In the case presented in Fig. 3.22, the transparency coefficient η is 0.4. The influence of refraction on the radiation intensity in the spot can be estimated from the quantity $\zeta(z) = \eta q_B(0, z)/q(0, z)$ depending on z. Below, we define the refraction broadening as $\zeta(z_f)$, and it is equal to 3 for the case under study. The COD parameters calculated in air and argon are presented in Figs. 3.25, 3.31, and 3.33. The geometrical parameters are the same as in Fig. 3.32. Figure 3.33 shows that the refraction broadening of the focal spot increases with increasing the radiation power. As in the case of laser beam attenuation caused by absorption, this effect is stronger in argon than in air. Calculations showed that the discharge thickness in argon greater (see l_1 in Fig. 3.25), while the distance from the lens to the plasma front, from which ζ depends, is smaller than in air. As a result, the radiation intensity in the focal spot is lower in the former medium. The refraction properties of the COD upon laser welding of steels were studied in [45]. Figure 3.34 presents he results of measurements of the refraction angle of a diagnostic CO_2 laser beam directed normally to the welding direction. Welding is performed by a 2 kW CO_2 laser in argon (30 l/min). The welding speed is 0.5 m/min. As expected, the refraction angle is maximal at the plasma plume edge, while at the centre it is close to zero. The maximum attenuation of the radiation intensity in the argon atmosphere compared to other gases was observed in laser welding experiments. In [46], the welding depth

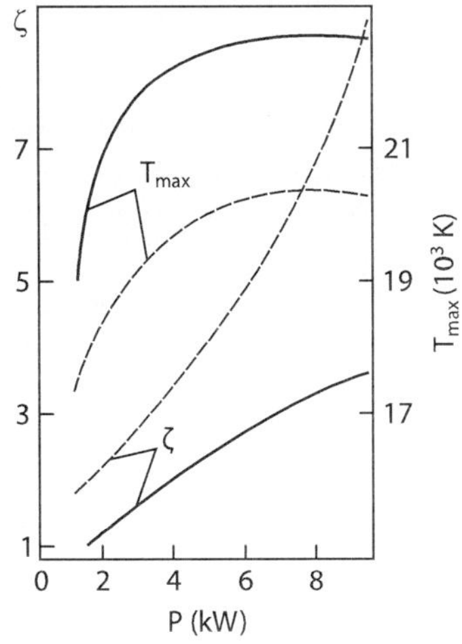

Fig. 3.33 Dependences of the refraction broadening of the laser beam and the maximal temperature in the discharge on the laser power in argon (*dotted curves*) and air (*solid curves*)

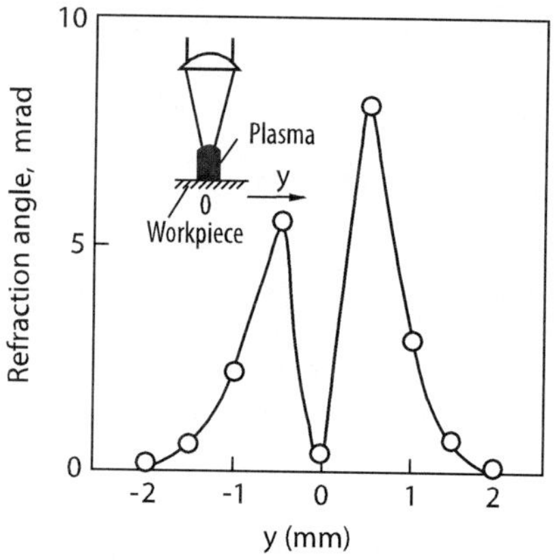

Fig. 3.34 Refraction angle of the probe CO_2 laser beam propagating through a plasma plume during laser welding of a 2 kW CO_2 laser at a speed of 50 cm/min. The argon flow rate was 30 l/min), [45]

in air for $P \sim 5\,kW$ proved to be five timed greater than in argon. Similar results were also obtained with a higher power CO_2 laser ($P = 12\,kW$) [47].

In [48], the laser radiation energy absorbed in a target was measured under conditions close to calculation parameters. A 5 kW cw laser was used to ignite a discharge near a refractory tungsten target, so that the target vapor had no effect on the COD. It was shown that the target heating in the helium atmosphere (where a plasma plume was very small) was five times greater than in the argon atmosphere under the same conditions.

3.4.5 The Radiative–Conductive COD Model

It was pointed out in the previous section that the role of natural convection in the COD balance should increase with increasing the laser power and, correspondingly, the COD size. In addition, such a discharge becomes optically thicker and the role of reabsorption of thermal radiation from plasma increases [14]. Consider the manifestation of these factors with the laser power (P =10–30 kW). The scheme of the calculation model is presented in Fig. 3.35. For $P = 10\,kW$, the COD was ignited in the lens focus at a distance of 2 cm from the sample surface, the focal spot radius being 2 mm. Calculations [39] showed that under these conditions the energy losses from a hot plasma due to the heat conduction and thermal radiation were 40% and 60%, respectively. The gas velocity due to natural convection achieved ~6 m/s, which affected the temperature field of the COD (see, e.g., Fig. 3.36). The necessity of the correct consideration of energy losses due to emission increases with further increasing the laser power and the COD size. Thus, for $P = 30\,kW$ and $r_f = 3\,mm$, the difference between the temperature distributions obtained for the optically thin plasma and plasma in which reabsorption is taken into account becomes considerable. When reabsorption is taken into account, photons emitted from the hot region are absorbed and heat peripheral layers. As a result, the COD size increases twice (Fig. 3.36). In this case, the maximum temperature decreases. This fact, which seems surprising at first glance, is explained by the fact that due to

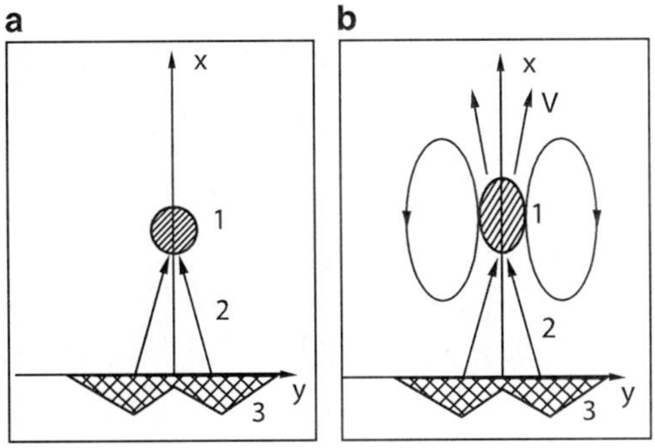

Fig. 3.35 Scheme for calculations of a COD; (**1**) plasma; (**2**) laser beam; (**3**) surface; convective gas flow in the discharge vicinity

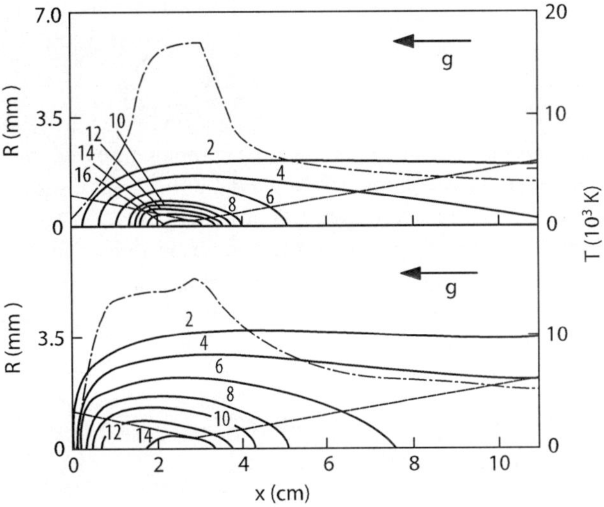

Fig. 3.36 Quasi-stationary temperature field of a COD maintained by a 3-kW laser taking into account convection in the gravitational field: (**a**) calculations by neglecting the reabsorption of thermal radiation and (**b**) taking reabsorption into account. The *solid curves* correspond to the temperature distribution in 1,000 K. The *dot-and-dash* curves correspond to the temperature distribution on the laser beam; the *dashed curves* show the boundary of the laser beam in vacuum. The laser beam is directed from left to right

the increase in the COD volume, the laser radiation of lower intensity reaches the focus, which leads to the decrease in the maximal temperature.

3.5 LCWs and a COD in a Gas Flow

As pointed out in the previous section, the ignited cw optical discharge is burning only under certain conditions. For example, for focusing angles $\varphi < \varphi_{\min}$, the discharge is not stabilized and is displaced from the place of its initiation. Under

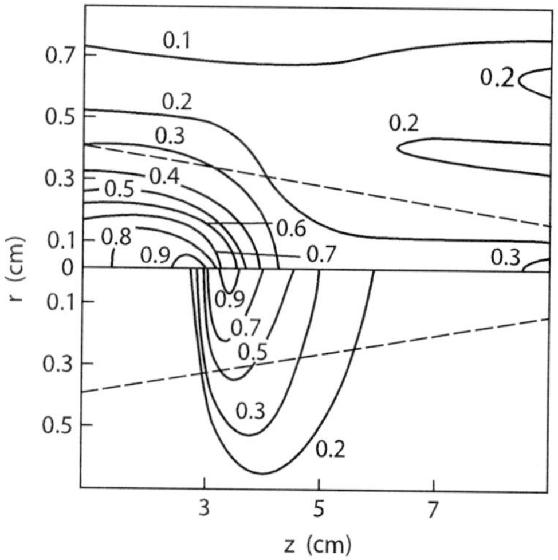

Fig. 3.37 Spatial intensity distribution (*top*) and the temperature field (*bottom*) in a light combustion wave at $t = 5.4$ ms: data are normalized to the maximum values $q_{max}=2.7 \cdot 10^4$ W/cm^2, $T_{max} = 22{,}000$ K; the *dashed line* is the laser beam boundary

such conditions, a LCW can be produced (see, e.g., Fig. 3.37). An optical discharge in a gas flow is analogous to the phenomenon of propagation of combustion waves. Indeed, a combustion wave "blown" by a counterpropagating gas flow can be stopped. However, a gas flow not always blows away an optical discharge. It can only display the discharge, by changing the combustion place. In the case of high laser powers, the optical discharge can considerably attenuate and refocus the laser beam. One of the methods used to reduce the plume influence is the quenching of the latter by a transverse gas flow (see section 4.3.4).

3.5.1 Light Combustion Wave

This wave was observed experimentally and was theoretically described for a cylindrical (unfocused) light beam [1]. Although in laser technologies, focused radiation is commonly used, to understand the general properties of the process, we consider first combustion waves in an unfocused (cylindrical) light beam.

A combustion wave in a laser beam is described by the same equations as a COD. However, in this case it is principally important to take into account the term with the time derivative in (3.56).

After the usual change of variables $\xi = x - ut$ and $\theta = \int \kappa \mathrm{d}T$, (3.56) takes the form

$$-u\rho_0 \frac{c_p}{\kappa} \frac{\mathrm{d}\theta}{\mathrm{d}\xi} - \frac{\mathrm{d}^2\theta}{\mathrm{d}\xi^2} = q\mu - \frac{\theta}{r_f^2}. \tag{3.58}$$

We will neglect the laser beam absorption in (3.58) for simplicity, and take side losses into account approximately, as in the case of the COD. This equation, as (3.46), can be integrated by representing $\mu(\theta)$ in the form of a "step" [1]. For clarity, we obtain approximate expressions for the wave parameters. The front width l_f can be found by equating terms in the left-hand side of (3.58):

$$l_f \approx \kappa / c_p \rho_0 u \tag{3.59}$$

By equating the second term in the left-hand side to the second term in the right-hand side of (3.58) and using (3.59), we obtain the expression for the velocity

$$u \approx \frac{\kappa}{c_p \rho_0} \sqrt{\frac{\mu q}{\theta *}} \tag{3.60}$$

The minimal radiation intensity at which the wave still exist is determined from the condition of vanishing of the right-hand side in (3.58) at $\theta = \theta^*$:

$$q_{\min} \approx \theta^* / \mu r_f^2 \tag{3.61}$$

This condition coincides with the lower combustion threshold (3.55) of the COD. If the absorption of the laser beam is taken into account, the temperature behind the front will be lower. The characteristic temperature decrease l_3 is determined by the maximal absorption coefficient: $l_3 \approx \mu^{-1}$. The convective flow behind the front appearing when μ is nonzero can be neglected if the velocity u is not too large:

$$u < u_\kappa = \kappa / c_p \rho_0 \mu r_f^2 \tag{3.62}$$

where the value of u_κ can be found from the equality of the thermal and convective flows behind the front. Condition (3.62) together with (3.60) restricts the value of q by the relation

$$q < q^* = q_{\min} (\mu r_f)^{-2} \tag{3.63}$$

Because usually the inequality $\mu r_f \ll 1$ is fulfilled, such a wave has a broad domain of existence. As we saw, the lower COD maintenance threshold, coinciding with the LCW maintenance threshold, amounts to a few kilowatts in air. Therefore, in the power range from a few kilowatts to a hundred kilowatts (for $\mu r_f \approx 1$), the combustion wave behaves as a wave with weak absorption (3.60).

The numerical calculations of the LCW were performed for a focused beam and geometry close to the real one. As for the COD, the system (3.56), (3.57) was solved. The distance between a lens and a target (which in this case also is not vaporized) was set equal to 10 cm. The focal distance of the lens was $F = 15$ cm, so that the focus was located behind the target. The beam radius r_0 on the lens was chosen comparatively small to provide a small focusing angle $\varphi = r_0/F$. The radiation intensity on the lens and target was $q_0 = 4.3 \cdot 10^4$ W/cm^2 and $q_0 = 4 \cdot 10^5$ W/cm^2, respectively. The calculations showed that under these conditions the combustion wave reaches the lens for the time ~7 ms, and then the discharge is burning near the lens stationary. The average velocity of waves in this case is ~20 m/s. One can see from Fig. 3.37 that the combustion wave is a solitary wave – a soliton of length along the beam greatly exceeding the beam diameter.

The temperature and length of a soliton slightly decrease during its movement. After "adhesion" of the soliton to the lens, the soliton temperature and length become noticeably smaller. Figure 3.37 shows that the transverse size of the soliton exceeds the local radius of the light beam. The temperature on the beam boundary is $T = 17{,}000$ K.

The velocity of a wave of any type at power noticeably exceeding the threshold is determined by the light intensity $q(z)$ [1]. Depending on the relation between r_0 and r_f, the local intensity in the wave decreases during its propagation. In the case under study, it decreased by 10–20 times. The processing of the results shows that the velocity u decreases with decreasing q approximately as $q^{1/2}$ (in fact, somewhat weaker). This corresponds to the combustion wave weakly absorbing light (3.60). The soliton length is determined by the absorption length: $l_a \approx \mu^{-1}$. The cooling length corresponding to thermal losses in side directions due to heat conduction is described by the expression $l_c = ur^2/\chi$ (where r is the local beam radius and χ is the thermal diffusivity.) Estimates made for typical LCW conditions show that a zone with $T \approx 20{,}000$ K, where $\chi \approx 10^3$ cm^2/s is coolest at the length 0.3 cm. However, for a zone with $T \approx 10{,}000$ K ($\chi \approx 10^2$ cm^2/s), we have $l_c \sim 3$ cm. Thus, the length of the hot zone of a soliton is approximately determined by the absorption length $l_a = 1.5$ cm, while the low-temperature tail is determined by the cooling length l_c. The soliton width $l_\perp$ can be easily estimated. The time t^* for which the soliton propagates over its length is $t^* = (\mu u)^{-1}$, where u is determined from (3.60).

During this time, the heat propagates a distance

$$l_\perp \approx 2\sqrt{\chi t^*} \approx 2(\theta^*/q\mu^3)^{1/4}$$

Under our conditions ($q \approx 30$ kW/cm^2, $\theta^* = 0.3$ kW/cm, and $\mu = 0.7$ cm^{-1}), $l_\perp \approx 0.6$ cm, which is consistent with calculations. Thus, calculations showed that, when the laser beam radius does not exceed 1 cm and the laser power amounts to a few tens of kilowatts, a quasi-stationary LCW with low absorption is produced. The transverse size of the LCW slightly decreases with increasing intensity. Note that the combustion front was stable in all cases. This can be explained by the fact that the front thickness was of the order of its width. But during the propagation of the wave from the target to the lens, the refraction broadening first increases and then decreases. The refraction broadening can be described approximately by the ratio of the cross section of the beam propagated through plasma to the initial beam cross section. We consider here the broadening of the beam on the target. From geometrical considerations, we have $\zeta \approx 1+\varphi^2 x^2/r_f^2$, where x is the distance from the target to the LCW. Because the refraction angle is $\varphi \approx \sqrt{\varepsilon'}$ (see Sect. 1.1), we have $\zeta \approx 1 + \varepsilon' x^2/r_f^2$. For the COD plasma and the LCW, we have $\varepsilon' \approx 10^{-2}$. For the data in Fig. 3.38, $r_f \approx 0.15$ cm and $\zeta \sim 10$, in

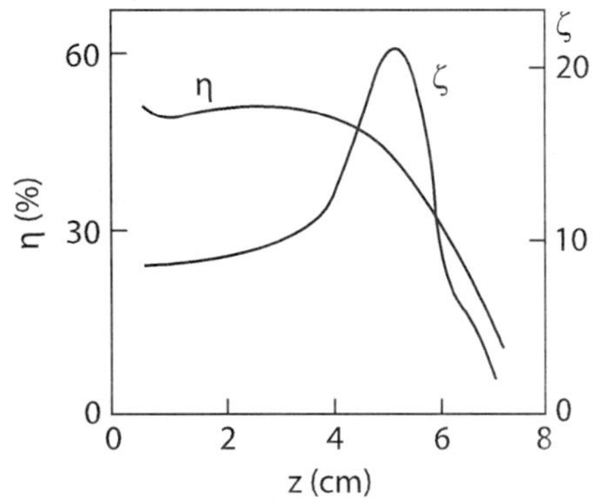

Fig. 3.38 Dependencies of the transparency coefficient η and spot broadening coefficient ζ on the distance from a plume to a mirror

accordance with the calculated value of ζ. The decrease of ζ observed in Fig. 3.38, is caused, as the increase in transparency, by the decrease in temperature, the density of charged particles and, hence, ε'.

3.5.2 Combustion Wave Supported Due to Thermal Radiation Transfer

Expression (3.60) and numerical calculations show that the wave velocity is determined by the laser radiation intensity. However, experiments with high-power lasers (<500 kW) showed that, in the case of the same intensities, the propagation velocity of a LCW in a larger-diameter beam was several times higher (Fig. 3.39 [50]). Laser radiation intensities used in [49] were low because laser beam diameters were rather large, from 5 to 21 mm. In [51], another mechanism of energy transfer in gas toward the laser beam was proposed. It was assumed that, instead of the heat conduction, the gas before LCW is heated by radiation emitted by a hot region in front of the wave. It is clear that such energy transfer will depend on the LCW size. Figure 3.40 presents the results of numerical calculations [50] of the LCW structure under experimental conditions [49]. The model proposed in [50] is more complete than that used in [51]. We will describe this model below, and here note only that the maximum velocity of gas flowing toward the beam (~10 m/s) is smaller than the wave propagation velocity (~25 m/s). Special analysis performed in [50] showed that the main mechanism of the LCW motion under these conditions is energy transfer due to thermal radiation. This model allows to consider combustion wave propagating in the vertical weakly focused beam with radius 1.06 cm [85]. The calculation show when beam is directed upward, the combustion wave propagates from top to bottom at velocities from 0.69 to 38.9 m/s upon variation of he laser radiation intensity from 0.017 up to 0.29 MW/cm^2. The heated gas moves upward due to the buoyancy force at the rate $v = 7.8$ m/s toward the wave. When the propagation direction of the laser beam is changed to the opposite, the convection rate coincides with the combustion wave velocity. In the case of low laser powers ($\leq$60 kW), the gas flow rate exceeds the combustion wave velocity. In this case,

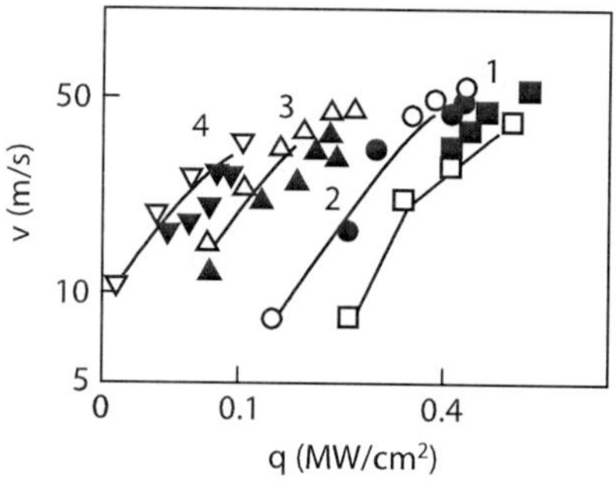

Fig. 3.39 Dependences of the propagation velocity of an optical discharge on the laser radiation intensity for laser-beam diameters 5 (**1**), 10 (**2**), 15.5 (**3**) and 21 mm (**4**). The *dark dots* are experiment, *light dots* are calculations

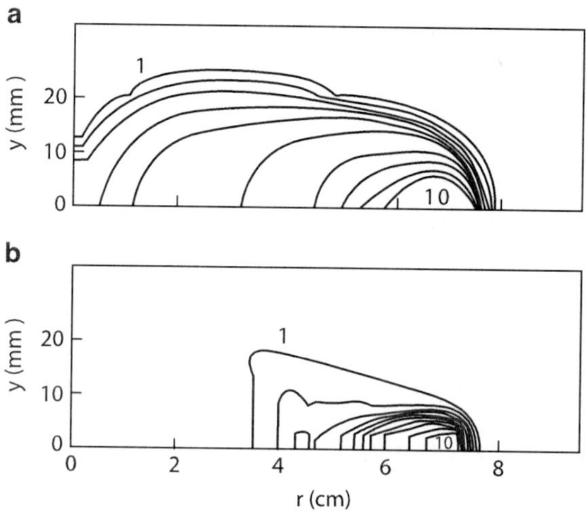

Fig. 3.40 Optical discharge structure 2.6 ms after discharge initiation by a 200-kW laser beam of radius 1.05 cm; (**a**) temperature field; (**b**) velocity field, $v_0 = 31.3\,\text{cm/s}^{-1}$ is the characteristic convection rate in the gravitational field. The beam is directed from right to left. Isotherm 10 corresponds to 20.5 K; then, approximately after 2,000 K, the velocity $v/v_0 = 36.8$ corresponds to line 10 and $v/v_0 = 3.7$ corresponds to line (I)

the vortex gas flow is excited. As the laser power is increased, the combustion wave velocity increases (Fig. 3.39) and exceeds the convection rate, and the gas flow becomes laminar.

3.5.3 Continuous Optical Discharge in a Gas Flow

We saw that, to obtain stationary combustion in an unfocused cylindrical laser beam, a certain laser power should correspond to a certain gas flow rate. If the laser beam is focused, this is not the case. The discharge can burn at different gas flow rates at the same laser power. The presence of the gas flow leads to the COD displacement depending on the rate v and change of its combustion thresholds. The influence of the COD gas flow can be taken into account, strictly speaking, only within the framework of a two-dimensional gas-dynamic model because a part of the gas will flow through the discharge region and another part will bend round it [52,76]. The same is true for the combustion wave [54]. A simple description is possible only within the framework of some hypothesis about gas flow. For example, we considered the LCW in the previous section by assuming that the mass flow density is preserved (the gas-dynamic analogue is the gas flow in a tube) and $\kappa/c_p \approx$ const. Before proceeding to the description of complex two-dimensional models of the COD in a gas flow, it is reasonable to consider the influence of the gas flow by using simplified models (3.44) and (3.46). This approach will offer at least the qualitative picture of the process and facilitate the understanding of complex models. The simplest treatment of a COD in a longitudinal flow can be given within the framework of a model taking into account only the heat conduction. We will assume for simplicity that the flow rate decreases with distance as $v = v_0 R/r$ (something average between gas flows in a cone and a tube). This assumption allows us to find easily the analytic solution of the equation. The presence of the gas flow is taken into account by adding the term describing convective cooling to the left-hand side of (3.46) under the condition that $\xi = 0$. Calculations [70] showed that all the parameters of the COD burning in the longitudinal flow depend on the dimensionless rate $(\kappa r)^{-1} c_p \rho v_0 R \mathrm{d}\theta/\mathrm{d}r$, which is $v = v_0 R c_p \rho/\kappa$ Peclet's number for this problem. At any direction of the gas flow, the lower combustion threshold increases. This is reasonable because the additional cooling of the discharge should lead to the increase in the laser power required for its support. When the plume is blown away along the laser beam, the stable combustion region ($\mathrm{d}\delta/\mathrm{d}P > 0$) increases. When the COD is blown in the direction opposite to the laser beam propagation, the stable combustion region decreases. These properties were also obtained in numerical calculations of one-dimensional equation (3.44). The inclusion of the term describing the convective heat removal [70] into (3.44) showed that the increase in the rate of gas flowing in the direction opposite to the laser beam propagation reduces the upper threshold down to a few kilowatts. This is consistent qualitatively with experiments [55]. In experiments with the counterpropagating gas flow [56], the COD could not be supported at all.

We presented above the results of numerical calculations for a discharge ignited by specifying the appropriate initial temperature distribution. It was shown in Sect. 3.2 that the optical breakdown in molecular gases near a refractory target has the thermal nature. In this case, the complete problem of the COD appearance and development can be solved by supplementing (3.44) with the boundary condition describing the temperature change at the focal spot center [57]:

$$T_M = \frac{\alpha R^2 \varphi^2}{\kappa_M} \sqrt{\frac{\chi_M}{\pi}} \int_0^t \frac{q_M(t-t')\mathrm{d}t'}{\sqrt{t'}(4\chi_M t' + R^2\varphi^2)}$$

where χ_M and κ_M are the thermal diffusivity and heat conduction of the target material, respectively, and q_M is the light intensity on the target. In addition, it

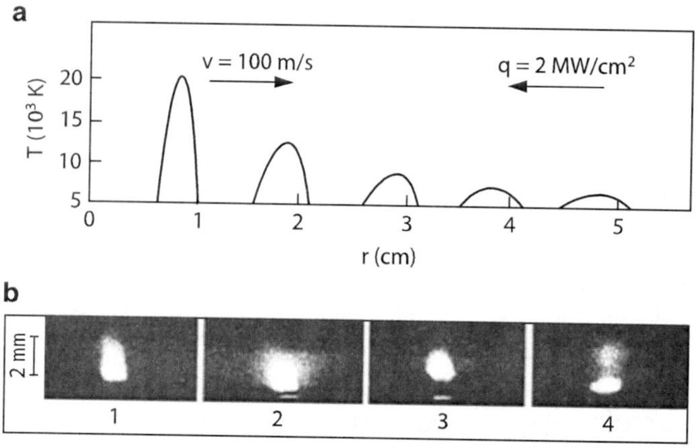

Fig. 3.41 Numerical simulation of optical discharge near tungsten target in air stream ($R \simeq 1$ mm, $\varphi \simeq 0.3$) (**a**) Dynamics of the oscillation combustion regime of an optical discharge near a steel target in argon. The CO_2 laser, $r_f = 0.1$ mm, $P = 0.6$ kW [82] (**b**)

should be taken into account that near the target surface the metal vapor always exists. We will assume that the vapor concentration is equal to the equilibrium concentration at temperature T_M: $n_M = N_s(T_M)$, $R < r < 2R$. We also assume for simplicity that the vapor concentration in the gas flow will change synchronously with the change in the concentration near the target. The absorption coefficient of radiation in this region is assumed equal to the sum of absorption coefficients in air and metal vapor. Calculations showed that, when the rate of gas flowing in the direction opposite to the laser beam propagation is high enough, the value of $P_{\max}$ (see Sect. 3.3) can be lower than the power P_{br} required for ignition of a discharge near the target. It was found that, under the condition $P_{\max} < P_{\text{br}} < P$, the COD was unstable and the system passed to the self-oscillation regime Fig. 3.41a. When q_M becomes higher than the radiation intensity required for optical breakdown at temperature T_M, a plasma bunch is produced near the target, which screens the target metal from radiation. As the plasma bunch moves in the direction opposite to the laser beam propagation, it is cooled and becomes more transparent. The thermal inertia prevents a rapid change in the surface temperature. Due to the combined action of these factors, the COD in the gas flow acquires a layer structure, and q_M oscillates in time. The oscillations of a plasma plume burning in gas vapor flows were observed in many experiments [2, 58, 59, 81]. Figure 3.41b presents the plasma cloud dynamics recorded by the method of high-speed photographing (5,000 frame/s) upon irradiation of a low-carbon steel sample in argon by a cw CO_2 laser. The maximum diameter of the focal spot was < 0.2 mm. The laser power density on the target was 2–4.5 MW/cm^2 [82]. The sample material was efficiently evaporated by a laser pulse. The vapor jet was ejected normally to the sample surface by the height up to 3 mm (frame 1 in Fig. 3.41b). Then, the optical breakdown occurred in the jet if $q > 2$ MW/cm^2 (frame 2). At this stage, the plasma plume absorbs laser radiation, which results in the decrease in the surface temperature in the focal zone and the decrease in the vapor density. Thereafter, the plasma bunch begins to decompose and move in the direction opposite to the laser beam propagation (frame 3). As the plasma is "bleached", the laser radiation intensity increases, resulting in the repeated heating of the target and generation of a vapor jet (frame 4). The process is repeated again. The pulsation frequency was 1 kHz.

High-speed frame-by-frame photographing at wavelengths 440 and 632.8 nm revealed the stratification of the plasma bunch with a spatial period of 0.3–0.9 mm [83].

We will describe a more complete COD model in the gas flow by calculating a discharge in an air flow [60] under conditions close to experimental [56]. The discharge was ignited in a CO_2 laser beam of power up to 6 kW. The beam was focused by a lens with the focal distance $F = 200$ mm. The beam divergence was 1.5–3 mrad. The beam convergence angle was varied from 1/8 to 1/5.

To describe correctly the gas flow in the optical discharge, it is necessary to solve the system of gas-dynamic equations: the continuity equation, the Navier–Stokes equation, and the equation of conservation of energy. The transfer of thermal energy plays an important role in the energy balance. This transfer was calculated by the method of multigroup diffusion approximation. In this complex model, laser radiation was calculated within the framework of geometrical optics by neglecting the refraction of beams. We saw earlier that the refraction of laser radiation can refocus a converging beam, increase the focal spot area and displace the focus by a few millimeters. It is possible that the latter effect will cause only some displacement of the COD, although probably the influence of refraction may not be restricted by only this factor.

$$\rho c_p \frac{\partial T}{\partial t} + \rho c_p \mathbf{V} gradT = div(\kappa gradT) + Q_L - Q_R \tag{3.64}$$

$$\frac{\partial \rho}{\partial t} + div(\rho \mathbf{V}) = 0 \tag{3.65}$$

$$\begin{aligned} \frac{\partial \rho u}{\partial t} + div(\rho u \mathbf{V}) = & -\frac{\partial p}{\partial x} - \frac{2}{3}\frac{\partial p}{\partial x}(\mu div \mathbf{V}) \\ & + \frac{1}{r}\frac{\partial}{\partial r}\left[r\mu\left(\frac{\partial u}{\partial r} + \frac{\partial v}{\partial x}\right)\right] + 2\frac{\partial}{\partial x}\left(\mu\frac{\partial u}{\partial r}\right), \end{aligned} \tag{3.66}$$

$$\begin{aligned} \frac{\partial \rho v}{\partial t} + div(\rho v \mathbf{V}) = & -\frac{\partial p}{\partial r} - \frac{2}{3}\frac{\partial}{\partial r}(\mu div \mathbf{V}) \\ & + \frac{\partial}{\partial x}\left[\mu\left(\frac{\partial u}{\partial r} + \frac{\partial v}{\partial x}\right)\right] + 2\frac{\partial}{\partial r}\left(\mu\frac{\partial v}{\partial r}\right) + 2\mu\frac{\partial}{\partial r}\left(\frac{v}{r}\right), \end{aligned} \tag{3.67}$$

$$Q = \int K_\omega (U_{b,\omega} - U_\omega) \mathrm{d}\omega \approx \sum_{k=1}^{N_k} K_k (U_{b,k} - U_k) \Delta\omega_k \tag{3.68}$$

$$div\left(\frac{1}{3K_k}gradU_k\right) = K_k(U_k - U_{b,k}), \quad k = 1, 2, \ldots, N_k \tag{3.69}$$

$$Q_L = K_{L,\omega}P_L \exp\left(-\frac{r^2}{R_L^2}\right)\exp\left(-\int K_{L,\omega}\mathrm{d}x\right)\frac{1}{\pi R_L^2} \tag{3.70}$$

where u and v are the projections of the velocity V on coordinated axes; ρ and c_p are the density and specific heat at a constant pressure, respectively; T is temperature; μ is the dynamic coefficient of viscosity; κ is the heat conductivity; K, U, and U_b are the volume absorption coefficient, the volume radiation density of the medium and perfectly black body, respectively; the subscripts ω and k refer to spectral and group characteristics, respectively, which are found by averaging in each of $N_k = 10$ spectral ranges $\Delta\omega_k$ and $K_{L,\infty}$ is the volume absorption coefficient for laser radiation. The boundary conditions for the solution of the system of equations (3.64)–(3.70) are: the velocity is $V_0 = (u_0, 0)$ and temperature is $T = T_0$ in the unperturbed approach stream ($x = 0$) and at a large distance from the symmetry axis ($r \to \infty$); on the symmetry axis, $v = 0$ and $\partial u/\partial r = \partial T/\partial r = 0$; and the first derivatives of the required functions are zero at the end of the calculation region downstream. The beam diameter at the input to the calculation region was 0.6 cm, the focus was located at a distance of 3 cm from the entrance to a chamber, i.e. the beam convergence angle was 1/5, in accordance with experimental conditions [56]. The radius of caustic was $r_f = 0.45$ mm, which also corresponded to the experimental value in [56]. The authors of [60] studied the COD structure in a longitudinal (along the beam) air flow and the domain of discharge existence in the power-flow rate coordinates. Figures 3.42 and 3.43 show the temperature and current line distributions in the COD at the same laser power and different velocities of the approach air flow, 2 and 8 m/s. The latter value of the velocity is limiting; when this velocity is exceeded, the discharge is quenched. At the smaller velocity $v = 2$ m/s, the discharge is displaced from the focus toward the beam (cf. Figs. 3.36 and 3.42). The isotherm lines are strongly thickened at the leading edge and are elongated at the trailing edge. The latter is explained by the increase in the gas flow rate behind the COD front, which was mentioned in the previous section. This increase in the gas flow rate does not correspond to the conservation of the flow density because the greater part of the approach gas flows around the hot region. No more than 1% of the gas flow comes into the high-temperature region of the COD (the 10 kK isotherm) [61]. To "provide" such flowing around, a narrow elevated-pressure region is established in front of the COD leading edge (Fig. 3.44 [61]). The passing part of the flow is accelerated on the rear slope of this region, the flow rate increasing approximately by an order of magnitude. The relation between these velocities is described by the phenomenological expression [62]

$$V_m \approx \sqrt{\frac{\rho_0}{2\rho_m}}V_0 \tag{3.71}$$

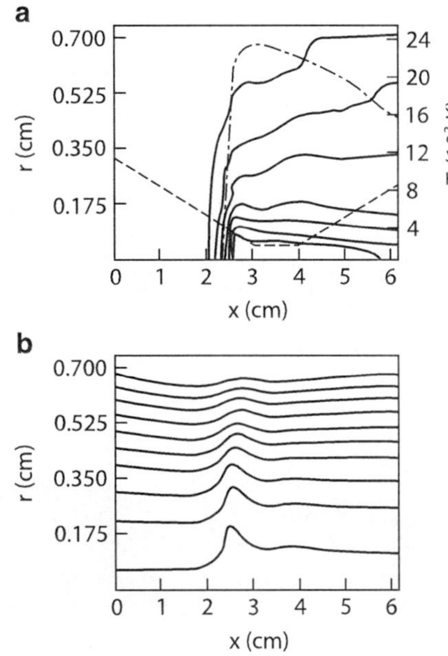

Fig. 3.42 Temperature (**a**) and flow function (**b**) distributions for $P = 5\,\mathrm{kW}$ and $u_0 = 2\,\mathrm{m/s}$; (**a**) the *dashed line* is the conventional boundary of the laser beam; the *dot-and-dash* curve is the temperature distribution along the symmetry axis (scale on the right); (**b**) each tenth isoline of the flow function is shown (from the maximum value of 3.6×10^{-2} g/s)

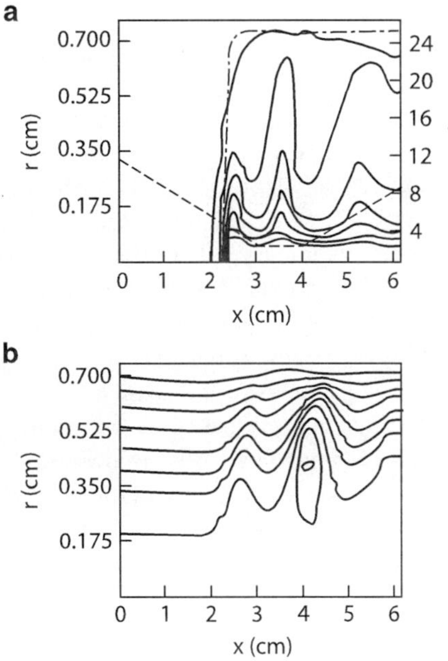

Fig. 3.43 Temperature (**a**) and flow function (**b**) distributions for $P = 5\,\mathrm{kW}$ and $u_0 = 8\,\mathrm{m/s}$; (**a**) the *dashed line* is the conventional boundary of the laser beam; the *dot-and-dash* curve is the temperature distribution along the symmetry axis (scale on the right); (**b**) each tenth isoline of the flow function is shown (from the maximum value of 1.44×10^{-1} g/s)

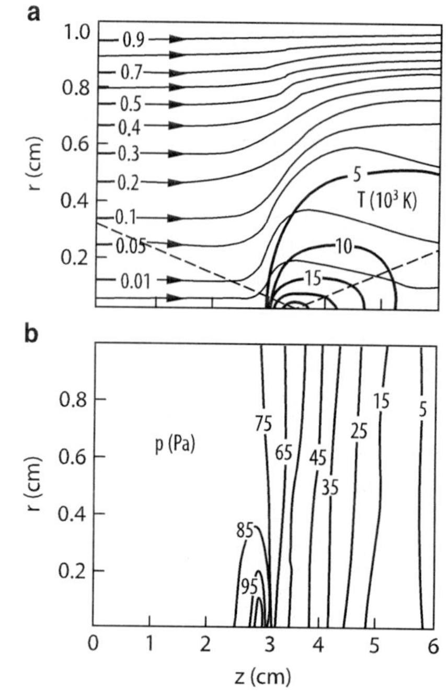

Fig. 3.44 Temperature and gas flow function (**a**), excess pressure (**b**) distributions; $P = 5\,\text{kW}$, $u_0 = 8\,\text{m/s}$ ($G_{\max} = 2.9\,\text{g/s}$), $\omega = 0$ (the *dashed line* is the conventional boundary of the laser beam)

The plasma begins to cool downstream, and the peripheral flow begins to come to the region of the gas flow trace. This is clearly seen from the current line structure in Fig. 3.42. As the critical flow rate is approached, the COD structure changes qualitatively. The COD front remains at its place, but periodic oscillations are excited downstream. Also, vortices are formed here (see closed current lines in Fig. 3.43). Such a nonstationary COD combustion regime near the discharge quenching boundary was observed experimentally [56, 63]. A comparison of the results of calculations with experiments [56] is presented in Fig. 3.45 [60]. It is shown that, as the gas flow rate increases, the region of the laminar gas flow in the COD changes to the region with regular oscillations. As the approach flow rate further increases, the artificially ignited discharge is quenched. It was found in experiments [63] that COD oscillations disappeared during the turbulization of the approach gas flow at high flow rates. The mechanism of this phenomenon was verified by studying the qualitative dependence of the maximal COD combustion rate on the gas turbulence degree [60]. For this purpose, the turbulence model [64] was used with the turbulent viscosity

$$\mu = \rho l^2 \left[\left(\frac{\partial u}{\partial r} \right)^2 + \left(\frac{\partial v}{\partial x} \right)^2 \right]^{1/2}$$

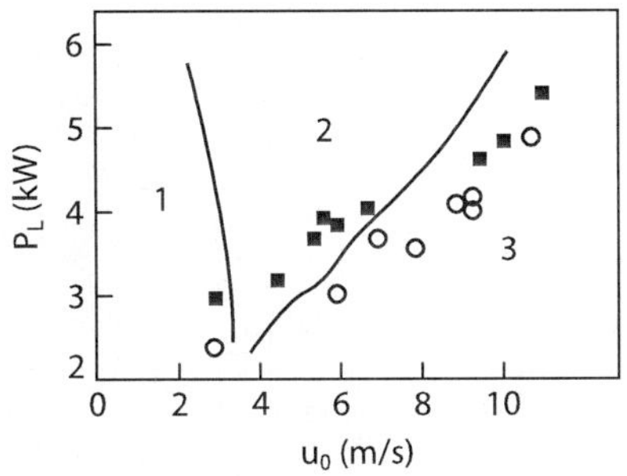

Fig. 3.45 Comparison of calculations with experimental data: (**1**) laminar flowing around region; (**2**) region of stable combustion with vortex and oscillatory processes; (**3**) instability region (discharge quenching). The *dark squares* and *circles* are limiting velocities obtained in experiments with longitudinal and transverse air flows

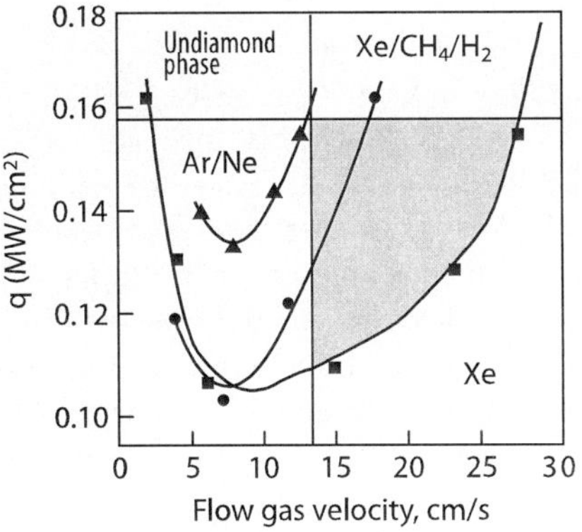

Fig. 3.46 Laser power thresholds vs gas flow rate at the atmospheric-pressure discharge in laser plasmotron. Dark region illustrate the possibilities of diamond formation

introduced into the hydrodynamic equation, where l is the Prandtl mixing length. Numerical calculations showed that the oscillation amplitude monotonically decreases with increasing l from 0.01 to 1.

It follows from the above discussion that the influence of turbulence on the COD stability is caused by the increase in the effective viscosity rather than the heat conduction, as would be assumed. Thus, the simplest models of a COD in the gas flow are in qualitative agreement with the results of complex radiative-gas-dynamic models. However, there exists a specific feature in the dependence $P_{\min}(V)$ at low flow rates. As V is increased from 0 to 10 cm/s, the value of $P_{\min}$ decreases approximately by a factor of 1.5 compared to $P_{\min}(V = 0)$ (Fig. 3.46). This is caused by the COD displacement to the focal spot, which alleviates the supporting of the COD and reduces the lower power threshold [65]. The temperature fields of the COD measured in argon show (Fig. 3.47) that a plasma bunch is displaced to the focus with increasing the gas flow rate and its temperature increases, although the

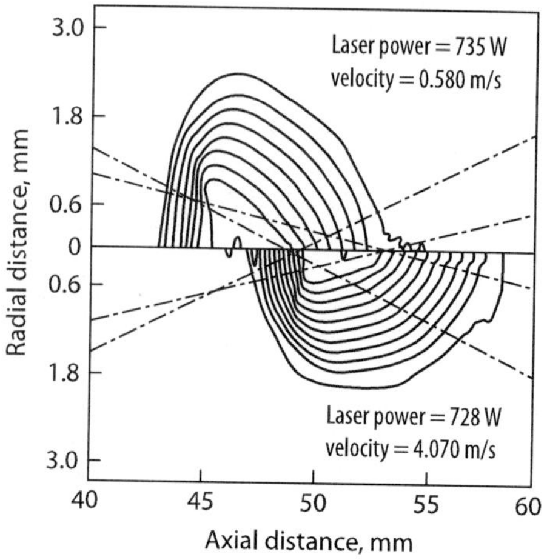

Fig. 3.47 Influence of the gas flow rate along the laser beam on the COD structure; the focal distance of a lens is 203 mm and the gas pressure is 2.5 atm. The last isotherm corresponds to 10,000 K, the next isotherms follows through 500 K

rate of the cooling gas flow increases almost by an order of magnitude [66, 77]. The analytic model does not describe this "delicate" effect. At such low rates, which do not exceed the natural convection rate, the spatial orientation of the laser beam becomes important. When the laser beam is directed vertically, the discharge supporting thresholds are smaller approximately by half than in the case of the horizontal orientation of the laser beam [67]. We will consider the influence of the transverse velocity of the COD in more detail because such a scheme is often used in practice, e.g., to quench the optical discharge during laser welding (see the next chapter). In this case, the term $vc_p\rho\theta/\varphi x$ (where v is a constant), which approximately describes the convective cooling of the discharge, should be added to (3.46). By solving equation (3.46) (where $v = 0$, plane light beam), as in Sect. 3.4, we obtain the expression for the COD combustion threshold in a transverse air flow:

$$\bar{P}_H = \mu\frac{r_f}{\varphi}\left[\frac{\mu r_f}{\varphi} + v^* + \sqrt{v*\left(\frac{\mu r_f}{\varphi} + v*\right)}\right] \tag{3.72}$$

where $v* = v\bar{\kappa}/\varphi\gamma\chi$ and other notations are as in Sect. 3.4. Figure 3.27 shows the dependence of the threshold discharge combustion power on the beam radius for $v = 20$ m/s, which was calculated by (3.72). The corresponding experimental dependence [27] is also presented.

Simulation of plasma blowing away during laser welding

The numerical study of the influence of the laser beam orientation within the framework of a complete gas-dynamic model involves additional difficulties because a COD in the gas flow oriented perpendicular to the laser beam should be described by using a three-dimensional model. The authors of [68] added to (3.64)–(3.70) the terms corresponding to the three-dimensional nature of the gas flow, thermal transfer, and thermal radiation. Laser radiation is described in the one-dimensional approximation, as in papers [60, 61] (Fig. 3.48). As in a longitudinal

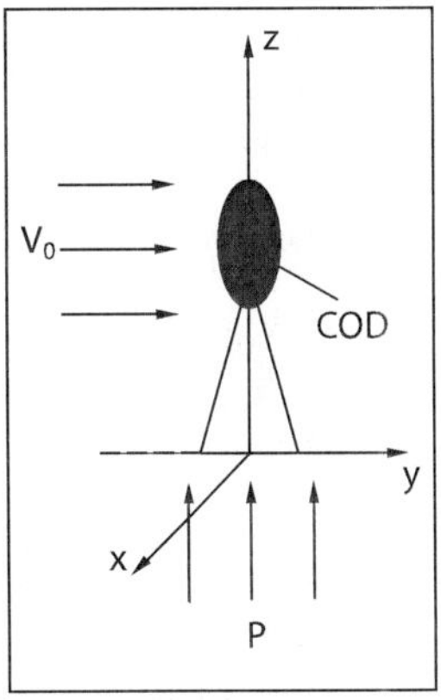

Fig. 3.48 Scheme for the COD support in a transverse gas flow

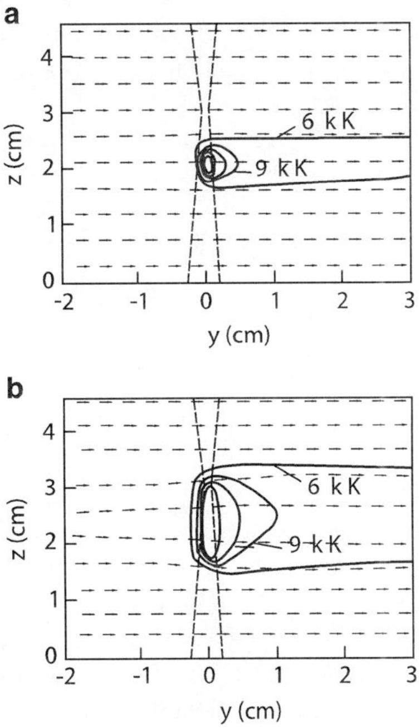

Fig. 3.49 Temperature distribution (isotherms with a step of 3,000 K) and the distribution of the flow direction field on the plane $x = 0$ for $V_0 = 8$ m/s (**a**) and 4 m/s (**b**) (the *dashed line* shows the conventional boundary of the laser beam)

discharge, steep thermal and gas-dynamic fronts are formed from the side of the approach transverse flow and long "tails" appear downstream (Fig. 3.49). Ahead of the front, the gas flow deviates perpendicular to the beam. In this case, the size of the hot plasma along the beam considerably decreases compared to the longitudinal discharge (cf. Figs. 3.49 and 3.42). Figure 3.49 also shows that the COD length along the beam considerably decreases with increasing the transverse flow rate. This circumstance probably explains the efficient action of the transverse gas flow on the COD during laser welding. Figure 3.50 shows the influence of the interaction of the

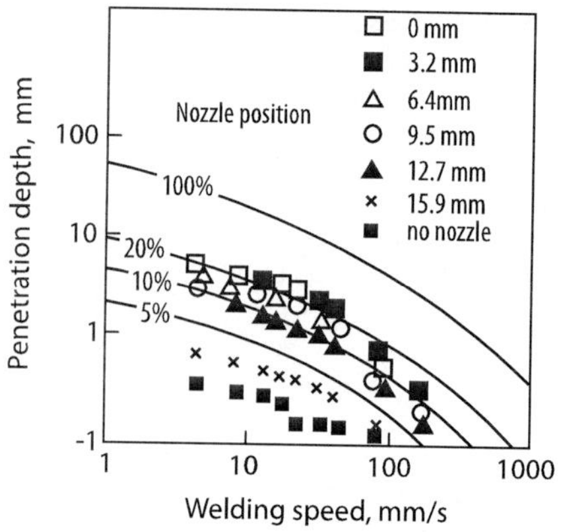

Fig. 3.50 Effect of the optical-discharge suppression on the welding depth. A gas nozzle supplied an argon jet perpendicular to the welding speed direction. The nozzle was located at different distances from a sample surface

transverse gas flow with the COD on the laser welding depth [69]. The maximum flow rate above which the COD is quenched weakly depends on the flow direction (along or perpendicular to the laser beam). This is demonstrated experimentally in Fig. 3.45.

It is of interest to consider the refraction of a laser beam in the plasma of an optical discharge burning in a transverse gas flow. This problem can be solved in the simplest way for a plane light beam. We assume, as before, that the gas flow rate is constant. Let us add to (3.56) the term $c_p \rho v \partial T/\partial x$ taking into account heat removal with the gas flow along the transverse coordinate x. Figure 3.51 presents the results of calculations by (3.56) and (3.57) performed for different flow rates at $P = 20\,\text{kW/cm}$ [70]. The focusing angle is $\varphi \approx 0.15$. One can see from Fig. 3.51a,b that the temperature distribution in the plume along the beam has two maxima. The optical discharge, as usual, moves in the direction opposite to the laser beam propagation. At the same time, the absorption of light in the plume is insignificant, and therefore the radiation intensity in the geometrical focus proves to be high enough to maintain another plasma bunch. However, as in three-dimensional calculations, the discharge size decreases and its stratification disappears with increasing the gas flow rate due to cooling. As expected, the light beam is deflected toward the gas flow Fig. 3.51c, which is analogous to the beam refraction in the atmospheric air (see Sect. 1.1). As the gas flow rate increases, the plume is bleached and the COD is quenched at $v = v_\kappa$. To describe correctly the refraction of the laser beam in the COD in the transverse gas flow, it is necessary to solve simultaneously equation (3.57) for the propagation of laser radiation and the complete system of equations of radiative-gas-dynamic model (3.64)–(3.70) in the three-dimensional approximation.

3.5.4 *Optical Discharge in a Gas–Vapor Keyhole*

In previous sections, we considered the COD burning either away from boundaries or its combustion was limited by a plane target. In some technological operations

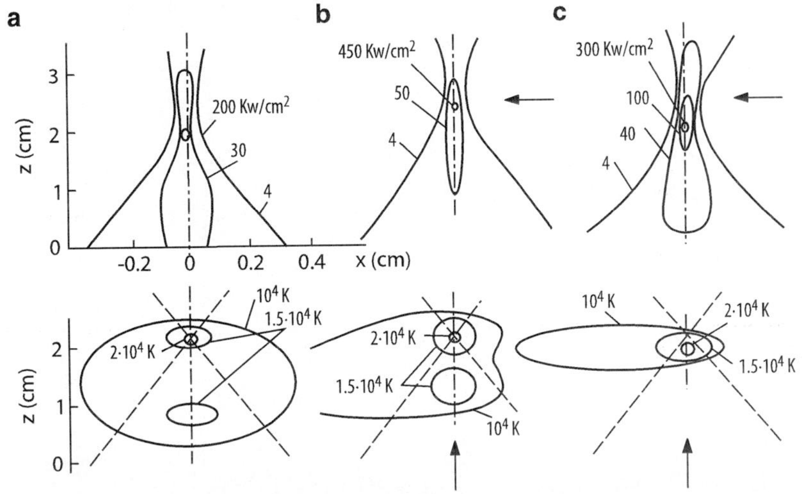

Fig. 3.51 Intensity distribution (*top*) and temperature field (*bottom*) in a COD in the transverse air flow for $v = 0$ (**a**), $v = 20\,\text{m/s}$ (**b**), and $v = 40\,\text{m/s}$ (**c**) ($P = 20\,\text{kW/cm}$, $\varphi = 0.15$); the *dashed line* is the beam boundary; the *arrows* show the gas flow direction (*top*) and beam direction (*bottom*)

such as welding and cutting, a plasma plume can probably appear also in a narrow channel produced in a sample. Therefore, it is of interest to consider this phenomenon. The experimental study of a COD in a keyhole, e.g., during welding is hindered because of a small size of the keyhole. For example, the authors of [?] measured the plasma temperature and density during the welding of steel in argon by a 3-kW CO_2 laser. The temperature and electron density at a point 0.5 mm above the steel surface were $T = 9{,}000$ K and $n_e = 2 \times 10^{16}\ \text{cm}^{-3}$, respectively. At a depth of 0.5 mm, temperature increased up to 13,000 K and the electron density by an order of magnitude. These values were already close to the case of a freely burning optical discharge. The influence of the plasma inside a keyhole on laser radiation was considered in experiments [71], where the attenuation of the CO_2 laser beam during welding thin (<1 mm) steel sheets was measured. The geometry of the experiment is shown in Fig. 3.52. It was found that during welding 0.12-mm-thick plates by a 12-kW laser at the speed of ~30 m/min, the laser beam propagated through a plate lost 50% of its power. This can be explained by the absorption of radiation in keyhole walls. As the welding speed was decreased, the attenuation of the beam in plasma became significant, corresponding to the absorption coefficient $\alpha \sim 1.5\ \text{cm}^{-1}$.

The theoretical study of an optical discharge in a keyhole is a complicated problem. We will see in the next chapter that the calculation of the parameters of a deep gas–vapor keyhole is an independent intricate problem. The consideration of plasma phenomena under these conditions will further complicate this problem. Estimates with the help of simplified modes are inefficient.

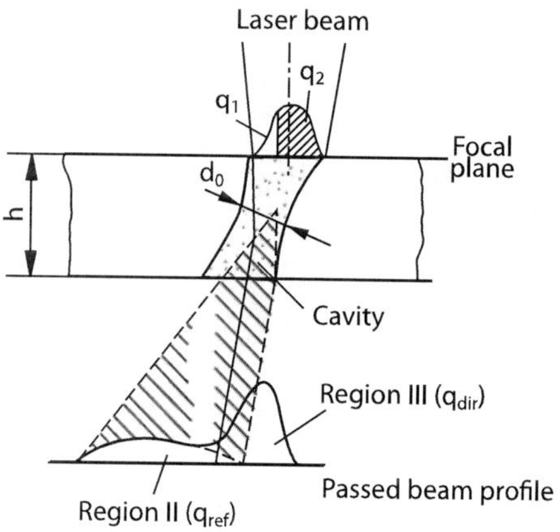

Fig. 3.52 Geometry of the experiment for measuring laser radiation transmitted through a steel plate during through welding

3.6 Laser Plasmatron and Deposition of Films

One can see from previous sections of this chapter that the investigation of an optical discharge is of interest from the practical point of view because the discharge plays a certain role in the laser processing of materials. In the last years a new technology for synthesis of diamond and diamond-like films by the laser-plasma method using a COD in the flow of special gas mixtures is being successfully developed [67, 72–74]. The importance of the development of efficient methods for deposition of such films is obvious. The method for deposition of such films with the help of a laser plasmatron offers a number of advantages compared to the methods developed earlier (high-frequency, microwave, glowing discharges, electric plasmatrons). The main advantages are:

1. The possibility to deposit films on parts without using a vacuum chamber.
2. The possibility of scanning and processing large parts.
3. The possibility to obtain pure films, not contaminated with electrode erosion products, as occurs, e.g., in electric plasmatrons, etc.

At present the record deposition rate for diamond films with the help of a laser plasmatron is $\sim$120 μm/h [73], which is close to the deposition rates of conventional methods.

3.6.1 Physical Processes in Optical Plasmatrons

Designs of a laser plasmatron for the deposition of diamond films can be different. For example, Fig. 3.53 shows the scheme of a setup which does not allow the scanning of the surface of parts being processed [67]. In [75], the scheme was proposed in which a plasma jet freely flows from a plasmatron nozzle on a substrate

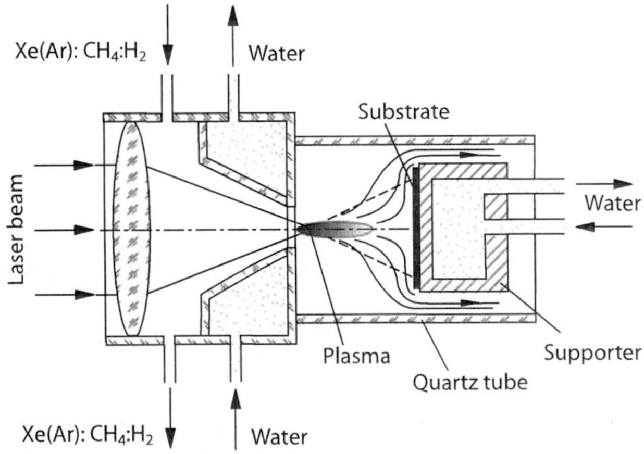

Fig. 3.53 Principal scheme of a plasma-chemical reactor based on a laser plasmatron for the deposition of diamond films

Fig. 3.54 Photography of a COD in a hydrocarbon gas mixture

(Fig. 3.54). This permits the displacement of the substrate and the deposition of films over large areas. Figure 3.55 shows a Raman spectrum indicating the deposition of a diamond film by using such a plasmatron. The deposition zone is protected from air (~0.1% of a nitrogen impurity suppresses the diamond film growth) by using the coaxial blowing with argon [75].

An optical discharge is ignited externally by introducing a wire to the focal volume or by a laser or electric spark and is supported near the nozzle cut. The region of COD existence in the gas flow was described above. Figure 3.46 shows that addition of molecular gases H_2 and CH_4, which are required for the deposition of a diamond film, to inert gas (argon or xenon) leads to the increase in the COD maintenance threshold. The formation of a diamond film is a "delicate" process,

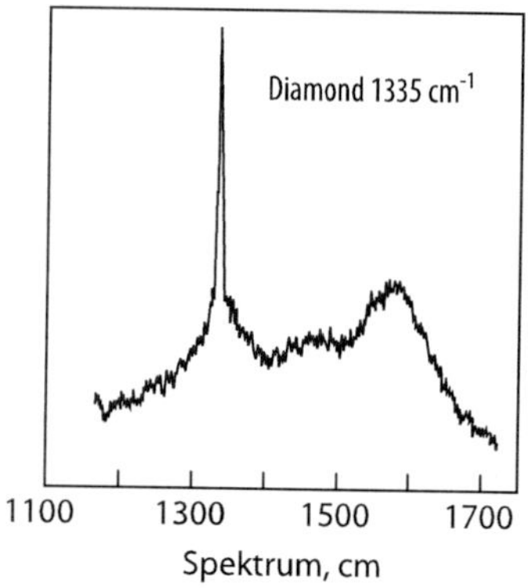

Fig. 3.55 Raman scattering spectrum confirming the formation of a diamond film

which depends on a careful selection of experimental parameters: the preparation of a Mo or W substrate, the substrate temperature, the chemical composition of a gas jet, its velocity, etc. [74]. The maximum diamond deposition rate is achieved at the substrate temperature 1,200° ± 100°C. As shown above, approximately half the energy of a CO_2 laser beam is spent to support the COD (Fig. 3.31). Despite a high reflection coefficient of the substrate, the transmitted radiation plays an important role in its thermal balance. The COD transparency decreases with increasing laser power (Figs. 3.31 and 3.56) and increases with increasing the amount of a molecular component in the gas jet (Fig. 3.56 [67]). The latter property is related to the increase in the COD maintenance threshold with increasing the amount of a molecular component in the gas flow. As the radiation power approaches the lower threshold, the absorption in the COD deceases due to a decrease in its size (Fig. 3.28). The molecular component is added to increase the film deposition rate. In this case, due to the decrease of absorption in the COD, its efficiency decreases. It is necessary either to increase the laser power or to use beams from two lasers focused on one spot [36]. It is also possible to add molecular gases to a plasma jet behind the COD (downstream). The latter method leads to a noticeable increase in the deposition rate (~60 μm/h), but here problems with the efficient mixing of gases appear (Fig. 3.57). It is also possible to increase the deposition rate by increasing the total gas pressure. In this case, the COD maintenance threshold (3.55) should decrease, and relatively cheap 2–3 kW lasers can be used. In this connection, it is interesting to consider the properties of a laser plasmatron at high pressures.

3.6.2 *High-Pressure Plasmatron*

To elucidate the basic properties of such a plasmatron, we consider the case of high-pressures ($p = 50$ atm) [76]. As follows, e.g., from (3.52), the lower COD maintenance threshold decreases with increasing pressure because the absorption coefficient μ increases by more than an order of magnitude compared to the case

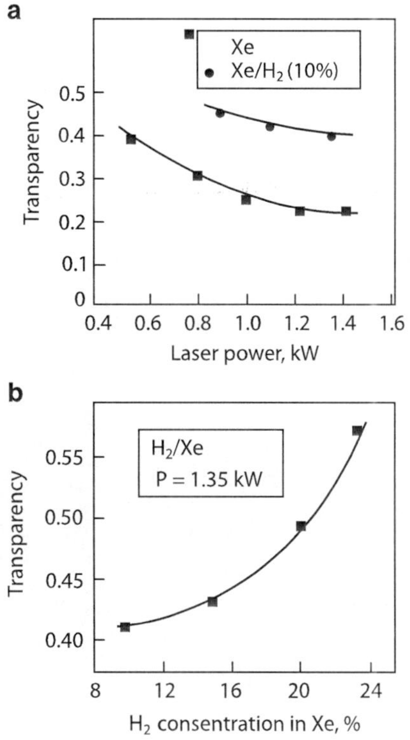

Fig. 3.56 Dependence of the transmission coefficient of plasma for laser radiation on the laser power (**a**) and hydrogen concentration in a plasma-carrying gas (**b**)

of atmospheric pressure. On the other hand, the heat conduction of gas increases with increasing pressure, which partially prevents a drastic decrease in P_{min}. It is interesting to select the laser plasmatron parameters so that the laser power would proportionally increase with increasing pressure. In this case, the laser radiation intensity in the focal volume remained close to that for the atmospheric plasmatron described above (Fig. 3.46). This leads to the power $P = 50$ kW and the beam waist radius $r_f \sim 0.4$ cm [76]. Gas flow rates are chosen typical for an optical plasmatron (10–50 cm/s). Under these conditions, the general picture of discharge combustion is similar to the atmospheric-pressure discharge (cf., e.g., Figs. 3.57 and 3.42). The main feature of the high-pressure COD is that the length of the high-temperature region is considerably (by 3–4 times) is smaller than the COD length at the atmospheric pressure. This is explained by the high absorption coefficient at high pressures. According to the numerical experiment, the relation between different channels of energy losses from a COD at high pressures is as follows: 50%, 30%, and 20% of energy are lost due to heat conduction, thermal radiation, and removed by the gas flow, respectively. The latter value determines the efficiency of a laser plasmatron at high pressures. As the air flow rate is increased above 0.6 m/s, the discharge is quenched. It means this regime of plasmatron is regime near lower laser power threshold, like at atmospheric-pressure (Fig. 3.46). As the laser power

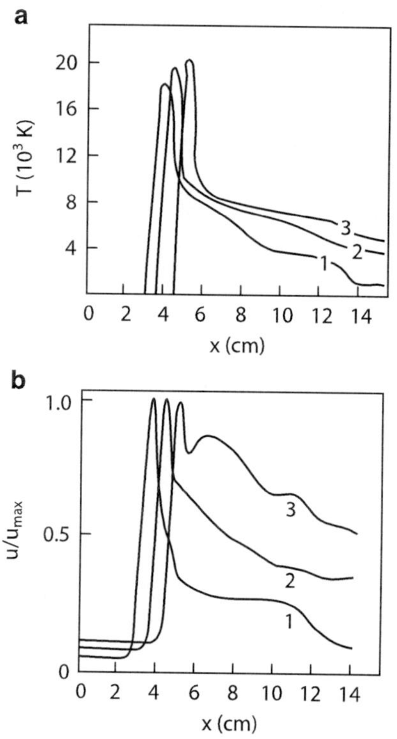

Fig. 3.57 Temperature (**a**) and velocity (**b**) distributions along the symmetry axis of an optical discharge in air for $P = 50\,\text{kW}$; (**1**) $u_0 = 0.1$ m/s, $u_{\max} = 1.87$ m/s; (**2**) $u_0 = 0.2$ m/s, $u_{\max} = 2.78$ m/s; (**3**) $u_0 = 0.3$ m/s, $u_{\max} = 3.08$ m/s

is doubled ($P = 100\,\text{kW}$), the discharge at $V = 0.3\,\text{m/s}$ becomes unstable. As the gas flow rate is increased up to 0.5 m/s, the discharge is stabilized. The latter factors illustrate the COD regime near the upper laser power threshold. Numerical calculations suggest that the power and efficiency of an optical plasmatron can be increased by increasing the gas pressure.

References

1. Yu.P. Raizer, *Lazernaya iskra i rasprostranenie razryadov (Laser Spark and Propagation of Discharges)* (Nauka, Moscow, 1974)
2. A.I. Barchukov, F.V. Bunkin, V.I. Konov, et al., Zh. Eksp. Teor. Fiz. **66**, 965 (1974)
3. E. Sturmer, M. Von Allmen, J. Appl. Phys. **49**, 5648 (1978)
4. E. Locke, E. Hoag, R. Hella, IEEE J. Quant. Electron. **2**, 132 (1972)
5. E.V. Dan'shchikov, V.A. Dymshakov, F.V. Lebedev, et al., Preprint Inst. Atom. Energy-3534 (1982)
6. N.N. Rykalin, A.A. Uglov, M.M. Nezametdinov, Dokl. Akad. Nauk SSSR **218**, 330 (1974)
7. R.S. Devoto, Phys. Fluids **16**, 616 (1973)
8. R.S. Devoto, J. Plasma Phys. **2**, 17 (1968)
9. Yu.P. Raizer, *Osnovy sovremennoi fiziki gazorazryadnykh lazerov (Fundamentals of the Modern Physics of Gas-discharge Processes)* (Nauka, Moscow, 1980)

10. G.I. Kozlov, V.A. Kuznetsov, V.A. Masyukov, Zh. Eksp. Teor. Fiz. **66**, 954 (1974)
11. A. Noiberger, Raket. Tekh. Kosm. **1**, 3 (1975)
12. I.V. Nemchinov, M.I. Popov, L.I. Shubadeeva, Inzh. Fiz. Zh. **4**, 577 (1982)
13. A.M. Bonch-Bruevich, V.I. Zinchenko, Ya.A. Imas, et al., Zh. Tekh. Fiz. **51**, 919 (1981)
14. Yu.N. Raizer, S.T. Surzhikov, Kvantovaya Elektron. **11**, 2301 (1984)
15. V.Yu. Baranov, I.V. Kochetov, V.G. Pevgov, et al., Preprint Inst. Atom. Energy-3080 (1979)
16. V. Autric, J.P. Carcssa, D. Duffrense, Acad. Sci. **B288**, 237 (1979)
17. V.A. Aleksandrov, N.L. Konchakov, E.E. Son, Fiz. Plasmy **4**, 1182 (1978)
18. B.Y. Moizhes, G.E. Pikus (eds.) *Termoemissionnye preobrazovateli I nizkotemperaturnay plazma (Thermal Emission Transducers and Low-temperature Plasma)* (Nauka, Moscow, 1973)
19. A.A. Vedenov, G.G. Gladush, Fizx. Khim. Obrzb. Mater. **1**, 142 (1979)
20. A.A. Vedenov, G.G. Gladush, A.N. Yavokhin, Fiz. Plazmy **9**, 434 (1983)
21. Yu.B. Zel'dovich, G.I. Barenblat, V.B. Lbrovich, et al., *Matematicheskaya teoriya goreniya i vzryva (Mathematical Theory of Combustion and Explosion)* (Nauka, Moscow, 1980)
22. B.M. Smirnov, *Iony i vozbuzhdennye atomy v plazme (Ions and Excited Atoms in Plasma)* (Atomizdat, Moscow, 1974)
23. A.A. Vedenov, G.G. Gladush, A.N. Yavokhin, Kvantovaya Elektron. **8**, 1485 (1981)
24. V.S. Vorob'ev, A.L. Khomkin, Pis'ma Zh. Tekh. Fiz. **9**, 1157 (1983)
25. M.I. Arzuev, A.I. Barchukov, F.K. Bunkin, Kvantovaya Elektron. **2**, 963 (1975)
26. Ya.B. Zel'dovich, Yu.P. Raizer, *Fizika udarnykh voln I vysokotemperagturnykh gidrodinamicheskikh yavlenii (Physics of Shock Waves and High-temperature Hydrodynamic Phenomena)* (Nauka, Moscow, 1966)
27. M.S. Fowler, D.C. Smith, J. Appl. Phys. **46**, 138 (1975)
28. M.I. Arzuev, A.I. Barchukov, F.K. Bunkin, Kvantovaya Elektron. **2**, 1717 (1975)
29. E.V. Dan'shchikov, F.V. Lebedev, A.V. Ryazanov, Kvantovaya Elektron. **11**, 1424 (1984)
30. B. Seidel, J. Beersiek, E. Beyer, Proc. SPIE **2207**, 279 (1994)
31. S. Tsukamoto, K. Hiraoka, Y. Asai, et al., Proc. ICALEO'96 **B76** (1996)
32. W.W. Duley, *Laser Welding* (Wiley, New York, 1999)
33. A. Poueyo-Verwaerde, R. Fabbro, G. Deshors, et al., J. Appl. Phys. **74**, 5773 (1993)
34. M. Schellhorn, A. Eichhorn, *ECLAT'96* **107** (1996)
35. G.I. Kozlov, I.K. Selezneva, Zh. Tekh. Fiz. **48**, 386 (1978)
36. E.B. Kulumbaev, V.M. Lelevkin, Vestn. Kyrgyz.-Ross. Slavyansk. Univ. **2** (2002)
37. D. Kefer, R. Welle, C. Peters, AIAA J. **24**, 1663 (1986)
38. C. Carloff, C. Gillet, E. Krametz, A. Muller, J.H. Schufer, J. Uhlenbusch, J. Arch. Mech. **36**, 473 (1984)
39. S.T. Surzhikov, Fluid Dynam. **32**(3) (1997)
40. G.G. Gladush, A.N. Yavokhin, Kvantovaya Elektron. **10**, 1399 (1983)
41. S.P. Babenko, M.A. Yakovlev, Fiz. Khim. Obrab. Mater. **1**, 44 (1988)
42. Yu.P. Raizer, Kvantiavaya Elektron. **11**, 64 (1984)
43. G.G. Gladush, A.F. Mamzer, A.N. Yavokhin, Fiz. Plazmy **11**, 236 (1985)
44. G.I. Kozlov, V.A. Kuznetsov, V.A. Masyukov, Zh. Tekh. Fiz. **49**, 2304 (1979)
45. I. Miyamoto, H. Maruo, Proc. LAMP, Osaka, **311** (1992)
46. A.G. Grigor'yants, S.F. Moryashchev, V.A. Fromm, Izv. Vyssh. Uchebn. Zaved., Ser. Mashin. **5**, 109 (1980)
47. T. Ishide, S. Shono, T. Ohmae, Proceeding LAMP'87, Oaka, 187 (1987)
48. G.G. Gladush, S.F. Moryashchev, A.A. Startsev, et al., Kvntovaya Elektron. **12**, 414 (1985)
49. E.I. Klosterman, S.R. Byron, J. Appl. Phys. **45**, 4751 (1974)
50. S.T. Surzhikov, Kvantovaya Elektron. **30**, 416 (2000)
51. J.P. Jackson, P.E. Nicolson, AIAA J. **12**, 1498 (1974)
52. K.G. Gus'kov, Yu.P. Raizer, S.T. Surzhikov, Kvantovaya Elektron. **17**, 937 (1990)
53. S.T. Surzhikov, Teplofiz. Vysok. Temper. **32**, 292 (1994)
54. I.A. Bufetov, A.M. Prokhorov, V.B. Fedorov, et al., Izv. Akad. Nauk SSSR, Ser.Fiz. **66**, 1141 (1982)

55. M.V. Gerasimenko, G.I. Kozlov, V.A. Kuznetsov, Kvantovaya Elektron. **10**, 709 (1983)
56. N.A. Generalov, A.M. Zakharov, V.D. Kosyn kin, M.Yu. Yakimov, Fiz. Goren. Vzryv. **22**, 91 (1986)
57. G. Karslou, Ya. Eger, *Teploprovodnost' tverdykh tel (Heat Conduction of Solids)* (Nauka, Moscow, 1964)
58. B.M. Zhirikov, A.I. Korotchenko, N.I. Popov, et al., Kvantovaya Elektron. **10**, 1190 (1983)
59. N.N. Rykalin, A.A. Uglov, M.B. Inat'ev, Dokl. Akad. Nauk SSSR **265**, 1117 (1982)
60. S.T. Surzhikov, A.A. Chentsov, Fiz. Plazmy **22**, 1054 (1996)
61. E.B. Kulumbaev, V.M. Lelevkin, Fiz. Plazmy **25**, 205 (1999)
62. V.Ts. Gurovich, E.B. Kulumbaev, V.M. Lelevkin, Fiz. Plazmy **24**, 1010 (1998)
63. A. Baranovsky, Z. Mukha, Z. Peradzyn'sky, Usp. Mekh. **1**, 125 (1978)
64. D.A. Anderson, J. Tannenhili, R. Pletner, *Vychislitel'naya gidromekhanika I teploobmen (Calculation Hydromechanics and Thermal Exchange)*, vol. 1 (Mir, Moscow, 1990)
65. Yu.P. Raizer, Usp. Fiz. Nauk **132**, 549 (1980)
66. R. Welle, D. Keefer, C. Peters, AIAA J. **25**, 1093 (1987)
67. S.A. Uglov, A.P. Bol'shakov, V.I. Konov, F. Dauzinger, B. Angstenberg, Laser-Inform **1**, 3 (1999)
68. E.B. Kulumbaev, V.M. Lelevkin, Fiz. Plazmy **25**, 568 (1999)
69. S. Chiang, C.E. Albright, *Proc. ICALEO'92* **491** (1992)
70. G.G. Gladush, A.F. Mamzer, A.P. Yavokhin, Abstracts of Papers, VI All-Union Conf. on Physics of a Low-temperature Plasma, vol. 2 (Leningrad, 1983), p. 176
71. I. Miyamoto, H. Marno, Y. Arata. Proceeding SPIE, 668, 11 (1986)
72. D. Dischler, C. Wild, *Low-Pressure Synthetic Diamond: Manufacturing and Aplications* (Springer, Heidelberg, 1998)
73. S. Metev, H. Brecht, J. Schwartz, G. Sepold, Diamond Relat. Mater. **11**, 472 (2002)
74. A.P. Bolshakov, V.I. Konov, A.M. Prokhorov, S.A. Ulov, F. Dausinger, Diamond Relat. Mater. **10**, 1559 (2001)
75. A.P. Bolshakov, V.G. Vostrikov, V.Yu. Dubrovsky, V.I. Konov, F.V. Kosyrev, V.G. Naumov, V.G. Ralchenko, Kvantovaya Elektron. 35, 385 (2005)
76. S.T. Surzhikov, Teplofiz. Vysok. Temp. **32**, 714 (1994)
77. D.R. Keefer, Laser-Induced Plasma and Applications. *Optical Engineering 21*, ed. by L.I. Radzieski, D.A. Gremers (Marcel Dekker, Inc., New York, 1989)
78. S. Fujinaga, H. Takenaka, T. Narikigo, S. Karayama, A. Masunam, J. Phys. **33**, 492 (2000)
79. S.V. Dresvin (ed.) *Fizika i tekhnika nizkotemeraturnoi plazmy (Physics and Techniques of Low-Temperature Plasma)* (Atomizdat, Moscow, 1972)
80. A.S. Predvoditelev, E.V. Stupchenko, A.S. Plemanov, et al., *Tablitsy termodinamicheskikh funktsii vozdukha (Tables of Thermodynamic Functions of Air)* (Computing Center, Acad. of Sciences, Moscow, USSR, 1962)
81. H.V. Fritsch, *Fundamental Aspects of Laser Material Processing, European Conference on Laser Treatment of Materials* (Oberursel, 1987), p. 19
82. A.E. Zaikin, A.V. Levin, A.L. Petrov, Kvanvaya Elektron. **18**, 1229 (1991)
83. A.E. Zaikin, A.V. Levin, A.L. Petrov, Kvanvaya Elektron. **22**, 145 (1995)
84. A.G. Gnedovets, E.B. Gnedovets, E.B. Kul'baiskii, I. Smurov, Appl. Surf. Sci. **96/98**, 272 (1996)
85. S.T. Surzhikov, Dokl. Ross. Akad. Nauk **420**, 64 (2008)
86. R. Ducharme, P. Kapadia, J. Dowden, K. Williams, W.M. Steen, *Integrate Mathematical Model of the Keyhole and Weld Pooll in theLaser Welding of Him Metal Sheets, ICALEO*, 177, 1992
87. E.V. Dan'shchikov, V.A. Dymshakov, F.V. Lebedev, V.D. Pis'mennyi, A.V. Ryazanov, Izv. Akad. Nauk SSSR, Ser. Fiz. **49**, 811 (1985)

Chapter 4
Properties and Mechanisms of Deep Melting of Materials by a cw Laser Beam

Abstract This chapter presents the models of deep melting of metals by laser radiation, which are based on the thermal balance and mechanical equilibrium of a gas–vapor keyhole. The models describe quantitatively the basic experimental properties of the process. More intricate approaches are developed which successively remove assumptions of the simplest models by taking into account the non-cylindrical shape of the keyhole, its finite width, stability to capillary waves, and the stability of the leading edge of the keyhole. The influence of the polarization and quality of laser radiation inherent in new laser systems are described.

The welding and cutting of materials by radiation beams from cw and repetitively pulsed CO_2 and Nd:YAG lasers have wide industrial applications (see, e.g., [1] and references therein).

Laser welding provides the fusion zone and weld of small sizes and a high welding speed. Unlike electron-beam welding, vacuum is not required and laser welding can be performed in gases. The weld depth produced by an electron beam in most cases only slightly exceeds that of the laser weld (for the same welding power and speed).

At the same time there exist a number of substantial differences. For example, a laser beam at a low welding speed penetrates to a smaller depth in a metal than an electron beam. Laser welding is sensitive to the type of gas in which welding is performed. There also exist other problems whose solution would expand industrial applications of laser welding. They include, e.g., the production of defect-free welds, the residual porosity, the expansion of a number of materials subjected to welding, etc.

Note that laser welding process is an intricate phenomenon which is determined by the interaction of a number of factors such as the dynamics of heat-and-mass exchange between the vapor, liquid, and solid phases in a weld pool, optical and plasma phenomena in a vapor keyhole and in the gas medium over the keyhole. Figure 4.1a presents the general schematic picture of the assumed processes which

G.G. Gladush and I. Smurov, *Physics of Laser Materials Processing*,
Springer Series in Materials Science 146, DOI 10.1007/978-3-642-19831-1_4,
© Springer-Verlag Berlin Heidelberg 2011

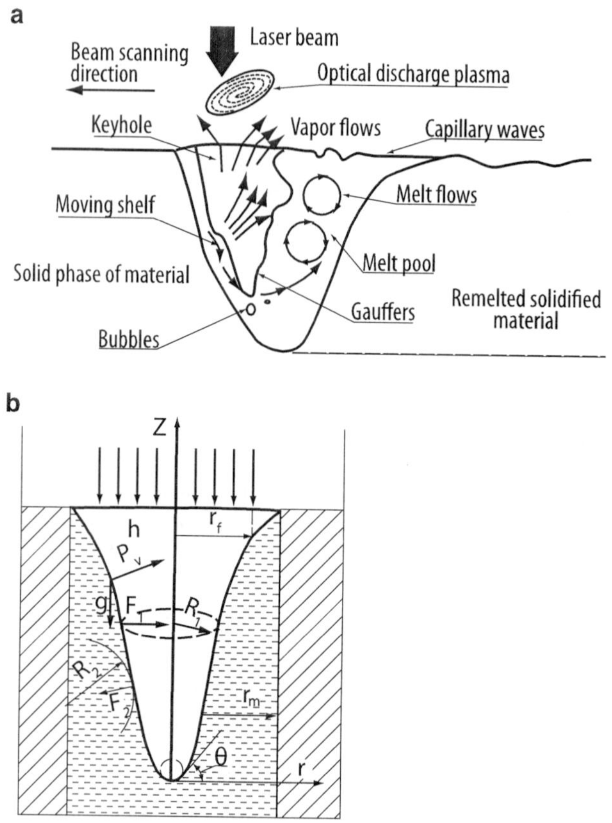

Fig. 4.1 Scheme of a keyhole (**a**) and melt pool geometry upon deep melting (**b**)

considerably affect each other [2]. For simplicity the propagation of a laser beam in the optical-discharge plasma and in a gas–vapor keyhole is not shown. All these processes determine eventually the main properties of the preparation of a weld pool such as the dependencies of the weld depth and width on the laser beam power, structure, and displacement speed. These processes will be considered to one or another extent in this chapter. Welding phenomena themselves, which determine the structure, mechanical, corrosion and other properties of the weld, are beyond the scope of our analysis. They are the subject of investigation in a special area of knowledge (see, e.g., [3]).

The optical, gas-dynamic, and hydrodynamical equations describing intricate laser welding process cannot be solved exactly (analytically or numerically). Therefore, it is necessary to construct simplified theoretical models and compare the results of calculations with experimental data. Unfortunately, the available experimental data and information on experimental conditions are too limited for obtaining reliable dependencies suitable for comparison with theoretical models. This circumstance complicates the elucidation of the role of individual factors in welding process and impedes the construction of the general picture of this process.

In the recent decade, quite sophisticated two- and three-dimensional numerical models of laser deep-penetration keyhole welding of metals have been developed (see, reviews [1, 4, 5]). However, they do not describe all the available experimental data and it is often difficult to understand from them the physical meaning of one or another aspect of the welding process. Therefore, to understand qualitatively the laser welding mechanism, it is reasonable to begin its consideration from simplest models.

4.1 Physical Processes Proceeding Upon Deep Melting of Fixed Samples

It is known from experiments that the weld depth increases with decreasing the sample travel speed. It can be expected from general considerations that for $V \to 0$ (fixed sample), the welding depth should achieve the maximum value for the specified power and the given material. However, such a regime proves to be unstable in practice. Consider the problem of laser welding for a fixed sample because it is substantially simpler than in the case of a moving sample due to the axial symmetry of the problem. The welding model for a moving sample will be presented in Sect. 4.2.

4.1.1 The Thermal Deep-Melting Model

As pointed out in Chap. 1, when the laser beam intensity exceeds a certain value [see (1.29)], the sample material begins to evaporate, resulting in the formation of a keyhole whose depth increases with time. The keyhole development dynamics will be discussed in Chap. 8; here we will consider the stationary state achieved after the end of the keyhole growth.

As the keyhole depth is made deeper, the amount of energy absorbed in the side walls of the keyhole will increase and, therefore, the radiation power incident on the keyhole bottom will decrease. For many metals used in technological applications, the vapor pressure at the melting point is lower than the atmospheric pressure. Upon keyhole welding of such metals, boiling occurs in a deep keyhole as well. Indeed, after termination of boiling, a melted metal formed around the keyhole fills it, the keyhole (channel) depth decreases, and boiling resumes. It seems that a stationary state appears when a small fraction of energy is spent for metal vaporization, which is required only to maintain the existence of the vapor keyhole and prevent its flooding with the melt (after switching off the laser beam, the liquid floods the channel, and because the melt volume considerably exceeds the vapor channel volume, only a small indentation is usually remained on the target). Consider in more details a quasi-stationary state when the laser beam penetrates to the depth h greatly exceeding the beam radius Fig. 4.1).

This state is called quasi-stationary because weak evaporation (maintaining the channel in liquid) causes only a slow lowering of the melt surface. Let us neglect this process and set the temperature of a sample equal to zero away from the interaction zone. This assumption is inessential for qualitative estimates. By solving the problem of the temperature distribution around a uniformly heated filament of radius r_f and length h by using the similarity between the stationary heat conduction problem and the electrostatic problem for a potential around a charged filament (see Sect. 2.2), we obtain approximately

$$T = \frac{P}{2\pi\kappa h}\ln\frac{h}{r}, \qquad r_f \leq r < h \tag{4.1}$$

where P is the laser power and κ is the heat conductivity of the sample material.

At a distance greatly exceeding the filament length, the temperature distribution is close to the temperature distribution from a point thermal source:

$$T = P/2\pi\kappa r, \qquad r \gg h \tag{4.2}$$

The radius is measured from the center of this source. It is assumed that all the energy is absorbed in the keyhole. This question was already discussed in Chap. 1. Experiments discussed below confirm this assumption. The depth of a stationary vapor channel and the fusion zone depth can be estimated from (4.1) to (4.2) taking into account that the temperature of the keyhole walls and bottom should be close to the boiling point T_b.

By setting $T = T_b$ in (4.1), we find

$$h \cong \frac{P}{2\pi\kappa T_b}\ln\left(\frac{P}{2\pi\kappa T_b}\frac{c}{r_f}\right) \tag{4.3}$$

where $c = \ln(h/r_f)$. Below, we will assume for estimates that $h/r_f \sim 10$ and $c \approx 2.3$. These values are typical for laser welding.

The fusion depth can be determined from the exact expression describing the propagation of heat from a source in the form of an ellipsoid of revolution with semiaxes r_f and h. This will lead to the appearance of the factor 2 in the logarithm in (4.1) [6]. However, such a refinement is not important for our purposes. We will show below that the half-width r_m of the fusion zone can be larger or smaller than the channel depth, depending on the laser radiation power. In the first case, the fusion zone size can be determined from expression (4.2)

$$r_m = P/2\pi\kappa T_m \tag{4.4}$$

In the second case, r_m is determined from (4.1) for $T = T_m$. The weld shape coefficient [keyhole aspect ratio (depth to width ratio) $h/2r_m$ [7] can be found in the first case for $r_m > h$ from (4.3) to (4.4) as

$$h/2r_m = (T_m/2T_b)\ln[cP/2\pi\kappa T_b r_f] \tag{4.5}$$

One can see from (4.5) that for low radiation powers, the weld shape coefficient for metals used in technological applications (T_b is 2–3 times larger than T_m) is smaller than or of the order of unity. As the laser power is increased, this coefficient weakly increases and, as follows from (4.5), achieves unity at the power

$$P_l = 2\pi\kappa T_m r_f \exp(2T_b/T_m) \tag{4.6}$$

By comparing (4.6) with the expression $P_b = \pi r_f T_b/\alpha$ for the critical power required to produce evaporation, we see that

$$P_l/P_b = 2\alpha(T_m/T_b)\exp(2T_b/T_m).$$

For many technological materials, this ratio exceeds unity, i.e. for $h \sim 2r_m$ the condition required for the laser beam penetration inside a metal is obviously fulfilled. As follows from (4.5), the weld shape coefficient weakly (logarithmically) increases with increasing power. If $h/2r_m \gg 1$, the value of r_m sh should be determined from (4.1), and the weld shape coefficient in this case is described by the expression

$$\frac{h}{2r_m} = \frac{1}{2}\left[\frac{P}{2\pi\kappa T_b r_f}\ln\frac{2,3P}{2\pi\kappa T_b r_f}\right]^{T_m/T_b} \tag{4.7}$$

Thus, at high laser powers, the keyhole depth and fusion zone width increase with increasing laser power; however, the keyhole depth increases somewhat faster, so that the aspect ratio increases. Note that, despite such a simplified deep fusion model, the estimates from (4.7) for steel ($r_f = 0.6$ mm, $\kappa = 0.7$ W/cm K, $T_b = 3{,}000$ K, $T_m = 1{,}800$ K) are close to experimental data (see Fig. 4.9b).

The expressions presented in this section were derived by assuming that the mechanical equilibrium of the vapor channel was provided by weak evaporation. Because the simultaneous consideration of the mechanical equilibrium and thermal balance is complicated, we consider first only the mechanical equilibrium.

4.1.2 Mechanical Limit of Laser Beam Penetration into Liquid

Let us assume that liquid occupies the region $z < h$ (h is the channel depth) and the coordinate origin is located at the center of the channel bottom (see Fig. 4.1b). The surface tension force and hydrostatic and atmospheric pressures of the liquid are equilibrated by the gas pressure inside the keyhole. In the case of a subsonic flow, the static gas pressure will be close to the atmospheric pressure, which considerably simplifies the picture of the gas-dynamic flow. We will restrict ourselves to the consideration of the surface tension $\sigma(1/R_1 + 1/R_2)$ ($R_{1,2}$ are the principal radii of curvature of the keyhole), the hydrostatic pressure $\rho g(h - z)$, and the recoil pressure $p_v = q^2 \cos^2\theta / L_b^2 \rho_a$ due to evaporation calculated from (1.31) (L_b is the vaporization enthalpy; θ is the angle between the incident light beam (vertical, in our case) and the normal to the surface; and ρ_a is the vapor density). We obtain the equation

$$\frac{Q^2}{1+\tilde{z}'^2} = \frac{\tilde{z}''}{(1+\tilde{z}'^2)^{\frac{3}{2}}} + \frac{1}{\tilde{r}} \frac{\tilde{z}'}{(1+\tilde{z}'^2)^{\frac{1}{2}}} + \xi^{-1}(\tilde{h} - \tilde{z}), \tag{4.8}$$

for the keyhole shape $z(r)$, where $Q = \sqrt{r_f / \rho_a \sigma L_b^2}$, $\xi = r_c^2 / r_f^2$, $z' = dz/dr$, $r_c^2 = \sigma/\rho g$ is the capillary constant [8]; ρ is the liquid density; g is the acceleration of gravity; $\tilde{r} = r/r_f$; and $\tilde{z} = z/r_f$. Equation (4.8) was derived assuming that the intensity distribution over the radius is constant for $r < r_f$. For simplicity, multiple reflections of beams in the keyhole are neglected. The boundary conditions at the bottom and surface of a liquid pool have the form

$$\tilde{z}(\tilde{r} = 0) = \tilde{z}'(\tilde{r} = 0) = 0, \qquad \tilde{z}(\tilde{r} = r_m) = \tilde{h} \tag{4.9}$$

Because we have three boundary conditions (instead of the two required for solving second-order equations), we can find additionally the value of $\tilde{h}$ from them. Note that the dimensionless parameter ξ for the conditions of the problem is quite large, being equal to 40 for $r_f = 0.5$ mm.

Consider first the simplest case for $\tilde{h} \ll \xi$, when the gravity force can be neglected. By omitting in (4.8) the term proportional to ξ^{-1}, we integrate approximately this equation. Near the liquid pool bottom, we have $\tilde{z}' \ll 1$. By neglecting $\tilde{z}'$ in (4.8) compared to unity, we obtain the simple equation

$$Q^2 = \tilde{z}'' + \tilde{z}'/\tilde{r}. \tag{4.10}$$

The solution satisfying the boundary conditions at the bottom is

$$\tilde{z} = Q^2 \tilde{r}^2 / 4.$$

On the side walls, we have $\tilde{z}' \gg 1$, and (4.8) is also simplified:

$$Q^2/\tilde{z}'^2 = 1/\tilde{r}.$$

By solving this equation, we obtain

$$\tilde{z} = \frac{2}{3} Q \tilde{r}^{\frac{3}{2}} + c,$$

where a constant c is chosen according to the condition of sewing with the solution near the bottom, while the sewing coordinate $\tilde{r}^*$ is determined from the condition of the equality of derivatives of these two solutions at the point r^*. Finally, we obtain

$$\tilde{z} = \frac{2}{3}Q\tilde{r}^{\frac{3}{2}} - \frac{4}{3}Q^{-2}; \quad \tilde{r} > \tilde{r}^* = 4Q^{-2}. \tag{4.11}$$

Note that the value of $\tilde{r}$ for which $\tilde{z} = 0$ is $\tilde{r}^{**} = 2^{2/3}Q^{-2}$, so that $\tilde{r}^{**} < \tilde{r}^*$ irrespective of Q. In the upper part of the channel, $Q = 0$ for $\tilde{r} > \tilde{r}_f$. The keyhole surface gradually comes to the horizontal plane, so that $\tilde{z}' \ll 1$ in this region, and equation (4.10) should be solved again for $Q = 0$.

$$\tilde{z} = c_1 + c_2 \ln \tilde{r}. \tag{4.12}$$

Constants c_1 and c_2 can be found from boundary condition (4.9) and conditions of sewing the function $\tilde{z}$ and its derivative $\tilde{z}'$ with the solution in the second region (4.11) for $\tilde{r} = 1$. The keyhole depth is determined from the same conditions:

$$\tilde{h} = \frac{2}{3}Q\left(1 + \frac{3}{2}\ln \tilde{r}_m\right) - \frac{4}{3}Q^{-2} \tag{4.13}$$

where r_m is determined from (4.1) or (4.4). The gravity force can be neglected if h is not too large. This restricts the radiation intensity by the relation $q < q_g$, where

$$q_g = \frac{3}{2}L_b\sqrt{\frac{\rho_a\sigma}{r_f}}\left(\frac{r_c}{r_f}\right)^2\left(1 + \frac{3}{2}\ln\frac{r_m}{r_f}\right)^{-1}$$

For $q = q_g$, we have $\tilde{h} = \xi$, i.e. $h = r_c^2/r_f$ (Fig. 4.2).

The vapor density ρ_α can be found from the condition of the equality of pressured inside and outside the keyhole:

$$\rho_a = Mp_A/kT_b,$$

where M is the mass of a vapor atom.

For example, for an iron target ($L_b \approx 10^4$ J/g, $\sigma = 1.5 \times 10^{-2}$ N/m, $r_c \approx$ 4 mm, $T_b \approx$ 3,000 K, $\rho = 7.5$ g/cm^3, atomic mass 57, $\rho_a = 2.6 \times 10^{-4}$ g/cm^3) at

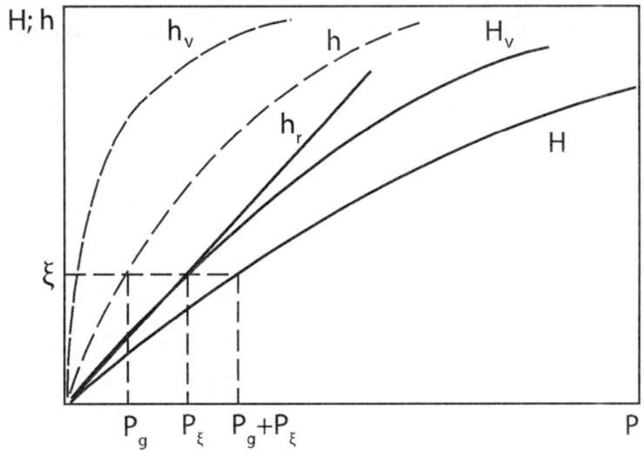

Fig. 4.2 Dependence of the welding depth in gas (h) and vacuum (h_v) on the laser radiation power: h_r is the welding depth determined by heat losses; H_v and H are depths determined by total heat losses in vacuum and gas; $P_g = \pi r_f^2 q_g$; P_ξ is determined from (4.3) for $h = \xi$

$r_f = 0.5$ mm, we obtain $h = r_c^2/r_f = 18$ mm. The power required for the thermal maintenance of such a channel is ~ 5 kW [see (4.3)] and the fusion radius estimated from (4.7) is about 3 mm. The power required for the mechanical maintenance of this channel, calculated from (4.13) is about 2 kW. Thus, 5 kW of a 7-kW laser beam is spent to the heat conduction and 2 kW is spent to evaporation. In this case, the channel depth is 18 mm and the fusion zone diameter is 6 mm.

At high powers, when $\tilde{h} \gg \xi$, the gravity force begins to play a role, and the surface tension in (4.8) can be neglected. In this case, we obtain

$$\tilde{z}' = \left[Q^2 \frac{\xi}{\tilde{h} - \tilde{z}} - 1\right]^{\frac{1}{2}}$$

For $\xi \ll \tilde{z} \ll \tilde{h}$, this equation has the approximate solution

$$\tilde{z} = Q\tilde{r}/\sqrt{\xi/\tilde{h}}. \tag{4.14}$$

Near the exit from the keyhole ($\tilde{h} \approx \tilde{z}$), the gravity force is again can be neglected, and solution (4.12) is valid in this region. The values of c_1, c_2, and $\tilde{h}$ can be found from the sewing conditions for the functions and derivatives of solutions (4.12) and (4.14) and boundary condition (4.9). As a result, we obtain

$$\tilde{h} = \left(Q\sqrt{\xi}\right)^{2/3} (1 + \ln \tilde{r}_m)^{2/3} \tag{4.15}$$

or in the dimensional units,

$$\frac{h}{r_f} = \left(\frac{q}{L_b\sqrt{p_a \rho g r_f}}\right)^{2/3} [1 + \ln(r_m/r_f)]^{2/3}. \tag{4.16}$$

Note that for $Q = \xi$, expression (4.13) is transformed to (4.15) with an accuracy to a numerical factor of the order of unity because the last term in (4.13) is small. Thus, for $q > q_g$, the dependence of the keyhole depth on power (4.16) becomes weaker than that at lower powers (4.13). The logarithmic additions to (4.13) and (4.16) are related to a finite width of the liquid pool. The numerical solution [6] of mechanical equilibrium equation (4.8) for forces in a keyhole in infinite liquid refines expression (4.16), by adding the factor 1/2 in the right-hand side of (4.16).

If laser welding is performed in vacuum (or for $p_v \gg p_A$), the recoil vapor pressure on the keyhole wall and bottom linearly depends on the radiation intensity [see (1.32)]:

$$p_v = q\frac{c_s}{L_b}\cos\theta, \tag{4.17}$$

where c_s is the speed (close to the sound speed) at which the metal vapor moves away from the surface. In this case, the left-hand side of (4.8) should be changed taking (4.17) into account. We restrict our consideration to a large fusion depth $h \gg \xi$. By performing calculations similarly to these presented above, we obtain the expression for the fusion depth h_v in vacuum

$$h_v = (qc_s r_f / L_b \rho g)^{1/2}[1 + ln(r_m/r_f)]^{1/2} \tag{4.18}$$

This expression is valid for sufficiently high intensities $q > q_{\text{gv}}$, where

$$q_{\text{gv}} = L_b \sigma^2 / c_s r_f^3 \rho g[1 + \ln(r_m/r_f)] \tag{4.19}$$

For $q = q_{\text{gv}}$, we obtain $h_v \approx \xi$. Let us find the intensity ratio at which the depths of the vapor channel for welding at atmospheric pressure and in vacuum (4.19) are equal:

$$q_g/q_{\text{gv}} = (\rho_a c_s^2 r_f/\sigma)^{1/2}.$$

For iron at $r_f = 0.5$ mm, this ratio is of the order of 10. Thus, for the same power, the channel depth in vacuum is much greater than at atmospheric pressure in air. For example, for $q = q_g$ [see (4.18)], the depth ratio for welding in vacuum and at atmospheric pressure is

$$h_v/h = \frac{3}{2}(\rho_a c_s^2 r_f/\sigma)^{1/4}.$$

For iron at $r_f = 0.5$ mm, we have $h_v/h = 4$ (see Fig. 4.2).

However, the difference between welding in vacuum and at atmospheric pressure proves to be not too large because in both cases the main part of laser energy is removed due to heat conduction. Figure 4.2 presents curves H, which are constructed by adding "inverse" functions $q(h)$ calculated from (4.3), (4.13), and (4.18). Curves H and H_v correspond to the total energy consumption to maintain the channel in the target (spent for heat conduction and vaporization). The maximum melting depth ratio for $h < r_c^2/r_f$ is determined by the relation

$$\beta = \frac{H_v}{H} = 1 + \frac{3}{8}\left(\frac{L_b^2 \rho_a \sigma r_f}{\kappa^2 T_b^2}\right)^{1/2} \ln\frac{r_c}{r_f}\left[1 + \frac{3}{2}\ln\frac{r_m}{r_f}\right].$$

For iron, $H_v \approx h_T$ for $h < \xi$, and $h/H \approx 1.7$ for $r_f = 0.5$ mm. As the radiation power is increased, β first increases and then decreases because at high powers the energy is mainly spent for vaporization, and these losses in vacuum and atmosphere are approximately equal.

Figure 4.3 demonstrates the increase in the welding depth with decreasing pressure for various metals [9].

This effect is especially noticeable at low welding speeds. It seems that the critical speed $v_{\min}$ decreases with decreasing the external pressure, so that the welding depth can be increased in vacuum. Note that the size of an erosion plume decreases with decreasing pressure. For $p < 5$ Torr, the plume almost does not rise over the sample surface (Fig. 4.4). In this case, the welding depth ceases to grow.

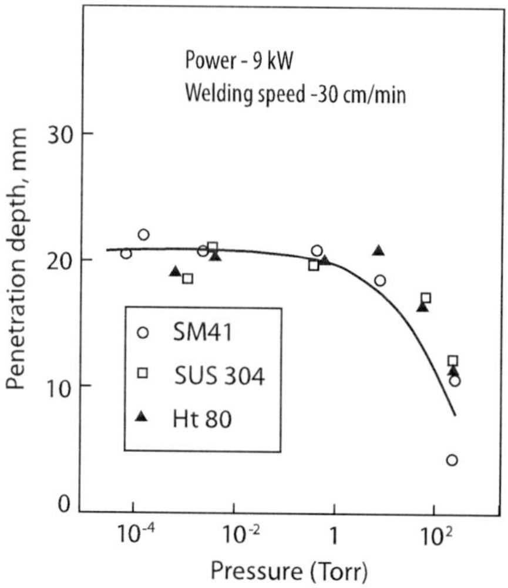

Fig. 4.3 Pressure dependence of the penetration depth

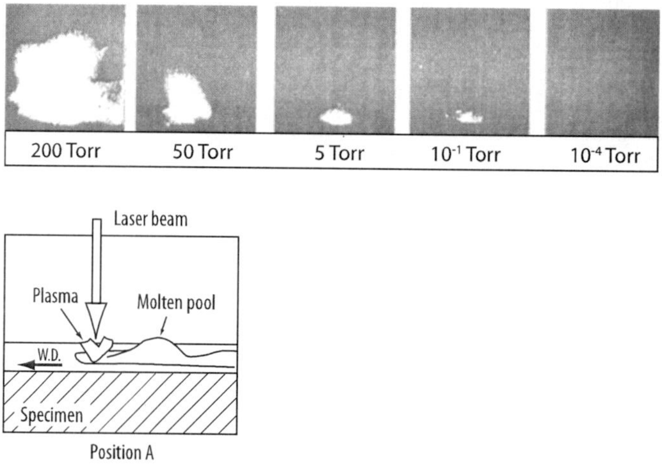

Fig. 4.4 Typical photographs of laser plasma at different pressures

We considered thermal loses in a sample at rest by neglecting the fact that the liquid melt moves to the vapor–liquid interface with a speed determined by the consumption of vapor leaving this surface. This liquid flow returns a part of heat to the keyhole, which is removed from it due to heat conduction in the absence of motion. The characteristic distance over which the heat is transferred from the source upstream of the liquid is

$$l = \chi/u, \tag{4.20}$$

where χ and u are the temperature conductivity and velocity of the liquid. If l greatly exceeds either the fusion zone width or the source height, the influence of the liquid on the thermal balance of the system is small. For vacuum welding, $l = \chi \rho r_f c_s \sigma^{-1}$ and for $r_f = 0.5\,\text{mm}$ in an iron target, we have $l \approx 5\,\text{cm}$, which greatly exceeds r_m and h, so that the heat conduction losses are determined by expression (4.1). In the case of welding at atmospheric pressure, $l = \chi \rho (r_f/\rho_a \sigma)^{1/2}$ and $l = 5\,\text{mm}$ for iron under the same conditions, which is comparable with the radius r_m of the fusion zone. In this case, the convective "return" of heat to the keyhole becomes considerable. As the radiation power is increased, the relative contribution of heat conduction losses decreases (because vaporization losses increase faster than heat conduction losses) and therefore the picture of the deep melting by a fixed laser beam described above remains valid.

At the beginning of this section we assumed that a keyhole in a massive metal target is quasi-stationary, i.e. the characteristic keyhole formation time is smaller than the liquid vaporization time from the fusion zone. Consider the example discussed above, when a 7-kW laser beam for $2r_f = 0.5\,\text{mm}$ penetrates into a steel target by a depth of 18 mm, the diameter of the fusion zone being 6 mm. Because in this case, 2 kW is spent for vaporization, the vaporization rate is $1/30\,\text{cm}^3/\text{s}$ (for steel $\rho L_b = 60\ \text{kJ/cm}^3$). Thus, all the material of the melted zone evaporates for the time of about 17 s. The maximum formation time of the keyhole (determined by the vaporization wave velocity $v = p/\pi r_f^2 L_b \rho$) is 0.1 s, so that the quasi-stationary condition is fulfilled. Thus, processes of heat transfer and mechanical maintenance of the gas–vapor keyhole play the dominant role in a complicated picture of laser welding in the limiting case of low welding speeds. By using the simplest models of these processes, we described approximately the basic properties of deep melting and analyzed the energy balance and the influence of atmospheric pressure.

4.1.3 Peculiarities of Deep Laser Beam Penetration into Liquid

The simplified thermal and mechanical models considered above have a number of disadvantages. In particular, the thermal model assumes that he faction of radiation energy spent for heat conduction is distributed over the keyhole height uniformly. However, experiments show that the heat release in the upper part of the keyhole is higher that in its lower part. The diameter of the gas–vapor keyhole is equal to the focal spot size. The dynamic equilibrium was considered by assuming that the two main forces – the static pressure of gas inside the keyhole and the external liquid pressure, completely compensate each other. The static vapor pressure on the keyhole walls during welding in vacuum was neglected. The consideration of this pressure on the keyhole walls and bottom substantially complicates the problem. The pressure can be found by solving three gas dynamics equations (continuity, motion, and energy) in the keyhole. Of course, the mass, momentum, and energy exchange between the vapor flow and liquid wall should be also taken into account. This requires the knowledge of the temperature distribution over the keyhole surface, which can be obtained by solving the thermal problem in the liquid surrounding the keyhole. The keyhole geometry suggests that all these problems are,

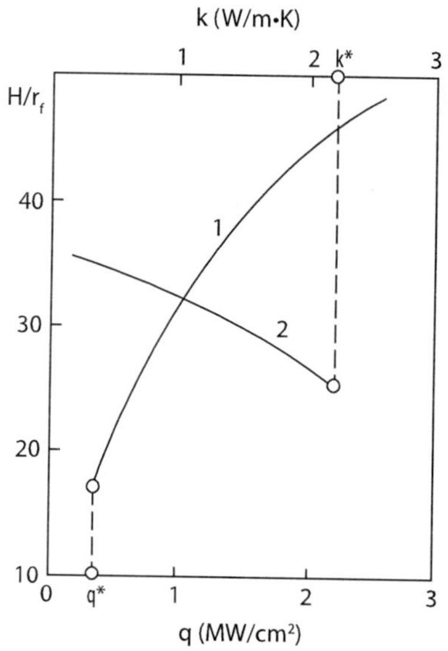

Fig. 4.5 Dependences of the relative welding depth in the subsonic flow regime on the absorbed radiation intensity for $\kappa = 2 \times 10^{-2}$ W/cmK (**1**) and on the heat conductivity for $q = 0.24$ MW/cm^2 (**2**)

strictly speaking, two-dimensional or even three-dimensional [1,4,5]. Some results of these models will be presented below.

This problem was solved numerically in the quasi-one-dimensional approximation in [10]. The solution for a deep keyhole was obtained only for materials with the low heat conduction.

Figure 4.5 shows the dependence of the keyhole depth on the light intensity in the focal spot [$\kappa = 2$ W/m K, $L_b \approx 2 \cdot 10^3$ J/g], in this case, $h \sim q^{2/3}$.

The specific feature of the maintenance of the keyhole in liquid in this model is the presence of the region of vapor condensation in the upper half of the keyhole. As the heat conductivity of the material increases, the keyhole depth weakly decreases (see Fig. 4.5) and for $\kappa = \kappa^*$, the solution corresponding to the deep keyhole ($h \gg r$) disappears. Thus, the vapor flow in the keyhole can be described in the quasi-one-dimensional approximation only for materials with the low heat conductivity (dielectrics).

Experiments on the laser beam penetration into fixed dielectric samples are scarce. In [11], a stable gas–vapor keyhole of depth 20 cm was obtained in glycerol by using a 1 kW cw CO_2 laser. However, a detailed comparison with the theory is complicated due to the lack of experimental data.

Experimental studies of the interaction of laser beams with transparent dielectrics are aimed at the elucidation of the mechanism of the maintenance of a vapor keyhole in the melt and liquid medium. The latter is of interest for medical applications of lasers [12]. Transparent dielectrics are convenient for studying both the shape of the vapor keyhole and the flow of liquid surrounding the keyhole. Similar X-ray studies of metals are quite time-consuming and have restrictions (see below). A deep keyhole in liquid at rest or flowing at a low velocity is almost always unstable [13, 14]. Its side surface is inhomogeneous and rapidly fluctuates with

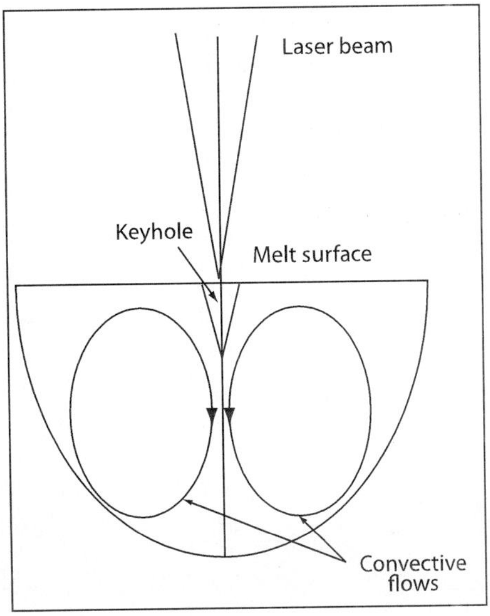

Fig. 4.6 Scheme of the melt motion

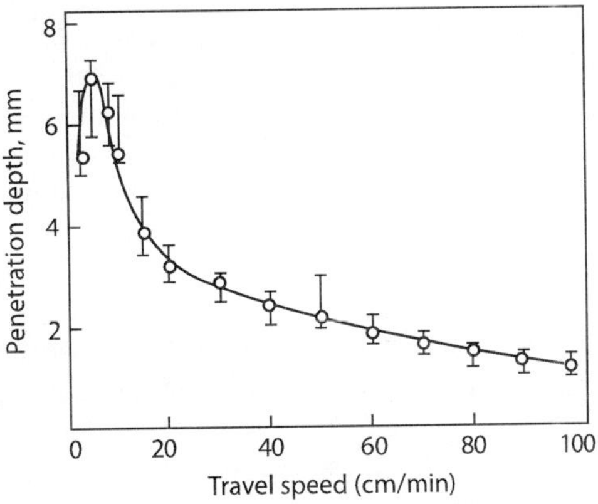

Fig. 4.7 Relation between the weld depth and welding speed (330 W)

time (see Fig. 4.1a). Fluctuations are maximal at the keyhole bottom. Here the liquid periodically floods a part of the channel and is periodically pushed out downwards by the laser beam [13]. Such fluctuations give rise to a toroidal vortex flow surrounding the keyhole. Near the keyhole walls, the liquid flows downwards rather then upwards, as if being carried away by vapor emerging from the keyhole Fig. 4.6. Such a keyhole behavior and convective cooling lead to a considerable additional heat removal from the keyhole. It seems that for this reason the keyhole depth ceases to grow or even decreases with decreasing the welding speed below V_{min} [14, 15] (see Fig. 4.7).

The value of v_{min} is determined by the keyhole stability and should be considered separately. It depends on the experimental conditions and is 1 cm/s for metals (see Fig. 4.8 and [15]) and $v_{min} \sim 1$ mm/s for quartz [14]. For $v < v_{min}$, a decrease in

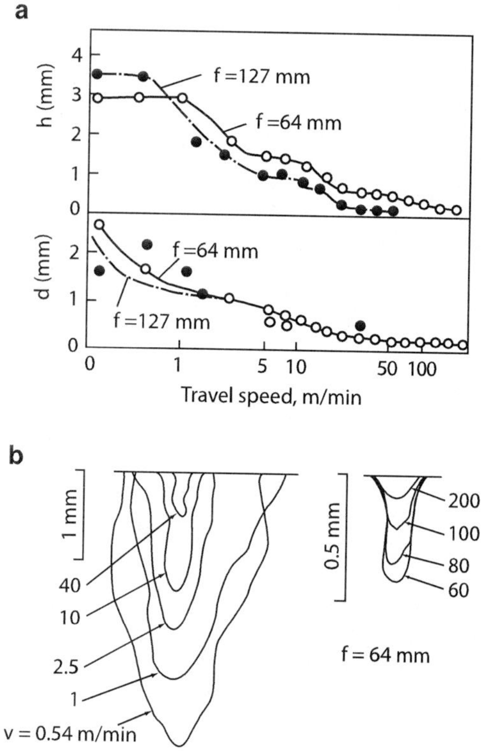

Fig. 4.8 Dependences of the parameters (**a**) and shape (**b**) of the weld in SS304 steel on the welding speed for $P = 1$ kW and different focal distances of a lens

the welding speed leads to the increase in the fusion zone width rather than in its depth (Fig. 4.8a).

As a result, the keyhole nature of laser welding, being one of the main of its advantages, disappears (see Fig. 4.8b).

The restriction of a convection current in water with the help of thin metal plates placed near the keyhole resulted in a considerable increase in the keyhole depth [13]. As the travel speed of a sample is increased to exceed $v_{\min}$, the width of a liquid layer in front of the keyhole decreases, which also stabilizes the keyhole [14]. We will consider later the reasons responsible for these instabilities and now will continue the study of the thermal model of a keyhole in a moving sample.

4.2 Thermal Deep Penetration Melting Model for a Moving Sample

4.2.1 Physical Processes in Welding of Materials

Laser welding is usually studied by performing deep melting of a thick sample moved in the horizontal direction with respect to the vertical laser beam. In the case of welding through the sample, the situation is more complicated because a part of laser radiation will pass through the sample; and vapor and maybe the metal melt will be ejected to the lower hole of the gas–vapor channel. We will analyze theoretically the laser beam penetration into a semi-infinite solid flat target and processes proceeding in this target. In Sect. 4.1, we presented the concept of physical processes in the case of a fixed target. It is obvious that the movement of a sample will cause its cooling and the sample temperature at each point should decrease. But because the temperature of the keyhole surface cannot be lower than T_b, the target motion will cause first of all a decrease in the vapor channel depth. More exactly, as mentioned above, this will occur for $v > v_{\min}$.

We will assume approximately that the temperature at a distance l [see (4.20)] from a cylindrical laser keyhole of radius r_f is $T_\infty = 0$ Then, h in the logarithm in (4.1) should be replaced by $r_f + \chi/V$, and we obtain

$$h = \frac{P}{2\pi\kappa T_b}\ln\frac{r_f + \chi/v}{r_f}, \tag{4.21}$$

instead of (4.3). This expression is valid at high enough welding speeds, when the width χ/v of the zone heated by the laser beam is far smaller than the keyhole depth h i.e. $v > \chi/h$. For many metals at the welding depth $h \sim 1$ cm the influence of sample movement becomes noticeable at the speed exceeding 1 mm/s. Laser welding is usually performed at welding speeds ~1–10 cm/s and therefore the weld depth strongly depends on the welding speed. According to (4.21), at low welding speeds, when $\chi/v > r_f$ the weld depth decreases logarithmically with increasing V, while for $\chi/v < r_f$ the weld depth is inversely proportional to the welding speed $h \sim v^{-1}$ (see Figs. 4.7 and 4.8). For $\chi/v \approx r_f$ some authors propose the empirical dependence $h \sim v^{-1/2}$ [16]. Thus, the heat consumption for maintaining the keyhole in the moving sample considerably increase compared to a fixed target. Note that (4.21) describes in fact the specific heat removal from a cylindrical source by a boiling metal flow. The heat removal from solid cylinder is described by semi-empirical formula [92]

$$\frac{P}{h} = \frac{2\pi\kappa T_b}{\ln(2h/r_f)}\left\{1 + \beta\left(\frac{rv}{\chi}\right)^{\gamma}\right\}$$

which is close to (4.21) for $\beta \approx 1, \gamma \approx 1$.

It follows from experiment [16] that the energy consumption for vaporization for $P \approx 1$ kW and $v > 1$ cm/s is small (see Fig. 4.9a). Let us find the laser power P_V above which this consumption becomes considerable and the dependence $h(P)$ begins to deviate from a linear one. This occurs when $h > r_c^2/r_f$ [see explanation of (4.16)]. By using (4.21) and assuming that $h = \xi$ we can easily obtain the expression for P_V:

$$P_V = \frac{2\pi\kappa T_b\sigma}{\rho g r_f}\ln^{-1}(1 + \chi/vr_f) \tag{4.22}$$

The value of P_V first increases weakly with the weld speed and then, for $v > \chi/r_f$, it increases strongly as $P_V = 2\pi c T_b\sigma v/g$ Thus, for $v = 1$ cm/s, we have

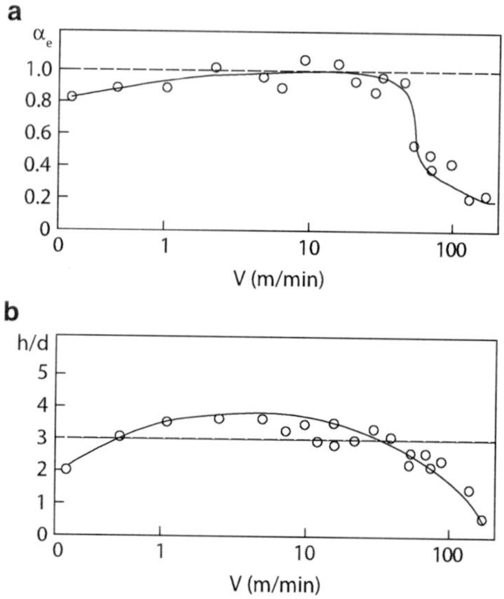

Fig. 4.9 Dependences of the effective absorption coefficient for CO_2 laser radiation (**a**) and the welding aspect ratio h/d (**b**) on the welding speed ($P = 3\,\text{kW}$, SS304 steel, $r_f = 0.085\,\text{mm}$)

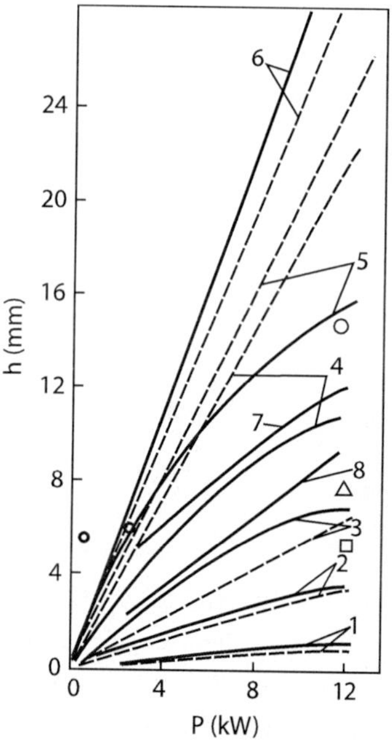

Fig. 4.10 Dependences of the welding depth on the laser radiation power at different welding speeds: (**1–5**) 304 steel, laser welding speeds are 42 (**1**), 17 (**2**), 8.4 (**3**), 4.2 (**4**), and 1.3 cm/s (**5**) [17]; (**6, 7**) NU-130 steel, electron-beam welding speeds are 1.1 (**6**) and 4.2 cm/s (**7**) [20]; (**8**) X18H10T steel, laser welding speed is 0.7 cm/s [18]; *solid curves* are experiment, dashed curves are calculated by (4.21) for $r_f = 0.25$ mm, = 0.25 W/(cm K); *points* are data for laser welding of aluminum at welding speeds 1.1 (*ring*), 2.6 (*triangle*), and 2.8 cm/s (*square*)

$P_V = 10\,\text{kW}$ for steel. Therefore, laser welding at high laser powers differs from electron-beam welding.

Indeed, the laser welding depth at high powers increases with P as $h \sim P^{2/3}$, while the electron-beam welding depth increases as $h \sim P$, according to (4.16) and (4.21). At lower weld speeds, this difference appears at lower powers. These tendencies are also manifested in experiments. Figure 4.10 shows the dependence

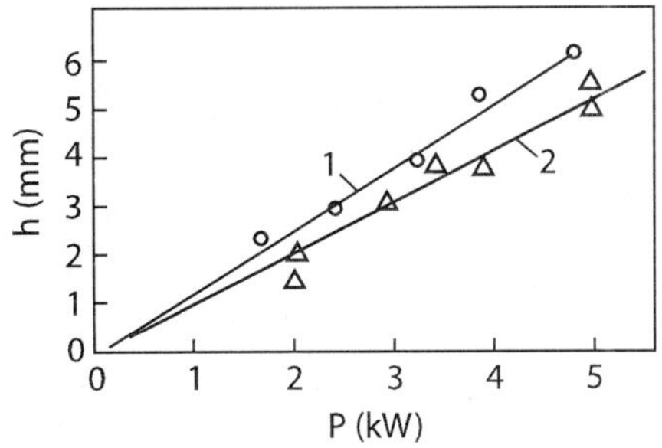

Fig. 4.11 Dependences of the welding depth for BT1 titanium alloy (**1**) and 17G2S steel (**2**) on the laser power ($v = 2.5$ cm/s [19])

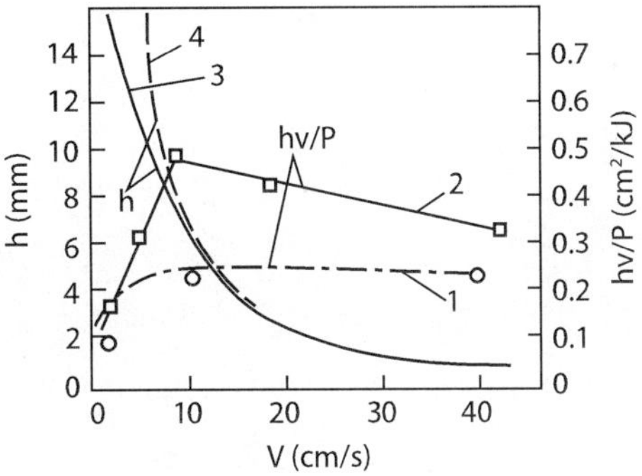

Fig. 4.12 Dependences of the welding depth and efficiency on the welding speed: (**1**) 17G2S steel, $P = 5\,\text{kW}$ [19]; (**2, 3**) 304 steel, $P = 12\,\text{kW}$ [17]; (**4**) calculation by (4.21) for $r_f = 0.25$ mm, $\kappa = 0.25$ W/cm K

of the welding depth on a CO_2 laser power for 304 stainless steel for different weld speeds. One can see that experimental data [17] are described by the linear dependence $h \sim P$ at high welding speeds. For low welding speeds ($v \sim 1$ cm/s), these dependences are linear only at low laser powers. The welding data [18] obtained for stainless steel showed that these dependences were also linear at higher powers $P > 10\,\text{kW}$. In this case, the focal size spot was rather large ($2r_f = 1$ mm) and the welding speed was low ($v \sim 0.7$ cm/s). The dependence of $h(P)$ for 17G2C steel [19] obtained for $P \leq 5\,\text{kW}$ and welding speed ~ 2.5 cm/s is also linear (see Fig. 4.11).

For comparison, Fig. 4.10 presents electron-beam welding data [20]. One can see that the electron-beam welding depth for the welding speed of the order of a few cm/s is close to the laser welding depth. However, for $v < 1$ cm/s, the electron beam penetrates much deeper than the laser beam and the electron-beam welding depth is consistent with estimates by (4.21). These tendencies are confirmed by data presented in review [21] (Fig. 4.13).

It is interesting to study the properties of deep melting of materials by laser radiation powers considerably exceeding 10 kW. It was pointed out in review [22] that $h \sim P$ for such powers at large welding speeds ($v > 10$ cm/s.) The power dependence of the maximum welding depth obtained at low welding speeds has the form $h \sim P^{0.7}$ (Fig. 4.14).

This dependence can be interpreted in different ways. First, as mentioned above, the deviation from linearity can be explained by the increase in vaporization losses. Second, a similar dependence can be obtained from (4.21) by considering the

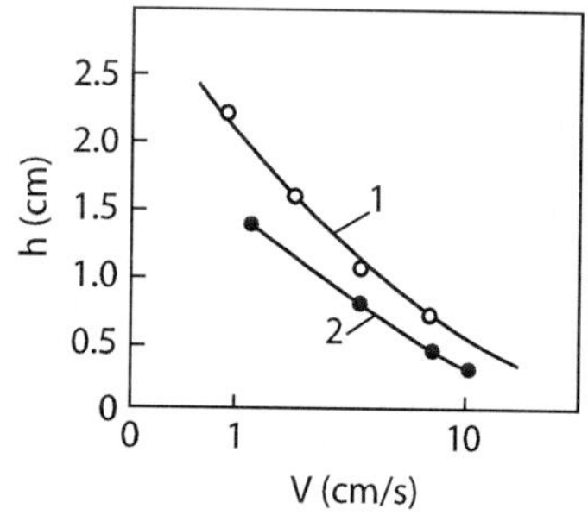

Fig. 4.13 Dependences of the welding depth on the welding speed for 304 steel ($P = 10\,\text{kW}$): (**1**) electron-beam welding; (**2**) laser welding

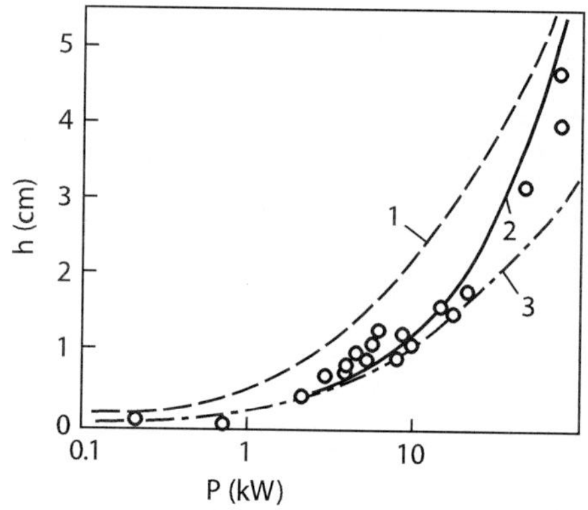

Fig. 4.14 Dependences of the maximum welding depth on the laser power: calculations for $v = 1$ cm/s (**1**) and 5 cm/s (**3**) [23] and experiment (**2**) [22]

waveguide propagation of light in a deep keyhole (see Sect. 1.3). By assuming that the keyhole diameter is determined by relation (1.22), where $\alpha_E^{-1} = h$ and by using (4.21), the dependences $h(P)$ can be plotted for different welding speeds. One can see from Fig. 4.14 that experimental data are in good agreement with calculations for welding speeds equal to 1 and 5 cm/s [23]. It seems that the penetration depth can be considerably increased by using polarized radiation with the electric field vector parallel to the lateral surface of the gas–vapor channel, which provides a weak decay of radiation in the channel [see (1.23)].

Let us find the conditions under which laser welding is most efficient, i.e. the value of hv/P is maximal. One can see from (4.21) that hv/P increases with increasing speed and for $v > \chi/r_f$ tends to the constant

$$hv/P = (2\pi c\rho T_b r_f)^{-1}. \tag{4.23}$$

This expression is valid when $P < P_V$ (where P_V is the power for $v = \chi/r_f$). One can see from (4.23) that the laser welding efficiency is inversely proportional to the beam radius and is independent of the heat conduction of a target. For example, $(hv/P)_{\max} \approx 0.2$ cm^2/kJ for steel for $r_f = 0.5$ mm (for $c\rho T_b = 15$ kJ/cm^3). This value is close to the experimental value [19] (see curve 1 in Fig. 4.12). This figure also presents the dependence of the welding efficiency on the welding speed [17].

The higher efficiency than in [19] is obviously explained by the use of the laser beam of a smaller radius.

Note that high-power laser welding also provides a high enough efficiency. For example, it was reported inside [22] that 3.7 cm-thick $HY - 80$ steel sheets were welded by a 90-kW laser beam at $v = 5$ cm/s with the efficiency $hv/P = 0.2$ cm^2/kJ ($r_f \sim 0.6$ mm). At the lower welding speed $v = 2$ cm/s, the welding depth was 5 cm for a 77-kW laser beam ($hv/P \sim 0.13$ cm^2/kJ).

4.2.2 Deep Melting of Various Materials

The welding depth for a fixed [see (4.3)] and moving [see (4.21)] sample depends on the boiling point and heat conduction of the sample material. To verify the correctness of the thermal welding model, we compare its results for materials with different thermal properties. Figure 4.11 shows the dependences of the melting depth on the laser beam power for $BT-1$ titanium alloy and low-alloy 17G2C steel. The room-temperature heat conductivity of steel is $\kappa = 0.57$ W/cmK [24], which is a few times higher than the heat conductivity of titanium $\kappa = 0.17$ W/cmK [24]. One can see from Fig. 4.11 that the difference between melting depth is small. This contradicts to the thermal model, which can be explained by the fact that in (4.21) the averaged value of κ was used. It is known [24] that the heat conductivity of steel decreases with temperature, while that of titanium increasers, so that their average values can be close.

The heat conductivity of pure aluminum and copper is 3–4 times higher than that of steel. However, for their alloys used in industry this difference is not so large. In addition, the heat conduction of aluminum strongly decreases with increasing temperature [24]. It is likely that for this reason the laser welding depth in aluminum only slightly differs from that in steel (see Figs. 4.10 and 4.15). It is not improbable that during aluminum welding a more energetically economical mechanism of maintaining the vapor channel is realized (described in Sect. 4.1), when the wall of the upper part of the keyhole does not boil but, on the contrary, absorbs vapor emerging from the keyhole bottom. Because the melting point of aluminum is not large ($T_m = 660°$C), the temperature of the upper part of the keyhole can be considerably lower than the boiling point.

The melting depth for copper is smaller than that for steel and aluminum. This is shown in Fig. 4.16, where electron-beam welding data are presented.

The electron-beam and laser welding depths obtained for aluminum alloys are compared in Fig. 4.15. These dependencies are similar to the corresponding dependencies obtained for steel samples (Fig. 4.13). Note that aluminum welding has a high efficiency, which, according to (4.23), should be close to the steel welding efficiency (the values of $c\rho T_b$ for these materials are close). However, the laser welding of aluminum, unlike steel, proves to be very unstable and many cavities are produced in the weld (Fig. 4.17).

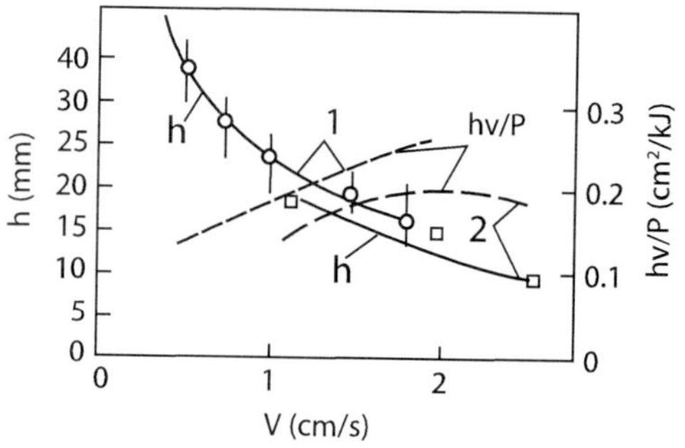

Fig. 4.15 Dependences of the welding depth and efficiency on the welding speed for AMG-6 aluminum alloy: (**1**) electron-beam welding for $P = 12$ kW and accelerating voltage 12 kV (vertical weld); (**2**) laser welding (horizontal weld)

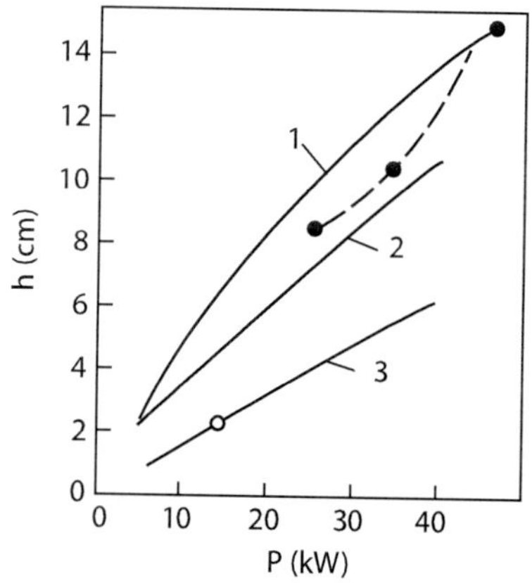

Fig. 4.16 Electron-beam welding of various materials: (**1**) D16 aluminum alloy; (**2**) 12X18H9E stainless steel; (**3**) M1 copper alloy; (**1–3**) vertical weld, accelerating voltage is 1,200 kV, $v = 0.4$ cm/s; (*dark ring*) 1201 aluminum alloy, horizontal weld, accelerating voltage is 100 kV, $v = 0.56$ cm/s; (*ring*) AMG-6 aluminum alloy, $v = 1.1$ cm/s, laser welding

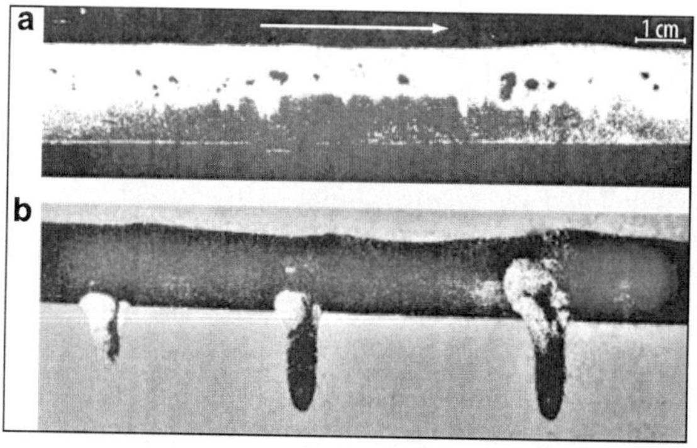

Fig. 4.17 Longitudinal section of a weld in AMG6 aluminum alloy for $P/h = 2\,\text{kW}\,\text{mm}^{-1}$ and $v = 2.8$ cm/s. The arrow shows the sample travel direction (**a**). Transverse section of welds in AMG6 aluminum alloy at different welding speeds for $P = 10$ kW (**b**) (photo by V.V. Fromm)

A great part of the melt is often ejected from the weld pool and solidifies outside Fig. 4.17b.

Such behavior can be explained by the low absorption coefficient of aluminum [22] [which is 3–4 times lower than that for steel (see Chap. 1)]. This explanation is also consistent with the fact that the quality of electron-beam welding of aluminum can be as high as that of steel [25].

4.2.3 Thermal Efficiency of Laser Welding

The efficiency of welding of any type is estimated by the thermal efficiency η_T, which is the ratio of the energy $(c\rho T_m + H_m\rho)hvd$ spent for sample melting to the total energy absorbed by the sample (d is the fusion zone width and H_m is the latent heat of fusion). In fact, this coefficient characterizes the role of heat conduction losses in the target because these losses usually play the main role in welding. At low welding speeds, the laser beam has time to heat a large volume around the weld in the sample. In this case, the value of η_T is small. The value of η_T increases with increasing welding speed.

In the case of a surface source (arc welding), the maximum value $\eta_T \approx \mathrm{e}^{-1} = 0.37$ was calculated in [7, 27]. Upon welding by using a volume source (laser or electron-beam welding), the value of η_T will be different. We assume, as before, that the heat release from the beam is distributed uniformly over height, and will neglect the influence of the melt flow (which will be considered in Sect. 4.3). In addition, to simplify the problem, we neglect the melting heat of the sample material compared to the difference of enthalpy at the melting point and at room temperature. Then, we can use the heat conduction equation. For a linear cylindrical source (of zero diameter) in a moving body, this equation has the solution [26] (see Fig. 4.18).

$$T(r,\varphi) = \frac{P}{2\pi\kappa h} K_0\left(\frac{vr}{2\chi}\right)\exp\left(-\frac{vr\cos\varphi}{2\chi}\right), \tag{4.24}$$

where r is the distance from the source (laser beam); φ is the polar angle measured from the welding direction (see Fig. 4.18); and K_0 is the modified Bessel function

$$K_0(\zeta) = \begin{cases} \ln(2/\gamma\zeta), & \zeta \ll 1,\ \gamma = 1.74; \\ \sqrt{2\pi/\zeta}\exp(-\zeta), & \zeta \gg 1 \end{cases}. \tag{4.25}$$

A model with the finite diameter of a gas–vapor keyhole will be considered below. If the heat conductivity depends on temperature, the value of κ in (4.24) should be

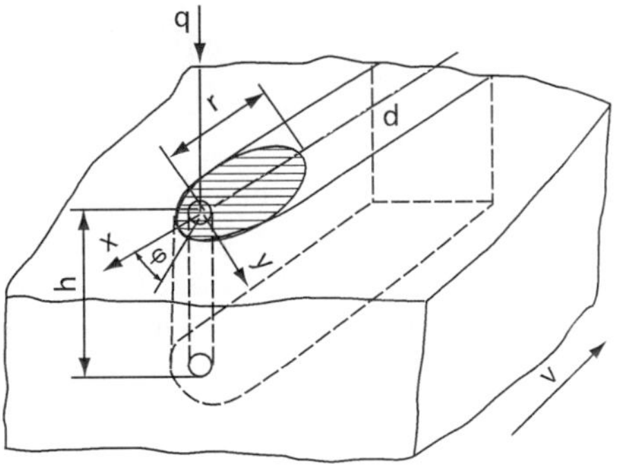

Fig. 4.18 Scheme of deep melting of a sample by a laser beam. The hatched region is melt pool

replaced by its average value $\bar{\kappa} = T_b^{-1} \int_0^{T_b} \kappa(T)\mathrm{d}T$. It follows from (4.24) to (4.25) that for distances $r < 2\chi/v$ at which the influence of the welding speed is absent, the temperature distribution coincides with (4.21) accurate to the factor of the order of unity under the same conditions. One can see from these expressions that isotherms are concentrated in front of the source and are strongly elongated behind it. As a result, the melting isotherm is displaced with respect to the laser beam (see the melting isotherm in Figs. 4.18 and 4.21).

The temperature in front of the source ($\varphi = 0$) decreases exponentially for $r > \chi/v$; behind the source ($\varphi = \pi$), temperature decreases slower as $T \sim r^{-1/2}$. Let us find the fusion zone width d, which enters into the expression for the thermal welding efficiency [26]. The melting isotherm $r(\varphi)$ can be found from (4.24) for $T = T_m$. It maximum distance from the x axis can be determined from the equality $\mathrm{d}y/\mathrm{d}x = 0$ (where $y = r\sin\varphi$ and $x = r\cos\varphi$), i.e. from the condition $\tan\varphi = r\mathrm{d}\varphi/\mathrm{d}r$.

By differentiating (4.24) at $T = T_m$, we find the relation between the coordinates r^* and φ^* of the point at which the fusion zone width $d = 2r^*\sin\varphi^*$ is maximal:

$$\cos\varphi^* = -K_0(\zeta^*)/K_0' \quad \zeta^* = r^*v/2\chi \tag{4.26}$$

The fusion zone width can be determined from

$$vd/2\chi = \zeta[1 - K_0^2(\zeta^*)/K_0'^2(\zeta^*)] \tag{4.27}$$

The values of ζ^* can be found from (4.24), where the angle φ^* should be excluded by using (4.26):

$$P/2\pi\kappa h T_m = \exp[\zeta^* K_0(\zeta^*)/K_0'(\zeta^*)] \tag{4.28}$$

Expressions (4.27) and (4.28) determine the dependence of the fusion zone width d on the power P (for the specified keyhole depth h). In the limiting cases corresponding to a slowly ($\zeta \ll 1$) and rapidly ($\zeta \gg 1$) moving target, by using expansion (4.25) we can obtained from (4.27) to (4.28)

$$\frac{vd}{\chi} = \begin{cases} \exp(-2\pi\kappa h T_m/P), & vd/2\chi \ll 1; \\ \sqrt{2/\pi e}P/\kappa h T, & vd/2\chi \gg 1. \end{cases} \tag{4.29}$$

The dependence of d on power is presented in Fig. 4.19, where experimental data [22] are also given.

Because we neglected the melting heat compared to the enthalpy difference, the thermal efficiency of laser welding can be written in the form $\eta_T = vdhc\rho T_m/P$. By substituting (4.29) into this expression, we find that η_T increases with increasing power and tends to the constant value $\sqrt{2/\pi e} \approx 0.48$ for $P/\kappa h T_m \gg 10$. The value of η_T can be found experimentally by drawing a straight line through the experimental point in Fig. 4.19 parallel to the dashed straight line. The dashed line is drawn at an angle of 45° to the abscissa. The ordinate of this straight line for

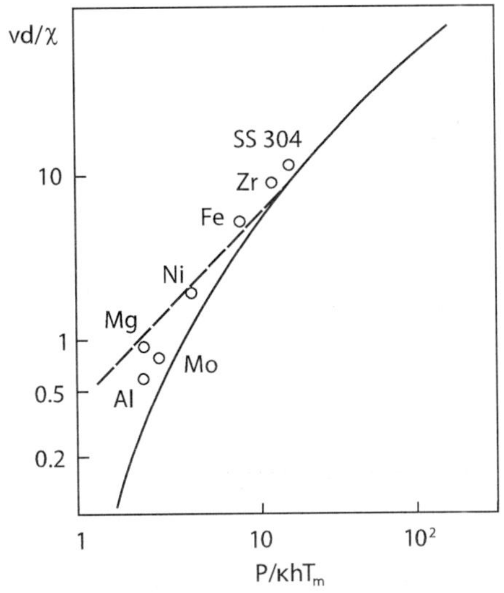

Fig. 4.19 Dependence of the dimensionless melting width on the dimensionless power

$P/\kappa h T_m = 1$ gives the value of η_T. One can see from Fig. 4.19 that the thermal welding efficiency for some metals (Mg, Fe, Ni) can achieve the limiting theoretical value. The thermal welding efficiency for aluminum and molybdenum is lower ($\eta_T = 20\%$) although this value is higher than the theoretical values corresponding to this power. The thermal welding efficiency for steel, titanium, and zirconium exceeds the limiting theoretical value equal to 0.48. It seems that the fusion zone width in this model is understated. It follows from (4.29) that the value of η_T for the specified P and h is independent of the welding speed. In the case of through-melting welding (when the welding depth is specified), the value of η_T weakly depends on the welding speed and is close to the limiting value Fig. 4.20 [28].

Upon welding without through melting, the welding depth decreases and, hence, the dimensionless power $P/\kappa h T_m$ increases, resulting in the increase in η_T.

The model (4.29) gives the relation between the four quantities P, v, d, and h. For the specified values of P and v, the values of d and h remain unknown. To determine them, it is necessary to use the additional relation, e.g., (4.21). Consider the dependence of the weld depth on the welding speed for the specified power P within the framework of the thermal model (4.21), (4.29). For the low speed $v < \chi/r_f$, the weld depth h weakly depends on the welding speed, according to (4.21); in this case, the power $P/\kappa h T_m$ changes weakly. Therefore, according to Fig. 4.19, the weld width d decreases with increasing the welding speed $d \sim v^{-1}$. The aspect ratio of the weld is $h/d \sim v$. When $v > \chi/r_f$, the weld depth $h = P/2\pi c\rho T_b r_f v$ decreases with increasing the welding speed. The parameter $P/\kappa h T_m$ increases with the welding speed, and this dependence becomes linear for $P/\kappa h T_m > 10$ (Fig. 4.19). In this case, the fusion zone width tends to the minimum value $d_{\min} = 2\pi \eta_T r_f T_b/T_m$ while the aspect ratio h/d of the weld is inversely proportional to the welding speed. These properties are observed experimentally: at low welding speeds, the fusion zone width decreases faster than its depth with increasing welding

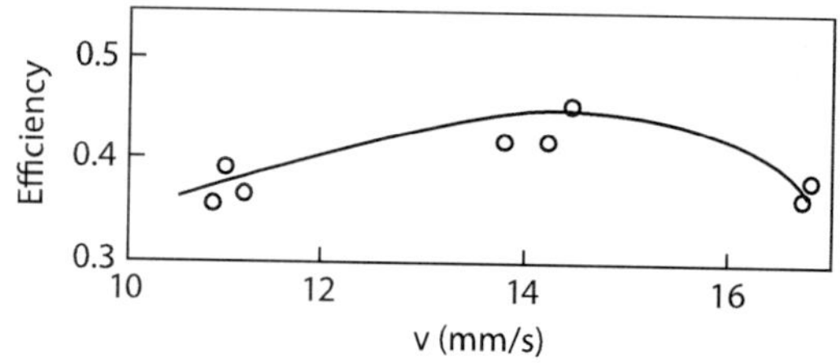

Fig. 4.20 Dependence of the welding efficiency on the welding speed

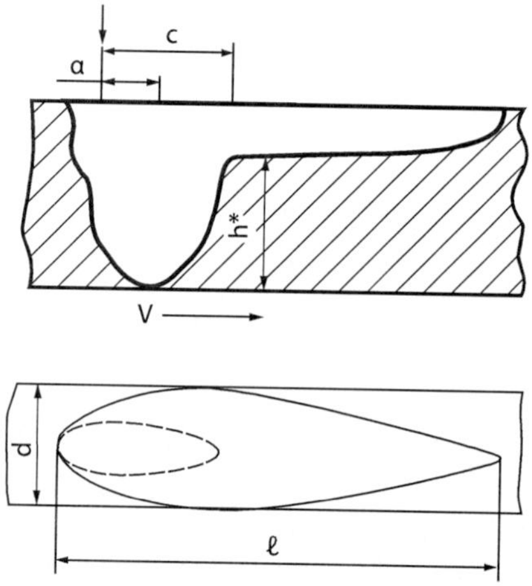

Fig. 4.21 Scheme of an empty weld pool: the hatched region is an unmelted sample

speed, and therefore the aspect ratio increases (Figs. 4.8 and 4.9). At high welding speeds, the aspect ratio decreases due to the decrease in the depth Fig. 4.9b. For some value of the welding speed $v = v_{cr}$, the aspect ratio becomes close to unity:

$$v_{cr} = PT_m/(2\pi)^2 c\rho T_b{}^2 r_f{}^2.$$

The condition $v < v_{cr}$ restricts the region of applicability of this model because it was assumed that $h \gg d$. For a steel target, $v_{cr} = 100\,\text{cm/s}$ for $P = 3\,\text{kW}$ and $r_f = 0.085\,\text{mm}$. This value is consistent with the experiment value [16] (see Fig. 4.9). The maximum aspect ratio of welding is achieved for $v \approx \chi/r_f$:

$$(h/d)_{max} \approx PT_m/2\pi^2 \kappa T_b{}^2 r_f. \tag{4.30}$$

For a steel sample at $P \sim 10\,\text{kW}$, we have $(h/d)_{max} \sim 10$. The experimental aspect ratio increases weaker with increasing power than it follows from (4.30). For example, for $P = 10\,\text{kW}$, the aspect ratio was $h/d \sim 5$ ($v = 3\,\text{cm/s}$) [29] and for $P = 90\,\text{kW}$, it was ~ 15($v = 5\,\text{cm/s}$) [22] (d was determined at the weld depth half-maximum).

In conclusion of this section, we consider the shape of the weld pool. One of the simple methods for determining the shape of the weld pool is the rapid ejection of a liquid metal (during welding) by knocking a sample [30] (Fig. 4.21).

A more laborious X-ray method will be considered below. This method can be used to study the weld pool dynamics.

The weld pool shape considerably differs from the elliptic cylinder shape predicted by the thermal model considered above. In reality, the fusion zone length changes with its depth, this length on the metal surface being greater than inside the metal sample. Thus, for $P = 3.6\,\mathrm{kW}$ the sample thickness 4 mm, and $v \sim 3\,\mathrm{cm/s}$, the parameters of the weld pool are: $l = 9.4\,\mathrm{mm}$, $h^* = 2.8\,\mathrm{mm}$, $d = 1.9\,\mathrm{mm}$, and $a = 1.1\,\mathrm{mm}$. The weld pool length in the simplified model can be found from (4.24) by assuming that $\varphi = \pi$:

$$l = P^2/\pi\chi v h^2 (c\rho T_m)^2. \tag{4.31}$$

Because in the case of through melting the weld depth is specified, it follows from (4.31) that l decreases with increasing welding speed ($l \sim v^{-1}$). In experiments, the weld pool length decreases with increasing welding speed weaker ($l \sim v^{-k}$, where $k \sim 0.3$–0.5) [30]. The discrepancy between the theory and experiment can be explained by the increase in the power absorbed by the sample with increasing welding speed during through welding: it is this part of the total laser power that enters (4.31). A weak decrease in the fusion zone width d with increasing v observed in these experiments can be also explained by this reason. It was shown in [30] that the distance a between the laser beam and the fusion zone center a increased with increasing welding speed (Fig. 4.21), which correlates with the "concentration" of isotherms in the theoretical model. Due to the increase in a the tilt of the front wall of the weld pool with respect to the laser beam increases (we will use this fact below).

The expression for the width d and length l of the fusion zone contains the laser power per unit length $P/h = \mathrm{d}P/\mathrm{d}z$. It was assumed in the theoretical model that P/h is constant along the keyhole depth. In reality, because the radius of the vapor channel increases with distance from its bottom, the specific power also $\mathrm{d}P/\mathrm{d}z$ should increase under the condition that the temperature on the keyhole surface is close to the boiling point T_b [see (4.21)]. For low welding speeds, $\mathrm{d}P/\mathrm{d}z$ increases logarithmically with r. For a high welding speed: $\mathrm{d}P/\mathrm{d}z \propto r$. By determining $\mathrm{d}P/\mathrm{d}z$ from (4.21) and substituting the result into (4.31), we obtain the distribution of the fusion zone length over its height:

$$l(z) = 4\pi \frac{v r^2(z)}{\chi} \left(\frac{T_b}{T_m}\right)^2. \tag{4.32}$$

Similarly, we can find from (4.29) the dependence of the weld width on z:

$$d(z) = \pi r(z) T_b / T_m \tag{4.33}$$

Because the vapor channel radius increases with distance from the bottom, the length and width of the weld pool increase as we approach the sample surface.

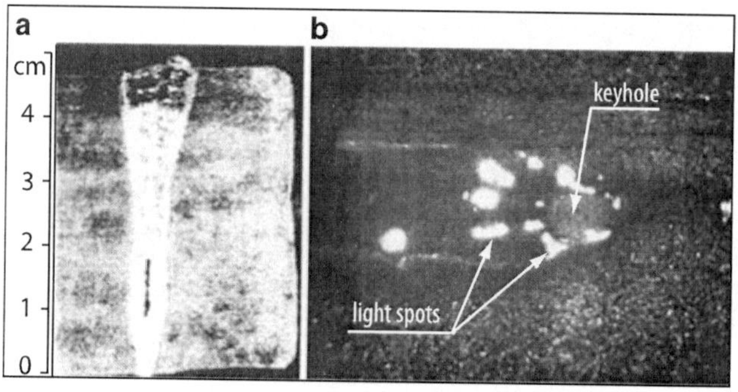

Fig. 4.22 Hot crack in a weld in St3 steel (photo by V.V. Fromm). The video image of weld pool and entrance in keyhole. The white spots are light spots on the molten metal

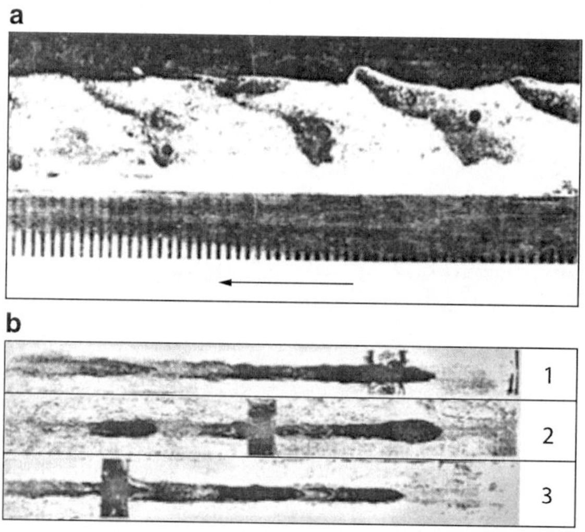

Fig. 4.23 Longitudinal section of a weld in carbon steel with copper marker inserts: (**a**) electron-beam welding, $P = 10\,\mathrm{kW}$, $v = 3\,\mathrm{cm/s}$ (photo by V.A. Lopota). (**b**) Distributions of a copper insert melt in different horizontal sections of a laser weld: in the upper, middle, and lower parts of the weld (photo by A. Grezev [33])

Such behavior is almost always observed (see Figs. 4.8, 4.21, and 4.22a). Similar reasoning is also valid for through melting if the target thickness is close to the welding depth in the case of non-through melting under the same conditions. Because the dependence $l(r)$ is stronger than $d(r)$, this leads to the corresponding cross section of the liquid pool (Fig. 4.21), whose length on the sample surface considerably exceeds the length near the weld root. This is also demonstrated in Fig. 4.23, which shows the longitudinal section of the weld for a steel sample into which copper pins are inserted at different heights. During the sample travel, a copper pin gets into the keyhole and is rapidly melted; due to the intense motion of liquid in the fusion pool, copper is rapidly spread over the entire pool and then the solution of copper and steel solidifies. In this way, the "copper coloring" reveals the shape of the liquid pool. Figure 4.23 shows that the pool shape determined by this method is similar to the shape obtained by the method of rapid metal spatter (see Fig. 4.21). The liquid pool not only determines the weld width but also affects the stability of deep melting and the weld quality. The flow of metals in the weld pool is discussed in the next section.

4.3 Hydrodynamical Processes During Deep Laser-Beam Penetration into Solids

We considered deep melting in previous sections by using approximate thermal models neglecting the motion of liquid in the weld pool. The liquid motion should appear for several reasons. In the case of a high welding speed, the boundary of the fusion zone with a solid metal is close to the vapor channel boundary. The melt boundary is located nearest to the vapor channel for $\varphi = 0$ (see Fig. 4.18). For usual welding parameters, when $P/2\pi\kappa h T_m > 1$, we can easily obtain from (4.24) the minimum melting radius:

$$\frac{r_m v}{\chi} \approx \ln\left(\frac{P}{2\pi\kappa h T_m}\sqrt{\frac{4\pi\chi}{v r_f}}\right). \tag{4.34}$$

(the distance from the laser beam to the melt boundary in the direction perpendicular to the speed is twice as large). One can see from (4.34) that, as the welding speed is increased, r_m decreases and tends to r_f. This is clearly seen by observing directly the weld pool by the method of ultrafast photography (see Fig. 4.22b) [31].

Upon stationary welding, the sample material should be vaporized from the front wall and condense on the rear wall and also flow in the form of liquid around the vapor channel over its lateral walls. In this case, the liquid velocity v^* should exceed the sample velocity: $v^*/v = r_m/(r_m - r_f) > 1$. The estimate shows that the mass transfer in the form of vapor across the channel is insignificant at least for high welding speeds. This follows from the energy considerations. The energy conservation law in the case of mass transfer in the form of vapor can be written as $P = \pi r_f h \rho L_b v$. [see (1.28)]. By substituting into this expression, e.g., the experimental parameters from Fig. 4.8 ($r_f = 0.1$ mm, $h = 1.3$ mm, $v = 16$ cm/s,) and $\rho L_b = 70$ kJ/cm^3, we obtain the laser power ~ 5 kW (in reality, $P = 1$ kW).

4.3.1 Experimental Study of Material Melt Flows

The flowing of a melt around a channel can be observed in model experiments on deep melting of transparent materials (glass [14] and quartz). For example, such experiments were performed [32] with a quartz sample into which molybdenum

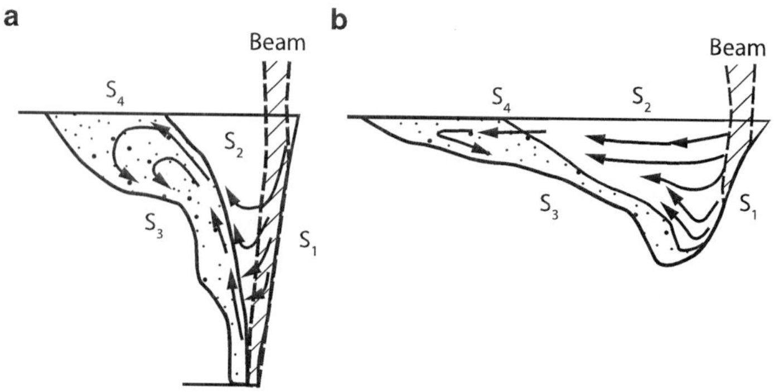

Fig. 4.24 Movement of a melt in a keyhole (300 W): (**a**) 20 cm/min; (**b**) 80 cm/min

powder was pressed. Apart from the flow along lateral walls across the channel, the motion of the melt along the channel was also observed. On the front wall the melt moved from top to bottom, while on the rear wall – from bottom to top. As we saw above, the liquid in a fixed keyhole flows along the channel walls from top to bottom. In a moving sample, due to the breaking of the axial symmetry, the melt from the front wall flows around the channel root and leaks into the weld pool from bottom to top (Fig. 4.24) [14]. One can see from Fig. 4.23 that the flow has the vortex nature in the entire volume of the weld pool. Finally, experiments show that there exits the enhanced-emission zone on the front wall of the keyhole (both upon laser and electron-beam welding), which is periodically created at the keyhole entrance and moves to the bottom. This effect was observed upon electron-beam welding by photographing the weld pool in its intrinsic X-rays [21].

The filming of laser welding in a quartz sample also revealed the periodic motion of the enhanced-emission zone on the front wall of the vapor channel [32]. This periodic motion is apparently responsible for the wavy structure of the weld surface and the formation of voids in the weld root. Note that the maintenance of the gas channel in liquid is an unstable process. Indeed, it is known that the gas bubbling through liquid causes a strong turbulent flow in the liquid [34]. The X-ray photographing of metal targets during electron-beam welding showed that the vapor channel change its shape with time. Because of a weak contrast of images, the authors of [35] did not publish experimental photographs but presented the corresponding drawings of the vapor channel and longitudinal weld at different instant of time Fig. 4.25.

The period of channel depth pulsations was 50–100 ms. One can see that the liquid periodically fills the gas channel, which leads to the formation of voids in the weld, while the melting depth periodically changes along the weld. Figures 4.17, 4.23 shows that similar pulsations of the channel depth can be observed upon laser welding as well.

How do hydrodynamical flows affect the welding process? On the one hand, deep melting is impossible without the corresponding structure of the flow in the

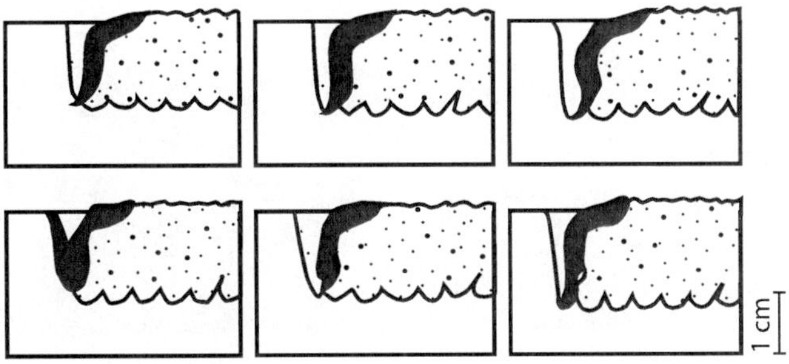

Fig. 4.25 Schematic X-ray images of the keyhole shape obtained at different instants (304 stainless steel, $P = 6\,\mathrm{kW}$, $v = 0.63$ cm/s)

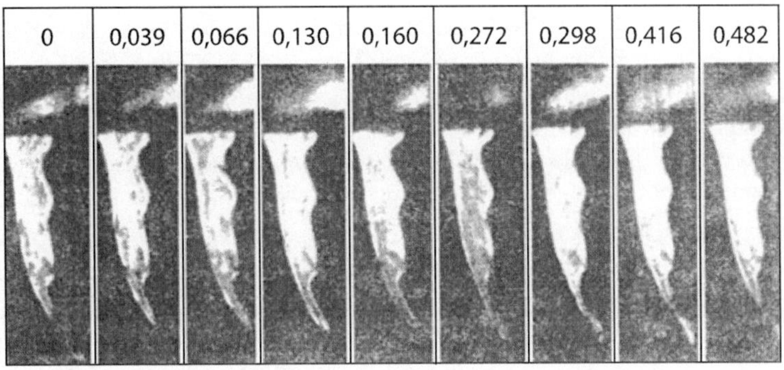

Fig. 4.26 Video images of the welding process in a quartz sample ($P = 0.8\,\mathrm{kW}$, $r_f = 0.8$ mm, $v = 1$ mm/s, $h = 21$ mm)

pool (e.g., the flowing around the vapor channel) because the vapor channel will be carried out from the beam by a moving sample. This means that the melt flow also can restrict the melting depth. On the other hand, it is clear that at a high welding speed, when the fusion zone–solid phase interface in the welding direction is close to the vapor channel, the liquid flow considerably affects the temperature distribution around the laser beam: the heat release zone is cooled more efficiently due to convective heat transfer and, therefore, the welding depth decreases.

Consider in more detail the observed picture of deep melting by a moving laser beam. Experiments [32] were performed with a quartz sample. One can see from Fig. 4.26 that the front wall of the channel is tilted with respect to the laser beam, which is directed vertically. The tilt angle increases with deepening of the channel. The rear wall of the channel is parallel to the laser beam and, therefore, direct beams do not fall on the weld root. The channel is strongly open in its upper part and its length is approximately 5 mm. The mean angle between the front edge of the channel and the sample plane depends on welding conditions: it increases with increasing the laser power and decreasing the weld speed (see Fig. 4.31. One can see from Fig. 4.26 that the shape of the lower part of the channel changes with a period

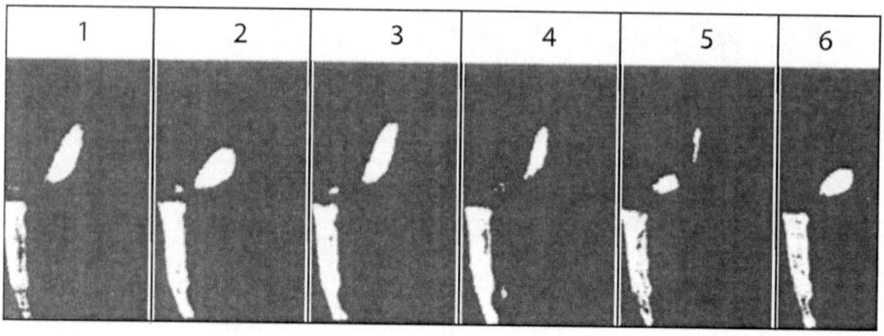

Fig. 4.27 Dynamics of vapor escape from a keyhole (time between frames is 0.3 s, $P = 1.3\,\text{kW}$, $v = 1\,\text{mm/s}$, $h = 24\,\text{mm}$, quartz)

of ~0.5 s. As a result, defects (voids) are formed in the weld root, which have the form of peaks, their frequency increases with increasing welding speed.

Because the beam diameter is much smaller than the keyhole length, only the beams reflected from the front wall can fall on the rear wall (see Fig. 4.26). For the same reason the front wall can be illuminated directly by the laser beam incompletely, its lower part being heated only by radiation reflected from the rear wall. The nonuniform illumination of the walls is demonstrated by photographs in Fig. 4.26. The nonuniform heating and the corresponding intensity are manifested in the wavy structure of the rear wall. It is clear that the pouring of the keyhole with the liquid can be prevented only when the liquid pool is small and liquid quartz has a high viscosity. A small size of the pool in quartz and a high viscosity of liquid quartz are confirmed by the fact that the vapor channel is not filled with liquid after switching off laser radiation [32].

By selecting different filters, the authors of [32] photographed simultaneously both the vapor channel and vapor jet emerging from the channel. Figure 4.27 shows that the angle at which vapor is ejected from the keyhole changes in the range 20–60°. The period of variations in the intensity and exit angle of vapor is 2–10 ms under these conditions and the mass removal rate is $\sim 10^{-2}$ g/s. A change in the exit angle of vapor is probably related to the movement of the enhanced heat release zone over the front wall of the keyhole, as mentioned above. The correlation between the vapor exit angle and the keyhole brightness is clearly demonstrated in Fig. 4.28, where the photographs of the sample welding process are presented (for quartz modification with a low viscosity close to that of steel).

Frames 1–5 show that the vapor exit angle with respect to the surface decreases with time and temperature in the lower part of the keyhole decreases. Thus, the heat release zone in this part of the period rises to the keyhole entrance. Frames 5–8 show that the vapor exit angle increases and the emission and temperature in the lower part of the keyhole increases. Thus, the heat release zone periodically descends deep into the keyhole (although this zone is not directly observed in these photographs).

Because the heat conduction of quartz glass is very low $\kappa = 1.4 \times 10^{-2}\,\text{W/(cm K)}$, the fusion zone behind the beam is small; and as a result, the keyhole is comparatively stable. The estimate by (4.21) for $P = 1\,\text{kW}$ gives the welding depth $\sim 6\,\text{cm}$

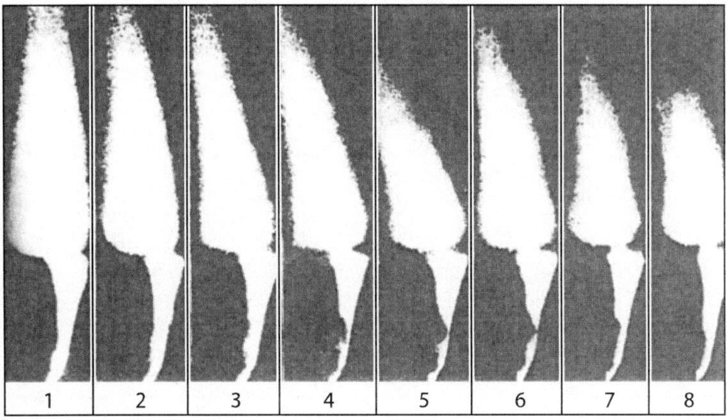

Fig. 4.28 Video images of melting of a quartz glass sample (time between frames is 7.5 ms, different filters for vapor and sample were used, $P = 0.65\,\mathrm{kW}$, $v = 1\,\mathrm{mm/s}$, $h = 11\,\mathrm{mm}$)

(the keyhole radius was set equal to 5 mm, in accordance with experiments, and the wall temperature was assumed 2,000 K). In this case, the amount of power expended for vaporization does not exceed 100 W.

The model experiment [32] on deep melting of a steel plate clamped from both sides by quartz plates is of interest for studying the welding process (Fig. 4.29).

One can see from the frames in Fig. 4.29 that, because a liquid metal is forced back from quartz by vapor, the interior of the keyhole glows weaker (a liquid metal has the higher luminosity than transparent quartz). The luminous metal zone is similar to the fusion zone in a massive sample (cf. Figs. 4.21 and 4.23). One can see from Fig. 4.29 that, as in a quartz sample, a vapor keyhole in the metal plate represents a hollow cone with the front wall tilted with respect to the laser beam and the rear wall virtually parallel to the beam. The keyhole shape, the brightness and uniformity of emission change periodically. It seems that liquid in the weld pool moves in the same way as in experiments with the quartz sample, only in this case there exists a broad heated (and probably melted) region behind the beam. This can be explained by a higher specific power P/h [see (4.31)]. The cooling due to heat conduction in the transverse direction (with respect to the travel direction) is absent because quartz plates clamped to the metal plate are poor heat conductors. This also explains a large welding depth (compared to a massive metal sample). Another feature of welding [36] is the occasional extrusion of a liquid metal on the front wall, which probably occurs at instants when the laser beam is incident on the lower part of the keyhole (frames 5 and 6 or 7 and 8). Note that a plasma plume in all the photographs presented above was either small (when melting was performed in the helium atmosphere) or emission was suppressed by optical filters. It seems that the luminous cloud observed in all the photographs is produced by hot vapor rather than by the plasma plume, which is confirmed by the fact that the cloud moved towards the beam and its emission enhanced with distance from the keyhole (probably due to the formation of small droplets upon vapor condensation).

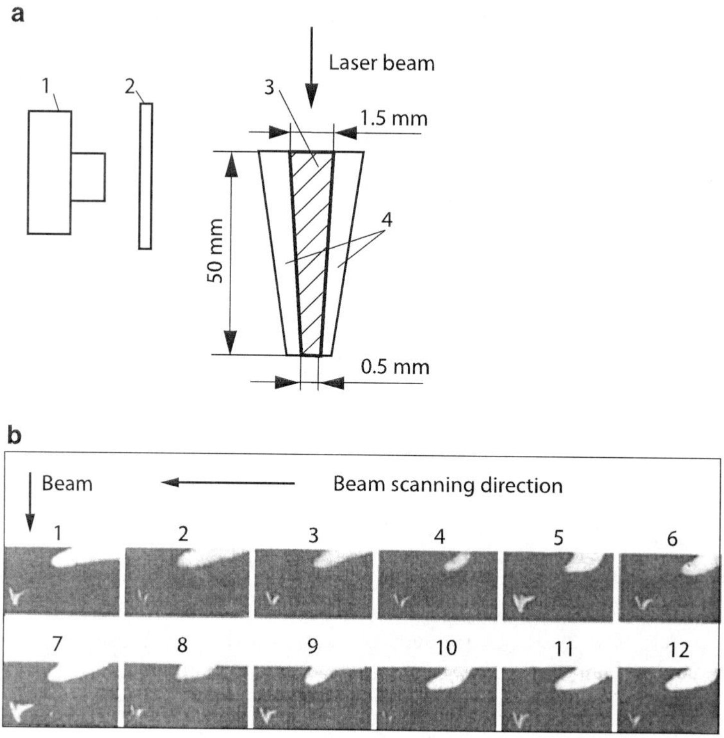

Fig. 4.29 Experimental scheme (**a**) and video images of a steel sample placed between quartz plates (**b**) obtained for $P = 4\,\text{kW}$, $v = 1\,\text{mm/s}$, $h = 22\,\text{mm}$ (time between frames is 1 ms) (photo by V.A. Lopota): (**1**) fast movie camera; (**2**) filter; (**3**) stainless steel plate; (**4**) quartz plates (the arrow indicates the sample travel direction)

Let us summarize the experimental studies of deep melting of quartz and a combined sample consisting of a metal and quartz. These investigations showed that the vapor channel width along welding direction could exceed the laser beam diameter by several times. The channel is strongly asymmetric due to the sample motion. The formation frequency of peaks in the weld root is close to the inverse time of the channel displacement over its width (Fig. 4.30).

At the same time, the illumination of the keyhole and vapor ejection change at the frequency which is approximately an order of magnitude higher. The formation of peaks in the weld root frequently occurs both in electron-beam and laser welding [20, 37]. This effect is sometimes explained either by the screening of a sample by vapor or by the filling of the lower part of the channel by liquid from the rear wall side. However, this effect also takes place upon welding of dielectrics both by high-power 3.5-kW laser radiation (when a plasma plume exists) and low-power 10 W laser radiation (when a plasma plume is absent) (see samples in Fig. 9.44. This suggests that the peaks are formed due to the keyhole instability. The stability problems will be considered below.

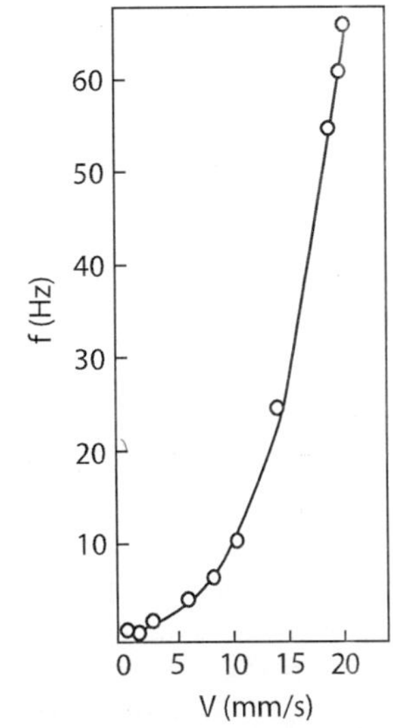

Fig. 4.30 Dependence of the frequency of formation of peaks on the travel speed of a quartz target irradiated by a 2.7 kW laser beam

4.3.2 Models of the Hydrodynamic Flow Upon Deep Melting

A variety of types of hydrodynamical flows and their intricate nature complicate the comprehensive theoretical description of this phenomenon. The problem is aggravated by the necessity of simultaneous consideration of the hydrodynamical flow and the propagation of heat in a sample because the forces producing the motion of liquid depend on the temperature distribution. We will assume that the temperature distribution on the vapor channel surface is asymmetric and the difference of temperatures of the hot front wall and rear wall is T_1.

To provide this condition, a thermal source should be nonuniform, emitting from the front wall a higher power than from the rear wall:

$$\tilde{P} = P + P_1 \cos\varphi, \qquad P_1 \ll P$$

where P is the average power released in the internal surface and P_1 is the half-difference of powers emitted from the front and rear walls.

The power asymmetry P_1 is determined by the channel shape – the number of beams incident on the front wall is greater than that incident on the rear wall (see Figs. 4.26, 4.27, 4.28, 4.29). It is clear that this asymmetry should increase with the welding speed (Figs. 4.31 and 4.24).

Consider briefly in the small speed and weak asymmetry approximation the flowing of liquid around the vapor channel caused by the difference between surface tensions on the front and rear walls as well as by the vapor flow in the keyhole [38]. This process can be caused by the difference between recoil vapor pressures on the front and rear walls.

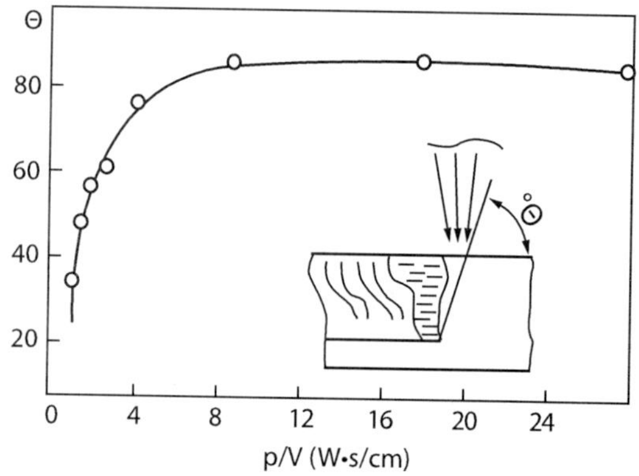

Fig. 4.31 Dependence of the tilt angle of the front face of a keyhole on the ratio P/v

In this case, the excessive momentum should be carried out from the target with vapors. Such a motion of vapor can be realized only in the upper part of the keyhole at the depth of the order of its radius (see Sect. 2.2). Vapors evaporated on the front wall at a large depth will condense on the rear wall by imparting to it the momentum that they carried away from the front wall. Nevertheless, in this case the liquid can overflow due to the friction of vapors on the lateral walls of the keyhole.

Theoretical studies [38] performed for $vr_f/2\chi \ll 1$ and $r_f/r_m \ll 1$ gave the following results. The overflow of a liquid metal due to surface tension forces is negligible in this case because metal vapors carry the condensation energy on the rear wall, thereby strongly reducing the temperature difference T_1.

The travel speed of the keyhole due to the friction of vapors on the lateral walls of the keyhole is

$$v = \frac{\nu_1}{\nu} \frac{\kappa}{2\rho L_b r_f} T_1 \tag{4.35}$$

where ν_1 and ν are the kinematic viscosities of vapor and liquid, respectively; κ is the heat conductivity of the liquid; and L_b, ρ, are the specific boiling enthalpy and density of the liquid, respectively. For example, for steel with $r_f \sim 0.5$ mm we have $v = T_1/40$ cm/s, which is quite acceptable value.

For $vr_f/2\chi < 1$, we can easily find from (4.21), (4.24), and (4.25) the relation between the fusion zone radius r_m and the vapor channel radius, which is equal to the laser beam radius r_f in this model:

$$r_m/r_f = (vr_f/2\chi)^{T_m/T_b - 1}. \tag{4.36}$$

One can see from (4.36) that, when the travel speed of the target is small ($vr_f/2\chi \ll 1$), the fusion zone radius greatly exceeds the channel radius, as assumed above. This expression also shows that the case $r_f \sim r_m$ is realized for $vr_f/2\chi \approx 1$, i.e. for a thin melt layer in front and on sides of the keyhole, convective heat transfer by a liquid metal flowing around the keyhole becomes important. It seems that the cavern has the form of an ellipsoid elongated along the speed direction. In this case, the study of the liquid flow is strongly complicated.

Finally, consider briefly the mechanical properties of weld joints determining the field of practical applications of laser welding [39, 41]. Weld joints often have defects in the form of hot cracks, which are produced upon crystallization of a liquid metal in the weld pool (Fig. 4.22a). It has been found experimentally that such defects are rarely produced in laser welding. Due to the formation of a fine-grained structure in the weld and near it, the probability of formation of cold cracks also decreases. Finally, the level of residual deformations upon laser welding is reduced by 3–5 times compared to arc welding [39, 40].

Note in the conclusion of Sects. 4.1 and 4.2 devoted to laser welding that most of the observed general welding properties can be described by models taking into account the absorption of light by keyhole walls, metal flow, and heat removal to the sample volume. Below, we will consider more accurate approaches in which the number of model elements will be reduced. Also, we will consider the models of nonstationary effects and the influence of various factors on the efficiency and quality of laser welding such as the polarization and quality of laser radiation, shield gases, etc.

4.3.3 *Influence of Laser Radiation Polarization and Shield Gas on Laser Welding Properties*

It was shown in the first chapter that, within the framework of geometrical optics, the effective absorption coefficient of laser radiation α_e in a keyhole increases with increasing its depth, more exactly h/d. The value of α_e for laser radiation polarized parallel to the plane of incidence achieves unity for $h/d \sim 6$. A laser beam polarized perpendicular to the plane of incidence is absorbed weaker by 1.5–2 times for the same value of h/d. These results were obtained for a plane symmetric or axially symmetric keyhole. In the latter case, it was assumed that radiation was polarized radially or circularly. It was shown above that the gas–vapor channel produced in laser welding has a more complicated three-dimensional shape. The propagation of plane-polarized radiation in such a keyhole will differ from simplest cases. However, a certain influence of polarization should be manifested. This was first demonstrated in laser cutting [59] (see Chap. 5). It was shown later [60] that the welding of carbon steel by a CO_2-laser occurred more efficiently when laser radiation was polarized parallel to the plane of incidence. Figure 4.32 compares four types of polarization: two linear polarizations ($\|, \perp$), circular and in the plane at an angle of 45° with respect to the welding direction [61].

One can see that, for the same expenditure of energy per 1 mm of the weld, radiation polarized parallel to the plane of incidence penetrates into a steel sample by the depth that is twice as large as that for radiation polarized perpendicular to the plane of incidence. In this geometry, the effective absorption coefficient probably weakly depends on polarization. This is confirmed by the fact that the weld width in the case of parallel polarization is approximately smaller by half than that for perpendicular polarization (see Fig. 4.33).

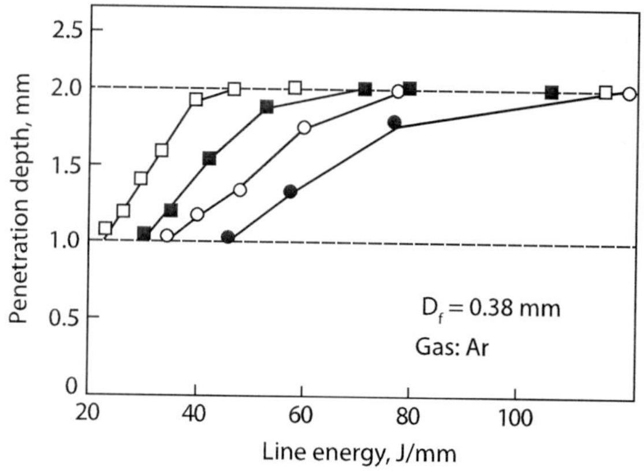

Fig. 4.32 Penetration depths obtained upon welding by laser beams with different polarizations. The depth 2 mm corresponds to through welding. (*square*) polarization parallel to the plane of incidence (3,900 W); (*dark square*) circular polarization (3,550 W); (*ring*) polarization in the plane at 45 to the welding direction (400 W); (*dark ring*) polarization perpendicular to the plane of incidence (3,850 W)

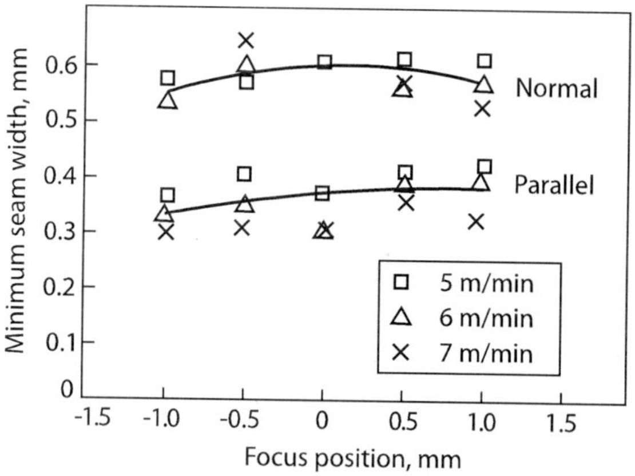

Fig. 4.33 Weld widths produced by laser radiation polarized parallel and perpendicular to the plane of incidence (1 mm thick steel, $P = 3.69\,\text{kW}$, $d_f = 0.36\,\text{mm}$, Ar)

It seems that the role of polarization is reduced to the creation of the favorable asymmetry of energy release on keyhole walls. Note that the influence of polarization is a "delicate" effect. Thus, the laser welding of aluminum [60] was independent of laser radiation polarization. However, the authors of paper [62] found the conditions under which the influence of polarization was observed. The important role was played by the gas in which welding was performed. When argon was replaced by helium, the advantage of parallel polarization was manifested in the same range of speeds and to the same extent as in welding of steel. Welding in the argon atmosphere was subjected to plasma effects, which were considered in the previous chapter. It is possible that these processes masked the influence of polarization [1].

Polarization can also lead to another interesting effect of the gas–vapor keyhole bending upon deep penetration of a laser beam into a material [63]. The mechanism of this effect is explained in Fig. 4.34. The necessary condition for such a bending is the asymmetric intensity distribution over the laser beam cross section and linear polarization. It is obvious that the keyhole profile at the initial state of its formation

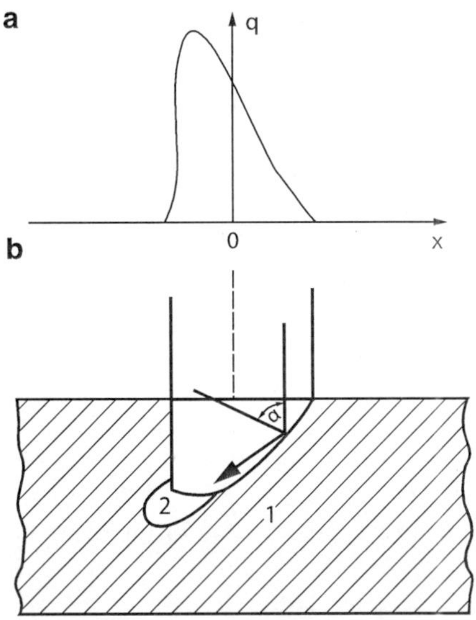

Fig. 4.34 Asymmetric distribution of the laser radiation intensity I along the beam diameter (along the X-axis) (**a**) and the corresponding longitudinal cross section of the keyhole at the initial stage of interaction of laser radiation with material (**b**): (**1**) metal; (**2**) metal zone that will be removed due to multiple reflection of beams

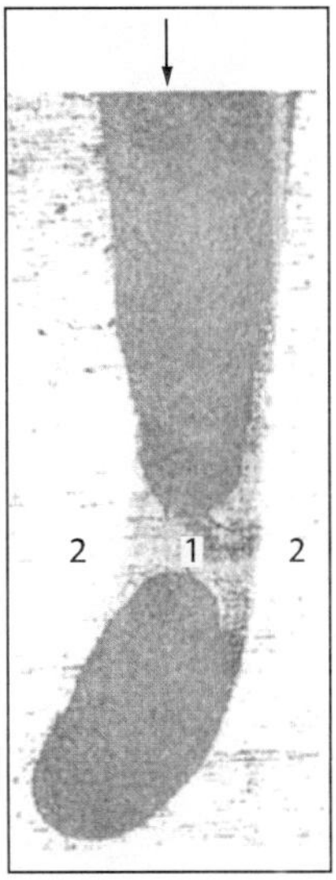

Fig. 4.35 Typical view of the longitudinal cross section of a laser damage channel in a plane perpendicular to the plane of linearly polarized radiation obtained for the asymmetric intensity distribution in the laser beam: (**1**) channel volume filled with melt; (**2**) metal; the arrow shows the incident beam direction; 30 magnification

will be also asymmetric. Beams incident on the flat slope of the keyhole will be reflected to the laser beam axis. If radiation is polarized in the plane perpendicular to the plane of incidence, the reflection coefficient will be large. This will lead to the absorption of these beams by the opposite wall, resulting finally in the keyhole bending. Figure 4.35 shows the cross section of the keyhole produced by the Nd laser beam focused to an aluminum sample. The bending effect disappears after the rotation of the polarization plane through 90°.

4.3.4 Role of Shield Gases in Deep Melting of Metals

In the previous chapter we considered plasma phenomena near the sample surface and inside the keyhole. The gas medium affects to a certain extent laser welding because it partially absorbs and scatters laser radiation and also can initiate chemical reaction with a hot metal. Based on technological requirements, shield gases are supplied through nozzles from different sides to the interaction zone of laser radiation with a sample. Consider the influence of gas jets on laser welding. First, note that if the jet is directed parallel to the surface and perpendicular to the laser beam, the melting depth will be minimal, although the nozzle is located close to the sample at a distance of ~1 mm. For the tilt angle of the nozzle to the surface plane ~30°, the welding depth increases by ~50% (see Fig. 4.36b [64]). To suppress the erosion plume, it is necessary to provide the sufficient shield-gas flow rate or, in other words, to provide the sufficient excess pressure. The scheme for measuring this quantity is presented in Fig. 4.36a [64]. The melting depth increases approximately

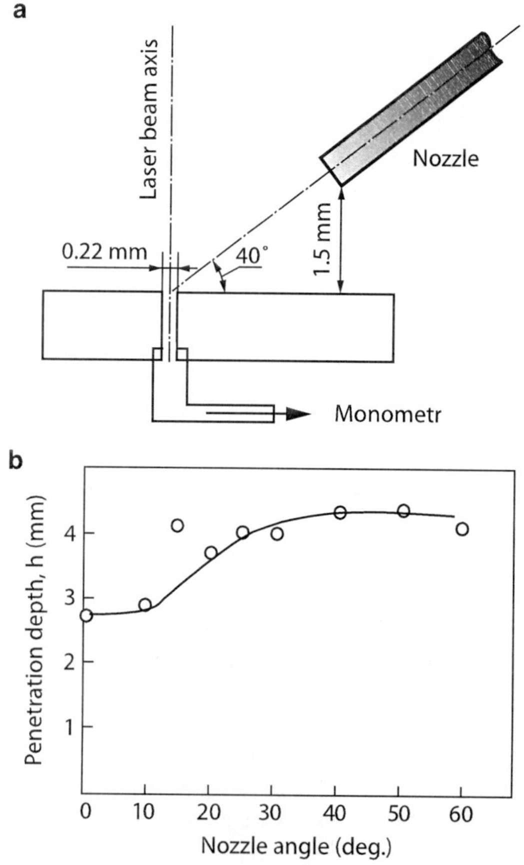

Fig. 4.36 Experimental setup. Measurement of operation pressure (**a**). Nozzle angle vs penetration depth (**b**)

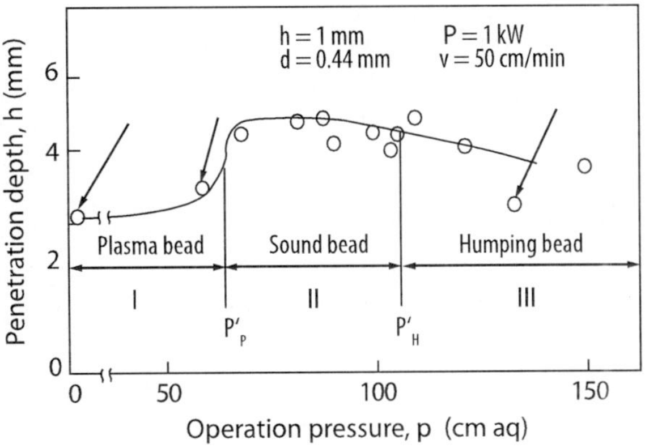

Fig. 4.37 Dependence of the melting depth on the shield gas pressure (1 kW, 50 cm/min)

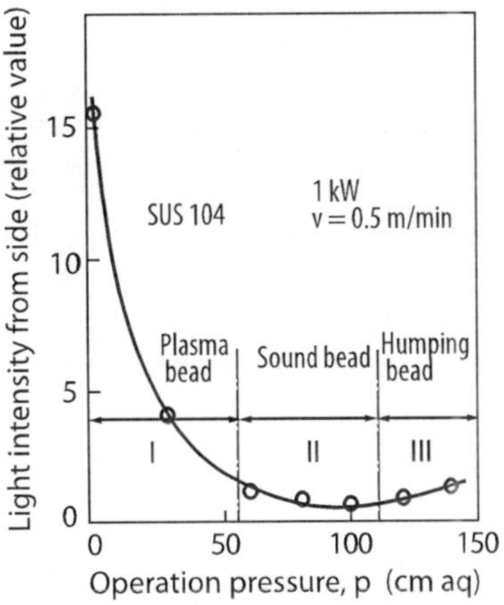

Fig. 4.38 Time-averaged light intensity measured with a phototransistor from the side (1 kW, 50 cm/min, $F = 5$ inch)

by 1.5–2 times (see Fig. 4.37 [64]). The increase in the melting depth is explained by the decrease in the influence of plasma processes, as is demonstrated in Fig. 4.38 [64]. In the gas pressure region, where the welding depth is smaller, the enhanced luminosity of the interaction zone of the laser beam with the sample is observed. The blowing of the plasma by the shied gas jet is typical for CO_2 lasers, while the plasma is virtually transparent upon Nd:YAG laser welding. The decrease in melting depth with further increase in gas pressure is explained by direct interaction of gas jet with molten metal. As a result, waves are excited in the weld pool, which flood the keyhole [64]. The excess vapor pressure in the keyhole was measured with a pressure gauge by using the scheme presented in Fig. 4.39. One can see that the vapor pressure is 20–25 (cm aq.) and is close to the surface tension pressure. When technological gas is supplied, the pressure in the keyhole increases correspondingly (cf. Figs. 4.39a,b). The action of the gas jet on the keyhole and weld pool is qualitatively illustrated by Fig. 4.40 [64].

The theoretical description of such intricate processes is a difficult problem. The interaction between the vapor and shied gas jets can be use to determine the dynamic

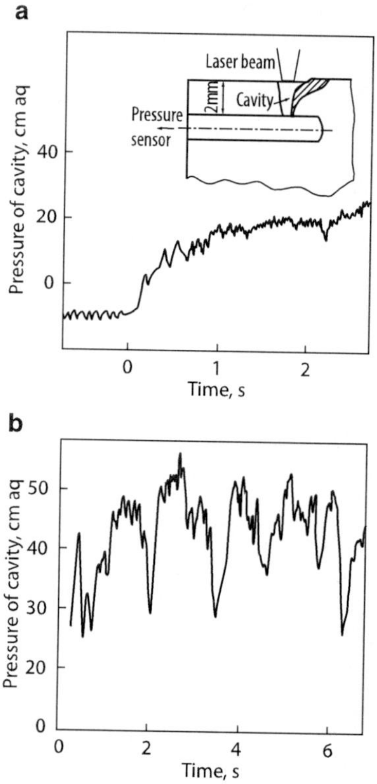

Fig. 4.39 Excessive pressure in the keyhole measures with a pressure gauge at $p = 0$ (**a**) and $p = 85$ cm aq. (**b**) (1 kW, 50 cm/min)

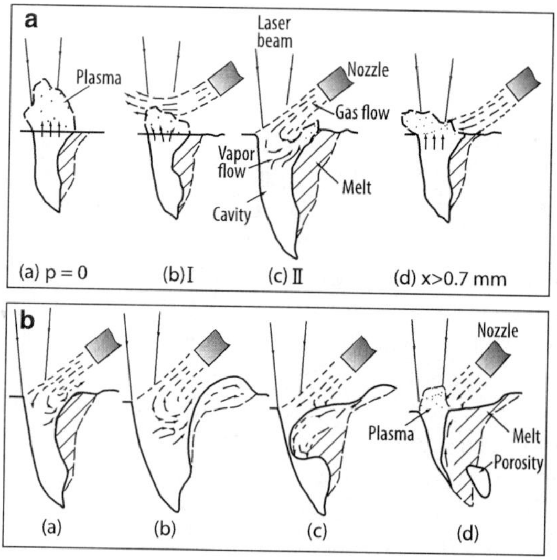

Fig. 4.40 (**a**) Scheme showing the role of the shield gas in regions I (a), (b), II (c), and at $x > 0.7$ mm (d) (see Fig. 4.37, regions I, II). (**b**) Scheme showing the interaction between the shield gas, plasma, keyhole, and molten metal in region III (see Fig. 4.37)

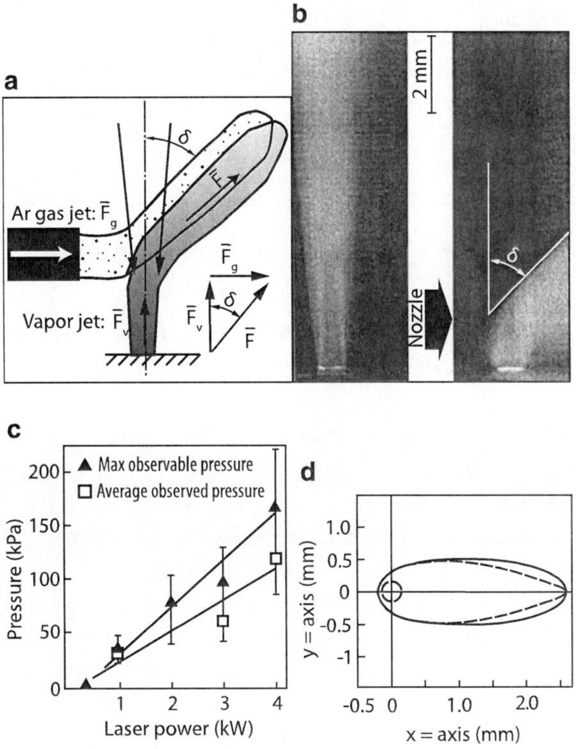

Fig. 4.41 (**a**) Scheme of the detection method: the Ar gas jet deflects the ejected vapor jet by an angle. The momentum F of the resulting jet is the sum of the gas jet momentum Fg and vapor jet momentum Fv. (**b**) Vapor jets produced upon irradiation of a 4 kW, 10 ms laser beam of diameter 0.6 mm with a fixed stainless steel sample in the absence (*left*) and presence (*right*) of a transverse Ar jet (the nozzle cross section is 4 × 2 mm and the flow rate is 30 l/min). The horizontal nozzle axis is located 3 mm above the sample surface. (**c**) Dependences of the dynamic pressure of a vapor jet on the laser power. Measurements are performed for a steel sample at different Ar flow rates. (**d**) Weld pool shape for a laser power of 2.0 kW and a welding speed of 10 mm/s. The *solid curve* and *dashed curves* are theoretical and experimental shapes, respectively. The *dashed circle* is a keyhole

pressure P_V of the vapor jet [65]. The method is based on measuring the deviation angle δ of the vapor jet upon blowing off the jet from the side by inert gas (see Fig. 4.41a,b [65]). It follows from the law of conservation of momentum that

$$\tan\delta = S_0\rho_0 v^2/S_V p_V,$$

where S_0 and S_V are cross-sectional areas of the shield gas and vapor jets, respectively; $\rho_0 v^2$ is the dynamic pressure of the shield gas jet; and p_V is the dynamic pressure of the vapor jet. Experiments were performed in the pulsed regime for $\tau < 10$ ms because the boundaries of the vapor jet are blurred at large times. Figure 4.41c shows that the dynamic pressure of the vapor jet increases linearly with the power of a Nd:YAG laser above the threshold $P = 0.5\,\mathrm{kW}$. A comparison of the dynamic pressure in Fig. 4.41c with pressure calculated by (1.31) gives the underestimated value of the absorption coefficient $\alpha \sim 0.1$. This is explained by the fact that not all the energy is spent for vaporization of the target material, which was observed both for welding and cutting (Chap. 6).

4.4 Models of a Gas–Vapor Keyhole of Finite Size

4.4.1 *Thermal Deep-Melting Model with a Gas–Vapor Keyhole of Finite Diameter*

The Swift–Hook–Gick weld pool model (4.24) is based on an idealized approach. A heat source is an infinitely thin heated filament in which heat of power P is released. As mentioned above, this model determines the relation between the weld depth and width. An additional condition for the thermal model can be the requirement that the temperature of the keyhole walls is equal to the boiling of a metal, as was assumed in the simplest model (4.21). We will attempt to combine these two approaches and try to construct a thermal model taking into account a finite size of a gas–vapor keyhole and the equality of the keyhole wall temperature to the boiling temperature. It is necessary to neglect in advance in this model the influence of the liquid motion on the temperature field of a sample. As shown above, this can be important at high welding speeds. The solution of the heat conduction equation in a moving sample in the case of a cylindrical source with walls at the boiling temperature has the complex form [42, 43]

$$T = 2(T_b - T_0)\exp\left(\frac{1}{2}\frac{Vr}{\chi}\cos\varphi\right)\left[\frac{I_0\left(\frac{1}{2}\frac{v}{\chi}r_k\right)K_0\left(\frac{1}{2}\frac{v}{\chi}r\right)}{2K_0\left(\frac{1}{2}\frac{v}{\chi}r_k\right)} + \sum_{n=1}^{\infty}(-1)^n\frac{I_n\left(\frac{1}{2}\frac{v}{\chi}r_k\right)K_n\left(\frac{1}{2}\frac{v}{\chi}r\right)}{K_n\left(\frac{1}{2}\frac{v}{\chi}r_k\right)}\cos n\varphi\right] + T_0. \tag{4.37}$$

Here, T_0 is the sample temperature away from the interaction zone; r_k is the radius of the gas–vapor keyhole; and I and K are the modified Bessel functions. Recall that I_n increases from zero to infinity and K_n decreases from infinity to zero for $\frac{1}{2}\frac{v}{\chi}r \geq 0$ [44].

Because a liquid metal flows around the keyhole, the energy supplied to the gas–vapor channel is removed by the heat conduction:

$$\frac{P}{hr_k} = -\kappa\int_0^{2\pi}\left.\frac{\partial T}{\partial r}\right|_{r=r_k}\mathrm{d}\varphi \tag{4.38}$$

By assuming that r_k is specified, e.g., is equal to the focal spot diameter, we can find from (4.37) the fusion zone width for the specified speed and, therefore, the weld width d. The weld depth h can be determined from (4.38) for the specified power P. Thus, as simplified model (4.21) and (4.29), this model allows us to determine welding parameters, although in more complicated but more rigorous way. This model can be used in the case of non-through and deep melting, when the total energy of a laser beam is absorbed in the keyhole. Otherwise, e.g., upon through welding, a part of radiation propagates through a sample. The model can be used to determine a fraction of absorbed radiation for a specified small thickness of a sample being welded. The authors of [43] used this model for measuring the diameter of a gas–vapor channel. As shown above, this diameter can differ from the focal spot diameter. The value of r_k can be determined from experimental data. For example, the melt pool length was determined by photographing a 1 mm-tick steel plate welded by a 2-kW CO_2 laser (the focal spot diameter was 0.4 mm). Figure 4.41d compares the calculated and experimental shapes of the weld pool. Also, the gas–vapor keyhole is shown whose radius was determined by using this model. Table 4.1 demonstrates that the theoretical value of the keyhole diameter decreases with increasing welding speed. As the keyhole size decreases, the temperature gradient near the keyhole increases and, therefore, power absorbed in the keyhole increases.

This is clearly demonstrated in Table 4.2 and is explained by the fact that in order to continue welding with a higher speed without decreasing the weld depth, it is necessary to increase the power absorbed by the sample. This situation is similar to remote cutting, where the fraction of radiation absorbed by the sample also increases with increasing the sample travel speed (see Fig. 5.12). The difference between diameters of the focal spot and gas–vapor keyhole is also demonstrated

Table 4.1 Geometry of weld pool produced by a continuous 2.0-kW carbon dioxide laser in a thin steel sheet of 1 mm thickness

Translation speed (mm/s)	Experimental weld length (mm)	Experimental weld width (mm)	Theoretical weld width (mm)	Theoretical keyhole radius (mm)
10	3.1	0.80	0.86	0.24
20	3.0	0.66	0.67	0.20
40	2.8	0.46	0.48	0.14
60	3.3	0.47	0.44	0.12

Table 4.2 Comparison of power absorption calculated from experimental weld pool shapes; (b) total bremsstrahlung and Fresnel absorption in keyhole produced by 2.0 kW laser beam of 0.2 mm radius in 1 mm thick sheet of steel

Translation speed (mm/s)	Power in weld pool (watts)	Power absorbed in keyhole (watts)
10	342	340
20	442	432
40	528	598
60	624	686

experimentally. For example, the diameter of a keyhole in a welded 3-mm-thick stainless steel plate exceeded the focal spot diameter [31] (Fig. 4.22b). The focal spot diameter for a 5.5 kW Rofin-Sinar CO_2 laser was 0.6 mm. For $v = 3$ m/min, the gas–vapor keyhole width fluctuated near 1.16 mm (13%) and length – near 1.57 mm (11%). These sizes decreased with increasing the welding speed or decreasing the laser power. It is interesting to compare the models of thermal sources in the form of a thin string (4.29) and a cylinder (4.37), (4.38). For large welding speeds $\frac{vd}{\chi} \gg 1$, when the welding efficiency in model (4.29) tends to the universal constant $\sqrt{2/\pi e}$, the welding efficiency in model (4.37), (4.38) is not the universal constant. As follows from (4.37), (4.38), it should depend on the boiling point T_b and melting point T_m. Calculations [42] within the framework of this model, taking into account that $d \to 2r_k$ and $T_0 \to 0$, give the expression

$$\eta \approx \frac{2T_m}{T_b + T_m}. \tag{4.39}$$

For example, (4.39) gives the welding efficiency 0.75 for iron, which exceeds by one and a half the welding efficiency according to model (4.29).

4.4.2 Self-Consistent Stationary Laser Welding Model

The previous models described the "global" properties of deep melting laser welding: the weld depth and width and their dependence on the welding speed. These approaches were not self-consistent and contained a number of model assumptions: the absorption of radiation was assumed uniform over the keyhole depth, absorption in the keyhole plasma was neglected, the keyhole shape was assumed cylindrical, and its diameter was assumed either equal to the focal spot diameter or was determined experimentally.

It is interesting to consider more complete models taking into account all these factors at least approximately. Our approach is based on the balance of heat from a linear source moving in an infinite sample. The temperature distribution from an infinitely thin source is described by expression (4.24). Let us separate any isotherm corresponding to a certain temperature T_0 and calculate the energy flow propagating from this isotherm outside

$$q_V(r,\varphi) = -\kappa \nabla T|_{T=T_0} \tag{4.40}$$

If we select a hole restricted by this isotherm and provide the heat flow described by $q_V(r,\varphi)$, the temperature field outside the isotherm will not change. Let us set $T_0 = T_b$ and calculate q at two points $\varphi = 0$ and $\varphi = \pi$ in the keyhole. Here, $\nabla T = \partial T/\partial r$ due to symmetry. Under these conditions, by substituting expression (4.24) for $T(r,\varphi)$ into (4.40), we obtain the heat flows q_{Vf} and q_{VB} at the leading and trailing edges of the gas–vapor keyhole [45]

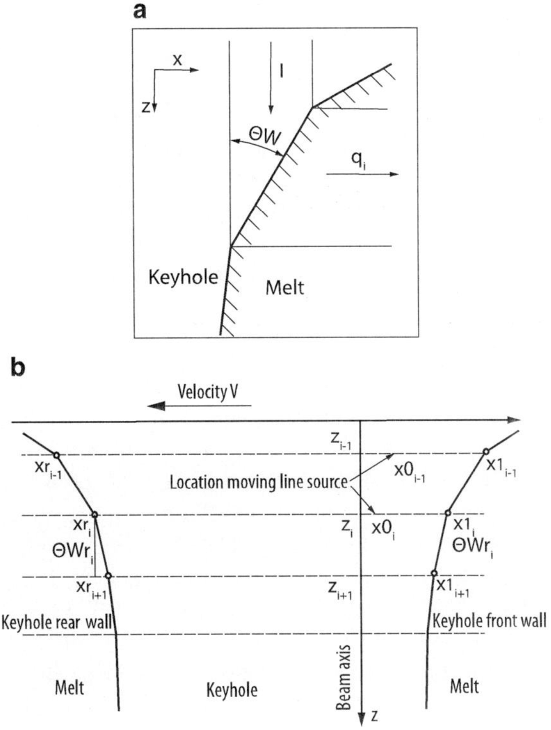

Fig. 4.42 (**a**) Resulting keyhole wall angle θ_W due to the relation of the incident beam intensity I to the head flow necessary for evaporation q_v; (**b**) pointwise wall calculation $x(z)$

$$q_{Vf} = (T_V - T\infty)\kappa P_e \left(1 + \frac{K_1(P_e x_f)}{K_0(P_e x_f)}\right), \tag{4.41}$$

$$q_{VB} = (T_v - T_\infty)\kappa P_e \left(-1 + \frac{K_1(P_e x_B)}{K_0(P_e x_B)}\right), \tag{4.42}$$

where x_f and x_B are the coordinates of the leading and trailing edges of the gas–vapor keyhole (Fig. 4.42), $P_e = v/2\chi$ and $K_1 = -K_0'(x)$ is the modified first-order Bessel function of the second kind [44]. The energy supplied by the laser beam should maintain these heat flow densities. Consider, e.g., the Gaussian distribution of laser radiation [46]

$$q(x,z) = I_0 \left(\frac{r_f}{r_{f0}}\right)^2 \exp\left(-\frac{2r^2}{r_f^2}\right), \tag{4.43}$$

where $r_{f0} = 2\lambda F M^2/\pi D$ is the focal spot radius, D is the beam aperture on a focusing lens with focal distance F, and M^2 is the beam quality parameter (the ratio of the focal beam diameter to the diffraction spot diameter). The laser beam radius is

$$r_f(z) = r_{f0}\left[1 + \left(\frac{z - z_0}{z_f}\right)^2\right]^{1/2}, \tag{4.44}$$

where z_f is the waist length (Rayleigh length) $z_f = 2r_{f0}F/D$. Under these conditions, the peak radiation intensity in the focal spot is $I_0 = 2P/\pi r_{f0}^2$ The energy balance at each point of the leading and trailing edges is described by the equality of the absorbed laser radiation flux and the flux q_V carried away due to heat conduction

$$\tan(\theta)\alpha(\theta)q(x,z) = q_V(x,z) - I_p - I_f, \tag{4.45}$$

where θ is the angle between the beam axis (vertical) and keyhole surface Fig. 4.42; I_p is the heat supply from the keyhole plasma; α is the absorption coefficient for laser radiation; and I_f is the radiation intensity caused by multiple reflection of laser radiation in the keyhole. It was shown in Chap. 3 that the heat flow from plasma is most considerable upon absorption of the CO_2 laser radiation. In the first approximation, the keyhole shape is calculated by neglecting multiple reflections of radiation and heat from plasma. Then, these quantities are calculated and the keyhole shape is calculated again taking them into account, etc. The shape of the front $x_f(z)$ and rear $x_b(z)$ keyhole walls is calculated by using (4.41)–(4.45). The calculation is ended when curves $x_f(z)$ and $x_b(z)$ intersect.

It was shown in Chap. 1 that multiply reflected radiation is usually calculated numerically. In the simplest case, when the absorption coefficient is independent of the angle of incidence, the analytic expression can be obtained. Let a keyhole has the shape of a turned over cone with the apex angle 2 θ_W. After n reflections, the beam intensity decreases to $I_n = I\rho_r^n$, where ρ_r is the reflection coefficient. The angle of reflection is $\theta_r = 2n\theta_W$. When $\theta_r > \pi/2$, the laser beam rapidly returns back and leaves the keyhole. This gives $n = \pi/4\theta_W$. Therefore, the average absorption coefficient for multiply reflected beams is

$$\alpha_e = 1 - \rho_r^{\pi/4\theta_W - 1}. \tag{4.46}$$

Figure 4.43 shows the dependence of the average absorption coefficient in the keyhole on the average angle between the leading and trailing edges for $\rho_r = 0.88$.

To calculate the absorption of laser radiation in the keyhole due to interaction with plasma, it is necessary to calculate the temperature of the latter and its pressure. The pressure decreases towards the exit due to a decrease in the surface tension forces (see Sect. 4.1):

$$p(z) = p_a + \frac{\sigma}{r(z)}. \tag{4.47}$$

Temperature is determined from the heat balance equation (Chap. 3). By assuming for simplicity that the keyhole has a cylindrical shape, we obtain the radial distribution of the plasma temperature

$$T(r) = T_V + (P^*/4\kappa)(r_w^2 - r^2), \tag{4.48}$$

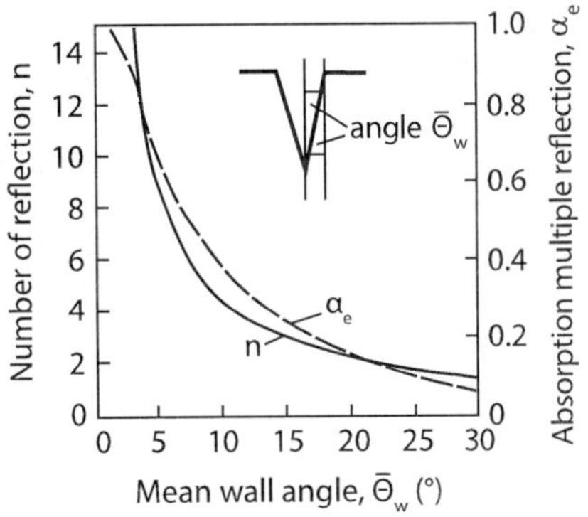

Fig. 4.43 Number of reflections n (*solid line*) and corresponding absorption coefficient α_e (*dashed line*) dependent on the mean wall angle $\bar{\theta}_W$

where P^* is the power absorbed in the unit plasma volume taking multiple absorption into account:

$$P^* = \frac{2\pi I_p r_w \Delta z}{\pi r_w^2 \Delta z} = 2I_p/r_w \tag{4.49}$$

where r_w is the mean radius of the keyhole. The integrated absorption coefficient for laser radiation in the keyhole plasma, taking into account multiple reflections from walls, is approximately determined by the average absorption coefficient $\alpha_p(T, P)$ and the keyhole depth h:

$$\alpha_{\text{pr}} = 1 - \exp(-\alpha_p 3h/2) \tag{4.50}$$

where α_p is the absorption coefficient of plasma (see Chaps. 1 and 3). The integration absorption coefficient for laser radiation in keyhole plasma upon single reflection is approximately determined by half of keyhole depth. By knowing all the absorption coefficients, we can obtain expressions for the intensity of energy fluxes incident on the keyhole surface:

$$I_F = (1 - \alpha_{p1})\alpha q(x, z) \tag{4.51}$$

$$I_f = (1 - \alpha_{p1})(1 - \alpha)\alpha_e q(x, r) \tag{4.52}$$

$$I_p = [\alpha_{p1} + \alpha_{\text{pr}}(1 - \alpha_{p1})](1 - \alpha)(1 - \alpha_e)q(x, z) \tag{4.53}$$

Expression (4.51) takes into account the probability of radiation propagation through the plasma upon single reflection. Expressions (4.52) and (4.53) are obtained similarly. For simplicity, we neglected in (4.51)–(4.53) the attenuation of laser radiation due to absorption in a plasma plume burning over the keyhole entrance. It will be shown below that this absorption does not exceed 10% of the laser beam power in a broad range of welding speeds.

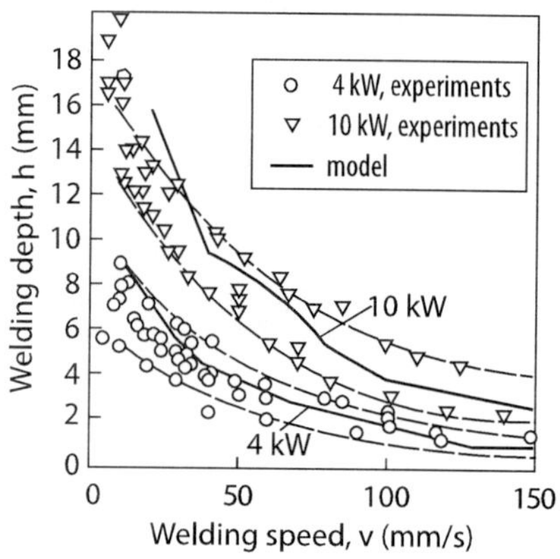

Fig. 4.44 Comparison of welding depth h versus welding speed v with experiments for 4 and 10 kW laser power

Table 4.3 Parameters used in the calculation

Laser power (cw)	P_l	4(10)	kW
Wavelength	λ	10.6	μm
Polarisation	–	Unpolarized	–
Beam quality	M^2	5.0	–
Beam mode	TEM	TEM_{00} TEM_{10*}	–
Focal length	f	200	mm
Beam diameter on optics	D_b	34	mm
Focusing number	$F = f/D_b$	6.0	–
Focal radius	R_{f0}	203	μm
Rayleigh length	z_r	±2.5	mm
Focal plane	z_0	Optimized (07)	mm
Welding speed	V_w	50	mms^{-1}
Material	–	Mild steel	–
Weld	–	Blind weld	–

By integrating equation (4.45), taking (4.51)–(4.53) into account, we can determine the shapes x_f and x_B of the leading and trailing edges of the keyhole and, therefore, the weld depth. Before considering the details of welding, it is reasonable to compare the calculated and experimental weld depths. Figure 4.44 shows weld depths calculated for cw CO_2 laser powers 4 and 10 kW. The 4 kW laser beam was assumed Gaussian (4.43), (4.44) and the 10-kW laser beam was assumed of a constant transverse intensity distribution. The calculation and experimental parameters are presented in Table 4.3.

One can see from the figure that for welding speeds above 1 cm/s the agreement between calculations and experiment is satisfactory. Note that simpler models also demonstrate agreement with experiments. For example, expression (4.21) is consistent with experimental data presented in Figs. 4.10 and 4.12. However, expression (4.21) was derived by using a number of assumptions (see the beginning of this section), which are absent in model (4.40)–(4.53). The latter model allows one to consider the important details of the welding process. Figure 4.45 shows a weld pool in the symmetry plane of the gas–vapor channel and the distribution of

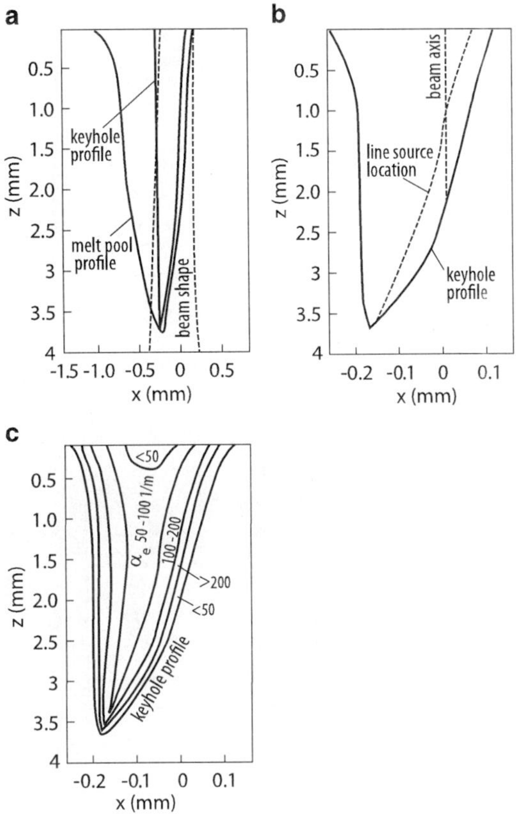

Fig. 4.45 Shape of beam, keyhole and molten pool (**a**), keyhole profile magnified in the x direction (**b**), distribution of plasma absorption coefficient α in the magnified keyhole (**c**) for 4 kW, $v_w = 50\,\text{mm}\,s^{-1}$ and steel

the absorption coefficient for laser radiation in the keyhole plasma. One can see that the absorption length is no less than 5 mm. For the keyhole depth ~3.5 mm the laser beam will be weakly absorbed in the plasma. This is confirmed by Fig. 4.46, which shows that the laser energy balance in the keyhole is mainly determined by the first and multiple absorption of the laser beam in keyhole walls. The loss of the laser radiation energy can be determined not only by the escape of beams after multiple reflection but also reflection from the sample surface outside the keyhole. The latter process can be more important at lower laser beam powers. The effective absorption coefficient measured in experiments with the 10-kW CO_2 laser [47] is in good agreement with calculations by expressions (4.40)–(4.53), although this model neglects a strong dependence of the Fresnel absorption coefficient on the angle of incidence. Note that in deep melting experiments at tighter focusing, the decrease in α_e with the welding speed is observed at considerably higher welding speeds ~ 30 cm/s (see Fig. 4.9a). Thus, despite the use of a number of model elements (the neglect of the dependence of the Fresnel absorption coefficient on the angle of incidence and the neglect of the melt flow around the keyhole), the model considered

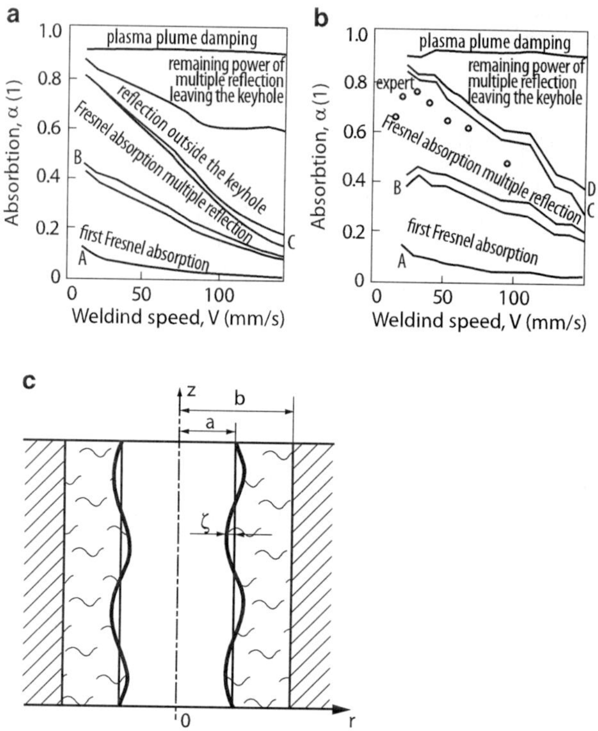

Fig. 4.46 Total energy balance of the deep welding process for blind welding of steel for 4 kW laser power (**a**) and for 10 kW (**b**). Zone *A* in the first plasma absorption; *B*, plasma absorption during multiple reflections; *C*, absorption outside the keyhole; *D*, reflection outside the keyhole; and expert., experimentally measured total absorption for 10 kW [47]. (**c**) Instability of a gas–vapor channel shape

above demonstrates the self-consistent approach to the description of the mechanism of deep melting of metals by laser radiation.

Along with the description of the general properties of deep-melting laser welding, it is also interesting to consider a number of other processes, which are not decisive but can affect some properties of welding. This concerns, e.g., the Marangoni convection. This process was considered in Chap. 2 as the main mechanism of the melt mixing in laser alloying of metal surfaces. It was shown that the appearance of new vortices in the melt can considerably affect the melt pool shape. The Marangoni flow can also appear in the weld pool upon laser welding. In this case, not the melt mixing itself becomes important but its influence on the pool shape and its important component – a gas–vapor keyhole. The vortex flow should be directed, as usual, from the hot region, i.e. from the keyhole to the distant boundary of the weld pool. This flow will cool the lower part of the pool and heat its upper part. This results in the narrowing of the lower part of the weld pool and extending of its upper part (Fig. 4.47). Calculations [48] have shown that this flow

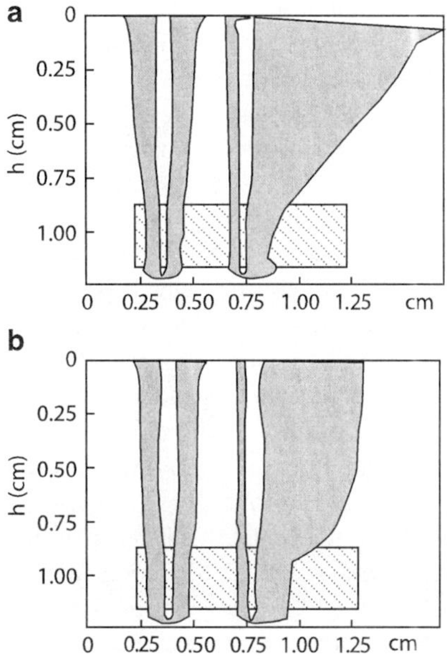

Fig. 4.47 Melt pool shape in the presence (**a**) and absence (**b**) of the Marangoni flow

is equivalent to the addition of $\sim$20% of the energy released in the keyhole to the energy distribution over vertical. The most important effect of the Marangoni flow in this case is the narrowing of the gas–vapor channel in the root region. As a result, the keyhole stability is impaired [48], which can deteriorate the welding quality.

4.4.3 Stability of a Cylindrical Gas–Vapor Keyhole

Nonstationary processes can affect laser welding in two ways. It is possible that the integrated properties and mechanisms of welding are the result of averaging of its parameters varying in time. This complicates the description of welding within the framework of simpler stationary models. On the other hand, nonstationary processes developing during laser welding are important in themselves because they determine the welding quality. Indeed, it is the transient nature of welding that leads to its "detrimental" inhomogeneities such as notches in the weld root, caverns in the weld, and spontaneous passing from welding to cutting due to sporadic ejection of the melt from the weld pool.

The study of nonstationary processes proceeding during a deep penetration of a laser beam into materials was started first of all from the analysis of instabilities produced upon the interaction of laser radiation with metals in the simplest case of a broad uniform beam and a homogeneous flat sample surface. These instabilities are classified in the first part of the book. Then, the instabilities were considered in the case of laser welding, which is, of course, a more laborious problem.

Almost all instabilities were considered in the linear approximation, which excluded the possibility of studying their influence on the integrated characteristics of the process and parameters of welding defects. The description of the influence of these instabilities on integrated characteristics was restricted by estimates. The correctness of the estimates can be verified either by a detailed comparison with experiments or by corresponding model calculations. In the latter case, approximate analytic calculations are used to elucidate the essence of processes proceeding during welding. On the other hand, linear theories determine the boundaries of various instabilities, which is undoubtedly important for determining the nature of nonstationary processes. In addition, linear theories give the characteristic oscillation frequencies of welding parameters. Therefore, it is reasonable to present in one figure the dependences of characteristic frequencies on the welding speed for various instabilities and the results of fragmentary experiments performed under different conditions. As for classical instabilities described in Chap. 1, they can be considered under welding conditions, strictly speaking, only for perturbation wavelengths that are smaller than the characteristic welding dimensions: the width and depth of the gas–vapor channel and the width and length of the weld pool. Of most interest are large-scale oscillations of the keyhole and melt pool because they most likely affect the welding quality and determine the dynamic equilibrium of the entire welding process. As shown in Chap. 1, under welding conditions, the capillary instability has the maximum increment. The maximum increment corresponds to very short perturbation waves. Therefore, the role of the capillary instability can be reduced to the generation of micro droplets from the surface of the leading edge of the keyhole. Because the mass loss upon welding is small, these droplets are probably evaporated in the laser beam, thereby increasing the metal vapor pressure in the keyhole [49].

Consider in more detail the development of this instability under conditions simulating laser welding. Such a restricted formulation of the problem is caused by the fact that no full-scale description of laser welding has been performed so far even in the absence of instabilities.

When a perturbation wavelength is comparable or larger than characteristic sizes of the welding process, it is necessary to solve the problem taking into account the real welding geometry at least to a minimal degree. Consider an infinite vertical hollow cylindrical nonviscous liquid layer of a finite thickness $b-a$, where a is the radius of the liquid surface and b is the radius of the boundary between the liquid and solid [50] (Fig. 4.46c). We assume that the energy αq is uniformly absorbed in the melt surface. The known temperature distribution with the gradient $G(r)$ will be established in the cylinder:

$$G = \frac{T_s - T_m}{r} \ln^{-1}(b/a) \equiv G_0/r \tag{4.54}$$

The temperature T_s of the melt surface is determined by the boundary condition

$$\kappa \partial T/\partial r|_{r=\alpha} = \rho L_b V(T_s) - \alpha q, \tag{4.55}$$

Where $V(T_s)$ is the vaporization rate. The geometry of the problem will change due to vaporization. However, because the development time of the instability is much shorter than the characteristic variation time of the gradient G, this change in the geometry can be neglected. Let us assume that small displacements ξ of the free surface of the liquid are axially symmetric:

$$\xi(r, z, t) = \xi_0 \exp(\gamma t + ikz)$$

The perturbations of the velocity potential can be obtained by solving the equation $\Delta\Phi = 0$ [8] for the potential flow for the zero velocity at the liquid boundary $r = b$.

$$\Phi(r, z, t) = \Phi_0 \exp(\gamma t + ikz)[I_0(kr) + AK_0(kr)] \tag{4.56}$$

where $A = I_1(kb)/K_l(kb)$, I_0, K_0, I_1, K_1 are modified Bessel functions of the second kind and Φ_0 is a constant. The liquid flow affects the perturbation δT of the melt temperature

$$(\gamma + \chi k^2)\delta T - \frac{\chi}{r}\frac{\mathrm{d}}{\mathrm{d}r}\left(r\frac{\mathrm{d}\delta T}{\mathrm{d}r}\right) = -G(r)u, \qquad u = \partial\Phi/\partial r \tag{4.57}$$

We will use for (4.57) the simplified boundary conditions

$$\left.\frac{\mathrm{d}}{\mathrm{d}r}\delta T\right|_{r=a} = 0, \qquad \delta T|_{r=b} = 0 \tag{4.58}$$

A more accurate condition at the free boundary [based on the linearization of (4.55)] considerably complicates calculations, but weakly affects the final result. The temperature perturbation on the melt surface is caused by the liquid flow and displacement ξ of the surface in the field of the temperature gradient $G(a)$

$$\delta T_s = \delta T(a) + G\xi, \tag{4.59}$$

where

$$\xi = \left.\frac{1}{\gamma}\frac{\partial\Phi}{\partial r}\right|_{r=a}$$

The perturbation of temperature on the melt surface leads to the perturbation of the recoil pressure

$$\delta p = \frac{\mathrm{d}p}{\mathrm{d}T}\delta T_s \exp(\gamma t + ikz) \tag{4.60}$$

Finally, all the perturbations are related by one balance equation for pressures on the liquid surface for $r = a$:

$$-\rho\gamma\Phi = (\sigma/a^2)(1 - a^2k^2)\xi + \delta T_s\frac{\mathrm{d}p}{\mathrm{d}T} \tag{4.61}$$

By substituting (4.56) and the solution of (4.57) into (4.61), we obtain the dispersion equation relating γ and the wave vector k. For short wavelengths $ak \gg 1$, the maximal increment $\gamma_{\max}$ has the form

$$\gamma_{\max}^2 = (\sigma/\rho)k_{\max}^3, \quad k_{\max} = \rho^{1/7}\sigma^{-5/7}\left[\frac{\mathrm{d}p}{\mathrm{d}T}G(a)\chi^{1/2}\right]^{4/7} \tag{4.62}$$

Numerical estimates by (4.62) for welding parameters for steel $\rho = 8$ g/cm^3, $\chi = 0.05\,\mathrm{cm^2/s}$, $\sigma = 10^3$ dyn/cm, $(\mathrm{d}p/\mathrm{d}T)G_0 = 2\times10^6\,\mathrm{dyn/cm^2}$ give $\gamma_{\max} \approx 3\times10^4\,\mathrm{s^{-1}}$, $k_{\max} = 200\,\mathrm{cm^{-1}}$. If vaporization is absent, the liquid cylindrical layer will be subjected to the Rayleigh instability. In the simplest case, this instability causes the decomposition of a water jet freely flowing from a tap into drops. For example, for long wavelengths, when the melt thickness $\Delta = b - a$ is smaller than the wavelength, we have

$$\gamma_R^2 = (\sigma/\rho a^2)k^2(1 - k^2a^2)\Delta \tag{4.63}$$

The maximal increment is achieved for $k_* = 0.7/a$

$$\gamma_{\max} \approx (\sigma\Delta k_*^4/\rho)^{1/2} \tag{4.64}$$

If the keyhole is maintained by the vaporization of the walls, $\mathrm{d}p/\mathrm{d}T \neq 0$, the term proportional to $\mathrm{d}p/\mathrm{d}T$ is added to (4.63). For high vapor pressures, when $\frac{\mathrm{d}p}{\mathrm{d}T}G_0 \gg \sigma/a$, the surface tension can be neglected,

$$\gamma \approx k\left(\frac{\partial p}{\partial T}G_0\Delta/\rho a\right)^{1/2} \tag{4.65}$$

For $(\mathrm{d}p/\mathrm{d}T)G_0 = 5\times10^6$ dyn/cm^2, $\rho = 8$ g/cm^3, $a = 0.25$ mm, $\Delta = 0.7a$, and $k = 45\,\mathrm{cm^{-1}}$, the estimate by (4.65) gives $\gamma = 3\times10^3\ \mathrm{s^{-1}}$. The linear theory shows that the Rayleigh instability develops faster for short wavelengths (4.63). This instability has time to develop before the keyhole collapse after switching off the laser or during keyhole displacement even at high speeds, up to $v = 1$ m/s. The smaller is the initial radius of the keyhole, the faster the keyhole is filled uniformly over z with liquid after switching off the laser:

$$\tau^{-1} = \sqrt{\sigma/a^3\rho} \tag{4.66}$$

The meaning of this expression is simple. According to the Bernoulli equation, the liquid velocity is

$$v = \sqrt{\frac{p}{\rho}} \equiv \sqrt{\frac{\sigma}{a\rho}}$$

Taking into account that the collapse time is $\tau = a/v$, we obtain (4.66). If the initial keyhole shape is close to cylindrical, we can expect that the keyhole collapse

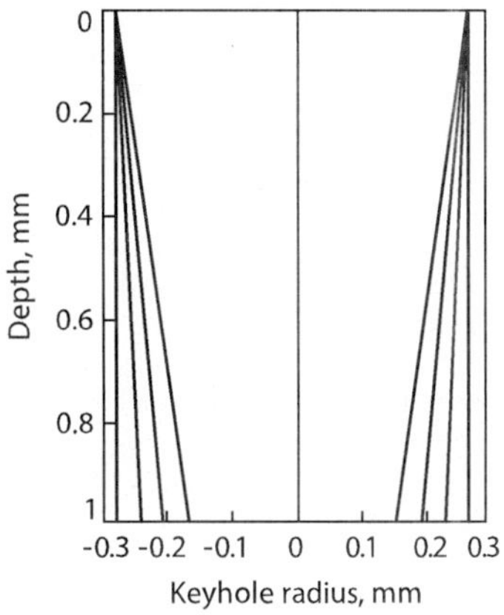

Fig. 4.48 The keyhole radius as a function of depth in a thin 1 mm sheet of steel for four different values of the weld translation speed: 10, 20, 40, 60 mm/s

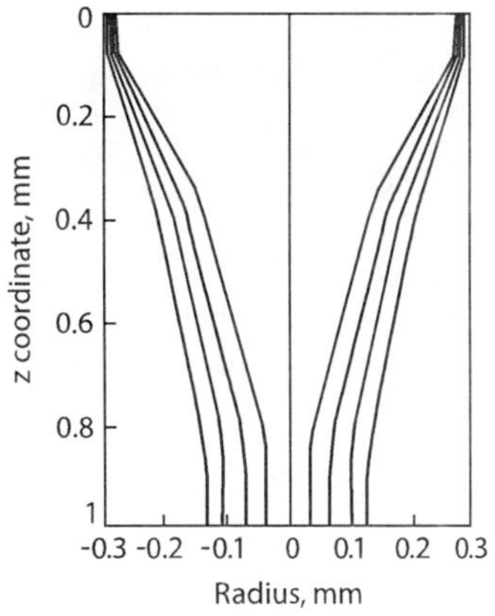

Fig. 4.49 The keyhole radius as a function of depth in a thin 1 mm sheet of steel at a weld translation speed of 60 mm/s at time $t = 0.1, 0.125, 0.15,$ and 0.175 ms following the extinction of the laser beam

dynamics will be determined by the development of instability, and the waist will be located in the middle of the keyhole depth. If the initial keyhole is a truncated cone, then, according to (4.66), it will collapse first of all in the initially narrow section. The numerical solutions of the nonstationary two-dimensional Navier–Stokes equation, describing the keyhole collapse in the weld pool after switching off laser radiation, correspond to this picture (see Figs. 4.48, 4.49 [52]).

These numerical calculations also describe the nonlinear stage of the Rayleigh instability. One can see from Fig. 4.49 that the instability amplitude is great, i.e. it is this process that determines the collapse time of the keyhole. The collapse occurs in the form of a waist. Figure 4.50 shows that the development rate of this instability increases with time, i.e. its development is explosive. It is clear that due to a strong nonuniformity of the keyhole collapse, discontinuities can be formed in the weld.

The development of the instability in the case of intense vaporization in the keyhole at the initial stage should be studied in the futures. This problem is important. At low welding speeds, welding passes to the oscillatory regime. The keyhole collapses periodically, then the laser beam again penetrates deep into the melt and recovers the keyhole. It seems that this process was observed in

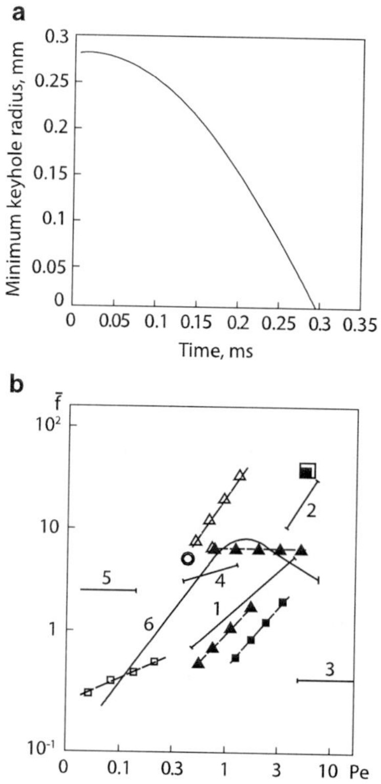

Fig. 4.50 The minimum radius of the keyhole in a thin 1 mm sheet of steel at a weld translation speed of 10 mm/s at function of time following the extinction of the laser beam (**a**). Frequencies on nonstationary phenomena in laser welding (**b**). Model: **1**. Moving step; **2**. Melt forcing over; **3**. Melt pool splash; **4**. Corrugations on the front wall; **5**. Keyhole collapse; **6**. Volume oscillations of the melt pool. Experiment: *square* [88]; *black square* [69]; *triangle* [54]; *black triangle* [54, 89]; *ring* [90]; *square with black square* [91]

experiments [17]. The periodicity of the process is caused by the thermal inertia of the keyhole recovery [51]. The oscillation period is determined by the collapse time and the penetration time of the laser beam. The collapse time can be taken from the results of calculations presented in Fig. 4.50a. The beam penetration time is determined by the ratio of the keyhole depth h to the penetration rate. For low welding speeds, we determine h from (4.21) and the penetration rate – from expressions (8.25) derived in Chap. 8 for the fountain regime of melt removal from the laser beam. For $P = 5\,\text{kW}$, $r_f = 2\times10^{-2}\,\text{cm}$, $v = 1\,\text{cm/s}$, $\chi = 0.2\,\text{cm}^2/\text{s}$, $\kappa = 0.76\,\text{W/cm K}$, $L_b = 0.7\,\text{kJ/g}$, we have

$$\bar{f} = f\frac{{r_f}^2}{\chi} = 3.3; \qquad Pe = r_f v/\chi = 0,1.$$

This quantity is plotted in Fig. 4.50b, where experimental data obtained under different conditions are summarized. The figure also presents model curves describing various instabilities observed upon laser welding. Most of the models give only tentative estimates and for this reason are not discussed here in detail. Nevertheless, when the sufficient amount of experimental data is available [49], a comparison with estimates can be methodologically useful. Below, we will describe numerical calculations of the amplitudes of inhomogeneities under nonstationary conditions by using a simplified welding model.

4.4.4 *Instability of the Leading Edge of a Keyhole*

The nonstationary uniform [53, 54] or nonuniform melt flow over the keyhole edge is often observed in experiments on welding and cutting of metals [55]. Several mechanisms explaining these inhomogeneities were proposed in the literature. In [49, 55], the washing of a liquid film by a step-like inhomogeneity moving from top to bottom was considered. Also, the simultaneous production of many inhomogeneities was discussed: frontal corrugations [49] or a chain of humps moving downwards along the leading edge of the keyhole [56, 57]. These effects were considered qualitatively and the authors mainly appealed to experimental data. As an example we consider the excitation and grow of humps on the front wall of a keyhole upon steel welding [56, 57]. This model also requires significant simplifications. First of all, a shallow and wide keyhole is considered with the leading edge completely illuminated by laser radiation. The rear wall of the keyhole and the weld pool are not considered in the model (see Fig. 4.51).

The melt on the keyhole edge moves to the side walls under the action of the vapor recoil pressure and then enters into the weld pool. The film thickness h is determined by the averaged balance equations for the mass and momentum

$$\frac{\partial h}{\partial t} + \frac{\partial (v_x h)}{\partial x} = -v_{\mathrm{dv}} + v_m \tag{4.67}$$

$$\frac{\partial}{\partial t}(v_x h) + \frac{\partial}{\partial x}\left(h v_x^2 + \frac{ph}{\rho}\right) = \frac{p}{\rho}\frac{\partial h}{\partial x} - (v_{\mathrm{dv}} - v_m)v_x - \frac{\mu v_x}{\rho h} \tag{4.68}$$

The coordinate system for (4.67), (4.68) is shown in Fig. 4.51. Here, v_{dv} is the rate of decrease in the film thickness due to vaporization; v_m is the rate of increase in h due to the moving of the melt boundary inside a sample; and μ is the melt viscosity.

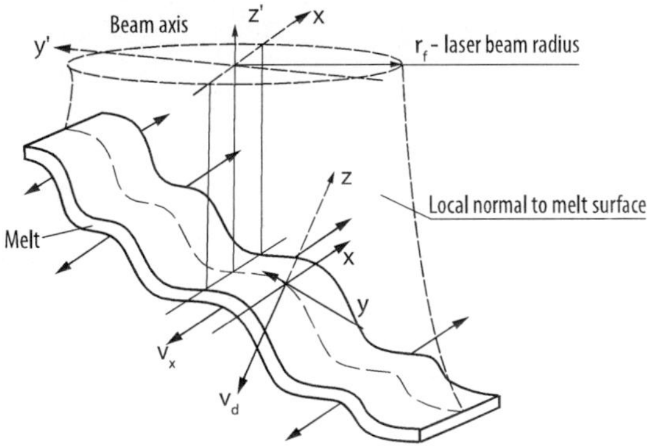

Fig. 4.51 Scheme of a front keyhole wall. The local coordinate system x, y, z is such that the z-axis is directed along the local normal lo the melt surface

The balance of heat flows along the normal to the liquid film surface is described by the equation

$$-\kappa_l \frac{\partial T}{\partial z}\bigg|_{z=0} + \rho v_{\mathrm{dv}} L_b = \alpha q \tag{4.69}$$

The vaporization rate is determined by the surface temperature T_s:

$$v_{\mathrm{dv}} = V_0 \exp(-U/KT_s) \tag{4.70}$$

where U is the vaporization energy for an atom of the sample material and V_0 is a constant of the order of the sound speed. The Stephen condition on the melt–solid interface has the form

$$\kappa_s \frac{\partial T_s}{\partial z}\bigg|_{z=z_m} - \kappa_l \frac{\partial T_L}{\partial z}\bigg|_{z=z_m} = \rho v_m H_m \tag{4.71}$$

where H_m is the specific melting energy and T_s and T_l are the temperatures of a solid and liquid, respectively. Due to a small film thickness, (4.71) takes into account only heat flows propagating normally to the film surface. The vapor recoil pressure p in (4.68) is determined by the approximate equation

$$p = A {T_s}^{-1/2} \exp(-U/kT_s) \tag{4.72}$$

where A is a coefficient depending on the pressure of the surrounding medium. The temperature distribution in the liquid film and solid is calculated taking into account transfer over all coordinates:

$$\frac{\partial T}{\partial t} + v_x \frac{\partial T}{\partial x} + v_d \frac{\partial T}{\partial z} = \frac{\kappa_l}{\rho_l c_l} \Delta T \tag{4.73}$$

$$\frac{\partial T}{\partial t} + v_d \frac{\partial T}{\partial z} = \frac{\kappa_s}{\rho_s c_s} \Delta T, \quad v_d = -\frac{\partial}{\partial x}(v_x h) - v_{dv}$$

where subscripts l and s refer to the liquid and solid phases, respectively. Calculations were performed at $\alpha(0) = 0.35$; for the angle of incidence above 70°, the reflectance drastically decreases. The system (4.67)–(4.73) was solved numerically taking into account the movement of a sample at a speed of v. After some time, as in the models considered above, the front profile becomes stationary and its height determines the welding depth. However, under certain conditions the instability develops on the front and the front profile becomes wavy (Fig. 4.52). Humps appear near the sample surface and come down, their amplitude increasing. Under conditions in Fig. 4.52, the hump length is about 20–50 μm, which gives for the speed 2 cm/s the oscillation frequency $f \sim 100$ Hz. In dimensionless units, $\mathbf{f} = f r_f^2/\chi = 1$ Hz for $Pe \sim 0,4$. This point is close to the experimental value [54] (see Fig. 4.50b). This instability is qualitatively close to the moving-step instability [49, 55], but here the numerical model predicts many such "steps" on the leading

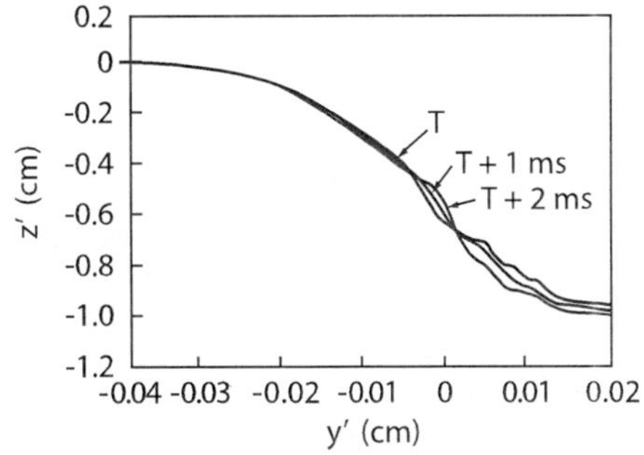

Fig. 4.52 Temporal evolution of the front pan of the keyhole for the following processing parameters: a Gaussian beam intensity distribution with a maximum of 3 MW cm^{-2} and radius 200 μm; a beam translation speed of 20 mm s^{-1}

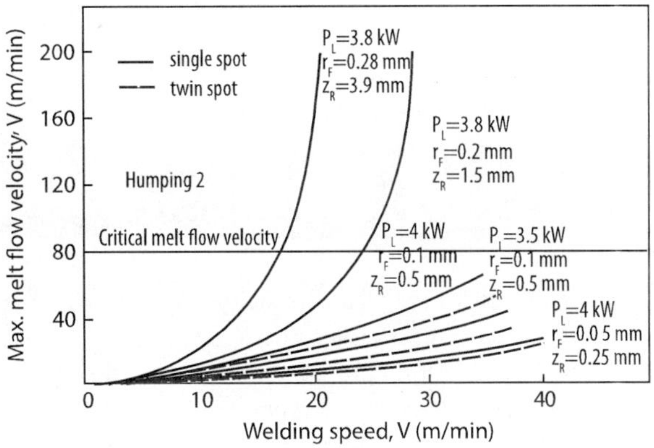

Fig. 4.53 Melt flow velocity versus welding speed

edge. From this point of view, the results of the numerical model are close to those of the moving-corrugation model [49], in which, however, the surface tension plays a decisive role.

Although the numerical model considered here is strongly simplified, it allows the study of oscillations with finite amplitudes and can be used for calculating the dynamic equilibrium in a gas–vapor keyhole upon welding of metals.

4.4.5 Melt Pool Instability

We considered nonstationarities directly in a gas–vapor keyhole. The welding quality can depend on processes proceeding in the weld pool, which are related to the properties of the gas–vapor keyhole. At high welding speeds $v > 30$ m/min, humps are formed on the weld surface. As shown above, the weld depth decreases with increasing weld speed, the fusion zone in front and on the sides of the gas–vapor keyhole becoming very thin [see (4.36)]. This leads to a considerable increase in the speed of the melt flowing around the keyhole. Observations showed [58] that this speed can exceed the welding speed by an order of magnitude, achieving $\sim$100 m/min. The focal spot size of the beam from a disc Yb:YAG laser was varied in a broad range by using a special optics. This provided variations in the rate of melt flowing around the keyhole. Figure 4.53 shows that the instability of humps appears under different conditions, but at the same flow rates. This demonstrates the hydrodynamical nature of this instability.

4.5 Remote and Hybrid Welding of Metals

The advances in the development of lasers stimulated the appearance of new directions in laser welding in the last decade. Attempts to use highly efficient diode lasers directly for laser processing of materials led to investigations of laser welding with a large focal spot. A new welding method – remote laser welding, which is promising for applications, e.g., in machine building, also belongs to this direction. Finally, interest in hybrid welding methods, which can favor industrial applications of laser welding, was revived again.

4.5.1 Features of Laser-Arc Welding of Metals

The use of an arc electric discharge in laser welding of metals is caused by the attempt to increase the total efficiency of the process. Indeed, because of the low electrooptical and total efficiencies of lasers, the use of an arc having a high electric energy–heat conversion efficiency seems quite promising [66]. In addition, the use of the arc can eliminate some disadvantages and problems in laser welding such as the formation of voids in the deep-melting weld, the necessity of the preparation of the edge of elements being welded and their careful positioning [67]. Hybrid welding uses mainly two types of arc discharges. The first type is a TIG arc with a refractory electrode directed to the exit of a gas–vapor keyhole to which a laser enters (see Fig. 4.54 [68]).

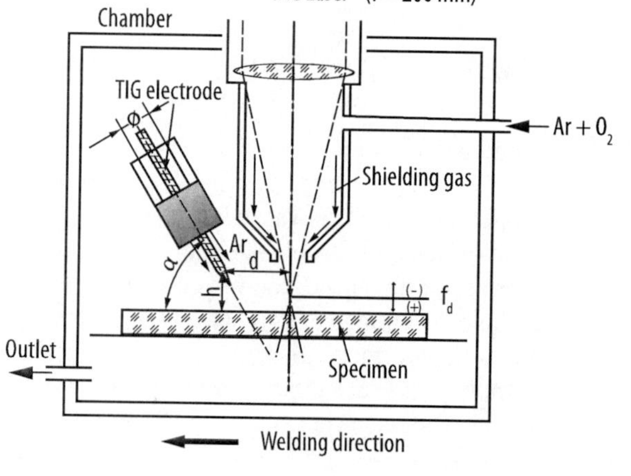

Fig. 4.54 Schematic experimental setup for various gas atmospheres in YAG-TIG hybrid welding

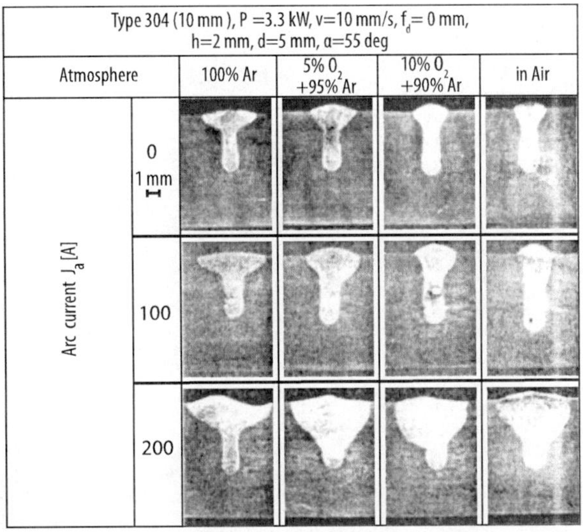

Fig. 4.55 Effects of arc current and environmental atmosphere on penetration geometry of laser and hybrid weld beads

Or the arc discharge is burning on the lower surface of a sample in the case of through melting. The second type is a MIG arc discharge, which is similar to that used in electrode welding. In this case, a metal wire is introduced to the interaction zone. These differences in the organization of hybrid welding lead to different properties. What can be expected when current is passed between the external electrode and the sample? It is clear that the arc will be "tied" to the hottest and nearest point to the electrode. This point in laser welding is the upper part of the gas–vapor keyhole. Thus, an additional virtually point radiation source appears upon hybrid welding. Although the laser beam passes through this heated zone, nevertheless it will lose the same amount of energy as in the absence of the arc. Therefore, it is difficult to expect that the keyhole depth and, hence, the melting depth will increase. The heat supplied by the arc is spent to increase the width of the upper part of the weld. This is clearly seen in Fig. 4.55 when the powers of the arc and Nd:YAG laser are close.

The depth and shape of the weld in stainless steel depend to a greater extent on the surrounding gas. As the arc power was doubled, the upper part of the weld considerably extended. The cross section of the weld becomes similar as a whole to its cross section upon arc welding. In this case, the root part retains the shape typical for laser welding; however, its internal structure changes. The X-ray transmission study shows that the efficiency of formation of pores is reduced upon hybrid welding (Fig. 4.56) [68].

The root part of the weld can be stronger affected in the case of through melting, when the arc electrode is directed to the exit of the gas–vapor keyhole on the lower surface of the sample. In this case, the melting depth also increases. The properties of hybrid welding change somewhat when the arc discharge is burning through a continuously supplied wire (MIG arc). In this case, along with the additional energy,

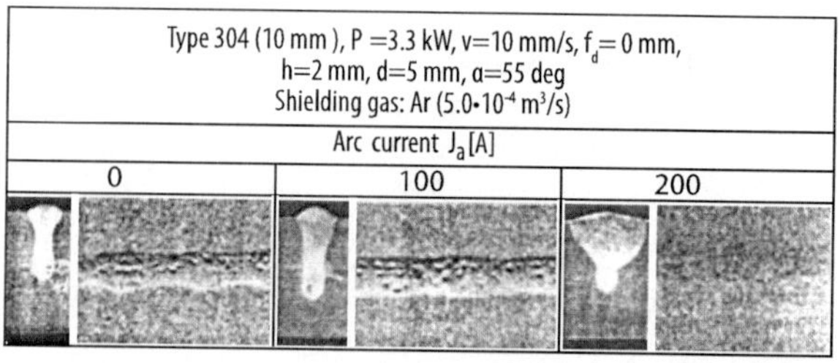

Fig. 4.56 Cross sections, X-ray inspection results and X-ray transmission in-situ observation results of Type 304 steel with low S content subjected to YAG laser only and hybrid welding at TIG arc currents of 100 and 200 A

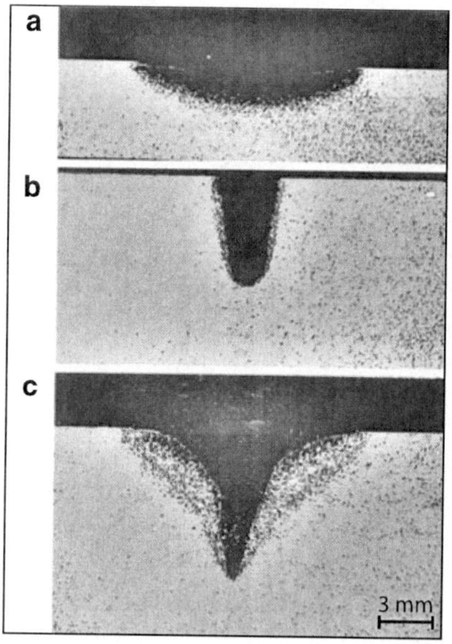

Fig. 4.57 (**a**) CMAW weld in type A-36 aluminium alloy. 830 J/mm, 270 mm/s, 1 m/min travel, $P_{\text{arc}} = 13$ kW; (**b**) Autogenous Nd: YAG Laser Weld in A-36, 4 kW, 240 J/mm, 1 m/min travel; (**c**) Hybrid Nd: YAG Laser Weld in A-36, 4 kW, 1060 J/mm, 5.7 kg/h wire: 1 m/min travel

the material of the melted wire also enters into the fusion zone. The combined action of the arc and laser in hybrid welding provides the increase in the welding depth (Figs. 4.57 and 4.58 [68]).

If there is a gap between samples being welded, it can be expected that the melt depth will increase. This occurs to a certain extent due to the flowing of the melt down under the action of gravity (Fig. 4.59).

This effect is observed when the gap between samples being welded is equal to or exceeds the size of the laser beam focal spot.

The X-ray data in Fig. 4.58 demonstrate the decrease in the amount of pores formed during welding with increasing the current (and power) of the arc discharge. The reason for this effect is not clear. It can be explained by the increase in the melt pool volume and the local lowering of the melt level under the action of the arc discharge. In this case, gas bubbles have time to leave the melt pool before its solidification. The scheme of this process is presented in Fig. 4.60.

This scheme is confirmed qualitatively by direct X-ray measurements of the weld pool geometry (Fig. 4.61 [68]).

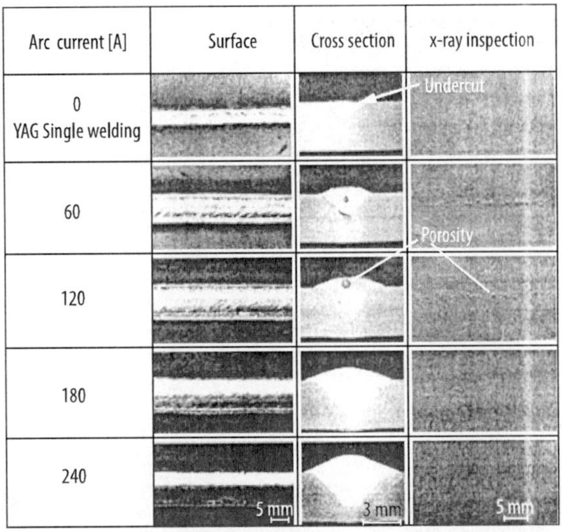

Fig. 4.58 Surface appearances, cross sections and X-ray inspection results of YAG laser and YAG-MIG hybrid weld beads

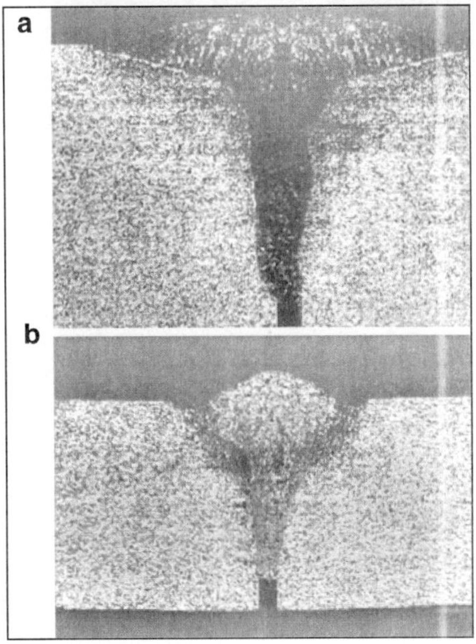

Fig. 4.59 (**a**) Hybrid with 0.89 mm gap and using 0.89 mm diameter wire at 1 m/min; (**b**) Hybrid with 1.14 mm gap and using 1.14 mm diameter wire at 1 m/min

The theoretical model of this effect is not developed yet; however, it concerns hybrid welding as a whole. Some aspects of this complicated process, e.g., a plasma plume attract the attention of theorists [70].

The use of hybrid technology for welding thick 2–25 mm steel samples is being actively studied by using both 7.5 kW Rofin Sinar Nd:YAG lasers and conventional 10–20 kW Trumpf CO_2 lasers. The high welding quality for such thick samples was achieved by selecting the optimal conditions of interaction of an arc discharge (MAG) and a laser beam [71].

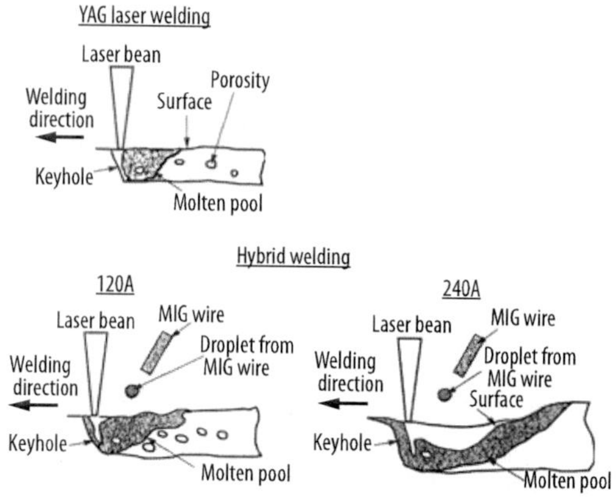

Fig. 4.60 Schematic representation of YAG-MIG hybrid welding phenomena, showing keyhole, bubble generation resulting in porosity, and concave surface of molten pool at 240 A

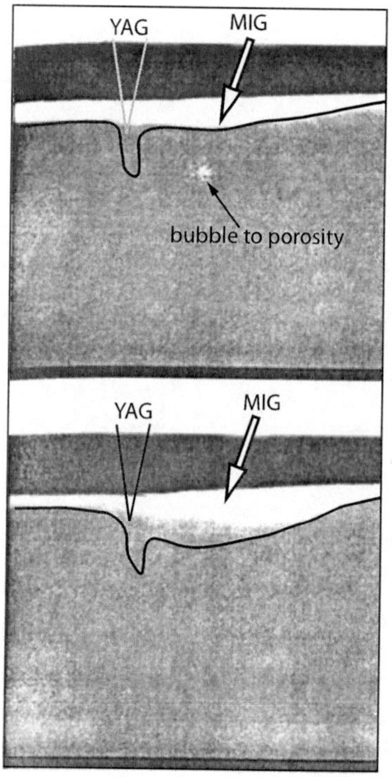

Fig. 4.61 X-ray transmission images during hybrid welding at 120 and 240 A, showing bubble generation and concave surface, respectively

4.5.2 *Remote Welding of Metals*

This type of welding is extensively being developed in the last decade. A laser beam is usually fixed during welding, while the sample is moved. The development of the scan technique and its computerization made it possible to perform laser welding of fixed samples by a laser beam which "finds" rapidly the required welding sites. The scheme of this process is presented in Fig. 4.62 [72].

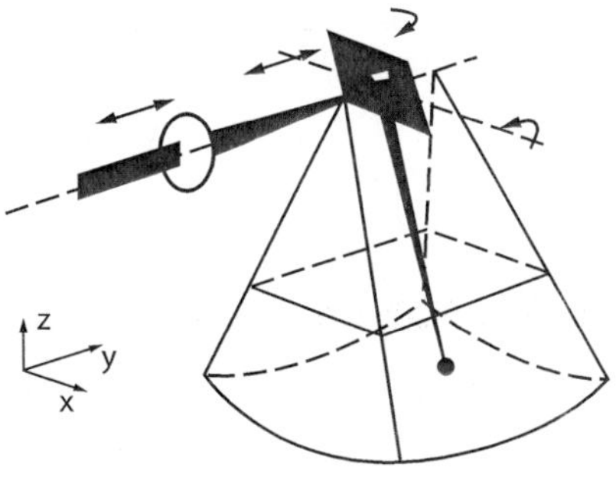

Fig. 4.62 Basic principle of the remote welding system

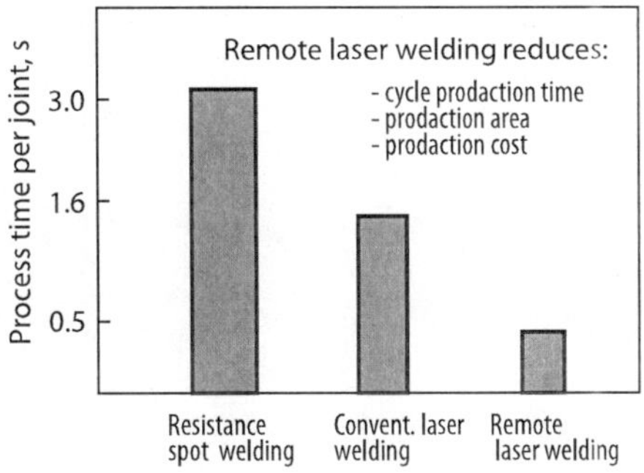

Fig. 4.63 Industrial motivation for remote laser welding (Source: Volkswagen AG)

The focal spot is displaced in the horizontal plane (coordinate x) by rotating a mirror and in the vertical direction (coordinate y) – by moving a focusing lens. For example, a system with a 6 kW CO_2 laser with a lens with the focal distance 1.6 m can displace the focus along the x axis by 1.5 m, along the y axis by 2.4 m, and along the z axis by 0.6 m [72]. Setups for remote laser welding with even better parameters [$4.5 \times 2.5 \times 0.9$ m (x, y, z), the beam positioning accuracy $\sim$0.2 mm] were described in [73]. Owing to the use of long-focus lenses in remote welding, the quality of laser radiation should be very high. Otherwise the focal spot will be large, which drastically deteriorates the quality and efficiency of laser welding.

What are the positive aspects of remote welding? The possibility of the fast displacement of a laser beam considerably reduces the duration of industrial welding (see Fig. 4.63 [74]).

In addition, this method reduces the investments by 30% and the industrial area by $\sim$50% [74]. The advantageous use of low-cost a CO_2 lasers involves some problems related to plasma processes proceeding during welding. These difficulties can be avoided by using more expensive diode-pumped Nd:YAG lasers. In remote welding, situations are typical when the laser beam is incident on the sample surface at an oblique angle. Experiments have shown that, if this angle does not exceed 15°, this does not impair considerably the welding quality [75]. We will discuss a similar process of remote cutting in the next chapter. However, along with a certain similarity, this process substantially differs from remote welding in that it does not use a gas jet, which is necessarily applied in laser cutting.

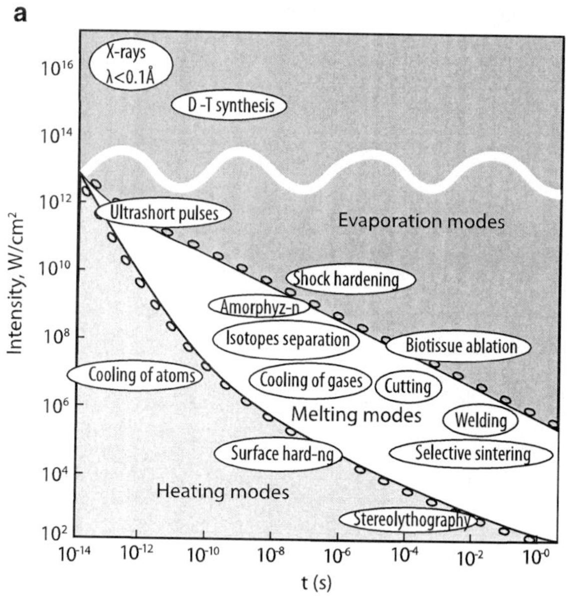

Fig. 4.64 Laser–matter interaction versus laser pulse duration

4.5.3 Influence of Laser Radiation Quality on Laser Welding

It was shown in the beginning of this chapter that the welding depth increases with decreasing the focal spot diameter [see, e.g., (4.3)]. As the focal spot is decreased, the radiation density increases and, therefore, the vaporization threshold will be exceeded (the necessary condition for deep-penetration keyhole melting) at a lower laser power (see Fig. 4.64). As shown in 1.1, to obtain the minimum size of the focal spot even in the case of ideal optics, it is necessary to use beams with a large convergence angle φ because $r_f \approx \lambda/\pi\varphi$. On the other hand, the increase in the convergence angle leads to the decrease in the beam waist (Rayleigh length): $L_f \approx 2\lambda/\pi\varphi^2$. It is clear that a long waist is preferable, providing a deeper penetration of the laser beam into material. If the possibility of keyhole welding (or laser drilling) is estimated from the point of view of the decay of the laser beam propagating in the keyhole, a small convergence angle is also preferable. Indeed, beams with a small convergence angle decay weaker, which is demonstrated by the results of special experiments presented in Figs. 1.8, 1.9. Thus, from the point of view of the diffraction approach the requirements to the convergence angle are mutually contradictive. In reality, the picture can be far from the diffraction one. The divergence $\Delta\varphi$ of a laser beam is caused by many other reasons. Because of this, the quality of laser radiation is characterized by two factors: the ratio of the diffraction divergence $\Delta\varphi_{\text{diff}}$ to the real divergence $\Delta\varphi$, i.e. $k = \Delta\varphi_{\text{diff}}/\Delta\varphi$ or by the beam parameters product $r_f\varphi$ (mm mrad, BPP-factor). The quality of radiation of various lasers is illustrated in Fig. 4.65 [2].

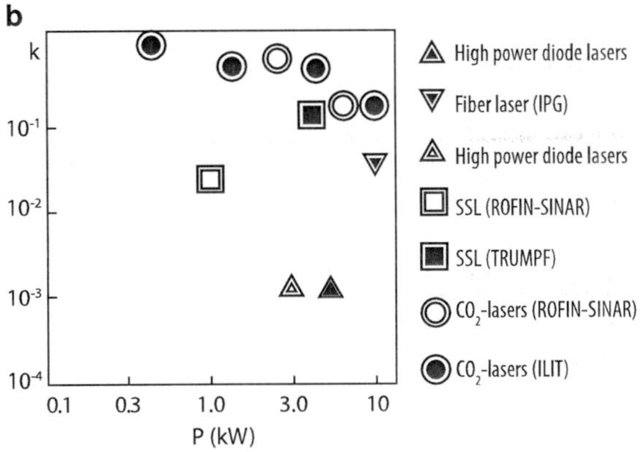

Fig. 4.65 Laser beam quality $k = \Delta\varphi_{diff}/\Delta\varphi$ versus laser power

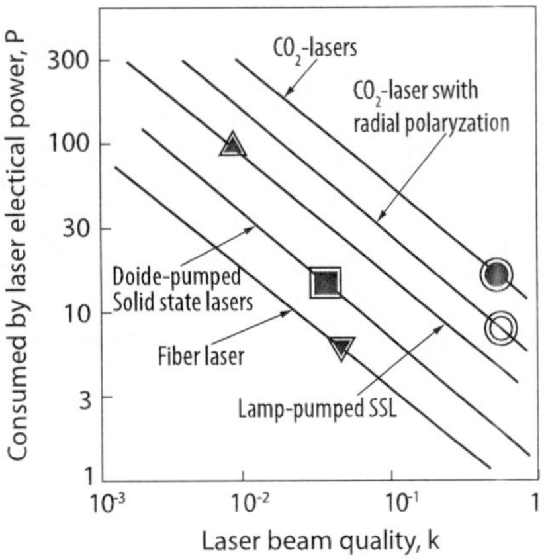

Fig. 4.66 Laser energy consumption inversely depends upon beam quality parameter (k). Points correspond to really achieved k values

Figure 4.66 demonstrates the influence of the radiation quality factor k on the efficiency of laser welding of materials [2]. The smaller is $r_f\varphi$, the small is r_f and the convergence angle, which also provides an increase in the welding efficiency. For example, the radiation quality of a 8 kW Nd:YAG laser with the active medium in the form of rods is 25 mm mrad [76]. A diode-pumped 4 kW disc Nd:YAG laser provides a higher radiation quality of 6–8 mm mrad [77, 78]. By using flashlamp-pumped lasers and diode-pumped disc lasers, it is possible, by varying the radiation quality in a broad range, to study its influence on laser welding properties [79].

Figures 4.67 and 4.68 show that, for the convergence angle $\theta = 11.31°$ and $P = 3$ kW, the welding depth increases as $1/d_f$ with decreasing the focal spot diameter [77]. However, for $d_f < 200$ μm the increase slows down and then the weld depth even decreases. In this case, the weld area almost does not change, as shown in Fig. 4.68a. This figure also presents the results of welding performed with a fiber laser [80], which has a higher radiation quality (2.5 mm mrad). As a result, at a small focal spot, the section broadens at the bottom (see Fig. 4.69).

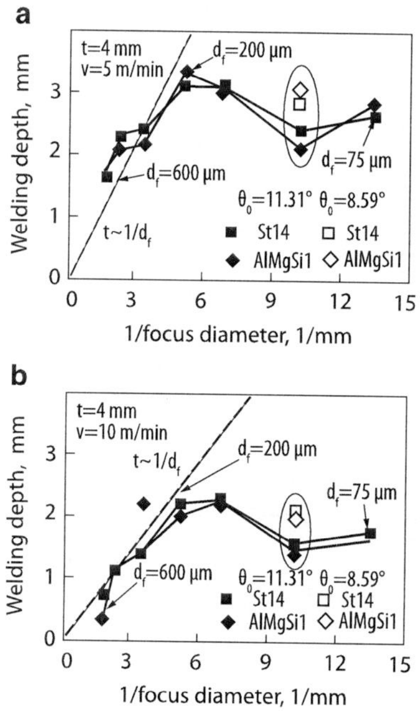

Fig. 4.67 (**a**) Welding depth for steel and aluminium as a function of the inverse focus diameter at a welding speed of 5 m/min (**a**) and 10 m/min (**b**) for $P = 3\,\text{kW}$

The focusing angle can be reduced by means of an optical scheme by retaining the focal spot size. This leads to the increase in the melting depth (Figs. 4.67, 4.68b,c) [80]. Thus, by controlling the quality of laser radiation, the welding depth can be increased. It follows from Fig. 4.68a that the thermal welding efficiency in this range of parameters is virtually independent of the radiation quality and is approximately 30%, which is lower than the ideal efficiency equal to 50% (see above the section on thermal efficiency of laser welding). It follows from Fig. 4.68c that the energy spent for melting the unit depth is large even for the high-quality radiation, amounting to 1–2 kW/mm, which exceeds this energy for usual CO_2 lasers (see Fig. 9.32). However, this specific melting is achieved at a high welding speed. As a result, the welding efficiency is $vh/P \approx 0.1\,\text{mm}^2/\text{J}$, which is 3–4 times higher than that for cw CO_2 lasers (see Figs. 4.12 and 4.15).

Due to a small focal spot, the aspect ratio of laser welds achieved great values ~60. From the point of view of industrial applications of high-power lasers, in particular, in welding, apart from the radiation quality other factors such as the conversion efficiency of electric energy to laser radiation, the weight and size, the ease of using, the necessity of cooling, and mobility are also important. In this connection, it is useful to compare various types of high-power industrial lasers (Table 4.4) [81]. One can see from the table that the best radiation quality can provide CO_2 lasers, but their efficiency is low, they have large dimensions, and are not very convenient for industrial applications. A new type of lasers – fiber

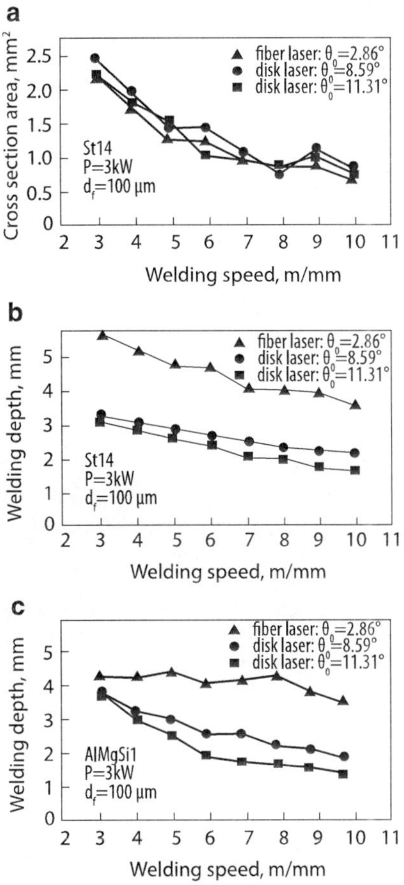

Fig. 4.68 Cross sectional area as a function of the welding speed for the three divergence angles in steel (**a**). Welding depth as a function of the welding speed for the three divergence angles in steel (**b**) and aluminium (**c**)

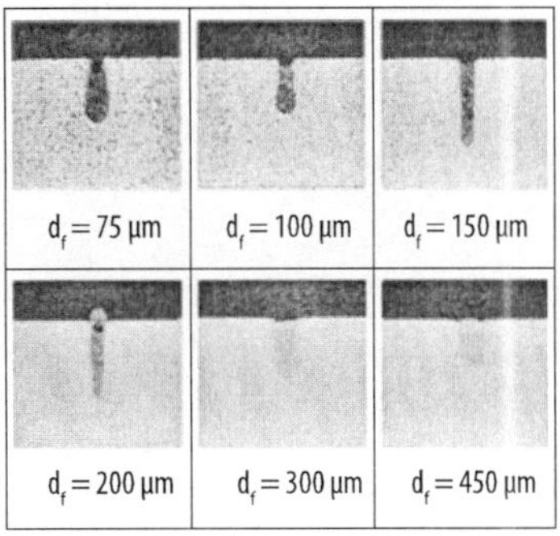

Fig. 4.69 Cross sections for the laser beam focus diameters up to 450 µm at welding speed of 9 m/min in 4 mm thick steel as bead on plate welds

laser can be distinguished. Active fibers used in these lasers combine the functions of lasing and the transport of radiation [82]. This provides the high efficiency and small weight, which in combination with the high radiation quality makes these lasers competitive and promising for industrial applications (see Fig. 4.70) [83, 84].

Table 4.4 Comparison of laser sources

	CO_2	Lamp-pumped Nd:YAG	Diode-pumped Nd:YAG	Yb-fibre (multi-mode)	Thin disc Yb-YAG
Lasing medium	Gas mixture	Crystal rod	Crystal rod	Doped fibre	Crystal disk
Wavelength, nm	10,600	1,060	1,060	1,070	1,030
Beam transmission	Mirror, lens	Fibre, lens	Fibre, lens	Fibre, lens	Fibre, lens
Typical delivery fibre ⊘, mm	–	0.6	0.4	0.1–0.2	0.15–0.2
Output powers[a], kW	Up to 15 kW	Up to 4 kW	Up to 6 kW	Up to 20 kW	Up to 4 kW
Typical beam quality[b], mm.mrad	3.7 3.7	25 12	12 <12	20 1.8	7 4
Maintenance interval, khrs	2	0.8–1	2.5	100[c]	2–5
Power efficiency, %	5–8	3–5	10–20	20–30	10–20
Approximate cost per kW, k$	6	130–150	150–180	130–150	130–150
Footprint of laser source	Large	Medium	Medium	Small	Medium
Laser mobility	Low	Low	Low	High	Low

[a] Commercially available.
[b] The top figures are for the max. available output powers, the bottom figures for the same type of laser but configured for optimum operation at 1 kW.
[c] Manufacturer's claim.

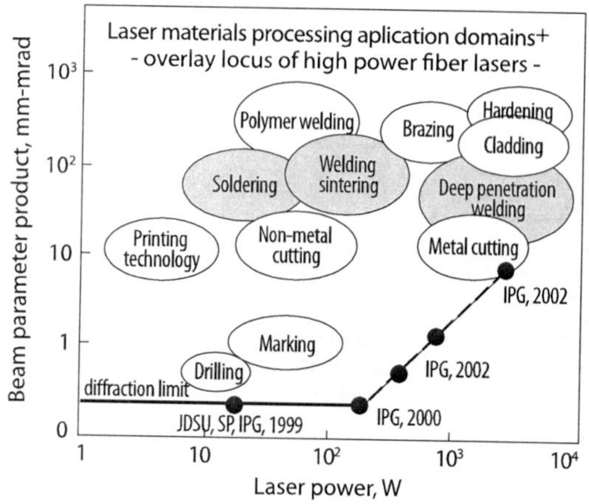

Fig. 4.70 Laser sources are also shown as points; all applications lying above and to the left of a source's *BPP* and power values can be addressed by that source

The parameters of modern high-power fiber lasers are presented in Table 4.5 [83]. The efficiency of welding aluminum alloys by using fiber lasers can be estimated from Fig. 4.71 [85].

One can see that the values of the specific melting h/P for the YLR 17000 and YLR 10000 lasers are close and noticeably exceed h/P obtained for the YLR 7000 laser. This is explained by the difference in the radiation quality of these lasers (Table 4.5).

Table 4.5 Comparison of basic beam parameters for three fiber laser systems measured in the BIAS lab

	Year	Max. beam power workpiece, kW	Fiber core diameter	Focal length, mm	Spot diameter, pm	BPP, mm·mrad
YLR-17000	2005	16.7	200	250	420	11.7
YLR-10000	2004	10.5	200	200	360	11.6
YLR-7000	2003	6.9	300	200	510	18.5

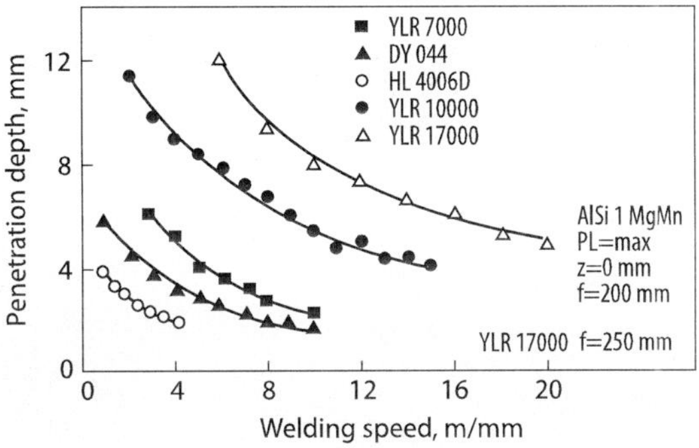

Fig. 4.71 Comparison of penetration for various lasers in AlSilMgMn

Solid-state Nd:YAG lasers and fiber lasers are pumped by diode lasers, which have the high efficiency of conversion of electricity to laser radiation. Due to the design of these lasers, their radiation has a high divergence and its focusing to a small spot is complicated. However, because of the high efficiency and small size of diode lasers it is reasonable to attempt to use their radiation directly for processing of materials, in particular, for welding of metals [86]. A DL025HQ diode laser emitting high-quality radiation at 0.8 and 0.94 μm was used for melting a 4-mm-thick stainless steel plate in the regime typical for laser welding. Argon was supplied at a rate of 60 l/min from the side to the interaction zone. The laser radiation was focused to a spot of size 1.2×1.2 mm by a lens with a focal distance of 9.5 cm. The maximum output power of the laser was 2.6 kW and the intensity in the focal spot was 1.56×10^5 W/cm^2 [87]. The experimental results were compared with model calculations assuming that the absorbed energy was transferred inside the plane due to heat conduction (Sect. 4.1). The shape of the plate was assumed invariable, and the absorption coefficient equal to 60% was obtained by fitting the experimental melting depth by the calculated depth. The experimental data and calculations are compared in Figs. 4.72 and 4.73 [87].

One can see that for $P < 1.6$ kW, the melting is caused only by the heat conduction, while for $P > 1.6$ kW, a gas–vapor keyhole begins to form and the deep-penetration keyhole melting regime appears. In Fig. 4.73, the curve 5 is calculated with help of (4.5), which describes the aspect ratio of magnitude about one. As expected, the transverse weld is characterized by a considerable thickening near the sample surface, Fig. 4.74. Note that the welding efficiency at the maximum depth is 0.22 ($v = 1$ m/min) and 0.27 cm^2/kJ ($v = 2$ m/min). These values are close

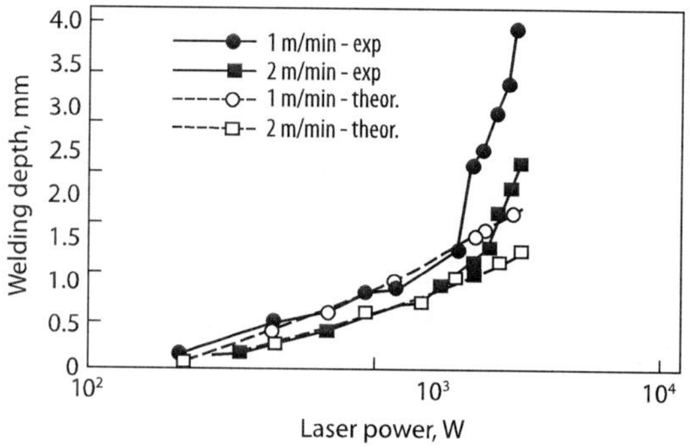

Fig. 4.72 Experimental and theoretical welding depths for welding speeds 1 m/min and 2 m/min. At 1 m/min the aspect ratio of 1 is reached as soon as the before threshold of 10^5 W/cm^2 is exceeded

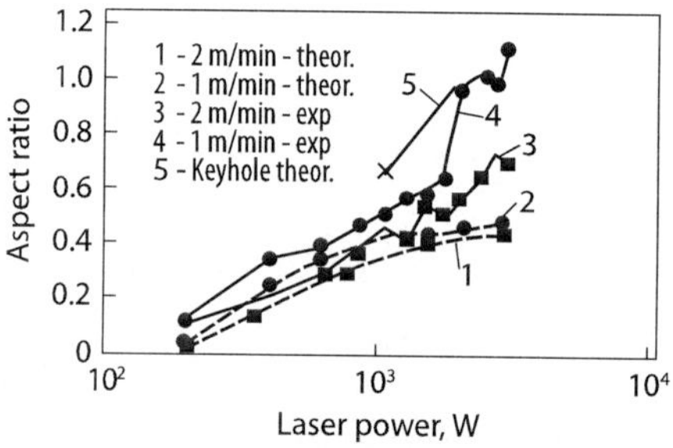

Fig. 4.73 Experimental and theoretical aspect ratios for welding speeds 1 m/min and 2 m/min. At maximum power the welding depth is 3.8 mm and the weld width is 3.3 mm, corresponding to the aspect ratio of 1.15

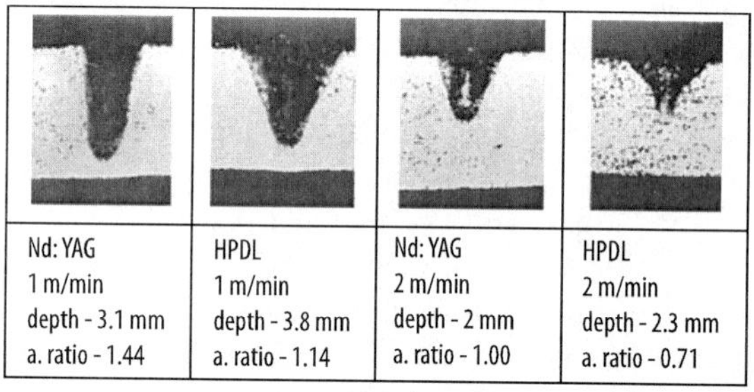

Fig. 4.74 Cross sections of Nd: YAG and HPDL welds at 2.6 kW and welding speeds 1 m/min and 2 m/min

to those for usual CO_2 laser welding (Figs. 4.12 and 4.15), which suggests that diode lasers are promising for welding of metals.

Note in the conclusion of this chapter that to construct a self-consistent physical model of laser welding, it is necessary, apart from theoretical studies, to perform systematic experimental investigations by fixing and varying the control parameters of experiments such as the spatial intensity distribution and power of laser beams, the fraction of absorbed energy, the radiation wavelength, and the welding speed. It is also necessary to carry out simultaneously "in vivo" measurements of the wavelengths of instabilities and their phase velocities, the oscillation frequencies of the vapor pressure, plasma emission, and "post mortem" measurements of the geometry of inhomogeneities in the solidified melt, and the fraction of laser radiation emerging from the opposite side during through welding. The difficulties of the theoretical (numerical) description of laser welding are caused by a variety of physical processes, which considerably affect each other. The study of the multidimensional nonstationary dynamics of heated liquids with the free surface is a difficult and tedious problem of computational mathematics.

References

1. W.W. Duley, *Laser Welding* (Wiley, New York, 1999)
2. V.S. Golubev, V.Ya. Panchenko, in *Proc. IX Int. Conf. Laser and Laser-Information Technologies. V Int. Symposium Laser Technologies and Lasers*, October 4–7 2006, Smollyan, Bulgaria
3. H. Cerjak (ed.) *Mathematical Modeling of Weld Phenomena*, vol. 3 (The Institute of Materials, London, UK, 1997)
4. M. Beck, A survey of modeling laser deep welding. ECLAT'96 **61**, (1996)
5. A.P. Mackwood, R.C. Crafer, Thermal modeling of laser welding and related processes: a literature review of Optics and Laser Technology **37**, 99 (2005)
6. J.G. Andrews, D.R. Atthey, J. Phys. D. Appl. Phys. **9**(15), 2181–2194 (1976)
7. N.N. Rykalin, A.A. Uglov, A.N. Kokora, *Lazernaya obrabotka materialov (Laser Machining of Materials)* (Mashinostroenie, Moscow, 1975)
8. L.D. Landau, E.M. Livshits, *Mekhanika sploshnykh sred (Mechanics of Continuous Media)* (Gostekhizdat, Moscow, 1953)
9. Y. Arata, N. Abe, T. Oda, *Proc. 5th Gas Flow and Chemical Laser Symp.*, Oxford, 20-24 August, 1984
10. A.A. Vedenov, G.G. Gladush, A.N. Yavokhin, Prikl. Mekh. Tekh. Fiz. **1**, 48 (1983)
11. V.A. Antonov, G.I. Kozlov. Kvantovaya Elektron. **4**, 1747 (1977)
12. E.D. Jansen, M. Frenz, K. Kadipasaoglu et al., OE Reports, Aug 1996, pp. 9–12
13. V.S. Golubev, A. N. Grezev, N. Kumar, G.L. Poltev, N.V. Grezev, M. Kupatadze. VIII Int. Conf. Laser and Laser-Information Technologies: Fundamental Problems and Application, Sept. 27-Oct. 1, 2003. Plovdiv-Smolyan, Bulgaria
14. Y. Arata, Plasma, *Electron and Laser Beam Technology* (American Society of Metals, Metals Park, OH, 1986)
15. A.V. Levin, Kvantivaya Elektron. **23**, 1037 (1996)
16. Y. Arata, Y. Maruo, I. Miyamoto, Tool. Engng. **25**, 24–32 (1981)
17. E.V. Locke, R.A. Hella, IEEE J. Quant. Electron. **10**, 179 (1974)
18. A.M. Belen'kii, G.M. Zuev, F.K. Kosyrev, et al., Svaroch. Proizvod. **11**, 23 (1997)
19. V.I. Lebedev, O.A. Velichko, F.K. Kosyrev, et al., Avtomat. Svarka **4**, 30 (1979)
20. N.N. Rykalin, I.V. Zuev, A.A. Uglov, *Osnovy elektronno-luchevoi obrabotki materiaov (Fundamentals of Electron-Beam Processing of Materials)* (Mashinostroenie, Moscow, 1978)
21. Y. Arata, I. Miyamoto, Technocrat, **11**, 33–42 (1978)
22. C. Banas, Opt. Engng 17, 210–216 (1978)
23. A.A. Vedenov, E.B. Levchenko. Kvantovaya Elektron **10**, 2107 (1983)
24. I.K. Kikoin (ed.) Tablitsy fizicheskikh velichin. *Spravochnik (Tables of Physical Quantities)* (Atomizdat, Moscow, 1976)
25. A.A. Bondarev, N.G. Tret'yak, Avtomat. Svarka. **4**, 1 (1981)
26. D.T. Swift-Hook, A.E.F. Gick, Weld J. **52**, 492S–499S (1973)
27. D. Rosenthal. Trans. Am. Soc. Mesh. Eng. **68**, 849 (1946)
28. A.G. Grigor'yants, F.K. Kosyrev, V.G. Fedotov, et al., Svaroch. Proizovod **9**, 1 (1980)
29. E.V. Locke, Opt. Eng. **17**, 192–197 (1978)
30. V.S. Gavrilyuk, A.G. Grior'yants, V.V. Ivanov, et al., Avtomat. Svarka. **6**, 27 (1983)
31. A. Salminen, A. Fellman, ICALEO 2006 Congress Proc. Laser Materials Proc. Conf. pp 18–2433
32. E.A. Mitkevich, V.A. Lopota, S.G. Gornyi, Avtmat. Svarka. **2**, 22 (1982)

33. A.N. Grezev, D.Sc. Thesis, Institute on Laser and Information Technologies, Moscow reg., Schatura, 2006
34. I.P. Ginsburg, V.A. Surin, A.L. Bagautdinov. Inzh.-Fiz. Zh. **33**, 213 (1977)
35. H. Tong, W.H. Giedt, Weld Res. Suppl. **6**, 259–266 (1970)
36. V.V. Bashenko, V.A. Lopota, E.A. Mitkevich, *Tezisy dokladov na Vsesoyuznom soveshchanii po primeneniyu lazerov v tekhnlogii mashinostroeniya, Zvenigorod, 1982 (Abstracts of Papers. All-Union Conference on Laser Applications in Machine Building Technologies, Zvenigorod, 1982)* (Nauka, Moscow, 1982), p. 75
37. R.W. Walker, Rapp. Ingenjorcvetenskapsakat **195**, 33–65 (1981)
38. A.L. Chernyakov. Cand. Sc. Dissertation, Institute of Theoretical Physics, Akad. Nauk SSSR, 1984
39. G.A. Nikolaev, A.G. Grigor'yants, Izv. Akad. Nauk SSSR, Ser. Fiz. **47**, 1458 (1983)
40. A.G. Grigor'yants, *Osnovy lazernoi obrabotki materialov (Fundamentals of Laser Machining of Materials)* (Mashinostroenie, Moscow, 1989)
41. A.G. Grigor'yants, I.N. Shiganov, A.I. Misyurov, *Tekhnologicheskie protsessy lazernoi obrabotki (Technological Laser Machining Processes)* (Izd. N.E. Bauman Moscow State Technical University, 2006)
42. R.V. Arutyunyan, V.Yu. Baranov, L.A. Bol'shov, D.D. Malyuta A.Yu. Sebrant, *Vozdeistvie lazernogo izlucheniya na materially (Interaction of Laser Radiation with Materials)* (Nauka, Moscow, 1989)
43. R. Ducharme, P. Kapadia, J.M. Dowden, K. Williams, W.M. Steen, *ICALEO 1992—177*, ed. by D.Farson, W. Steen, I. Migamoto, Orlando, Florida, USA
44. E. Jahnke, F. Emde, F. Losch, *Tables of Higher Functions*, 6th edn. (McGraw-Hill, New York, 1960; Nauka, Moscow, 1977)
45. A. Kaplan, J. Phys D Appl. Phys. **27**, 1805 (1994)
46. H. Hugel, *Strahlwerkzeng Laser* (Teubner, Stuttgart)
47. V. Del Bello, C. Rivela, M. Cantello, M. Penasa, SPIE Proc. On Indusrial and Scientific Uses of High-Power Laser 1502, 104–116, 1991
48. V. Lopota, G. Turichin, E. Valdaitseva, E. Beyer, S. Voellmar, *Laser-Assisted Micro-and Nanjtechnoligies 2003*, ed. by V. Veiko. Proceeding of SPIE, vol. 5399 (SPIE, Bellingham, WA 2004), p. 204
49. V.S. Golubev, *Analiz modelei dinamiki glubokogo proplavleniya materialov lazernym izlucheniem (Analysis of Models of Dynamics of Deep Melting of Materials by Laser Radiation)* Preprint no.83, (Institute of Applied Laser Studies and Technologics, RAS, Shatura, 1999)
50. F.Kh. Mirzoev. Kvantovaya Elektron. **21**, 147 (1994)
51. V. Golubev, F. Mirzoev, Pis'ma v Zhurnal Tekhnicheskoi Fiziki **29**, 30 (2003)
52. R. Ducharme, P. Kapadca, J.M. Dowden, *ICALEO'93/177* ed. by P. Denney, I. Miyamoto, B.L. Mordike, Orlando, Florido USA
53. S.G. Gornyi, V.A. Lopota, I.V. Matyushin, et al., Kvantovaya Elektron. **16**, 1643 (1989)
54. V.V. Bashenko, S.G. Gornyi, V.A. Lopota, *Proc. LAMP?87*, 1987
55. V.S. Kovalenko, V.V. Romanenko, L.M. Oleshuk, *Malootkhodnye protsessy rezki luchom lazera (Small-waste Laser Cutting Priocesses)* (Tekhnika, Kiev, 1987)
56. Akira Matsunawa, Vlad Semak. J. Phys. D Appl. Phys. **30**, 798 (1997)
57. V. Semak, W.D. Bragg, B. Damkroger, S. Kempka, J. Phys. D Appl. Phys. **D32**, L61 (1999)
58. K. Behler, P. Schafer, Proc. Of Third Int. WLT-Conf on Laser in Manuficturing 2005, Munich, June 2005
59. F.O. Olson, Cutting with polarized laser beams. DVS-Berichte **63**, 197 (1980)
60. E. Beyer, K. Behler, G. Herziger, Proc. LIM **233** (1988)
61. H.Hugel, F. Dausinger, *Laser Applications for Mechanical Industry* 207–218 (Kluwer Academic Publisher, 1993)
62. S. Sato, K. Takahashi, B. Mehmetli, J. Appl. Phys. **79**, 8917 (1996)
63. G.M. Mikheev, V.S. Idiatulin. Kvantovaya Elektron. **24**, 1007 (1997)
64. T. Matsumoto, N. Fukuda, Y. Kondo, A. Ohmori, K. Inoue, Y. Arata, J. Laser Appl. **11**, 258 (1999)

65. R. Fabbro, S. Slimani, F. Coste, F. Briand, B. Dlubak, G. Loisel, ICALEO 2006 Congress Proceeding. Laser Materials Processing Conference
66. M. Eboo, W.M. Steen, J. Clarck. Proc. Conf. Advances in Welding Processes. Harrogate, UK, 9–11 May 1978, pp. 257–265
67. E. Schubert, B. Wedel, G. Kohler. Proc. ICALEO 2002
68. S. Katayama, Y. Naito, S. Uchiumi, M. Mizutani, Proc. Third. Int. WLT-Conf. on Laser in Manufacturing 2005, Munich, June 2005
69. A.M. Grezev, Private Communication, 20.04.93
70. G. Turichin, E. Valdaytseva, E. Zemliakov, E. Pozdeeva, U. Dilthey, A. Gumenyuk, Proc. Third Int. WLT - Conf. on Lasers in Manufacturing 2005, Munich, June 2005
71. C. Fuhrman, D. Petring, R. Poprawe, Proc. Third Int. WLT - Conf. on Lasers in Manufacturing 2005, Munich, June 2005
72. W. Rath, Proc. Third Int. WLT-Conf. on Lasers in Manufacturing 2005, Munich, June 2005
73. R. Menin, Proc. of Automotive Circle International Conference: 6th European Conf., Bad Nauheim, Febr. 2005, 331–349
74. C. Emmelman, Proc. Third Int. WLT-Conf. on Lasers in Manufacturing 2005, Munich, June 2005
75. M. Grupp, T. Seefeld, F. Vollertsen, Int. Proc. of 2th WLT-Conf. on Lasers in Manufacturing 2003, Munich, June 2003
76. H. Schlueter, Proc. SPIE **5777**, 8–15 (2005)
77. F. Dausinger, A. Rus, Proc. of the 1th Pacific Int. Conf. on Application of Laser and Optics, 2004
78. H. Hugel, et al., Proc. SPIE **5958** (2005)
79. J. Weberpals, A, Rus, F.Dausinger, H. Hugel, Proc of the Third Int. WLT-Conf. on Laser in Manufacturing 2005, Munich, June 2005
80. ICALEO 2006 Congress Proc. Laser Material Processing Conf. p. 553.
81. G. Verhaeghe, P. Hilton, Proc. of the Third Int. WLT-Conf. on Laser in Manufacturing 2005, Munich, June 2005
82. A. Babushkin, N.S. Platonov, V.P. Gapontsev, Proc SPIE **5709**, 98–102 (2005)
83. P. Loosen, et al., Proc. Laser in Modern Manufacturing; Industrial Laser User, 13, November, pp. 15–19, 1998
84. V. Gapontsev, W. Krupke, *Laser Focus World* 83–87 (2002)
85. C. Thomy, T. Seefeld, F. Vollertseen, Proc. of the Third Int. WLT-Conf. on Laser in Manufacturing 2005, Munich, June 2005
86. D. Petring, C. Benter, R.Poprawe, ICALEO 2001, Jacksonville, USA
87. C. Benter, D. Petring, R. Poprawe, Proc. of the Third Int. WLT-Conf. on Laser in Manufacturing 2005, Munich, June 2005
88. A.M. Zabelin, Private Communication, 28.03.96
89. K. Williams, W. O'Neil, W.M. Steen, Melt-pool and keyhole dynamics during thin plate laser welding of steel, Proc. SPIE "Gas Flow and Chemical Lasers" **1810**, 594–597 (1992)
90. S. Woelker, L. Mazumber, Visualization and dimensional measurements of the laser weld pool // ICALEO'90, Laser Material Processing, 4–9, 1990, Boston MA, Proc. SPIE. **1601**, 422–429 (1990)
91. H. Gu, W.W. Duley, Acoustic emission and optimization of CO_2-LW of steel // Proc. SPIE **2500**, 77–85 (1994)
92. S.S. Kutateladze, Fundamentals of Heat Exchange, (Moscow, Atomizdat, 1979)

Chapter 5
Physics of Remote and Gas-Assisted Cutting with Lasers

Abstract This chapter is devoted to the physical mechanisms of cutting of metal. Simple process of cutting without gas-assisted can be used for remote cutting of metal constructions at large distance. The mechanisms of this process through melt removal by gravity force or recoil pressure of metal vapour are presented. Numerical results are in accordance with experiment. In gas-assisted cutting the peculiarities of the gas streams through a narrow cut determine the properties and qualities of process. The melt flow classification and calculation of influence of beam polarization on cutting rate are described. The instability of laser cutting in the oxygen atmosphere is considered.

In physics the whole matter of laser welding is intimately related to another important technological process. This process is laser cutting. Here, just like it works for welding techniques, a laser beam penetrates deep into a workpiece and melts the material. The basic difference is that the molten substance must be immediately removed or blown out by a gas jet to create a narrow kerf. Otherwise the "cutting" operation will only produce deep fusion penetration. This is why most laser cutting methods are usually called gas- or gas-jet-assisted. However, the action of the gas jet on the kerf region makes the overall process too complicated for analysis. The interaction between the gas stream and the liquid surface is a very complex problem alone. The theoretical and numerical models of this processes are presented which qualitatively describe the basic properties of gas-assisted and remote cutting metals. So at first it is worth to view the cutting process in a very simple arrangement to mark out the most distinctive properties and regularities. Having derived the basics we will later consider a number of sophisticated cases.

To begin with, let us examine the cutting technique with no gas jet involved when the melt is removed by either gravity or the recoil pressure of laser vaporized material. The latter is employed for cutting with pulsed laser beams and will be discussed later in Chap. 9. It must be noted that laser cutting of metals without a gas jet may find various industrial applications. In practical situations it often becomes

G.G. Gladush and I. Smurov, *Physics of Laser Materials Processing*,
Springer Series in Materials Science 146, DOI 10.1007/978-3-642-19831-1_5,
© Springer-Verlag Berlin Heidelberg 2011

necessary to perform cutting of metalware positioned a long distance away from the laser system. The task could be even trickier if installing the jet nozzle and focusing optics appears impossible in the required vicinity to the work object. Numerous projects are described in the literature suggesting the use of high power lasers to perform partition cutting of radiation-contaminated machinery and equipment, clearing the well heads from damaged constructions while fighting fires on oil and gas fields and other operations [1]. Further, we will identify this type of cutting technique as remote cutting.

5.1 Mechanism of Remote Cutting with cw Lasers

The gas-jet-assisted cutting is often described as punching of consecutive holes (see [2], for example). Leaving aside the validity of this approach we also begin our description of remote cutting with the simplest case of cutting a fixed workpiece by repetitive perforation. Here, we also use the limit of a weakly focused cutting beam which makes the focal spot intensity insufficient to heat the workpiece to the boiling point. On the other hand the spot size is assumed to be smaller than that of the drilled apertures. Moreover, for the sake of simplicity we will study the cutting of a thermally thin plate when the temperature disturbance along its depth is negligible. This is possible when the penetration time of the heat conduction wave deep into the workpiece is shorter than the other characteristic times including the drilling time. It is already clear from a general consideration that various processes may be responsible for creating the holes in the plate. These are thermocapillary flows in the molten material, oxidation of metal, recoil pressure, and gravity. Further in this chapter we describe some experiments where it is shown that if the plate orientation to the vertical axis is altered the regularities of the drilling process change considerably. This fact speaks for the priority role played by gravity in laser drilling of metal plates.

5.1.1 Physics of Melt Removal in Drilling of Vertical Plates

A thin plate can be fused all the way through its depth. The fusion spot in this case grows in time according to the function $\sqrt{t}$. The molten material is pulled down by gravity. However, the surface tension blocks it from leaking out completely. This is well seen from the cross-section photograph of a plate treated with the beam power weaker than needed to break through (Fig. 5.1). It is clear that with the greater diameter of the fusion zone the hydrostatic pressure should rise sufficiently to overcome the surface tension and build a situation similar to that when a water drop breaks off the tap. In case of a thin workpiece which is mounted vertically the hole can be drilled faster than it is done in the thick plate because the cavities created by descending liquid can join each other and cause the rapture of the liquid film.

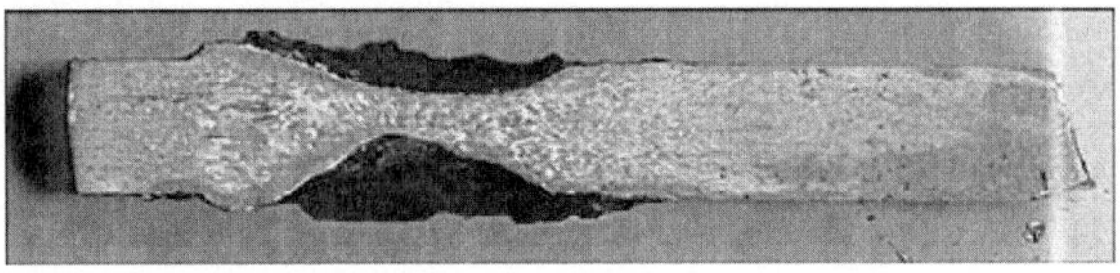

Fig. 5.1 Transverse slice of St3 after laser treatment prior to through melting, thickness $h = 3\,\text{mm}$, power $P = 3\,\text{kW}$

Let us now find the surface shape of the liquid [4, 10]. It is known that the static equilibrium of liquids is described by the Laplace equation [3]:

$$\frac{1}{R_1} + \frac{1}{R_2} + \frac{\rho g}{\sigma} y = const, \tag{5.1}$$

where R_1 and R_2 are the main radii of curvatures and y is the coordinate in which the gravity force is directed. The values of R_1 and R_2 here represent the nonlinear functions of coordinates. For the liquid area stretched out in a stripe along the x-axis the surface distortion must be only y-dependent:

$$-\frac{z''}{\{1 + (z')^2\}^{3/2}} + y = c. \tag{5.2}$$

Here, both y and z are made dimensionless with the capillary radius $r_c = \sqrt{\sigma/\rho g}$. In order to solve this equation, the boundary conditions must be set such that $y_{1,2} = 0, d$.

The behaviour of the boundary points is determined by wettability and the character of their possible displacement. The top point in case of some surface distortion can shift down the surface which has just been freed from the melt. The bottom point can shift down to the workpiece face plane. The latter one is normally dirty and, thus, is not wettable to the melt. Therefore, it is more likely (just as observed in experiments) that the bottom point should never be found displaced. As for the top point, the situation here is not that obvious so both variants deserve consideration.

We will start with the case of the top edge being free and the bottom one firmly fixed, i.e. $z(0) = 0$. We assume that $z'' = 0$ at $y = d$. This condition provides $c = d$ in (5.2). The nonlinear equation can only be solved numerically unless it is simplified for minor displacements:

$$-z'' = d - y. \tag{5.3}$$

Now it can be solved easily. The integration constants follow from the boundary condition and the invariance of the liquid volume. As a result we come to

$$z = \frac{1}{6}y^3 - \frac{\mathrm{d}}{2}y^2 + \frac{1}{4}\mathrm{d}^2 y. \tag{5.4}$$

We are particularly interested in finding $z(y = d)$, i.e. the maximum deepening of the crater δ:

$$\delta = \frac{1}{12}d^3, \tag{5.5}$$

The solution (5.5) is valid when $(z')^2 \ll 1$ or, more precisely, at either $1/16d^4 \ll 1$ or for dimension d $1/16(d/r_c)^4 \ll 1$.

With large values of d (5.2) must be solved numerically. These calculations demonstrated that enlargement of the fusion zone height makes δ grow faster giving $\delta \to \infty$ at $d \to 1$. Consequently, the maximum height of the melt pool is $d = r_c$. In this case it becomes apparent that if $y = r_c$ we get $z \to -\infty$ and $z'' \to 0$. This fact confirms the correct selection of the constant values in (5.2).

Now we can move on to a more complicated problem where both curvatures are important. Just like in the previous case we again restrict ourselves to the linear approximation. Thus, the differential operator in (5.2) is substituted with the Laplace operator [3]. In cylindrical coordinates it is written as

$$-\frac{1}{r}\frac{\mathrm{d}}{\mathrm{d}r}r\frac{\mathrm{d}z}{\mathrm{d}r} - \frac{1}{r^2}\frac{\mathrm{d}^2 z}{\mathrm{d}\varphi^2} + z\sin\varphi = c. \tag{5.6}$$

The boundary conditions read $z(r = R) = 0$ where R is the melt zone radius. The homogeneous equation has a solution bounded at $r = 0$:

$$z = c_1 r \sin\varphi. \tag{5.7}$$

The specific solution is

$$z = -\frac{c}{4}r^2 + \frac{r^3}{8}\sin\varphi. \tag{5.8}$$

The constants are derived from the boundary condition

$$\left(c_1 R + \frac{R^3}{8}\right)\sin\varphi - \frac{c}{4}R^2 = 0,$$

which is met if we take $c_1 = -R^2/8$ and $c = 0$. Finally, we come to

$$z = \frac{1}{8}r\left(r^2 - R^2\right)\sin\varphi. \tag{5.9}$$

From (5.9) it is seen that under this approximation the shape of the liquid surface is antisymmetric with respect to the horizontal diameter. This fact also follows from the form of (5.6). Such antisymmetry testifies that the height of the hummock is the same value as the crater depth.

These extremums are found at $r = R/\sqrt{3}$ and the maximum deepening is

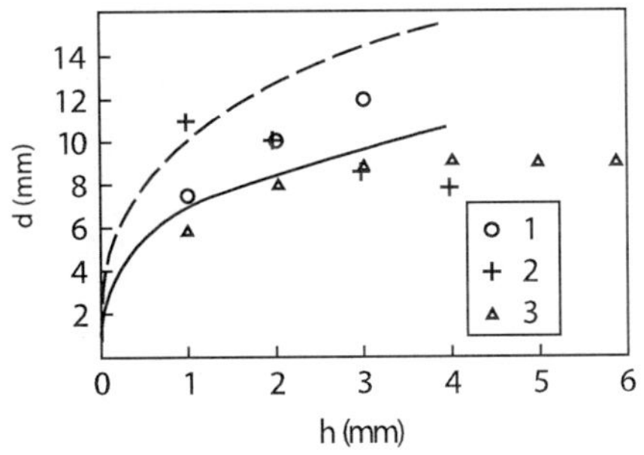

Fig. 5.2 The drilling diameter in metal plates located in different position (vertical or horizontal) vs its thickness. (**1**) St 3 plate in vertical position at $P/h = 1$ kW/cm, *dashed curve* is the theory for the same conditions. (**2**) Pb plate in horizontal position at $P = 1$ kW. (**3**) Pb plate in vertical position at $P = 1$ kW, *solid curve* is the theory

$$\delta = \frac{1}{12\sqrt{3}} R^3. \tag{5.10}$$

Comparison of (5.5) and (5.10) shows, when the height of molten strip equals height of ring circular crater is less in 13.5 times. This is primarily conditioned by the fact that the melt is simultaneously retarded from leaving the top and the bottom regions. Indeed, the equilibrium state analysis of the strip-shaped fusion zone with two anchored points demonstrates that the crater depth appears only 1.4 times as small compared to the "circular" fusion zone. The comparison with the experiment shows that the maximum height of the fusion zone obtained using the model (5.5) is significantly different from the measured one (see Fig. 5.2). Thus, the model (5.10) is more preferable for describing the thin plates. It is clear that the melt will leave a 2δ–thick plate if the fusion zone diameter in accordance with (5.10) exceeded its height (5.5) by the factor of 2.4.

Prior to studying melt ejection it is reasonable to compare the estimation (5.10) with the experimental data as this equation determines the calculation techniques to find the critical size of the fusion zone. Figure 5.3 shows theoretical and experimental curves obtained under assumption that the capillary radius was 2 mm. It follows from this figure that some experimental points lie outside the area where the linear theory is sufficiently accurate. While using the model (5.2) we could see that the nonlinear effects can increase δ sufficiently at large values of R. It is now clear that the melt leaves the plate completely if the doubled crater deepening δ matches the plate thickness h. This immediately gives the critical value for the size of the fusion zone d_{cr}:

$$d_{\mathrm{cr}} = 4.37 h^{1/3}. \tag{5.11}$$

The maximum diameter of the fusion zone can be found for a thick plate as well. Similar to the problem (5.2) we need to solve the nonlinear, but yet two-dimensional, (5.1) for the round-shaped melt pool. This difficulty can be overcome the following way. Both analytical and numerical solutions of (5.2) obtained for

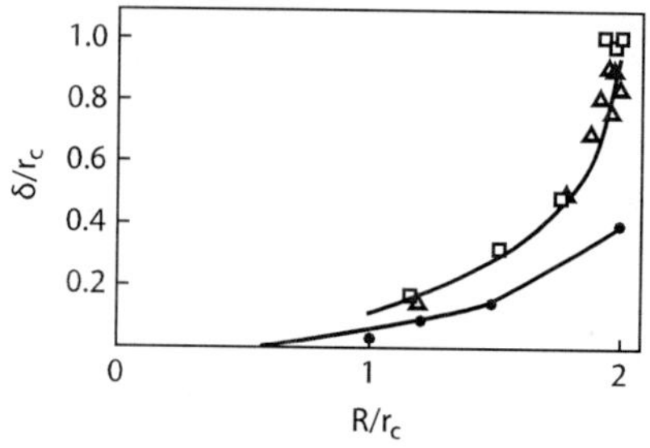

Fig. 5.3 Maximal height of the hummocks (*square*) and crater depth (*triangle*) in lead plate vs melt radius, black rings is numerical results

a fusion strip with boundary conditions the same as for the round pool (5.6) give different results with respect to the plate thickness limit. The thin plates enjoy

$$d_{cr} = 3.97h^{1/3}, \tag{5.12}$$

while the thick plates have [4]

$$d_{cr} = 3.6. \tag{5.13}$$

These solutions can join at $h = 0.75$. If we compare (5.12) to (5.11) we see that the critical value of the round zone is greater than that of the strip by the factor of 1.1. Finally, the critical size of the fusion zone for the vertical plates as function the plate thickness can be summarized as the following

$$d_{cr} = \begin{cases} 4.37\, h^{1/3}, & h < 0.75 \\ 3.96\,, & h > 0.75 \end{cases}. \tag{5.14}$$

In Fig. 5.2, this dependence is compared with the experimental data measured for led and steel plates. In cutting the steel the saturation of d_{cr} was never obtained due to limited power of the CO_2-laser system used in the experiments [10]. The led plates were perforated with the beam powers of 0.5 and 1.0 kW and $d_f = 2$ mm. All these facts prove that the diameter of the hole is determined by the workpiece thickness and if the plate is thicker than the capillary radius it is governed by the maximum value (5.14).

5.1.2 *Drilling of Horizontal Plates*

Now the situation is qualitatively different from the previous case. When perforating thick metal plates mounted vertically the larger thickness must be compensated with a greater diameter of the fusion spot. However, since thicker workpieces are heavier, keeping the molten cylinder still in a horizontal plate requires smaller curvature radius of the liquid surface. At the same time it can not be smaller than the radius of the fusion spot. The roughest approach to estimate the balance of forces gives the limiting value for the fusion diameter

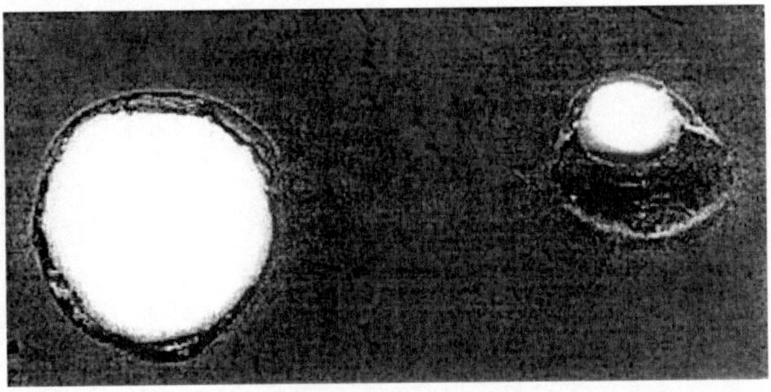

Fig. 5.4 Hole size against workpiece position: *left* – horizontal, *right* – vertical

$$D \approx 8/h. \tag{5.15}$$

Equation (5.1) for the top and the bottom surfaces of the molten zone can be written as one-dimensional. The equation that refers to the top surface is the same as the one used to find the surface shape of the threshold keyhole (see Chap. 4) at $Q = 0$. For the bottom surface it must be taken into account that the liquid pressure is the sum of the external pressure and surface tension. The numeric calculations of (5.1) together with the tangency condition for the top and bottom surfaces give critical diameter as a function of plate thickness [4] (Fig. 5.2). It is shown in the figure that (5.15) is valid, at least qualitatively, despite the numerical factor in (5.15) has been set almost twice as high. Figure 5.2 also shows that for thin workpieces the holes drilled in horizontal plates may substantially exceed the holes in vertical plates. This is very clearly illustrated in Fig. 5.4. If $h \sim r_c$ the diameters become equal. Thus, we can conclude that it is advantageous to drill or cut thin plates when they are fixed vertically in contrast to the thick plates which are easier to be cut if positioned horizontally.

5.1.3 Self-Consistent Drilling Model

From the above analysis we have learnt that growing of the fusion zone changes its shape. This, in turn, changes the size of the heat affected zone in the workpiece. For thin plates this change must be negligible and the temperature can be calculated with no account for possible deviations from the perfectly plane surface. The expansion time of the fusion edge in such as a case is described by the well known equation [5]:

$$\tau = \frac{r_m^2}{\chi} \exp \frac{4\pi\kappa T_m}{P/h}. \tag{5.16}$$

This concept was checked for validity in experiments on thin steel plates (steel St3). Figure 5.5 shows the expansion dynamics of the fusion zone for the plates of

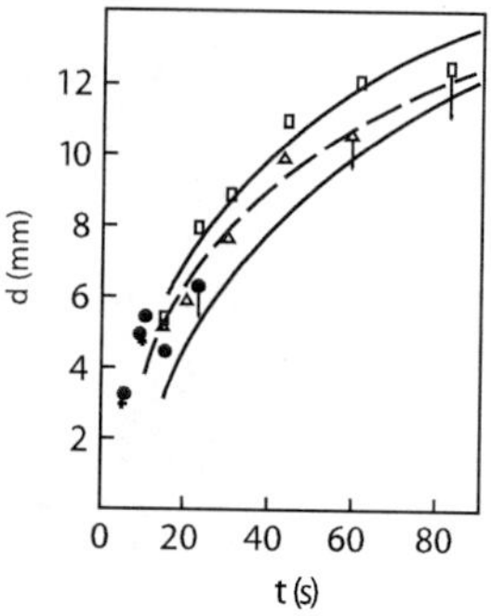

Fig. 5.5 Dynamics of the fusion zone expansion in thin steel plates of various thickness at $P/h = 1\,\text{kW}$, $d_f = 2\,\text{mm}$. Vertical arrows indicate piercing points; *dashed curve* represents theory for $\alpha = 28\%$. *box* – $h = 3\,\text{mm}$, $\triangle$ – $h = 2\,\text{mm}$, $\bullet$ – $h = 1\,\text{mm}$

various thicknesses. The beam power was increased proportionally to the workpiece thickness keeping $P/h = const$. That was done because this relation is the only one relevant to heating of a thermally thin material. In other words, the plates of different thickness should enjoy the same heat scenario given the same P/h ratio. Figure 5.5 proves that it is really so to a certain precision. It is necessary to note that the beam intensity was proportional to the thickness of workpieces. The arrows indicate the moments when a hole is created. One can see that the hole diameter grows with the thickness despite the fact all pieces are identical in terms of heat conduction. Let us finally note that this elementary model for heating a thermally thin plate (5.16) at constant parameters is generally found in good agreement with the experiment when $\alpha \approx 28\%$.

5.1.4 Thermally Thick Limit

For the sake of simplicity we will study an excessively thick workpiece approximated as a semispace. In this case the heating problem may have a steady-state solution (see Chap. 1). The temperature distribution over the depth at $r = 0$ is described analytically [6] as

$$T = \frac{P}{\pi r_f^2 \kappa}\left(\sqrt{z^2 + r_f^2} - z\right), \quad r_f < r_m. \tag{5.17}$$

This equation can be successfully used to estimate the power being absorbed to allow the fusion isotherm to reach the critical depth. However, the critical power is unlikely to be found using the critical diameter of the fusion zone (5.14) as the fusion diameter is now found via the integral [6]:

$$P_{\text{cr}} = \frac{\pi \kappa T_m r_c \xi^2}{\sqrt{1 + \xi^2} - 1}, \quad \xi = \frac{r_f}{r_c}. \tag{5.18}$$

Here, it is taken into account that in agreement with Fig. 5.2 the critical depth is the maximum deepening of the crater δ and $\delta_{\max} \approx r_c$. On the other hand the beam power in this model is bounded from above. It is necessary that the maximum temperature never reached the boiling point. This limitation can be derived from (5.17) at $z = 0$:

$$P < \pi \kappa r_f T_b. \tag{5.19}$$

Substituting (5.18) into (5.19) we get the lower bound for focal spot radius r_f:

$$\frac{\xi}{\sqrt{1+\xi^2}-1} < \frac{T_b}{T_m}. \tag{5.20}$$

For $\xi \ll 1$ (5.20) gives

$$\frac{r_f}{2r_c} > \frac{T_m}{T_b}. \tag{5.21}$$

Since the critical fusion radius r_m is close to $2r_c$ as follows from (5.14), we can rewrite (5.21) this way:

$$1 > \frac{r_f}{2r_c} > \frac{T_m}{T_b}. \tag{5.22}$$

Now it is clear that these inequalities are in full agreement to each other and at $P > P_{\text{cr}}$ determined by (5.18) we can expect a droplet of molten material to flow out of the vertical workpiece (see Fig. 5.6) followed by steady dripping. This should constitute the drilling process with no vaporization involved. Without going into the details of this drilling technique we need to note that the bore is usually found significantly inclined from the beam axis to up to 30°. This is conditioned by the necessity for the melt to leave the bore and the beam reflection from the bottom

Fig. 5.6 The fusion zone structure in a lead workpiece (side view)

Fig. 5.7 The cut made in a lead workpiece: $h = 30\,\mathrm{mm}$, $P = 1\,\mathrm{kW}$ (side view)

wall (see Fig. 5.7). From what we have studied now the drilling of metals without a gas jet is fairly described in a simple case of thermally thin plates. Further in this chapter we will use these fundamentals to study a more sophisticated case when the workpiece moves relative to the beam. We classify the latter as remote cutting.

5.1.5 Remote Cutting

Cutting process with a cw laser and the lack of gas jet assistance to remove the melt will be studied using numerical results because it poses a two-dimensional nonstationary problem of heat distribution in a flat metallic workpiece. The workpiece geometry is being altered in time because of the melt leaving it. The latter consequently affects the beam absorption. The most complex issues here are the mass transfer and melt removal. Broadly speaking, it must be looked at as a three-dimensional flow of a molten liquid. For the sake of simplicity we can assume that, similar to the drilling process, the melt is removed when the vertical size of the pool meets some critical value. Of course, the shape of the fusion zone may vary considerably from a half-moon to an ellipse stretched in the direction of the beam route. For this model, we assume that the critical size does not depend on the pool shape and is utterly determined by (5.14). The heat conduction equation here is supposed to have only constant coefficients although this simplification is not necessary if it is solved numerically [11]:

$$c\frac{\partial T}{\partial t} - \kappa\left(\frac{\partial^2 T}{\partial x^2} + \frac{\partial^2 T}{\partial y^2}\right)$$
$$= \frac{q_0}{h}\exp\left(-\frac{(x-vt)^2+y^2}{r_f^2}\right) - \varepsilon\frac{\sigma_B T^4}{h} - \frac{q_u}{h}\theta(T-T_b) - \frac{q_k}{h} \quad (5.23)$$

where c and κ are the heat capacity and heat conduction coefficient respectively, h denotes the workpiece thickness, q_0 is the intensity of absorbed radiation, σ_B is the Stefan–Boltzmann constant and ε marks greyness. The laser beam travels along the x-axis at with speed v. The value $\theta(\xi) = 0$ when $\xi < 0$ and $\theta(\xi) = 1$ while $\xi > 0$. The last three terms in the right-hand-side of (5.23) describe the energy losses due to radiation, vaporization of the material and natural convection in the air.

The heat escape via vaporization will be described using the simplest available technique because a qualitative account is sufficient in our case. The energy flux carried away with the vapor is calculated as the product of the boiling enthalpy L_b, the vapor density n_s and its thermal velocity:

$$q_u = 0.2 n_s \bar{V}_T L_b = \frac{2.56\times 10^3}{\sqrt{T}}(p_s - p_A)\ (\text{W/cm}^2), \quad (5.24)$$

where $p_A = 760\,\text{mm Hg}$. The vapor density is conditioned by the saturation pressure [20], i.e.

$$p_s = \frac{4}{T^{1.27}} \times 10^{(13.27-19710/T)}. \quad (5.25)$$

The values of these coefficients are taken for iron which is the material treated in the experiments. In (5.24) and (5.25) the pressure is expressed in mms of mercury column and the temperature in the K scale.

Intensive vaporization starts when the surface temperature exceeds the boiling point of the material T_b, i.e. $p_s > p_A$.

Calculation of the heat losses due to natural air convection q_k is accompanied with serious difficulties as this process is a non-local function of surface temperature [21]. Since its exact form is not important for this problem we can just stay restricted to the local temperature dependence of the losses found from

$$q_k = \kappa\frac{T - T_a}{H^{1/4}}\left(\frac{g}{\nu^2}\right)^{1/4}. \quad (5.26)$$

Here, κ and ν are the heat conduction coefficient and kinematic viscosity of air, H describes the characteristic vertical size of the workpiece, T_a is the air temperature and g is gravity acceleration. For $H \sim 5\,\text{cm}$, $T \sim 1{,}000\,^\circ\text{C}$, and $q_k \approx 1.67\,\text{W/cm}^2$ the equation recommended in [5] for similar condition gives approximately $1\,\text{W/cm}^2$. Despite the small value of q_k at slow cutting speeds the workpiece is heated substantially to make the convectional cooling noticeable. The workpiece plate edges as well as the cut edges require the use of the second type conditions as these edges are free, i.e. $\partial T/\partial n = 0$, where n is the normal to the boundaries. As for the initial condition we assume that the workpiece temperature is that of the environmental air T_a.

Equation (5.23) is written in a dimensionless form using the conventional technique [22]:

$$\theta = \frac{4\pi k T}{P/h}, \quad \bar{x} = x/r_f, \quad \bar{y} = y/r_f, \quad \bar{t} = t/(r_f^2/2\chi),$$

(the explicit dimensionless form is not written because of its bulkiness). If the heat losses are disregarded this equation has the only parameter $r_f/(4\chi/v)$ which is the ratio of the focal spot to the width of the heat front. It must be noted that these losses become notable mainly in case of cutting in the pulsed mode. In order to simplify (5.23), like it is often done to study laser treatment methods, the phase transition energy in the fusion process is omitted as for iron is never beyond 30%.

The laser beam travels near the center of the workpiece along its long side. The workpiece dimensions are 5×10 cm. The calculation algorithm foresees that vertical size of melt pool riches the critical value d_{cr} (5.14) all melt is removed from a workpiece. The thermal physics parameters were simply assumed constant and close to steel St35, i.e. $\kappa = 0.3$ W/cm K, $\chi = 0.13$ cm^2/s. The absorption factor α was set equal to 13% while the thermal emittance factor was $\varepsilon = 0.3$.

Besides, it was important to calculate respectively the fraction of radiation passed through the workpiece and delivered to its surface β.

5.1.6 Experimental Techniques and Results

A horizontal beam from a cw CO_2 laser was guided to a workpiece. The beam power could be as high as 6 kW [23]. The workpiece was fixed in a vertical position allowing the beam to hit its plane surface at right angle. This arrangement was used to make cuts in several steel plates (St3) of about 5 cm tall and 10 cm across with their thickness ranging from 1 to 4 mm. The CO_2 laser performance was controlled with a power gauge. The amount of radiation that broke through the workpiece was measured as well so the value of β was at hand. The cutting speed was varied between 0.3 and 1.0 mm/s while the spot size was ranged from 0.5 to 5.0 mm. The results of these experiments are shown in Figs. 5.8, 5.9, 5.10.

The aim of this experiment was to determine the process responsible for the mass transfer from the interaction zone. Thus, the cuts were made using a broad range of beam intensities which was realized by changing the spot size at constant power. The spot size was altered from a reasonably small and typical for laser welding, $d_f \sim 0.5$ mm, to some big ones, i.e. $d_f \sim 5 - 7$ mm. This trick allowed the focal spot intensities to differ by over two orders of magnitude.

It is known from the literature [20,24,25] that for welding speeds below a certain value ($\sim$1 mm/s) the weld depth stops growing while the weld width increases

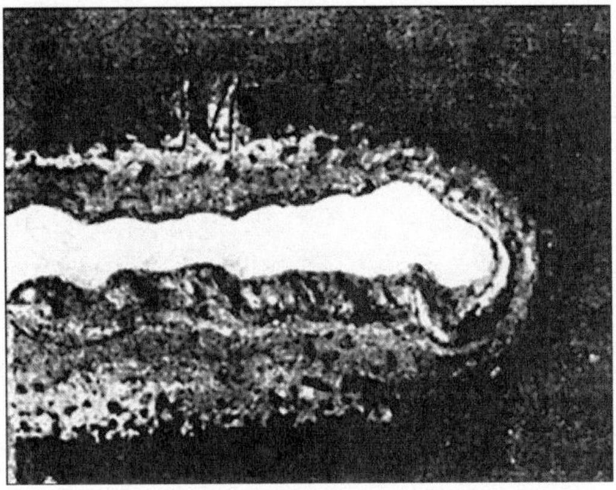

Fig. 5.8 The general view of a cut made with a laser in a 2 mm St.3 plate at $P = 3.2\,\text{kW}$ and $V = 0.6\,\text{mm/s}$

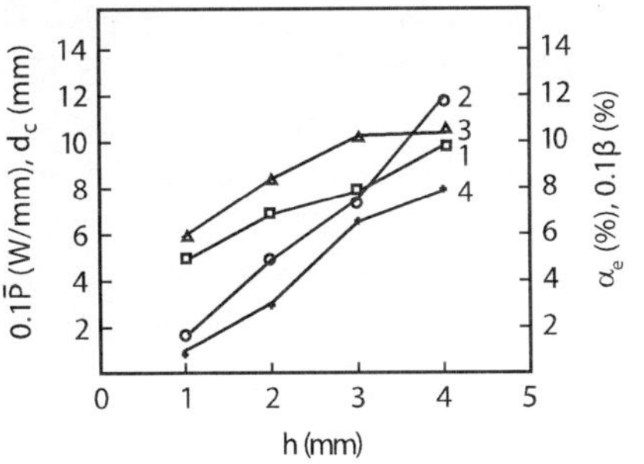

Fig. 5.9 Cutting process parameters for St.3 plates as functions of workpiece thickness at cutting speed $V = 0.6\,\text{mm/s}$, $P = 4\,\text{kW}$ ($h = 1\,\text{mm}$), $P = 3.2\,\text{kW}$ ($h = 2\,\text{mm}$), $P = 4\,\text{kW}$ ($h = 3\,\text{mm}$), $P = 3.6\,\text{kW}$ ($h = 4\,\text{mm}$); **1** – cut width d_c; **2** – effective absorption α_e; **3** – fraction of beam incident to work surface β; **4** – specific power absorbed by workpiece $\widetilde{P} = \alpha_e P/h$

sharply (see Chap. 4). Apparently, this property can be quite beneficial for cutting metal sheets. In fact, the widening of the fusion zone makes it possible for the molten material to leave and create a kerf. The experiments carried out with the beams sharply focused ($d_f \sim 0.5\,\text{mm}$) and the workpieces lowly moving relative to the beam fully confirmed this hypothesis. Indeed, at slow delivery rate of about 0.5 mm/s the welding process is known to turn into cutting process because the molten metal fails to freeze once the beam has passed further. Instead, it flows down to the bottom edge to create a frozen wavy buildup (see Fig. 5.8). The video footage and examination of the workpieces both prove undoubtedly that the remote cutting is nothing similar to any of the previously practised laser processing techniques. First of all, the kerf width produced now is substantially greater than the size of the focus spot (ranging from factor 2 to 10 as seen from Fig. 5.9). Secondly, the entire operation is clearly seen as an oscillation-like process. Indeed, the interaction zone

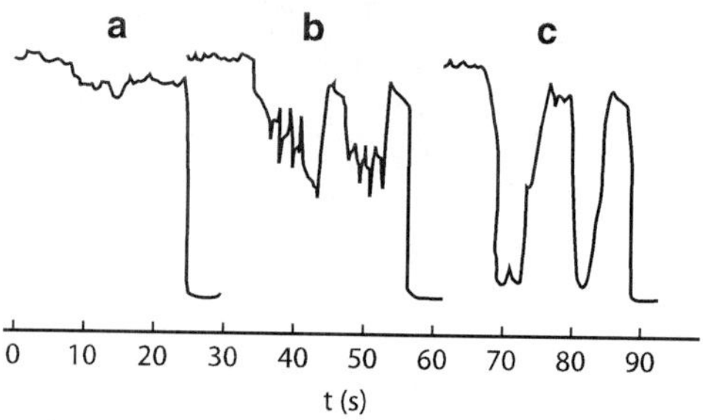

Fig. 5.10 Oscillogram traces of the power lost through the kerf, recorded for St.3 plates; $V = 0.3$ mm/s; $P = 4$ kW; workpiece thickness $h = 1$ mm – (**a**), 2 mm – (**b**), 3 mm – (**c**)

is heated, the melt leaves it and freezes. Then the situation is replicated again and again. Another difference is that the loss of the workpiece mass in comparison to the mass of remelted material is rather small. Besides, it is highly interesting to note that at a constant specific value P/h, where P is the beam power and h is the workpiece thickness, the kerf appears to be wider in thicker plates.

Figure 5.9 shows the growth of the kerf width for thicker samples even at constant beam power. This fact, being surprising at first glance, is easily explained if one notes that the effective absorption α_e becomes stronger as well. High values of α_e are conditioned by a larger fraction of radiation β falling on the workpiece. Moreover, α_e increases with h quite rapidly making the specific power of absorbed radiation $\alpha_e P/h$ a growing function of the workpiece thickness. Thanks to this fact we now understand why the growing thickness of workpieces makes it possible observe wider kerfs.

5.1.7 Oscillatory Type of Remote Cutting

Let us now discuss the details of α_e growth for thick workpieces. As we showed above the remote cutting of metals is an oscillatory process. For typical cutting conditions, i.e. $d_c \approx$ 7–10 mm, $V = 0.5$ mm/s, the oscillation period can be evaluated as $T \sim d_c/(2V) =$ 7–10 s. Figure 5.10 shows cutting of a thin metal plate and it is seen that the overwhelming part of radiation travels freely through the kerf. In other words only a small overlapping of the beam spot and the workpiece surface is already sufficient to create a narrow melt region. The workpiece material is removed as the liquid metal periodically runs off out of the kerf. As soon as the melt has left the interaction zone the entire beam power is freely transmitted behind the workpiece. This is in full agreement with the fact that the fusion zone exceeds the beam focal spot. With greater h the target screens the beam well enough to let only a tiny fraction of power behind it. In this case, the melt removal is also followed by free passing of the beam through the created aperture. But as soon as the beam meets the unfused metal it becomes screened again. This picture can be successfully used to derive an approximate condition for cutting maintenance. The beam power must be such that the fusion zone could reach the critical size before the workpiece is moved a distance of the size d_{cr}. The expansion time for the fusion edge can be estimated from the nonstationary model of heat conduction in thermally thin samples (5.16). This time must be shorter than the spot travel time $\tau = d_{cr}/(2V)$.

Taking into account (5.16) we can obtain the cut initiation condition

$$\frac{Vd_{cr}}{2\chi}\exp\left(\frac{4\pi\kappa T_m}{\alpha P/h}\right) < 1. \tag{5.27}$$

Within the numerical factors this equation agrees with the Swift–Hook and Gick theory. In order to describe the initial stage of the cutting process starting from the workpiece edge (5.27) should be supplemented with $2P$ (instead of P) as the heat in this case is applied to a semiplane.

5.1.8 Comparison of Calculated and Experimental Results

Figure 5.11 demonstrates the cut shapes calculated for steel plates with different thicknesses h [11]. The cut edges have wavy forms produced by the melt leaving the cut zone periodically. The cuts made in experiments with steel plates of thicknesses larger than 1 mm have very similar shapes as it is very well seen in Fig. 5.8. From Fig. 5.11 it follows that for thicker plates and, therefore, lower specific power P/h the spatial period increases notably. The energy conservation law makes it evident that the size of the fusion zone (its length and width) should decrease with lower P/h or with the higher beam travel speed. Since in this case $d_{cr}(h) \sim h^{1/3}$ the kerf width must be found greater on thicker plates. In order to "provide" a wider fusion zone one has to make the plate "takes" more energy from the beam which is exactly what happens if the fraction of the beam energy incident to the workpiece is increased. This is well seen from Fig. 5.12 where the calculations as well as the experiments prove that β is higher for thicker samples.

The cutting behavior is better revealed by varying the cutting speed since the only changing parameter in this case is $r_f V/4\chi$. It describes the size ratio of the focal spot to the heat front. The experiments show that the fraction of the beam

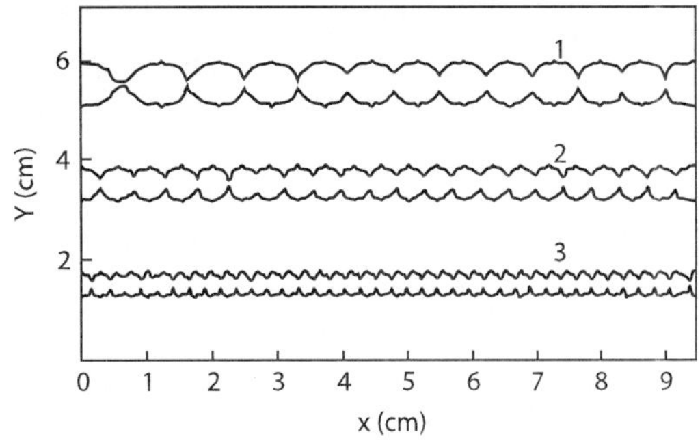

Fig. 5.11 The form of cut a steel (St.3) plate at $P = 4\,\text{kW}$ and $V = 0.6\,\text{mm/s}$: **(1)** $h = 3\,\text{mm}$, **(2)** $h = 2\,\text{mm}$, **(3)** $h = 1\,\text{mm}$

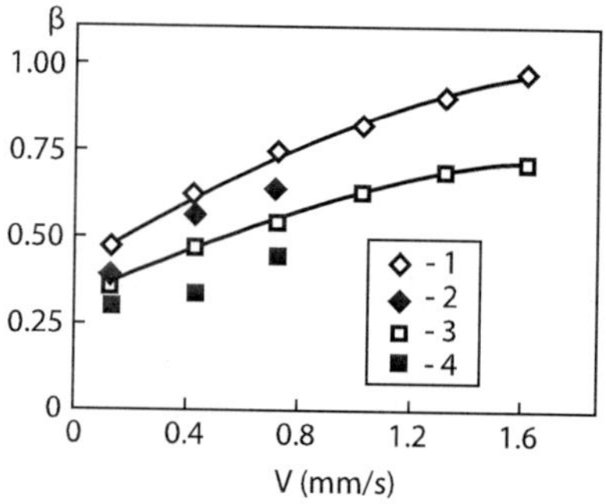

Fig. 5.12 The fraction β of laser radiation incident on a workpiece versus the rate of laser beam travel v at $P = 4\,\mathrm{kW}$: (**1**) $h = 3\,\mathrm{mm}$, calculation; (**2**) $h = 3\,\mathrm{mm}$, experiment; (**3**) $h = 2\,\mathrm{mm}$, calculation; (**4**) $h = 2\,\mathrm{mm}$, experiment

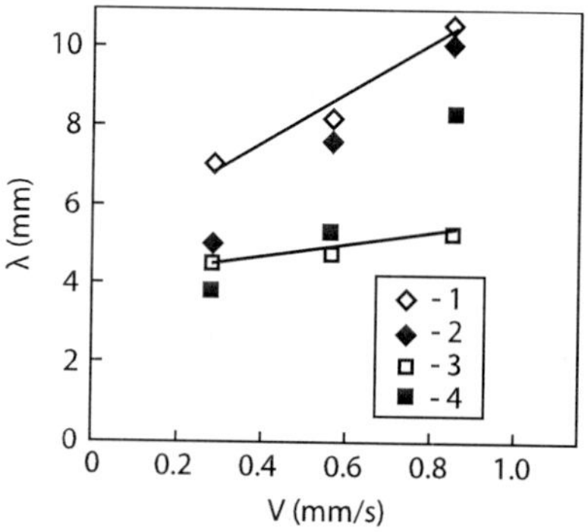

Fig. 5.13 The oscillation wavelength of the cut shape versus the cutting speed v for different values of the plat thickness. The notation is the same as in Fig. 5.12

which illuminates the workpiece grows monotonously with rising cutting speed. It must be clear from a general consideration that when a faster speed causes process failure one finds $\beta = 1$. Naturally, we understand the cut is no longer produced when the specific power is insufficient. This explains why with higher speed the curves in Fig. 5.12 tend monotonically to unity and a curve corresponding to lower specific power would lie above that for higher P/h. It is evident that this simple model already gives a satisfactory qualitative agreement between the experimental measurements and calculated data.

As it was mentioned above the conservation of energy provides shorter fusion zone if the beam is moved faster. Since the molten material is removed by dripping out periodically the size of the cut heterogeneities λ should seem to decrease with higher speeds. However, as it follows from Fig. 5.13, both the experiments and calculations show that λ increases with higher v. Such behavior is as well conditioned by growing β. Indeed, according to the Swift–Hook and Gick model the fusion zone length can be stretched by the adding extra power ($\sim P^2$ or stronger) and reduced in reverse proportion to the process speed (see Chap. 4, (4.31)). Consequently, even a small rise in β at growing v will lengthen the fusion zone at high speed. This is exactly what conditions elongation of the spatial cutting period with rising speed. However, here the agreement between the measurements and calculations is rather qualitative which is explained, most likely, by a more complicated character of melt removal. So the model under discussion describes the averaged characteristics only.

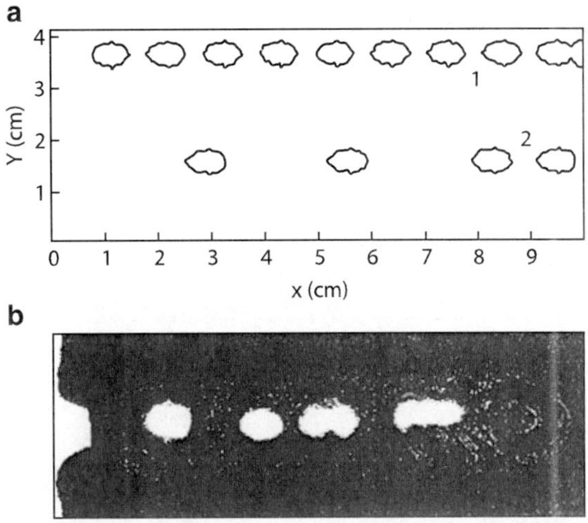

Fig. 5.14 (**a**) The form of cuts for unstable cutting modes (St.3 steel) at $P = 2\,\text{kW}$ and $h = 1\,\text{mm}$: (**1**) $v = 4\,\text{mm/s}$; (**2**) $v = 4.5\,\text{mm/s}$; (**b**) typical form of cut in the case of unstable cutting

5.1.9 Disruption of Cutting Operation

It would seem that the Swift-Hooke-Jeek model could be used to calculate the threshold of disruption of cutting with increasing cutting speed (or with decreasing specific power). The speed at which the melt width becomes smaller than d_{cr} will be the critical cutting speed. However, when the cutting ceases, the value of β increases jump-wise to unity. Because of this, the cutting operation will resume, β will decrease, and the cutting will again cease. Thus, the cutting-fusion transition passes through an unstable stage which reduces to intermittent cutting [11]. The relative duration of pauses increases with the cutting rate (see Fig. 5.14). Finally, when the cutting speed reaches such a value that the melt width will be smaller than d_{cr} if even at $\beta = l$, the cutting will cease completely. This is evident from Fig. 5.15, where the results of calculation of threshold cutting rates are compared to the critical cutting rate calculated by the Swift-Hooke-Jeek model. It is difficult to estimate the critical speed at which the interruptions of cutting begin, because we do not know the value of β_{cr}. This may be done only by numerical calculation (see Table 5.1, where v is the critical cutting speed, above which the cut is not produced).

The regularity of cutting in the transition (unstable) region largely depends on the thermal front width $\sim 4\chi/v$. At low cutting speeds, it is comparable to the plate dimensions ($\chi \simeq 0.13\,\text{cm}^2/\text{s}$, $v = 0.5\,\text{mm/s}$, $4\chi/v \simeq 10\,\text{cm}$). Therefore, in the process of cutting, the entire plate is heated, pauses grow ever shorter as the laser beam moves toward the plate edge, and, finally, cutting becomes continuous.

At high cutting speeds, when $4\chi/v \ll r_f$ one can see in Fig. 5.14a that pauses alternate with cutting regions. Figure 5.14b shows a photograph of a typical unstable cut of a steel plate.

Figure 5.15 gives the experimental parameters of remote cutting in the predicted region. One can also see that the cutting speed may be substantially increased at high values of the P/h ratio (about 4 kW/mm).

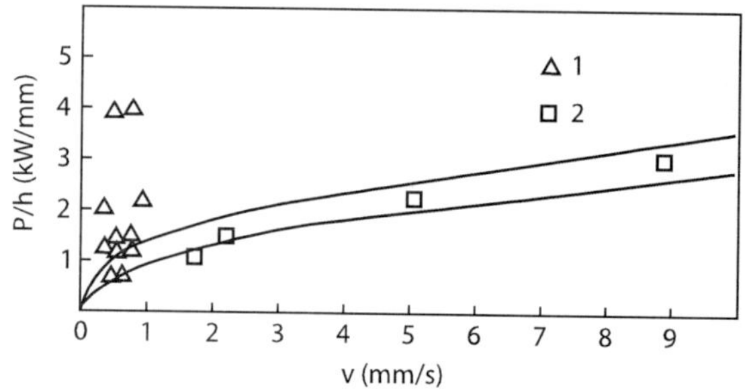

Fig. 5.15 Pattern of cutting (St.3 steel) versus specific power and speed. The stable cutting region is above the upper curve, and the unstable cutting region is below it; the fusion region is below the lower curve; (**1**) experiment and (**2**) theory

Table 5.1 Critical parameters of remote cutting

v, mm/s	0.015	0.9	3	13
β_{cr}	0.68	0.75	0.82	0.9
$\eta(\%)$	3	15	28	44

It is of interest to estimate the efficiency of remote cutting: $\eta = vdcT_m/(\alpha P/h)$. It is clear that its maximum value will be found at the boundary of the cutting region. These values are presented in the table. Here it is convenient to recall that the fusion efficiency in the case of deep-melting welding does not exceed 50%.

The kerf width varies as observed when cutting thin titanium foils under water without a gas jet as reported in [7] with continuous Nd laser. Decreasing power or increasing the scanning speed the cutting was not created continuous and produced a series of independent apertures. Oscillations of the cut width were observed in the air environment too. It is possible that the mechanism of this phenomena is identical to the mechanism described earlier for cutting steel plates.

5.2 Properties of Gas-Assisted Cutting

In the previous section we showed that lasers can be successfully used to cut metals without a gas jet involved. However this technique produces a wide kerf especially in plates thicker than the capillary radius. The situation is sharply different if the melt is blown away by jets of air or some other gases. First of all, this gas assistance considerably reduces the cut width. It can be as narrow as fractions of a millimeter. Besides, using air or oxygen can give rise to an exothermic reaction of the workpiece material oxidation which can substantially (up to factor 2) increase the depth or speed of cutting. From general analysis it becomes clear that the positive action of the gas jet is conditioned by the efficient removal of the molten material from the cut zone. It occurs either because of the static pressure difference created by the gas flow or the friction force between the streaming gas and molten material or both. In any case one has to consider the properties of the gas flow in such a complex geometry of the cut produced jointly by a laser beam and a gas jet.

It could first seem that a higher gas pressure in a nozzle would ensure better efficiency of cutting. But we can immediately ask how far we can go on increasing the gas pressure. Besides, under certain conditions a higher pressure might cause an reverse effect of a shallow cut depth and slow cutting speed. Now we will take a better look at some regularities of gas flow.

5.2.1 Gas Dynamics in Laser Cutting

As soon as the gas jet leaves the nozzle it starts interacting with the environment or the atmospheric air. As we showed in Chap. 1 it is the ratio of gas velocity to the sonic speed which governs the jet properties in the air. A subsonic jet dissipates within nine nozzle diameters from the exit section ($\sim 9d$, where d is the jet diameter when entering atmosphere or the nozzle diameter). In practice the distance between the nozzle exit section and workpiece surface is several nozzle diameters. So the gas jet hits the workpiece and diverges in different directions. As a result a gas flow with so called stagnation zone is created [12]. Such subsonic flow is formed only if the nozzle pressure is not above the twice atmospheric pressure. In this case the pressure induced to the workpiece surface is not greater than twice the atmospheric pressure. Since the cut width is several times smaller than the nozzle diameter the kerf weakly affects the gas-jet dynamics above the work surface. Consequently, as the cut itself is nothing but a flat nozzle, it forms a subsonic gas flow as well.

The situation becomes significantly different if the nozzle pressure is higher than twice the atmospheric pressure. In case of a free gas jet, as we could see in Chap. 1, moving away from the entrance point (nozzle exit section) the gas velocity rises continuously above the sonic level at the nozzle exit section. At the distance $x/d \simeq 0.6\sqrt{p/p_A}$ from the nozzle a standing shock wave or the Mach disk is created as shown in Fig. 1.13. When cutting the nozzle is moved to the work surface as close as nearly a nozzle diameter. Consequently, the shock disk is found very close to the surface. Here rises a question of the pressure following the shock disk as it governs the flow properties inside the kerf. Now our first task is to determine the pressure preceding the Mach disk P_1. It can be estimated by using the well–known equations from the gas dynamics of an isentropic flow:

$$\frac{T_0}{T} = 1 + \frac{\gamma_a - 1}{2} M^2 \cong 1 + 0.2M^2 \qquad (\gamma_a = 1.4), \tag{5.28}$$

$$\frac{P_0}{P} = \left(\frac{T_0}{T}\right)^{\frac{\gamma_a}{\gamma_a - 1}} = \left(\frac{T_0}{T}\right)^{3.5}, \tag{5.29}$$

$$\frac{\rho_0}{\rho} = \left(\frac{T_0}{T}\right)^{\frac{1}{\gamma_a - 1}} = \left(\frac{T_0}{T}\right)^{2.5}. \tag{5.30}$$

Here, "0" subscripts refer to the values inside the nozzle. From this equations one can see that the gas velocity can reach the sonic limit ($M = 1$) at the nozzle exit section if

$$\frac{P_0}{P_A} = \left(\frac{\gamma+1}{2}\right)^{\frac{\gamma}{\gamma-1}} \simeq 2 \quad (\gamma = 1.4). \tag{5.31}$$

This point has already been discussed above. Now the system of equations needs to be completed with a useful relation between the cross-section area and the Mach number, M_1, as in (5.28)–(5.30) all the quantities are functions of the latter parameter:

$$\frac{A}{A^*} \simeq \frac{0.6}{M}\left(1 + 0.2M^2\right)^3 \quad (\gamma = 1.4), \tag{5.32}$$

where A^* is the minimum cross-section area of the canal, i.e., found exactly at the nozzle exit section. It is worth to estimate the value of A/A^* for a freely expanding jet. If the opening angle is about 1 we get

$$\frac{A}{A^*} \simeq \left(1 + \frac{h}{r}\right)^2,$$

where h denotes the spacing between the nozzle and the work surface and r is radius of the nozzle aperture. For the ratio, $h/r \sim 1$, we have $A/A^* \sim 4$ and from (5.32) we find $M_1 \sim 3$. Knowing the Mach number before the jump in (5.28) and using (5.29) we can immediately evaluate P_1:

$$P_1 = P_0\left(1 + 0.2M_1^2\right)^{-3.5} \simeq 3 \times 10^{-2} P_0. \tag{5.33}$$

Following the jump the values of pressure, P_2, and the Mach number, M_2, are found from the well-known relations at the shock front [3]:

$$\begin{aligned}
\frac{P_2}{P_1} &= \frac{2\gamma}{\gamma+1}M_1^2 - \frac{\gamma-1}{\gamma+1} \cong 1.15M_1^2 - 0.15, \\
\frac{\rho_2}{\rho_1} &= \frac{u_2}{u_1} = \frac{(\gamma+1)M_1^2}{(\gamma-1)M_1^2 + 2} \approx \frac{0.7M_1^2}{0.2M_1^2 + 1}, \\
M_2^2 &= \frac{2 + (\gamma-1)M_1^2}{2\gamma M_1^2 - (\gamma-1)} \cong \frac{M_1^2 + 5}{7M_1^2 - 1}.
\end{aligned} \tag{5.34}$$

From (5.34) we get

$$\begin{aligned}
P_2 &= P_1(1.15M_1^2 - 0.15) = 0.5P_0, \\
M_2 &\approx 0.5.
\end{aligned} \tag{5.35}$$

Thus, from (5.35) it is clear that if $P_0 \approx 2$ atm the pressure at the kerf entrance region must be close to atmospheric pressure giving conditions for the subsonic gas flow. As the nozzle pressure is increased up to 4 atm the pressure at the kerf entrance should exceed 2 atm and make it possible for a supersonic flow to develop. Since the kerf geometry is rather complex and very different from geometry of the

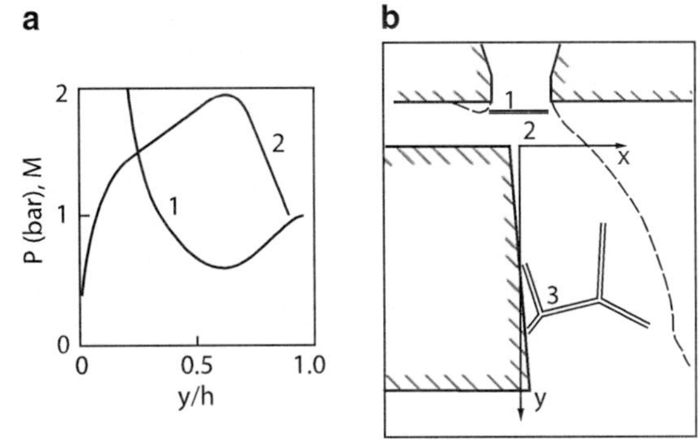

Fig. 5.16 The profiles of static pressure (**1**) and Max number M (**2**) at $d/2$ distance from kerf front, $h/d = 12$, $d_c/d = 3.5$; $p_0 = 5.4$ at, nozzle hight over plate surface is $1.2d$. (**a**) The structure of assisted gas flow: (**1**) axe symmetric supersonic jet, (**2**) subsonic gas flow, (**3**) plate supersonic flow. Twice lines are shock waves, dashed line is free boundary (**b**)

flat nozzle the study of gas dynamics for such conditions must include numeric calculations together with laboratory observations [26] or simulation experiments [8]. Let us start with the latter. Estimations of the gas flow parameters show that the Reynolds number, Re, is approximately 7,000 and the thickness of the turbulent boundary layer $\delta^* \sim 0.1\text{Re}^{-0.2} \approx 5 \times 10^{-3}\,\text{cm} \ll d$. Consequently, one would observe only an inviscid flow inside the kerf. This allows simulation of gas dynamics in a real cutting operation by using enlarged geometrically similar models in a wide range of nozzle pressures and geometrical relations. The enlarged dimensions of the model experiment were crucial to ensure measurements of the basic gas-dynamic parameters such as pressure and velocity. Geometries similar to those typical for cutting assemblies provide three regions where flow patterns have different types (see Fig. 5.16b). In the region between the nozzle exit section and the Mach disk settled above the work surface the gas flow forms a supersonic jet with inhomogeneous velocity distribution along the radius. The lower region has a complicated three-dimensional structure showing sub- and near-sonic gas flows within the space of about the kerf width. The third region is similar to a plane supersonic flow and occupies the most part of the cut section. Here the gas jet is accelerated up to $M = 2$ and the pressure falls down to atmospheric (Fig. 5.16a). As it follows from the figure the acceleration stage is followed by slowing–down on the system of oblique jumps. If both P_0 and the nozzle aperture are increased the supersonic acceleration region may expand and slowing-down may occur outside the cut zone. It would be appropriate now to recall the experiments [26] in which the pressure in the gas jet was measured near the cut bottom by using a Pitot tube (see Fig. 5.17). It was demonstrated that the pressure within the kerf first rises as the nozzle pressure is increased until at $P_0 \sim 3$–3.5 atm it drops down suddenly to become 2–3 times as small (see Fig. 5.18). Apparently, in this case the compression shock sank down beneath the Pitot tube so it measured the low pressure typical for a supersonic flow discussed earlier. It is clear that the compression shock within the

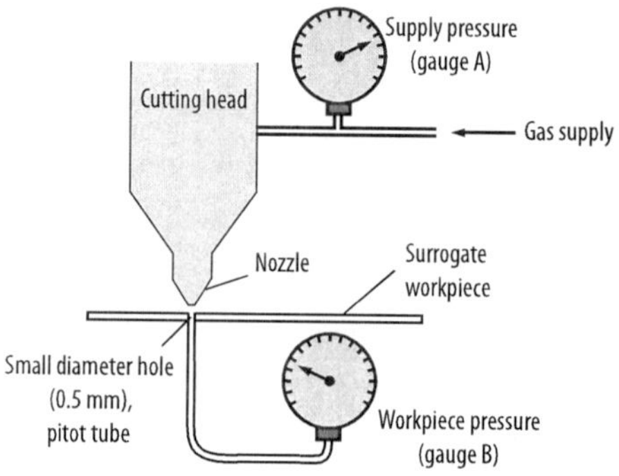

Fig. 5.17 A schematic of the experimental setup for monitoring the output pressure of a cutting nozzle for various supply pressures

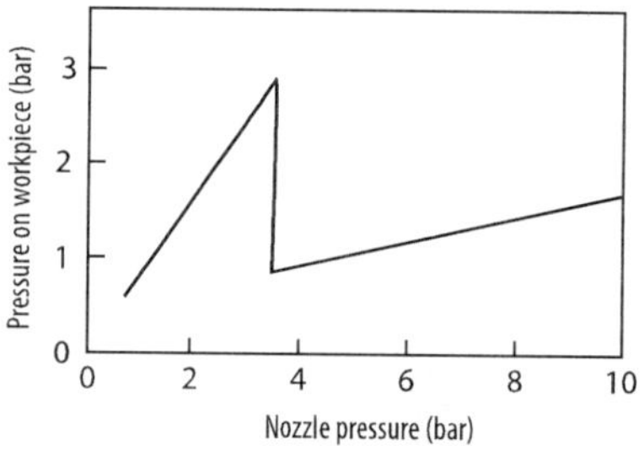

Fig. 5.18 The pressure on the workpiece exerted by a typical cutting nozzle as a function of supply pressure. Note the sudden decrease in workpiece pressure as the supply pressure is increased from 3.0 to 3.5 atm. In this case the gas used was air through a 1.5 mm diameter nozzle with a nozzle–workpiece distance of 3.5 mm

kerf and the pressure rise along (downward) the cut front can work to slow down the melt outflow and, consequently, to reduce the maximum cutting speed. The latter was also observed in experiments [26] (see Fig. 5.19). From this figure it is evident that further increasing of the nozzle pressure would cause the maximum cutting speed to grow up again approaching a saturation point. At the same time the pressure inside the kerf continues to rise with the supply nozzle pressure. It should mean that the melt removal has reached the maximum and no longer affects the cutting speed. The fact that the maximum cutting speed comes to a saturation plateau when the pressure inside the kerf appears significantly low (compare Figs. 5.18 and 5.19) is likely to prove the changing of melt removal mechanisms. At low supply pressure the melt is pushed out from the cut by the static pressure. In the opposite case when the nozzle pressure is high the melt is removed by the friction force which increases with the gas flow velocity passing through the kerf. The mechanisms of melt removal are discussed later in this chapter. Meanwhile, we move on to review the numerical methods used to study the gas dynamics in laser cutting.

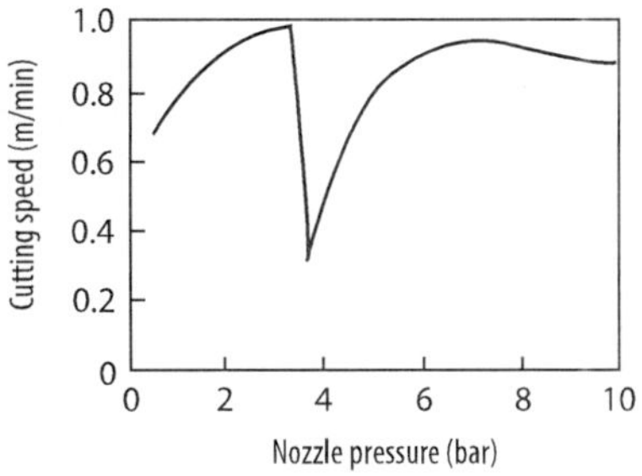

Fig. 5.19 The variation of maximum cutting speed with increasing supply pressure. The gas and nozzle characteristics were the same as and the close correlation is obvious. The laser power was 830 W and the material was 2.6 mm thick PE16 alloy

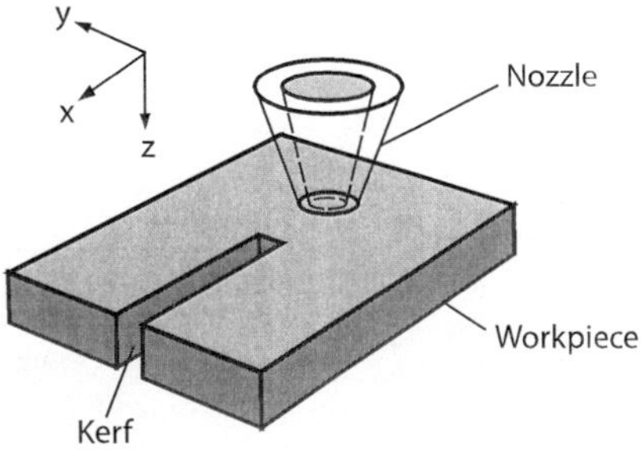

Fig. 5.20 The schematic representation of the cut zone used in numerical studies

5.2.2 *Numerical Studies of Gas Dynamics*

Numerical studies of gas flow dynamics are usually carried out for geometries similar to the shape of a cut as shown in Fig. 5.20. The geometry parameters are assumed independent of laser cutting and varied around the values known from real experiments. The nozzle construction is chosen to be of two basic types. The first type is a conical nozzle where the gas velocity at the exit section is a function of gas pressure and reaches the sonic limit. The other type is a de Laval nozzle where the gas velocity becomes sonic at the "throat" or minimum cross-sectional area and later increases to supersonic values at the exit section. The flow of a gas through the nozzles and the kerfs was assumed viscous and laminar. The complete Navier-Stokes equations are solved using the pseudoviscosity method in a three-dimensional geometry (Fig. 5.20) [42]. Velocity boundary conditions at the solid walls is set to be zeros. The heat flux to the walls is also assumed null. The density on this boundary is found from the continuity equation and the velocity conditions. Normal derivatives are set equal to zero at the symbolic permeable boundary [42]. Just as we did to introduce the analytical treatment we now start with moderate supply pressures in the nozzle, i.e., $P \leq 2$ atm. Figure 5.21 illustrates numerical results which describe oxygen flow through the cut zone at $P = 1.5$ atm and the channel width comparable to the nozzle exit diameter. As one would expect the flow of oxygen through the kerf channel remains subsonic with $M \approx 0.5$ which is close to the estimate made for a free gas jet as follows from (5.35). The lines of the flow are directed along the solid wall. The flow separation usually possible at high pressures is not observed here. It must be noted that the gas flow at the kerf outlet demonstrates

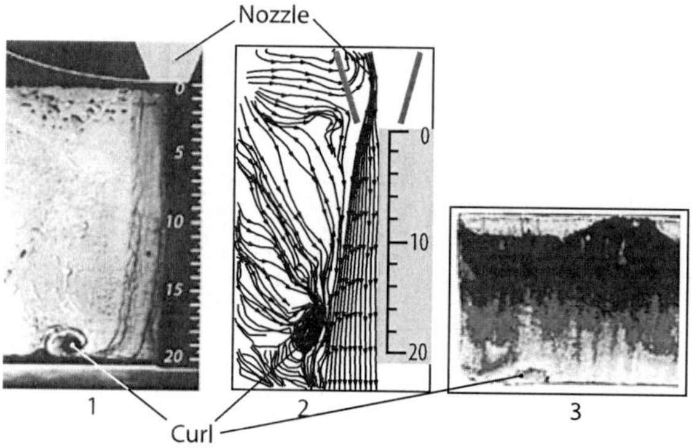

Fig. 5.21 Studying of gas dynamics at laser cutting with assisted gas of low pressure cone nozzle, p = 1.5 atm, plate thickness: 20 mm, Kerf width: 0.7 mm. (**1**) The flow visualization with a liquid coat of work piece walls. (**2**) Numerical results of the streamlines. (**3**) Longitudinal kerf section after laser cutting of steel at low pressure oxygen

Fig. 5.22 Polarization types of a laser beam

Laser beam polarization type	Schematic view
Plane polarization, an electrical vector is parallel to the beam velocity vector (P)	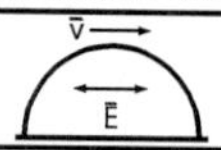
Plane polarization, an electrical vector is perpendicular to the beam velocity vector (S)	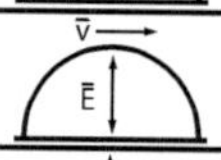
Circular polarization (C)	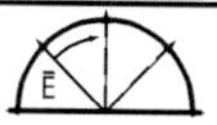
Azimuthally polarization (A)	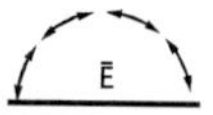
Radial polarization (R)	

a distinctive property. There rises a vortex which in conditions of a real cutting operation would capture the molten material to produce the dross on the bottom edge of the cut. The same figure shows a flow visualization produced by applying a liquid coat to the workpiece and which contains a clearly seen vortex pattern.

Recalling the case of a free gas jet we can conclude that as the pressure is raised inside the conical nozzle the gas flow between the nozzle and the work surface increases to supersonic velocities. As soon as the pressure behind the first Mach disk jumps to overcome 2 atm a supersonic flow of gas is expected to form throughout the cut kerf. For the nozzle pressure of 8 atm the Mach number is rather small, i.e., about 1.2. Consequently, when the gas is passing through a 16 mm workpiece at about 10 mm distance the supersonic flow is decelerated to subsonic speeds (Fig. 5.23). At the same point the pressure rises and the boundary layer becomes separated. The streamlines divert away from the cut front to form a reverse flow as shown in Fig. 5.23. This picture has been confirmed in numerous model and on-site experiments. For example, when cutting of a stainless steel workpiece was assisted with a nitrogen jet the bottom edge of the cut was found blackened (Fig. 5.34). It meant the reverse flows carried the atmospheric oxygen to the workpiece surface.

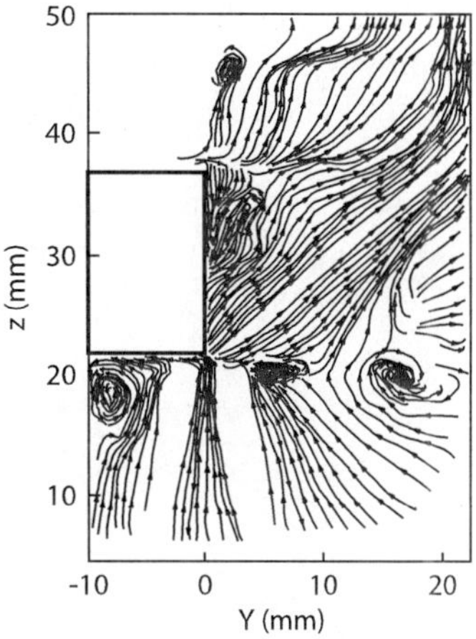

Fig. 5.23 Numerical simulation of gas dynamics during laser cutting. Konical nozzle, neutral assisted gas pressure $p = 7$ atm. Kerf width: 0.5 mm, plate thickness: 16 mm

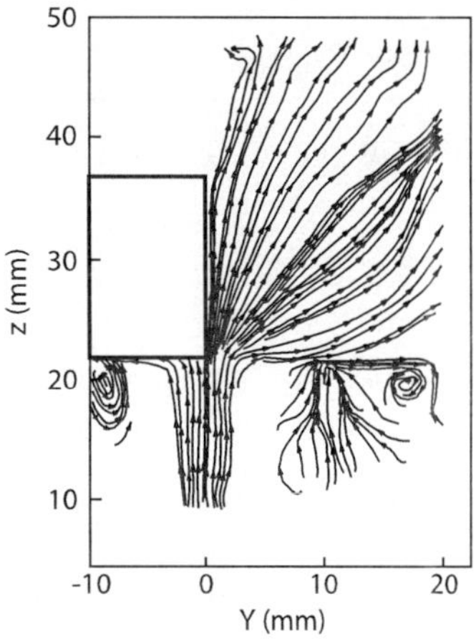

Fig. 5.24 Numerical simulation of supersonic gas dynamics with Laval nozzle. Kerf width: 0.5 mm, assisted gas pressure: 7 atm.; plate thickness: 16 mm

The changes of the flow patterns increase the roughness of the cut walls and spoil the cut quality. These negative effects are very well seen in Fig. 5.34.

Numerical calculations show that raising gas pressure causes the separation zone to move down to the bottom of the workpiece. For instance, in order to push the separation zone entirely out of kerf at workpiece thickness of 30 mm the pressure must be as high as 20 atm.

Since a narrow kerf would decelerate the gas flow and cause its separation a wider kerf is expected to reduce this deceleration effect and diminish the separation zone. Calculations showed that if the kerf width is kept close to the nozzle outlet diameter the separation does not occur at all.

When a supersonic nozzle is used in a similar cutting arrangement one should observe formation of a supersonic flow all the way throughout the kerf depth with the Mach number, M, being approximately within the range of 1.2–2. Figure 5.24

shows that in this case the flow is not separated from the cutting front. The cut quality is high. However, when cutting thicker plates with $h \approx$20–30 mm are cutting the separation is natural and can be fought by increasing the kerf width.

Let us now apply the simplest model to describe effects of the pressure within the kerf on cutting speed. The first step is to write down the conservation law equations considering the fluxes of mass, momentum and energy in a liquid layer with thickness Δl:

$$vh = \Delta l v_{\text{liq}}, \tag{5.36}$$

$$\Delta p = \rho v_{\text{liq}}^2/2, \tag{5.37}$$

$$\eta P = (c\rho T + H_m \rho) v_{\text{liq}} d \Delta l, \tag{5.38}$$

where Δp is the pressure drop along the cutting front. As long as the liquid layer is not too thin its thickness can be evaluated from the heat balance equation as

$$\Delta l \simeq \sqrt{4\chi h / v_{\text{liq}}}. \tag{5.39}$$

Substituting v_{liq} and Δl derived from (5.37) to (5.39) respectively into (5.36) we get

$$h = \frac{\chi}{v^2}\left(\frac{2\Delta p}{\rho}\right)^{1/2}. \tag{5.40}$$

Knowing the cut depth, h, it is possible to find from (5.38) the metal surface temperature:

$$T = \frac{\eta P v}{d c \rho \chi}\left(\frac{\rho}{2\Delta p}\right)^{1/2} - \frac{H_m}{c}. \tag{5.41}$$

As Δp increases the temperature of the molten material drops down but never below the melting point. This limitation is the very condition to find the maximum value of Δp. If the pressure gets higher the cutting speed can no longer be accelerated.

$$\Delta p_{\max} = \frac{1}{2}\frac{\eta^2 P^2 v^2 \rho}{d^2 (c\rho\chi)^2 (T_m + H_m/c)}. \tag{5.42}$$

Using $\Delta p_{\max}$ from (5.42) to substitute Δp in (5.40) we come to a conclusion that the cutting depth at $\Delta p = \Delta p_{\max}$ obtains the value limited by the energy conservation law (5.38). From (5.40) it follows that cutting speed increases smoothly, $v \propto (\Delta p)^{1/4}$, and must show saturation at $\Delta p > \Delta p_{\max}$. These regularities have been also observed in experiments (see Fig. 5.25) [51]. Figure 5.25 shows the discharge rate of the assist gas is increased by the rising pressure which approximately fits to $v \propto (\Delta p)^{1/4}$.

From (5.42) it follows that at a higher power the gas pressure must be raised considerably if one wants a deeper cut. For instance, from (5.42) we see that cutting

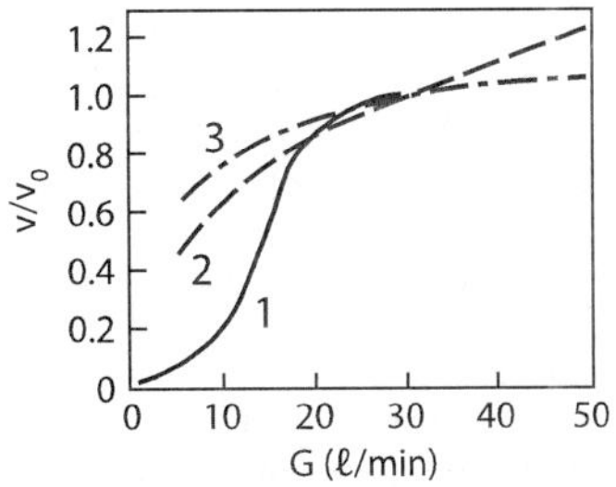

Fig. 5.25 Cutting speed versus assist gas discharge (G). V_0 denotes cutting speed at $G = 30$ l/min.
1 – experimental data;
2 – calculated data (oxygen);
3 – calculated data (inert gas)

steel with $d \sim 1$ mm, $\eta \approx 0.5$, $v \sim 10$ cm/s, and $P \sim 5$ kW would require $\Delta p \sim 0.1$ MPa. The latter is in good agreement with the experiments described earlier.

The need to ensure a large pressure gradient along the entire cutting front while keeping the kerf narrow can turn out a limiting factor for cutting thick metal pieces. In order to conclude the topic of the flow patterns made by assist gases in laser cutting we need to note that their effect goes way beyond changing the cutting speed and depth by dragging the melt out. Laser cutting of steel with an oxygen jet demonstrated that cutting performance depends crucially on the purity of the assist gas [27]. Moreover, it was claimed that the required purity had to be over 99.7%. For example, entrainment of the environmental air by an oxygen flow can under certain conditions seriously reduce the cutting performance. Some fine effects of gas dynamics are discussed later in Sect. 5.3 in connection to laser cutting with oxygen. Now it is advantageous to consider the picture of melt removal in metals cut with a gas jet.

5.2.3 Mechanisms of Melt Removal

First of all we need to note that the mechanisms of melt removal during gas assisted cutting can be grouped into two major classes:

- Even removal of molten material along the entire cut.
- Nonstationary removal of molten material (humps and shelves).

The steady-state mechanism can be maintained either by the difference in static pressure of the gas flow (this option was discussed above) or due to viscous friction forces. Later, we show that the latter mechanism can dominate over the first one because of the pressure gradient within a narrow range of conditions: sufficiently low beam powers (~100 W) and high cutting speeds (~10 cm/s), i.e., small depth of cuts (<1 mm). This is caused primarily by development of a flow pattern in the boundary liquid layer during its interaction with the gas jet [9]. In this case the motion driven by friction forces is maintained only within a thin boundary layer near the interface between liquid metal and assist gas. When deeper cuts are made at low cutting speeds the molten material can be removed due to formation of shelves. When we were describing the properties of a gas jet we took into accounts its

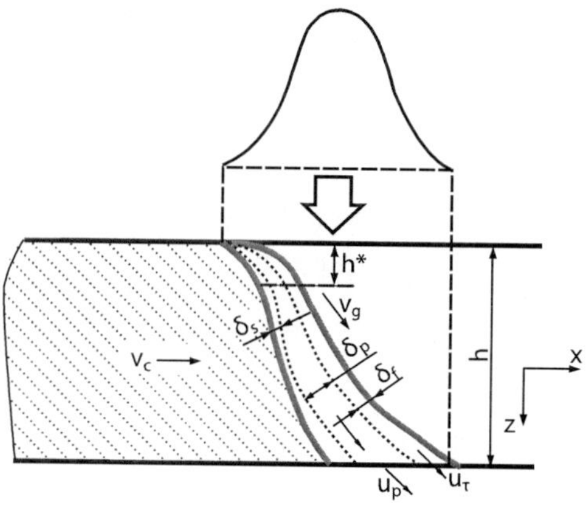

Fig. 5.26 The structure of steady – state removal of molten material: laser cutting velocity V_c, gas velocity V_g, the thickness and liquid velocity in boundary layer in liquid – gas interface δ_f, u_τ; the thickness and liquid velocity driven by pressure gradient δ_p, u_p; the thickness of boundary layer near solid metal surface δ_s; the liquid flow to be multilayered below h^*; the plate thickness h

turbulent character. According to [12] the tangential stress τ_f, i.e., the force per unit area of either solid or liquid surface in this case is

$$\tau_f = 0.03\mathrm{Re}^{-1/5}\rho_g V_g^2, \tag{5.43}$$

where $\mathrm{Re} = zV_g/\nu_k$, V_g is the gas velocity in the flow core, ν_k is the kinematic gas viscosity, and z denotes the vertical coordinate (see Fig. 5.26). The estimates of typical cutting parameters are the following: $V_g \sim 3 \times 10^2$ m/s, $h \sim 0.1$–1.0 cm, $\rho_g \sim 10^{-3}$ g $\cdot$ cm^{-3}, $\nu_k \sim (0.3 - 1) \times 10^{-1}$ cm^2/s, $\mathrm{Re} \approx 10^5$–10^6, $\tau_f = 3 \times 10^{-3}\rho_g V_g^2$. For an isentropic flow the value of $\rho_g V_g^2$ depends on the Mach number, M, according to [3]:

$$\rho_g V_g^2 = p_0 \gamma M^2 \left[1 + \frac{\gamma - 1}{2} M^2\right]^{-\frac{\gamma}{\gamma-1}},$$

where p_0 is the pressure at the entrance to the kerf. In a supersonic flow, $M \sim 2.3$, the tangential stress is estimated as $\tau_f \approx 2 \times 10^{-3} p_0$, whereas for transonic velocities, $M \sim 1$, it is $\tau_f \approx 2.5 \times 10^{-3} p_0$. It is obvious that the force p_0, acting normally on the liquid surface is several orders of magnitude stronger that the tangential stress τ_f. Nevertheless, as it was mentioned above, there are conditions under which this stress becomes the dominant acting force.

Now, we will formulate the conditions needed for the boundary layer flow to occur. First of all, let us consider a flow driven by viscosity forces at the very top of the cut front, i.e., within the region $0 < z < h^*$, where h^* marks its boundary and is defined below (see Fig. 5.26). While the velocity is zero, $V_z = 0$, at the liquid–solid interface there is τ_f force acting at the liquid–gas boundary:

$$\tau_f = \mu \frac{V_z}{\delta}, \tag{5.44}$$

where V_z is the velocity of the liquid. By using the mass balance equation we can write

$$V_z\delta = V_c z, \tag{5.45}$$

and from (5.44) immediately get

$$V_z = (\tau_f V_c z/\mu)^{1/2}, \tag{5.46}$$

$$\delta = (\mu V_c z/\tau_f)^{1/2}. \tag{5.47}$$

We need to take into account that in the layer δ the liquid has two velocity components, V_z and V_c. The latter one is directed against the viscous momentum diffusion which propagates from the liquid–gas interface with the speed of v/δ. As soon as the two rates become equal the viscosity can no longer accelerate the near-bottom layers of the molten material. Thus, the liquid flow from the bottom would push away the liquid layer driven by the viscosity force. This is how the surface boundary layer, δ_f, is formed. There remains a layer of liquid between the bottom and layer δ_f which can be accelerated only by a static pressure gradient. From (5.46) to (5.47) together with the equality condition for cutting speed and viscous momentum diffusion, $V_c = v/\delta$, we get the depth at which such flow is formed:

$$h^* = v\tau_f/\rho V_c^3. \tag{5.48}$$

If the cut thickness $h > h^*$ we can expect the liquid flow to be multilayered below this point. This should limit either the cut depth or cutting speed as follows from

$$V_c h^{1/3} > (v\tau_f/\rho). \tag{5.49}$$

The cutting speed at which such flow is formed can be estimated by substituting τ_f from (5.43) into (5.49). For example, for $h = 1 - 10\,\text{mm}$ and nozzle pressure of about 1 atm $V_c > 1 - 4\,\text{cm/s}$. Earlier we assumed that in the layer where the flow was not separated this flow existed due to the friction force (see (5.44)), i.e., it was assumed stronger than the volume force $\mathrm{d}p/\mathrm{d}z \approx p_0/h$:

$$\tau_f > p_0 \frac{\delta}{h}. \tag{5.50}$$

Substituting δ from (5.47) into (5.50) we derive the condition for the viscous flow in the separation-free layer $z < h^*$:

$$z < z^* = \tau_f^3 h^2/\mu V_c P_0^2. \tag{5.51}$$

From condition $z^* < h^*$ using (5.48) and (5.51) we finally come to

$$V_c h < \frac{P_0}{\tau_f} v. \tag{5.52}$$

Substitution of $\tau_f = 2 \times 10^{-3} p_0$ (see above) to (5.52) gives

$$hV_c < 0.5 \times 10^3 \nu \simeq 3\text{cm}^2/\text{s}, \quad (\nu \simeq 6 \times 10^{-3}\text{cm}^2/\text{s}). \tag{5.53}$$

At faster speeds, following from this inequality, the flow without the separation should be driven by the pressure force. Under these conditions its length is different. The vertical component, V_z, is determined by the pressure gradient and reads

$$V_z(z) = V_0\sqrt{z/h}\,, \quad V_0 = (p_0/\rho)^{1/2}. \tag{5.54}$$

From (5.54) to (5.45) and using condition $\nu/\delta = V_c$ we can conclude that separation of the surface boundary layer occurs when $z > z^*$, where

$$z^* = (\nu V_0)^2/hV_c^4$$

Consequently, condition for the layer $h < z^*$ is different from (5.49) and reads

$$V_c h^{1/2} > (\nu V_0)^{1/2} \simeq (1.5 - 2.5)\text{cm}^{3/2}\text{s}$$

The numerical estimate was obtained for $\nu = 6\times10^{-3}$ cm^2/s, $V_0 = 3\times10^2$–10^3 cm/s.

So if condition (5.52) is met the flow in the upper part of the cut is driven by friction while in the opposite case it is done by gas pressure. As it was shown in [9] even when the flow in the upper part is maintained by friction between the gas and molten metal in the lower section, where the boundary layer is formed, most melt is, after all, moved because of the pressure gradient in the gas jet.

It was shown in [9] that along with the liquid–gas boundary layer near the interface between the solid wall and liquid there is a natural boundary layer (see Fig. 5.26) which, however, has little effect on the overall transfer of molten material. It is convenient to draw the regions with different modes of melt removal on a two-dimensional diagram showing the cut depth versus cutting speed (Fig. 5.27). For the sake of generality this diagram must be presented in dimensionless coordinates: $\bar{h} = h/d$, Pe $= V_c d/\chi$. In this case the condition allowing ejection of melt by friction forces (5.53) is the following:

$$\bar{h}Pe < 0.5 \times 10^3\text{Pr},$$

where Pr $= \nu/\chi$ is the Prandtl number. For metals the number is small, Pr $\sim$0.05–0.1. Below, this figure will also contain the regions showing the other unsteady mechanisms of melt removal. Finishing our description of steady mechanisms we need to note that at sufficiently high pressures the cutting speed becomes independent of pressure and, consequently, invariant to the mechanism of the material removal. Nevertheless, studying this mechanisms is crucial for understanding what causes roughness of the cut edges and dross clinging to the bottom part of the cut. Purity of the cut edges is of great importance in practical use of laser cutting machines in industry. A great number of works is dedicated to this

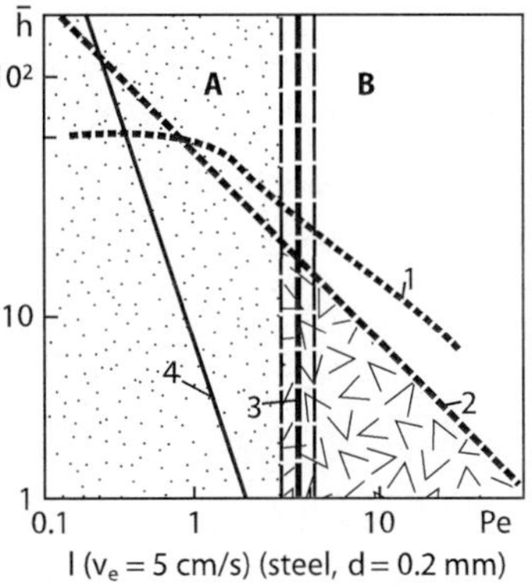

Fig. 5.27 The two-dimensional diagram of character of melt removal. $Pe = V_c b/\chi$, d is cut width; $d = 0.2\,\mathrm{mm}$, $V_c = 5\,\mathrm{cm/s}$. The line 1 is boundary of full cutting at $P = 4\,\mathrm{kW}$; line 2 is boundary of regime of melt removal by turbulent gas flow (left from line 2); line 3 is boundary between regimes with shelves (**a**) and humps (**b**); line 4 is boundary between regimes with surface flow (*right*) and the regimes with homogeneous melt oscillations on cut front

issue. The edge roughness is linked to numerous instabilities which develop while a gas jet interacts with liquid molten material. To some extent these processes in laser cutting are similar to interactions between the vapor jets and the melt in laser welding described in the previous chapter.

5.2.4 Instabilities and Nonstationary Mechanisms of Melt Removal

In this part of the section we describe some instabilities that are most likely to contribute considerably to our understanding of mechanisms that produce roughness. One of the classic approaches to deal with the problem was the hypothesis that some induced periodic variations of melt thickness became possible even in a steady-state layer accelerated by a density gradient if some damping oscillations were resonant to the turbulent pulsations in the gas jet. These oscillations according to [14] may be the very process responsible for the roughness of the cut edge. Moreover, some type of instability can as well be observed for the boundary layer between the melt and the gas jet. However, as suggested by the authors of this model [9] its role in the melt transfer is quite small. Under the condition there is a gas flow above the liquid surface it is natural to surmise a possibility for the so-called "wind" or Kelvin–Helmholtz instability [3]. This type of instability has a simple interpretation: the gas velocity is greater above a liquid hump, then, as it follows from the Bernoulli equation, the gas pressure above the hump is weaker and, therefore, the hump continues to grow. In the elementary case of infinite media and the velocity jump at the gas-liquid interface the increment for this instability [3] is written simply as:

$$\gamma \cong kV_g\sqrt{\frac{\rho_g}{\rho_l}}, \tag{5.55}$$

where $k = 2\pi/\lambda$ and λ is the perturbation wavelength. For smaller wavelengths the increment is limited by the surface tensions. Besides, the finiteness of melt thickness δ is expected to contribute to some extent as well. Finally, from [9, 13] it is known to be:

$$\gamma = \sqrt{k\delta}\left(kV_g^2\frac{\rho_g}{\rho_l} - k^3\frac{\sigma}{\rho_l}\right)^{1/2}. \tag{5.56}$$

The increment reaches its maximum at $k = k_{\max} = (2/3)\rho_g V_g^2/\sigma$ and becomes

$$\gamma = \gamma_{\max} = k_{\max}V_g\sqrt{\frac{\rho_g}{\rho_l}}\sqrt{k_{\max}\delta/3}.$$

The conditions for gas laser cutting ($\rho_g V_g^2 \sim 5\times10^6$ dn/cm^2, $\sigma \sim 2\times10^3$ dn/cm, $\delta \simeq 10\,\mu$m) if estimated numerically provide the characteristic perturbation wavelength of approximately 40 μm and the increment $\gamma_{\max}$ of the order of 10^6 s^{-1}. In general, for most types of instabilities, their influence is characterized by their nonlinear stages. In terms of the qualitative analysis of this stage the hump growth time before the droplet detachment is only enough for the workpiece to travel a small distance of

$$\Delta l \simeq V_c/\gamma_{\max} \simeq 10^{-6}\text{cm}.$$

Consequently, the wind instability seems unlikely to form notable grooves on the cut edge though its contribution into the melt removal can be quite considerable [9]. Some other models are known in the literature to be similar in formulation to the wind instability. For instance in [15] they study the viscous flow without the boundary layer described above. The melt flow was ensured by the static pressure difference and the jet friction. The surface tension was taken into account as well. Such conditions were shown to provide a steady-state flow under dominance of the viscous forces. On the other hand when the pressure effects prevail the flow is unsteady. However, similarly to the case of the wind instability, the maximum increment corresponds to the wavelengths being too small, i.e. of about 20 μm, which is approximately an order of magnitude shorter than the scale of striations in the experiments [15].

In addition to the instabilities that develop on the melt front discussed earlier some other alternative mechanisms of melt removal are actively discussed. They are inhomogeneous and of unsteady character. In the early works it was noted that the melt seemed to be "scrubbed" from the cut front. Usually, the melt humps were seen to move downward periodically. They were brighter and observed very clearly in special experiments (see, for instance, Fig. 5.28, the review [27]). This motion is often assumed to form the cut edge roughness. Later in this section we show from a qualitative picture that this situation is more complicated and requires further thorough analysis using numerical techniques. Such studies have

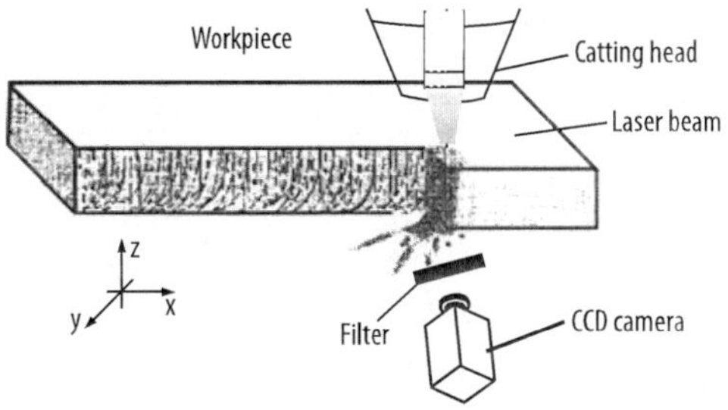

Fig. 5.28 The scheme of the experiment for observation of bright spot moving along the cut front from top to down

not been performed so far. Some researchers do not go beyond the qualitative and fragmentary models to be validated by comparison with the experimental data (see [9]). Taking into account these notes we just give one example of such models. We consider a possible origin of periodic shelves described above.

The experiments show that they originate at the top edge of the cut where the film flow starts. We will show that there is also a limit for the flow onset. The situation is similar to condition for the melt leakage in remote cutting (see Sect. 5.1). The liquid starts leaking as soon as the gravity pressure is no longer compensated by the surface tension. In our case, the action of the gravity pressure is represented by the pressure difference in the gas jet along the bead size r_0. This pressure difference we evaluate from the variations of the flow cross-section $\delta S/S \simeq 2r_0/d$. In this case the variation of the pressure at the cut inlet we define using (5.34) as

$$\delta p = -p\gamma \frac{M^2}{1-M^2}\frac{2r_0}{d}. \tag{5.57}$$

When the surface tension becomes smaller than δp, the bead starts "traveling" down the cut front.

$$\frac{\sigma}{r_0} < \delta p. \tag{5.58}$$

Substituting δp from (5.57) we get the maximum size of the bead:

$$r_{cr} = \left[\frac{\sigma d}{p\gamma}\frac{1-M^2}{M^2}\right]^{1/2}. \tag{5.59}$$

Substituting the typical cutting parameters ($M \approx 0.5$; $\gamma = 1.4$; $p \approx 2$ atm) we get $r_{cr} \approx 60\,\mu$m, this value is significantly less than the cut width $d \approx 200 - 300\,\mu$m, which is exactly as assumed in the model. The time of the bead drain along the cut front is estimated as $h/\sqrt{2p/\rho_l} \sim 0.3\,\mu$s. It is substantially smaller than the workpiece shift time in the drop scale: $r_{cr}/V_C \sim 3$ ms ($V_C \sim 2$ cm/s). Consequently, the time prior the bead formation determines the periodicity of the process and, as the result, the width of roughness bands:

$$\lambda = \frac{hV_C}{\sqrt{2p/\rho_l}} + r_{cr}. \tag{5.60}$$

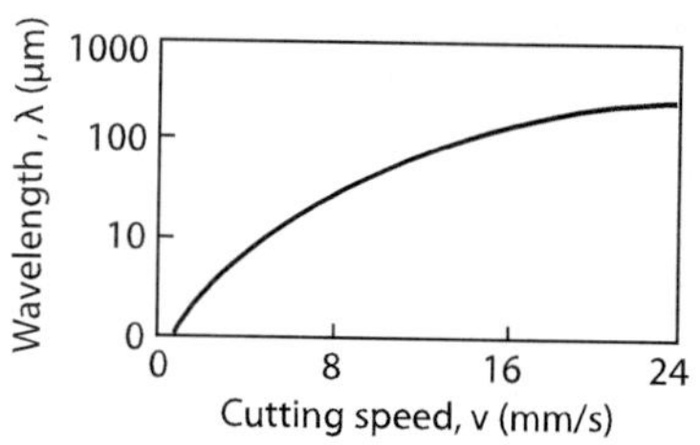

Fig. 5.29 Wavelength of the periodic striations of the cut edges vs. cutting speed (mild steel, 5 mm thickness, beam power 1,200 W, oxygen cutting)

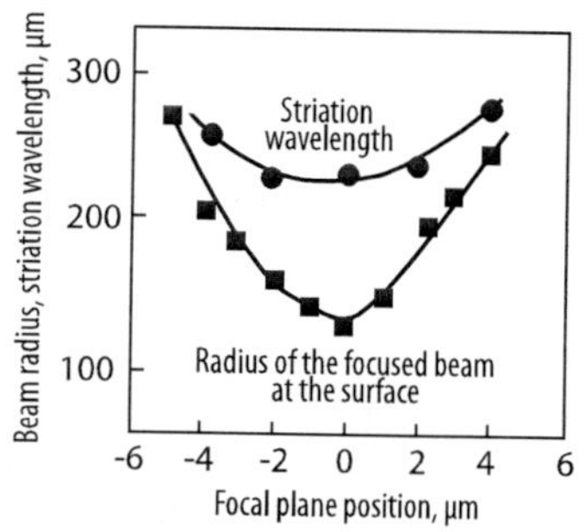

Fig. 5.30 Correlation between the wavelength of the striations and the beam radius at the top surface in dependence of the position of the focal plane (6 mm mild steel, beam power 1.5 kW, oxygen gas pressure 0.3 MPa, cutting speed 2 m/min)

From (5.60) it is well seen that the roughness (striations) wavelengths increase with the cutting speed (see Fig. 5.29 [17]) and the workpiece thickness. The latter was observed in experiment [28]. As it follows from Fig. 5.30 [28] the wavelength is not straightforwardly dependent on the size of the focal spot.

So the unsteady removal of the molten material can be realized through either sputtering a liquid film or "scrubbing" it by a hump sliding down the cut front. The model experiments show that these two mechanisms can work simultaneously.

5.2.5 *Modelling of Melting Front and Melt Removal in Gas-Assisted Cutting of Metals*

In experiments on cutting steel sheets, it was noticed that when the sheets were thicker than 5 mm ejection of the melt was accompanied by light emission. This bright glowing was due to the melt droplets entrained by the gas flowing from the kerf [43]. In case the condensed phase carries small particles the cut quality is satisfactory. If the particles are large the cut quality is poor. In other words, the cut quality depends strongly on how the melt is sputtered. For example, in cutting a stainless steel plate of 2 mm thickness with a CO_2 laser system providing the power of 1.5 kW, scanning speed of 5 cm/s and air pressure of 8 atm, the size of the particles would be around 80 μm [43]. This size is substantially smaller than the kerf width. Such droplets can easily travel with the gas jet through the kerf zone.

Model experiments have been implemented to make sure the melt can be removed in such a way. In this model metal is substituted with paraffin which is

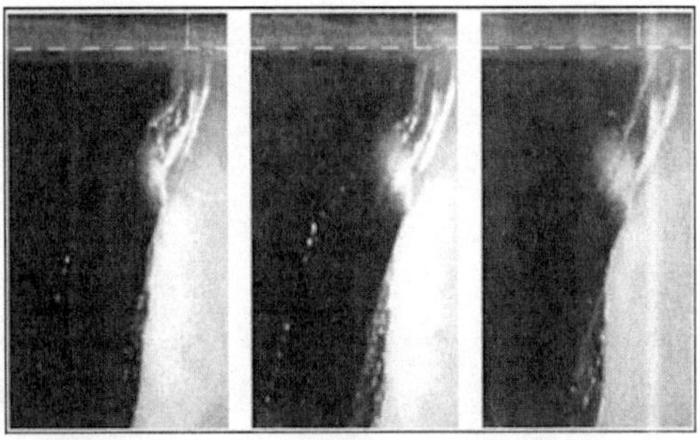

Fig. 5.31 The model experiment to study the melt removal at laser cutting. Metal is substituted with paraffin between two glass plates. The nozzle pressure is about 5 atm. The time interval between frames is 0.125 s. The *dot and dash line* is axe of nozzle, and *dashed line* is the boundary of glass plate

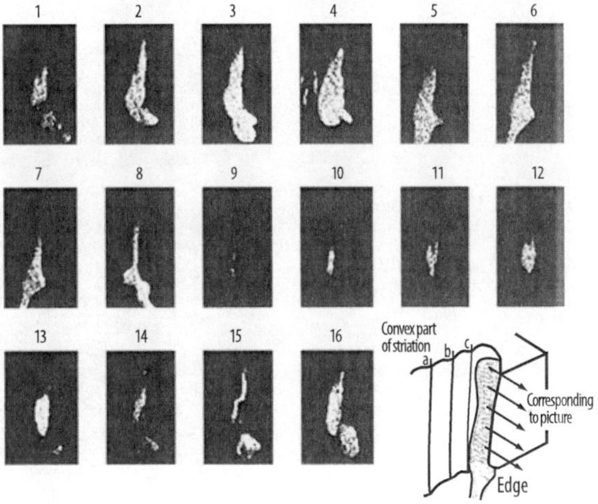

Fig. 5.32 High speed movie picture of edge cutting at 80 cm/min (thickness: 1.2 mm. laser power: 200 W). Time interval of each picture is 1 ms

placed between two glass plates. This substitution allows observation of melting process and removal of paraffin with a jet of hot air. If the nozzle pressure is moderate, about 3 atm, excitation of surface waves is well observed along with the film flow. At the same time the film is not sputtered. If the air pressure is increased up to 5 atm there rises a shelf in the melt front for the melt droplets to slide down from as shown in Fig. 5.31. Further increase of pressure causes the shelf to grow and shift down to the bottom allowing the film to be dispersed in a cloud of tiny droplets. However, the model experiments reviewed here do not meet the strict modelling requirements to describe melt removal in laser cutting of metals. For better results it is necessary to find model materials with properties similar to metallic.

Generally, the generation of striations on the cut edge as well as the frozen drops of molten material beneath the bottom edge known as burrs dross, broadening of the kerf width near the bottom comprise a collection of complicated unsteady and inhomogeneous processes. In particular this is well seen from the footage of the front glowing performed from different angles (see Figs. 5.32 and 5.33 [29]) and from the photographs of the cut walls (see Fig. 5.34 [42]) and the bottom part (Fig. 5.35 [43]). It is therefore not surprising that different model are suggested to explain this phenomenon (see review [9]). However, they are rather

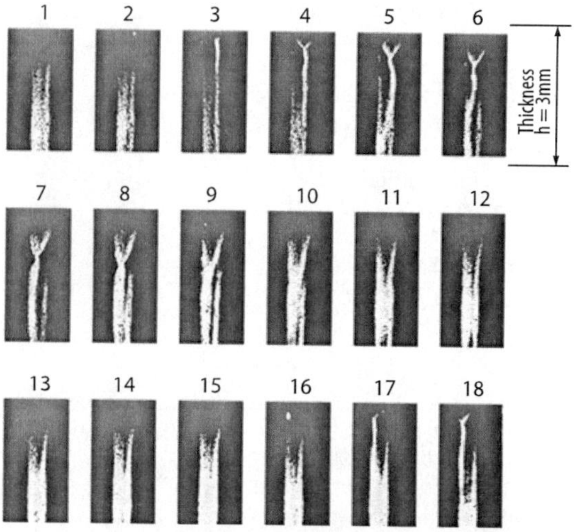

Fig. 5.33 High speed movie pictures of normal cutting at 100 cm/min (thickness: 3 mm. laser power: 1 KW). Time interval of each picture is 1 ms

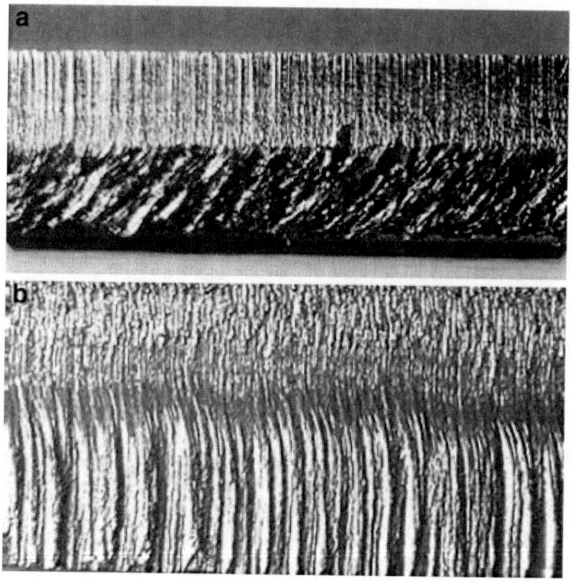

Fig. 5.34 The photograph of cut walls (**upper**) St. steel plate, thickness is 7 mm, (**down**) Ti plate, thickness is 20 mm

qualitative, approximate and have never been brought to quantitative comparison with experiment.

The amplitudes of these striations are of special attention and pose a more complicated research topic. It determines the roughness and, correspondingly, its quality. In this aspect, one has to apply to experiment. We now briefly outline the mechanisms waiting to be described theoretically.

Roughly, the quality of the gas laser cutting may be described by three parameters: cut broadening near the bottom F, the height of the solid drops formed at

Fig. 5.35 The photograph of plate bottom after laser cutting (steel plate thickness $h = 0.5\,\text{mm}$, $P = 800\,\text{W}$, $V_c = 9\,\text{m/s}$, $p = 6\,\text{atm}$, nitrogen)

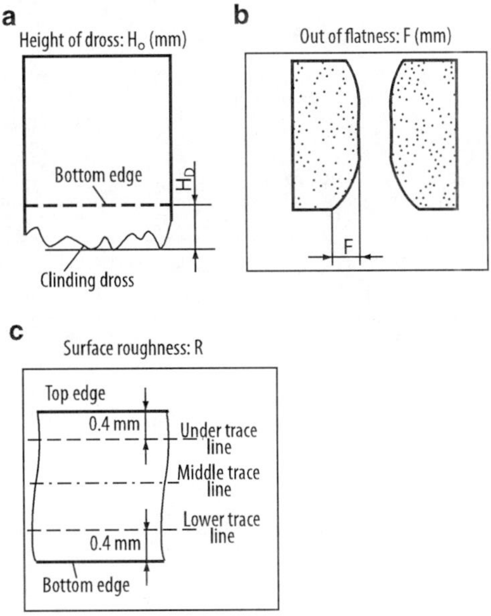

Fig. 5.36 Evaluation factors of cut quality and those method

the bottom edge (burrs dross) H_0 and the height of the inhomogeneities on the cut edge R. The latter usually grows along the cut depth and thus is measured at three points of the cut, i.e., at the upper, the middle, and lower near the bottom (see Fig. 5.36 [30]). All these magnitudes depend on the cutting conditions such as the laser power, nozzle pressure, the focal length and the cutting speed (Fig. 5.37). All three magnitudes have the minimum as functions of the cutting speed (Fig. 5.38). The fact that the defect dimensions become smaller for faster cutting speeds seems quite natural since for a rapidly moving workpiece the perturbation in the melt has shorter time to effect the cut edge. Further increasing of the speed at constant laser power and workpiece thickness the cut depth approaches its limit of incomplete cutting. In such a case the cut front near the bottom edge is notably deviated from the vertical axis, leaves the beam and is maintained by the hot melt flowing from the upper layers. It is well observed from the high-rate footage (Fig. 5.39) and follows from temperature measurements along the cut front (Fig. 5.40 [29]). To finalize the

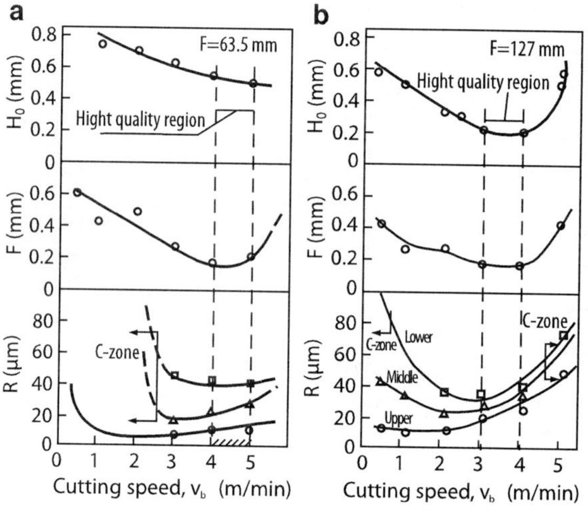

Fig. 5.37 Effect of focal length and speed on cut qualities, (thickness: 2 mm, oxygen pressure: 1.5 kg/cm2, laser power: 1 kW) (**a**) $F = 63.5$ mm (**b**) $F = 127$ mm

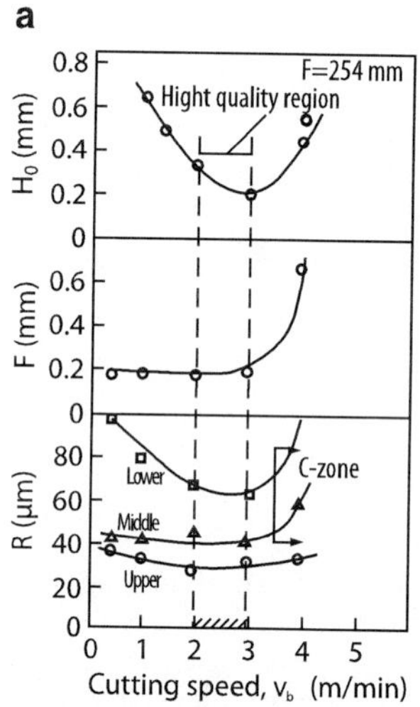

Fig. 5.38 Effect of focal length and speed on cut qualities, (thickness: 2 mm, oxygen pressure: 1.5 kg/cm^2, laser power: 1 kW. $F = 254$ mm)

section on melt removal it is worth noting out the importance of the gas jet influence on the cut quality. The authors of [30] suggest the cut quality can be improved by introducing a second jet (tandem nozzle cutting) directed into the cut from the angle to the vertical axis of abut $\pi/4$ (Fig. 5.41). Figure 5.42 shows that the additional jet decreases the striation size and ensures the vertical edges. It is evident that the cut quality is a strong function of the gas jet structure and is not conditioned by extra gas introduction. Indeed, as shown in Figs. 5.37, 5.38 the gas consumption beyond a certain limit in a conventional cutting setup has no effect on the cut quality. Having reviewed the mechanisms of the melt removal we will later consider the general picture allowing for the heat processes responsible for the initial stage.

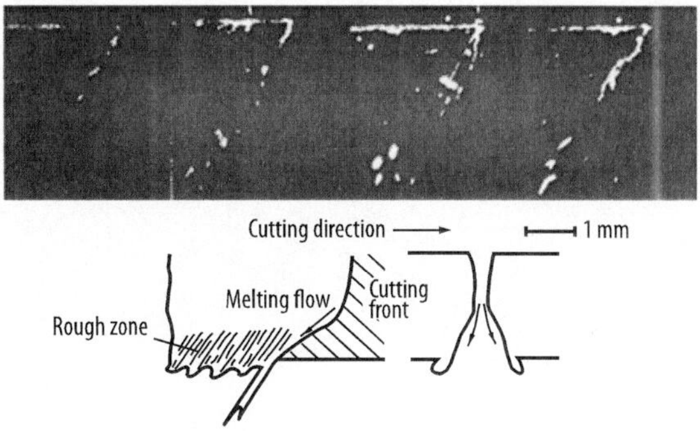

Fig. 5.39 High speed movie pictures of cutting accompanied with rough kerf zone. (cutting speed: 5 m/min)

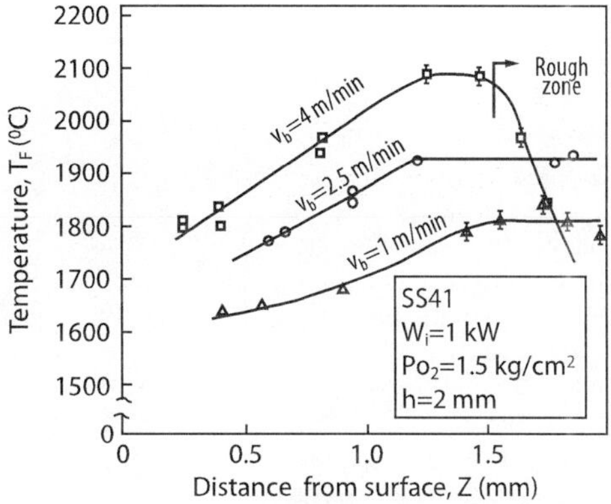

Fig. 5.40 Relationship between temperature of cutting front and distance from top surface for various cutting speeds

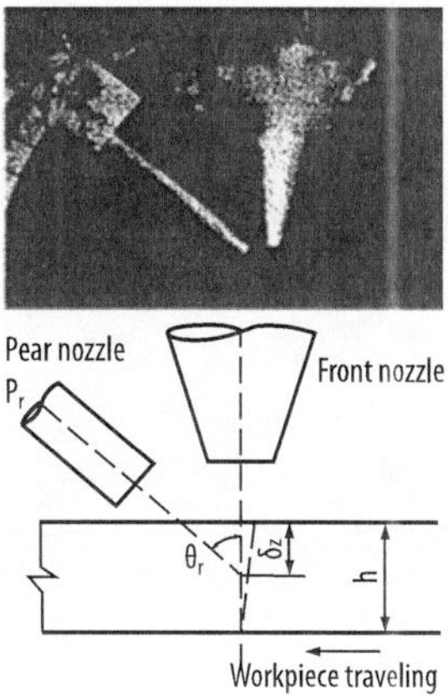

Fig. 5.41 Schematic diagram of tandem nozzle cutting method

5.2.6 Properties and Efficiency of Gas-Assisted Cutting

Basic properties and regularities of laser cutting are qualitatively described by the Swift–Hook and Gick theory. If cutting performance is high, i.e., when heat conduction losses are considerably small compared to what is spent on heating and

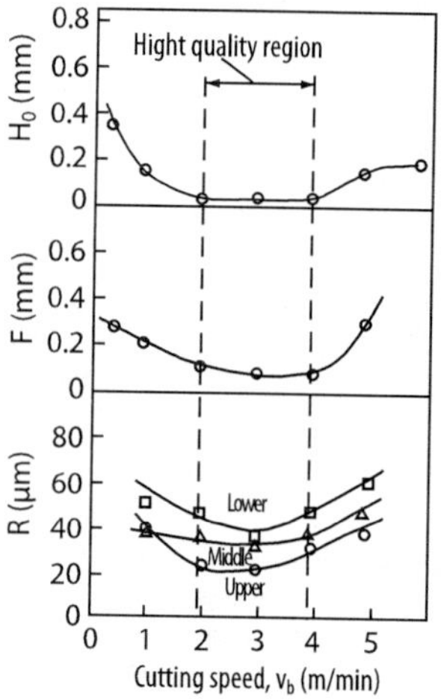

Fig. 5.42 Evaluation of tandem nozzle cut qualities, (thickness: 2 mm)

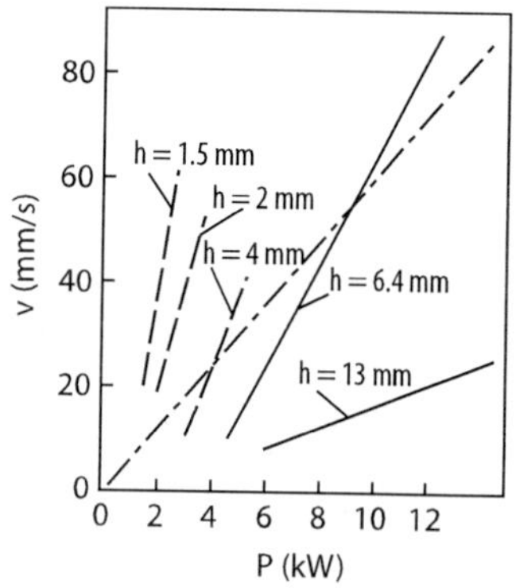

Fig. 5.43 The maximal cutting velocity vs CO_2-laser power at different thickness of workpieces. Solid curve is experiment [40], dot and dash curve is calculation on 5.61 for steel (St304) in noble gas at $\eta = 0.3$, $d = 0.5$ mm; dashed curves are experiment [41], stainless steel in oxygen

melting the material, this theory gives a simple equation for energy balance:

$$hVd(c\rho T_{\text{melt}} + H_{\text{melt}}\rho) = \eta P, \tag{5.61}$$

where P is the summed power of the absorbed beam and the exothermic oxidation reaction, h and d denote depth and width of the cut, V is the rate at which the workpiece is moved relative to the beam, $\eta = \alpha_e \eta_t$ is cutting efficiency where η_t and α_e are the thermal efficiency factor and effective light absorption coefficient respectively. Let us remind that in laser welding $\alpha_e \approx 1$ while $\eta_t = 0.48$ is the maximum value provided by theory (see Sect. 4.2). If the kerf width is assumed to be the same as the beam diameter then from (5.61) it follows that $V \sim P$ at constant h and $h \sim V^{-1}$ at constant P. These relations are in good qualitative agreement with the experimental results shown in Figs. 5.43 and 5.44. These figures also show theoretical curves drawn using (5.61) for different values of η. The theoretical curve associated with cutting steel 304 (*dash-dot* line) was drawn for $h = 6.4$ mm, $d \sim 0.5$ mm, and $\eta = 0.3$.

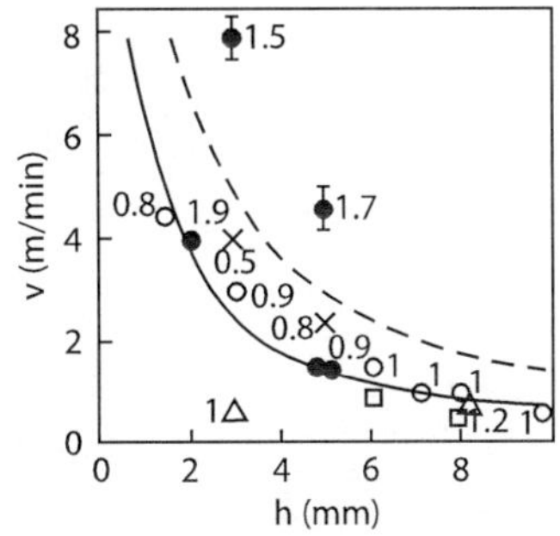

Fig. 5.44 The maximal cutting velocity vs work piece thickness h, CO_2-laser power $P = 5\,\mathrm{kW}$ [41], numbers near points are kerf width d (mm), calculation result (5.61) at $\eta = 0.5$, $d = 0.5\,\mathrm{mm}$ (*dashed curve*), and $d = 1\,\mathrm{mm}$ (*solid curve*). Points are experimental results [41]. Stainless steel in oxygen and argon (*black circles with stick and crises*), tungsten in argon (*triangles*), middle steel in air (*open circles*), stainless steel in air (*black circles*) Cr steel in air (*squares*)

Comparison of this curve with the experimental data demonstrates that efficiency, η, should grow with increasing cutting speed. In this respect, laser cutting of metals is identical to the welding operation. On the contrary, unlike welding, the cutting efficiency is highly dependent on η_t and α_e. The value of η_t in a cutting process must be greater than a similar parameter in welding. This is because in cutting there is no necessity to heat metal up to the boiling point and, consequently, the heat conduction losses should be smaller. Effective beam absorption, α_e, can be weaker in cutting as compared to welding. A keyhole analogy observed in cutting has no bottom and no back wall so the beam from a CO_2 laser is not absorbed as effectively as in a weld pool (see Sect. 1.2).

Theoretical curves in Fig. 5.44 were plotted using (5.61) with $\eta \approx 0.5$ for cut widths $d = 1\,\mathrm{mm}$ and $d = 0.5\,\mathrm{mm}$. These curves fit qualitatively to empirical results. However, the simplest form of (5.61) has no account for the effects of beam focusing conditions on the workpiece surface (see Figs. 5.45 [31]). Besides, the situation when cut depths becomes smaller with bigger spots cannot be explained by a subsequent increase of the kerf width (compare Figs. 5.45 and 5.46 [31]).

Importance of energy use efficiency is determined by both the economic reasons and cut quality. If the thermal efficiency factor, η_t, is small heating of the workpiece material becomes substantial. This causes irreversible deformations and spoils the cut shape. It is worth noting right away that according to experiments the cutting process is generally a complex three-dimensional phenomenon. Thus, the cutting efficiency can be evaluated using approximate calculation methods only. For example, complexity of a cutting operation follows from Fig. 5.46 where the kerf width, d, as a parameter in the energy balance equation, is shown to change appreciably with cutting speed. Moreover, it is clear that d fluctuations are of different character at each point of kerf. As the cutting speed is approaching its critical value the width at the bottom part of the kerf is increasing (see Fig. 5.36) which is rather surprising from the view of energy balance. This situation is quite

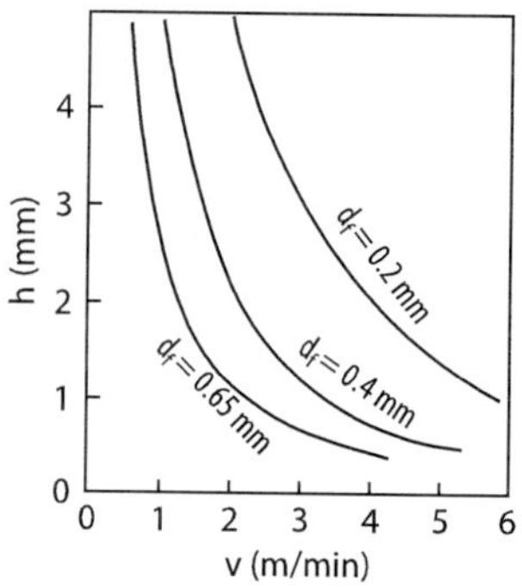

Fig. 5.45 Variations of cutting deepness with cutting speed at different spot diameters (middle steel)

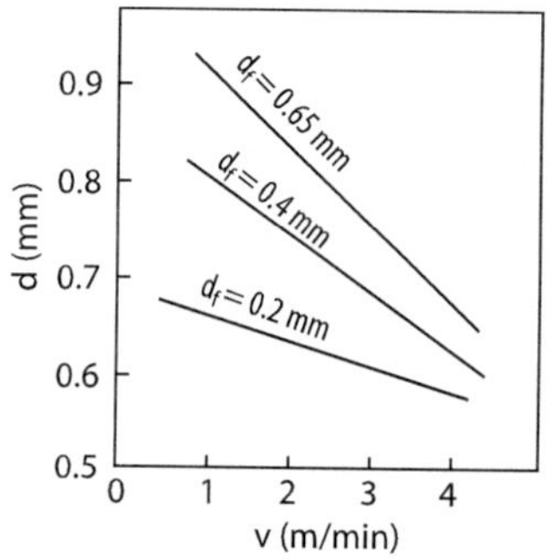

Fig. 5.46 Variations of kerf width with cutting speed at different spot diameters

similar to remote cutting where the kerf tends to be wider in thicker workpieces whiles the power and scanning speed are kept constant. The latter was explained by the fact that the actual incident power was greater for thicker plates (see Sect. 5.1). On the other hand, Fig. 5.42 shows that a higher cutting speed can widen the kerf by a factor of 1.5 at most. This is about how much the kerf width is greater than the beam spot diameter. These are the factors that determine precision in calculated cutting efficiency.

For the sake of generality we will define the total power of heat delivered which comprises the power of the beam and the exothermic oxidation process as $P_{\text{total}} = \alpha_e P$. According to [16] the use of oxygen as an assist gas gives $\alpha_e \approx 2$. In [32] the thermal efficiency factor was introduced in the following form

$$\eta_T^{-1} = 1 + (\chi\tau_m/d^2)Pe^2 + \left(\frac{1}{2Pe} + \frac{1}{Pe^2}\right)^{1/2}, \tag{5.62}$$

where τ_m is the time spent to remove the melt and Pe $= dV_c/\chi$. This equation defines the maximum efficiency, i.e., it uses the value V_c for the maximum allowed cutting speed at which the cutting process is kept continuous while no beam power is lost through the kerf. As long as the molten material is removed by either the static pressure gradient or the viscous friction of the gas the time is defined as $\tau_m = h/V_m$. If the melt is removed because of instabilities, h has to be substituted with the instability wavelength: $\tau_m = \lambda/V_m$. The value of η_T is a nonmonotonic function of the Peclet number (Pe). For $\mathrm{Pe} = (0.45d^2/\chi\tau_m)^{0.37}$ we get

$$\eta_T = \eta_{\mathrm{Tmax}} \simeq \left[1 + 1.6(\chi\tau_m/d^2)^{0.25}\right]^{-1}.$$

For example, at $d \simeq 2 \times 10^{-2}\,\mathrm{cm}$, $\chi = 0.1\,\mathrm{cm}^2/\mathrm{c}$, $\tau_m \simeq 10^{-3}\,\mathrm{s}$, $\mathrm{Pe}_m = 0.85$, $\eta_{\mathrm{Tmax}} \simeq 0.4$. If the work is done by instabilities $\tau_m = 10^{-5}\,\mathrm{s}$, $\mathrm{Pe}_m \approx 4.5$, and $\eta_{\mathrm{Tmax}} \approx 0.7$. From these estimates it is clear that a characteristic efficiency under such conditions is independent of the beam power.

Earlier, when we were discussing the mechanisms of melt removal, it was mentioned that if the Peclet number $\mathrm{Pe} \leq 3$ the material was pushed out by migration of shelves. The task to describe this type of cutting process is yet to be solved so we will restrict ourselves to finding an approximate estimate. For example, in [9] it is suggested to use the following interpolation of η_T:

$$\eta_T = \left\{(1 + 1.2\mathrm{Pe}^{-0.75})\left[1 + 0.3(q/q_1)^{0.6}\right]\right\}^{-1}. \tag{5.63}$$

where q is the beam power density. q_1 is the power density carried away by material vapors and defined as $q_1 = (L_b/\kappa T_b p c_s)$ with c_s denoting the speed of sound through the vapor and L_b being vaporization enthalpy. It is apparent from (5.63) that η_T is a function of beam power. Figure 5.47 [9] demonstrates how η_T evaluated using (5.63) varies with cutting speed. For that occasion the experimental data were recalculated using $d = 0.2\,\mathrm{mm}$ and $\alpha_e = 2$. The latter parameter describes the heat contributed by the oxidation reaction. This value of α_e has been reported in the literature [16].

5.2.7 *Beam Polarization*

Another effect that contributes to cutting performance is the beam polarization. As it was mentioned earlier, in a cutting process the character of beam propagation is different from what happens in welding because there is no back wall in the

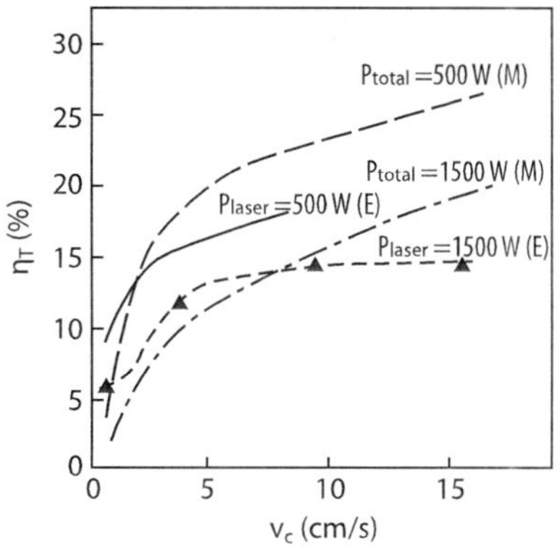

Fig. 5.47 The thermal efficiency of laser cutting at two CO_2-laser powers. Curves (M) describe model (5.62), curves (E) describe experimental data

beam-generated keyhole. In the keyhole welding mode the beam penetrates down the channel undergoing multiple reflections until it is completely absorbed. In order to provide a good cutting performance it is necessary that most of the beam power is absorbed at the cutting front where the workpiece material is subject to heating, melting and removal. At a constant beam power the cutting depth has to increase as the kerf width is narrowed. However, if the beam was passing through a very narrow kerf it would be effectively absorbed on the cut walls which would, in turn, considerably limit the cut depth. It means there is some optimum value for the kerf width which corresponds to maximum depth of the laser cut. The optimum width depends on beam polarization as it determines radiation absorption inside the cavity (see Sect. 1.2).

We will study effects of beam polarization on cutting speed in a wide cutting region where the cut walls are well separated and do not affect propagation of light. We will use the arrangement illustrated in Fig. 5.48 and take into account that the liquid material is to be removed from the workpiece. In this case the conservation condition for mass flow gives

$$Vh = V_N 2r_f / \cos\varphi, \quad \tan\varphi = h/2r_f,$$

where V denotes cutting speed and V_N is the rate at which the liquid level is lowered relative to the surface normal as determined by the actual removal mechanism. For the sake of simplicity we can assume that the melt is removed by vaporization. Then the lowering rate V_N can be found from (1.31) in which q is multiplied by $\cos\varphi$ and function $\alpha(\varphi)$ is introduced [see (1.19 and (1.20)]. In case $V \gg V_V = q\alpha/L_b\rho$ for both polarizations $h/2r_f = V_V/V$. For slow cutting, when $V \ll V_V$, the normal polarization $\alpha_\perp = (\alpha\cos\varphi)$ would give a different dependence of the cut depth $h_\perp$ on cutting speed:

$$\frac{h_\perp}{2r_f} = \sqrt{\frac{V_V}{V}} \gg 1. \tag{5.64}$$

For parallel polarization at the incidence angle close to $\pi/2$ from (1.21) we can get $\alpha_\| = \alpha\cos\varphi/|\zeta|$. In this case the cut depth is found in a similar manner

$$\frac{h_\perp}{2r_f} = \sqrt{\frac{V_V}{V}\frac{1}{|\zeta|}} \gg 1. \tag{5.65}$$

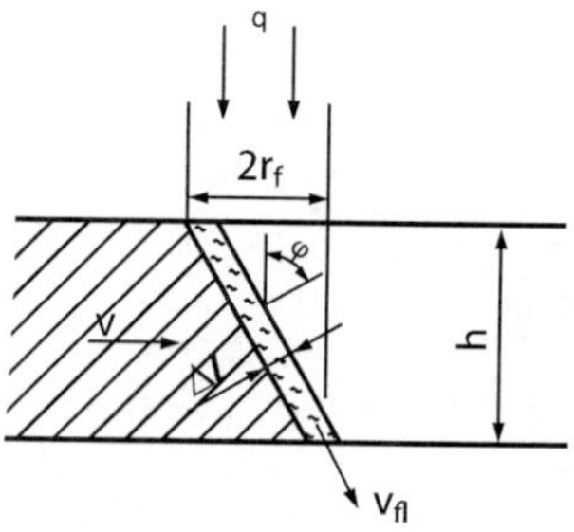

Fig. 5.48 The scheme of homogeneous laser cutting

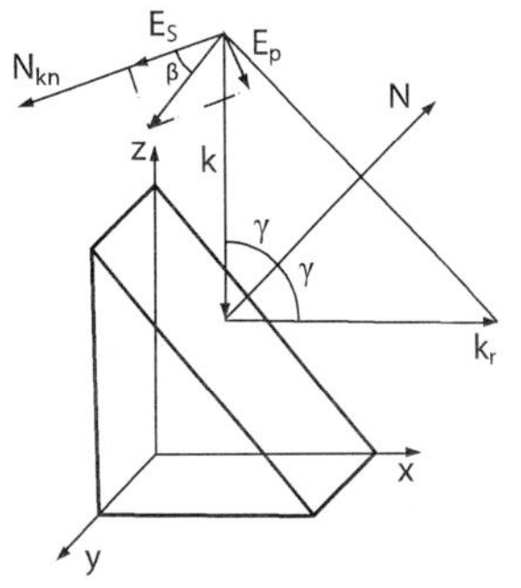

Fig. 5.49 The scheme of interaction of laser beam with cut surface

where ζ is defined in Sect. 1.2. Since $|\zeta| \ll 1$ these inequalities show that for the same beam power parallel polarization would work more efficiently during cutting than normal polarization. This advantageous effect of polarization has been confirmed in experiments [33]. The same result was obtained in [18] by applying a rigorous approach to study the problem.

In the model described above as well as in [18] the problem was treated as one-dimensional, i.e., the distance between the side walls of the slot was assumed sufficient to allow the beam to interact with the cut front only. However, as we could see earlier, in practical situations the kerf width is usually comparable to the spot diameter which makes it necessary to take into account the effects caused by the cut walls. In this case the problem becomes two-dimensional and the cut surface must be treated as a function of x and y coordinates [9] (see Fig. 5.51). For illustrative purposes the table lists the various types of beam polarization used during laser cutting (Fig. 5.22). Just as in [18] it is assumed that melt thickness is negligibly small while absorption is independent of the surface temperature. Besides, only a single action of the laser beam is taken into account. It is also assumed that locally there is a balance of absorbed energy and removed material. The gas jet is expected to blow the melt out instantly. Under such simplified conditions coordinate z of the beam affected zone obeys the equation:

$$\frac{\partial z}{\partial t} - V_c \frac{\partial z}{\partial x} = -V_N \left[1 + \left(\frac{\partial z}{\partial x} \right)^2 + \left(\frac{\partial z}{\partial y} \right)^2 \right]^{1/2}, \tag{5.66}$$

where V_c is the cutting speed along the x-axis and V_N is the melt lowering rate along the surface normal. The term in the square brackets describes the relationship between shifting along the z-axis and the normal line. Using the local condition for the energy and mass balance we can write the equation for V_N as

$$V_N = \alpha q(x, y, z) \cos \gamma \left[\rho_m H_m + c_s^0 \rho_s^0 (T_m - T_0) \right]^{-1}, \tag{5.67}$$

where ρ_m and H_m are, respectively, the material density and the latent heat of fusion at melting temperature T_m. Notations c_s^0 and ρ_s^0 refer to specific heat capacity and material density at room temperature T_0. Finally, γ is the angle at which the beam is incident to the work zone and $q(x, y, z)$ is the beam intensity found from

$$q(x, y, z) = \frac{2P}{\pi r_b^2} \exp\left(-2r^2/r_b^2\right),$$

$$r_b = \left\{ r_f^2 + \left[\frac{(z - z_f)\lambda}{\pi r_f} \right]^2 \right\}^{1/2},$$

where r_b and r_f are the current beam radius and the spot radius respectively, z_f measures the distance from the plane surface of the workpiece, $z = 0$, to the focal plane, and P is the beam power. Let the laser beam be linearly polarized. In this case projection E_p to the plane of incidence is corresponded with absorption coefficient $\alpha_{\parallel}$ while projection E_s orthogonal to the plane is affected by coefficient $\alpha_{\perp}$. Let the angle β be between the vector E and the normal to the plane of incidence N, then $E_p = E \sin\beta$ and $E_s = E \cos\beta$. Hence, the absorption coefficient of radiation with an arbitrary linear polarization is

$$\alpha(\gamma, \beta) = 1 - R_s(\gamma)\cos^2\beta - R_p(\gamma)\sin^2\beta, \tag{5.68}$$

where R_s and R_p are the reflection coefficients of beams with different polarization (see Chap. 1). At $\beta = \pi/2$ we have a linear P-wave, and $\beta = 0$ corresponds with this linear S-wave. Since the cut surface is a function of x and y, then the normal N to it also has two projections, N_x and N_y. As the equations for the shape of cut surface depends on x, y, it is necessary to express the dependence of the absorption coefficient on the same coordinates. Consider the case where the elliptical polarization of a beam takes place at which the vector end of the elliptical field in the plane x, y describes an ellipsis with half-axes a and b parallel to x and y, respectively. As assumed $a^2 + b^2 = 1$. Here, the absorption coefficient $\alpha = \alpha_x a^2 + \alpha_y b^2$, where α_x and α_y are the absorption coefficients of radiation that is linearly polarized to the axes x and y. Upon easy transformations from (5.68) we obtain:

$$\alpha(\gamma, \mathbf{N}) = 1 - R_s\left(a^2 N_y^2 + b^2 N_x^2\right) - R_p\left[a^2\left(1 - N_y^2\right) + b^2\left(1 - N_x^2\right)\right]. \tag{5.69}$$

That is, the coefficient of absorption depends on the angle of incidence γ, orientation of the vector of the normal N and polarization of radiation, the latter is given by the ratio of the ellipsis axes b/a. In the case of the circular polarization $b/a = 1$ (see Fig. 5.22), for thin sheets, when the cut walls are close to vertical, from (5.69) we have

$$\alpha_c = 1 - 0.5\left(R_s + R_p\right). \tag{5.70}$$

This is a well-known expression used to evaluate the absorption coefficient in the case of circular polarization. At $a = 1$, $b = 0$ from (5.69) we have the wave polarized in the direction of cutting:

$$\alpha_p = 1 - R_p + N_y^2(R_p - R_s). \tag{5.71}$$

In the center of the cut front, where $N_y = 0$, expression (5.71) goes to the known coefficient of absorption (see Chap. 1). Similarly, from (5.69) the absorption coefficient for the wave with normal polarization is as follows:

$$\alpha_s = 1 - R_p + N_x^2(R_p - R_s). \tag{5.72}$$

Knowing the coefficient of absorption at different wave polarizations, one can solve (5.66) with boundary conditions, i.e. the equality to zero of derivatives $\frac{\partial z}{\partial x}$ and $\frac{\partial z}{\partial y}$ at the infinity. The upper surface of the sample was located at $z = 0$, and the cut grew to a limiting value which was called the depth of cutting. This depth depends on two dimensionless parameters, i.e. the b/a polarization and the relation of the maximum rate of substance fracture to the velocity of cutting $\sigma = 2P[\pi r_f^2 V_c(\rho_m H_m + \rho_s^0 c_s^0(T_m - T_0))]^{-1}$. In the given calculations this value was obtained for the following parameters of stainless steel: $T_m = 1{,}700\,\text{K}$, $c_s^0 = 477\,\text{J/kg·K}$, $\rho_s^0 = 7{,}870\,\text{kg/m}^3$, $\rho_m = 6{,}610\,\text{kg/m}^3$, $H_m = 276\,\text{kJ/kg}$ and $N = 17.8 + i4.2$ is the complex index of refraction. The calculation results for the depth of cutting vs. the type of polarization and parameter σ are shown in Fig. 5.50. As seen, the circular polarization produces the deepest cut. The advantage of the linear polarization coincident with the direction of cutting ($b = 0$) persists in this case as well. However, this advantage is not too great when compared with the case where the walls are far apart from each other (see (5.64), (5.65)). It is of interest that the depth maximum is attained at the elliptical type of polarization: $b/a \simeq 0.8$. A typical shape of the cut obtained at $\sigma = 300$ (that corresponds to $P = 2.2\,\text{kW}$ and $V_c = 4.5\,\text{cm/s}$) is presented in Fig. 5.51. Figure 5.52 demonstrates an absorption index on cutting proceeded at two different types of polarization. When cutting with S-polarized radiation ($\mathbf{E} \perp \mathbf{V}_c$, the maximum absorption index falls at the side walls. The cut turns out to be wide and not deep. With P-polarization ($\mathbf{E} \parallel \mathbf{V}_c$) the α maximum falls at the cut front. The cut turns out to be narrow and deeper. Then a question arises: why is the P-polarized radiation unable to cut metal to a maximum depth? Obviously, it is the result of closing side walls with each other. For them this radiation is S-polarized and its absorption index, at steep walls, proves to be small. Such a phenomenon can be interpreted as beam "sticking" between

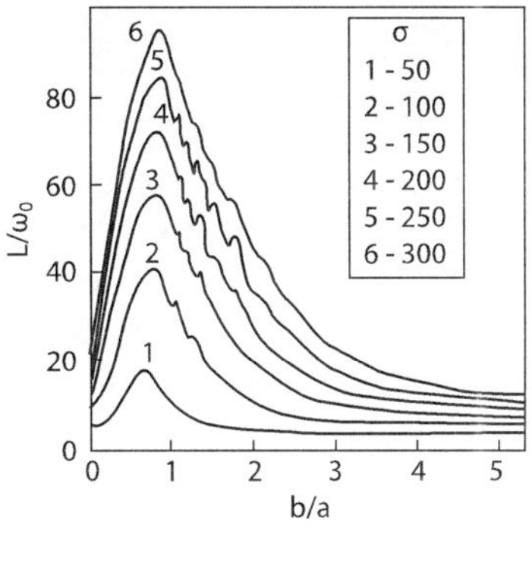

Fig. 5.50 Maximal cut thickness L/ω_0 VS ratio of the ellipsis axes b/a at different values σ, $\omega_0 = r_f = 100\,\mu\text{m}$

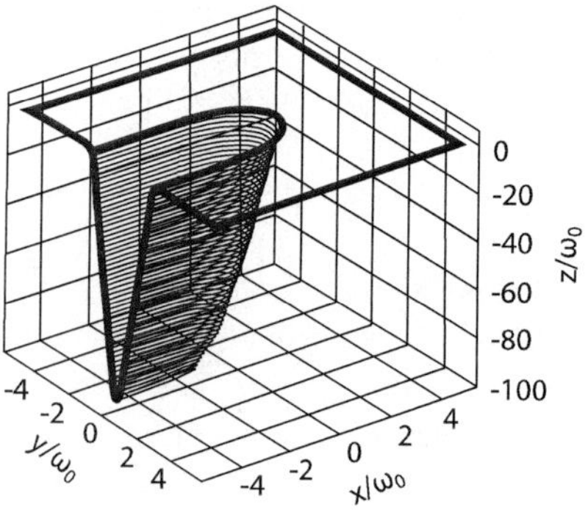

Fig. 5.51 The typical kerf surface at $\sigma = 300$, $b/a = 0.75$, $\omega_0 = r_f = 100\,\mu\text{m}$

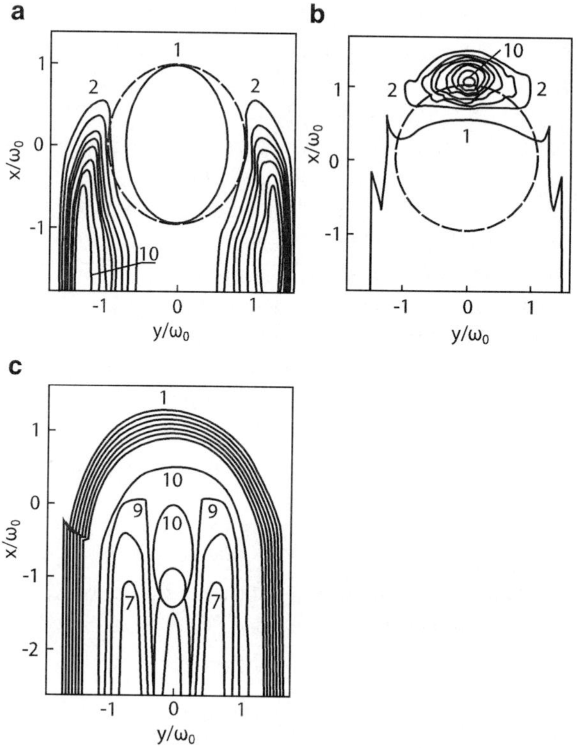

Fig. 5.52 Local absorption coefficients $\alpha(x, y, z_m(x, y))$ on the kerf surface for three kind of polarizations (**a**) α_s) (linear s-polarization), (**b**) α_p (linear p-polarization), (**c**) α_c (circular c-polarization). The *dashed line* is the minimal cross section of laser beam, $r_f = \omega_0 = 100\,\mu\text{m}$. (**a**) $\alpha_c = 0.05$ (1), 0.12 (2), 0.65 (10); (**b**) $\alpha_p = 0.04$ (1), 0.11 (2), 0.65 (10); (**c**) $\alpha_c = 0.09$ (1), 0.24 (7), 0.3 (9), 0.32 (10)

weakly failing side walls. In Fig. 5.53 are shown calculated profiles of the cut over the plane of cutting symmetry $y = 0$ (see Fig. 5.51) for P- and S-polarizations [46]. In the case of P-polarization without regard to the effect of absorption on the walls the cut depth, first, dominates over the depth obtained with S-polarization. This is associated with the fact that the index of S-wave absorption at the cut front is smaller by a factor of 2 than that of P-waves. However, with due account of the absorption on the side walls starting from a certain depth there occur limitations on the cut depth. This can be attributed to the development of the channel configuration

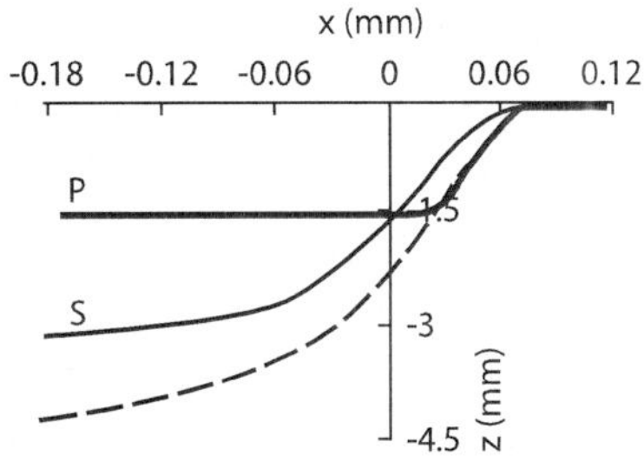

Fig. 5.53 The kerf profiles in a plane $y = 0$ for p and s polarization. The *dashed line* is the kerf profile with p-polarization beam without interaction with sides

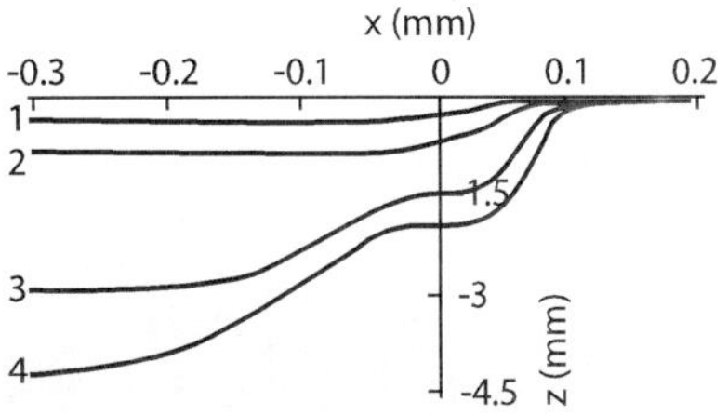

Fig. 5.54 The kerf profiles in a plane $y = 0$ for c-polarization (1.3) and R-polarization (2.4). The ratio of intensity to cutting velocity q/v for curves (1.2) is smaller compare with curves in ten times

which is commensurate with a dramatic decrease in the absorption of P-radiation. Sometimes the process of cutting performed by polarized radiation is compared with that of metal cutting done by a side milling cutter. The cutter shall be so sharpened that it can cut metal both at the front and sides, thus providing its penetration deep into metal [46]. The axisymmetric types of polarization C and R (see Fig. 5.22) are distinguished from each other by their absorption indices: $\alpha_c = 1/2(\alpha_p + \alpha_s)$, and $\alpha_R = \alpha_p$. Since under conditions of cutting $\alpha_s \ll \alpha_p$, then $\alpha_R = 2\alpha_c$. Such a difference in absorption is responsible for difference in cutting depths. The quantitative relationship of ultimate depths depends on the power of radiation and the velocity of cutting, Fig. 5.54 [46]. Besides, the circular and radial polarization offers some technical advantage as it produces a cut of good quality at a complicated configuration of cutting.

5.2.8 Multiple Reflections

As it can be seen in Chaps. 1 and 4, at a deep penetration of laser radiation inside the matter the aspect of multiple reflection of radiation from the channel walls becomes of importance. The propagation of radiation passes into to a waveguide mode. This is most significant for radiation of a CO_2 laser as here at small angles of incidence the absorption index is also small (Chap. 1). The phenomenon of multiple reflections is essential to the case where the channel is closed, i.e. no side losses exist. This is realized in laser welding and drilling (Chaps. 4 and 8). On cutting one half of the channel is open, so part of the beams reflected falls at the side walls far from the area of their initial incidence. Such beams do not act in the process of metal cutting and this very fact has been presumed in the models considered above [46].

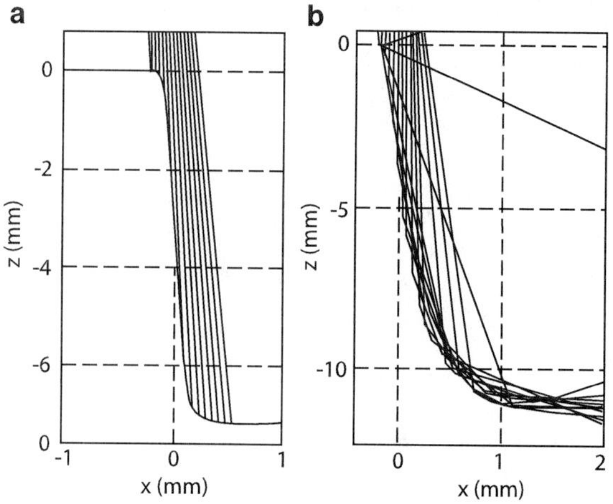

Fig. 5.55 The cut shape in the case of laser cutting without (**a**) and with (**b**) allowance for reflection

However when cutting samples of large thickness (>10 mm) multiple reflection of laser radiation was observed [47]. The need to take account of the reflected beams follows also from simple models which considered the propagation of radiation in the cut of the given configuration [48, 49].

Let us consider the effect of multiple reflections in terms of the self-consistent model described above. With this goal, it is necessary to add in (5.66) the energy transferred by the reflected beams which fall in the point of observation. The energy loss from the preceding reflection needs to be taken into account. Absorption indices are described by Frenel formulas. For example, it was assumed that the polarization of the incident radiation was circular and it remained unchanged during the process of reflections. The laser beam was splitted into 10,000 elementary rays with their energy being consistent with the Gauss distribution [50]. Figure 5.55 shows trajectories of main beams and a configuration of the cut surface with and without regard to (*b*) and (*a*) reflections. It is apparent that a part of the reflected beams contribute significantly to the process of cutting. They fall in the cut bottom and thereby increase the depth of cutting approximately 50%. In conclusion, it should be noted, that the above-given results were obtained in terms of a simplified local model and are rather euristic in nature. Results comparable with those from experiment require calculations performed in a broader formulation of the problem.

5.3 Physical Processes in Laser Cutting with an Oxygen Jet

The use of oxygen for gas laser cutting seems to be natural, since, from one hand, oxides are formed at the cut front with their absorption index being several times higher than that of metals. On the other hand, the process of metal oxidation is an exothermic reaction followed by heat release. As could be seen earlier, accounting of these factors is equivalent to an increase in the absorption coefficient by a factor of two. The advantage gained using oxygen to cut carbon steels proves to be very significant. In [31], the case of cutting carbon steel with a laser beam (CO_2-laser) of its power $P = 0.9\,\text{kW}$ at a velocity $V = 1$ m/min, and an oxygen pressure of 0.15 MPa is presented, the cut depth was found to be 5.3 mm. With the replacement of oxygen by argon the depth of cutting reduced to 0.85 mm. But such effectiveness of oxygen cutting is not unique to many other materials. On describing the process of oxygen cutting, there appear some added difficulties. In this case it is imperative that the dynamics of an oxide film at the melt-gas boundary be considered. Removing this film in due time enables the oxygen to directly undergo a reaction with hot metal and hold a high rate of burning. Processes which proceed at burning on the fuel surface are rather complicated and present a subject of study for a special direction of research. So the models of laser oxygen cutting are confined to the simplest approaches and aimed at clarifying the qualitative features of the process. As the process of burning in these conditions is limited by diffusion of reagents (oxygen) through a liquid medium of the oxide film, the rate of metal burning, in its simplest case, is presented in [34]:

$$\mathrm{d}m/\mathrm{d}t = k/\delta_0,$$

where k is a constant of chemical reaction (g/cm · s) and δ_0 is the thickness of the oxide film. The value of the latter is determined from the mass balance equation of the oxide film:

$$\frac{k}{\delta_0(z)} = \frac{1}{\cos\alpha(z)}\frac{\partial}{\partial z}[\rho_0\delta_0(z)V\sin\alpha(z)] \tag{5.73}$$

where α is the angle between the normal to the film surface and the velocity of cutting V (see Fig. 5.56). This angle is related to the shape of the solid body-melt surface $h(z)$

$$\cos\alpha = \frac{1}{\sqrt{1+(\mathrm{d}L/\mathrm{d}z)^2}}, \tag{5.74}$$

Knowing the value of α, we find the normal and longitudinal components of velocity at the boundary of the melt: $V_0(x) = V\cos\alpha$, $u_0(x) = V\sin\alpha$. The plane flow

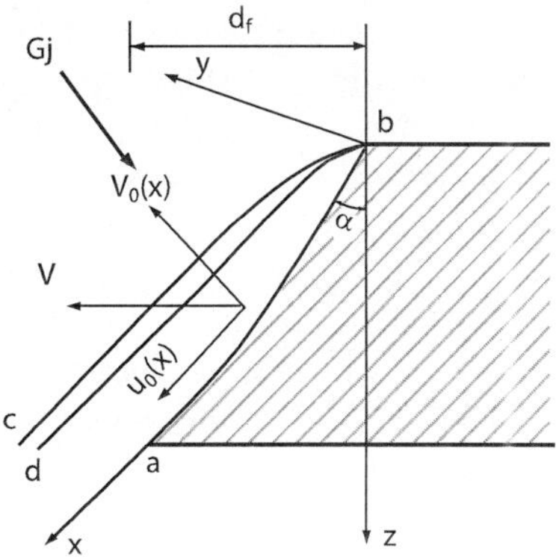

Fig. 5.56 The scheme of computer simulation. The line ab is melting border, bc is cutting surface, abc is material melt, d_f is a laser diameter, Gj is gas jet and bcd is oxide film

of the melt is described in the approximation of the boundary layer. The melting surface $L(z)$ and the normal to it are chosen as x and y coordinates. At the boundary $y = 0$, $\Psi_x = V_0(x)$, $\Psi_y = -u_0(x)$, where $\Psi(x, y)$ are the current function. At a free boundary of the melt at $y = \delta(x)$ the liquid is carried away by the frictional force $\tau(x)$:

$$\mu \frac{\partial}{\partial n} \sqrt{\Psi_x^2 + \Psi_y^2} = \tau(x). \tag{5.75}$$

The equation for the current function $\Psi(x, y)$ is found from Navier–Stokes equation. The heat transport equation shall be added to that for the hydrodynamics of liquid, here it is believed that the temperature along the liquid film changes rather slightly:

$$c\rho V_0(x) \frac{\partial T}{\partial y} - \kappa \frac{\partial^2 T}{\partial y^2} = 0. \tag{5.76}$$

On the free liquid surface, i.e. at $y = \delta(x)$ the condition of heat flow conservation must be

$$-\kappa \frac{\partial T}{\partial y} = q + q_o - q_t, \quad T = T_s, \tag{5.77}$$

where q is the density of the heat absorbed by the laser radiation flow.

$$q = (4P\alpha/\pi d_f h) \cos \alpha$$

$q_o = (\mathrm{d}m/\mathrm{d}t) Q_o$ is the heat release from the oxidation reaction per unit area, Q_o is the specific heat of combustion reaction and $q_t \simeq \kappa(T_s - T_0)/d_f$ is an approximate expression for the heat loss on the side walls due to heat conduction. At the boundary of melting, the difference in the heat flows goes to melting the solid target:

$$-\kappa_1 \frac{\mathrm{d}T_1}{\mathrm{d}y} = -\kappa_2 \frac{\mathrm{d}T_2}{\mathrm{d}y} + \rho V_0 H_m, \quad T_1 = T_2 = T_m, \tag{5.78}$$

where T_1 and T_2 are the temperatures of the liquid and the solid body, respectively. The system of equations with boundary conditions was analyzed by way of automodel solutions with the use of some simplifications. The description of this analysis is omitted because of its cumbersomness and results are displayed in figures. Figure 5.25 shows (in relative units) a dependence of cutting upon the gas flow rate (i.e. inert gas and oxygen). As pointed out earlier, in terms of simple estimations, at a certain flow rate the velocity of cutting by an inert gas attains its saturation. But this is not the case for the velocity of cutting in the situation of an increase in the oxygen flow rate. As the velocity of gas increases the both films (of melt and oxide) diminish. A decrease in the oxide thickness leads to an increase in the rate of oxidation and heat release reactions. This allows the velocity of cutting to be raised. The adequacy of the model describing physical reality is proved by Fig. 5.57 [34].

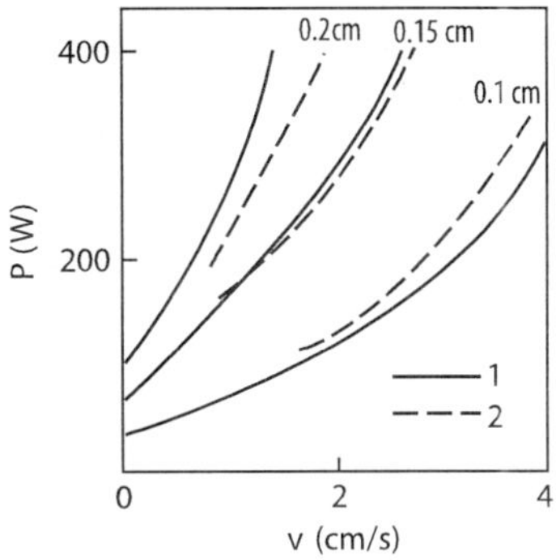

Fig. 5.57 The cutting speed vs laser power and plate thickness, experiment (**1**) [51], numerical simulation (**2**)

5.3.1 Model of Stationary Cutting of Steel in an Oxygen Jet

5.3.1.1 Thick steel plates

Physics of metal sheet laser cutting with an oxygen jet has been discussed earlier. It was seen that in the case of cutting with inert gases at an increase in the sheet of thickness the gas jet run into a problem of penetration into a narrow and deep slit of the cut. In addition, when applying a subsonic oxygen jet to cut thick plates, we have to reduce the pressure of oxygen [44]. In the opposite case the process of cutting goes to an unsteady (uncontrollable) regime with a very low quality of cutting (the cause of this phenomenon will be discussed later). These problems can be overcome, for instance, by a sharp increase of the cut width and the use of a supersonic jet of oxygen. The authors of [45] call this method LASOX. The major factor here is in shaping a narrow supersonic jet which penetrates into the cut slit and keeps a uniform distribution of cutting parameters. The appropriate parameters of the nozzle were chosen on the basis of three-dimensional numerical analyses of jet flows over the entire gas dynamic path: the nozzle and plane slit channel geometrically similar to the laser cut. The analyses showed [44] that, for instance, at an oxygen pressure of 9atm in the nozzle volume the most uniform distributions of the jet velocity in the slit were observed at d_{out}/d_c =1.5–1.7; l/d_c =5–7, see Fig. 5.58. The channel width was set equal to the outlet diameter of the nozzle, i.e. d_{out} (d_c is the nozzle throat diameter and l is the distance between the throat and the nozzle outlet section). Providing for the results of the treatments the parameters of the field experiments were chosen as follows: $d_c = 1.9$ mm, $d_{\text{out}} = 3$ mm, $l = 8$ mm. The CO_2 laser power was 2 kW. The diameter of the focal spot on the plate surface was taken equal to 4 mm. This stems from the concern to obtain a wide cut. The distance between the nozzle and the sample surface was from 4 to 5 mm that is also substantially large in comparison with that used in the "classical" case of laser cutting. Under these conditions the plates of low-carbon steel to 50 mm thick were cut. The cut width was 3.4 mm and the height of roughness was ~20 μm, the latter is close to the results of "classical" cutting, see Fig. 5.37. The experiments [44] have shown that such method of cutting proved to be rather effective and stable in a sufficiently

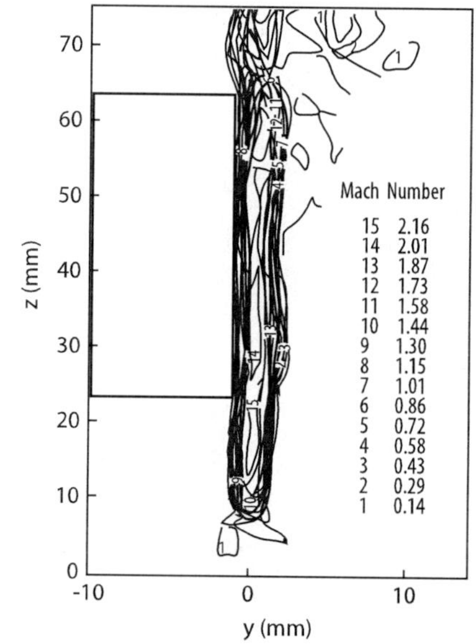

Fig. 5.58 The structure of supersonic oxygen jet at LASOX regime

broad range of parameters. Besides, it is less sensitive to impurities in oxygen when compared to the cutting at a low oxygen pressure that has been noted above.

5.3.2 *Instability of Laser Cutting in the Oxygen Atmosphere*

When cutting metals with large values of specific heat Q_0 (in reaction of carbon steels and titan burning) in the mode of low cutting speed rates there occurs transition to the so-called uncontrolled mode. The latter mode is accompanied by a significant increase in the cut width and the appearance of irregular edges. To study this phenomenon it is necessary to resolve the problem of the cutting stability in the jet of oxygen. As seen from previous divisions, the laser cutting of metals is a complex three-dimensional phenomenon and any full-scale investigation of its stability does not appear feasible. In order to understand the qualitative side of the transition into the "uncontrolled mode," we limit ourselves to a simple model which permits conducting study of stability [35]. The model is concentrated on the burning rate being a function of temperature. The laser beam role is confined to the sustaining of this burning process. When the laser is switched off, the process stops and so the physics of non-self-sustained burning of metal is investigated. To describe this model used are one-dimensional equations of the theory [36] of burning of condensed media with a laser source of heat $Q(x)$ added:

$$\frac{\partial}{\partial t}T + V\frac{\partial T}{\partial x} = \chi\frac{\partial^2 T}{\partial x^2} + T_{\text{bur}}\phi(T, c) + Q(x) + \varepsilon(T_0 - T), \qquad (5.79)$$

$$\frac{\partial}{\partial t} c + V \frac{\partial c}{\partial x} = \phi(T, c). \tag{5.80}$$

Here, χ is the thermal diffusivity, T_0 is the temperature far from the zone of interaction, ε is the rate of heat loss (compare with 5.26), T_{bur} is the temperature of the substance at its complete combustion in the absence of heat loss, $\phi(T, c)$ is the rate of burning reaction and $\phi = k_0 \exp(-E/T)$.

As seen from these equations, they describe the burning of a combustible additive with the concentration of $c \leq 1$ dissolved in a thin sheet (dimensionless units). The latter is heated by the laser and the burning reaction, but remains intact. Though this model does not describe the melt removal, it describes the instability of non-self-sustained burning to which some non-stationary processes occurred at the laser cutting in an oxygen atmosphere are related. In view of the system non-linearity (5.79 – 5.80), one fails to get a solution in a general form. With regard to the fact that the heat sources, due to the processes of burning and absorption of laser radiation, occupy narrow zones, they can be replaced by δ-functions. In this occasion, the condition of non-self-sustained burning of metal is following.

$$E/2\ln(V_0/V) - T_0 - T_{\text{bur}}/\beta > 0, \tag{5.81}$$

$$T_l/\beta > E/2\ln(V_0/V) - T_0 - T_{\text{bur}}/\beta, \tag{5.82}$$

where T_l is the temperature of the laser-heated substance in the absence of heat loss, $\beta = (1 + 4\chi\varepsilon/V^2)^{1/2}$, $V_0 = \sqrt{k_0\chi}[\ln(k_0/4\varepsilon) - \ln\ln(k_0/4\varepsilon)]^{-1/2}$ If the upper inequality is violated, in case the intensity of burning reaction is high, then a self-sustained wave of burning may propagate. If the upper inequality is satisfied, but this is not the case with the lower one, the laser radiation is not sufficient to sustain the burning front before the laser beam. In this case, no stationary solutions exist.

The non-self-sustained burning was examined for its stability by a perturbation method. Figure 5.59 shows an area of instability for a non-self-sustained wave of burning in the coordinates: a dimensionless velocity of the plate motion and $\Theta = 2T_{\text{bur}}/E$. It is apparent that with a decrease in the rate and intensity of the burning reaction the non-self-sustained burning becomes unstable. Such trends were noted in experiments [2]. The non-linear stage of this instability is demonstrated in Fig. 5.60. The figure was obtained by way of the numerical solution of the original system of equations. It is clear that the temperature profile fluctuates in time, the burning wave runs away from the laser beam, then it stops so that the laser beam approaches to the boundary of the non-burnt substance and the process is repeated. Thus, the simplest model shows that fluctuations in the process of cutting in an uncontrolled mode are possible, apparently, due to peculiarities of the balance of heat coming from the laser and the oxidation reaction. It is clear that such model consideration is of a qualitative nature and for comparison with experiment a full-scale simulation is required.

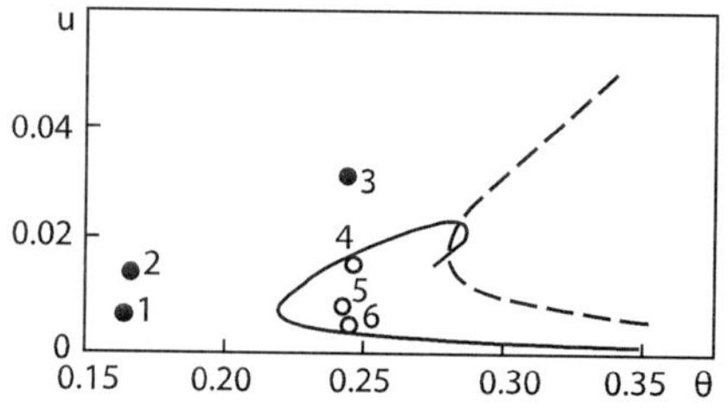

Fig. 5.59 The regions of non-self-sustained regimes. The *solid line* is the border of stability, *light points* (**4, 5, 6**) demonstrate the fluctuation regimes, *dark points* (**1, 2, 3**) are regimes without oscillation. The *dashed line* is the border of self-sustained regimes

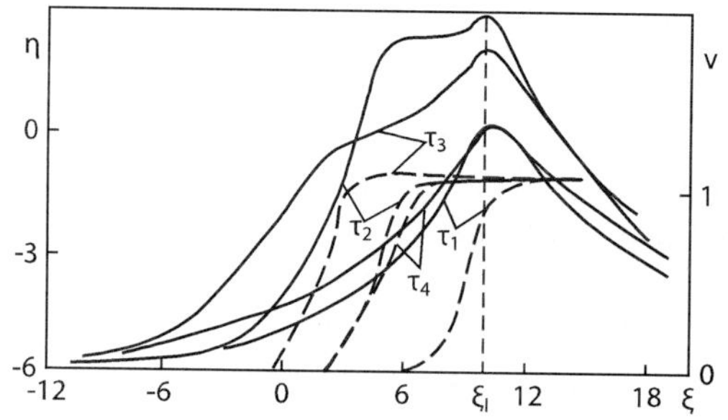

Fig. 5.60 The non-linear stage of instability obtained by way of numerical solution. Solid lines are temperature profiles η (relative units), dashed lines are combustible additive profiles $v = 1 - c$ at successive moments of times ($\tau_1 < \tau_2 < \tau_3 < \tau_4$). ξ_l is laser coordinate (relative units)

5.3.3 Experimental Studies of High-Quality Laser Cutting of Thick Mild Steels with Oxygen Assist Gas

The simple model described above illustrated the possible mechanism producing the instability of oxygen-assisted laser cutting of thin steel samples. Let us consider the results of some experiments to find the optimal conditions for laser cutting with an assist gas of a mild steel sheet with a thickness of 5–25 mm [52]. The working gas was oxygen of high purity: 99.999%. Mild steel was cut using CO2 laser radiation with circular polarization. Experiments have shown that minimum roughness of laser cutting is achieved under conditions of constant deposited energy per unit volume: $P/hdv = 19.4\,\text{J/mm}^3$. This condition must be complemented with relations between P [W] and V [mm/s] with the sample thickness h [mm]: $P = 194\,\text{h}$, $V = 11/(0.35 + 0.02\,\text{h})$ [53].

After determining the parameters for the oxygen-assisted laser cutting with minimal surface roughness it is interesting to consider the energy balance for such a case. The most difficult task here is to define the power released in the exothermic oxidation reaction, referred to above. This quantity could be determined from the energy balance as following: the power expended on melting of steel, plus the power lost by the thermal conductivity, less laser power absorbed in the cut.

The first value was calculated while the other two were found experimentally. The losses on thermal conductivity were determined using the calorimetric method. The absorption coefficient was measured as the difference between the laser power and radiation power at the exit of the cut. In the experiment, this value was a slightly increasing function of sample thickness: at $3 < h < 18\,\text{mm}$ $0.6 < \alpha < 0.8$. As a result, the specific capacity of all four components of energy balance were similar in the interval of 130–170 W/mm. Knowing these values, we can easily determine the effective absorption coefficient, taking into account heat emission due to oxidation α_e and heat absorption η_T. The magnitude of α_e increases weakly with increasing thickness h: $1.5 < \alpha_e < 1.7$. The heat absorption coefficient is $\eta_T \approx 0.46$. These values were determined for the minimal surface roughness on the walls of the cut and may differ from the values in the general case as it was discussed in previous sections.

References

1. A.G. Krasyukov, F.K. Kosyrev, V.G. Naumov, V.M. Shashkov, XII Int. Symp. Jn Gas Flow and Chemical Laser. 31 August-5 Sept. 1998. St. Petersburg, Russia
2. A.G. Grigor'yants, A.A. Sokolov, *Lazernaya reska metallov (Laser cutting of metals)* (Vys'shaya shkola, Moskva, 1988)
3. L.D. Landau, E.M. Lifshitz, *Gidrodinamika (Hydrodynamics)* (Nauka, Moskva, 1986)
4. V.V. Likhanskiy, A.I. Loboiko, G.F. Antonova, et al., Kvantovaya Elekronika **26**, 2 (1999)
5. J.F. Ready, *Action of powerful laser radiation* (Mir, Moscow, 1974)
6. H.S. Carslow, J.C. Jaeger, *Conduction of heat in solids* (Oxford, 1964)
7. M.I. Arzuov, Zh.I. Dzhumabekov, V.I. Konov, et al., Phys. Chem. Mater. Process. **3**, 136 (1989)
8. N.K. Makashov, E.C. Asmolov, V.V. Blinkov, et al., Kvantovaya Elektronika **19**, 910 (1992)
9. V.S. Golubev, *Mechanisms of Melt Removal at Laser Cutting of Materials. The Deep Penetration and Filamentation of Power Laser Radiation in Matter*, ed. by V.Ya. Pahchenko, M. Interkontakt, Science, 266c.2009
10. G.F. Antonova, G.G. Gladush, A.G. Krasyukov, et al., High Temp. Phys. **37**, 865 (1999)
11. G.F. Antonova, G.G. Gladush, A.G. Krasyukov, et al., High Temp. Phys. **38**, 501 (2000)
12. H. Schlichting, Grenzschlicht-Teorie, Verlag G. Braun, Karlsruhe (1967)
13. R.L. Johnson, J.D. OKeefe, AIAA J. **12**, 1108 (1974)
14. D. Schuosker, *Ind. Las. Handbook*, ed. by D. Belforte, M. Levitt, Penwell, Tulsa, OK, pp. 65–79, 1987
15. M. Vicanek, G. Simon, H.M. Urbassek, I. Decker, J. Phys. D Appl. Phys. **20**, 140–145 (1987)
16. J.F. Ready, D.F. Forson (eds.), *LIA Handbook of laser Material Processing*, vol. 20 (Laser Institute of America, 2001)
17. D. Schuocker, *Hight Power Laser in Production Engineeting* (ICP, London, 1999)
18. A.A. Vedenov, O.P. Ivanov, A.L. Chernyakov, Kvanovaya Elekronika **11**, 2397 (1984)
19. A.V. Zaytsev, O.B. Kovalev, A.M. Orishich, I.M. Fomin, Kvanovaya Elekronika **35**, 200 (2005)
20. R.V. Arutyunyan, V.Yu. Baranov, L.A. Bol'shov, D.D. Malyuta, A.Yu. Sebrant, *Vozdeistvie lazernogo izlucheniya na materially (Interaction of Laser Radiation with Materials)* (Nauka, Moscow, 1989)
21. B. Gebhart, I. Dzhalurija, R. Mahadzhan, B. Sammakija, *Convectiv Flows in Heat-Mass Exchange* (Mir, Moscow, 1991)

22. J. Ready, *Industrial Applications of Lasers*, (Academic Press, New York, London, Tokyo, 1997)
23. F.K. Kosyrev, N.P. Kosyreva, A.P. Leonov, V.A. Timofeev, Avtomaticheskaya svarka **10**, 51 (1978)
24. A.A. Vedenov, G.G. Gladush, *Physical processes at laser material processing*, (Energoatomizdat, Moscow, 1985)
25. N.N. Rykalin, A.A. Uglov, I.V. Zuev, A.N. Kokora, *Laser Material Processing*, (Mashinostroenie, Moscow, 1985), p. 496
26. J. Powell, *CO_2-Laser Cutting*, (Springer, 1998)
27. T. Fushimi, H. Horisawa, S. Yamaguchi, et al., Proc. SPIE. **2888**, 90–95 (2000)
28. S.M. Shariff, G. Sundararajan, S.V. Joshi, J. Laser Appl. **11–12**, 54–63 (1999)
29. Y. Arata, Dynamic behavior in laser gas cutting of midl steel, in *The book: Plasma, Electron and Laser Beam Technology* (American Society of Metals, Metals Park, OH, 1986)
30. Y. Arata, New laser-gas-cutting technology for stainless steel, in *Book: Plasma, Electron and Laser Beam Technology* (American Society of Metals, Metals Park, OH, 1986)
31. V.S. Kovalenko, V.V. Romanenko, L.M. Oleschuk, *Without Waste Processes of Laser Cutting* (Technika, Kiev, 1987)
32. I.O. Bazyleva, M.G. Galushin, V.S. Golubev, E.A. Dubrovina, V.A. Karasev, Proc. SPIE **4644**, 77–78 (2001)
33. F.D. Olsen, in Optoelectron. Techn. Vort. 5 Intern. Kong. Laser-81, Munchen, 1981, Berlin, 1982, p.227
34. T.V. Bystrova, V.B. Librovich, V.I. Lisitsyn, Phys. Burning Explosion **9**, 725 (1973)
35. O.P. Ivanov, A.L. Chernyakov, Kvantovaya Elektronika **13**, 2287 (1986)
36. G.M. Makhviladze, G.V. Novozhilov, Appl. Mech. Tech. Phys. **5**, 51 (1981)
37. P. Sforza, V. Santecesaria, Proc. SPIE **2207**, 836–847 (1994)
38. O. Elemming, F.D. Olsen, Proc. SPIE **2207**, 402–413 (1994)
39. V.A. Karasev, V.I. Ledenev, V.P. Yakunin, Proc. SPIE **3688**, 163–168 (1999)
40. E.V. Locke, R.A. Hella, IEEE J. Quant. Electron. **QE-10**, 179 (1974)
41. A.I. Tikhomirov, A.Yu. Mikheev, N.V. Evdokushin, et al., Automat. Weld. **3**, 68–70 (1982)
42. A.V. Zaytsev, O.B. Kovalev, A.M. Orishich, V.M. Fomin, V.B. Shulyatyev, Comput. Technol. **11**, 75 (2006)
43. O.B. Kovalev, A.M. Orishich, A.P. Petrov, V.M. Fomin, P.V. Yudin, et al., Appl. Mech. Tech. Phys. **45**, 162 (2004)
44. A.V. Zaytsev, O.B. Kovalev, A.G. Malikov, A.M. Orishich, V.B. Shulyatyev, Kvantovaya Elektronika **37**, 6 (2007)
45. W. O'Neil, J.T. Gabzdyl. Opt. Laser. Eng. **34**, 355 (2000)
46. V.G. Niziev, A.V. Nesterov. J. Phys. D Appl. Phys. **32**, 1455–1461 (1999)
47. A.N. Malov, N.A. Malov, A.M. Orishich, V.B. Shulyatev. Proc. IV Int. Conf. Laser Technologies and Their Implementation, St. Petersburg, Sept. 24–26, 2003
48. J. Milewski, E. Sklar, Model. Simul. Mater. Sci. Eng. **4**, 305 (1996)
49. J. Duan, H.C. Man, T.M. Yue, J. Phys. D Appl. Phys. **34**, 2127 (2001)
50. O.B. Kovalev, A.V.Zaitsev, J. Appl. Mech. Techn. Phys. **46**, 9–13 (2005)
51. R.P. Babenko, V.P. Tychinsky, Kvantovaya Electronika **5**, N11 (1972)
52. A.G. Malikov, A.M. Orishich, V.B. Schulyatev, Kvantovaya Elektronika **39**, 547 (2009)
53. V.M. Fomin, A.G. Malikov, A.M. Orishich, V.B. Schulyatev, Doklady Akademii Nauk **428**, 325 (2009)

Chapter 6
Interaction of Pulsed Laser Radiation with Materials

Abstract In this chapter, the possibilities of precision processing of surfaces for microelectronics and to the laser deposition of special films are considered. Using a very short radiation pulse can form the regime in which a material is removed due to vaporization with the formation of liquid phase in minimal amount. The physics of laser ablation is considered in the two regimes: ablation in vacuum and buffer gas. The results of comparison of numerical simulations with experiment are described.

We considered in previous chapters cw CO_2 laser. In recent years, along with repetitively pulsed high-average power CO_2 lasers, repetitively pulsed UV lasers with considerable average powers have been developed. These lasers are of interest for use in laser technologies [1, 2]. To study the interaction of such lasers with targets, it is necessary to know the mechanisms of pulsed interaction of light with materials. Note that the interaction of pulsed laser radiation with matter has comparatively long been studied because the first high-power lasers were pulsed. The physical and technological aspects of this interaction are considered in [3, 4], where processes proceeding during material processing by radiation from neodymium and ruby lasers are discussed.

CO_2 and UV lasers were created later. Because the radiation wavelengths of these lasers considerably differ from that of Nd lasers, the interaction of their radiation with matter has a number of specific features.

In this chapter, we consider the interaction of millisecond solid-state laser pulses with materials only briefly because this question was already discussed in monographs (see, e.g., [1, 3–6]). The same concerns CO_2 lasers. We discuss this issue for the generality of consideration of this problem. In the recent decade, the attention of researchers was shifted to the consideration of shorter pulses from Nd and UV lasers. This is related to the search for the possibilities of precision processing of surfaces for microelectronics and to the laser deposition of special films and coatings. The matter is that by using a very short radiation pulse, researchers try to find a regime in which a material is removed from a substrate

G.G. Gladush and I. Smurov, *Physics of Laser Materials Processing*,
Springer Series in Materials Science 146, DOI 10.1007/978-3-642-19831-1_6,
© Springer-Verlag Berlin Heidelberg 2011

due to vaporization with the formation of a liquid phase in minimal amounts. This is required both for the precision processing of surfaces and to prevent the formation of drops during film deposition.

Experiments have shown that short laser pulses can provide conditions when plasma "has no time" to be produced. Therefore, we begin this chapter with the discussion of the kinetics and gas dynamics of pulsed vaporization of materials without plasma formation. Then, we will consider the conditions of the appearance of an optical discharge, different types of its development and its influence on surface processing (Chap. 7). In the next sections, we will describe the mechanisms of the melt removal from a shallow pool induced by vapors, as well as the features of deep penetration of radiation from pulsed Nd and CO_2 lasers into a target (Chap. 8).

6.1 Physics of Pulsed Laser Ablation and Deposition of Films

To understand the direction of investigations of the ablation physics, we consider briefly the properties of the method of pulsed laser deposition (PLD) of films (see, e.g., review [7] and references therein). The advantages of this method are (a) the universal nature, allowing the deposition of dielectrics, superconductors, semiconductors, etc.; (b) the pulsed type of the process, allowing good control of the film thickness; and (c) the high kinetic energy of atoms and molecules during pulsed ablation, providing their greater mobility, which is important for growing high-quality films. A disadvantage of PLD is the inhomogeneous film deposition caused by the nonuniform angular directivity of a vapor jet ($\sim \cos^n \theta$) and by the possible deviation from the stoichiometric composition. The latter is especially important for superconducting films.

The stoichiometry can be in principle controlled by varying the pressure and composition of a residual gas. In such a way, the energy of a vapor cloud can be also controlled because too hot vapor can damage a film. The basic mechanisms of vaporization of a plane surface exposed to laser radiation are considered in a simple case in Chap. 1. These mechanisms cannot be used to describe pulsed ablation in our case. This is explained first of all by the two-dimensional and nonstationary vapor flow in our case. In addition, during irradiation by nanosecond laser pulses, the vaporization time is considerably shorter than the propagation time of a vapor cloud, and vapors can also flow back to the target and condense on its surface. As usual, this field of science developed gradually – from simple to complex concepts. Numerical and analytic models were proposed which considered individual aspects of this phenomenon such as one-dimensional plane nonstationary flows [13–16] and one-dimensional nonstationary flows occurring simultaneously in gas and condensed media [17]. The lateral expansion of a vapor cloud was simulated by considering the one-dimensional spherically symmetric flow of vapors [18]. The two-dimensional model of the nonstationary flow of vapors was developed in papers [19–21] taking plasma phenomena into account. We will consider it in the next sections. Note here that this model neglected kinetic effects, which are important for laser film deposition.

These questions were also considered by the numerical Monte-Carlo method [22–24]. However, the authors studied only the case of low vapor densities, when collisions between vapor atoms can be neglected compared to collisions of vapor atoms with surrounding gas atoms.

By restricting ourselves to a brief description of the development of the laser ablation mechanism, studies we consider the model in which the different aspects of this process are interrelated. As mentioned above, PLD consists of the two stages of vaporization and propagation of vapors. For 10 ns pulses and the typical rate of ablated atoms $v_v \sim 10^4$ m/s, the thickness of a vapor cloud is 0.1 mm. We assume that the radiation intensity in a focal spot of radius r_f is spatially constant but depends on time. The typical size of a focal spot on a target during PLD is chosen ~1 mm. This means that the vaporization process can be considered in the one-dimensional approximation. For the same reason, the heating of the target surface, its melting and vaporization can be also considered in the one-dimensional approximation. The duration of this stage should not exceed the lateral expansion time of vapors. It can be set equal to a few light pulse durations. In this case, this ablation stage can be uses as the initial condition for a problem of the two-dimensional expansion of a vapor cloud, Fig. 6.1a.

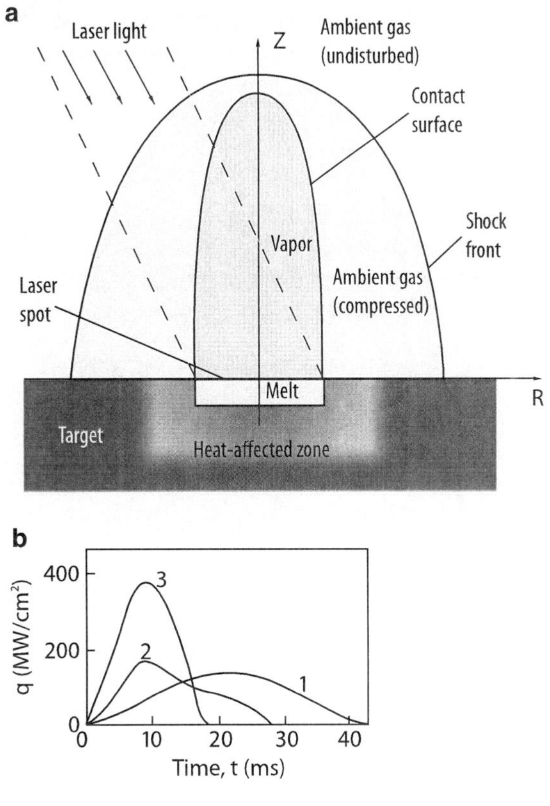

Fig. 6.1 The scheme of numerical simulation (**a**). Pulse shapes of Nd:YAG (**1**) and KrF (**2**) lasers and the second harmonic of the Nd:YAG laser (**3**) (**b**)

6.1.1 Initial Stage

The heating and melting of a sample surface were described, taking into account the phase transition (melting) temperature, by using the approximation in which this quantity was included into the heat capacity for $T > T_m$:

$$(\partial/\partial t - U_e \partial/\partial z_c)(N_c L_c) = \partial/\partial z_c(\kappa_c \partial T_c/\partial z_c) \tag{6.1}$$

where U_e is the vaporization front velocity, L_c is the enthalpy per atom, N_c is the atomic density, T_c is temperature, κ_c is the heat conductivity, and z is the coordinate directed normally to the sample surface inside it. The heat conductivities are assumed different for solid and liquid phases:

$$\begin{aligned} \kappa_c &= \kappa_s \quad T_c < T_m \\ \kappa_c &= \kappa_l \quad T_c > T_m \end{aligned} \tag{6.2}$$

where T_m is the melting temperature. The heat capacities of the solid and liquid phases are C_s and C_l, respectively.

The enthalpy of a target atom is defined as $L_c = T_c C_s$, when $L_c \leq T_m C_s$. When $L_c > C_s T_m + H_m$, we have $L_c = C_l(T_c - T_m) + C_s T_m + H_m$. When $C_s T_m + H_m > L_l > T_m C_s$, temperature is equal to the melting temperature. By using this approach, the Stefan problem is solved, the motion of the melt boundary being described by the point displacement, where $T = T_m$.

The gas phase consists of monoatomic vapor of density N_v with atoms of mass m_v. The surrounding inert gas is characterized by the density N_g and atomic mass m_g with the polytropic index γ.

In the simplest case of nonviscous gas, by neglecting the heat conduction, the behavior of vapors and surrounding gas is described by usual equations of conservation of the number of particles, momentum, and energy:

$$\begin{aligned} &\partial N_v/\partial t + \partial(N_v U_z)/\partial z = 0 \\ &\partial N_g/\partial t + \partial(N_g U_z)/\partial z = 0 \\ &\partial(\rho U_z)/\partial t + \partial(\rho U_z^2 + P)/\partial z = 0 \\ &\partial(E + \rho U_z^2/2)/\partial t + \partial(U_z(E + \rho U_z^2/2 + P))/\partial z = 0 \end{aligned} \tag{6.3}$$

where U_z is the velocity, $\rho = m_v N_v + m_g N_g$ is the density, $P = (N_v + N_g)kT$ is pressure, and $E = (3/2N_v + 1/(\gamma - 1)N_g)kT$ is the internal energy. At the condensed phase–gas interface, mass and energy flows should be conserved:

$$N_l U_l = N_v U_z \tag{6.4}$$

$$U_z(N_v H_v + E + \rho U_z^2/2 + P) - \kappa_l \partial T_l/\partial z = q(t), \quad z = 0 \tag{6.5}$$

Here, H_v is the vaporization energy for a substrate atom and $q(t)$ is the absorbed radiation intensity. Equations (6.1)–(6.5) should be supplemented with the boundary conditions

$$T_c \to T_a, \ z_c \to \infty$$

where T_a is the surrounding gas temperature. The initial conditions are determined by the surrounding gas parameters:

$$T = T_c = T_a, \quad P = P_a, \quad N_v = 0, \quad U_z = 0, \quad t = 0$$

The boundary conditions for the one-dimensional problem have the form

$$P(z \to -\infty) \to P_a$$
$$N_g U_z = 0, \quad z = 0$$

As for conditions for vapor at the boundary $z = 0$, we will use here the relation between hydrodynamic parameters at the Knudsen layer boundary and on the condensed phase surface, which we discussed in detail in Chap. 1.

After the radiation pulse end, the surface is rapidly cooled, and vaporization changes to condensation. The mass flow to the surface in described in this case by the usual Hertz–Knudsen equation

$$N_k U_k = (P_s - P_k)/(2\pi m k T_s)^{1/2} \tag{6.6}$$

Here, N_k, U_k, and P_k are the density, velocity, and pressure at the Knudsen layer boundary, T_s and P_s are temperature and pressure of the saturated vapor over the surface heated to the temperature T_s.

The one-dimensional problems formulated above are required for determining the parameters of a vapor layer, which are in turn necessary to solve the problem of the vapor expansion into the surrounding gas. This problem is described by the same equations (6.3) supplemented with the terms describing the lateral expansion of vapors in the radial direction. The problem is considered axially symmetric:

$$\begin{aligned}
&\frac{\partial N_v}{\partial t} + \frac{\partial (N_v U_z)}{\partial z} + \frac{1}{r}\frac{\partial (r N_v U_r)}{\partial r} = 0 \\
&\frac{\partial N_g}{\partial t} + \frac{\partial N_g U_z}{\partial z} + \frac{1}{r}\frac{\partial (r N_g U_R)}{\partial r} = 0 \\
&\frac{\partial (\rho U_z)}{\partial t} + \frac{\partial \rho U_z^2 + P}{\partial z} + \frac{1}{r}\frac{\partial (r \rho U_z U_r)}{\partial r} = 0 \\
&\frac{\partial (\rho U_r)}{\partial t} + \frac{\partial (\rho U_z U_r)}{\partial z} + \frac{1}{r}\frac{\partial [r(\rho U_r^2 + P)]}{\partial r} = \frac{P}{r} \\
&\frac{\partial (E + \rho U^2/2)}{\partial t} + \frac{\partial [U_z(E + \rho U^2/2 + P)]}{\partial z} + \frac{1}{r}\frac{\partial [r U_r (E + \rho U^2/2 + P)]}{\partial r} = 0
\end{aligned} \tag{6.7}$$

where U_r is the radial velocity, and $U^2 = U_r^2 + U_z^2$. The boundary conditions are

$$P \to P_a \quad \{r, Z\} \to \infty$$

On the condensed phase surface outside the focal spot, we have $N_v = 0$ and $U_z = 0$. Within the focal spot, we have $N_g = 0$ and the Jouguet condition for vapor

$$N_g = 0, \quad U_z = C = (5kT/3m_v)^{1/2}, \quad z = 0$$

The three systems of equations were solved numerically for ablation of graphite and gold targets irradiated by nanosecond pulses from different lasers in vacuum and inert gas atmosphere.

Figure 6.1b shows the shapes of 1.06 μm pulses from a Nd:YAG laser [25], 248 nm pulses from an excimer KrF laser [26], and 532 nm second-harmonic pulses from a Nd:YAG laser [27]. The properties of materials used in calculations are presented in Table 6.1. Notes: (a) for $\lambda = 248$ nm; (b) for $\lambda = 532$ nm; (c) average value for $T > 2{,}000$ K; (d) average value for $T < 2{,}000$ K. The data for gold are close to these presented in the table in Chap. 1 and are presented here for generality. Consider the behavior of a condensed medium by using Fig. 6.2. The intensities and durations of pulses from Nd:YAG and KrF lasers incident on graphite and gold are close. According to estimates, the Au surface, by neglecting vaporization during the pulse, should be heated up to 7.5–10 kK (for different values of the heat conductivity). Due to a great difference between the thermal coefficients of irradiated materials, carbon should be heated approximately ten times greater. Clearly, the heating process was restricted by the vaporization of these materials. Although the vaporization energy and density of carbon atoms are considerable greater than these for gold, the gold target was vaporized weaker Fig. 6.2b than the carbon target Fig. 6.2a. This occurs because the interaction under study has the

Table 6.1 Properties of materials used in calculations

	Parameter	Au	Graphite
Reflectance	r	0.33^a; 0.67^b	0
Density of particles	N_c	5.91×10^{28} m^{-3}	1.13×10^{29} m^{-3}
Temperature of			
melting	T_m	1,336 K	...
boiling	T_m	3,150 K	5,100 K^c
Heat of			
melting	H_m	0.131 eV	...
boiling	H_b	3.45 eV	9.13 eV^c
Specific heat for			
solid	C_s	28.8 J/(mol K)	24.9 J/(mol K)
liquid	C_l	31.3 J/(mol K)	...
Heat conductivity for			
solid	K_s	284 W/(m K)	1.6 W/(m K)^d
liquid	K_s	155 W/(m K)	0.81 W/(m K)^c

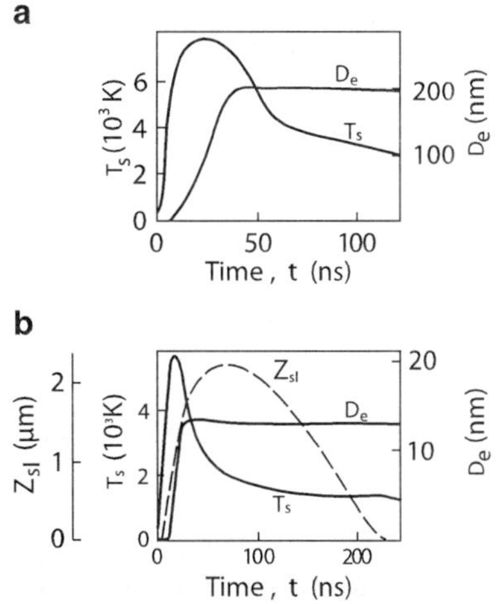

Fig. 6.2 Time dependences of the surface parameters of carbon (**a**) and a gold (**b**) targets irradiated by Nd:YAG laser pulses. Curves T_s and D_e describe the surface temperature and the evaporated layer thickness, respectively; Z_{sl} is the melting zone

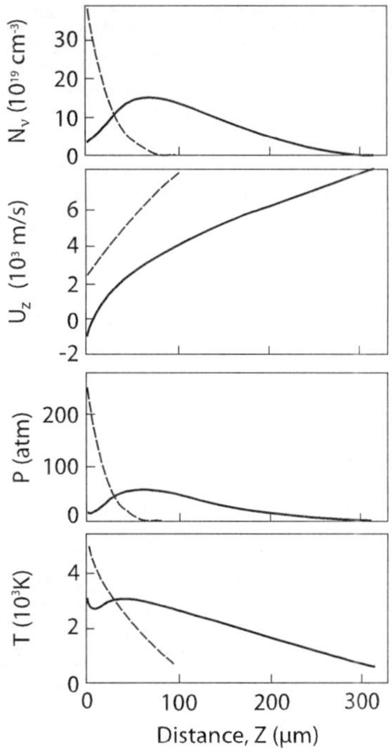

Fig. 6.3 Distribution of the vapor parameters near a carbon target after the pulse end ($\tau = 20$ ns, *dashed curve*) and after $\tau = 40$ ns (*solid curve*)

pulsed nature - the gold target temperature increases slower due to the higher heat conduction of gold.

Figure 6.3 shows the parameters of a vapor layer by the 20 ns pulse end and after 40 ns. One can see that by the end of the pulse, all the parameters of vapor monotonically decrease with distance from the surface due to adiabatic expansion.

The vapor temperature on the surface is lower than the surface temperature (cf. Figs. 6.2 and 6.3). Figure 6.2 shows that the surface temperature decreases after the laser pulse end, vaporization ceases and changes to condensation. As a result, all the parameters of the vapor layer near the target decrease and the velocity of vapors changes its sign, Fig. 6.3. During PLD, the pressure of the residual gas in a chamber is usually chosen low ($\leq$100 Pa). One can see from Fig. 6.3 that the vapor pressure is many orders of magnitude higher, so that the surrounding gas has no effect on the initial stage.

6.1.2 Ablation to Vacuum

As mentioned above, the one-dimensional stage of vapor expansion for t = 50 ns was used as the initial condition of the two-dimensional problem. After the "switching on" of the lateral expansion, an expansion wave propagates along a vapor layer from the edge of the vapor cloud for $r = r_f$ Fig. 6.4a. The motion along the normal remains virtually one-dimensional. The velocity of the expansion wave is equal to the local speed of sound C; according to Fig. 6.3, $T_v \approx 3{,}000$ K and $C = 1{,}800$ m/s. In this case, the expansion wave reaches the focal spot center by $t = 500$ ns (Fig. 6.4b). After that, the expansion of vapors acquires the two-dimensional nature. By this instant, due to the adiabatic expansion, the vapor temperature becomes lower than the surrounding gas temperature. The vapor density decreases so by the time $\sim$2 μs that the mean free path of atoms becomes larger than the vapor cloud size. From this moment, all the atoms move by inertia:

$$r = U_r t, \quad z = U_z t, \quad N(r,z,t) = N(r_0 + U_r t, \ z_0 + U_z t) \tag{6.8}$$

Here, we come to the question of the angular distribution of vapor atoms, which is important for PLD. The distribution function of atoms in the hydrodynamic approximation is assumed Maxwellian, with the temperature, density, and hydrodynamic velocity depending on coordinates. It is clear that the angular distribution is determined by the temperature and relation of the transverse and longitudinal velocities. It is reasonable to introduce the distribution function for particles flying at an angle θ with energy E from all the points of the volume occupied by the vapor cloud. For this purpose, we should pass from Cartesian coordinates in the velocity space to spherical coordinates and integrate over the entire volume. Figure 6.5a shows the energy distribution function of particles flying normally to the surface. One can see that the particle energy increases up to $t = 500$ ns due to the work of pressure forces. Then, the energy does not virtually change. As for the angular distribution for particles of any energies (Fig. 6.5b), the angular scatter first decreases up to $t = 500$ ns, which is caused by the decease in the vapor temperature. Then, when the whole cloud "acquires" the transverse velocity, the angular scatter slightly increases. This explains the decrease in the angular scatter with increasing the focal spot size Fig. 6.6.

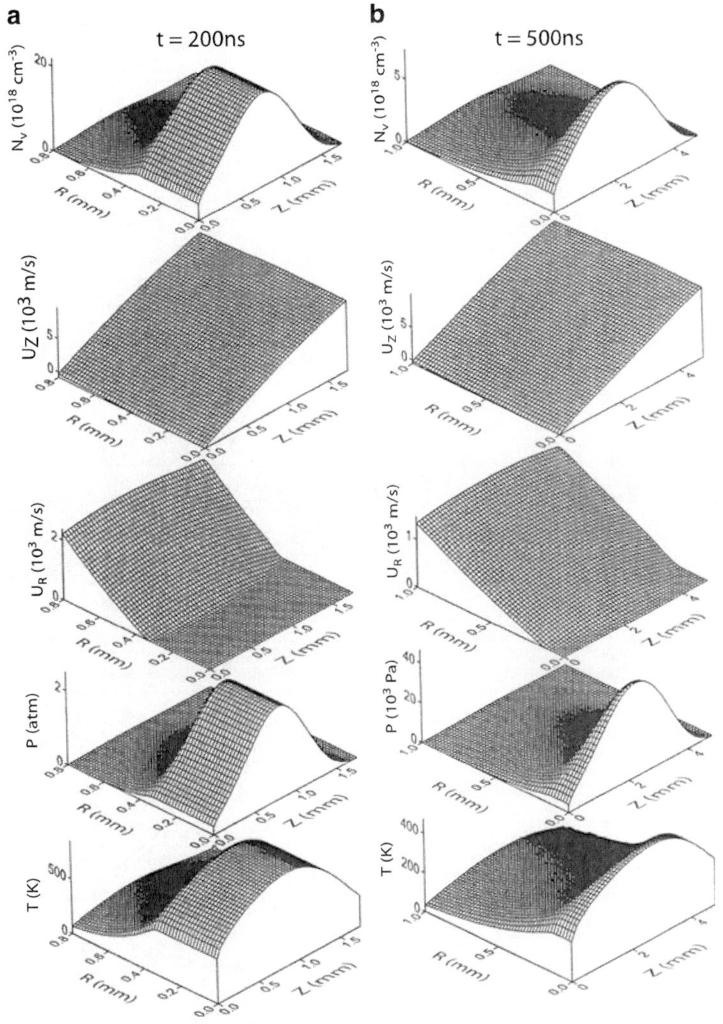

Fig. 6.4 Two-dimensional distributions of the carbon vapor density at successive instants of time after the laser pulse end

6.1.3 *Ablation to Buffer Gas*

In this case, the picture of the vapor flow considerably changes. However, this occurs at later stages for $t > 500$ ns. At the early stage ($t < 500$ ns), the vapor pressure is so high that the influence of the residual gas is insignificant (Fig. 6.7a). For $t \approx 500$ ns, when the vapor pressure decreased and became equal to that of the buffer gas, the latter begins prevent the expansion of vapors. Vapors are accumulated near the contact surface, especially near the front part of the vapor cloud. A shock

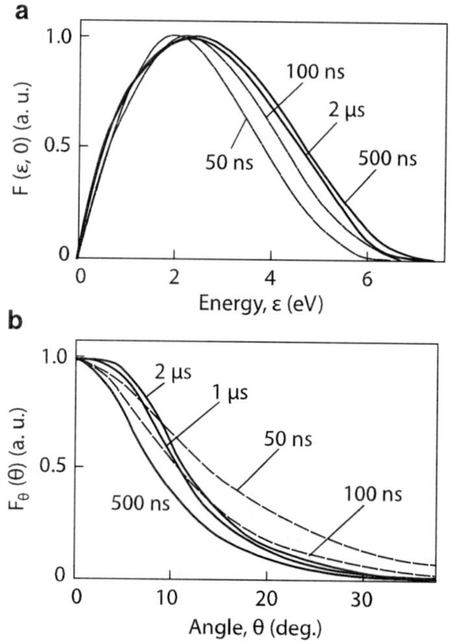

Fig. 6.5 Energy distribution function of particles flying normally to the surface (**a**) and the angular distribution function of particles (energy-averaged) (**b**)

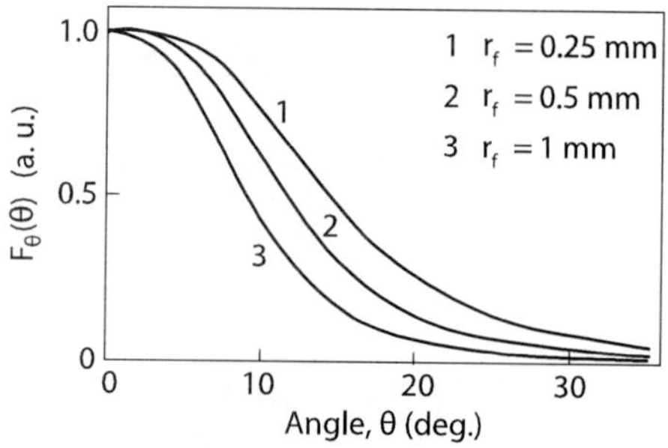

Fig. 6.6 Angular distribution functions of carbon particles in vacuum for different focal spot radii ($\tau = 25$ ns, $\varepsilon = 4\,\mathrm{J/cm^2}$, $t = 5\,\mu$s)

weave is formed in the buffer gas, which propagates ahead the contact boundary (Figs. 6.7b,c). The region occupied by vapor decreases with increasing the mass of gas molecules (Fig. 6.8).

The angular scatter decreases with increasing the residual gas pressure (Fig. 6.9). This is explained by the fact that the resistance to vapor expansion proves to be greater from sides. This is caused by a greater dynamic pressure of vapors in the axial direction than in the lateral one. This effect was also observed in experiments [28].

Figure 6.9a also shows the change in the velocity distribution of vapor particles flying normally to the surface. This distribution in vacuum can be approximated by the Boltzmann distribution with the effective temperature. As the gas pressure increases, the distribution becomes a triangle one. The tail of fast particles is cut because these particles are concentrated at the vapor cloud front and are faster

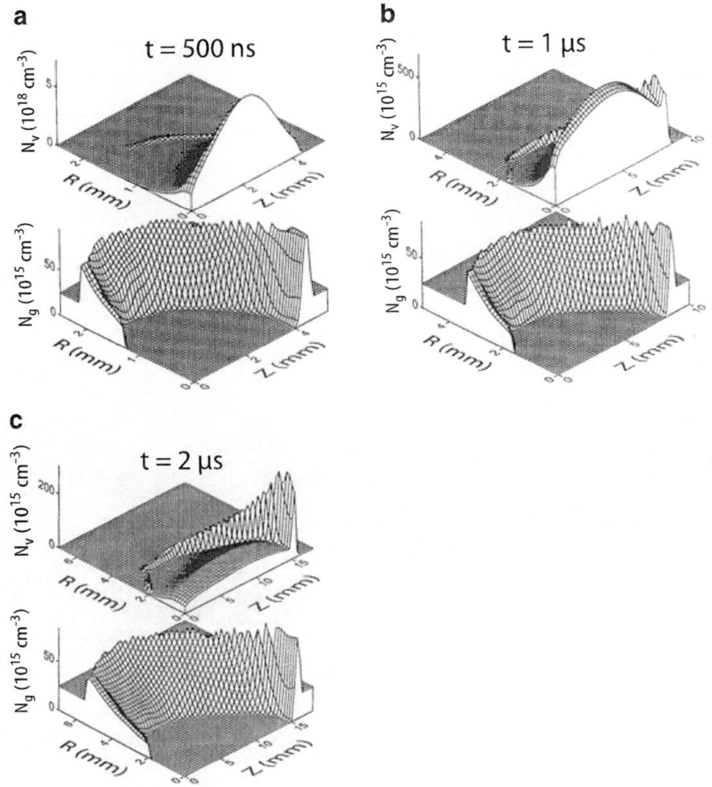

Fig. 6.7 Distributions of vapor density (*upper row*) and surrounding gas (*lower row*) at successive instants of time

decelerated by the surrounding gas. The same effect is observed with increasing the gas mass (Fig. 6.10). The latter effect depends on the relation between masses of target atoms and surrounding gas. If the target atom mass exceeds the gas atom mass, this dependence is inverse, the angular scatter decreasing with increasing the gas molecule mass (Fig. 6.11).

6.1.4 Comparison with Experiments

Numerical calculations were performed with parameters close to experimental ones. Figure 6.12 demonstrates a comparison of the calculated propagation of the contact boundary with experimental data. The position of the contact boundary was measured in experiments by the emission of carbon atoms [25]. One can see that the vapor expansion velocity decreases with increasing the surrounding gas

Fig. 6.8 Influence of the surrounding gas type on the target vapor distribution

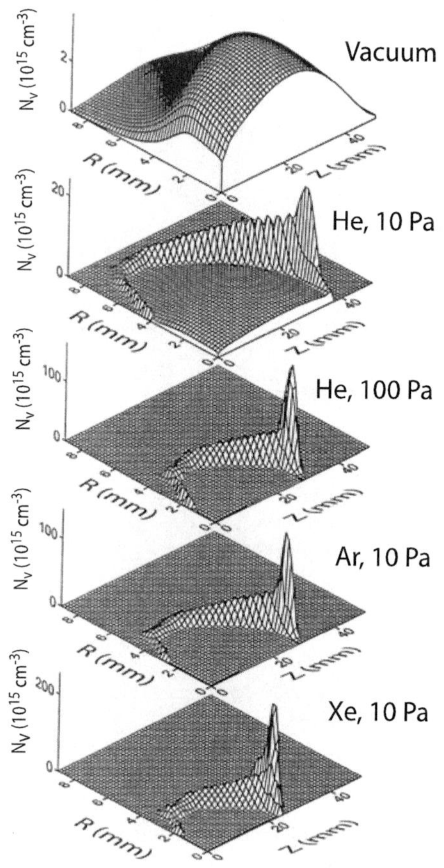

pressure. As the gas pressure increases, the results of calculations better correspond to experiments.

The discrepancy at low pressures is probably caused by fast particles, which were neglected in calculations.

By using the time-of-flight method, the function $\sqrt{\varepsilon}F(\varepsilon,0)$ was found in experiments [26] (see Fig. 6.14). This function can be calculated with the help of the numerical model discussed here. Although the calculations correlate qualitatively with experiments, the quantitative discrepancy is obvious. This can be explained by the fact that targets were irradiated by many pulses. It is known that the target surface changes after irradiation, which was neglected in calculations.

In experiments [27], the function $v^2F(mv^2/2)$ was measured, which can also be compared with numerical calculations. Figure 6.13 demonstrates rather good agreement. The study of the ablation mechanism is still far from complete. For example, it is assumed [7] that the vaporization, formation of clusters and molecules in vapor, etc. can be governed by a nonthermal mechanism.

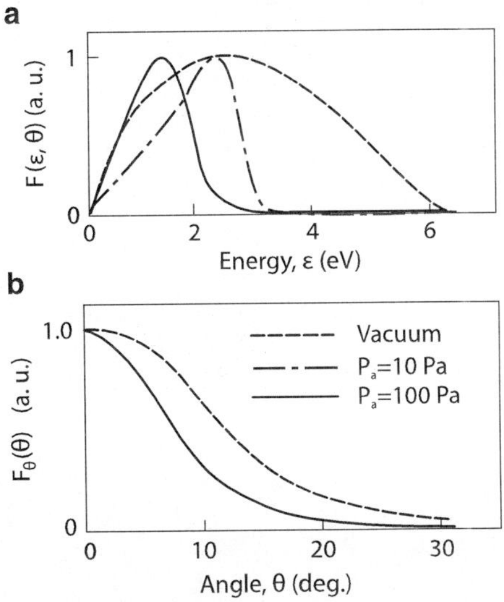

Fig. 6.9 Velocity distributions of vapor particles flying normally to the surface (**a**) and angular distributions (**b**) at different surrounding gas pressures

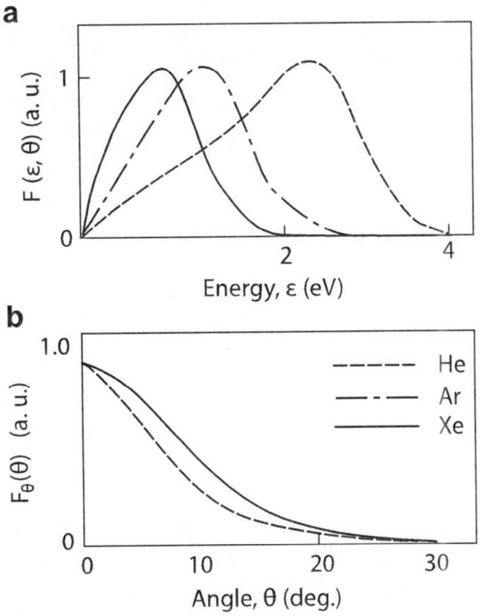

Fig. 6.10 Same parameters of the carbon vapor as in Fig. 6.9 as functions of the atomic mass of the surrounding gas

It is interesting to compare the properties of ablation under different conditions. For example, the ablation of an AlN target by a KrF in the N_2 atmosphere was described in [7]. The ablation time was ~30 ns. The focal spot area in [7] was smaller ($0.2 \times 0.8\,\text{mm}^2$) and the energy density in the focal spot was higher ($20\,\text{J/cm}^2$). Because the experimental conditions in [7] were different, we can perform only qualitative comparison. Figure 6.15 shows the dynamics of the fluorescence front at different pressures of the surrounding gas. As in Fig. 6.12, the monotonic decrease in the velocity and length of a vapor cloud is observed

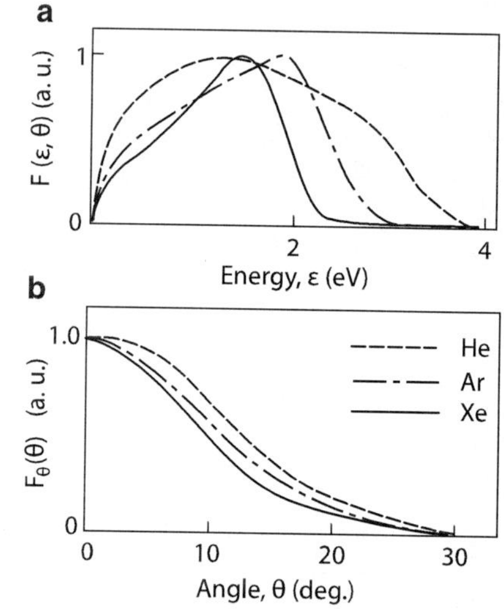

Fig. 6.11 Same parameters as in Fig. 6.10 for the gold vapor

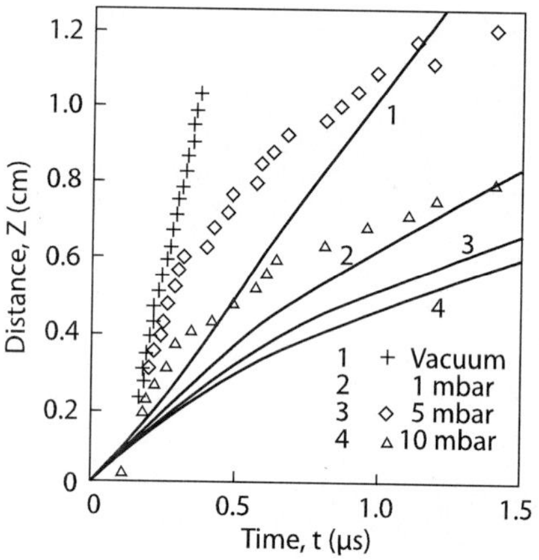

Fig. 6.12 Dynamics of the contact interface between the carbon vapor and external gas at different pressures

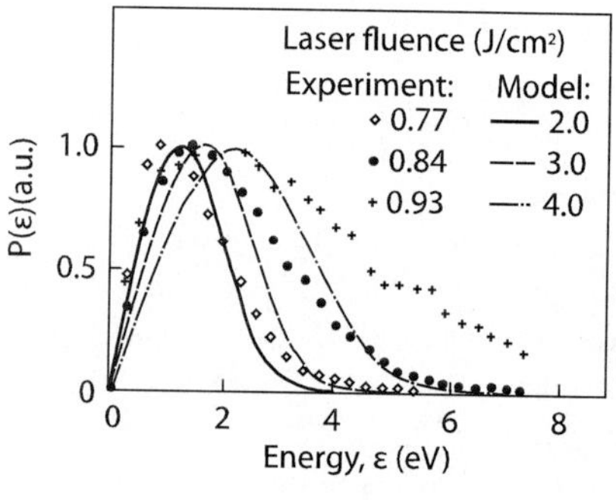

Fig. 6.13 Energy distribution of Au vapor particle for different energy densities ($\lambda = 248$ nm, $r_f = 0.5$ mm, $\tau = 30$ ns, $t = 5\,\mu$s)

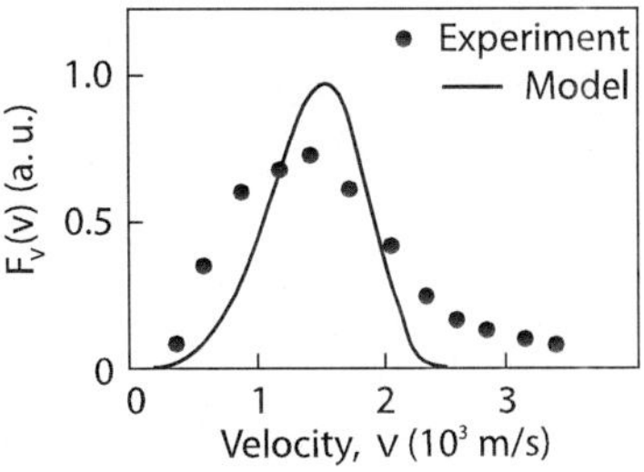

Fig. 6.14 Velocity distribution of Au vapor particles ($\tau = 11$ ns, $\lambda = 532$ nm), $r_f = 25\,\mu$m, $\varepsilon = 4.9$ J/cm^2, $t = 500$ ns

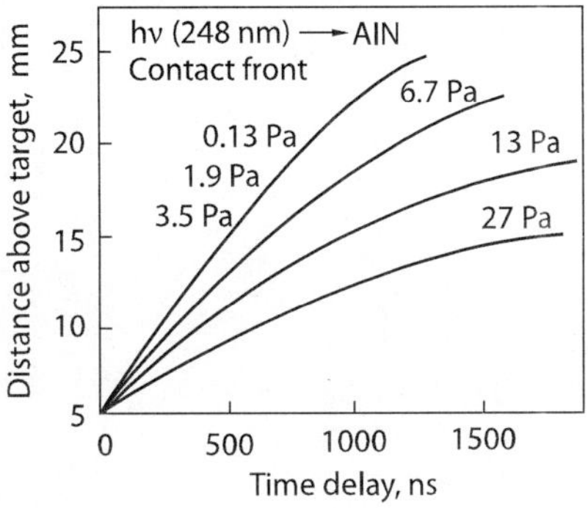

Fig. 6.15 Dynamics of the fluorescence front of the AlN vapor in the nitrogen atmosphere at different nitrogen pressures (KF laser, $\tau = 30$ ns)

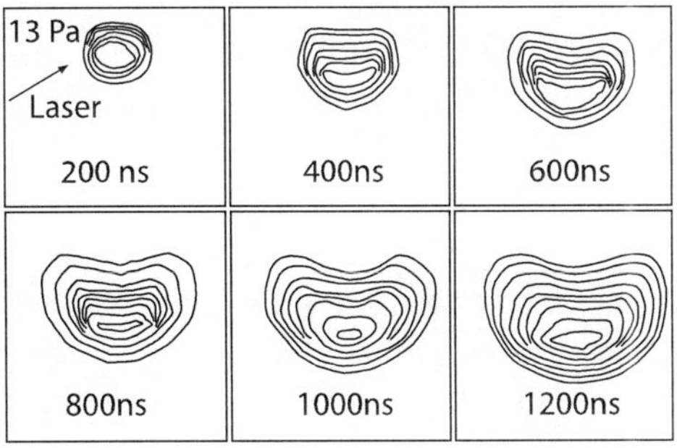

Fig. 6.16 Expansion dynamics of the vapor of a target irradiated by a laser pulse in gas

with increasing pressure. As for the internal structure of the cloud, the results of calculations are also in qualitative agreement with experiments. Figure 6.16 shows that, as in calculations, vapors are concentrated near the front part of the contact boundary. It is important to note the vapor cloud front is comparatively smooth within the spatial ($\sim$150 μm) and temporal ($\sim$5 ns) resolution of an ICCD camera. The question of the stability of the contact vapor-surrounding gas interface was actively discussed in the literature [29]. Here, the situation can be favorable for the development of the Rayleigh-Taylor instability, when light liquid (vapor) pushes heavy liquid (cold surrounding gas).

Consider again the question of the angular divergence of the vapor cloud. The calculations show that the angular divergence is smaller in the case of large focal spots Fig. 6.6. In experiments with a copper target in vacuum irradiated by 532 nm, $\tau = 6$ ns, $\varepsilon = 2.7$ J/cm^2 second-harmonic pulses from a Nd:YAG laser,

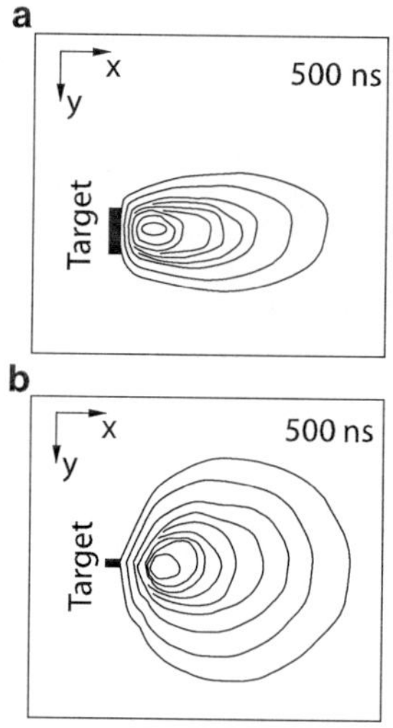

Fig. 6.17 Expansion dynamics of a vapor cloud in vacuum at different focal spot sizes

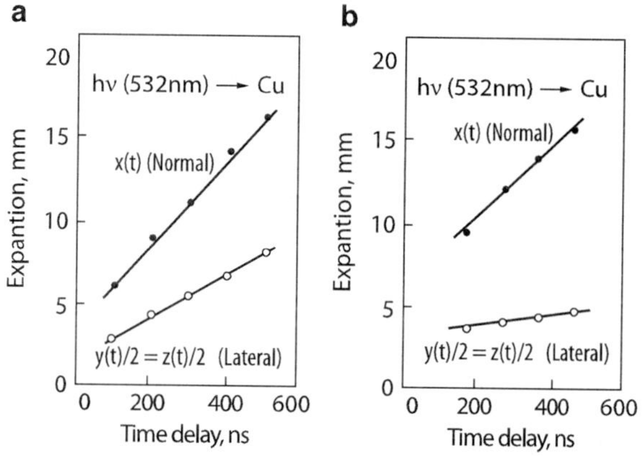

Fig. 6.18 Expansion dynamics of a copper target vapor cloud normally to the target surface and in the lateral direction, (**a**) small focal spot, (**b**) large focal spot

a vapor cloud expands in the lateral direction faster when the focal spot radius is smaller Fig. 6.17, Fig. 6.18. One can see from the same figures that the expansion velocity along the normal is independent of the focal spot size, in accordance with calculations. The weaker lateral expansion in the case of large focal spots in calculations is explained by the fact that the lateral expansion begins later because the unloading wave arrives later to the cloud center.

In conclusion, consider an interesting effect observed upon ablation produced by radiation focused into a noncircular spot. This is the so-called rotation effect [7] which lies in the fact that if the focal spot is elongated to one side, the film deposited on a substrate proves to be elongated in the perpendicular direction Fig. 6.19.

This effect can be qualitatively explained by using the dependence of the lateral expansion velocity on the focal spot size considered above. Indeed, the focal spot along the y axis is smaller (Fig. 6.19a), and hence the expansion velocity of the

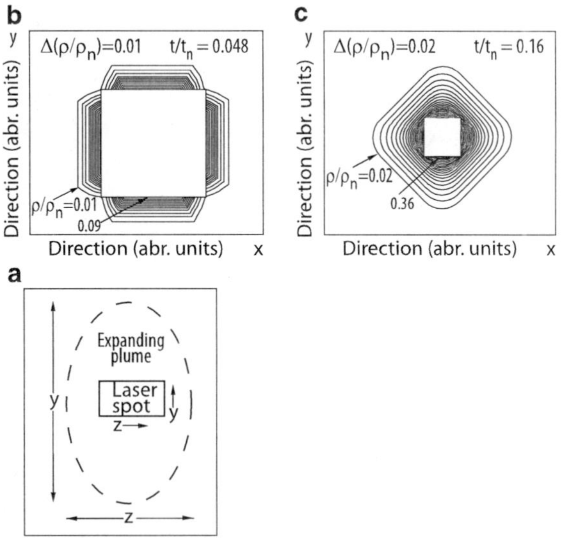

Fig. 6.19 Scheme of a target ablation for a nonuniform irradiation spot (**a**). The numerical calculation of the expansion dynamics of a gas column during the instant removal of walls: (**b**) the beginning of the process, (**c**) the final stage

vapor cloud along the y axis is greater than that along the z-axis. This leads to a greater expansion of the cloud along the y axis and, hence, to the rotation of the greater transverse size by 90° with respect to the greater size of the focal spot. The mechanism of this effect was investigated in detail by using numerical simulations [7]. Generally speaking, this problem is three-dimensional, but we can restrict our consideration to a simplified two-dimensional model. The system of equations (6.7) was considered in the Cartesian coordinate system in the transverse direction. In thc dircction normal to the target surface, the problem was assumed uniform. Such a tube with a rectangular cross section was filled with gas, and the walls were rapidly removed at the instant $t = 0$. The initial gas density and temperature were selected close to their experimental values presented in Fig. 6.18. The density and time are presented in relative units. Figures 6.19b and 6.19c correspond to the time intervals 200–700 ns and 600–2,000 ns, respectively. The first figure shows the beginning of gas expansion after the removal of the walls. One can see that the vapor expansion along diagonals lags behind the vapor expansion normally to the edges. Later, this leads to the rotation of the vapor cloud (Fig. 6.19c). Thus, it becomes clear that this effect has a pure gas-dynamic nature. In detail this phenomenon was analyzed in the review [8] and the references therein.

6.1.5 Ablation Efficiency

We considered above comparatively long nanosecond pulses of duration up to 30 ns, when two-dimensional gas dynamics becomes important. The main attention was given to this case. The initial ablation stage, when the lateral expansion of

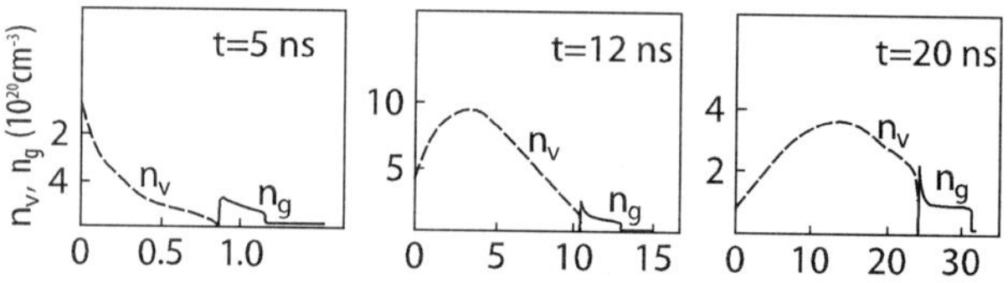

Fig. 6.20 Vapor profile n_v and ambient gas profile n_g

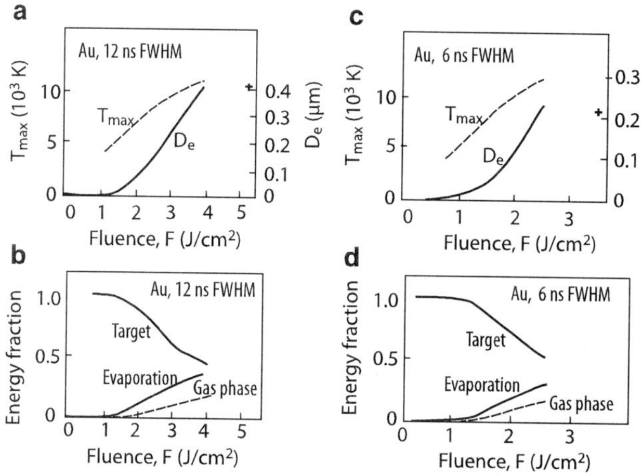

Fig. 6.21 Dependences of the temperature and thickness of the evaporated layer on the energy density absorbed in a target for laser pulse durations 12 ns (**a**) and 6 ns (**c**). The energy spent for vaporization and energy distributions in targets and vapors for different energy densities and different pulse durations (**b, d**) (crosses are experimental data)

vapors still remained unimportant, was treated as the initial condition for the two-dimensional problem of vapor expansion. This problem was studied in the one-dimensional approximation. For shorter pulses of duration 6–17 ns, the one-dimensional approximation will be sufficient to describe the dynamics during the entire pulse. In this case, calculations can be performed with more accurate boundary conditions, described in Chap. 1. All this allows us to consider the important question about the ablation efficiency.

Figure 6.20 presents the density dynamics of gold vapors evaporated by a 266 nm, 6 ns, 2 J/cm^2 laser pulse. One can see that gold vapors are flying away from the target surface during the pulse and compress the surrounding Ar gas at pressure of 1 atm in front of them. When laser radiation is switched off, the target is cooled, and the saturated vapor pressure becomes lower than the vapor pressure in a bunch. Condensation begins near the target surface, and density decreases here. At later times ($t = 20$ ns), the expansion of the vapor bunch also plays a role. Figure 6.21 presents the generalized picture of ablation for different fluences. One can see that, as the fluence is increased, the vaporization efficiency increases, achieving 30%. The beam energy is mainly spent to heat the target. In the case of longer pulses lighter metals, the vaporization efficiency increases and achieves ~50%, for example for Al [50]. In the latter case, the increase in the vaporization efficiency is caused by the increase in the sound speed and a faster expansion of the vapor cloud. A comparison with experiments, which can be often performed in laser technologies, is complicated because the absorption coefficient for laser radiation is unknown under experimental conditions. The absorption coefficient of gold at 193 nm at room temperature is 0.75. Taking this value into account, the agreement between experimental and calculated keyhole depths is satisfactory. For high fluences, the approach of the target temperature to the critical point and the formation of plasma in vapors can play a role ([41]; see also Chap. 7).

In the beginning of this chapter it was noted that the ablation techniques may be used in microelectronics to deposit thin films on a solid surface. The energy spectrum of the deposited particles determines the important properties of such films [9]. However, this method of film deposition suffers from dripping of the target material onto the substrate. One way to solving this problem is to use the scheme of two intersecting beams, resulting from the exposure of two targets [10]. In this method, since the droplets do not dissipate, they fly past the substrate. The plasma ions, due to the scattering effect, do fall partially on the substrate. The greatest deviation is observed for the ions for which the product of their density to the scattering cross section is maximum ($\lambda = 1.05\,\mu\text{m}$, $\tau = 15\,\text{ns}$, $\varepsilon = 15\,\text{J/cm}^2$, Si) [11].

Another way to remove the droplets may be their mechanical separation. Rotation of the separator dist and the laser startup are synchronized so that the disk hole is located along the torch axis at the time of ablation. Because of the difference between the velocities of the ion-atomic mixture and the droplets the latter ones are cut off [12].

6.1.6 Ablation of Materials Irradiated by Ultrashort Laser Pulses

We considered above the ablation of materials under equilibrium conditions, when the characteristic times of processes exceed the relaxation times during which different equilibrium states are established. This is first of all the establishment of the equal electron and lattice temperatures in a target and the establishment of the Maxwell distribution in the atomic and electron subsystems in the gas cloud. Clearly, under nonequilibrium conditions, ablation can reveal new properties. The time of energy transfer from the electron subsystem of the lattice is the shortest. The estimate by Coulomb collisions gives the value of $\tau_e \sim 1$ ps for metals. For example, $\tau_e \sim 5$ ps for Ni [30]. For semiconductors, in particular, for silicon, $\tau_e \sim 1$ ps [31]. Thus, if the laser pulse duration is shorter than 1 ps, the whole absorbed laser radiation energy is released only in the electron subsystem. The lattice heating begins only within the time τ_e after the laser pulse end. The laser radiation energy is absorbed in a narrow layer of width $\sqrt{\varepsilon}$ times smaller than the wavelength (see

Sect. 1.2, Chap. 1). Because $\varepsilon \sim 10^4$ for metals, this layer, for example, for Ni is ~100 nm and 400 nm for Si [32]. The sound speed in Ni is 5,000 m/s and 2,000 m/s in Si. This gives the relaxation time of thermoelastic stresses ~20 ps for Ni and 200 ps for Si. Because the lattice heating time is considerably shorter than the thermoelastic stress time, the overheated layer can be in a metastable state. After relaxation, this layer should expand adiabatically into vacuum, cool off and produce a cloud of nanoparticles. Because the heating time in both cases is shorter than the relaxation time, the formation of nanoparticles should not depend considerably on the laser pulse duration. One can see from Fig. 6.22 that the shape and parameters of the size distribution of particles are not strongly different for laser pulses of different durations [33]. A photograph of the expansion of an ablation cloud is presented in Fig. 6.23. Figure 6.24 shows the velocity distribution of nanoparticles. Their average velocity is of the order of the sound speed. The same figure shows that upon ablation, along with the cloud of nanoparticles, a cloud of atoms and ions is formed, which flies from the surface at the velocity 100 times greater than the velocity of nanoparticles. Such high velocities and energies of atoms and ions can be caused by the intense heating of electrons in strong electric fields of laser radiation followed by the acceleration of ions in spatial charge layers [33].

The problem of ablation is very broad, and we considered here only the most general features caused by gas dynamics. A great part of this issue is the interaction of vapors with surrounding gases when they are mixed, take part in chemical reactions, etc. Here, we will consider only the simplest case of mixing due to the diffusion of atomic vapors, without chemical reactions. Such a situation is realized,

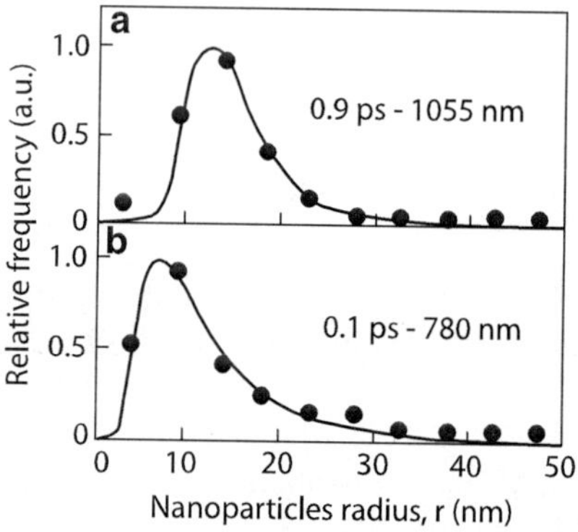

Fig. 6.22 Size distributions of nanoparticles for different laser pulse durations

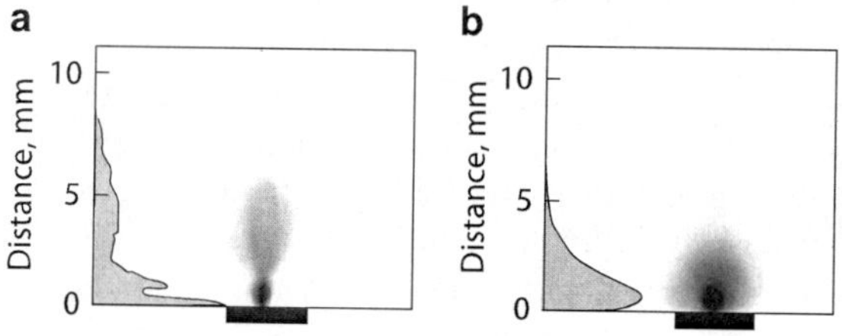

Fig. 6.23 Photographs of an ablation cloud and the vapor density distribution along the symmetry axis

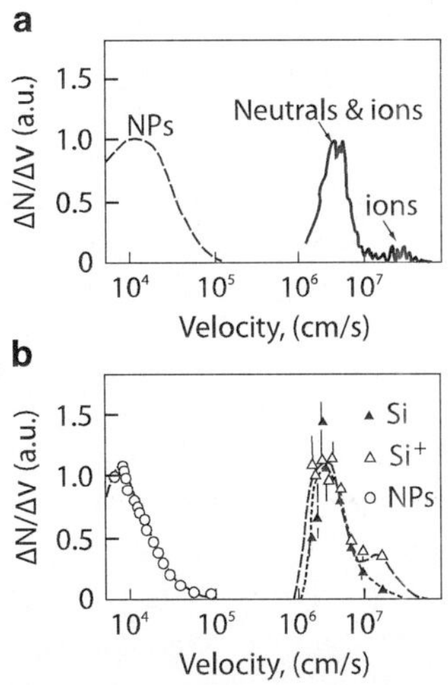

Fig. 6.24 Velocity distributions of nanoparticles and vapor atoms and ions. Theory (*upper curves*), experiment (*lower curves*)

e.g., in a practically important case of synthesis of nanoparticles during pulsed vaporization.

6.2 Modelling of Synthesis of Nanoparticles Upon Pulse Laser Vaporization

Nanocrystal materials obtained from ultradispersed powders attract increasing recent attention due to their specific physical and mechanical properties. Modern technologies for synthesis of nanoparticles are based on homogeneous nucleation in gases. They include synthesis in an arc plasma, high-frequency plasma discharges, magnetron sputtering, vaporization of materials in a vacuum chamber, electron-beam methods, etc. [34–36]. The typical yield of a product, provided by different methods of condensation in gases and plasma, is a few grams per minute. Considerable efforts were directed to the synthesis of monodisperse particles because a narrow size distribution is necessary for the production of high-quality nanomaterials. The plasma technology of condensation in a gas phase [37, 38] is used to produce monodisperse Cr and Ni particles with an average size of $\sim$10 nm. However, the deposition rates are very small ($\sim$0.1 nm/s). This method is based on the spatial separation of nucleation and condensation growth. As a result, condensation centers have the same conditions for their growth. The method of vaporization to a gas phase [39] and synthesis in a high-frequency discharge provide a greater yield (the deposition rate is $\sim$1 μm/s for Fe particles), but the size distribution is broader.

Laser ablation is a promising direction for the synthesis of nanoparticles because it provides a simple control of the vaporization process [39–43]. The laser ablation technology does not require a chamber for the evaporated substance and allows the synthesis of nanoparticles of refractory or chemically active materials. The maximum mass removal speed achieved in such experiments is 1 g/s [42]. We have shown in the previous section that at high energy flux densities or at low pressures of the surrounding gas, a supersonic vapor jet is formed. The adiabatic expansion becomes the main cooling mechanism. The overcooled vapor condenses partially on the target surface and in the gas phase in the form of nanoparticles [40, 41]. As shown in Chap. 1, a subsonic vapor flow is formed at lower energy flux flow densities. In this case, the heat conduction and diffuse mixing with the surrounding gas become the main cooling mechanisms [42, 43].

It was shown experimentally that the most suitable regime for the laser synthesis of particles is the conditions when the vapor pressure near the irradiated target surface slightly exceeds the surrounding gas pressure and a submerged erosion jet is formed. For example, the irradiation of a WO_3 and MoO_3target in the H_2 atmosphere at pressures 1–15 atm by 1.06 μm, 1 ms, 3×10^5 W/cm^2 laser pulses results in the formation of tungsten particles of size 10–100 nm [42], Figs. 6.25, 6.26. Note that there is no point in increasing the power density above the value indicated above because this leads to the removal of large drops from the surface. The condensation conditions (the overcooling degree, etc.) are substantially different at different points in the flow, and as a result, the size of produced particles varies in a broad range.

A decrease in the laser power density reduces the yield of particles, but it can be used for obtaining high-quality ultradispersed powders without inclusions of drops. It is interesting to study laser vaporization at low power densities, when the saturated vapor pressure does not exceed the external pressure and an erosion jet is not formed. In such diffusion vaporization regime, vapor propagates into gas exclusively due to diffusion to the gas phase [44]. The erosion jet is considered later in this chapter.

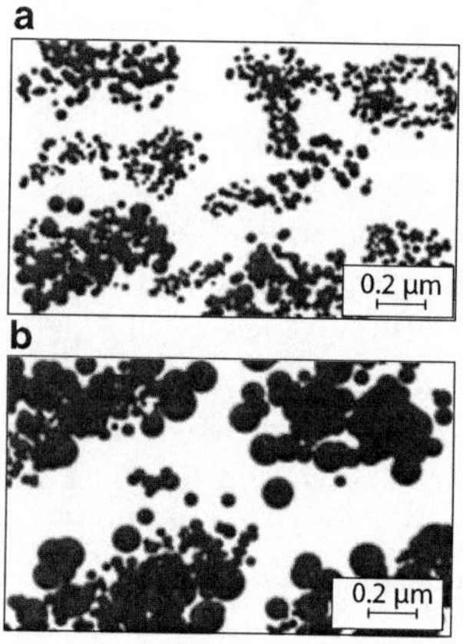

Fig. 6.25 Tungsten parcels deposited at different hydrogen pressures: (**a**) $p = 5 \cdot 10^5$ Pa, (**b**) $p = 1.5 \cdot 10^6$ Pa

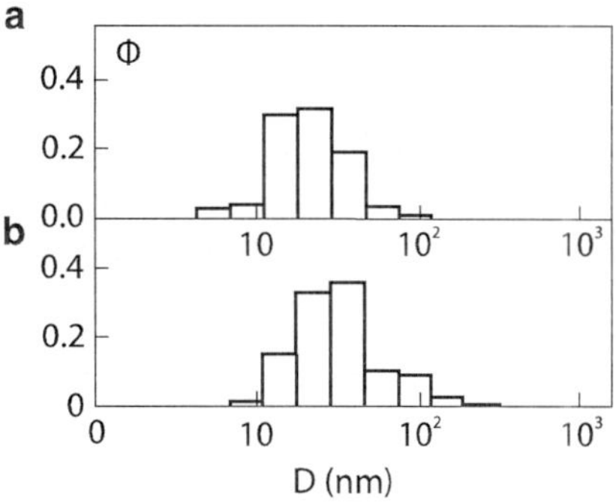

Fig. 6.26 Size distributions of molybdenum particles at different hydrogen pressures: (**a**) $p = 2 \cdot 10^5$ Pa, (**b**) $p = 5 \cdot 10^5$ Pa

6.2.1 *Diffusion Model*

Laser radiation is partially reflected and partially absorbed by a target. It is assumed that the absorption depth in the target is much smaller than the characteristic size of a zone heated due to heat conduction, which is turn much smaller than the laser beam diameter. Therefore, the heat propagation inside the target can be described, as during PLD, by the one-dimensional heat conduction equation

$$\frac{\partial T_c}{\partial t} = \frac{1}{N_c C_{\text{pc}}} \frac{\partial}{\partial z_c} \left(\kappa_c \frac{\partial T_c}{\partial z_c} \right) \tag{6.9}$$

where C_{pc} is the heat capacity of a condensed phase, N_c is the numerical atomic density, T_c is temperature, κ_c is the heat conductivity, t is time, and z_c is the spatial coordinate along the internal normal to the surface. Equation (6.9) is valid inside solid and liquid phases, and the Stefan conditions at the phase interface have the form

$$T_c = T_m \quad z_c = z_i \tag{6.10}$$

$$H_m N_c \frac{\partial z_i}{\partial t} = \left(\kappa_c \frac{\partial T_c}{\partial z_c} \right) \bigg|_{z_c = z_i + 0} - \left(\kappa_c \frac{\partial T_c}{\partial z_c} \right) \bigg|_{z_c = z_i - 0} \tag{6.11}$$

where T_m is the melting temperature, H_m is the melting heat, and z_i is the melting front coordinate. The initial condition for (6.9) is $T_c = T_a$, where T_a is the external temperature. The boundary condition at $z_c = 0$ describes a heat source produced by pulsed laser radiation:

$$-\kappa_c \frac{\partial T_c}{\partial z_c} = \begin{cases} q_\alpha & at \ 0 \le t \le \tau \\ 0 & at \ t > \tau \end{cases} \tag{6.12}$$

where q_α is the absorbed energy flux and τ is the pulse duration. Thermal losses caused by vaporization and radiation from the surface are neglected because they are very small compared to the laser energy in the irradiation regime under study.

It is assumed that vapor is sufficiently cold and remains transparent for laser radiation, while the diffusion length of vapor molecules is much smaller than the beam diameter. Processes under study are characterized by high temperatures and high temperature gradients. Therefore, we should analyze the heat conduction, diffusion and thermal diffusion processes. The hydrodynamic continuity, vapor mass, and energy transfer equations for a binary gas mixture [45] in the one-dimensional case have the form

$$\frac{\partial N}{\partial t} + \frac{\partial (NU)}{\partial z} = 0 \tag{6.13}$$

$$\frac{\partial (Nx)}{\partial t} + \frac{\partial}{\partial z}\left[NxU - ND\left(\frac{\partial x}{\partial z} + \gamma \frac{x(1-x)}{T}\frac{\partial T}{\partial z}\right)\right] = -I \tag{6.14}$$

$$\frac{\partial (NE)}{\partial t} + \frac{\partial}{\partial z}\left[NEU - \kappa \frac{\partial T}{\partial z} - \gamma PD\left(\frac{\partial x}{\partial z} + \gamma \frac{x(1-x)}{T}\frac{\partial T}{\partial z}\right)\right] = H_b I \tag{6.15}$$

where $z = -z_c$ is the coordinate along the external normal to the surface, N is the numerical density of the mixture, U is the average molecular velocity, $x = N_v/N$ is the vapor concentration, E is the average energy per molecule, T is temperature, D is the diffusion coefficient, κ is the heat conductivity, γ is the thermal diffusion factor, and H_b is the latent heat of evaporation. The equation for momenta can be omitted because pressure $P = NkT$ is virtually constant in subsonic flows. The formation and growth of particles in the condensed phase is described by the flow in (6.14) and by the corresponding energy source in (6.15).

In the case of a low vapor concentration ($x \ll 1$) considered here, energy equation (6.15) can be transformed to

$$\frac{\partial (NE)}{\partial t} + \frac{\partial}{\partial z}\left(NEU - \kappa \frac{\partial T}{\partial z}\right) = H_b I \tag{6.16}$$

Thermal diffusion terms are proportional to x and, therefore, can be neglected in (6.16). On the contrary, thermal diffusion remains significant in vapor transfer processes, and all the terms in diffusion equation (6.14) are proportional to x.

The condensation rate (the number of molecules condensed in the unit volume per unit time) is described by the expression

$$I = 4\pi J \int_0^\infty F(R)R^2 \mathrm{d}R + \frac{4}{3}\pi R_*^3 N_c G \tag{6.17}$$

with the condensation flow determined by the Knudsen formula

$$J = \frac{Px - P_s}{(2\pi m_v kT)^{1/2}} \tag{6.18}$$

Here, the saturated vapor pressure

$$P_s = P_0 \exp\left(\frac{H_b}{kT_b} - \frac{H_b}{kT}\right) \tag{6.19}$$

can be determined from Clausius–Clapeyron's equation assuming that H_b is independent of T [46]. (T_b is the boiling temperature at the normal pressure $P_0 = 1$ atm). The nucleation rate (the number of condensation centers produced in the unit volume per unit time) is found in the theory of homogeneous nucleation [46, 47]:

$$G = \frac{2(Nx)^2}{N_c}\left(\frac{\sigma}{2\pi m_v}\right)^{1/2} \exp\left(-\frac{b}{\theta^2}\right)$$

where σ is the surface tension, and the coefficient b is

$$b = \frac{16\pi}{3}\frac{\sigma^3}{N_c^2 H_b^2 kT}.$$

The overcooling degree is defined as $\theta = (T_e - T)/T_e$ with the equilibrium temperature for the specified vapor concentration

$$T_e = \left[\frac{1}{T_b} - \frac{k}{H_b}\ln\left(\frac{Px}{P_0}\right)\right]^{-1}$$

determined from (6.19). The critical radius of a nucleus is found from $R_* = 2\sigma/(N_c H_b \theta)$. The size distribution $F(R, z, t)$ of condensate particles (FdR is the number of particles with radii from R to $R + \mathrm{d}R$ in the unit volume) is determined from the transport equation

$$\frac{\partial F}{\partial t} + \frac{\partial FU}{\partial z} + \frac{\partial FL}{\partial R} = 0 \tag{6.20}$$

where $L = J/N_c$ is the linear growth rate of a particle.

The initial conditions for the system of equations (6.13), (6.14), (6.16), and (6.20) are $U = x = F = 0$, $T = T_a = 298$ K. The boundary conditions for hydrodynamic equations (6.13), (6.14), and (6.16) on the target surface $Z = 0$ describe the liquid–vapor equilibrium: $U = 0$, $x = P_s/P$, $T = T_l(0)$. These conditions are valid both for vaporization and condensation. If the saturated vapor pressure is higher than the partial pressure xP for $z > 0$, vaporization occurs, otherwise vapor condenses on the surface. The vaporization and condensation rates are determined by diffusion processes in the gas phase. It is reasonable to define the boundary condition for (6.20) as a balance between nucleation and growth of particles in the $R = 0$ plane: $FL = G$. Here, FL can be treated as the R-component of the distribution function flux vector in the $R - z$ space.

The properties of the target material used in calculations are presented in Table 6.2 [48]:

Table 6.2 Material properties

	Zn	Sb	Cr	Zr	W
Numerical density $N_c (10^{28}\ \mathrm{m}^{-3})$	6.57	3.31	8.34	4.26	6.34
Heat capacity C_{pc} (J kg^{-1} K^{-1})					
Solid phase	430	220	540	340	160
Liquid phase	480	260	600	380	180
Heat conduction κ_c (W m^{-1} K^{-1})					
Solid phase	110	18	70	23	110
Liquid phase	60	27	70	23	110
Melting temperature T_m (K)	693	904	2,150	2,125	3,693
Boiling temperature T_b (K)	1,179	1,907	2,945	4,613	5,953
Melting heat H_m (kJ mol^{-1})	7.2	20.1	16.3	14.6	35.1
Vaporization heat H_v (kJ mol^{-1})	115	124	338	558	770
Surface tension σ (N m^{-1})	0.77	0.35	1.54	1.40	2.32

The dependences of the kinetic transfer coefficients on the gas temperature were calculated from the model of solid elastic spheres [49] as

$$\kappa = \kappa_0 (T/T_b)^{1/2} \tag{6.21}$$

$$D = D_0 \frac{(T/T_b)^{1/2}}{N/N_0}$$

where $N_0 = P_0/kT_b$, κ_0 and D_0 are the gas density, heat conductivity, and the mutual diffusion coefficient, respectively, at the normal pressure $P_0 = 1$ atm and the boiling temperature T_b of the target material. The values of κ_0 were calculated by interpolating of extrapolating experimental data [48] by using (6.21). The diffusion coefficient and heat conductivity were calculated by molecular masses m_i and diameters d_i based on the Chapman–Enscoge theory. In the first approximation, they are described by the expressions

$$D_0 = \frac{3}{2N_0(d_v + D_g)^2} \left(\frac{kT_b(m_v + m_g)}{2\pi m_v m_g} \right)^{1/2}$$

$$\gamma = \frac{5}{2} \frac{2 + 2\mu - \{(1+s)^2 \mu^{3/2}/4[2(\mu+1)]^{1/2}\}(15\mu - 7)}{30\mu^2 + 16\mu + 13}$$

where $\mu = m_v/m_g$ is the mass ratio and $s = d_v/d_g$ is the ratio of molecular diameters. The kinetic coefficients are presented in Table 6.3.

6.2.2 Results and Discussion

The results of numerical calculations for a Zn target irradiated by $\tau = 1$ ms, $q = 4 \times 10^4$ W/cm^{-2} laser pulses in the surrounding Ar gas are presented in

Table 6.3 Vapor kinetic coefficients

	H_2	Ar	Zn	Sb	Cr	Zr	W
Molecular mass (a.e.)	2.0	40.0	65.4	121.8	52.0	91.2	183.9
Molecular diameter d (nm)	0.241	0.316	0.228	0.287	0.211	0.264	0.231
Heat conductivity κ_0 (10^{-2} W m^{-1} K^{-1})							
in Ar			4.76	6.05	7.52	9.41	10.69
in H_2			44.9	N/A	70.9	N/A	N/A
Diffusion coefficient D_0 (10^{-4} m^2 s^{-1})							
in Ar			1.90	2.91	8.42	12.20	18.60
in H_2			9.19	N/A	38.60	N/A	N/A
Thermal diffusion factor γ							
in Ar			0.14	0.42	0.04	0.30	0.44
in H_2			0.80	N/A	0.73	N/A	N/A

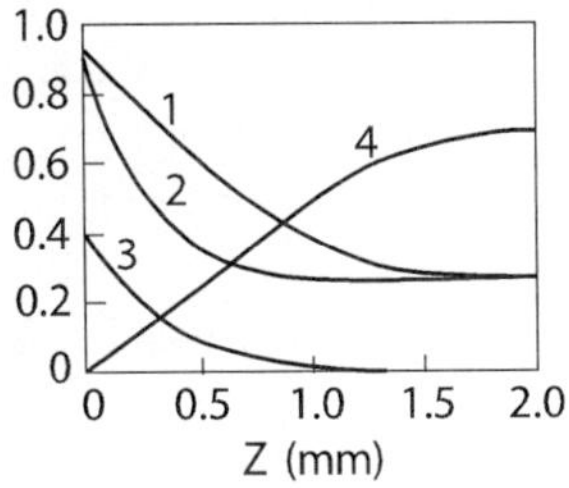

Fig. 6.27 Spatial distributions of dimensionless hydrodynamic parameters in Ar for $P = 1$ atm and $t/\tau = 0.9$. (**1**) gas temperature T_g; (**2**) temperature $T(-z_c)$ of a Zn target; (**3**) vapor concentration x; (**4**) velocity u

Figs. 6.27, 6.28, 6.29, 6.30. A change in the surface temperature is determined by the absorption of laser radiation and thermal losses due to heat removal into a target and surrounding gas. The competition of these processes determines the growth of the target and gas temperature during pulsed irradiation (see curves 1 and 2 in Fig. 6.27). The thermal expansion of gas near the target produces a hydrodynamic flow (curve 4 in Fig. 6.27). The material evaporated from the target surface diffuses into gas (curve 3 in Fig. 6.27).

Vapor passes to the oversaturated state due to mixing with cold gas, and the homogeneous nucleation begins (dotted curve in Fig. 6.28). After the end of the laser pulse, the surface temperature begins to drop. Vaporization changes to vapor condensation on a cold surface. The homogeneous nucleation still continues (solid curve 1 in Fig. 6.28) and vapor passes to nuclei, resulting in the formation of nanoparticles.

The dependence of the homogeneous nucleation rate on the coordinate has a sharp maximum (Fig. 6.28). Therefore, the nucleation process can be considered as the motion of a thin layer inside which nuclei are formed (the nucleation front).

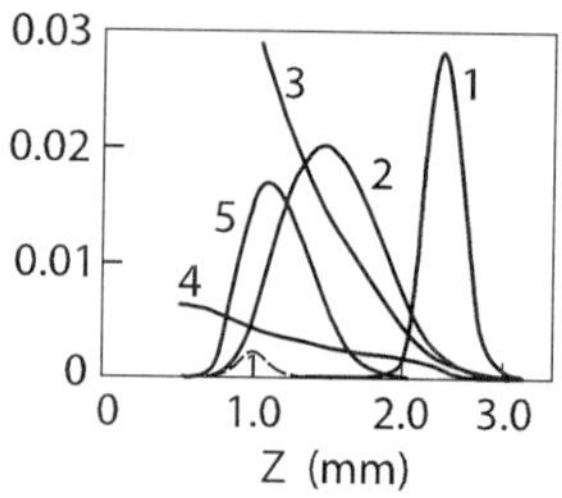

Fig. 6.28 Dimensionless characteristics of the particle formation in the Zn–Ar system for $P =$ 1 atm and $t/\tau = 0.9$ (*dashed curves*) and $t/\tau = 2.0$ (*solid curves*); (**1**) nucleation rate $g \cdot 10^{-4}$; (**2**) particle concentration $f_0 \cdot 10^{-5}$; (**3**) average particle size; (**4**) standard deviation (dispersion); (**5**) mass concentration of particles

Particles grow behind the nucleation front (see curve 5 describing the spatial distribution of the condensate mass). Thus, nucleation and condensation processes are spatially separated. Note that particles do not grow near the surface because vapor condenses predominantly on the target surface rather than on nuclei. Usually, about half the evaporated material condenses back on the target, while the fest of the material is spent to form particles. The average size of particles (curve 3 in Fig. 6.28) decreases with distance from the target surface. The standard deviation (curve 4) proves to be much smaller than the average size of produced particles.

The total amount of vapor per target surface unit is defined as

$$W_v = \int_0^{\infty} N x \mathrm{d}z$$

The amount of condensate atoms per target surface unit is W_c. The sum of these two integrals gives the amount of the evaporated material: W_v and W_c.

The dimensionless values of these integrals (see Table 6.4) as functions of time are presented in Fig. 6.29. The amount of the evaporated material achieves a maximum at the laser pulse end. After the end of the laser pulse, nucleation begins, the surface temperature drops, and as a result, vapor intensely condenses back on the surface. The noticeable condensation of vapor on nuclei begins at $t/\tau \approx 1.5$. By this moment, about half the evaporated material condenses on the surface and nucleation is virtually completed. Thus, the nucleation and condensation particle growth processes are strongly spatially and temporally separated.

The calculated dependences of the average size of condensing particles on the working parameters of the process are presented in Fig. 6.30. The laser power density was selected so for different target materials that the maximum vapor pressure was slightly lower than the specified external pressure of the surrounding gas. It was found that the average size of synthesized particles decreased with increasing pressure and boiling temperature of the target material and increased with increasing the laser pulse duration. One can see that the properties of the

Table 6.4 Dimensionless variables

Variable	Normalization unity
Temperature T/T_b	T_b
Coordinate in gas $z = Z/Z_0$	$Z_0 = (D_0\tau)^{1/2}$
Coordinate in target $z = Zc/Z_{c0}$	$Z_{c0} = (\lambda_c\tau/N_c C_{pc})^{1/2}$
Velocity $u = U/U_0$	$U_0 = Z_0/\tau$
Particle radius $r = R/R_0$	$R_0 = (N/N_c)\,\tau(kT_b/2\pi m_v)^{1/2}$
Momentum $f_k = F_k/F_k^0$	$F_k^0 = (R_0)^{k-3}N_0/N_c$
Nucleation rate $g = G/G_0$	$G_0 = F_0^0/\tau$
Vapor and condensate density $(i = v)$	
$(i = c) w_i = W_i/W_i^0$ (per surface unity)	$W_i^0 = N_0 Z_0$
Density of condensation centers	
$w_n = W_n/W_n^0$ (per surface unity)	$W_n^0 = F_0^0 Z_0$
Size distribution $h = H/H_0$	$H_0 = F_{-1}^0 Z_0$

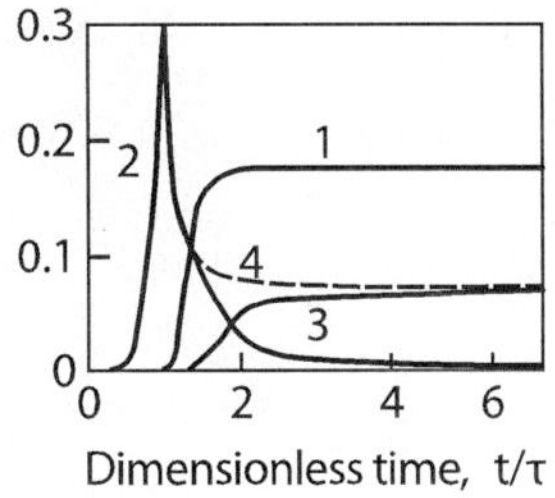

Fig. 6.29 Vaporization, nucleation, and condensation kinetics in the Zn–Ar system for $P = 1$ atm and $\tau = 1$ ms. Dimensionless densities of condensation centers ($w_n \cdot 10^{-4}$) (**1**); vapor (w_v) (**2**); condensate (w_c) (**3**); and the target material in the gas–condensate mixture ($w_v + w_c$) (**4**)

process in the diffusion regime of particle synthesis can differ qualitatively from these properties during synthesis with the help of a submerged vapor jet (cf. Figs. 6.25, 6.26 and 6.33a). This is explained by the different mechanisms of these technologies.

6.2.3 *Erosion Jet*

We will consider a submerged jet in the quasi-stationary regime [51]. For this purpose, it is necessary to select a pulse of duration exceeding the pressure equalization time across ($\tau \gg r_f/c_s$) and along ($\tau \gg l/V$) the jet (V is the vapor speed). It was shown in Chap. 1 that a subsonic jet penetrates into the surrounding gas by the depth equal to ten calibers. Therefore, the fulfillment of the second inequality provides the quasi-stationary condition. Under experimental conditions ($r_f \approx 0.5$ mm, $\alpha q = 5\times10^5$ W/cm^2, $V \sim 50$ m/s [42]) compared with calculations,

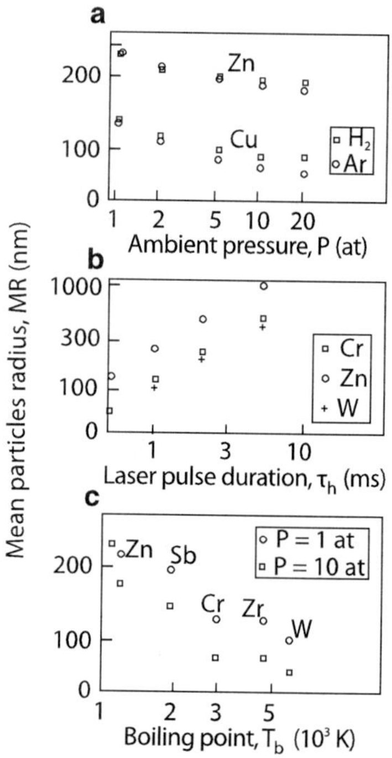

Fig. 6.30 Influence of the parameters of the laser vaporization process on the average particle size: (**a**) surrounding gas pressure ($t_h = 1$ ms); (**b**) laser pulse duration (Ar, $P = 1$ atm); (**c**) material (Ar, $t_h = 1$ ms)

a pulse of duration ~1 ms is sufficient to produce a quasi-stationary jet. In gas-dynamic equations (6.13)–(6.15), in the stationary case, the nonstationary terms should be replaced by the terms describing transfer across the flow:

$$\frac{\partial N}{\partial t} \to \frac{1}{r}\frac{\partial}{\partial r}(rNU_r) \tag{6.22}$$

$$\frac{\partial}{\partial t}(Nx) \to \frac{1}{r}\frac{\partial}{\partial r}(rNxU_r) \tag{6.23}$$

where U_r is the transverse velocity. In addition, the diffusion vapor transfer should be taken only across the flow because the jet length is considerably greater than its width. Therefore, the replacement should be made in (6.14).

$$\frac{\partial}{\partial z}(ND)\frac{\partial x}{\partial z} \to \frac{1}{r}\frac{\partial}{\partial r}\left(rD\frac{\partial x}{\partial r}\right) \tag{6.24}$$

It is necessary to take into account the momentum transfer across the jet:

$$\frac{1}{r}\frac{\partial}{\partial r}(rNUU_r) + \frac{\partial}{\partial z}(NU^2) = \frac{1}{r}\frac{\partial}{\partial r}\left(r\nu N\frac{\partial U}{\partial r}\right) \tag{6.25}$$

In the equation for the conservation of the jet energy, we will take into account the subsonic nature of the flow, i.e. $p = const$:

$$\frac{1}{r}\frac{\partial}{\partial r}(rU_r) + \frac{\partial}{\partial r}U = \frac{1}{p}\left[\frac{1}{r}\frac{\partial}{\partial r}\left(rN\chi\frac{\partial T}{\partial r}\right) + \frac{H_b}{c_p}I\right] \tag{6.26}$$

Gas-dynamic equations should be supplemented with the boundary conditions. On the substrate surface, the energy balance

$$xNUL_b = \alpha q - q_1, \quad r < r_f \tag{6.27}$$

should be fulfilled, where q_l is the energy loss due to heat conduction and radiation. We assume for simplicity that the velocity of vapors flying away from the surface quadratically depends on the distance from the focal spot center,

$$U = U_m(1 - r^2/r_f^2), \quad r \leq r_f$$

and is zero for $r > r_f$, where U_m is a parameter of the problem. Within the focal spot on the surface, temperature is close to the boiling temperature T_b and a gas-vapor mixture consists of vapors ($x = 1$). Outside the focal spot, $T = T_a$, $x = 0$, where T_a is the surrounding gas temperature. At infinity, we have $U = U_r = 0$, $T = T_a$, $x = 0$. We will determine the condensation rate I as in the diffusion model (6.16)–(6.20).

The detailed structure of the erosion jet near a Cr target surface in argon is shown in Fig. 6.31a. One can see that inert gas penetrates into the vapor jet near the target. The maximum of the nucleation rate is located here. The condensation front, as in the diffusion model, does not coincide with the nucleation front. They are formed on the gas–vapor jet-cold Ar gas interface. Condensate particles are transferred by

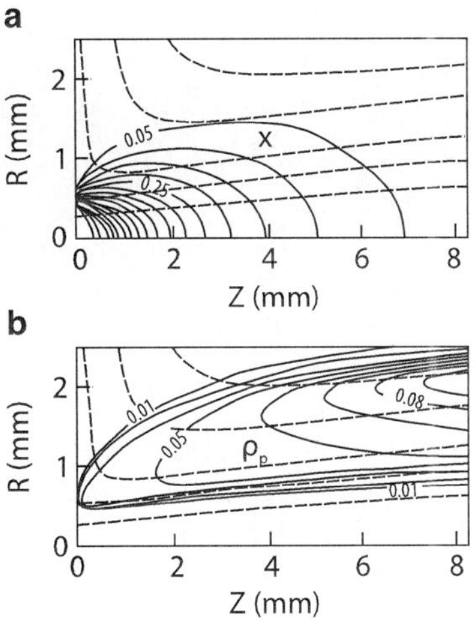

Fig. 6.31 Distributions of the vapor density of chromium x (*solid curves*) and argon atoms (*dashed curves*) near a target surface (**a**), particle-mass density ρ_p (solid curves) (**b**)

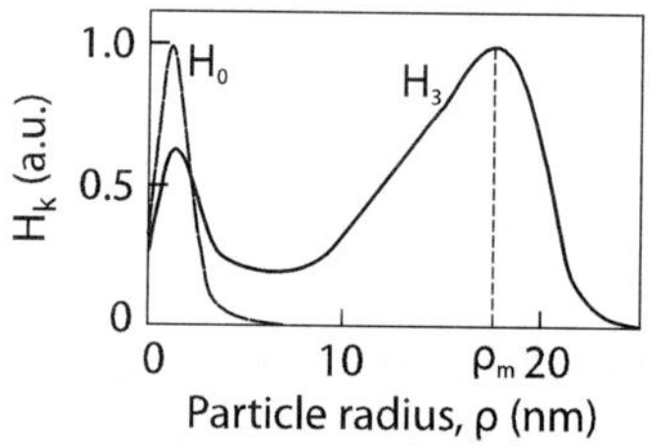

Fig. 6.32 Size distributions of the condensate particle-number near the symmetry axis H_0 and particle-mass away from the symmetry axis H_3 (evaporation of Cr in Ar, $p = 1$ atm, $r_f = 0.5$ mm)

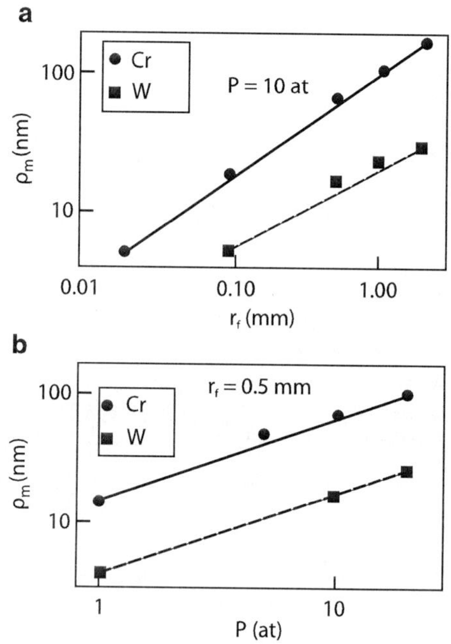

Fig. 6.33 Dependences of the size of chromium and tungsten particles on the focal spot size (**a**) and external gas pressure (**b**)

the flow the periphery to the region nearer to axis. Here, the maximum density of condensate particles is observed. However, the condensate density has a maximum away from the system axis (Fig. 6.31b). This is explained by the fact that the number of particles is smaller here, but they are larger, which is well demonstrated in Fig. 6.32.

Figure 6.33b shows that small particles can be obtained by reducing the focal spot radius or decreasing the external gas pressure Fig. 6.33a. Here, it is reasonable to compare the results of calculations with experiments. The particle size in experiments was $\rho_m \propto p^{0.8}$ [42], while in numerical calculations it was $\rho_m \propto p^{0.77}$ (Fig. 6.33b).

In conclusion of this section, we summarize the results of the study of the ablation process. In the simplest case, the ablation mechanism during the deposition of films is investigated quite thoroughly and is in good qualitative agreement with experiments. At the same time, taking into account a great variety of real technologies, it is necessary to study many processes such as plasma formation, mixing and chemical reactions with the surrounding gas, the ejection of drops of the target melt, etc. As for the synthesis of nanoparticles, it is desirable to consider the two-dimensional flow and mutual diffusion of vapors and surrounding gases, chemical reactions during ablation into the active external medium. The next chapter is devoted to plasma formation processes, which often accompany ablation.

References

1. R.V. Arutyunyan, V.Yu. Baranov, L.A. Bol'shov, et al., *Vozdeistvie lazernogo izlucheniya na materially (Interaction of Laser Radiation with Materials)* (Nauka, Moscow, 1989)
2. W.W. Duley, *UV Lasers: Effects and Applications in Materials Science* (Cambridge University Press, Cambridge, 1996)
3. N.N. Anisimov, Ya.A. Imas, G.S. Romanov, et al., *Deistvie izlucheniya bol'shoi moshchnosti na metally (Interaction of High-power Radiation with Metals)* (Nauka, Moscow, 1970)
4. N.N. Rykalin, A.A. Uglov, A.N. Kokora, *Lazernaya obrabotka metallov (Laser Metal Processing)* (Mashinostroenie, Moscow, 1975)
5. J.F. Ready, *Lasers in Modern Industry* (Society of Manufacturing Engineers, Dearborn, Michigan, 1979; Mir, Moscow, 1981)
6. A.A. Vedenov, G.G. Gladush, *Fizicheskie oprotsessy pri lazernoi obrabotke materialov (Physical Processes in Laser Material Processing)* (Energoatomizdat, Moscow, 1985)
7. R. Kelly, A. Miotello, A. Melle, A. Guidoni, in *Plume Formation and Characterization in Laser-Surface Interactions. Experimental Methods in the Physical Science*, vol. 30, *Laser Ablation and Desorption*, ed. by J.C. Miller, R.E. Haglund (Academic Press, San Diego, 1998)
8. S.I. Anisimov, B.S. Lukyanchuk, Uspekhi *172*, 301 (2002)
9. K.L. Saenger. J. Appl. Phys. **70**, 5629 (1991)
10. A. Gorbunov, A. Tselev, W. Pompe, Proc. SPIE **3688**, 351 (1999)
11. E.V. Khaydukov, O.A. Novodvorsky, A.A. Lotin, et al., J. Tech. Phys. **80**, 59 (2010)
12. O.A. Novodvorsky, A.A. Lotin, E.V. Khaydukov, Patent 89906 RF, Inst. of Laser and Information Technology of Russian Academy of Sciences, 2009
13. C.J. Knight, AIAA j. **17**, 519 (1979)
14. K.R. Chen, J.N. Leboeuf, R.F. Wood, D.B. Geoghegan, J.M. Donato, C.L. Liu, A.A. Puretzky, J. Vac. Sci. Technol. **A14**, 1111 (1996)
15. R. Bellantone, Y. Hahn, J. Appl. Phys. **76**, 1436 (1994)
16. R. Bellantone, Y. Hahn, J. Appl. Phys. **76**, 1447 (1994)
17. P.V. Breslavsky, V.I. Mazhukin, *Mathematical Modeling of the Process of Surface Vaporization by Laser Radiation in the Atmosphere with Back Pressure* (Institute of Mathematical Modeling, Russian Academy of Sciences, Moscow, 1992)
18. A.V. Bulgakov, N.M. Bulgakova, J. Phys. D **28**, 1710 (1995)
19. V. Mazhukin, I. Smurov, G. Flamant, J. Comput. Phys. **112**, 78 (1994)
20. V.I. Mazhukin, I. Smurov, G. Flamant, Appl. Surf. Sci. **96–98**, 89 (1996)
21. V. Mazhukin, I. Gusev, I. Smurov, G. Flamant, Microchem. J. **50**, 413 (1994)
22. D. Sibold, H.M. Urbassek, J. Appl. Phys. **73**, 8544 (1993)
23. T.E. Itina, A.A. Katassonov, W. Marine, M. Autric, J. Appl. Phys. **82**, 3526 (1997)
24. T.E. Itina, A.A. Katassonov, W. Marine, M. Autric, J. Appl. Phys. **77**, 849 (1995)
25. T. Kerdja, S. Abdelli, I. Madaiaga, J. Appl. Phys. **80**, 5365 (1996)
26. T.D. Bennett, C.P. Grigoropoulos, D.J. Krainovich, J. Appl. Phys. **77**, 849 (1995)
27. J.W. Elam, D.H. Levy, J.Appl. Phys. **81**, 539 (1997)
28. D.J. Lichtenwalner, O. Auciello, R. Dat, A.I. Kingon, J. Appl. Phys. **74**, 7497 (1993)
29. S.I. Anisimov, Proc. FIAMN 2000, St.Petersburg 2000.

30. D.S. Ivanov, L.V. Zhigilei, Phys. Rev. **B68**, 064114 (2003)
31. D.R. Lide, *CRC Handbook of Chemestry and Physics*, vol. 75 (CRC Press, Boca Raton, 1994)
32. D.S. Ivanov, L.V. Zhigilei, Appl. Phys. A **79**, 977, (2004)
33. S. Amoruso, G. Ausanio, R. Bruzzese, et al., E-MRS, Strasburg France, 2005
34. H. Haberland, *Clusters of Atoms and Molecules*, vol. 1 (Springer, Berlin, 1994), p. 207
35. R.P. Andres, R.S. Averback, W.L. Brown, L.E. Brus, W.A. Goddard, A. Kador, S.G. Louie, M. Moscovits, P.S. Peercy, S.J. Riley, R.W. Siegel, F. Spaepen, Y. Wang, J. Mater. Res. **4**, 704 (1989)
36. E. Pfender, Plasm. Chem. Plasm. Process. **19**, 1 (1999)
37. S. Yamamuro, K. Sumiyama, K. Suzuki, J. Appl. Phys. **85**, 483 (1999)
38. T. Hihara and K. Sumiyama, J. Appl. Phys. **84**, 5270 (1998)
39. Y. Sasaki, M. Hyakkai, E. Kita, A. Tasaki, H. Tanimoto, Y. Iwamoto, J. Appl. Phys. **81**, 4736 (1997)
40. S.V. Gaponov, A.A. Gudkov, A.A. Fraerman, Zh. Tekh. Fiz. **52**, 1843 (1982)
41. A. Kar, J. Mazumder, Phys. Rev. E **49**, 410 (1994)
42. A.G. Gnedovets, E.B. Kul'batskii, I. Smurov, G. Flamant, Appl. Surf. Sci. **96/98**, 272 (1996)
43. A.G. Gnedovets, A.V. Gusarov, I. Smurov, G. Flamant, Appl. Surf. Sci. **109/110**, 74 (1997)
44. A.G. Gnedovets, A.V. Gusarov, E.B. Kul'batskii, I. Smurov, G. Flamant, *Thermal Plasma Torches and Technologies*, vol. 2 (International Science Publishing, Cambridge, 1998), p. 430
45. J.H. Ferziger, H.G. Kaper, *Mathematical Theory of Transport Processes in Gases* (North-Holland, Amsterdam, 1972)
46. Y.B. Zel'dovich, Y.P. Raiser, *Physics of Shock Waves and High-Temperature Hydrodynamic Phenomena*, vol. 2 (Academic Press, New York, 1967)
47. Y.P. Raiser, Zh. Exp. Teor. Fiz. **37**, 1741 (1959)
48. I.S. Grigor'ev, E.Z. Melikhov (eds.), *Spravochnik fizicheskikh velichin (Handbook of Physical Quantities)* (Energoatomizdat, Moscow, 1991)
49. S. Chapman, T.G. Cowing, *The Mathematical Theory of Nonuniform Gases* (University Press, Cambridge, 1970)
50. A. Gusarov, I. Smurov, J. Appl. Phys. **97**, 014307 (2005)
51. A. Gusarov, A.G. Gnedovets, I. Smurov, G. Flamant, Appl. Surf. Sci. **154-155**, 331 (2000)

Chapter 7
Pulsed Surface Plasma

Abstract In Chap. 7, the pulsed optical breakdown near surface and dynamics of plasma plume are considered. The laser radiation reflection by plasma plume, plasma transparency and plasma phenomena during the deep penetration of laser beam into a sample are described. The results of 2D numerical simulation of plasma dynamics and mass removal are comparised with experiment.

7.1 Pulsed Optical Breakdown Near a Surface

Stationary plasma phenomena – an optical plume and breakdown, were considered in previous chapters. Even if the breakdown condition $q > q_{\mathrm{br}}$ is fulfilled, some finite time is required for the discharge development. The higher is the excess of the radiation intensity over the breakdown threshold q_{br}, the shorter is this time. By solving the problem of a pulsed optical breakdown, it is necessary to find the dependence of the breakdown time on the radiation intensity. As in the case of stationary breakdown, two different cases can be realized here. If a vapor flow is absent before the breakdown (refractory materials, high pressure of the surrounding gas), a thermal breakdown can occur. When a vapor flow develops before the breakdown (low-boiling materials, low pressure), the breakdown can have a different nature (we will call it a pulsed nonequilibrium breakdown). We will consider the optical breakdown near a target by assuming that the radiation intensity is lower that the breakdown threshold of pure gases (without a target). Because the breakdown occurs directly in the gas surrounding the target, the surface of a material irradiated by microsecond pulses from a CO_2 laser even is not partially melted, although the radiation intensity and energy is quite sufficient damage considerable the target ($q \approx 1{,}000\,\mathrm{MW/cm^2}$, $\varepsilon \approx 1\,\mathrm{kJ/cm^2}$) [1].

G.G. Gladush and I. Smurov, *Physics of Laser Materials Processing*,
Springer Series in Materials Science 146, DOI 10.1007/978-3-642-19831-1_7,
© Springer-Verlag Berlin Heidelberg 2011

7.1.1 Nonstationary Thermal Breakdown

The total time of a pulsed thermal breakdown τ_{br} is the sum of the target heating time τ_h, the vapor diffusion time τ_d, and the plasma generation time τ_g [2, 3]. In the case of a pulsed thermal breakdown, the time τ_q is determined by (3.27) and (3.28). It is clear that the value of τ_g will be small at high radiation intensities, and the breakdown time will be determined by the sum $\tau_h + \tau_d$. The minimal value of τ_g can be easily estimated when the energy loss from plasma is small. If the temperature of the wall is not very high, so that $\sigma_{\text{ea}} N > \sigma_c(T_s)n$, we have for atomic gas

$$\tau_g = \frac{cm\omega^2 I}{4\pi e^2 v_e} \frac{1}{N\sigma_{\text{ea}} q} \ln \frac{n_c}{n_0} \approx \frac{I}{240} \frac{T_s}{300} \frac{1}{qp}, \tag{7.1}$$

where τ_g is the breakdown time (in ms); n_c is the density at which Coulomb collisions begin to dominate; q is the radiation intensity (in MW/cm^2); n_0 is the initial plasma density; and p is pressure (in MPa).

The specific heat of a molecular gas is higher than that for an atomic gas (because in the former case, along with the heating of electrons, the heavy component of plasma is also heated). The plasma generation time for a molecular gas breakdown is

$$\tau_g = \frac{c_p T_s cm\omega^2}{4\pi e^2 v_e \sigma_{\text{ea}} n_0 N q}. \tag{7.2}$$

One can see from (7.1) to (7.2) that the breakdown time ratio for atomic and molecular gases is $In_0/T_m c_p$ and depends on the ration of the initial plasma concentration and the neutral gas density. In both cases, we have $\tau_{\text{br}} \approx q^{-1}$. If the surface temperature is high enough, so that Coulomb collisions dominate from the very beginning of the breakdown development, the breakdown time is shortened and becomes the same for atomic and molecular gases:

$$\tau_{\text{br}} = \frac{cm\omega^2 I}{4\pi e^2 v_e \sigma_c n_0 q}, \quad n_0 > n_c \tag{7.3}$$

For example, for a tungsten target in argon at $T_s = 5{,}000$ K and $p = 0.1$ MPa and $q = 5$ MW/cm^2, we have $\tau_{\text{br}} = 0.1$ ms ($n_0 = 1.2 \times 10^{15}$ cm^{-3}). One can see from (7.1), (7.2), and (7.3) that the breakdown time depends significantly on the surface temperature and radiation intensity. The surface temperature in pulsed breakdown experiments [4] is a function of the radiation intensity and time because the breakdown occurs during the increase of temperature (Fig. 7.1). The usual inequality $\tau_{\text{br}} < \tau$ is not the sufficient breakdown condition because the surface temperature is unknown.

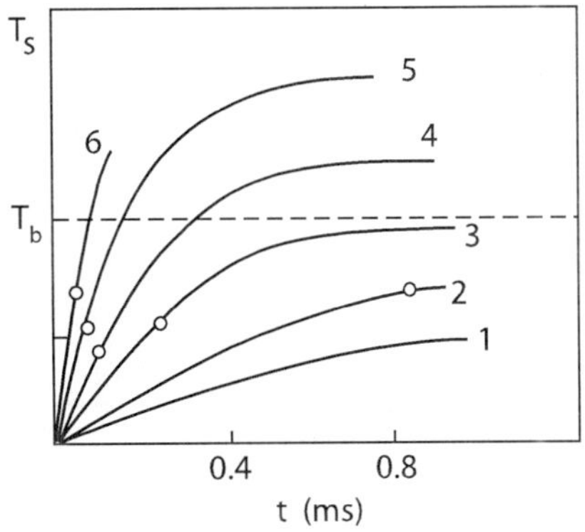

Fig. 7.1 Time dependences of the surface temperature T_s of a refractory metal (tantalum) for different radiation intensities (increasing from curve 1 to curve 6). Circles indicate the instants of an optical breakdown

7.1.2 Quasi-Stationary Breakdown

As pointed out in Sect. 3.2, the breakdown threshold decreases with increasing the surface temperature. If by the end of a pulse of intensity q the surface temperature did not achieve the temperature at which the breakdown threshold q_{br} became lower than q, the breakdown will not occur (see Fig. 7.1). As q is increased, a high temperature will be achieved earlier and when the breakdown threshold will become lower than q, the breakdown will occur (see curve 2 in Fig. 7.1). The higher is q, the earlier the breakdown occurs and the lower is the surface temperature during the breakdown (see curves 2 and 4 in Fig. 7.1). Thus, the quasi-stationary stage of the breakdown is realized, when the stationary theory can be applied in the nonstationary case. At high q, the time during which the surface temperature achieves the breakdown temperature becomes small. Gas near the focal spot has no time to be heated and the sample material vapor has no time to diffuse far enough. Therefore, beginning from some value of q, the breakdown temperature begins to increase with increasing q (curves 5 and 6 in Fig. 7.1). In other words, because the irradiation time is too short, to obtain the breakdown, it is necessary to increase the surface temperature. Thus, the increase in the breakdown intensity with decreasing the breakdown time is caused by the influence of the heated surface. In this case, the quasi-stationary theory also can be used, however, the focal spot radius in expressions (3.34) and (3.35) for the stationary breakdown threshold should be replaced by the characteristic diffusion size or the characteristic size $\sqrt{\chi\tau_h}$ related to heat propagation, where $\tau_h = T^{*2}\kappa_M^2/q^2\chi_M$ is the heating time up to the breakdown temperature T^*. By substituting $\sqrt{\chi\tau_h}$ instead of r_f into (3.34), we obtain that q_{br} increases with T^*:

$$q_{\mathrm{br}} \sim T^* \exp(-I/2T^*)$$

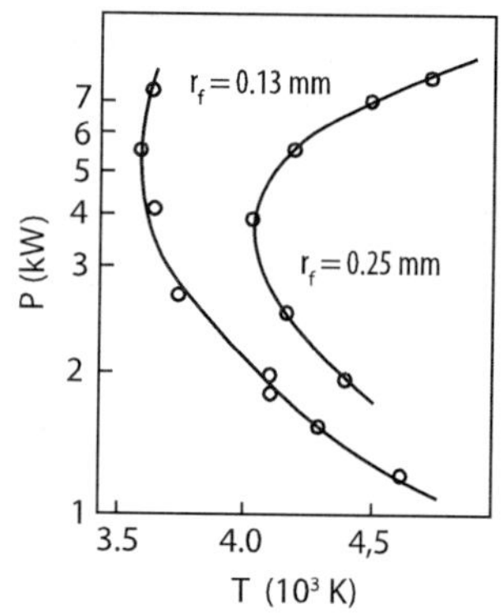

Fig. 7.2 Dependences of the air breakdown power near a tantalum target surface on the surface temperature for two focal spot radii

which is observed in experiments with molecular gases [5] at high breakdown powers (Fig. 7.2).

As the value of q is further increased, the surface temperature can achieve the boiling temperature. In this case, the breakdown in vapor at rest changes to the breakdown in the vapor jet.

7.1.3 Optical Breakdown in a Target Vapor Jet

We can distinguish a breakdown of two types: a pulse breakdown, which occurs simultaneously with the development of the nonstationary vapor flow, and a slow quasi-stationary breakdown, when the vapor flow has time to be established before the breakdown. The first case is realized, e.g., when a sample is irradiated by a high-intensity ($q \approx 100$–$1{,}000\,\mathrm{MW/cm^2}$) microsecond laser pulse. The analytic theory of such a breakdown is absent at present, and so we will restrict ourselves to the description of corresponding numerical calculations. In [6], the development of vaporization and plasma formation on graphite and aluminum targets in vacuum as studied in the one-dimensional approximation. The latter assumption is quite significant and restricts a comparison with experiments by the case of a large focal spot area, short pulses and, hence, a high pulse intensity. In [6], three gas-dynamic equations were solved numerically (conservation of mass, momentum and energy) together with the laser radiation transfer equation:

$$\frac{\mathrm{d}\rho}{\mathrm{d}t} + \frac{\mathrm{d}\rho v}{\mathrm{d}x} = 0, \quad \frac{\mathrm{d}v}{\mathrm{d}t} + v\frac{\mathrm{d}v}{\mathrm{d}x} = \frac{\nabla p}{\rho},$$

$$\frac{\mathrm{d}}{\mathrm{d}t}\rho(\varepsilon + v^2/2) + \frac{\mathrm{d}}{\mathrm{d}x}\rho v(\varepsilon + v^2/2 + p/\rho) = \mu q,$$

$$\frac{\mathrm{d}q}{\mathrm{d}x} = \mu q. \tag{7.4}$$

On the target surface the Jouguet conditions for the gas velocity were taken into account (the gas velocity is equal to the local sound speed) and the law of conservation of energy in the vaporization wave. The plasma density, which determines the value of the absorption coefficient μ, was assumed equilibrium and was determined from the Saha formula (3.4). The calculation was performed for $\tau = 2\,\mu\text{s}$, $q = 320\,\text{MW/cm}^2$ pulses corresponding to radiation from a free-running ruby laser.

The results of the calculation and experiments on the breakdown dynamics and measurements of the recoil momentum are in goof agreement for some metals (e.g., Bi [7]). The optical breakdown in experiments with Al occurs earlier than predicted by calculations. In [8], this was explained by the inhomogeneity of the target surface and the presence of "flakings" having a poor thermal contact with a sample mass. In [9], calculations were performed, similar to [6], by assuming that a target consisted of isolated 4-μm thick layers. The calculations are in good agreement with experiments with $\tau = 5\,\mu\text{s}$ pulses for $q_{\text{br}} = 22\,\text{MW/cm}^2$ [10]. In experiments with $\tau = 1$ ms pulses for $q_{\text{br}} = 0.1\,\text{MW/cm}^2$ [8], the breakdown was observed for $t = 400\,\mu\text{s}$. The calculations showed that the target should not be heated up the melting temperature for this time. The theory can be matched with the experiment assuming the presence of 4-μm thick flakings on the target.

Figure 7.3 presents the dependences of the boiling onset time τ_b on q and the breakdown time τ_{br} calculated for $\lambda = 10.6$ and $1.06\,\mu\text{m}$ assuming that a thermally isolated Al layer has a thickness of $1\,\mu\text{m}$ [9]. In addition, the experimental data obtained for $\lambda = 1.06\,\mu\text{m}$ [11] are also presented. One can see from this figure that the difference in the breakdown times for different laser wavelengths is not large. This can be explained as follows. The absorption coefficient for the CO_2 laser radiation in an aluminum target is twice as small that for the neodymium laser radiation (7 and 15%, respectively, Fig. 1.2). This leads to a high value of τ_b; at the same time, the absorption coefficient for the $\lambda = 10.6\,\mu\text{m}$ radiation by vapor is approximately 100 times higher than that for the $\lambda = 1.06\,\mu\text{m}$ radiation. It seems

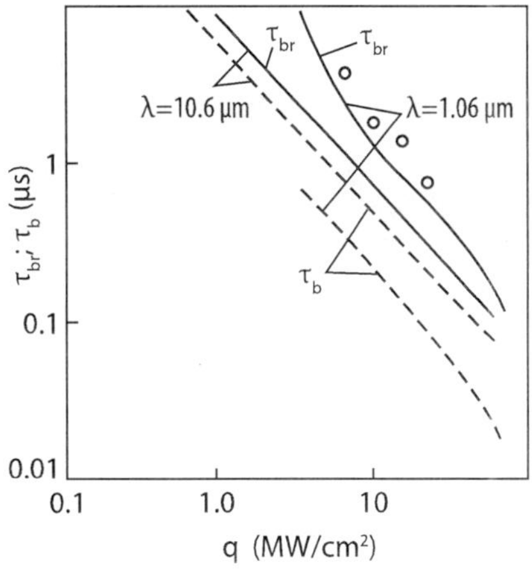

Fig. 7.3 Dependences of the boiling onset time τ_b and the breakdown time τ_{br} on the laser radiation intensity

that the combination of these two factors leads to a small difference in breakdown times at different wavelengths.

The dependence of the breakdown threshold on the radiation wavelength can be found by substituting the vapor density near the target into (7.1) [see (1.31)]:

$$N = \alpha q / L_b M c_s \tag{7.5}$$

where L_b is the vaporization enthalpy; c_s is the sound speed; α is the surface absorption coefficient; and M is the mass of molecules of the target material.

Taking into account that, according to (1.18), $\alpha \approx \sqrt{\omega}$, we obtain for (7.1) and (7.5)

$$\tau_{\mathrm{br}} \sim \omega^{3/2}/q^2$$

or $q \approx \sqrt{\tau_{\mathrm{br}}}$, which is consistent with experimental data generalized in [12]. For the specified pulse duration, the breakdown threshold q_{br} increases with increasing frequency:

$$q_{\mathrm{br}} \approx \omega^{3/4}/\sqrt{\tau}. \tag{7.6}$$

One can see from Table 7.1 that the experimental dependence $q_{\mathrm{br}} \sim \lambda^{-1}$ [13] agrees with theoretical dependence (7.6). A comparison of the breakdown thresholds for $\tau = 1.5$ and $\tau = 0.5\,\mu\mathrm{s}$ [13] showed that $q \sim \tau^{-1/2}$. The same dependence was observed in experiments with a brass target in air, where the breakdown threshold was (2–3)$\cdot 10^7$ W/cm^2 for $\tau = 1\,\mu\mathrm{s}$ and $6 \cdot 10^6$ W/cm^2 for $\tau = 20\,\mu\mathrm{s}$.

Thus, the total breakdown time in the case of microsecond laser pulses is determined by the plasma generation time rather than the vaporization time (otherwise the dependence $q_{\mathrm{br}} \sim \sqrt{\lambda}$ should be observed). Equations (7.1)–(7.3) for the pulsed breakdown time were derived by neglecting the energy loss from the interaction zone. If this loss is not very small, the breakdown threshold will increase with decreasing the focal spot radius due to the increase in the side thermal loss (Fig. 7.4) (air, Al). As a whole, the threshold intensity weakly depends on the focal spot radius for not too small spots, and therefore the threshold power of a laser increases with increasing the focal spot area (Fig. 7.5 [14]). This property is retained for very large spots. For $S = 133\,\mathrm{cm}^2$, the air breakdown threshold near aluminum and steel targets was ~100 MW/cm^2 [14].

Table 7.1 Breakdown intensity q_{br} (in MW/cm^2) near a target as a function of λ for $\tau = 1.5\,\mu\mathrm{s}$

	λ, μm			
Target	1.3	2.9	4	10.6
Copper mirror	50	55	38	13
Steel mirror	N/A	26	14	7
Aluminum mirror	N/A	24	15	6
Unpolished aluminum	12	11	11	5

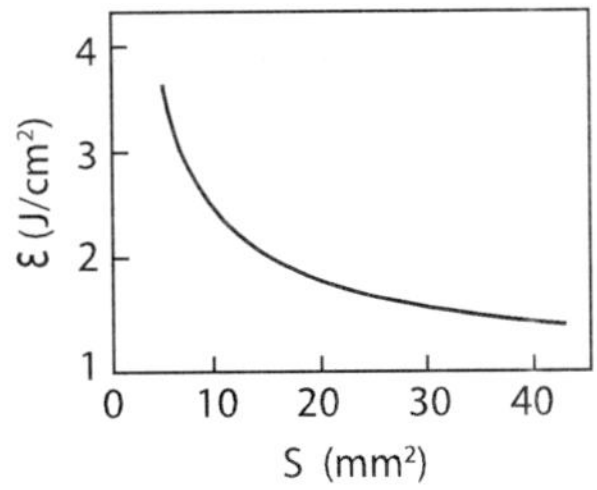

Fig. 7.4 Dependence of the CO_2 laser energy density at which an optical breakdown occurs near a target on the focal spot area

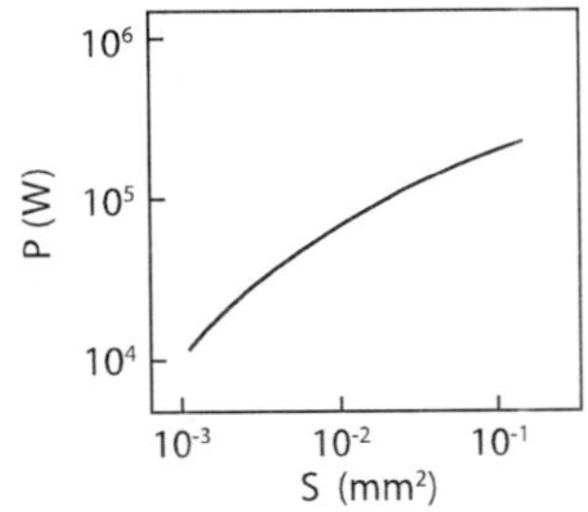

Fig. 7.5 Dependence of the air breakdown power near an aluminum target on the focal spot area

7.1.4 Two-Dimensional and Nonequilibrium Effects in the Pulsed Breakdown

The attempts to take into account the two-dimensional nature of a flow existing in real experiments were made already in the first calculations of the interaction of radiation with metals within the mechanism of developed vaporization. This concerns the interaction both in vacuum [6] and air [15]. Two-dimensional effects are especially important when samples are irradiated by long light pulses and the vapor cloud size exceeds the focal spot size. The side expansion was taken into account in [15] by assuming that the flow had a spherically symmetric character; at the same time, the propagation of laser radiation was also considered within the framework of a one-dimensional problem. Calculations were performed for $\tau \approx$ 1–5 ms, $q \approx$0.1–1 MW/cm^2 triangular pulses irradiating a 16–32 mm thick titanium sample. The focal spot diameter was set equal to 0.5–1.3 cm [12]. The calculations were performed using the scheme similar to [6]. Table 7.2 presents a comparison of calculations with experimental data. One can see that the experimental data agree with calculations performed for a homogeneous target. It is interesting to analyze the plasma formation dynamics. Consider the third line of Table 7.2.

After evaporation for $\tau_b = 0.4$ ms, due to the rapid expansion of vapor a weak shock wave is produced in air in which temperature is slightly higher than that of the surrounding air (Fig. 7.6). At the same time, the vapor temperature increases, a breakdown occurs in the vapor, and its temperature increases up to 20,000 K. The shock wave at the moment $t = 0.49$ ms propagated from vapor to air and its pressure achieves 0.8 MPa. According to the Hugoniot relation, such a compression in a shock wave should result in the increase in temperature only up to 700 K, whereas the temperature in the wave is much higher (20,000 K, see Fig. 7.6). This suggests

Table 7.2 Dependence of the breakdown and boiling onset times on the intensity of a 5-ms laser pulse

Intensity, MW/cm^2	Target thickness, mm	Spot diameter, cm	τ_{br}, ms (experiment)	τ_{br}, ms (calculation)	τ_b, ms (calculation)
0.3	32	1.0	1.80	1.80	1.46
0.4	16	1.0	1.00	1.20	1.01
1.7	16	0.5	0.57	0.49	0.40

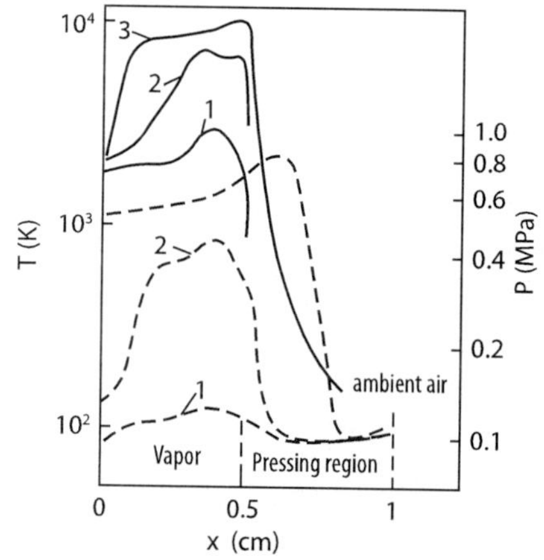

Fig. 7.6 Vapor temperature (*solid curves*) and pressure (*dashed curves*) distributions for a titanium target vaporized in air upon irradiation by a CO_2 laser. $t = 0.475$ (**1**), 0.477 (**2**), and 0.49 μs (**3**)

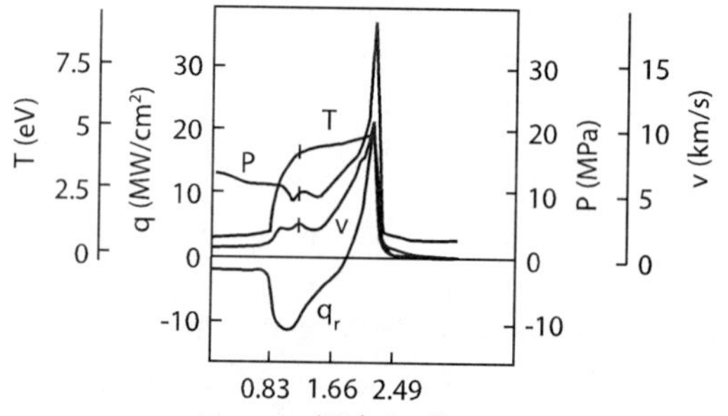

Fig. 7.7 Distributions of gas-dynamic parameters during irradiation of an aluminum target in air by a neodymium laser for $t = 0.62$ μs and $q = 100$ MW/cm^2. The *dashed line* is the vapor–air interface, the air pressure is 0.01 MPa

that a radiation wave is produced in gas due the absorption of laser radiation because this process is manly determined by the absorption of radiation from plasma by cold air in front of the wave. The same picture is observed when microsecond pulses are used (Fig. 7.7). The development of a plasma bunch after breakdown will be considered in greater detail in next sections.

The authors of [16] analyzed an optical breakdown produced near a target by microsecond pulses by taking into account inhomogeneities on the target surface. However, unlike [9], where inhomogeneities were simulated by replacing the target by a thin layer and the gas flow was assumed flat, in [16] a decisive role of the two-dimensional nature of the gas flow appearing during the evaporation of thermally isolated grains on the target was emphasized. The grains were simulated by small

$\delta = 0.2\,\mu\text{m}$ thick discs of radius $R \approx 6\,\mu\text{m}$. Due to their small volume, grains are rapidly evaporated, the flow becomes two-dimensional after a very short time, and a vapor microjet appears in the surrounding gas. When the radiation intensity absorbed in the target is of the order of 10 MW/cm^2, the jet will be supersonic. The initial electrons are produced due to thermal ionization near the surface, i.e. their density is determined by the Saha formula with the surface temperature T_s. Because the laser radiation intensity considerably exceeds the stationary breakdown intensity, the balance equation for the plasma concentration and the concentration of Al atoms excited to the first level was written by neglecting losses. Therefore, it was assumed that after the establishment of a gas-dynamic flow, the concentration of excited atoms and electrons begins at once increase against its background behind the Mach disc (see Sect. 1.4):

$$\frac{\mathrm{d}N_1}{\mathrm{d}t} = k_1^* n_e N_M, \quad \frac{\mathrm{d}n_e}{\mathrm{d}t} = k_i n_e N_1^*,$$

$$\mu q = \acute{\varepsilon}_1 \frac{\mathrm{d}N_1^*}{\mathrm{d}t} + (I - \acute{\varepsilon}_1)\frac{\mathrm{d}n_e}{\mathrm{d}t}$$

where $\acute{\varepsilon}_1$ is the energy of the first excited level. One can see from these equations that plasma is generated due to the ionization of the first excited level. The authors assume further that $a = k_1^*/k_1$ is a constant. This allows them, not knowing the ionization and excitation rates, to calculate the increase in concentrations N_1 and n_e with time and the breakdown time τ_{br}. Figure 7.8 presents the dependence of the boiling onset time τ_b for an aluminum grain for $\alpha = 0.05$, $\delta = 0.2\,\mu\text{m}$, $T_b = 3{,}000\,\text{K}$, $T_0 = 300\,\text{K}$, $\rho = 27\,\text{g/cm}^3$, and $a = 1$. For $q > 10\,\text{MW/cm}^2$, the total plasma formation time is determined by the vaporization time. For $q < 10\,\text{MW/cm}^2$, the vaporization time becomes smaller than the breakdown time τ_{br}. Therefore, the plasma formation time will be $\tau_b + \tau_{\text{br}}$. For comparison, the dot-and-dash curve in Fig. 7.8 presents the results of one-dimensional calculations [9] (see the dependence τ_{br} for $\lambda = 10.6\,\mu\text{m}$ in Fig. 7.3 and experimental data obtained in different works [17]). One can see that, by varying the grain size, the agreement with experiments can be obtained. A disadvantage of this approach is the assumption that the ionization coefficients for the excited level and the excitation coefficient for this

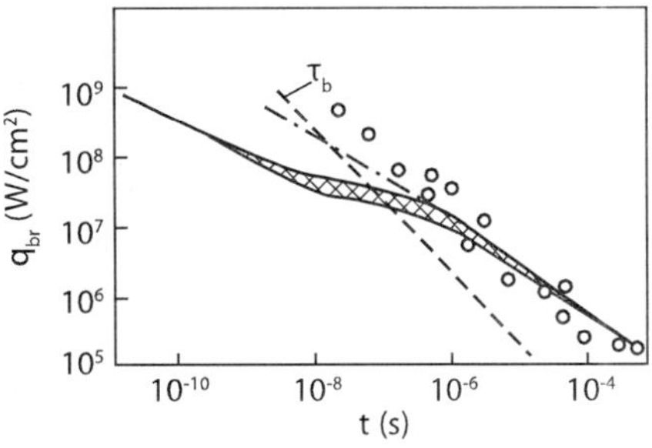

Fig. 7.8 Dependence of the air breakdown intensity near an aluminum target on a CO_2 laser pulse duration

level are constant. This approach also uses the assumption that a breakdown in a small jet of diameter about 20 μm is transformed to the breakdown of the entire focal volume of a laser beam of diameter ~500 μm.

In summary, we can say that the breakdown of gases near a target by millisecond pulses can be described as a whole by theoretical considering a uniformly evaporating target. Upon irradiation by millisecond laser pulses, plasma is produced after the beginning of vaporization and the damage of the target.

In the case of microsecond pulses, theoretical models can be correlated with experiments by assuming the presence of thermal grain inhomogeneities on the target surface. In this case (as for millisecond pulses and refractory targets), plasma screens the target before the beginning of its damage. It is clear that in the case of long enough pulses, the produced plasma can escape from the focal spot region, be reducing screening, which does occur for $\tau > 5$–$20\,\mu$s (see below).

7.2 Nonequilibrium Mechanisms of the Pulsed Breakdown

The consideration of the breakdown mechanism presented above is based on the assumption of the equilibrium ionization of material vapor. Such a situation can take place for refractory materials. This is not the case when we are dealing with a breakdown near low-melting (low-boiling) material such as aluminum, copper, steel, etc. Here, at the boiling temperature not exceeding 3,000°, the plasma density calculated by the Saha formula is low ($<10^{13}$ cm^{-3}). This results in a considerable deviation from the local thermal equilibrium. Indeed, according to estimates presented in Chap. 3 [see (3.43)], the step ionization frequency becomes lower than the inverse diffusion time and lower than the inverse light pulse duration which we will consider below. Thus, a breakdown should be described under these conditions by considering a nonequilibrium plasma taking into account processes violating the equilibrium such as the escape of radiation from plasma, diffusion of charges, vapor gas dynamics (in the case of a pulsed breakdown), etc. To calculate ionization correctly, it is necessary to calculate the population kinetics of atomic levels. The solution of this problem becomes too cumbersome if plasma inhomogeneities are taken into account. We consider first the influence of the nonequilibrium plasma on the breakdown mechanism for the simplest case of homogeneous plasma, when the nonequilibrium nature is caused by the radiation escape from plasma. We will study the qualitative features of the influence of the structure of atomic levels on the breakdown mechanism for two different materials: aluminum and copper [18]. They have close thermal properties (boiling temperatures are $T_b(Al) = 2{,}793$ K and $T_b(Cu) = 2{,}816$ K and vaporization heats $H(Al) = 292$ kJ/mol and $H(Cu) = 300$ kJ/mol) and ionization potentials, but different electronic energy level diagrams. In particular, the first excited level of Al is located considerably higher than the similar level of the copper atom ($E_1(Al) = 3.14$ eV, $E_i(Al) = 5.98$ eV, $E_1(Cu) = 1.39$ eV, $E_i(Cu) = 7.72$ eV, Fig. 7.9). We will consider the breakdown of the vapor of these materials under typical experimental conditions. The Nd laser radiation intensity

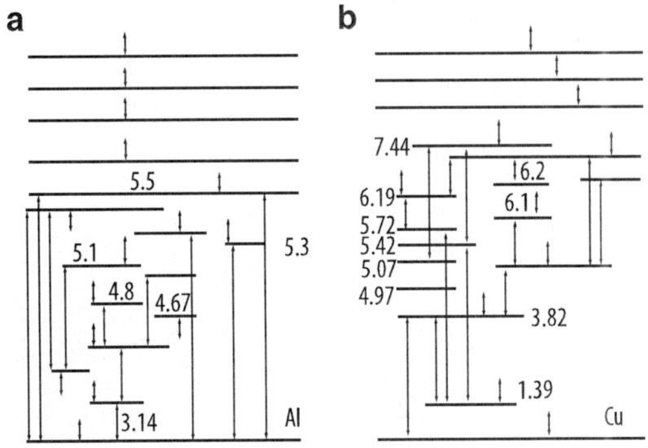

Fig. 7.9 Energy level diagrams for aluminum (**a**) and copper (**b**) atoms

will be varied from 10^6 to 10^9 W/cm^2 for Al vapor and from 10^7 to 10^{10} W/cm^2 for copper vapor. We will calculate the dependence of the breakdown time on the laser radiation intensity, which is measured in experiments. The results of calculations will strongly depend on the initial temperature of vapor. Because this quantity is not calculated in the given model, the calculations were performed for different values of this parameter close to the boiling temperature. We will begin from the case

$$T_e(0) = T_g(0) = 0.2\,\text{eV},$$
$$N_0^0(Al) = 6 \times 10^{18}\,\text{cm}^{-3}, \quad N_0^0(Cu) = 8,3 \times 10^{17}\,\text{cm}^3,$$
$$N_e^0(Al) = 4 \times 10^{13}\,\text{cm}^{-3}, \quad N_e^0(Cu) = 5 \times 10^{11}\,\text{cm}^{-3}.$$

The vapor and plasma densities are determined based on the equilibrium expressions for saturated vapor (6.19) and the Saha equation [see (3.4)]. The equilibrium population of the considered levels for $t = 0$ were calculated from the Boltzmann distribution.

Analysis of the energy diagrams of these atoms showed that in the case of aluminum it is sufficient to take into account 13 atomic levels, 9 ion levels, and 11 levels of a doubly ionized atom. For copper, 11 atomic levels and 6 ion levels were considered. Direct and reverse bound-bound and bound-free transitions were taken into account. To calculate the rate of these processes, it is necessary to know the electron temperature. The latter was determined from the energy balance equation for the electron subsystem, which took into account the absorption of laser radiation due to bremsstrahlung backscattering (3.2) and energy transfer from electrons to atoms and ions in elastic and inelastic collisions. Ions and atoms receive their energy from electrons and do not lose it. Thus, the temperature of heavy particles either increases with time or, if $T_e = T_a$, tends to a constant. We do not present here all the balance and rate equations because they are too cumbersome.

Our calculations showed that the plasma formation process in vapors has a threshold, i.e. if the laser radiation intensity is lower than a certain value, then the electron temperature in plasma increases nonuniformly with time and tends to a constant, which only slightly exceeds the initial temperature (~30%) (Fig. 7.10a),

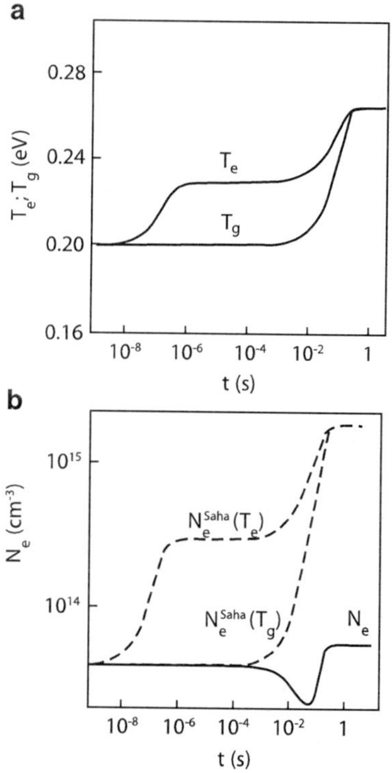

Fig. 7.10 Time dependences of the aluminum vapor plasma parameters in the pre-breakdown regime: (**a**) electron and heavy particle temperatures and (**b**) plasma density

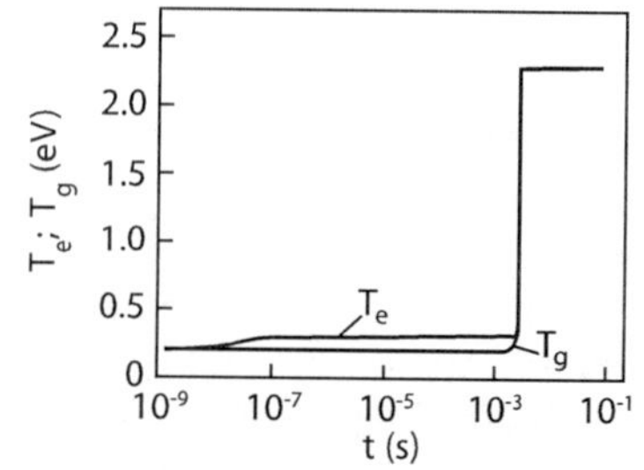

Fig. 7.11 Time dependences of the electron temperature T_e and heavy particle temperature T_g during the breakdown of Al vapor

remaining constant up to a few tens of seconds. The plasma density also increases not very strongly (Fig. 7.10b). One can see from these figures that the plasma concentration, calculated by assuming the existence of equilibrium (according to Saha), exceeds by several orders of magnitude the plasma concentration obtained in the nonequilibrium model. This suggests that nonequilibrium processes are important under these conditions.

For $q = 10^8$ W/cm^2 for Al and $q = 5 \times 10^9$ W/cm^2 for Cu, the situation considerably changes. The electron temperature increases by an order of magnitude for the time ~1 ms (Figs. 7.11, 7.12) and vapor is completely ionized (Fig. 7.13). It is interesting that in both cases the electron temperature increases with time by two steps. The real plasma concentration is lower by several orders of magnitude than

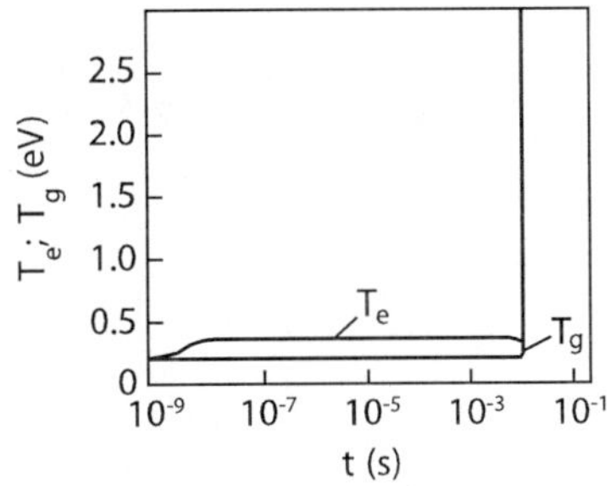

Fig. 7.12 Same as in Fig. 7.11 for *Cu* vapor

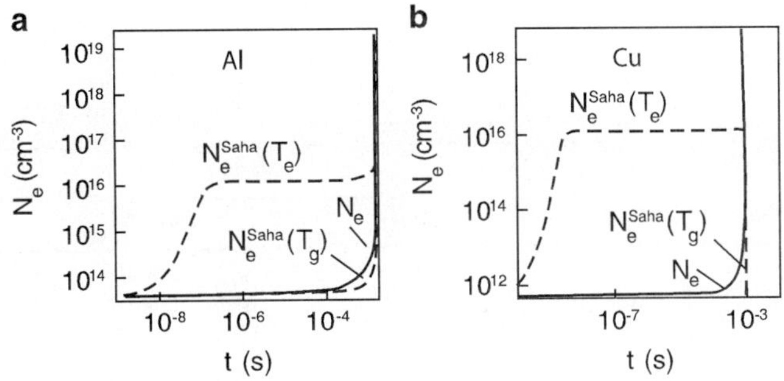

Fig. 7.13 Time dependences of the Al (**a**) and *Cu* (**b**) vapor plasma density during breakdown

its equilibrium value. This indicates that the nonequilibrium nature of plasma during the breakdown becomes even more considerable.

The fact that the breakdown intensity is higher for atoms with a greater ionization potential is reasonable. But what is the influence of the atomic energy level diagram discussed in the beginning of this section? Special calculations have shown that the lower is the first exited level, the higher is the breakdown threshold.

Finally, we consider the summary results of the model: the dependence of the breakdown time on the radiation intensity (see Fig. 7.14). This dependence is usually determined experimentally. One can see from Fig. 7.14 that the breakdown time strongly depends on the initial temperature and vapor density near a target. This circumstance emphasizes once more the importance of a more general formulation of the problem taking into account the heating, vaporization, and vapor expansion, as was done in the previous chapter in the consideration of ablation. Nevertheless, it is interesting to compare the results of calculations with experiments. The last experimental point in Fig. 7.3 corresponds to the intensity $q = 2 \times 10^7$ W/cm^2. One can see from Fig. 7.14 that, to match the experimental breakdown time $\tau_{br} \sim 1\,\mu$s with the calculated time, it is necessary to select the initial vapor temperature $T_0 =$ 0.4 eV. As we saw in the previous chapter, the target surface temperature at such radiation intensities can rise up to 5–6 kK.

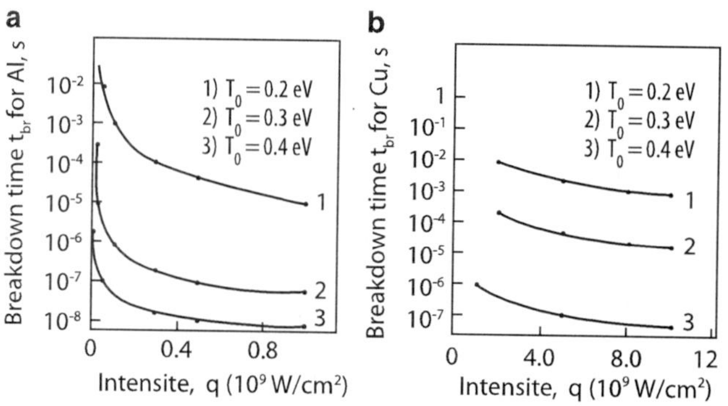

Fig. 7.14 Dependences of the Al (**a**) and *Cu* (**b**) vapor breakdown times on the Nd laser radiation intensity for different initial temperatures

Table 7.3 Breakdown intensity versus pulse duration

	τ, s			
q^*_{exp}, q^*_{teor}	10^{-3}	10^{-4}	10^{-5}	10^{-6}
q^*_{exp}, W/cm^2	$\sim 3 \times 10^6$	4.5×10^6	6×10^6	$\sim 3 \times 10^6$
q^*_{teor}, W/cm^2	8.6×10^5	1.6×10^6	4.5×10^6	2.1×10^7

A similar situation is also observed upon matching calculations [74] with experiments [75] with nanosecond pulses. For τ = 4–7 ns, Al vapors were broken for $q_{exp} = (1.5–2.5)$ GW/cm^2. When the initial temperature is increased up to T_0 = 0.38 eV, the calculated threshold intensity for τ = 7 ns is 2.5 GW/cm^2.

As for a breakdown produced by the lower-frequency CO_2 laser radiation, its threshold within the framework of the given model is determined by the frequency dependence of the absorption coefficient $\mu \sim \omega^{-2}$ (3.2). As follows from calculations [19] that we do have $q_{br} \sim \omega^2$. In experiments, $q \sim \omega$, and the reasons for such a discrepancy were considered above. The general notion about correlation of calculations and experiments is presented in Table 7.3 (T_0 = 0.2 eV, Al, CO_2-laser). Table data correlate with Fig. 7.7. What will be the result of calculations of the breakdown voltage for radiation frequencies exceeding the frequency of the $\lambda = 1.06\,\mu$m radiation? The optical breakdown produced by radiation from excimer lasers near a target [$\lambda = 0.355\,\mu$m ($\hbar\omega$ = 3.5 eV) pulses from a XeF laser and $\lambda = 0.248\,\mu$m ($\hbar\omega$ = 4.99 eV) pulses from a KrF laser] was studied in papers [20–22]. Pulses of different durations and different shapes were used in experiments. It was found that first a material was evaporated and then a plasma bunch was formed in the vapor. This allowed the authors to develop a simplified breakdown model, which is similar to that considered above for the plasma formation near a target irradiated by a pulse from a neodymium laser. As in this model, the initial parameters of vapor and plasma were not calculated but were set equal close to their experimental values:

$$T_0 = 0.2\,\text{eV},\ N_0^0 = 6 \times 10^{18}\,\text{cm}^{-3},\ N_e^0 = 3 \times 10^{14}\,\text{cm}^{-3}. \tag{7.7}$$

Unlike the situation with the breakdown of vapor by IR radiation, when $h\nu \leq 1\,\text{eV}$, in the case of breakdown by UV radiation, photoprocesses became important. This is first of all the optical excitation of electronic levels due to the resonance and nonresonance absorption of laser radiation by atoms. The latter process is caused by the broadening of the absorption line due to collisions and heating of vapor. These processes, although do not cause directly ionization, are quite important because their cross section $\sigma = 10^{-12}\,\text{cm}^2$ is very high. In this case, photoprocesses caused by cw radiation prove to be insignificant because of the low temperature of a medium.

Finally, it is also necessary to take into account photoionisation from the ground level and excited levels. The latter processes can additionally "supply" energy into the electronic subsystem, by increasing its temperature. It is also assumed that photons emitted in radiative decays leave the system. This is, as in the case of IR radiation, the main energy loss channel, which causes the deviation from equilibrium. All photoprocesses were additionally taken into account in the system of kinetic equations for the case of IR radiation.

The results of numerical investigations of the mechanism of breakdown by UV radiation are generalized in Fig. 7.15. Curve 1 in all figures corresponds to the dependence $q_{\text{br}} \sim \omega^2$. Figure 7.15a presents the results of calculation of the breakdown intensity taking into account all the processes mentioned above. Curve 3 in Fig. 7.15b shows the breakdown intensity when only photoionisation is taken into account among optical processes. One can see that the influence of photoionisation begins already for $\hbar\omega \geq 1\,\text{eV}$. The ionization rate increases due to photoionisation of upper excited levels. The first large decrease in the threshold by several orders of magnitude occurs when the photon energy achieves the ionization energy of the atom. Even more remarkable result is achieved when resonance excitation is "switched on" (Fig. 7.15c). The resonance excitation of the first level reduces the breakdown intensity by more than an order of magnitude compared to photoionisation from the ground level. As mentioned above, this is caused by the high photoexcitation cross section compared to the photoionisation cross section. This analysis explains the mechanism of nonequilibrium breakdown of vapors near a target in the uniform model (see curve 2 in Fig. 7.15a).

The role of photoprocesses in the plasma formation near the surface is illustrated in Fig. 7.16. This figure presents the dynamics of plasma concentration and temperature observed upon irradiation of Al by $q = 2 \times 10^7\,\text{W/cm}^2$, $\tau = 60\,\text{ns}$ pulses from a KrF laser. One can see that even when the electron temperature decreases, the plasma concentration continues to increase rapidly.

As for a comparison with experiments, a direct quantitative comparison is complicated because the model is too restricted (the target heating and vaporization process are neglected). This situation is similar to the picture of plasma formation upon irradiation of a target by a Nd laser. In any case, the calculated breakdown time should be shorter than the experimental time for the same pulse intensity and shape.

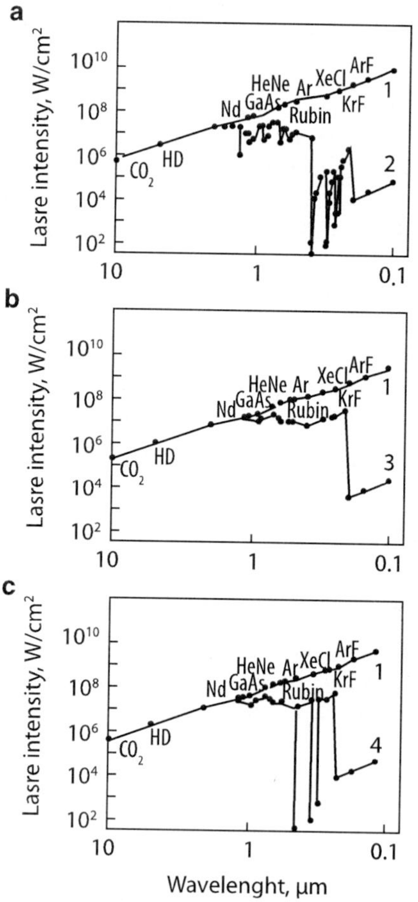

Fig. 7.15 Wavelength dependence of the threshold intensities with consideration for all the processes (**a**), photoionisation (**b**), and resonance photoexcitation (**c**)

For example, for $q = 1.6\times10^9$ W/cm^2 the experimental breakdown time is $\tau = 12$ ns [20–22], while the calculated breakdown time is shorter that this value. As for a XeF laser, because the energy of photons emitted by this laser is smaller ($\hbar\omega = 3.5$ eV), the difference between the photon energy and the energy of the transition from the ground state to the first excited state (3.14 eV) is smaller than that for a KrF laser. As a result, the calculated and experimental breakdown intensity and time for this laser are lower than those for the KrF laser. As should be, the calculated breakdown time for $q = 1.8 \times 10^8$ W/cm^2 is smaller than the experimental value $\tau = 4.5$ ns.

Table 7.4 shows the generalized comparison of experimental data with modelling [77].

Thus, this model agrees qualitatively with experiments. To obtain the quantitative agreement, it is necessary to increase the number of considered processes, first of all taking into account the target heating and vaporization.

Fig. 7.16 Time dependences of the electron (T_e) and heavy particle (T_g) temperatures (**a**), and the electron density N_e upon irradiation by 0.248 μm, 60 ns laser pulses

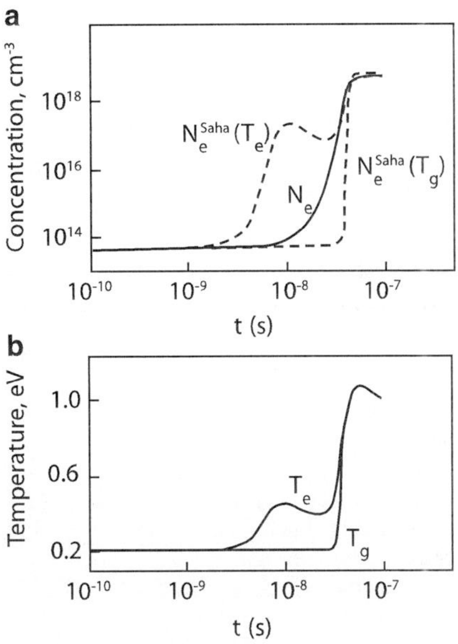

Table 7.4 Comparison of the experimentally measured threshold intensities for the vapor plasma formation with the numerically predicted values

Wavelength (μm)	1.06	0.532	0.355	0.308	0.308	0.248	0.248
Reference	[75]	[80]	[80]	[81]	[82]	[83]	[79]
Pulse duration (ns)	7	6	6	20	20	15	17
Threshold intensity (experiment) (10^8 W/cm^2)	14	3.8	3.0	2.0	2.2	1.15	1.85
Threshold intensity (modelling) (10^8 W/cm^2)	10	4.3	2.9	1.1	1.1	1.2	1.25

7.3 Dynamics of a Plasma Plume and its Interaction with a Laser Beam

In the previous section we considered the mechanisms of plasma production by irradiating a target by laser pulses. Below, we will describe the properties of pulsed laser plasma, which of interest for some laser technologies.

It is known that the CO_2 laser beam energy is more efficiently transferred to metal targets with the help of plasma. When plasma is produced, the mechanical action of the target increases. However, the thermal action on the material decreases during a breakdown. In technological operations such as cutting and welding, the most important factor is the laser radiation intensity. Therefore, we will consider the influence of plasma formation on the screening, refraction, and reflection of a laser beam.

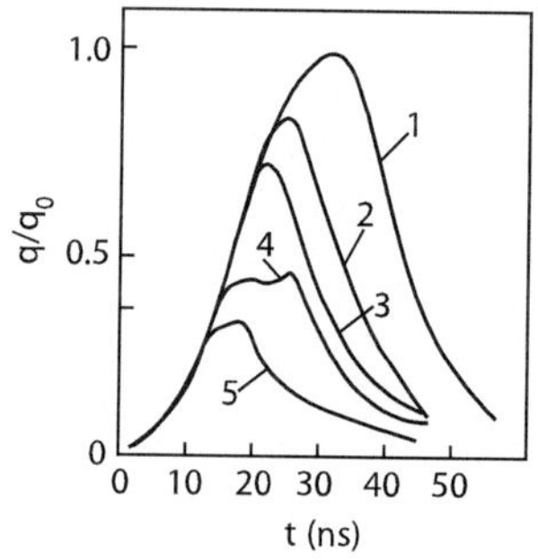

Fig. 7.17 Oscillograms of the incident neodymium laser pulse and laser pulse transmitted through a hole in an aluminum target in vacuum for different radiation intensities (the results are normalized to the total intensity): (**1**) incident beam; $q = 30$ (**2**), 110 (**3**), 250 (**4**), and 580 MW/cm^2 (**5**)

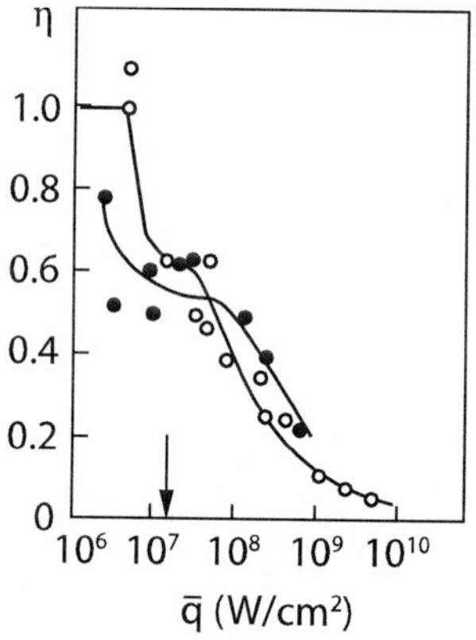

Fig. 7.18 Dependences of the integrated (averaged per pulse) transmission coefficient of a plasma plume on the radiation intensity for an aluminum target. Laser beam is defocused (*black ring*) and focused (*ring*). The *arrow* shows the breakdown threshold

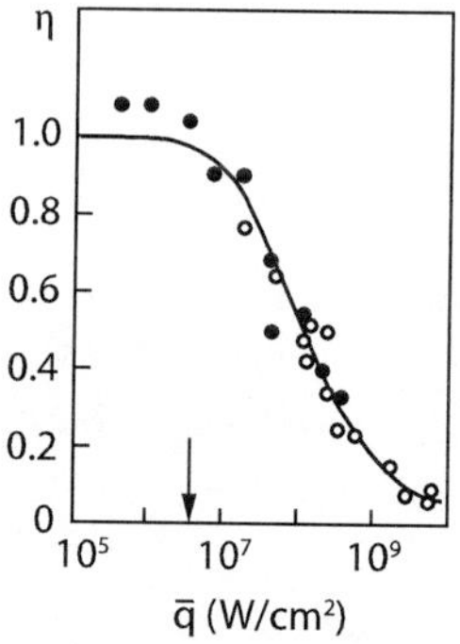

Fig. 7.19 Same as in Fig. 7.18 for a lead target

7.3.1 Propagation Mechanisms of the Surface Plasma

Figures 7.17 and 7.18 present the results of experimental measurements of the propagation of neodymium laser radiation through the plasma of the target vapor in vacuum [23]. The measurements were performed through a small hole in the target. One can see from Fig. 7.17 that the screening onset time is displaced to the light pulse onset with increasing the radiation intensity, the fraction of light propagated on average per pulse being decreased (Figs. 7.17–7.19). The corresponding one-dimensional calculations [7] are in qualitative agreement with experiments. If the target is located in gas at the atmospheric pressure (e.g., in air), the situation changes considerably. While in vacuum a shock wave appearing after the optical breakdown decays at the edge of a vapor cloud, the propagation of a plasma bunch in gas depends on the laser radiation intensity. For $q < 10\,\mathrm{MW/cm^2}$, after the formation of plasma in vapor, pressure drastically increases and a weak shock wave runs away forward in air. The plasma propagates due to absorption of its thermal radiation by cold air (see the previous section and Fig. 7.6). This plasma motion is called a subsonic radiative wave (SRW), which is typical in the case of millisecond laser pulses. At higher laser beam intensities $q \geq 10\,\mathrm{MW/cm^2}$, which are typical for microsecond pulses, the shock wave absorbs laser radiation and passes to air. In air a laser-supported detonation wave (LDW) is produced [24, 25]. At even higher intensities, plasma propagates in the form of a supersonic radiative wave [26, 27]. The latter is formed in a nearly plane geometry. The main difference of the radiative wave from the LDW is that the regions in front of the wave, as in the case of the SRW, are heated due to absorption of radiation leaving the plasma. Below, we will consider this phenomenon by calculating the propagation of the surface plasma upon irradiation by nanosecond, gigawatt pulses.

Generally speaking, the dynamics of pulsed laser plasma is not restricted only by the types of motion mentioned above. These mechanisms can change each other or can exist simultaneously. In some cases (see below), it is difficult to separate one or another mechanism. Thus, after the optical breakdown near a target in air, as follows from one-dimensional calculations [28], the intense laser radiation ($\sim 100\,\mathrm{MW/cm^2}$) is completely screened by the LDW. However, radiation from the wave front in plasma can fall on the target. But this radiation almost does not reach the target because it is absorbed in vapor appearing at the very beginning of laser irradiation Fig. 7.7. One can see that pressure and temperature jumps are located in air. Curve q_r describes the radiation flux intensity distribution emitted by a strongly heated plasma layer. This radiation is absorbed both in front of the wave (in cold air layers) and behind it. Due to this absorption, pressure near the target surface can increase by several times, thereby increasing the mechanical action on the target.

It is interesting to compare calculations and experimental data on energy transfer to a target (within the focal spot) in the presence of the breakdown. Figure 7.20 presents the dependence of the energy absorbed in the target per $\tau = 1\,\mu\mathrm{s}$ pulse on the pulse power or energy, which was calculated for the one-dimensional target (solid curve) and taking into account flakes of thickness $4\,\mu\mathrm{m}$ (dashed curve). The calculations ignored energy transfer to the target due to radiation from plasma because the focal spot in experiments [10] was small and the two-dimensional nature of the experiment was manifested. One can see that no more than 10% of the laser beam energy is incident on the target. Note that plasma phenomena near a target can be considered in the one-dimensional approximation only at the beginning of the process. Indeed, the plasma moving at the speed of $\sim 10^6$ cm/s Fig. 7.21 is displaced from a target by 1 cm for $1\,\mu\mathrm{s}$, which is usually strongly exceeds the focal spot size, i.e. the problem becomes two-dimensional. Below, we will consider "two-dimensional" effects as well.

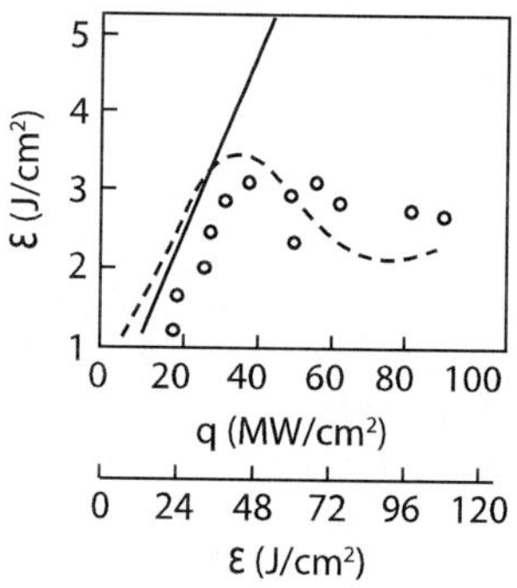

Fig. 7.20 Dependences of the energy density absorbed in a target within the focal spot on the incident radiation intensity and energy density

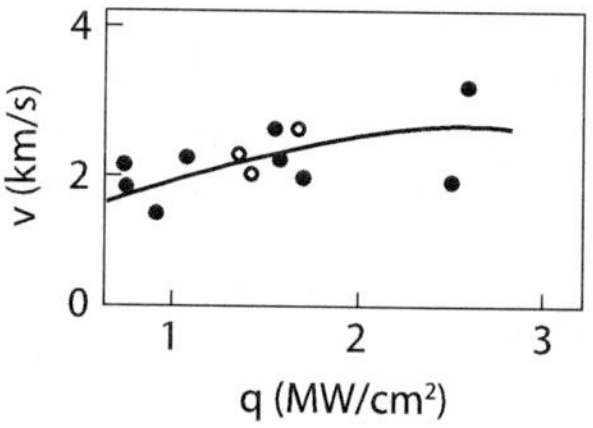

Fig. 7.21 Dependence of the plume propagation speed on the radiation intensity: (*ring*) graphite target; (*black ring*) duralumin target

Note that upon the oblique incidence of an intense laser beam, plasma propagates along the beam [24]. In this case, the plasma properties are independent of the target material. Figure 7.21 shows the dependences of the propagation velocity of a plasma plume in air for different targets [29].

Thus, the plasma front moves at the supersonic speed in the form of an LDW produced near the target (such a wave is observed after the gas breakdown in the absence of the target as well [25]). The sweep of the plasma emission near the target in the LDW regime is shown in Fig. 7.22 [30]. The shape of the CO_2 laser pulse is shown in Fig. 7.23. If $q < q_{SRW}$ and q_{LDW}, where q_{SRW} and q_{LDW} are the thresholds of the SRW and LDW, the plasma remains near the target and the optical discharge is burning in the mixture of vapors and surrounding gas [24]. This reminds an optical discharge burning in a gas flow (laser plasmatron, see Chap. 3). To realize a laser plasmatron, the vaporization of a target is needed, i.e. $\alpha q > \kappa T_b / r_f$ [see (1.29)]. If the target is not evaporated (e.g., a target is made of transparent LiF), then for $q < q_{SRW}$ and q_{LDW}, an LCW can appear in the surrounding gas (see Chap. 3). In the two last cases, plasma can transmit a great part of radiation to the target. When the LDW is formed (Fig. 7.22), only a small part of radiation is incident on the target (Fig. 7.23). The wave is decomposed by reaching the end of the beam caustic because the radiation intensity behind the caustic end decreases and $q < q_{LDW}$. At this instant the plasma plume becomes transparent. If a pulse is not ended by this time, the light can reach the target, but a new detonation wave may not be formed. This is explained by the fact that after the propagation of the wave, the gas near the target remains strongly heated and rarefied, while the gas breakdown threshold near the target and the LDW threshold at a low pressure increase with decreasing density [17]. The initial gas density will be restored after some time

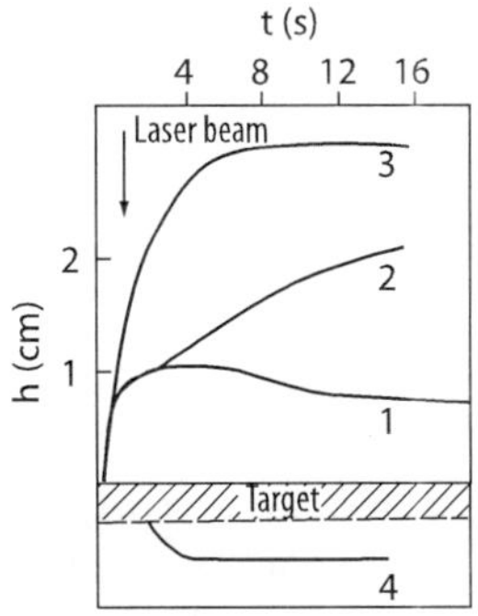

Fig. 7.22 Photoscan of a plasma plume formation neat a steel plate for $\tau = 30\,\mu s$: (**1**) a target has a through hole in it, $E = 4$ J; (**2**) no hole, $E = 4$ J; (**3**) no hole, $E = 10$ J; (**4**) a through hole exists, $E = 4$ J, the plume is located behind the target

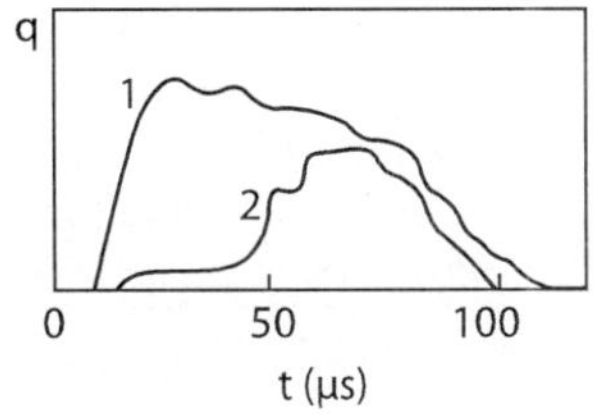

Fig. 7.23 Shapes of the incident CO_2 laser pulse (**1**) and CO_2 laser pulse transmitted through plasma ($E = 4$ J) [30]

(which is determined by cooling due to radiation, heat conduction, and convection). This time is usually quite large. For example, a heated zone is cooled due to the heat conduction for the time $\tau = d^2/\chi \approx 199\,\mu s$ ($d \approx 0.1$ cm, $\chi = 10^2\,cm^2s^{-1}$). The efficiency of the action of two laser beams focused at closely separated spots on a target was studied experimentally in [32]. It was found that if the delay between laser pulses is small, an LDW is not formed during the second pulse and the target is damaged stronger than during the first pulse. However, if the radiation intensity is very high, LDWs can be excited again [24].

7.3.2 Propagation of a Laser-Supported Detonation Wave in the Surrounding Gas

Consider the approximate quantitative properties of laser irradiation under LDW excitation conditions (radiative waves are described by numerical calculations). Because we are interested in the screening time and the LDW excitation threshold, we recall the theory of this phenomenon [25]. It follows from the definition of the LDW that it represents a usual hydrodynamic discontinuity with the absorption of light in this discontinuity. The propagation of a plane hydrodynamic discontinuity in gas is described by the three equations of conservation of mass, momentum and energy and by the Jouguet condition behind the wave front. Calculations give expressions for the LDW speed

$$v_0 = [2(\gamma^2 - 1)q_0/\rho_0]^{1/3} \tag{7.8}$$

and the internal energy of a material behind the wave front

$$\varepsilon = \frac{2^{2/3}\gamma}{(\gamma^2-1)(\gamma+1)}\left(\frac{q_0}{\rho_0}\right)^{2/3}. \tag{7.9}$$

These relations were obtained by assuming that the width of the LDW front is considerably smaller than the radius of a light channel. Otherwise, it is necessary to take into account the energy loss related to the lateral expansion of gas, and the LDW may not appear at all in the case of a small diameter of the LDW beam. The wave front thickness is determined by the thickness of the heat release zone, i.e. by the light absorption length l_a. The length $l_a = \mu^{-1}$ increases with decreasing temperature behind the wave front. This temperature decreases with decreasing the radiation intensity [see (7.9)]. Therefore, the condition $l_a \approx r_f$ determines the minimal intensity q_d at which the LDW exists. Because the temperature at the wave support threshold proves to be low, we can assume that $\mu \sim \exp(-I/kT)$. Therefore, the temperature T_c behind the wave front at the wave support threshold weakly depends on the gas density:

$$T_c = \frac{I}{k}\ln^{-1}\left\{\frac{r_f 4\pi e^2 v_e \sigma_{0c} N}{mc\omega^2 (T/10^4)^2}\frac{2g_+}{g_a}\left(\frac{2\pi mkT}{h^2}\right)^{3/2}\right\} \tag{7.10}$$

whereas q_d depends linearly on the density:

$$q_d = \frac{T_c^{9/4}}{2}\left(\frac{\alpha(\gamma^2-1)(\gamma+1)}{\gamma}\right)^{3/2} MN \tag{7.11}$$

where M is the mass of an atom or a molecule; α is the proportionality coefficient in the approximate relation $\varepsilon = \alpha T^{3/2}$ [25]; and $\sigma_{0c} = 10^{-13}\,\mathrm{cm}^2 \cdot \mathrm{eV}^2$.

It follows from (7.10) to (7.11) that the wave support threshold decreases with decreasing the gas density. But at some small value of N [when the logarithm in (7.10) tends to zero], q_d begins to grow with decreasing density. There exists the density threshold below which an LDW cannot be formed at any intensity. The existence of this threshold can be easily explained by considering the expression for the photon path l_μ in air [see (3.5)]:

$$l_\mu = \frac{mc\omega^2}{4\pi e^2 v_e}\frac{(T/10^4)^2}{\sigma_{0c}}\frac{1}{z^3 N^2 x_e^2} \tag{7.12}$$

where z is the mean ion charge and x_e is the single ionization degree. It follows from (7.12) that at low temperature the value of l_μ decreases with increasing T due to the increase in the ionization degree. In the temperature range from 2 to 25 eV, when single ionization is completed, a further decrease in l_μ with increasing T is caused by the increase in z. As follows from [25], for air in this range, $z = T$, where T is expressed in eV. Therefore, $l_\mu \sim T^{-3/2}$. For $T > 25\,\mathrm{eV}$, z increases weakly [25], and therefore l_μ begins to grow with increasing T. Thus, there exists a minimum of the absorption length. It is clear that if this minimum exceeds the beam radius, it is impossible to support the LDW at any intensity. The minimal photon path in air is achieved at $T = 25\,\text{eV}$:

$$l_\mu \approx 10^{35}/N^2.$$

For $r_f \approx 0.01$ cm, we have $N_{\text{th}} = 10^{18}\,\text{cm}^{-3}$ ($p \sim 0.01$ MPa). This value agrees by an order of magnitude with the experimental value 0.01 MPa obtained in air [33]. Because l_μ is a nonmonotonic function of temperature, the equation $l_\mu(T) = r_f$ has two solutions: T_{c1} and T_{c2} ($T_{c1} < T_{c2}$). Correspondingly, there exist two solutions of q for the same value of N, i.e. there exist not only the lower but also upper the LDW support threshold. The lower threshold q_d is determined by expressions (7.10, (7.11). To determine the upper threshold, it is necessary to find the critical temperature T_{c2}, which is calculated from (7.12) for $T > 25\,\text{eV}$. For air,

$$T_{c2} = r_f^{2/3} N^{4/3} A \tag{7.13}$$

where $A = [24\pi e^2 \sigma_{0c} z_0^3 \times 10^{13}/mc\omega^2]^{2/3}$, z_0 is the nuclear charge. By substituting (7.13) into (7.11), we obtain the expression for the upper limit of the LDW existence:

$$q_u = r_f^{3/2} N^4 A^{9/4} M[\alpha(\gamma^2 - 1)(\gamma + 1)/\gamma]^{3/4}. \tag{7.14}$$

For $N = N_{\text{th}}$, the thresholds are equal: $q_d = q_u$. Note that q_u strongly increases with increasing density and is very high at the atmospheric pressure. There exists an analogy between the LDW and a continuous optical discharge (COD). In both cases, both the lower and upper optical discharge support thresholds in the radiation intensity exist. For the LDW, only the lower pressure threshold exists, while for the COD – only the upper pressure threshold (Fig. 3.29).

By knowing the LDW support limit, we can determine the screening time τ_s during which light does not fall on the target. For this purpose, we will calculate the length L propagated by the wave before its decay. The laser beam intensity decreases with increasing distance z from the focal plane [see (1.6)]:

$$q = q_0 \left[1 + \frac{z^2\varphi^2}{r_f^2}\right]^{-1}$$

where 2φ is the focusing angle.

The length L can be determined from the equality $q = q_d$:

$$L = \frac{r_f}{\varphi}\sqrt{q_0/q_d - 1}. \tag{7.15}$$

The minimal LDW speed in the beam is determined from (7.8) for $q = q_d$. By dividing the length by speed, we obtain the maximal screening time τ_s:

$$\tau_s = [P - P_d]^{1/2} \pi^{-1/2} \varphi^{-1} \rho_0^{1/3} q_d^{-5/6} (2\gamma^2 - 1)^{-1/3}. \tag{7.16}$$

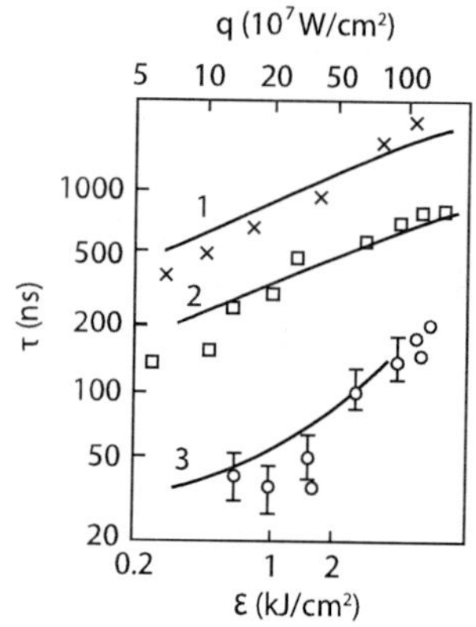

Fig. 7.24 Dependence of the LDW lifetime on the CO_2 laser radiation intensity: (**1**) argon; (**2**) air; (**3**) helium

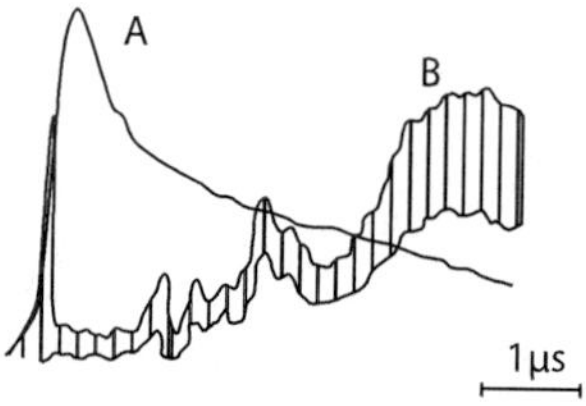

Fig. 7.25 Shapes of the incident CO_2 laser pulse (**A**) and CO_2 laser pulse reflected after the breakdown (**B**) (the scale for B is magnified by 15 times compared to the scale for A)

It follows from (7.16) and (7.11) that τ_s weakly depends on the gas density ($\tau_s \sim \rho_0^{-1/2}$). The value of τ_s depends on the ionization potential of gas: $\tau_s \sim I^{-5/6}$. Figure 7.24 presents experimental screening times and these times calculated by (7.16). Experiments [34] were performed by focusing 5 μs pulses from a CO_2 laser to a very small spot of diameter 65 μm. The energy density in the spot was varied from 0.2 to 10 kJ/cm^2. The screening time was determined from the shape of incident and reflected light pulses (Fig. 7.25). As expected, the oscillogram of the reflected light exhibits three phases. During the first phase, the light experiences specular reflection from a metal, this phase duration being 150–200 ns. Then, the reflected signal drastically decreases for approximately 10 ns and the second, screening phase appears, during which an LDW exists. At these stages, the picture resembles externally the picture described above for the interaction of a short pulse from a neodymium laser with a target in vacuum (Fig. 7.17). In the case of the interaction of longer pulses from a CO_2 laser with a target considered here, the third phase exists, which appears after the LDW decay (Fig. 7.25). The light is incident on the target again. Figure 7.24 shows that the LDW lifetime depends on the gas type and radiation intensity, which is consistent with the theory. By knowing the screening time, we can calculate the transparency coefficient of the laser plume in a pulsed optical discharge. It is clear that the longer is the screening time, the smaller is the transparency coefficient (Fig. 7.26a). The argon plasma screens a sample most strongly. Figure 7.26b presents the dependence of the sample damage efficiency Y on the radiation intensity. One can see that the damage efficiency correlates with the plume transparency (the damage efficiency Y is the ratio of the damage depth to the radiation energy).

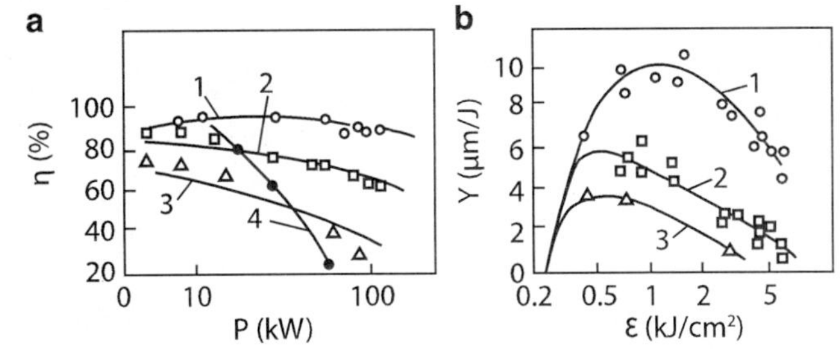

Fig. 7.26 Transparency coefficients η of a plasma plume (**a**) and destruction efficiencies Y of a copper target (**b**): (**1**) helium [34]; (**2**) air [34]; (**3**) argon [34]; (**4**) air [36]

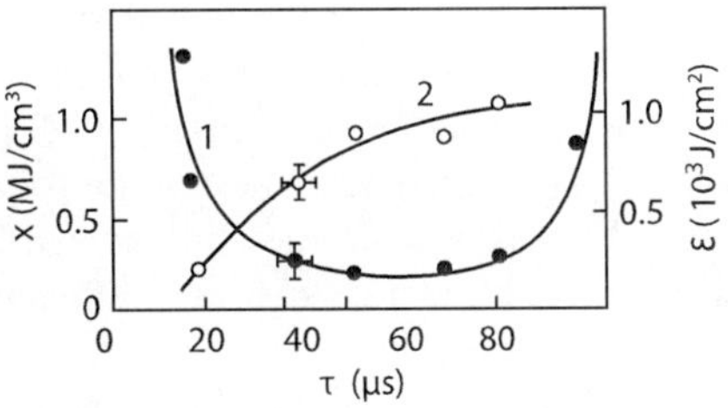

Fig. 7.27 Dependences of the specific destruction energy (**1**) and the radiation energy density transmitted through plasma (**2**) on the pulse duration at the constant pulse energy $E = 1.3$ J

It follows from the above discussion that a short intense light pulse is strongly screened by a plasma plume. The reduction of the pulse energy ε is not always advantageous. For example, the average power of a repetitively pulsed laser $\bar{P} = \varepsilon f$ will be reduced. To increase the pulse repetition rate f with decreasing the output energy is technically difficult. The pulse repetition rate of the first technological lasers was restricted by ~1 kHz [35]. At present the pulse repetition rate of these lasers is restricted by ~10–100 kHz. The high average power of a laser provides the high efficiency of a technological process. The efficient interaction of laser radiation with a target can be achieved by increasing simultaneously the CO_2 laser pulse energy and duration. Self-sustained discharge pulsed CO_2 lasers can emit pulses of duration from 30 to 800 μs with energies from 1 to 20 J [35]. In this case, the screening time is a small fraction of the pulse duration, and thus the radiation energy loss due to screening will be small. Figure 7.27 shows the increase in the energy transmitted through a laser plume with increasing the CO_2 laser pulse duration for the constant pulse energy [36]. The transparency increases both due to the light pulse elongation and screening time shortening because power decreases with increasing τ [see (7.16)]. The energy transmitted through the plume was measured through a hole in a sample, as in Fig. 7.18. The plume length increases with the laser power (Fig. 7.28).

By substituting experimental data [36] ($r_f = 0.4$ mm $\varphi \approx 1/8$, and $q_d = 8\,\mathrm{MW/cm^2}$ into (7.15), we obtain for a 2 J, $\tau = 20\,\mu$s pulse the plume length

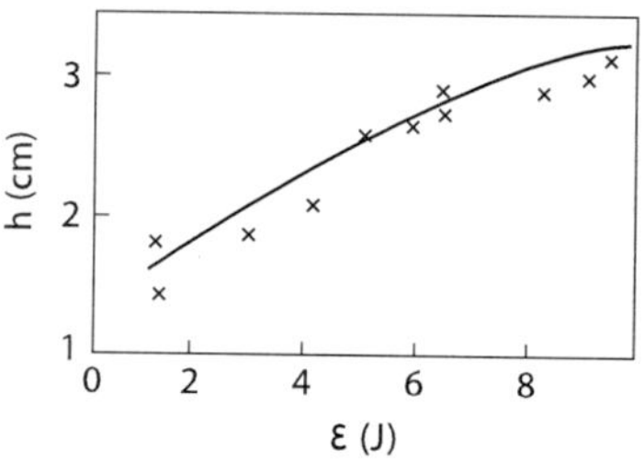

Fig. 7.28 Dependence of the plasma plume length on the CO_2 laser pulse energy ($d_f = 0.5$ mm, $\tau = 70\,\mu$s)

$L \sim 5$ mm, which is somewhat smaller than experimental values. The plasma plume length can also depend of a material vapor jet. Indeed, the plasma plume is larger in the absence of a through hole, when a developed vapor jet exists (Fig. 7.22). After the formation of a hole, the jet length begins to decrease. After 5–10 pulses, only the trace of the LDW remains in the diagram (Fig. 7.22 [30]).

Let us compare the results of experiments performed for different diameters of a focal spot on a target and the same laser radiation power. It follows from (7.16) that the screening time and, hence, the transmission coefficient are determined only by the laser beam power. For example, for $P = 65$ kW and a small spot with $d = 65\,\mu$m, we have $\eta \approx 0.7$ (curve 2 in Fig. 7.26), while for a large spot with $d \approx 600\,\mu$m, we have $\eta = 0.25$ (curve 4 in Fig. 7.26 and curve 2 in Fig. 7.27, $\tau = 20\,\mu$s). In the case of a small spot, the screening duration was $\sim$0.5 μs (curve 2 in Fig. 7.24). It is clear that if η was determined by only τ_s, the value of η for a large spot would be close to unity because $\tau = 20\,\mu$s. This contradiction can be explained as follows. First, transmission coefficients η are measured in these cases differently and describe, generally speaking, different processes. The transmission coefficient of the plume in Fig. 7.26 [34] is calculated by the laser power, while in Fig. 7.27 it is calculated by the laser beam intensity.

The radiation intensity on a target is determined not only by the absorption but also by refraction of radiation in the plume. The effect of the latter can be considerable, as is shown in Sects. 1.2 and 3.4. The possibility of a strong influence of refraction is also confirmed by the fact that, despite a weak decrease in the transmission coefficient with increasing power, the drilling efficiency drops almost by an order of magnitude (Fig. 7.26a,b). This process was specially studied in [64], where the refractive index of plasma at the laser radiation wavelength was calculated by measuring the plasma density distributions at different instants of time. Knowing the refractive index distribution and solving the problem of radiation propagation within the framework of geometrical optics, the authors calculated the beam trajectories at different instants of time (Fig. 7.29). One can see that at the initial stage of plasma development, radiation can be focused after the breakdown due to refraction. At the later stages of the discharge, when pressure is equalized and the plasma density decreases at the boundaries, the refraction effect leads to the defocusing of beams, as in the case of a COD. A comparison with experiments showed that the time dependence of the plasma transparency, measured through a hole in a target, can be explained only taking refraction into account. We will

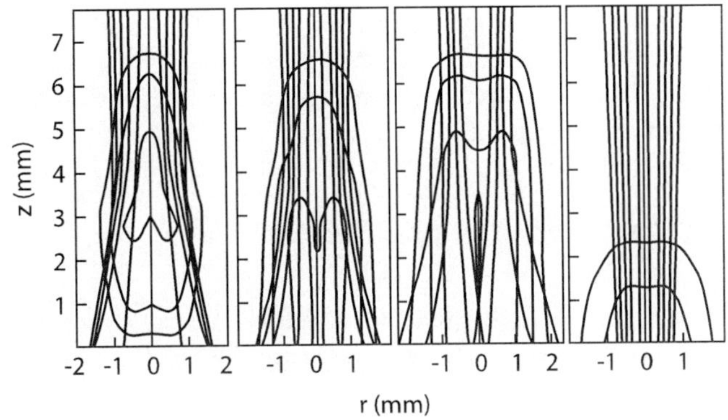

Fig. 7.29 Beam trajectories calculated during the propagation of a laser pulse through a plasma plume upon the surface breakdown of air. Figures in the left-to-right sequence correspond to instants 0.88, 1.32, 1.98, 3.74 μs after the breakdown. The isolines n_e are also presented

Fig. 7.30 Dependences of the laser radiation intensity on a target q_M on the incident radiation intensity; (**1**) without plasma; (**2**) $\tau = 80\,\mu\text{s}$, $E = 1.3\,\text{J}$ [36]; (**3**) $\tau = 1\,\mu\text{s}$ [17]

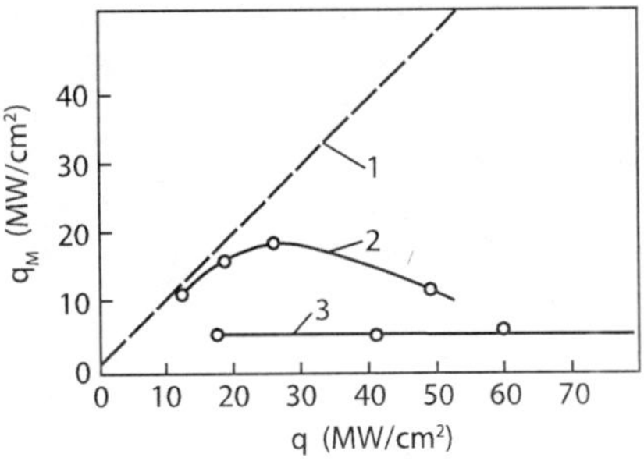

show below that the laser radiation intensity distribution on a target can be strongly inhomogeneous due to the influence of two-dimensional boundary effects as well.

If the laser beam diameter is large, the radiation intensity can be noticeably decreased due to a large screening time caused by a slow expansion and cooling of the heated compressed gas remained after the LDW propagation. Indeed, the ratio of beam diameters in experiments [34, 36] is equal to ten. A comparison of the slit scan, frame photography and intensity oscillograms of light transmitted through the plasma plume showed that complete screening occurs only during the LDW existence (2 μs). At the rest of the time, the beam was partially attenuated in the decomposing LDW and a plasma jet of the material vapor (Figs. 7.22 and 7.23) or in COD in gas [30].

Note in conclusion that in the case of a large spot ($d = 0.6\,\text{mm}$), the plasma plume so strongly screens radiation that the radiation intensity on the target decreases with increasing the initial beam intensity. This is demonstrated in Fig. 7.30, where the transparency of the plume is compared for the same intensity and size of a focal spot but different irradiation times. The irradiation time of the target for $\tau \sim 1\,\mu\text{s}$ (curve 3) is considerably smaller than for $\tau \approx 20$–$80\,\mu\text{s}$ (curve 2). The run of these curves can be explained by different influence of screening times.

Thus, a pulsed optical discharge considerably differs from a stationary discharge, in which the transmission coefficient also decreases with increasing radiation intensity, but the radiation power on a target always increases with increasing the incident beam power (see Fig. 3.31).

7.3.3 *Reflecting Properties of a Plasma Plume*

In previous sections, we considered the radiation intensity transmitted through plasma to a target. Now we will study the energy losses due to which a part of radiation does not reach the target. The radiation energy can be either absorbed in plasma or be reflected from it. The latter is possible because the plasma density in the LDW, as we saw, can exceed the air density at the atmospheric pressure equal to $\sim 3 \times 10^{19}\,\mathrm{cm}^{-3}$. The critical plasma density n_c (see Sect. 1.1.2) at which the CO_2 laser radiation does not penetrate into plasma is not too high ($\sim 10^{19}\,\mathrm{cm}^{-3}$). The reflection of laser radiation from a plasma plume depends on the conditions of interaction of radiation with a target, i.e. is determined by the radiation intensity, the focal spot size, and the pulse duration. In the case of usual pulsed TEA CO_2 lasers emitting $\tau \approx 1$–$3\,\mu\mathrm{s}$, $q = 10$–$100\,\mathrm{MW/cm^2}$ pulses, the reflection of light from plasma is insignificant [38]. Figure 7.31 presents the radiation pattern for laser radiation reflected from the zone of interaction of the laser beam with a brass target. In the case of low intensity, the reflected light propagates virtually isotropically. As the incident light intensity is increased, a greater part of radiation is reflected backward. This is explained by the fact that an LDW is generated faster at higher intensities, and it has time to propagate over a large distance. Thus, in the case of high radiation intensities, plasma has the form of a cylinder elongated along the laser beam, its front end reflecting light. In the case of low-intensity laser radiation, the surface of a heated gas region is a hemisphere, which provides isotropic scattering. As a whole the total energy of reflected light in these experiments is low, being from 3.5% to 15% of the incident beam energy (Fig. 7.31).

In the case of shorter light pulses ($\tau \leq 0.1\,\mu\mathrm{s}$) and a large focal spot $S \approx 1\,\mathrm{cm^2}$) [39], the form of a plasma plume produced near the target is close to a flat layer ($q = 100\,\mathrm{MW/cm^2}$), which is confirmed in experiments with a high-speed

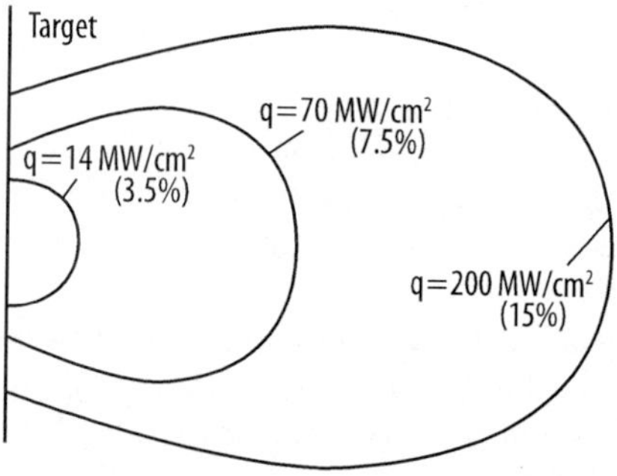

Fig. 7.31 Reflected CO_2 laser radiation pattern ($\tau = 1\,\mu\mathrm{s}$)

photodetector [39]. However, the angular distribution of reflected light did not correspond to reflection from a flat mirror. The total reflected light energy was 53% of the incident energy, while only 3.3% of the energy was reflected to the aperture of a focusing mirror ($q = 100\,\mathrm{MW/cm^2}$).

7.3.4 Numerical Modelling of a Pulsed Optical Discharge

As is shown in previous sections, plasma phenomena represent a combination of complicated dynamic processes. These phenomena can be explained with the help of numerical calculations. It was shown in Sects. 7.1 and 7.2 that these calculations agree with experiments on the optical breakdown near a target. However, one-dimensional calculations discussed in upper sections, as was already pointed out, can be only applied at short irradiation times and large focal spots, which is rarely realized in practice. Experiments show that a plasma bunch departs from a target during a light pulse by a distance considerably exceeding the focal spot size (Figs. 7.28, 7.29). Thus, a pulsed optical discharge is usually two-dimensional.

Numerical calculations are necessary not only for the description of the lateral expansion on the propagation of an optical discharge. As mentioned above, some types of optical discharges propagate due to radiation emitted by the initial plasma near a target and absorbed by a cold gas, i.e. this mechanism of the discharge propagation is determined by the radiative gas-dynamics mainly described by numerical methods.

As shown above, laser radiation is incident on a target either before screening or after the decomposition the screening discharge. In the first case, it is better to use a short light pulse, whereas in the second one – a long pulse. The latter case can be qualitatively analyzed, while the analysis of the case with a short pulse, when the duration of a propagated pulse is determined by the formation of the screening discharge, is very complicated. Consider first optical discharges in the simplest case, when a discharge develops at once in the surrounding gas, e.g., in air [40].

Radiation transfer is taken into account within the framework of radiative gas-dynamics by adding to the energy balance equation the terms describing the release or decrease of energy due to the absorption of the photon energy flux **W**:

$$\frac{1}{2}\frac{\mathrm{d}}{\mathrm{d}r}(rW_r) + \frac{\mathrm{d}}{\mathrm{d}z}W_z \tag{7.17}$$

The photon energy flux is described in the simplified diffusion approximation by the equations [41,42].

$$divW_\nu + c\kappa_\nu U_\nu = c\kappa_\nu U_{\nu,\mathrm{eq}} \tag{7.18}$$

$$\frac{1}{3}cgradU_\nu + \kappa_\nu W_\nu = 0$$

where U_ν is the radiation energy density and $U_{\nu,\mathrm{eq}}$ is the equilibrium energy density. The frequency dependence of κ_ν and U_ν is taken into account in the usual multigroup approximation [43, 44].

The absorption coefficient $\kappa_\nu = \kappa_\nu(\nu, p, T)$ is a function of three variables and its frequency dependence is complicated due to the involvement of a few radiation mechanisms. This coefficient was determined by using the quantum-mechanical Hartree–Fock–Slater model [45, 46]. The results of calculations were tabulated in the form of three-dimensional tables. The temperature, density, and energy ranges were $T \in [0.01 - 200\,\mathrm{eV}]$, $p \in [5 \times 10^{-6} - 10^{-2}\,\mathrm{g/cm^3}]$, $h\nu \in [0.01 - 250\,\mathrm{eV}]$, while the number of groups was set equal to 10. The absorption coefficient κ_ν in each group $\nu \in [\nu_{k-1}, \nu_k]$ was averaged over the Planck and Rosseland distributions:

$$\kappa_k = \int_{k-1}^{k} \kappa_\nu U_{\nu,\mathrm{eq}} \mathrm{d}\nu \Big/ \int_{k-1}^{k} U_{\nu,\mathrm{eq}} \mathrm{d}\nu,$$

$$\kappa_k^{-1} = \int_{k-1}^{k} \kappa_\nu^{-1} (\mathrm{d}U_{\nu,\mathrm{eq}}/\mathrm{d}T) \mathrm{d}\nu \Big/ \int_{k-1}^{k} (\mathrm{d}U_{\nu,\mathrm{eq}}/\mathrm{d}T) \mathrm{d}\nu.$$

The equations of states were determined by the method proposed in [46] and were represented in the form of two-dimensional tables.

7.3.5 *Modeling Results*

7.3.5.1 Beginning of the Process and Initial Data

A region $0 \le (r \times z) \le (L_r \times L_z)$ filled with motionless cold air of density $\rho_{pd}(0, r, z) = \rho_0$ at temperature $T(0, r, z) = T_0$ is considered. The plane $z = 0$ corresponds to the free surface of a condensed medium. The medium was irradiated by single $\tau = 1.7 \times 10^{-8}$ s laser pulses at $\lambda = 1.06\,\mu\mathrm{m}$ or $\lambda = 0.353\,\mu\mathrm{m}$ with the peak intensity q_0 =(4–17) GW/cm^2.

The initial stage of the laser radiation action, corresponding to the surface evaporation, the optical breakdown of evaporation products and the appearance of a plasma zone, was modeled by a motionless thin hot air layer of size $0 \le (r \times z) \le (r_{\mathrm{hot}} \times z_{\mathrm{hot}})$ and density $\rho(0, r, z) = \rho_{\mathrm{hot}}$ at temperature $T(0, r, z) = T_{\mathrm{hot}}$ located near the surface. Here, $r_{\mathrm{hot}} < L_r$, $z_{\mathrm{hot}} \ll L_z$, $T_{\mathrm{hot}} \gg T_0$, $\rho_{\mathrm{hot}} = \rho_0$. The values of parameters used in calculations were $L_r = 5\,\mathrm{cm}$, $L_z = 10\,\mathrm{cm}$, $r_{\mathrm{hot}} = 2R$, $R = 0.15\,\mathrm{cm}$, $T_0 = 0.03\,\mathrm{eV}$, and $\rho_0 = 1.25 \times 10^{-3}\,\mathrm{g/cm^{-3}}$.

The value of a pair $(T_{\mathrm{hot}}, z_{\mathrm{hot}})$ for a $\lambda = 1.06\,\mu\mathrm{m}$ laser was determined by comparing the calculated and experimental transmission coefficients [47, 48]. For a fixed intensity G_0, a pair of values of $(T_{\mathrm{hot}}, z_{\mathrm{hot}})$ was selected for which the calculated transmission coefficient became equal to 0.5 at the same time as in experiments. The corresponding results are presented in Table 7.5. It follows from these data that the transmission of plasma strongly depends on the temperature

Table 7.5 Specified parameters of the initial region

Radiation power density q_0, GW/cm^2	Screening time t_s, ns	Parameters of the initial region: $z_{hot} = 50\,\mu m$ T_{hot}, eV	$z_{hot} = 100\,\mu m$ T_{hot}, eV
6	10–15	1.05	0.98
10	0	1.05	0.98
13	−13	1.10	1.02

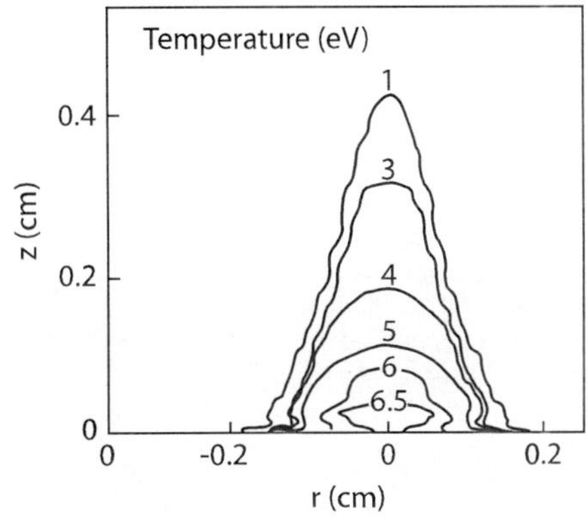

Fig. 7.32 Plasma temperature distribution by the end of the 34 ns, 1.06 μm pulse

($\sim$1 eV) of the initial hot region. The influence of the region thickness is much weaker, and its doubling from 50 to 100 μm is compensated by the decrease of temperature approximately by 5%. When the specified region temperature T_{hot} was below 0.9 eV, plasma was not formed at any radiation intensities within the range studied. For $\lambda = 1.06\,\mu m$, $q_0 = 6$, 10, and 13 GW/cm^2, Table 7.5 presents experimental screening times t_s and parameters of the initial hot region, temperature T_{hot} and thickness z_{hot} for which the calculated screening time coincides with the experimental value.

Because experimental data on the plasma transparency in air in the UV range are not available, the initial values of (T_{hot}, z_{hot}) were determined by a different method. Indirect estimates based on the observation of pressure signals [48, 49] suggest that the screening of radiation at the laser pulse peak $t_s \approx 0$ ns occurs at the radiation intensity of 6 GW/cm^2. These data correspond in calculations to $T_{hot} = 2$ eV and $z_{hot} = 100\,\mu m$.

7.3.6 *Expansion Mechanisms of Plasmas*

The features of the evolution of plasmas produced by IR and UV laser pulses can be analyzed by the example of calculations performed for wavelengths $\lambda = 1.06\,\mu m$ and $\lambda = 0.353\,\mu m$ at the same radiation intensity $q_0 = 10$ GW/cm^2 (Figs. 7.32–7.39). The intense absorption of laser radiation in the initial region causes a rapid

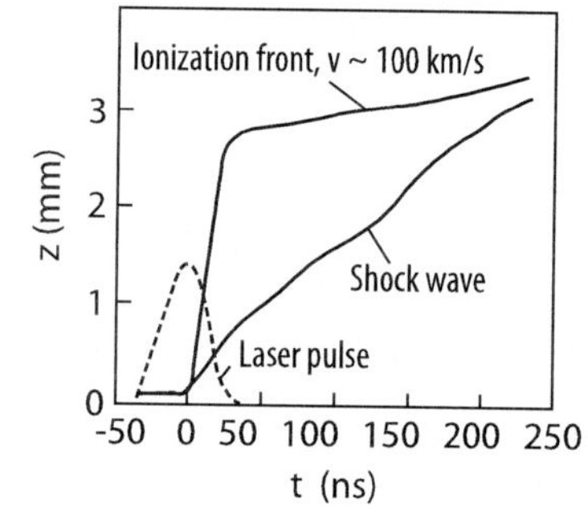

Fig. 7.33 Positions of the shock wave and ionization front upon irradiation by a 1.06-μm pulse

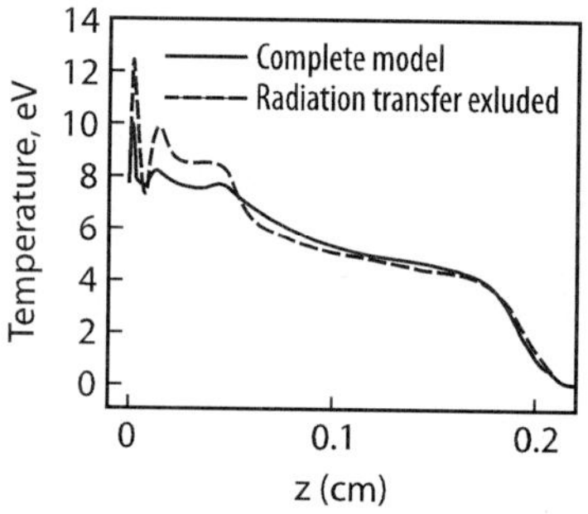

Fig. 7.34 Temperature distributions $T(z, r = 0)$ at $t = 17$ ns calculated taking into account plasma radiation transfer (*solid curve*) and neglecting it (*dashed curve*) upon irradiation by a 1.06 μm pulse

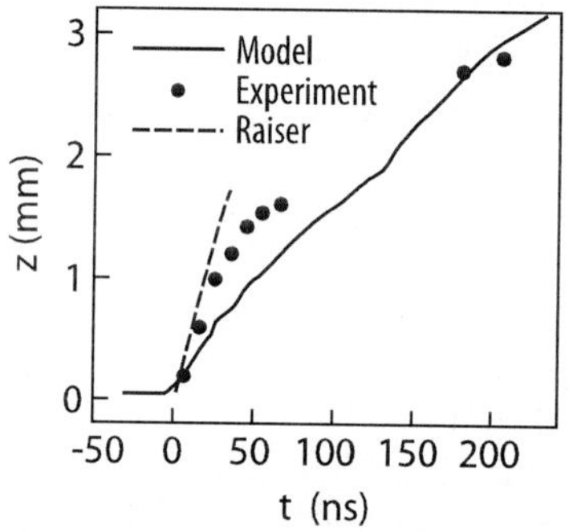

Fig. 7.35 Calculated (*solid curve*), experimental (*dots*) positions of the shock wave front, and these positions calculated by the Raiser formula

increase in the plasma temperature, exceeding 9 eV. The excess pressure in the initial hot region leads to the formation of a shock wave along its interface with the unperturbed gas. The maximum density on the shock wave front achieves 3×10^{-3} g/cm^3, exceeding more than twice the background density. The equilibrium electron concentration, which is determined by the plasma temperature and density, also increases with a maximum of $n_e \approx 10^{20}$ cm^{-3} in the shock wave at the region center. In the case of irradiation by IR pulses, the absorbed laser energy proves to be sufficient to form an ionization wave [50] corresponding to the high-temperature region with a high electron density (Fig. 7.33), which rapidly propagates toward the laser with a virtually invariable gas density. The motion of the ionization front and shock wave is shown in Fig. 7.32. The position of both boundaries is determined on the laser beam axis (straight line $r = 0$), while the coordinate of the ionization front is found from the condition $\alpha = N_e/N_0 = 0.01$, where N_e and N_0 are the electron

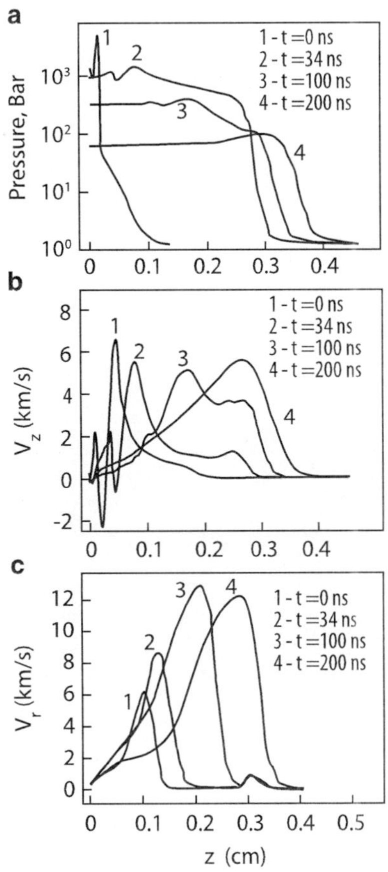

Fig. 7.36 Pressure $P(z, r = 0)$ (**a**), velocity $v_z(z, r = 0)$ (**b**), and $v_r(r, z = 55\ \mu\text{m})$ (**c**) distributions at different instants upon irradiation by a 1.06 μm pulse

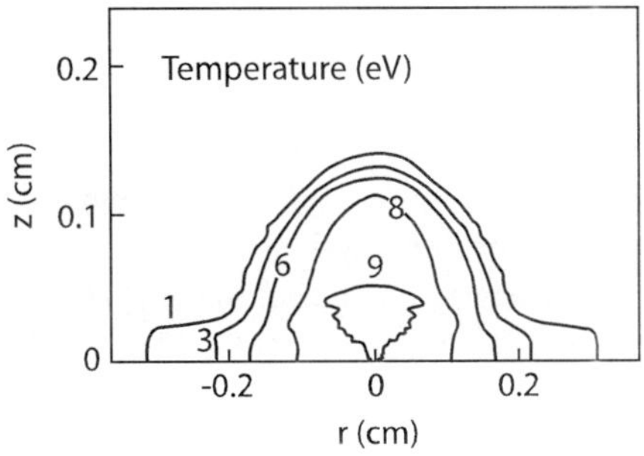

Fig. 7.37 Plasma temperature distribution by the 0.35 μm pulse end at $t = 34$ ns

and neutral atom concentrations, respectively, and the coordinate of the shock wave is determined by the pressure maximum. The ionization wave, as mentioned above, is the fastest type of plasma expansion. Figure 7.33 shows that its speed achieves 100 km/s, considerably exceeding the sound speed. The shock wave, which is determined by slower hydrodynamic processes, propagates at a considerably smaller

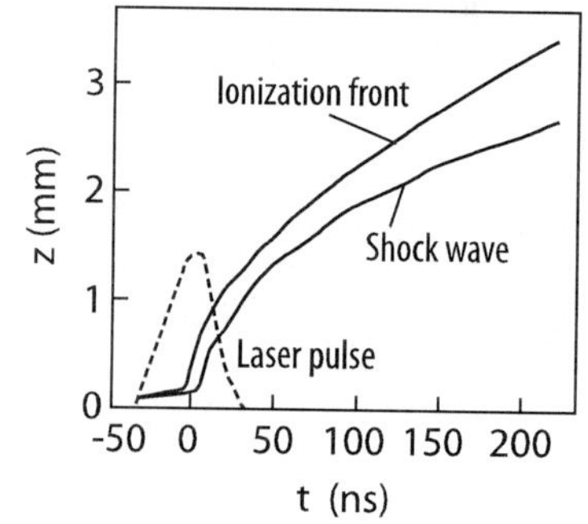

Fig. 7.38 Positions of the shock wave and ionization front upon irradiation by a 0.353 μm laser pulse

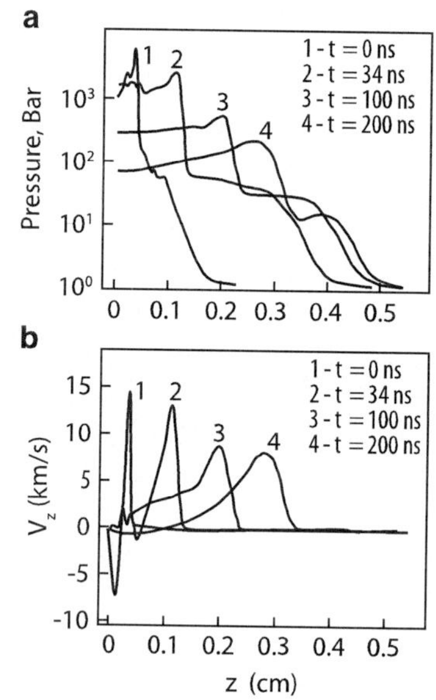

Fig. 7.39 Pressure $P(z, r = 0)$ (**a**) and velocity $v_z(z, r = 0)$ (**b**) distributions at different times upon irradiation by $\lambda = 0.353$ μm pulses

speed. After the end of a laser pulse, the ionization front drastically slows down, the shock wave catches up with it, and beginning from the moment $t \approx 200$ ns, the plasma expansion is characterized only by one front.

The energy balance of the plasma includes three components: the laser radiation energy, the plasma emission energy, and the work of pressure forces. The balance of these components determines one or another mechanism of plasma expansion. Upon irradiation by IR pulses, the release of the laser energy dominates in the region of the ionization front, while the work of pressure forces dominates in the shock wave region. The role of thermal radiation is illustrated by two temperature profiles $T(z)$ in Fig. 7.34 for $r = 0$ at $t = 17$ ns. The solid curve is calculated by using the complete model, while the dashed curve is calculated by neglecting plasma radiation transfer. It follows from this figure that radiation efficiently redistributes energy in the hottest region near the surface and in the region of the shock wave front. In the absence of radiation, the shock wave front becomes steeper and the maximum temperature increases approximately by 2 eV. The contribution of radiation in the region of the ionization front is negligible. The maximum of the emission spectrum of this region at temperatures 2–3 eV is located in the range 6–8 eV. This emission is weakly absorbed by cold air, and as a result, the emitted energy freely escapes from the interaction region.

The calculated plot of the shock wave motion (solid curve in Fig. 7.35) can be compared with the corresponding experimental curve obtained by high-speed photography (dotted curve in Fig. 7.35) [48] and the curve calculated analytically by the expression $V_0 = [2(\gamma^2 - 1)q_0/\rho_0]^{1/3}$ (dashed curve), where V_0 is the shock wave speed and $\gamma = 1.4$ is the adiabatic index. The figure shows that calculated and experimental values are in good agreement. At the initial stage, the analytic dependence describes the front dynamics with a good accuracy.

The gas-dynamic parameters of the process – the distributions of pressure $P(z, r = 0)$, azimuthal $V_z(z, r = 0)$ and radial components of the velocity at several successive instants are presented in Fig. 7.36. The initial moment of the plasma formation is characterized by a drastic increase in pressure, achieving a few kilobars (curve 1 in Fig. 7.36a). Then, the high-pressure region propagates together with the ionization wave, the maximum pressure being achieved in the shock wave (curve 2). Pressure rapidly decreases during plasma cooling (curves 3 and 4 in Fig. 7.36a). The profiles of the azimuthal velocity (curves 1 and 2 in Fig. 7.36b) are bipolar: the positive branch corresponds to the shock wave, while the negative branch describes the compression wave propagating toward the surface. The motion of the shock wave is restrained by the high pressure in the region in which the ionization front has propagated, and therefore the characteristic velocity of this wave (6 km/s) is almost half the expansion velocity in the radial direction (Fig. 7.36c). At the instant of time 200 ns, a plasma cloud takes the form of a hemisphere of radius ~0.3 cm (Fig. 7.36b,c).

In the case of irradiation by UV pulses, the plasma expansion has a number of qualitative differences (Fig. 7.37). The absorption coefficient of plasma, which decreases with decreasing the wavelength, proves to be insufficient to support the ionization wave. Laser radiation is screened predominantly on the shock wave front, resulting in the formation of a laser-supported detonation wave, which provides the propagation of plasma initiated by UV laser pulses. The boundary of the ionization region moves slightly ahead of the shock wave (Fig. 7.38). Because an LRW is not excited by UV pulses, pressure and temperature in front of a shock wave are lower than in the case of irradiation by IR pulses (cf. Figs. 7.36a and 7.39a). As a result, the laser radiation intensity on the LDW front in the former case is higher than in the latter. This leads to the increase in pressure and the LDW speed (cf. Figs. 7.36a and 7.39a; Figs. 7.36b and 7.39b).

7.3.7 Plasma Transparency and Transmission Coefficient

Due to the absorption of laser radiation in plasma, the real intensity $q_{tr}(r, t)$ and duration τ_{tr} of a pulse incident on the surface of a condensed medium are smaller than the intensity $q(r,t)$ and duration τ of the initial pulse. Consider the dependence of the parameters of the transmitted pulse on the peak intensity q_0 in the IR and UV regions. Figures 7.40 and 7.41 present the time dependences of the incident $[q(0,t)]$ and transmitted $[q_{tr}(0,t)]$ radiation intensities at the focal spot center $r = 0$ obtained upon irradiation by IR and UV pulses of peak intensities 6 and 13 GW/cm^2.

In the case of irradiation by $q_0 = 6\,\text{GW/cm}^2$ IR pulses (Fig. 7.40a), plasma appears at the end of the trailing edge of a laser pulse ($t \approx 12$ ns) and its screening action is insignificant. As a result, the intensity and FWHM of the incident and transmitted pulses coincide (Fig. 7.40a). In the case of irradiation by $q_0 = 13\,\text{GW/cm}^2$ IR pulses, plasma appears at the onset of the leading edge of a laser pulse ($t \approx -17$ ns), its screening action is manifested completely, and the transmitted pulse considerably differs from the incident pulse. The maximum intensity achieves $q_{tr} = 6\,\text{GW/cm}^2$, being half that of the incident beam, while the pulse duration $\tau_{tr} \approx 10$ ns is three times smaller than that of the incident beam (Fig. 7.40b). This means that, as the laser beam intensity is further increased, the intensity of the laser pulse acting on the condensed medium will not increase, while its duration will be reduced.

Upon irradiation by $q_0 = 6\,\text{GW/cm}^2$ and $q_0 = 13\,\text{GW/cm}^2$ UV pulses, the screening effect appears beginning from the leading edge of the laser pulse Fig. 7.41. However, unlike irradiation by IR pulses, the screening action of plasma is manifested no so drastically. And although the transmitted pulses differ "in details", their

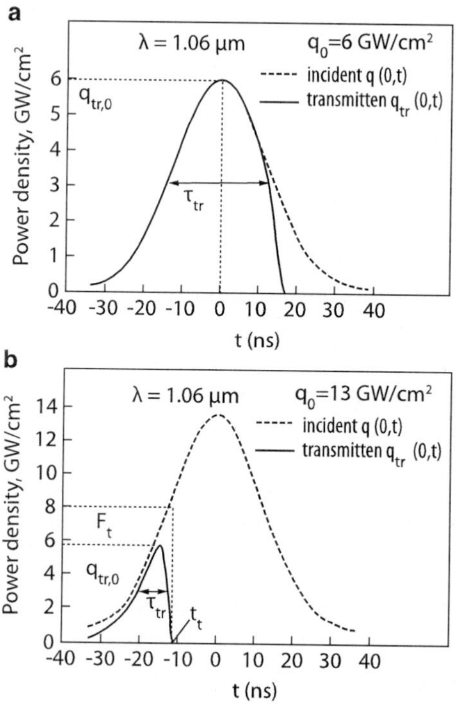

Fig. 7.40 Transmitted $[q_{tr}(0,t)]$ and incident $[q(0,t)]$ 1.06 μm laser radiation intensities. $q_0 = 6$ (**a**) and 13 GW/cm^2 (**b**)

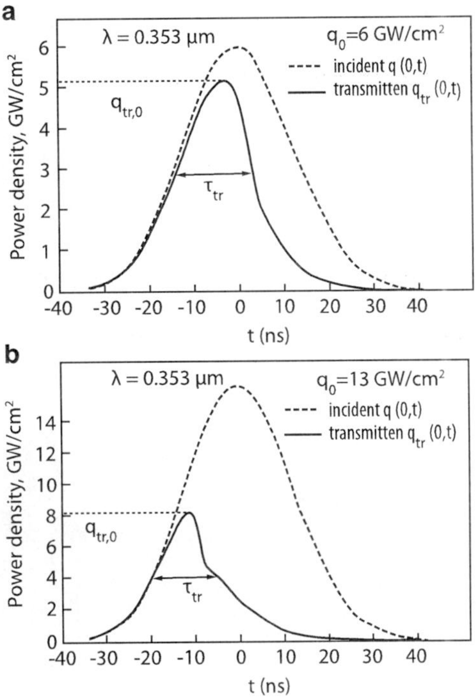

Fig. 7.41 Transmitted $[q_{\mathrm{tr}}(0,t)]$ and incident $[q(0,t)]$ 0.353 μm laser radiation intensities. $q_0 = 6$ (**a**) and 13 GW/cm^2 (**b**)

average parameters are close: $q_{\mathrm{tr},0} \approx 6\,\mathrm{GW/cm^2}$ and $\tau_{\mathrm{tr}} \approx 18\,\mathrm{ns}$ for $q_0 = 6\,\mathrm{GW/cm^2}$ and $q_{\mathrm{tr},0} \approx 7\,\mathrm{GW/cm^2}$ and $\tau_{\mathrm{tr}} \approx 15\,\mathrm{ns}$ for $q_0 = 13\,\mathrm{GW/cm^2}$.

Note that the screening of laser radiation occurs not simultaneously over the entire focal spot area (Fig. 7.42). Although the laser pulse is short, a plasma cloud is inhomogeneous along the surface due to the inhomogeneous Gaussian intensity distribution over the radius of the incident laser beam. Absorption begins at the focal spot center and propagates to the edges. Here, the question again arises of the validity of the integrated measurements of the plasma transparency with the help of a hole in a target (see Figs. 7.17, 7.18, 7.27). It seems that such measurements are of qualitative nature.

7.3.8 Comparison with Experiments

Figures 7.43 and 7.44 present the results of a series of calculations performed for irradiation by IR pulses. The radiation intensity q_0 was varied from 6 to 17 GW/cm^2 with a step of 1 GW/cm^2, and $q_{\mathrm{tr},0}$ and τ_{tr} were determined in each case. One can see from Fig. 7.43 (solid curve) that the maximum intensity of the transmitted pulse first increases, passes through the maximum and then becomes constant at a level of ~6 GW/cm^2. Experimental estimates (dotted curve) also presented in this figure

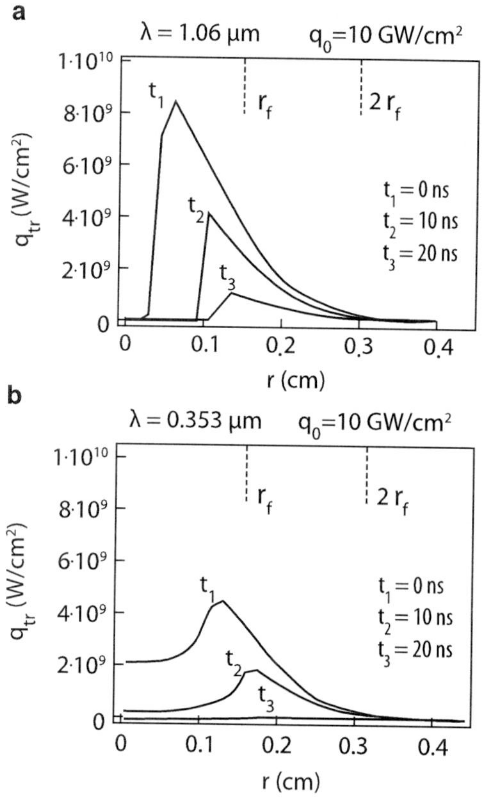

Fig. 7.42 Spatial profiles $q_{\rm tr}$ of laser radiation on a surface at different instants for $\lambda = 1.06\,\mu$m (**a**) and $\lambda = 0.353\,\mu$m (**b**)

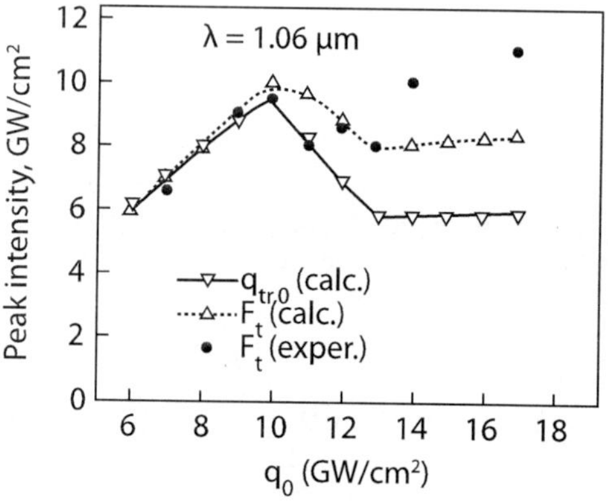

Fig. 7.43 Dependences of the maximum intensity of the transmitted 1.06 μm laser radiation on the incident radiation intensity: calculated values $q_{\rm tr.0}(q_0)$, experimental estimate $F_t(q_0)$ [47] and the calculated estimate $F_t(q_0)$

well agree with calculations in the region $q_0 < 12\,\rm GW/cm^2$. However, at higher intensities, they become constant at a level of $\approx 10\,\rm GW/cm^2$ with an experimental accuracy of $\pm 2\,\rm GW/cm^2$ [47].

Figure 7.44 presents the calculated (solid curve) and experimental (dotted curve) dependences of the transmitted pulse width on the radiation intensity $\tau_{\rm tr}(q_0)$ [47].

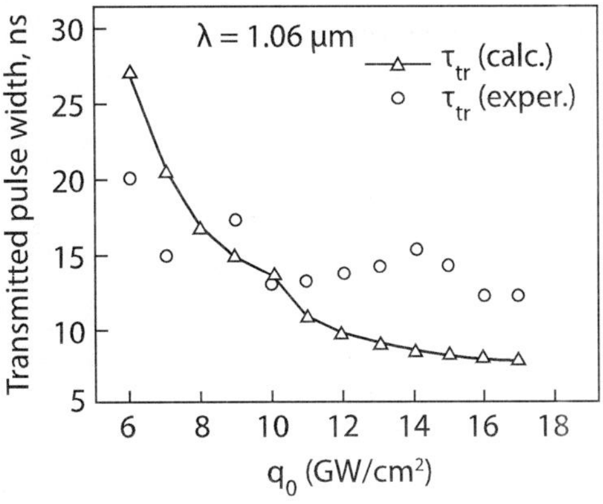

Fig. 7.44 Dependences of the width of a $\lambda = 1.06\,\mu$m transmitted laser pulse on the incident radiation intensity (calculated values $\tau_{tr}(q_0)$ and experimental estimates [47])

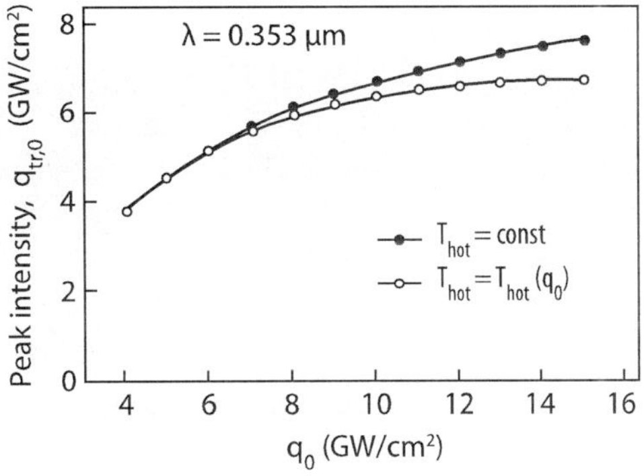

Fig. 7.45 Dependences of the maximum intensity of transmitted $\lambda = 0.535\,\mu$m laser radiation on the incident radiation intensity $q_{tr.0}(q_0)$ obtained for different initial data

The pulse width decreases from its maximum value 25–30 ns down to constant level of 8 ns in calculations and 10–15 ns in experiments.

Figures 7.45 and 7.46 show the dependences $q_{tr,0}(q_0)$ and $\tau_{tr}(q_0)$ upon irradiation by $q_0 \in$4–15 GW/cm^2 UV pulses. Beginning from $q_0 = 6$ GW/cm^2, the transmitted radiation intensity becomes lower than the incident radiation intensity (curve $T_{hot} =$ *const* in Fig. 7.45). In this case, unlike irradiation by IR pulses, the radiation intensity increases monotonically. If the initial conditions are specified taking into account the increase in the temperature $T_{hot}(q_0)$ of the initial hot region with increasing q_0, then q_{tr} achieves $\approx$6 GW/cm^2 (Fig. 7.45). The values of $T_{hot} = T_{hot}(q_0)$ presented in Fig. 7.46 show that the duration of the transmitted UV pulse is $\approx$15 ns, being twice that of the IR pulse.

The fraction of laser energy transmitted to a target can be estimated from the dependences of the maximum intensity and duration of the transmitted pulse on the radiation intensity (Figs. 7.43 and 7.44). It amounts to $\sim$50% for $q_0 = 6$ GW/cm^2. However, this estimate will be somewhat understated because of the spatial inhomogeneity of transmitted radiation. The matter is that the data presented in these figures correspond to the central region of the focal spot, which is, as we saw, is most strongly screened (Fig. 7.42). Integration over time and target surface gives the fraction of radiation transmitted to the target surface equal to $\sim$85% of the incident

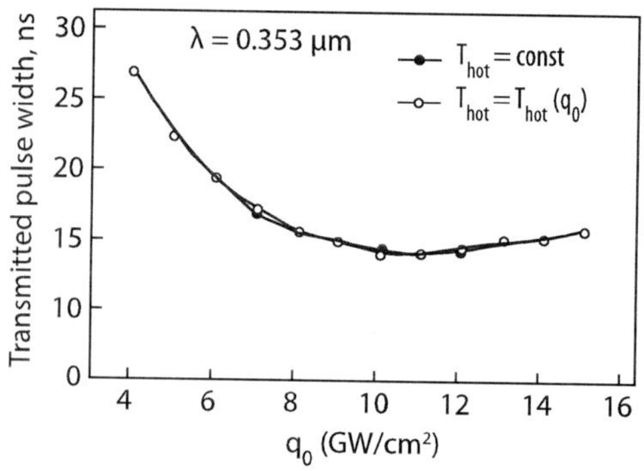

Fig. 7.46 Dependences of the width of transmitted $\lambda = 0.535\ \mu$m laser radiation on the incident radiation intensity $\tau(q_0)$ obtained for different initial data

radiation. The calculation of radiation energy emitted by plasma in the direction of the target shows that this energy amounts to 10% of the energy absorbed by plasma. As a result, 87% of the laser pulse energy is supplied to the target. Approximately, the same part of UV radiation achieves the target. Figures 7.43–7.46 show that the fraction of the transmitted IR and UV radiation energy can decrease down to 1% and 25%, respectively, with increasing the laser radiation intensity. Thus UV radiation is weaker attenuated in plasma.

7.4 Plasma Processes in Material Vapors

In the previous paragraph, we considered pulsed plasma phenomena in air initiated by IR and UV laser radiation. Plasma initiation processes were simulated by specifying a narrow air layer near a target, the layer temperature being sufficient for initiating a plasma plume. This artificial procedure allowed the separation of vaporization processes from other processes proceeding in the environment in order to focus attention on the latter. As for the relation of these calculations to the physical reality, such a situation is possible, e.g., during irradiation of refractory metals, when the surface temperature is high but vapor is still absent (see part 3.2). It is interesting to consider a more realistic case, when plasma appears "spontaneously". For this purpose, we can analyze plasma processes proceeding over a low-melting metal target. In this case, a discharge will develop in the vapor of a metal whose ionization potential is considerably lower than that in air. Therefore, the conditions can be realized when the breakdown can be described in the equilibrium case, i.e. when the plasma density is described by the Saha equation. We described the "equilibrium" breakdown earlier (see part 3.2) under stationary conditions or in the case of long laser pulses. The equilibrium mechanism of plasma formation allows a unified description of plasma and gas-dynamic processes. In this case, gas-dynamic equations for a vaporizing target (7.4) should be supplemented with plasma processes such as laser radiation transfer (7.4) and hot vapor plasma radiation transfer (7.17) and (7.18). We consider plasma processes in the target vapor

Table 7.6 Thermo-physical and optical properties of aluminium

Variable	Value	Variable	Temperature dependences:		
			$T_0 = 300$ K	$T_m = 933$ K	$T_{cr} = 8{,}000$ K
T_b, K	2720	κ, W/cm K	2.37	0.75	0.16
p_b, Bar	1	ρ, g/cm^3	2.70	2.33	0.64
L_v, J/g	1.1×10^4	C_p, J/gK	0.95	1.20	1.20
A_m, g/mol	27	α	0.1	0.1	0.3

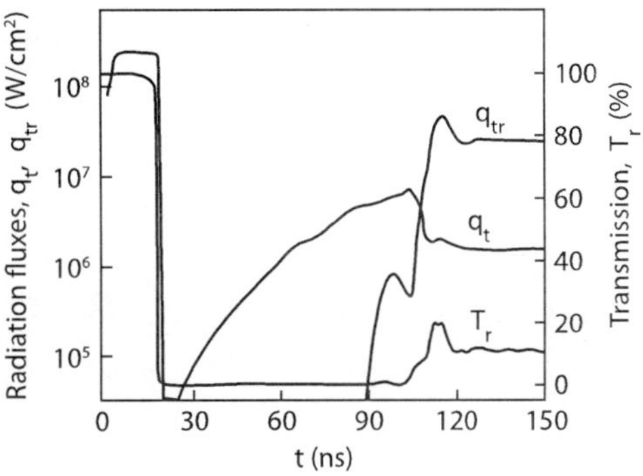

Fig. 7.47 Transmitted laser intensity q_{tr}, transmission coefficient T_r, and thermal radiation flux q_t vs time for $r = 0$ and $z = 0$

by the example of practical interest when an aluminum target is irradiated in vacuum by a 150-ns, 0.8 GW/cm^2 "rectangular" pulse from a Nd laser ($r_f = 0.25$ mm) [52]. Because plasma is initiated in fact by a target, the target surface temperature should be calculated with good accuracy taking into account the dependence of thermal coefficients on temperature and phase transitions, as was done in the PLD case (Chap. 6). The relevant parameters are indicated in Table 7.6.

Figure 7.47 demonstrates that the plasma has appeared 20 ns after the irradiation beginning, its density being so high that the plasma absorbs all incident radiation. Correspondingly, the transparency coefficient drops down to zero. Because of screening, the target surface temperature drastically decreases and vaporization not only ceases but changes to condensation (Fig. 7.48). The first stage of the plasma process ends by this moment. A hot plasma layer begins to expand and is further heated. It expands to both sides (towards the target and from it). When hot vapor again touches the target surface, the surface temperature begins to grow rapidly and condensation weakens. This is the long second (condensation) stage. The vapor continues to expand towards the target and to the sides, its density decreases (Fig. 7.49). As a result, the plasma begins to transmit partially laser radiation (Fig. 7.50), the surface temperature increases and condensation changes to subsonic vaporization (Fig. 7.48). Thus, a quasi-stationary plasma jet of the material vapor is produced (stage III) with a small Mach number $M \approx 0.03$. However, the vapor density in this case is rather high, being almost equal to the saturated vapor density ρ_s. Recall that $M = 1$, $\rho = 0.32\rho_s$. Therefore, this jet resembles a subsonic vapor jet flowing with a counterpressure to the surrounding gas. Such is the plasma formation effect during vaporization to vacuum!

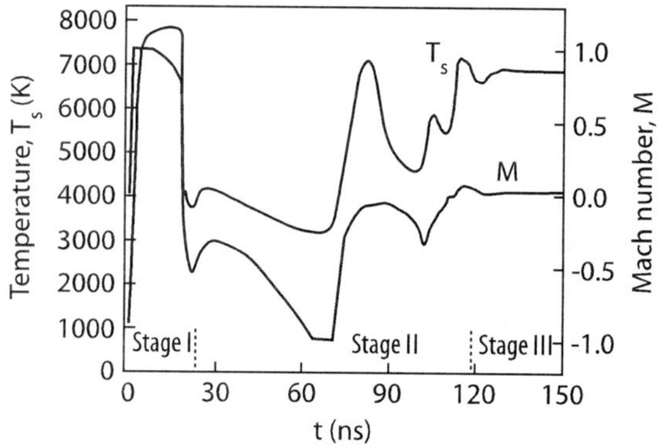

Fig. 7.48 Time dependences of the target surface temperature and Mach number for $r = z = 0$

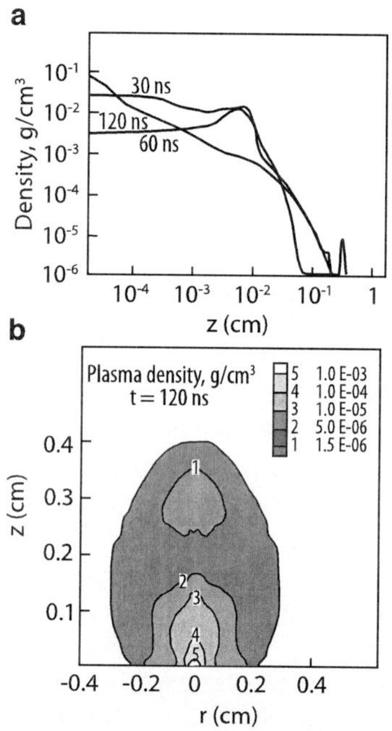

Fig. 7.49 Plasma density profiles at the beam axis (**a**) and contour plot at $t = 120$ ns (**b**)

7.4.1 Plasma Processes on a Target Surface

A plasma vapor cloud expands not only forward but also in the radial direction. Such a motion occurs both during evaporation in vacuum without plasma upon ablation (see Chap. 6) and in air upon generation of a plasma bunch (see above). During the propagation of a plasma bunch in air, the maximum intensity of transversely inhomogeneous laser radiation was observed at the edges of the focal spot on the target surface. The same occurs, but for different reason, in our case as well. This is well illustrated in Fig. 7.50. During the propagation of a rarefaction wave from the edge to center, the plasma begins clear at the edges. The laser radiation

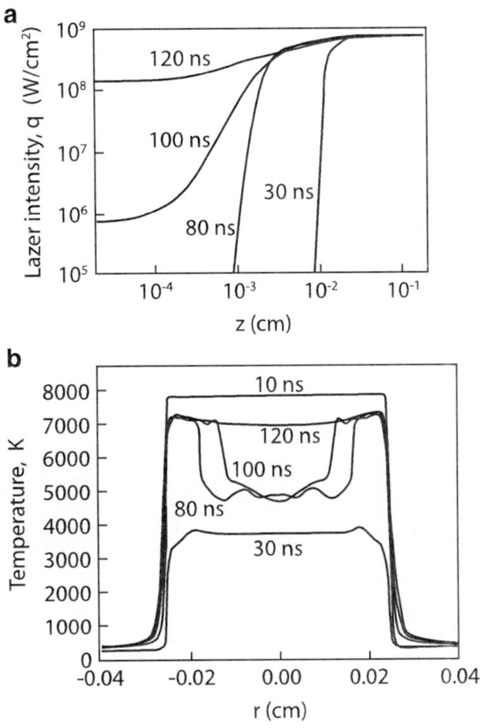

Fig. 7.50 Incident laser intensity profiles on the beam axis (**a**) and surface temperature profiles vs the beam radius (**b**)

intensity on the target proves to be maximal at the focal spot edges as well. The surface temperature rapidly responds to laser radiation. One can see that the plasma itself, more exactly its thermal radiation only weakly affects the surface temperature profile. Indeed, the plasma temperature is maximal at the bunch center, whereas the surface temperature on the axis is minimal. Finally, when the quasi-stationary state is established, the plasma bunch becomes considerably greater than the focal spot (Fig. 7.49b) and the plasma density weakly changes within the focal spot. Because of this, laser radiation is attenuated uniformly, and the target temperature in the focal spot is also virtually homogeneous. Because vaporization strongly depends on temperature, the material removal profiles also demonstrate edge effects. The material amount removed at the focal spot edges is several times greater than that removed at the focal spot center (Fig. 7.51).

The above discussion suggests the question of how the material removal rate will change with decreasing the focal spot size, when edge effects become decisive. The answer to this question is demonstrate in Fig. 7.52a. When the focal spot size becomes comparable with the characteristic size of the edge inhomogeneity, the plasma is bleached stronger and the bleaching begins considerably earlier. Because of this, the material removal at the focal spot center proves to be considerably greater in the case of a small focal spot compared to a large spot for the same laser radiation intensity, Fig. 7.52b.

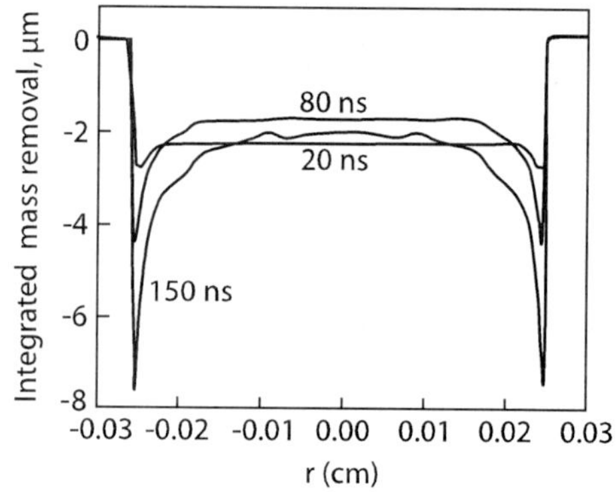

Fig. 7.51 Integrated mass removal vs radius

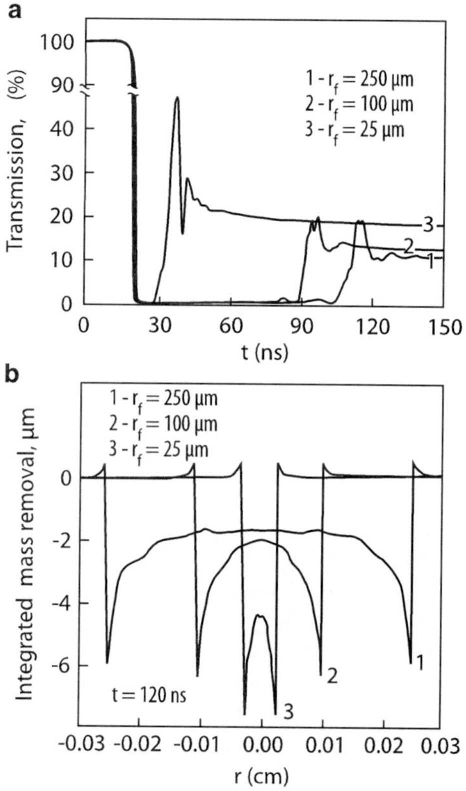

Fig. 7.52 Transmission coefficients upon irradiation by laser beams with radii $r_f = 250$, 100, and 25 μm (**a**) and the integrated mass removal vs r for these beam radii (**b**)

Before comparing calculations with experiments, it is interesting to consider the interaction of higher frequency laser radiation with a target. This radiation is weaker absorbed in the plasma and, therefore, is assumed more promising for the ablation technology. Figure 7.53a presents calculations of the removal of an Al target material per 25 ns pulse at 1.06 and 0.248 μm [76]. One can see that due to the lower breakdown threshold for UV radiation, ablation is followed, in the

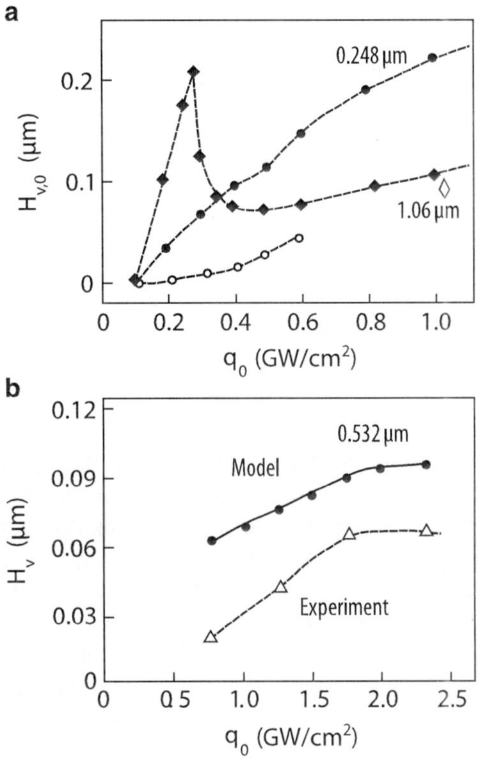

Fig. 7.53 (**a**) Dependences of the removed layer thickness on the $\lambda = 1.06\,\mu$m and $\lambda = 0.248\,\mu$m laser radiation intensities. *Dark symbols* are calculations, *open symbols* are experiment. (**b**) Dependences of the removed layer thickness on the laser radiation intensity calculated taking into account the influence of plasma (*solid curve*); triangles are experimental data, $\lambda = 0.532\,\mu$m, $r_f = 0.025$ cm

intensity range under study, by the plasma screening of the target. Upon irradiation by low-intensity $1.06\,\mu$m pulses, a breakdown does not occur and the removal of the target material increases with increasing q. Above the vapor breakdown threshold (0.25 GW/cm^2), the ablation efficiency drastically decreases and then weakly increases with increasing the radiation intensity. As a whole, because of a higher absorption coefficient of plasma in the IR region, the ablation efficiency of IR radiation is considerably lower than that of UV radiation. However, the dependence of the ablation efficiency on the radiation frequency is not so strong as could be expected based on the expression for the absorption coefficient. This is explained by the complicated nature of ablation controlled by the optical discharge. The same figure presents experimental data for radiation at $1.06\,\mu$m [78] and $0.248\,\mu$m [79]. In the case of visible radiation, experimental data better agree with the model (Fig. 7.53b [31]). As a whole, the theory is in qualitative agreement with experiments, the quantitative difference is probably explained by incomplete information about the properties of a target at high temperatures (7,000–8,000 K) obtained in calculations.

Thus, we see that ablation during initiation of the optical breakdown differs qualitatively from ablation without plasma. The removal of the material of target decreases due to screening, the jet becomes subsonic, and the keyhole depth increases near focal spot edges. The latter circumstance is important for laser applications in microelectronics, where vertical walls can be required for laser etching.

7.4.2 *Plasma Processes During Vaporization of Metals in Air*

In previous sections, we considered plasma formation near a target either in a surrounding gas or material vapor in vacuum. The plasma dynamics during the vaporization of a target in a gas medium attracts natural attention. The vaporization of a target to a medium with a counterpressure was considered in the study of ablation without plasma formation (see Sect. 6.1). The numerical solution of such problems involves a difficulty of describing the vapor–air interface. While this difficulty was surmounted in one-dimensional models, in two-dimensional cases special methods are required to do this. The examples of plasma formation during the vaporization of a target into a surrounding gas by nanosecond and microsecond pulses are reported in the literature. Note at once that these earlier calculations neglect processes which were taken into account in recent more detailed models. This concerns first of all the description of the change of vaporization to vapor condensation. Let us begin with the case that is most close to the irradiation of an aluminum target in vacuum by a nanosecond pulse, which we considered above.

Consider the irradiation of an aluminum target in air by 100 ns pulses from a Nd laser [53]. The pulse intensity is 380 MW/cm^2, which is almost three times lower than that in the previous case. Thus, we should solve the same equations as for vaporization in vacuum. Because the radiation intensity in our case is lower than in the case of vacuum, plasma processes will develop slower. The total screening will occur at $t = 50$ ns ($t = 25$ ns in the compared variant). The high-pressure region passes into air, where an LDW propagates (Fig. 7.54a). As a result, plasma is not "clarified". After the end of the laser pulse at $t = 100$ ns the shock wave continues to run in air and is gradually attenuated due to expansion. By the end of the laser pulse almost all energy remains in plasma (Fig. 7.54b). Figure 7.54b also shows the dynamics of the thermal radiation of plasma to the environment and towards the target. One can see that this part of energy in the thermal balance is small at all. Only 4% of the laser pulse energy reaches the target for the time exceeding the laser pulse duration by two orders of magnitude. The values of the same order of magnitude are also typical for the propagation of a plasma cloud in vapor or air (see Fig. 7.47). Numerical calculations agree with experiments [51], where it was found that 6% of the laser pulse energy was incident on an aluminum target due to emission from plasma.

Note here calculations [54] on the irradiation of an aluminum target by 26 ns, $\lambda = 248$ nm, $q \cong 1$ GW/cm^2 pulses from a KF laser. The target was irradiated in air at a low pressure of $\approx 10^{-3}$ atm (100 Pa). The authors of [54] compared the expansion of a vapor cloud without a breakdown and with the plasma formation (Fig. 7.55). One can see that plasma formation due to the absorption of laser radiation leads to a considerable increase in the bunch size and changes its structure. Note that

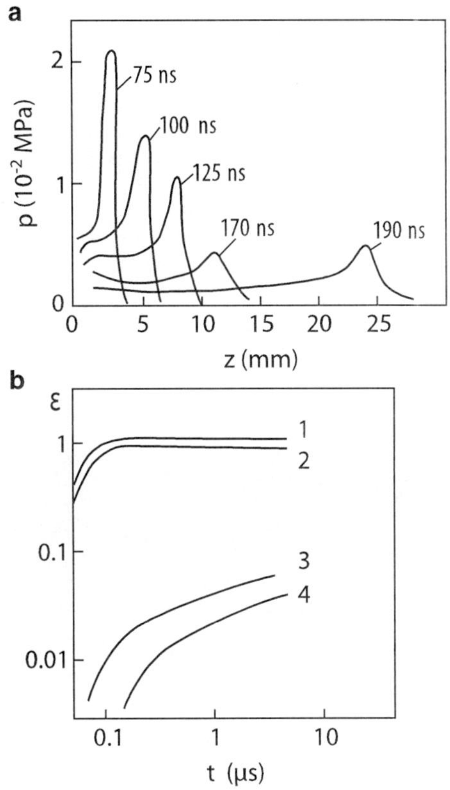

Fig. 7.54 (**a**) Distributions of gas-dynamic parameters along the symmetry axis for an aluminum target irradiated by a neodymium laser in air ($\tau = 0.1\,\mu$s, $q = 380\,$MW/cm^2. (**b**) Time dependences of the total (**1**) and thermal (**2**) energy of a plasma plume and energy losses due to radiation to the environment (**3**) and towards a target (**4**) (data are normalized to the total radiation energy)

ablation without plasma formation for other materials has a similar structure, which is described in the beginning of Chap. 6, Fig. 6.8.

As mentioned above, LDWs are not formed at intensities below $\sim$10 MW/cm^2, which is demonstrated by one-dimensional calculations (Fig. 7.6). The same result follows from two-dimensional calculations [55]. The $q = 2\,$MW/cm^2 beam of a neodymium laser illuminates an aluminum target in air for 20 μs. After transient processes, a quasi-stationary supersonic flow of the aluminum vapor is established near the target. The passage to the subsonic flow occurs through a standing shock wave with a pressure jump twice the atmospheric pressure. The vapor temperature does not exceed 10 kK, which weakly attenuates the neodymium laser radiation. Thus, in the case of a low-intensity laser beam, an erosion jet with weakly attenuated laser radiation can be formed. We discussed such a flow of vapors in Chap. 1. A similar picture was also observed in other experiments.

Numerical methods for studying physical processes were developed simultaneously with the development of laser technologies. Recently a method was developed which allows a more detailed description of a complex gas-dynamics during irradiation of metal targets by laser pulses. The effect of processes involved in this case requires quite accurate calculations. This concerns, in particular, the interaction

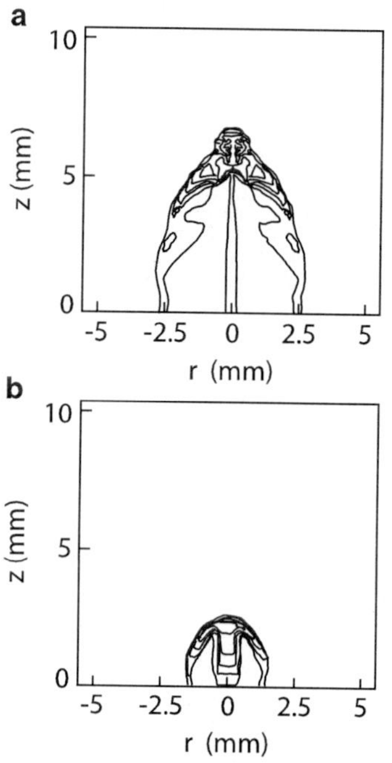

Fig. 7.55 Dimensionless pressure contours calculated for an aluminum target irradiated by $\tau = 26$ ns, 25 J/cm^2 pulses from an excimer laser for $t = 500$ ns taking plasma into account (**a**) and neglecting plasma (**b**). The air pressure is $P = 10^{-3}$ atm, $r_f = 0.5$ mm

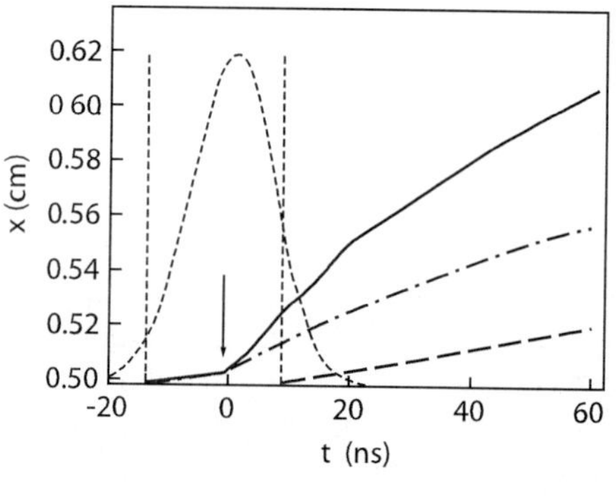

Fig. 7.56 Motion trajectories of phase interfaces: the solid curve corresponds to a shock wave in air, the dot-and dash curve corresponds to the contact air–vapor interface, and the dashed curve corresponds to a contact discontinuity in vapor. $q_0 = 7 \times 10^9$ W/cm^2 $\tau = 10^{-8}$ s

of shock waves in a vapor–air atmosphere, which is typical for these conditions. It is known that when one wave catches up and "absorbs" another, a flow with a sharp temperature peak appears [56]. This effect is very important for our conditions because it can lead to a rapid formation of an LDW, which completely screens the target. The dynamic adaptation method [57–59] can be used to describe the motion of many interfaces, i.e. discontinuities running in different directions such as shock waves, the vapor–air interface, and contact discontinuities in a homogeneous matter (vapor). This method allows one to distinguish the fine structure of a plasma cloud propagating into the surrounding gas. Figure 7.56 shows the trajectories of interfaces obtained in one-dimensional calculations of the vaporization of an aluminum target in air irradiated by 10 ns, $q = 7$ GW/cm^2 pulses from a Nd laser [56].

Figure 7.56 shows the time dependences of the spatial position of a shock wave and two contact discontinuities (vapor–vapor and vapor–air interfaces). The vertical dashed straight lines indicate the instants of the evaporation and formation of a shock wave and the instant of formation of a contact discontinuity in vapor. The arrows shows the instant corresponding to the passage of the shock wave from plasma through the air–vapor interface followed by the absorption of the shock wave in air. This is demonstrated by a drastic change in the slope of the trajectories of the shock-wave and contact interface. In this case, the shock wave in air is transformed after amplification to an LDW and escapes from the air–vapor interface. The speed of the shock wave in air increases from 2.7×10^3 m/s up to $\approx 2.6 \times 10^4$ m/s, and temperature increases from $\approx 5.5 \times 10^3$ K to 6×10^5 K. A contact discontinuity in vapor is produced during the repeated vaporization process, which we observed earlier in the two-dimensional model of vaporization into vacuum (Fig. 7.48). In a more sophisticated model [59], the ratio of vapor densities and temperatures in a contact discontinuity in the case of a uniform pressure can be quite significant (2–16).

7.4.3 Plasma Phenomena During the Deep Penetration of a Laser Beam into a Sample and Breakdown on Microdroplets

During pulsed or repetitively pulsed making holes by intense laser radiation, a plasma plume is formed both over the surface of a sample and in its depth in a gas–vapor channel. It is clear that an important factor such as the side discharging of plasma is absent in deep channels. From this point of view, the situation is similar to the one-dimensional case. However, thermal radiation can be strongly absorbed in the channel, and friction and heat exchange with side walls can occur. Therefore, the development of a plasma plume under such conditions requires a special consideration. Optical discharges were studied in narrow quartz capillaries [60–62] and small copper tubes with narrow side slits [63, 64] simulating gas–vapor channels. Plasma plumes were also investigated directly in gas–vapor channels produced by a laser beam in a transparent material [65]. The restriction of the side expansion of a plasma plume ignited by a millisecond pulse reduces the SRW support threshold in air down to 2.5 MW/cm^2. For the $\lambda = 5.2\,\mu$m radiation from a CO-laser, this threshold is reduced down to 0.5 MW/cm^2 [61]. Studies with microsecond pulses from a CO_2 laser with $q > 50$ MW/cm^2 [37, 63], showed that a plasma bunch propagates in the beginning of a pulse in the LDW regime at a speed of $\sim$10 km/s. Then, the speed drastically decreases and a luminous zone fills the entire channel from its bottom to the shock-wave front, as after the LDW decomposition (see Fig. 7.57(1)). As q is decreased, the discharge propagates in the SRW regime, as in previous cases. The question arises: How do the side walls affect the screening properties of plasma plumes in capillaries? Because the LDW support threshold is

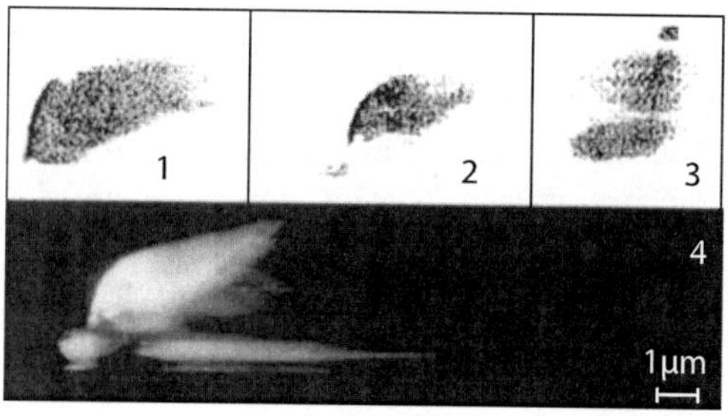

Fig. 7.57 Photographs demonstrating the development of a breakdown on the surface of a quartz plate (frame 1) for $h = 1$ mm (2) and 3 mm (3). $E = 2$ J, $\tau = 100\,\mu$s. 1, 2, 3 are negatives; 4 is 2 at the magnified scale (positive)

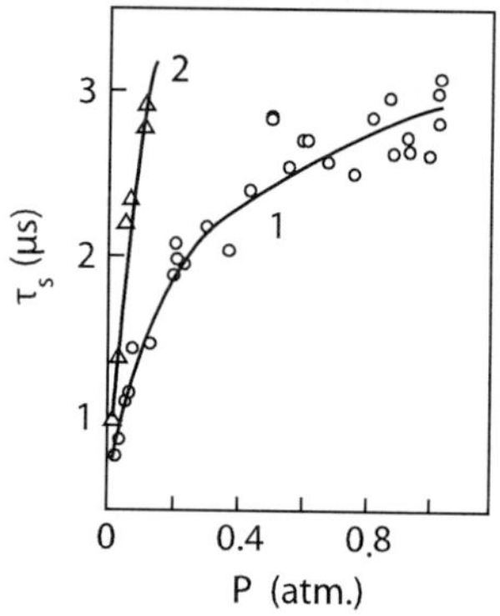

Fig. 7.58 Dependences of the complete screening time on the air pressure: (**1**) free surface; (**2**) tube of diameter 1.9 mm and length 12 mm

reduced, this wave will decompose later, and thus the screening time will be longer. This is well illustrated in Fig. 7.58 [64].

A plasma plume in a real gas–vapor keyhole can be affected not only by walls but also by the laser radiation distribution. As shown in Chap. 1, this distribution strongly depends on the channel shape. The plasma plume dynamics can be studied under these conditions by using frame photography or photochronography of an optical discharge and a keyhole in a quartz plate. Figure 7.57 presents photographs of the breakdown development in the atmospheric air on the surface and in the keyhole produced by repetitively pulsed CO_2 laser drilling [65] ($E = 2$ J, $\tau = 100\,\mu$s, and $q_0 = 5\,\mathrm{MW/cm^2}$). One can see that the propagation speeds of plasma over the surface and in the channel are close. The discharge propagates during the first microseconds along a narrow front, which indicated to the LDW regime. One can see that the keyhole emission disappears due to LDW screening. Then, after the LDW decomposition, laser radiation again penetrates into the keyhole. This is demonstrated by the regenerated quartz emission.

This emission also shows that the breakdown appears within some delay τ_d after the laser pulse onset. In the case of shallow keyholes, a plume is produced at the keyhole bottom. As the keyhole depth increases, the delay time increases (Fig. 7.59). When a certain depth of the keyhole is achieved, the depth at which the plume is generated ceases to grow (Fig. 7.57(3)). As the keyhole depth further increases, the breakdown does not occur at all. This effect demonstrates the influence of the keyhole depth on the laser radiation intensity on the keyhole walls. The

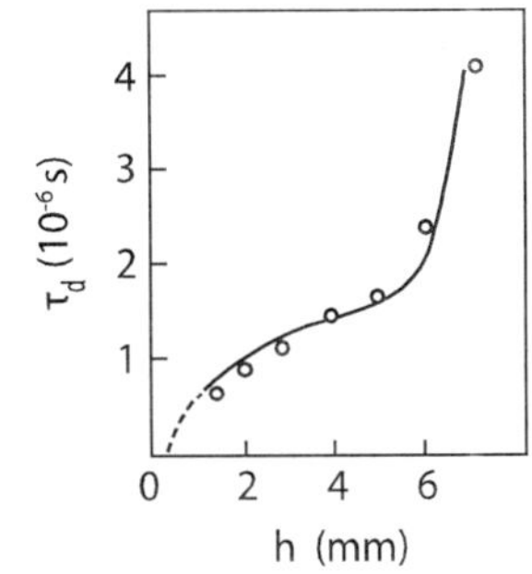

Fig. 7.59 Typical dependence of delay time τ_d on keyhole depth h for $E = 4\,\mathrm{J}$ and $\tau = 30\,\mu\mathrm{s}$

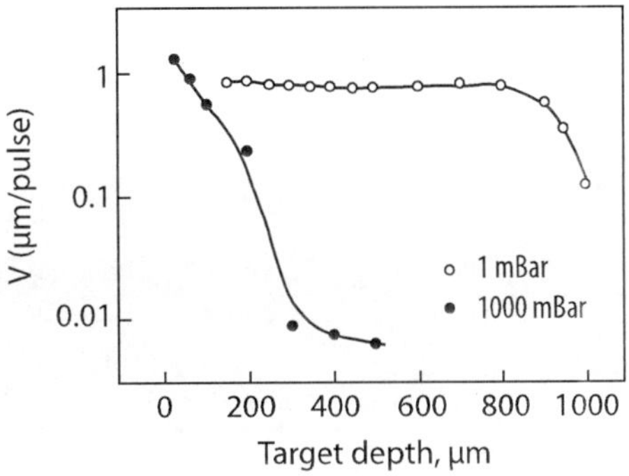

Fig. 7.60 Comparison of average ablation rates for channels in steel irradiated by 300 ps, 80 J/cm^2 laser pulses

calculations of the light intensity in a keyhole performed above by the method of geometrical optics showed that the fraction of beams reflected from the keyhole bottom decreases with increasing the keyhole depth. This leads to the attenuation of the light intensity on the walls. The effect of plume quenching with increasing the keyhole depth can be important in material processing.

Plasma processes in the case of deep penetration of laser radiation into a sample can have a different nature. This is confirmed by the fact that plasma processes are enhanced on the contrary with increasing the channel depth. This leads to a drastic decrease in the ablation rate of steel irradiated by picosecond pulses (Fig. 7.60 [66]). The plasma screening during pulsed laser ablation is one of the key problems preventing the production of high-quality narrow and deep channels. The plasma produced upon irradiation interacts with the side walls of keyholes, which considerably reduces the efficiency and accuracy of the technological process [67, 68]. Plasma processes during ablation drilling are most intense in air at the atmospheric pressure (Fig. 7.60). A strong screening of the plasma occurs due to formation of microparticles or clusters in a keyhole. If a successive pulse in the repetitively pulsed drilling regime acts before the settling or removal time of these microparticles, the breakdown of air occurs (Fig. 7.61). Figure 7.61 shows that plasma produced by a single picosecond pulse weakly screens a laser beam despite the high beam intensity $\sim$300 GW/cm^2. Let us compare this with nanosecond pulses, when screening appears at a lower intensity (Fig. 7.44). The fact that microparticles do initiate the breakdown of air is confirmed by direct experiments

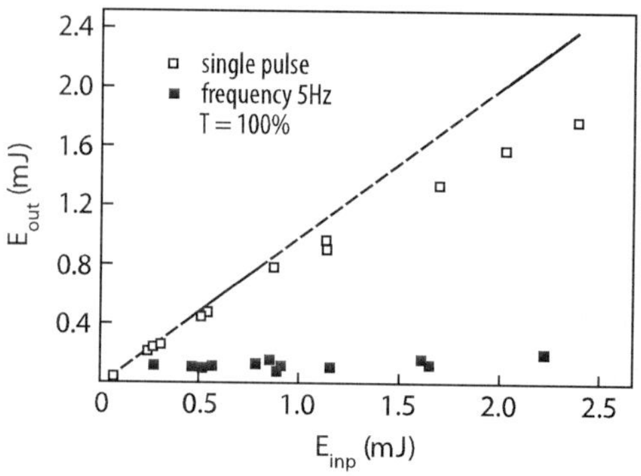

Fig. 7.61 Dependences of the laser pulse energy at the output of a broad through ablated channel ($d \gg d_w$) of length 500 μm on the input energy upon irradiation by single 300 ps, 80 J/cm^2 laser pulses (*open squares*) and these pulses with a repetition rate of 5 Hz (*dark squares*). The dashed straight line corresponds to the 100% transmission of the channel

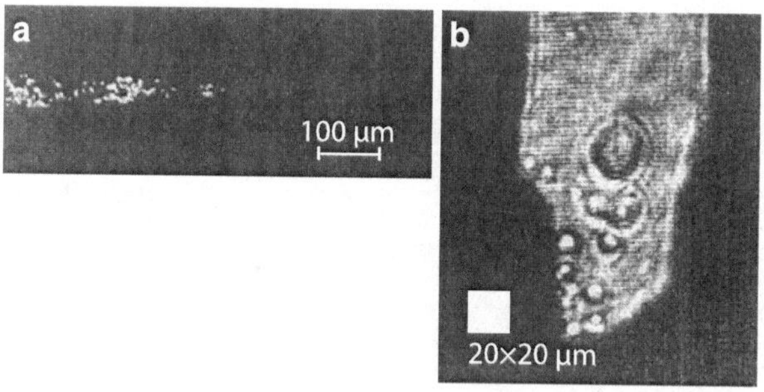

Fig. 7.62 Formation of multiple sources of optical breakdown plasma on ablated microparticles in air. (**a**) Shadow photograph of the formation process of a through channel in a 100-μm thick steel plate; (**b**) interferometric image of the channel bottom formation process in a 100-μm thick steel plate

(Fig. 7.62 [69, 70]). For a pulse repetition rate of ablating pulses equal to 2 Hz, breakdown sources are distinctly observed inside an ablated channel in the form of light dots on a shadow photograph or spherical phase objects upon interferometry. This plasma cloud, unlike plasma regularly appearing at the channel bottom, is located away from the bottom. This plasma absorbs up to 90% of laser energy, resulting in a considerable increase in the diameter at the upper part of the channel (Fig. 7.63).

Figure 7.60 shows that the breakdown on microparticles occurs at the atmospheric pressure. When pressure is decreased by 2–3 times, the breakdown cannot occur. It can be expected that, when the average power is increased (with increasing a pulse repetition rate), the density of air decreases due to its heating in the channel, which will lead only to rare breakdowns. A similar effect was considered earlier upon irradiation by successive microsecond pulses. If the second pulse was supplied

Fig. 7.63 Cross section of a channel produced in steel by 300 ps, 75 J/cm^2 laser pulses

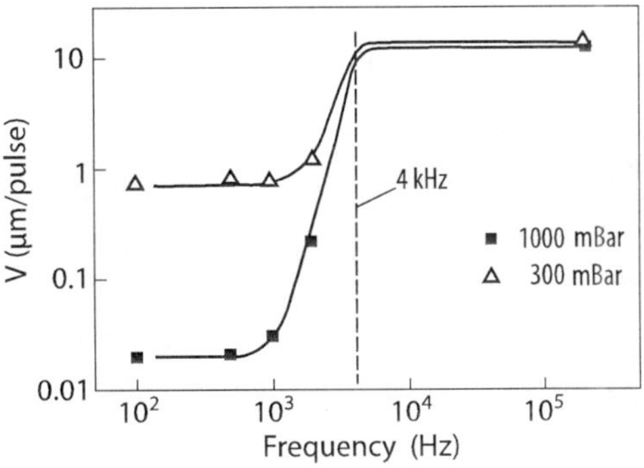

Fig. 7.64 Dependences of the ablation rate of 500-μm thick steel plates on the 100 J/cm^2 nanosecond pulse repetition rate

with a small delay after the first one, no plasma screening appeared. A similar effect is illustrated in our case in Fig. 7.64. This allows the use of lasers with high pulse repetition rates to obtain ablation conditions close to ablation in vacuum. This will considerably simplify the technological microprocessing process because there is no need to use vacuum chambers and pumps.

Note in conclusion that we have considered not all the aspects of the physics of an erosion plume. We discussed this phenomenon mainly from the point of view of its influence on material processing. We paid no attention to the interesting properties of this process such as the appearance of high-energy particles in the laser plasma, nonequilibrium effects, the instabilities of vapor–air interfaces, etc. (see monograph [71] and references therein).

Consider here in more detail one of the properties of a plasma plume, namely, energy transfer from a laser to a target with the help of a laser plume.

Although the effect of a drastic increase in the absorption coefficient for radiation from a CO_2 laser during a breakdown near a surface has been known almost for 30 years, it was not adequately theoretically described so far. Figure 7.65 [17] shows that the effect of the breakdown plasma is quite considerable, but the intensity of this energy flux is low. This energy is supplied to a target too slowly (for the time 100 times longer than the pulse duration (Fig. 7.66 [72]) and to a large area (10 times larger than the focal spot size (Fig. 7.67 [73]). As shown above, the radiation

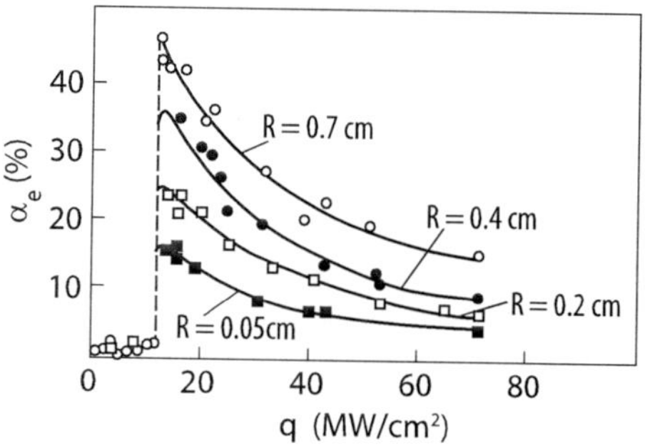

Fig. 7.65 Dependences of the effective absorption coefficient of a copper target of different radii in air on the pulsed CO_2 laser radiation intensity ($r_f = 0.5$ mm, $\tau = 1\,\mu$s)

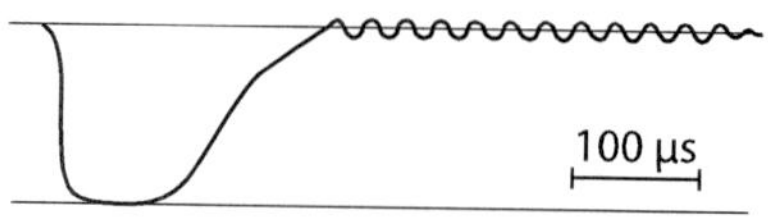

Fig. 7.66 Microwave radiation signal propagating through the breakdown plasma

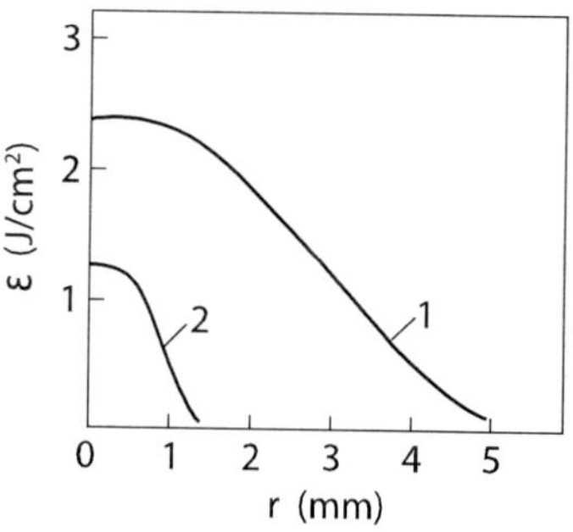

Fig. 7.67 Heat release profiles on an aluminum target of diameter 5 cm and thickness 0.3 mm for $\tau \sim 7\,\mu$s and $2r_f \sim 2.5$ mm: (**1**) with plasma, $E = 9.3$ J, $\alpha = 18\%$; (**2**) without plasma, $E = 1.4$ J, $\alpha = 2.3\%$

intensity is most important for laser processing. Nevertheless, e.g., in the case of repetitively pulsed processing, this effect can increase the average temperature of a target, thereby increasing the processing efficiency.

References

1. G.A. Abil'sitov, V.S. Golubev, Vestn. Akad. Nauk SSSR **11**, 37 (1982)
2. A.A. Vedenov, G.G. Gladush, A.N. Yavokhin, Kvantovaya Elektron. **8**, 1485 (1981)
3. A.S. Kovalev, A.M. Popov, Zh. Tekh. Fiz. **51**, 73 (1981)
4. E.V. Dan'shchikov, V.A. Dymshakov, F.V. Lebedev, et al., Preprint. Atomic Energy Inst.-3534 (1982)
5. E.V. Dan'shchikov, V.A. Dymshakov, F.V. Lebedev, Kvantovaya Elektron. **9**, 1703 (1982)
6. G.G. Vilenskaya, I.V. Nemchinov, Prikl. Mekhan. Tekh. Fiz. **6**, 3 (1969)
7. A.P. Golub', I.E. Markovich, I.V. Nemchinov, Dep. VINITI, no. 3300-79 (1979)
8. C.T. Walter, R.N. Barnes, R.E. Beverly, J. Appl. Phys. **48**, 2973 (1978)
9. A.P. Golub', I.V. Nemchinov, Kvantovaya Elektron. **7**, 1831 (1980)
10. J.A. Mckay, R.D. Bleach, D.J. Nagel, J. Appl. Phys. **50**, 3231 (1979)
11. A.V. Bessarab, G.V. Dolgaleva, N.V. Zhidkov, et al., Fiz. Plazmy **5**, 558 (1979)
12. E.A. Berchenko, A.V. Koshkin, A.P. Sobolev, et al. Kvantovaya Elektron. **8**, 1582 (1981)

13. A.V. Bessarab, V.N. Novikov, D.V. Pavlov, et al., Zh. Tekh. Fiz. **50**, 886 (1980)
14. V.P. Ageev, S.G. Burdin, I.N. Goncharov, et al., Kvantovaya Elektron. **10**, 774 (1983)
15. P.D. Thomas, AIAA J. **13**, 1279 (1975)
16. G. Weyl, A. Pirri, R. Root, AIAA J. **19**, 460 (1981)
17. A.M. Prokhorov, *Fizicheskie printsypy obrabotki materialov impul'snoi lazernoi plazmoi (Physical Principles of Material Processing by Pulsed Laser Plasma)*, Nauchnye osnovy progressivnoi tekhnologii (Scientific Foundations of Progressive Technology) (Mashinostroenie, Moscow, 1982), pp. 97–126
18. V.I. Mazhukin, V.V. Nosov, I. Smurov, G. Flamant, Surv. Math. Ind. **10**, 45 (2001)
19. V.I. Mazhukin, V.V. Nosov, M.G. Nikiforov, I. Smurov, J. Appl. Phys. **93**, 56 (2003)
20. H. Schitenhelm, G. Calliees, P. Berger, H. Hugel, J. Phys. D **29**, 1564 (1996)
21. H. Schitenhelm, G. Callies, P. Berger, H. Hugel, Appl. Surf. Sci. **109/110**, 494 (1997)
22. D.I. Roden, J. Miteldorf, G. Kothandaraman, A.N. Pirri, E.R. Pugh, J. Appl. Phys. **53**, 3190 (1982)
23. Yu.I. Dymshits, V.G. Neverov, Pis'ma Zh. Tekh. Fiz. **1**, 558 (1975)
24. A.I. Barchukov, F.V. Bunkin, V.I. Konov, et al., Zh. Eksp. Teor. Fiz. **66**, 965 (1974)
25. Yu.P. Raizer, *Lazernaya iskra i rasprostranenie razryadov (Laser Spark and Propagation of Discharges)* (Nauka, Moscow, 1974)
26. I.V. Nemchinov, Izv. Akad. Nauk SSSR, Ser. Fiz. **46**, 1026 (1982)
27. V.A.Danilychev, B.D. Zvorykin, *High Power Lasers and Laser Matter Interaction* (Nauka, Moscow, 1983)
28. I.V. Nemchinov, T.I. Orlova, Fiz. Plazmy **4**, 952 (1978)
29. A.A. Bakeev, L.A. Vasil'ev, L.I. Nikolashina, Kvantovaya Elektron. **2**, 1278 (1975)
30. A.G. Borkin, G.G. Gladush, S.V. Drobyazko, et al., *Vzaimodeistvie izlucheniya, plazmennykh i elektronnykh potokov s veshchestvom (Interaction of Radiation, Plasma and Electron Flows with Matter)* (Izd. TsNIIatominform, Moscow, 1984), pp. 97–98
31. A. Mele, Giardini Guidoni, R. Kelly, C. Flamini, C. Orlando, Appl. Surf. Sci. **109-110**, 584 (1997)
32. W.E. Maher, R.B. Hall, J. Appl. Phys. **47**, 2486 (1976)
33. A.I. Barchukov, F.V. Bunkin, V.I. Konov, et al., Zh. Eksp. Teor. Fiz. **66**, 965 (1974)
34. E. Sturmer, M. Von Allmen, J. Appl. Phys. **49**, 5648 (1978)
35. A.A. Vedenov, *Fizika elektrorazryadnykh lazerov (Physics of Electric-Discharge Lasers)* (Energoizdat, Moscow, 1982)
36. A.A. Vedenov, G.G. Gladush, S.V. Drobyazko, et al., Kvantivaya Elektron. **8**, 2154 (1981)
37. B.N. Anisimov, R.V. Arutyunyan, L.A. Bolshov, et al., Preprint of Atomic Energy Ins.-4586/7, M. 1988
38. V.P. Ageev, V.I. Konov, A.S. Silenok, et al. Abstracts of Papers. *V Vsesoyuznoe soveshchanie po nerezonansnomu vzaimodeistviyu opticheskogo izlucheniya s veshchestvom (V All-Union Conference on Nonresonance Interaction of Optical Radiation with Matter)* (Izd. GOI, Leningrad, 1981), p. 221
39. V.A. Baiko, V.A. Danilychev, B.N. Duvanov, et al., Kvantovaya Elektron. **6**, 1323 (1979)
40. V.I. Mazhukin, V.V. Nossov, I. Smurov, Mathematical Model **15**, 23 (2003)
41. M.F. Modest, *Radiative Heat Transdfer* (McGraw-Hill, New York, 1993)
42. Ya.B. Zeldovich, Yu.P. Raizer, *The physics of Shock Waves and High Temperature Hydrodinamics Phenomena* (Academic Press, New York, 1967)
43. B.N. Chetverushkin, *The Dynamics of Radiating Gas* (Nauka, Moscow, 1992)
44. V.I. Mazhukin, A.A. Uglov, B.N. Chetverushkin, Doklady Akademii Nauk **256**, 1100 (1981)
45. V.A. Kamenshikov, Zh.A.Plastinin, B.M. Nikolaev, L.A. Novitsky, *Radiation Gases Properties at High Temperature*, (Mashinostroenie, Moscow, 1971)
46. G.S. Romanov, Yu.A. Stankevich, L.K. Stankevich, K.L. Stepanov, Thermodynamic and optical properties of gases in a wide range of parameters // Int. J. Heat Mass Transfer **38**(3), 545–556 (1995)
47. L. Berthe, R. Fabbro, P. Peyre, E. Bartnicki, Experimental study of the transmission of breakdown plasma generated during laser shock processing // Eur. Phys. J. Appl. Phys. **3**(2), 215–218 (1998)

48. R. Fabbro, L. Berthe, Plasma observation under lazer shock processing // Private communication (1998)
49. R. Fabbro, P. Peyre, L. Berthe, A. Sollier, E. Bartnicki, Physics of application of laser shock processing of materials // SPIE Proc. **3888**, 155–164 (2000)
50. Yu.P. Raizer, *Gas Discharge Physics*, (Nauka, Moscow, 1987)
51. R.W. Mitchel, R.W. Conrad, E.L. Roy, J. QRST **20**, 519 (1978)
52. V.I. Mazhukin, V.V. Nossov, I. Smurov, Thin Solid Films **453-454**, 353 (2004)
53. A.M. Bonch-Bruevich, V.I. Zinchenko, Ya.A. Imas, et al., Zh. Tekh. Fiz. **51**, 919 (1981)
54. J.R. Ho, C.P. Grigoropulos, G.A.C. Humphrey, J. Appl. Phys. **79**, 7205 (1996)
55. M.A. Ilyashevich, L.Ya. Minko, G.S. Romanov, Yu.A. Stankevich, Ya.A. Chivel, A.N. Chymakov, Izvestiya Akademii Nauk. Seriya Phizicheskaya **49**, 1132 (1985)
56. A.A. Demin, The method of dynamical adoptation at burn problem and laser matter interaction, PhD Thesis, Lomonosov University, Moscow, 2004
57. V.I. Mazhukin, A.A. Samarskii, Surv. Math. Ind. **4**, 85 (1994)
58. V.I. Mazhukin, I.Yu. Smurov, C. Duruy, D.J. Jeandel, Num. Heat. Transf. A **26**, 587 (1994)
59. V.I. Mazhukin, A.A. Samarskii, M.M. Chuiko, Dokl. Ross. Akad. Nauk **368**, 307 (1999)
60. I.A. Bufetov, A.M. Prokhorov, V.E. Fedorov, et al., Pis'ma Zh. Eksp. Teor. Fiz. **39**, 216 (1984)
61. A.A. Bakeev, L.A. Vasilev, M.E. Zemskov, et al. Kvantovaya Electron. **10**, 1812 (1983)
62. A.A. Bakeev, L.I. Nikolashina, N.V. Prokopenko, et al., Phizika i chimiya obrabotki materialov **6**, 17 (1985)
63. V.N. Anisimov, R.V. Arutyunyan, L.A. Bol'shov, Kvantovaya Elektron. **14**, 1485 (1987)
64. V.N. Anisimov, O.N. Derkach, D.D. Malyuta, et al., Abstracts of Papers. *VII Vsesoyuznoe soveshchanie po vzaimodeistviyu opticheskogo izlucheniya s veshchestvom (VII All-Union Conference on Interaction of Optical Radiation with Matter)* (GOI, Leningrad, 1988), p. 228
65. A.G. Borkin, G.G. Gladush, S.V. Drobyazko, et al., in *Trudy Vsesoyuznoi Konferentsii "Primenenie Lazerov v Narodnom Khozyastve" (Proceedings of the All-Union Conference on Applications of Lasers in National Economy)* (Nauka, Moscow, 1986), p. 119
66. S.M. Klimentov, T.V. Kononenko, P.A. Pivovarov, et al., Kvantovaya Electron. **31**, 278 (2001)
67. S.M. Klimentov, T.V. Kononenko, S.V. Garnov, et al., Izvestiya RAN, Seriya Phizicheskaya 65 (2002)
68. B.N. Chichkov, C. Motta, S. Nolte, von F. Alvensleben, A. Tunnermann, Appl. Phys. A. **63**, 109 (1996)
69. S.V. Garnov, V.I. Konov, A.A. Mallyutin, O.G. Tsarkova, I.S. Yatskovsky, F. Dausinger, Laser Phys. **13**(3), 386–396 (2003)
70. S.V. Garnov, V.I. Konov, A.A. Mallyutin, et al., Kvantovaya Electron. **33**, 758 (2003)
71. R.V. Arutyunyan, V Yu. Baranov, L.A. Bol'shov's, D.D. Malyuta, A.Yu. Sebrant, *Laser Materials Interaction* (Nauka, Moscow, 1989)
72. V.P. Ageev, V.I. Konov, T.M. Murina, et al., Kratkie Soobshch. Fiz. **5**, 6 (1978)
73. S. Marcus, et al., IEEE J. Quantum Electron. **11**, 49D (1975)
74. V.I. Mazhukin, M.V. Mazhukin, P. Berger, Math. Model. **17**, 27 (2005)
75. M. Ignachacichus, E. Khazakacichus, G. Orishevsky, V. Danyunas, Kvantovaya Electron. **18**, 1325 (1991)
76. V.I. Mazhukin, V.V. Nossov, I. Smurov, Appl. Suf. Sci. **253**, 7686 (2007)
77. V.I. Mazhukin, V.V. Nossov, I. Smurov, Appl. Phys. **101**, 1 (2007)
78. A.D. Aksakhalyan, S.V. Gaponov, V.I. Luchin, A.P. Cherimanov, Sov. Phys. Tech. Phys. **33**, 1146 (1988)
79. R. Timm, P.R. Wellmatt, J.R. Huber, J. Appl. Phys. **80**, 1794 (1996)
80. S. Amoruso, A. Amodeo, V. Berardi, R. Bruzzesse, N. Spinelli, R. Velotta, J. Appl. Surf. Sci. **96-98**, 175 (1996)
81. V.I. Mazhukin, P.A. Prudkovskii, A.A. Samokhin, Zh. Matem. Modelirovanie **5**, 3 (1993)
82. D.V. Gaidarenko, A.G. Leonov, D.I. Chekhov, Sov. J. Plasma Phys. **17**, 534 (1991)
83. J.E. Rothenberg, G. Koren, Appl. Phys. Lett. **44**, 664 (1984)

Chapter 8
Physics of the Damage and Deep Melting of Metals by Laser Pulses

Abstract This chapter is devoted the physical mechanisms of damage of metals by laser pulses in different cases. In simplest case of shallow pool, the classification of melt removal is presented at different magnitudes of pulls duration and energy density. Analytical calculations are compared with numerical modeling and experiment. The specific mechanisms of removal of target material at short and ultra short laser pulses are considered. The importance of nonlinear conversion of laser radiation in cone radiation for material processing by femtoseconds pulses is described. The models of drilling of metals in waveguide regime of laser radiation are presented. The mechanism of control of pulsed metal welding are considered.

8.1 Qualitative Hydrodynamical Model of Laser-Induced Melt Removal

In Chaps. 6 and 7, we have already concerned the problem of the damage of solids by laser pulses. However, pulse durations were so short that the mass of a material removed from a target was small and the thickness of the removed layer was only a few microns (Fig. 7.52).

The material removal mechanism was purely evaporative. It is clear that in the case of longer, millisecond and microsecond pulses, the situation should change qualitatively. Indeed, almost all the photographs taken during irradiation of metals by laser pulses clearly demonstrate the tracks of particles of liquid metal. This feature has been pointed out already in the first monograph on laser irradiation of materials [1]. Since then the problems of the damage hydrodynamics of materials have received permanent attention. Note that hydrodynamics of laser melting is extremely intricate because it is determined by many interrelated processes. In this case, the three, or more exactly, four states of material – plasma, gas, liquid, and solid – interact with each other. The motion of a melt is a two-dimensional flow with a free and evaporating boundary. To overcome these difficulties, researchers

G.G. Gladush and I. Smurov, *Physics of Laser Materials Processing*,
Springer Series in Materials Science 146, DOI 10.1007/978-3-642-19831-1_8,
© Springer-Verlag Berlin Heidelberg 2011

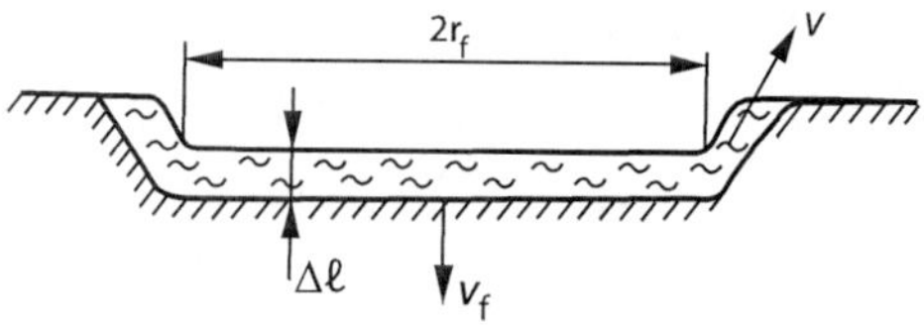

Fig. 8.1 Scheme of the damage of a metal target by ejecting a liquid melt

try to separate all the complex of problems into individual problems and to study individual aspects of the total process. We start with the simplest case when a melt fills a wide and shallow pool.

A shallow melt pool can appear either during perforating holes of diameter exceeding a plate thickness or at the initial stage of perforating deep holes. In any case, the removal of the melt occurs by neglecting the influence of side walls.

8.1.1 Removal on a Melt from a Shallow Pool

Consider a laser beam of intensity q incident on a material and absorbed in the material surface, the absorption coefficient being α.[1] Below, we will consider shallow melt pools, when $2\sqrt{\chi\tau} \ll r_f$ (Fig. 8.1) [2]. In this case, the surface temperature increases with time as

$$T \cong \frac{q}{c\rho}\sqrt{\frac{t}{\chi}} \tag{8.1}$$

where c, ρ, χ are the specific heat, density, and thermal diffusivity of matter, and r_f is the beam radius. The surface temperature achieves the boiling temperature T_b for $t = \tau_b$, where

$$\tau_b = c^2\rho^2T_b^2\chi q^{-2}. \tag{8.2}$$

Because the boiling temperature is 2–3 times higher than the melting temperature (Table 1.3), the material will be melted by this instant approximately by the depth $\sqrt{\chi\tau_b} = \kappa T_b/q$ (all thermal coefficient are assumed independent of temperature). Vapor produced during surface boiling will exert pressure on liquid. Because the laser beam diameter is limited, the melt can be either extruded or evaporated. It is apparent that the first method of damage is more efficient (requires the lower energy) because the specific boiling enthalpy greatly exceeds the energy required to heat the material up to the boiling temperature. Which of these methods is realized depends on the value of q. The liquid velocity v at the edge of the melt pool can be found from the Bernoulli equation (the Reynolds number in this problem is rather large):

$$p = \rho v^2/2. \tag{8.3}$$

[1]Below, for simplicity q is the absorbed radiation intensity.

The pressure p in a supersonic vapor flow is determined by the recoil pressure (see Sect. 1.4):

$$p = \rho_v c_s^2 = \rho v_u c_s. \tag{8.4}$$

The condition of conservation of mass flow $\rho_v c_s = \rho v_u$ is used, where v_u is the rate of liquid level lowering caused by evaporation, and ρ and ρ_v are the liquid and vapor densities, respectively. The rate v_f of liquid level lowering caused by its extrusion on pool edges is determined from the condition of conservation of flow

$$v_f = 2\Delta l v / r_f \tag{8.5}$$

where Δl is the effective width of a "gap" near the pool edge through which liquid flows. The value of Δl depends on whether the flow steady or not

8.1.2 Fountain Wave Regime

In the case of a steady flow, a ejection occurs in the fountain regime [3]. The width l_f of the front of this wave (the liquid film thickness) is determined by the balance of the heat conduction flow and convective heat removal towards the laser beam:

$$l_f \approx \chi / v_f. \tag{8.6}$$

The front propagation time is $t_f \approx \chi / v_f^2$. During this time, a thermal wave will propagated in the side direction by the distance $\Delta l \approx \sqrt{\chi t_f} = l_f$. By substituting the expression for Δl into (8.5) and taking into account (8.3) and (8.4), we obtain

$$v_f = (2 v_u c_s)^{1/4} \sqrt{2\chi / r_f}. \tag{8.7}$$

Taking into account that the laser energy is spent for evaporation and heating of the flowing liquid, we obtain

$$\rho L_b v_u = q - c\rho T_b v_f, \tag{8.8}$$

where L_b is the specific evaporation enthalpy. As will be shown below, energy is mainly spent in this case for evaporation, and therefore the last term in (8.8) can be neglected. We obtain from (8.7) to (8.8)

$$v_f = \left(\frac{2 q c_s}{\rho L_b}\right)^{1/4} \sqrt{2\chi / r_f}. \tag{8.9}$$

The validity of this expression for the damage rate is confirmed experimentally in [3]. The condition of realization of the hydrodynamic melt removal $v_f > v_u$ bounds the value of q from above:

$$q < q_1 = \rho L_b [c_s \chi^2 / r_f^2]^{1/3}. \tag{8.10}$$

Fig. 8.2 Regions of metal damage mechanisms

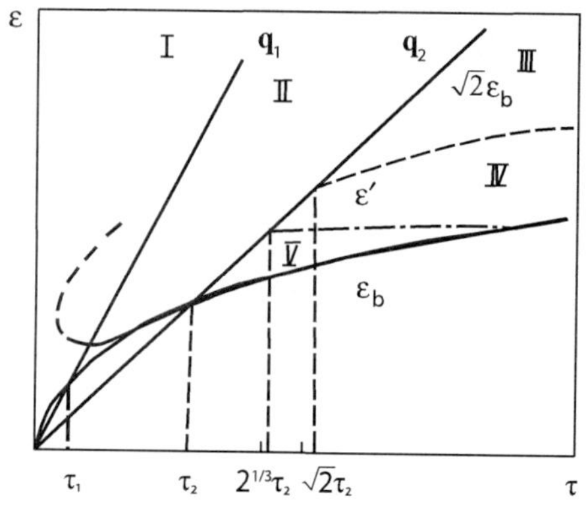

For steel, we have $\rho L_b = 70\,\text{kJ/cm}^3$, $\chi = 0.05\,\text{cm}^2/\text{s}$, $r_f = 0.1\,\text{cm}$, $q_1 \approx 10^6\,\text{W/cm}^2$ (see Fig. 8.2). To obtain the fountain regime, the initial depth $\kappa T_b q^{-1}$ of the melt should not exceed the front size l_f. This condition bounds the value of q from below:

$$q > q_2 = \left[c\rho T_b \sqrt{\frac{2\chi}{r_f}} \right]^{4/3} \left(\frac{2c_s}{\rho L_b} \right)^{1/3} \tag{8.11}$$

For steel, we have $q_2 = 3 \times 10^5\,\text{W/cm}^2$ $r_f \approx 0.1\,\text{cm}$ ($c\rho T_b = 12\,\text{kJ/cm}^3$) (Fig. 8.2). Inequalities (8.10) and (8.11) do not contradict each other because the ratio $q_1/q_2 = (\rho L_b/c\rho T_b)^{4/3} > 1$. It is easy to verify that, if condition (8.11) is fulfilled, the second term in (8.8) can be neglected. Taking into account that the gap width is l_f, we obtain

$$l_f \approx \sqrt{\chi r_f / v}. \tag{8.12}$$

This expression coincides with expression for the thickness of a thermal boundary layer at the pool edge. For this reason, the melt bottom in the fountain wave regime is not flat in general, its shape depending on the velocity distribution over the pool radius, which is determined by the laser radiation intensity over the pool surface. Because the liquid flow is inhomogeneous, it is difficult to calculate accurately the bottom shape in this case. For metals, a viscous boundary layer is smaller than a thermal layer because of a small Prandtl number, and therefore the flow will be nonviscous (as assumed above). For dielectrics, the viscous and thermal layers have close thicknesses. The velocity v_f, calculated in the limiting case of the viscous flow, weakly differs from that calculated from (8.9):

$$v_f = \left[\frac{\chi^3}{\rho \nu r_f^2} \frac{q c_s}{L_b} \right]^{1/4} \tag{8.13}$$

where ν is the kinematic viscosity of the melt.

8.1.3 Liquid Splash Regime

At lower laser radiation intensities, the initial depth of the melt (the depth at which liquid starts boiling) exceeds the front size (8.6), and therefore the law of liquid flowing out of the pool caused by evaporation will be initially different. Consider in more detail the regime $q < q_2$, which we will call the ejection regime [4]. If we assume that the size of a "gas piston" is close to the laser beam diameter, then the effective width of a "gap" through which the liquid is ejected will be approximately equal to the initial depth $\kappa T_b q^{-1}$ of the melt. Taking this into account, we can obtain from (8.3) to (8.5) and (8.8) the expression for the rate of liquid level lowering

$$v_f = \frac{\kappa T_b}{r_f}\sqrt{\frac{2c_s}{qL_b\rho}}. \tag{8.14}$$

Expression (8.14) was derived by assuming that energy is still completely spent for evaporation. In this case, the ejection of the liquid over pool edges does not affect the energy balance and does not reduce the liquid temperature. This takes place because the lowering of the pool bottom (i.e. melting) can be neglected in this regime because of the rapid liquid level lowering. No new (cold) liquid enters the melt during the ejection time $\tau_s = \kappa T_b/qv_f$.

Because of this, the motion of liquid along the pool bottom does not cause the radial redistribution of temperature, the temperature remaining approximately constant and close to the boiling temperature. This circumstance significantly simplifies the problem, allowing the exact solution to be obtained in a particular case of the intensity distribution in a laser spot.

The motion of a cylindrically symmetric thin liquid layer is described by the equations

$$v_r = -r\dot{h}/h \tag{8.15}$$

$$\frac{1}{\rho}p_0 = -\frac{r_f^2}{4}\frac{\mathrm{d}}{\mathrm{d}t}\left(\frac{\dot{h}}{h}\right) + \frac{r_f^2}{8}\left(\frac{\dot{h}}{h}\right)^2, \tag{8.16}$$

where $h(t)$ is the liquid layer thickness and p_0 is pressure at the liquid layer center (which is assumed time-independent).

By integrating (8.15) and (8.16) with respect to time, we obtain [4]

$$\frac{h(t)}{h_0} = ch^{-2}\frac{t}{t_0}, \quad t_0 = \sqrt{\frac{\rho r_f^2}{2p_0}}, \quad h_0 = \frac{\kappa T_b}{q}. \tag{8.17}$$

The radial distribution of pressure can be determined from (8.16):

$$p(r)/p = 1 - r^2/r_f^2. \tag{8.18}$$

In the case of a Gaussian beam $[q/q_0 = \exp(-r^2/r_f^2)]$, the pressure distribution near the pool center will be nearly parabolic (8.18). One can see from (8.17) that the level lowering rate v_f first increases with time (liquid is accelerated) and then decreases for $t > t_0$ (the gap width decreases). The maximum rate coincides with the value calculated from (8.14).

If light is switched off at the instant $\tau < t_0 + \tau_b$, the liquid flows out from this point on by inertia and its level lowers slower:

$$\frac{h}{h_1} = \left[1 + \frac{t}{t_0} th \frac{\tau - \tau_b}{t_0}\right]^{-2}, \quad t > \tau. \tag{8.19}$$

The inertial flowing out of the liquid metal is restricted by cooling accompanied by solidification. This restricts the flowing out time t_s:

$$t_s < \kappa^2 T_b^2 / q^2 \chi = \tau. \tag{8.20}$$

Thus, metal melts during the time τ_b, the it is extruded from the pool by the vapor pressure during the time $\tau - \tau_b$ and flows out by inertia during the time τ_b. If $\tau - \tau_b = t_0$, the inertial stage is absent. Because τ_b depends on ε, the condition of boiling onset $\tau_b < \tau$ bounds the value of ε from above:

$$\varepsilon > \varepsilon_b = c\rho T_b \sqrt{\chi \tau} \tag{8.21}$$

Condition (8.21) is illustrated by curve ε_b in Fig. 8.2. The intersection points of curve ε_b with lines q_1 and q_2 correspond to the critical values of duration and energy. For $\tau < \tau_1$, the removal of the target material in the liquid form is impossible at any energy. For $\tau_1 < \tau < \tau_2$, the fountain regime is possible, while for $\tau > \tau_2$, depending on the pulse energy, all the three regimes are possible: the liquid ejection regime at low energies (regions III and IV), the fountain regime at intermediate energies (region II), and the evaporation regime at high energies (region I).

The times τ_1 and τ_2 are described by the expressions

$$\tau_1 = \left(\frac{cT_b}{L_b}\right)^2 \left(\frac{r_f^4}{\chi c_s^2}\right)^{1/3}, \quad \tau_2 = \left(\frac{L_b}{cT_b}\right)^{2/3} \left(\frac{r_f^4}{\chi c_s^2}\right)^{1/3} \tag{8.22}$$

For steel, e.g., we have ($r_f = 0.1$ cm), $\tau_1 = 5\,\mu$ s, $\tau_2 = 100\,\mu$s.

Let us plot curve $\varepsilon\prime$ in Fig. 8.2 for which the condition of liquid ejection from a pool by the pulse end is fulfilled:

$$\tau - \tau_b = t_0. \tag{8.23}$$

At large times curve ε' approaches curve ε_b and it intersects line q_2 at the point $\tau_2' = 2^{1/3}\tau_2$. If a point corresponding to the pulse energy ε and duration τ lies below curve ε', this means that the inertial regime is realized (region V). If $\varepsilon > \varepsilon'$, the inertialless ejection regime is realized (region IV). It is appropriate to demonstrate the condition of applicability of the approach considered above for laser processing in the atmospheric air. This condition is fulfilled if the intensity of absorbed laser radiation exceeds the radiation intensity at which the vapor flow velocity on the target is equal to the sound speed (see Chap. 1):

$$q > q_3 = 2c_s L_b \rho_a \tag{8.24}$$

where ρ_a is the environmental gas density. For typical thermal parameters of steel machining, we have $q_3/q_2 \sim 0.2$. This means that the region of ejection removal of the melt, when the vapor jet is supersonic, is quite large. The intersection of curves (8.24) and (8.21) gives the pulse duration below which processing is performed in the supersonic regime of vapor flowing into the environmental gas atmosphere:

$$\tau_3 = \left(\frac{\rho}{\rho_a}\right)^2 \left(\frac{cT_b}{L_b}\right)^2 \frac{\chi}{c_s^2}. \tag{8.25}$$

The estimate by this expression gives $\tau_3 \approx 300\,\mu\text{s}$ for steel. For $q < q_3$, the vapor flow will be subsonic and vapor pressure will decrease with deceasing intensity not linearly but faster – quadratically with q: $P = (q/q_3)^2 2c_s^2 \rho_a$. Therefore, the efficiency of material removal for $q < q_3$ considerably decreases.

8.1.4 Specific Damage Energy

Consider the dependence of the specific damage energy X on the light pulse energy density ε for different pulse durations τ. The damage mechanism depends on the laser pulse duration, which finally determines the value of the specific damage energy. For generality we neglect the influence of inhomogeneities because the description of inhomogeneous damage mechanisms is complicated. Let $\tau > \tau_2$; if ε lies on curve ε', the liquid will be completely removed by the end of the pulse, so that

$$X = \varepsilon^2/\kappa T_b \tau. \tag{8.26}$$

Because ε' tends to ε_b with increasing τ, X tends to the minimum possible value [5]:

$$X_{\min} = c\rho T_b. \tag{8.27}$$

The specific energy increases with increasing $\varepsilon(\varepsilon > \varepsilon')$ because laser-induced damage will occur in the time interval $(\tau - \tau_b - t_0)$ in the fountain regime:

$$X = \varepsilon[\kappa T_b/q + v_f(\tau - \tau_b - t_0)]^{-1} \tag{8.28}$$

for $\varepsilon = \sqrt{2}\varepsilon_b$

$$X = \frac{2\rho c T_b}{1 + v_f \tau / 2\sqrt{\chi\tau}} \tag{8.29}$$

where v_f is determined by (8.9). Later on, $X \approx (\varepsilon/\tau)^{3/4}$ for $\varepsilon > \tau q_2$ and $X = \rho L_b$ for $\varepsilon = q_1\tau$, i.e. the passage to the evaporation regime takes place. For $\varepsilon < \varepsilon'$, by using (8.17), (8.19), and (8.20) and taking into account that the boiling time $(\tau - \tau_b)$ is small compared to the ejection time t_0, we obtain the volume of the removed material:

$$\frac{H}{h_0} = \left(\frac{\tau - \tau_b}{t_0}\right)^2 \left[1 + \frac{2\tau_b}{\tau - \tau_b}\right]. \tag{8.30}$$

The second term in brackets is related to the inertial flowing out of the liquid. One can see from (8.30) that the removal of material near the damage threshold ($\tau \approx \tau_b$) mainly occurs after the shutdown of the light pulse:

$$X = \frac{\varepsilon^3 r_f^2 L_b}{4c^3\rho^2 T_b^3 \chi^2 \tau^3 c_s (1 - \varepsilon_b^2/\varepsilon^2)}. \tag{8.31}$$

Expression (8.31) has a minimum for $\varepsilon = \sqrt{5/3}\varepsilon_b$. For $\tau = 2\tau_2$, the value of $X_{\min}$ is close to $c\rho T_b$. Thus, the value of $X_{\min}$ decreases somewhat in the inertial ejection regime.

For $\tau \le \tau_2$, the material is always removed in the fountain wave regime:

$$X = \varepsilon / v_f(\tau - \tau_b) \tag{8.32}$$

where v_f is determined by (8.9). Expression (8.32) for $\varepsilon = \sqrt{3}\varepsilon_b$ has a minimum $X = 1.5c\rho T_b$ for $\tau = \tau_2$. The value of $X_{\min}$ increases with decreasing τ.

Thus, as the laser pulse duration τ is increased, the minimum damage energy decreases and curves $X(\varepsilon)$ are flattened with increasing ε.

It is easy to verify that for $q = q_2$ the simmer time τ_b and the ejection time t_0 are equal. In curve ε' $\tau - \tau_b = t_0$; at the intersection point of dependences q_2 and ε' the simmer and ejection times are equal to half the pulse duration τ. In this case, the melted metal outflow velocity is $v = 2r_f/\tau$. Knowing the liquid ejection velocity, we can easily obtain the upper limit of the depth of a hole imposed by the gravitational force:

$$h < 2r_f^2/\tau^2 g$$

where g is the acceleration of gravity. This restriction is insignificant in practice, and the limiting depth is determined by other factors. It seems that the depth of a hole in the case of low pulse repetition rates is manly restricted by the cooling of the melted metal on side walls. If this hole is perforated after irradiation by many pulses, then

$$h < (\Delta l)^2 v/\chi.$$

However, this restriction is absent in the case of a single pulse because side walls are hot in this case. In the fountain regime, when the liquid film thickness is small,

$\Delta l = l_f$, by using (8.5) and (8.6), we obtain the limiting depth of the hole $h < r_f$. In the liquid ejection regime upon irradiation by many pulses, a deeper hole can be produced:

$$h < \tau_b r_f / t_0 \approx r_f q^{-3/2}$$

$\tau_b = t_0$ for $q = q_2$; for $q < q_2$, $\tau_b > t_0$. Thus, the limiting drilling depth upon repetitively pulsed irradiation increases with increasing the laser pulse duration and energy (but with decreasing the pulse intensity). All the restrictions imposed on the hole depth are alleviated with increasing the pulse repetition rate and also when multiple reflections of light in the hole are taken into account. This question will be considered in detail below.

In conclusion of this section, we will analyze the influence of the surface tension of the liquid on the phenomena under study. To initiate the motion of the liquid under the action of the recoil vapor, the vapor pressure should exceed the surface tension force:

$$p > 2\sigma / \Delta l.$$

By substituting (8.4), (8.6), and (8.8) into this expression, we see that the role of the surface tension is insignificant. Indeed, the minimum value is $p\Delta l / 2\sigma = c_s \kappa T_b (2 L_b \sigma)^{-1}$. This ratio for iron is 3.

8.1.5 *Numerical Modelling of Metal Removal from a Shallow Melt Pool*

Numerical studies were performed by using a simplified model as well. It was assumed that the Peckle number Pe = RePr was smaller than unity. Because the Prandtl number for metals is $Pr \ll 1$, we can neglect the convective heat transfer during motion of the melt even in the case of comparatively high melt motion velocities (when $Re > 1$). Because we consider, as above, a shallow melt pool, the influence of variations in the free surface shape on the thermal balance can be also neglected. This allows us to describe separately the propagation of heat and hydrodynamic flows of the melt, which considerably simplifies the problem, for example, compared to the description of the Marangoni flow for $Re > 1$ (see Sect. 2.2). Such a simplification allows us to take into account the deformation of the free surface caused by the surface tension and recoil vapor pressure [36]. The latter quantity is most simply described during evaporation to vacuum (Chap. 1). In this case, vapor flies away from the melt surface at the supersonic speed, the Mach number being $M = 1$. The recoil pressure is $p_r = p_1(1 + \gamma)$, where the gas pressure p_1 is expressed in terms of the saturated vapor pressure p_s as $p_1 = 0.2 p_s$ (1.46). This gives $p_r \approx 0.5 p_s$. The saturated vapor pressure is described by the Hertz–Knudsen expression

$$p_s = p_0 \exp[(H_v/k)(1/T_b - 1/T)]. \tag{8.33}$$

Here, H_v is the heat of evaporation and T_b is the boiling temperature. The Laplace pressure can be obtained from (4.8) for the mechanical equilibrium for a gas–vapor keyhole produced upon laser welding taking into account that $\partial h/\partial r \ll 1$. As a result, the total pressure will depend on the temperature and shape of the free surface:

$$p = 0.5p_M - \sigma\left(\frac{1}{r}\frac{\partial h}{\partial r} + \frac{\partial^2 h}{\partial r^2}\right). \tag{8.34}$$

The equation for the free surface shape is determined by the melt motion velocity on this surface:

$$\frac{\partial h}{\partial t} = \left(v_z - v_r\frac{\partial h}{\partial r}\right)_{z=h(r,t)}. \tag{8.35}$$

The melt motion is described by the nonstationary Navier–Stokes equation, in which nonlinear terns can be omitted (Sect. 2.2):

$$\frac{\partial v_r}{\partial t} = -\frac{1}{\rho}\frac{\partial p}{\partial r} + \nu\frac{\partial^2 v_r}{\partial z^2} \tag{8.36}$$

$$\frac{\partial v_z}{\partial t} = -\frac{1}{\rho}\frac{\partial p}{\partial z} + \nu\frac{\partial^2 v_z}{\partial z^2} \tag{8.37}$$

$$\frac{1}{r}\frac{\partial}{\partial r}(rv_r) + \frac{\partial v_z}{\partial z} = 0. \tag{8.38}$$

The flow velocity at the melt–solid interface is set equal to zero. The friction force at the free boundary is determined by the thermocapillary force (Sect. 2.2):

$$\eta\frac{\partial v_r}{\partial z} = \left.\frac{\partial \sigma}{\partial T}\frac{\partial T}{\partial r}\right|_{z=h} \tag{8.39}$$

The temperature distribution in the melt and solid is determined by the conductive heat conduction equations. Because a shallow pool is considered, where the melt depth is $S(z) \ll r_f$, we can restrict our consideration to the one-dimensional Stefan problem (Sect. 2.2)

$$\frac{\partial T_l}{\partial t} = \kappa_l\frac{\partial^2 T_l}{\partial z^2}, \quad \frac{\partial T_s}{\partial t} = \kappa_s\frac{\partial^2 T_s}{\partial z^2} \tag{8.40}$$

$$-\kappa_l\left.\frac{\partial T_l}{\partial z}\right|_{z=S(r,t)} + \kappa_s\left.\frac{\partial T_s}{\partial z}\right|_{z=S(r,t)} = H_m\rho\frac{\partial S/\partial t}{1+(\partial S/\partial r)^2} \tag{8.41}$$

$T_l(r, z = S, t) = T_s(r, z = S, t) = T_m$.

The dependence of $S(r,t)$ is determined by the inhomogeneous nature of the absorbed laser radiation:

$$-\kappa_l\left.\frac{\partial T_l}{\partial z}\right|_{z=0} = q_0\exp(-r^2/r_f^2) \tag{8.42}$$

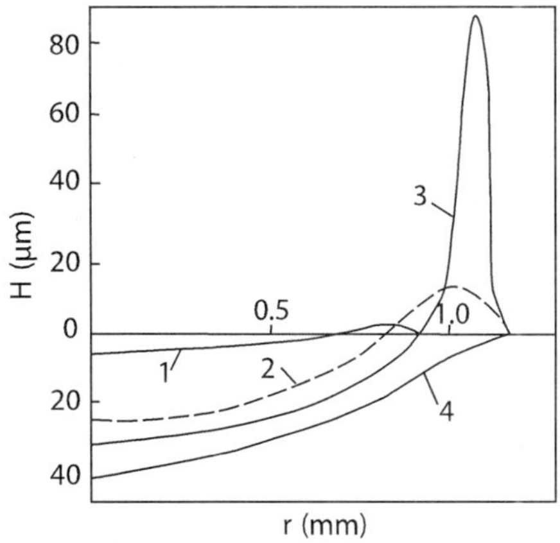

Fig. 8.3 Free surface shape versus radius at moments $t = 1.2$ ms (curve 1), 1.8 ms (curve 2), and 2.0 ms (curve 3). Curve 4 is the solid–liquid interface at $t = 2$ ms. (Ti target, $q_0 = 3.5 \cdot 10^4$ W/cm^2, Gaussian laser beam, $r_f = 1.4$ mm)

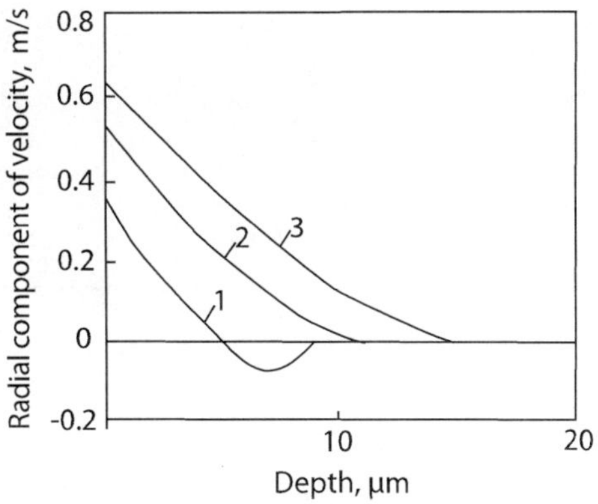

Fig. 8.4 Dependence of the radial component of the melt velocity on the pool depth; $t = 1.0$ ms (curve 1), 1.2 ms (curve 2), 1.6 ms (curve 3). Curve 1 corresponds to the vortex convection caused by thermocapillary forces (the velocity is negative near the pool bottom). Curves 2 and 3 correspond to the melt movement only in one direction, from the center towards the pool boundary, i.e. to the melt removal

By solving one-dimensional equations (8.40) with boundary conditions (8.41) and (8.42), we can obtain the shape of the melt pool at different instants (Fig. 8.3) [6]. Knowing the melt pool shape and solving the system of equations (8.33)–(8.39), we can obtain the velocity field for a liquid metal. We took into account that the pool bottom descended at the velocity that was smaller than the velocities of other processes, i.e. the pool shape was quasi-stationary. Figure 8.4 shows the dependences of the radial velocity over the pool depth at the successive instants. One can see that this velocity changes its sign at the beginning of melting, i.e. the flow has the vortex nature, as should be in the case of the Marangoni effect. The melt surface temperature increases with time. Because of the drastic temperature dependence of the vapor pressure, the recoil pressure is abruptly "switched on", resulting in the

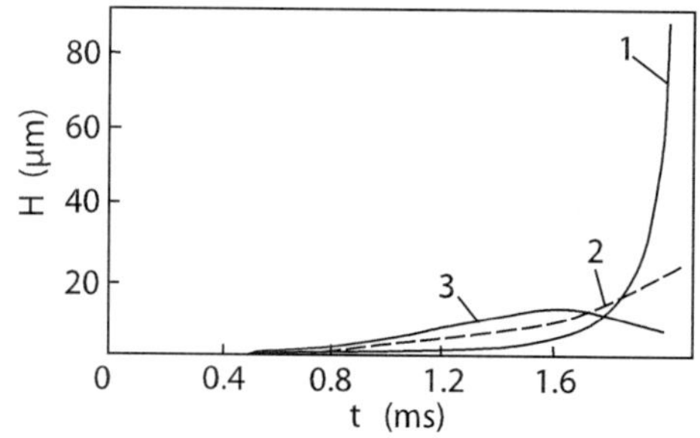

Fig. 8.5 Time dependences of the melt maximum rise on the pool boundary (curve 1), the maximum sink (curve 2) and melt thickness (curve 3) at the pool center. (Ti target, $q_0 = 3.5 \cdot 10^4$ W/cm^2)

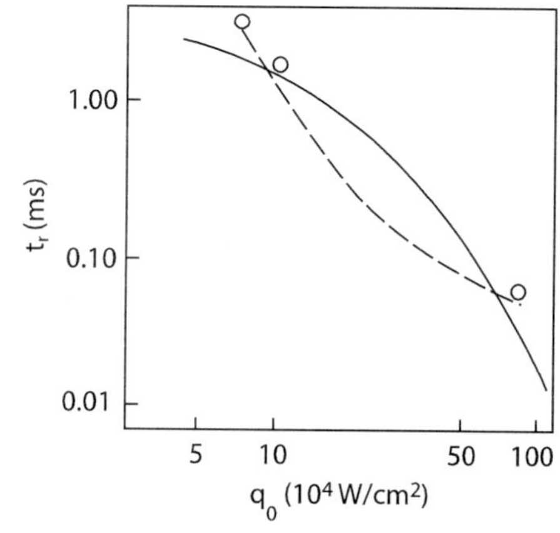

Fig. 8.6 Dependence of the metal removal threshold time on the absorbed energy density. A Ti sample; solid and dashed curves correspond to numerical and analytic calculations, respectively

ejection of the melt to the pool boundary (Figs. 8.3, 8.4). Because of the Rayleigh instability, the melt should be further removed in the form of drops. Although a melt roller was also observed in experiments at the edge of the irradiated region (see Sect. 8.2). The radial velocity becomes positive over the entire pool depth (Fig. 8.4). One can see from Fig. 8.5 that the melt thickness at the pool center first increases with time and then decreases due to the melt flowing out to the boundary. This demonstrates the transient regime of melt removal, which corresponds to the ejection regime considered analytically above. In this connection, it is interesting to compare the melt removal times obtained by using these two models. According to the analytic analysis, the melt removal time τ_r is the sum of the time τ_b (8.2) of heating up to the boiling temperature and the ejection time t_0 (8.17):

$$\tau_r = \left(\frac{c\rho T_b}{q}\right)^2 \chi + r_f \sqrt{\frac{\rho L_b}{2c_s q}}. \tag{8.43}$$

A comparison of numerical calculations with theoretical expression (8.43) ($\chi = 0.06\,\text{cm}^2/\text{s}$, $c_p = 0.7$ J/gK, $L_b = 9.8$ kJ/g) shows that they well agree as a whole (Fig. 8.6). The discrepancy appears at high radiation intensities. This is explained by the fact that the numerical model neglects the evaporation energy loss. As a result, the surface temperature and vapor recoil pressure become overstated at high radiation intensities, and therefore the calculated melt removal time is understated.

8.2 Experimental Studies of the Interaction of Millisecond Laser Pulses with Materials

8.2.1 Experimental Study of Shallow Damage of Materials

Targets were damaged by radiation from neodymium and ruby lasers by using either high-power short $\tau \leq 0.1\,\mu s$ pulses in the giant pulse regime or low-intensity ($q \approx 1$) MW/cm^2 long ($\tau \sim 1$ ms) pulses in the free running regime. As shown in the previous section, the damage efficiency depends not only on the pulse intensity but also on its duration.

Let us assume that the pulse duration is small and the keyhole depth increases during irradiation by the value not exceeding the keyhole width (this condition provides the one-dimensionality of the problem). Figure 8.7 shows light pulses

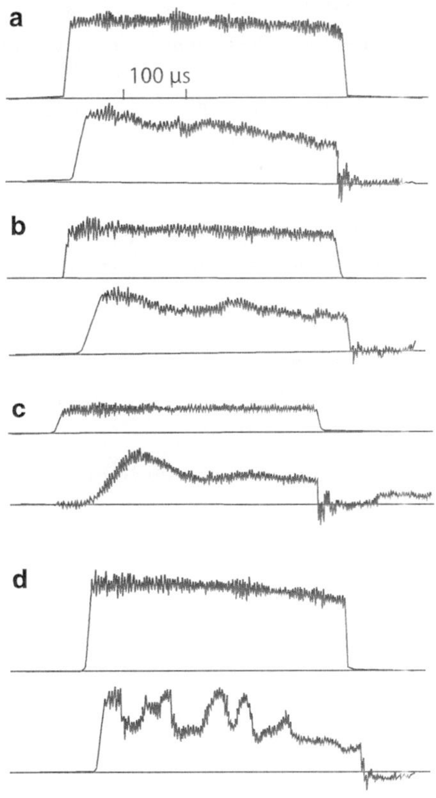

Fig. 8.7 Oscillograms of neodymium laser pulses (*top*) and piezoelectric sensor signals (*bottom*) on a target for different pulse intensities and different pressures: (**a**) 0.9 MW/cm^2, 0.9 MPa; (**b**) 0.6 MW/cm^2, 0.6 MPa; (**c**) 0.3 MW/cm^2, 0.09 MPa; (**a–c**) bismuth target; (**d**) lead target, 1.1 MW/cm^2, 1 MPa

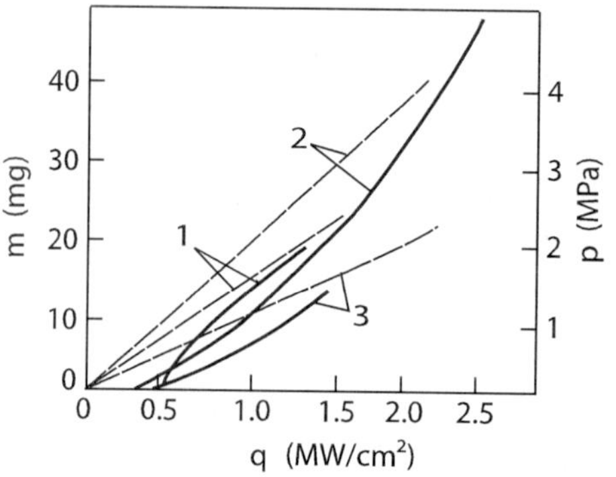

Fig. 8.8 Dependences of the mass removal (**1**) and recoil pressure (**2, 3**) on the radiation intensity for bismuth and lead targets: the dashed straight lines are theoretical dependences, the solid curves are experimental dependences; (**1**) ruby laser, lead, experiment [8] and calculation by (8.48) for $\alpha = 0.35$; (**2**) neodymium laser, bismuth, experiment [7] and calculation by (8.44) for $\alpha = 0.5$; (**3**) neodymium laser, lead, experiment [7] and calculation by (8.44) for $\alpha = 0.5$

and recoil pressure pulses appearing after irradiation of a bismuth target [7]. The recoil pressure pulse repeats the light pulse profile at comparatively high intensities (Fig. 8.7a). As the radiation intensity is reduced, the shape of the pressure pulse changes: pressure increases slower, achieves a maximum and then considerably decreases (Fig. 8.7b,c). A similar picture is observed for a lead target. Figure 8.8 presents the dependence of the maximum recoil pressure on the laser radiation intensity for bismuth and lead. In the latter case, the recoil pressure is affected for $q > 1.2\,\mathrm{MW/cm^2}$ by a periodically appearing plasma plume (Fig. 8.7d). A plasma plume also appeared upon irradiation of the bismuth target, but it did not affect noticeably processes under study. Note that experiments with a lead target irradiated by a ruby laser, in which no plasma was produced in the range of laser intensities used, gave similar results. One can see from Fig. 8.8 that the dependence $p(q)$ is essentially nonlinear for both targets. Both dependences pass below the limiting theoretical curves:

$$p_1 = c_s q / L_b. \tag{8.44}$$

Photographs of the irradiated region show that the target material is melted and partially ejected by the middle of the laser pulse (Fig. 8.9a). By the end of the pulse, a melted metal roller is distinctly observed on the pool edges (Fig. 8.9b), which is later spattered from the target. The conditions approximately correspond to Fig. 8.7c. Thus, one can see from Fig. 8.9 that under these conditions the hydrodynamic mechanism of the target damage is realized. It was shown in the previous section that for $q > q_2$ the fountain mechanism is realized, when the total energy of the laser beam is spent for evaporation. In this case, the mass consumption is a nonlinear function of q:

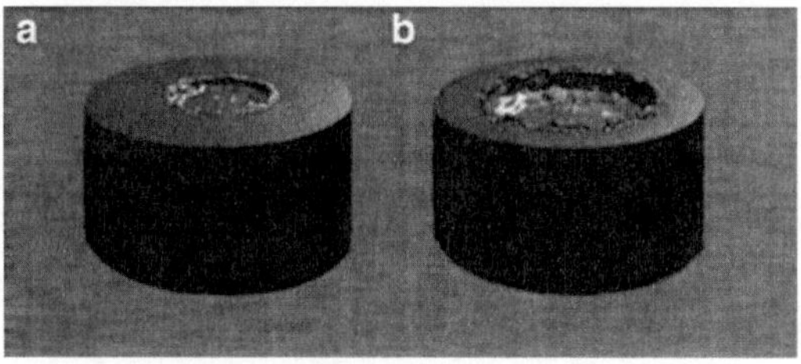

Fig. 8.9 Lead target during irradiation by a laser pulse ($q \simeq 0.4\,\mathrm{W/cm^2}$, (**a**) $t = 160\,\mu\mathrm{s}$, (**b**) $t = 400\,\mu\mathrm{s}$)

$$m = S\tau\rho\sqrt{\frac{2\chi}{r_f}}\left(\frac{2c_s q}{\rho L_b}\right)^{1/4}, \tag{8.45}$$

and the recoil pressure increases with q [see (8.44)]. Expression (8.45) was obtained from (8.9). The laser radiation intensity q was changed in these experiments by varying energy, the rest of the parameters being constant. Therefore, we can use below the scheme in Fig. 8.2. For $\varepsilon'/\tau < q < q_2$, the liquid is removed from the pool during the first part of the pulse in the ejection regime, and then – during the second part of the pulse, in the fountain regime. As shown above, the laser beam energy in the ejection regime is completely spent for evaporation, i.e. $p = c_s q/L_b$ in this case as well. However, the energy balance in the fountain regime is considerably determined by fountain energy losses [see (8.8)]. In this case,

$$v_f \approx q/c\rho T_b. \tag{8.46}$$

Taking (8.46) into account, we obtain from (8.4) to (8.7) the expression for the recoil pressure:

$$p_2 = \left(\frac{q}{c\rho T_b}\right)^4\left(\frac{r_f}{2\chi}\right)^2\frac{\rho}{2} = p_1(q/q_2)^3, \quad q < q_2. \tag{8.47}$$

The mass consumption in this case linearly depends on q [see (8.46)]:

$$m = S\tau q/cT_b. \tag{8.48}$$

Figure 8.8 presents the experimental dependence of the mass removal on the ruby laser radiation intensity. Curve 1 is plotted by expression (8.48) ($S = 10^{-2}\,\mathrm{cm^2}$, $\tau \approx 1$ ms, the absorption coefficient is $\alpha \approx 0.35$, and $cT_b = 217$ J/g). The theory and experiment are in good agreement. The numerical estimate of q_2 [see (8.11)] for lead gives the value $0.2\,\mathrm{MW/cm^2}$.

Because $\alpha \sim 0.3$, we can assume that the value of q_2 lies in the middle of the region of radiation intensities studied. Because $q_3 \sim 0.29q_2$, this means that the "working point" lies in the region of supersonic flows of vapor [see (8.24)]. This is confirmed by the direct observation of the vapor jet. Near the target a Mach disc is distinctly observed [9]. An additional correlation with experiment can be made by estimating the melt removal velocity from expressions (8.3) and (8.44). For $q = 0.3$–$0.9\,\mathrm{MW/cm^2}$ and $\alpha = 0.3$, we obtain $v \approx 6-10\,\mathrm{m/s}$. These values agree with the velocity of drops observed experimentally [10]. Thus, the model described above can be used to explain experimental data. For small values of q during the first part of the pulse, when the laser beam damages the target in the ejection regime, pressures is determined by expression (8.44). During the second part of the beam the fountain regime is realized and pressure decreases according to (8.47).

Thus, by the example of lead and bismuth targets we can say that upon irradiation by low-intensity pulses from neodymium and ruby lasers, when the depth of the hole does not exceeds its width, the average damage energy can be small – equal to the heat content at the boiling temperature of the target material. This fact is consistent with theoretical concepts presented in the beginning of this section. Experimental and theoretical investigation of liquid removal from a allow bath under action of CO_2 laser radiation pulse focused to a spot of size substantially smaller than the bathlength are presented in [37]

8.2.2 *Deep-Penetration Keyhole Damage by a Single Pulse from a Neodymium Laser*

In the case of a high-intensity pulse, when the depth of the hole produced by the pulse greatly exceeds the hole diameter, the damage mechanism changes. In this case, the damage produced by a single pulse can be described in the following way. Material is removed from the bottom of the hole due to evaporation, while from the sides of the hole the melt is removed by the vapor jet. In this case, not only the depth h of the hole increases with time, but also its width, the mass removal mainly occurring due to an increase in the hole diameter. Measurements of the average damage energy of metals during perforating deep dead-end holes showed that its value lies between L_b and L_m, more exactly, between L_b and $L_b/2$ (Fig. 8.10) [11].

Physical processes proceeding during the damage of metals by the CO_2 laser radiation, which will be considered in the next section, are even more diverse. The

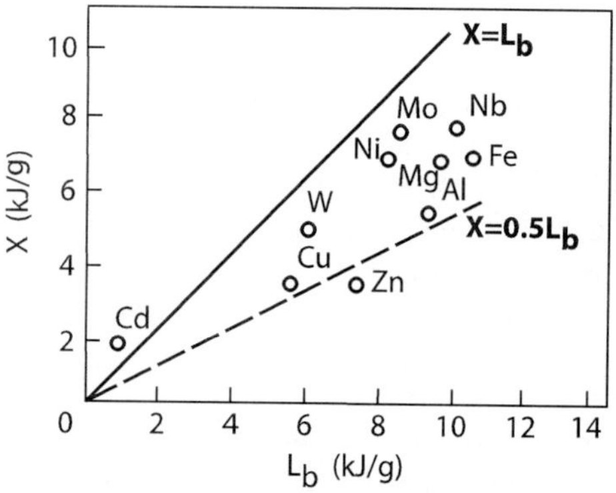

Fig. 8.10 Dependences of damage energy X on evaporation enthalpy for different materials

picture is complicated by the appearance of a plasma plume, which, as shown in Sect. 7.3, significantly affects the radiation intensity incident on a target. Another specific feature of this process is related to a lower absorption coefficient for radiation compared to radiation from a neodymium laser.

8.3 Damage of Materials by Microsecond and Ultrashort Laser Pulses

8.3.1 Experimental Studies of the Damage of Metals by Pulsed CO_2 Laser Radiation

We already discussed the efficiency Y of metal damage by pulses of a CO_2 laser by considering a pulsed optical discharge near a target in Sect. 7.3 [12]. Figure 7.26b presents the dependence of Y on the energy density ε for different gases surrounding the target (the pulse duration was $\tau \approx 5\,\mu s$ and the focal spot diameter was $d = 60\,\mu m$). The damage efficiency in the given gas is related to the transparency of a plasma plume. This is demonstrated by the position of curves for the transmission coefficient η (Fig. 7.26a) and damage efficiency Y (Fig. 7.26b). For ε exceeding a threshold, which is the same for all gases, the damage efficiency increases with increasing ε, achieves a maximum and then decreases much faster than the plasma transmission does. It seems that the damage features are observed here, which we discussed at the beginning of section. The typical values of the interaction parameters [see Fig. 8.2 and expressions (8.10), (8.11), (8.21), and (8.22)] for experimental conditions in Fig. 7.26 are: $q_1 \approx 70\,\mathrm{MW/cm^2}$, $q_2 \approx 15\,\mathrm{MW/cm^2}$, $\tau_2 = 1\,\mu s$, $\varepsilon_b = 13\,\mathrm{J/cm^2}$ ($L_b\rho = 52\,\mathrm{kJ/cm^3}$, $c\rho T_b = 8.4\,\mathrm{kJ/cm^3}$, $\chi = 0.5\,\mathrm{cm^2/s}$, $r_f = 30\,\mu m$ and $c_s = 7.5 \cdot 10^4\,\mathrm{cm/s}$). By assuming that the absorption coefficient of copper at a high temperature close to the boiling temperature is 5%, we can find from Fig. 7.26 the damage threshold, which is equal to $12\,\mathrm{J/cm^2}$, in accordance with the theoretical value $\varepsilon_b = 13\,\mathrm{J/cm^2}$. Because the pulse duration is $\tau > \tau_2 \approx 1\,\mu s$, for $\varepsilon > \varepsilon_b$ the most efficient damage mechanisms, the ejection and fountain regimes, can be realized. One can see from Fig. 7.26 that the damage efficiency maximum is achieved for $\varepsilon \approx 1\,\mathrm{kJ/cm^2}$, which corresponds to $q = 200\,\mathrm{MW/cm^2}$. If the absorption coefficient is 0.05, the absorbed power is $10\,\mathrm{MW/cm^2}$, which is lower than q_1 and q_2. The specific damage energy in this case (taking the absorption coefficient into account) is $5\,\mathrm{kJ/cm^3}$, which is lower than $c\rho T_b = 8.4\,\mathrm{kJ/cm^3}$. The minimum of Y corresponds to $\varepsilon = 7\,\mathrm{kJ/cm^2}$ and the absorbed power density $70\,\mathrm{MW/cm^2}$. The latter value coincides with q_1.

The experiments described above were performed by using very tight laser beam focusing (the focal spot diameter was $d = 65\,\mu m$. The maintenance of such focusing conditions in practice is quite complicated. Because of this, we consider here the most typical case of irradiation of targets by laser beams of greater

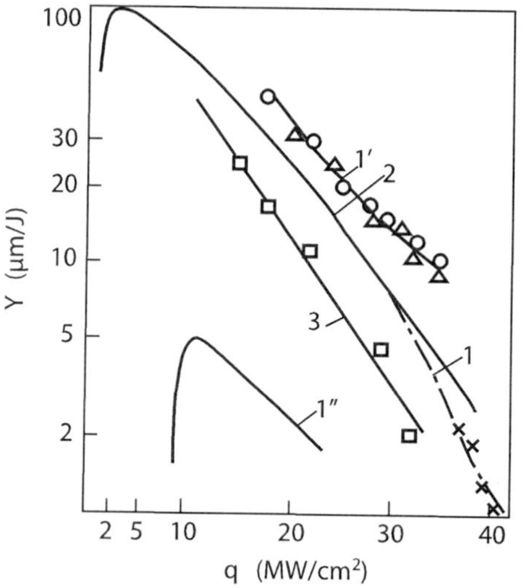

Fig. 8.11 Dependences of the stainless steel damage efficiency on the CO_2 laser radiation intensity in different gases: (**1**, 1′, 1″) air, (1′) data from [13], (1″ data from Fig. 7.26 (curve 2); (**2**) helium; (**3**) argon)

diameters $d \approx 0.3$–0.4 mm [5, 13]. In this case, the value of $\tau_2 \sim r_f^{4/3}$ increases correspondingly [see (8.22)]. To process a metal most efficiently (in the ejection regime), it is necessary to increase the laser pulse duration so that $\tau > \tau_2$. This condition together with (8.21) results in the increase in the laser pulse energy. The dependence of the average damage efficiency $Y = h/nE$ on the maximum radiation intensity was experimentally studied in [13] ($E \leq 1.2$ J, $\tau = 50\,\mu$s, $d = 300\,\mu$m). Here, the depth h is equal to the plate thickness; nE is the total energy spent to produce a hole in a plate; and n is the number of pulses.

Because of a peculiar shape of a laser pulse (a short spike of duration 1 μs followed by a long tail), the maximum power can strongly differ from the average power. For conditions in Fig. 8.11, the ratio of these powers is 25. One can see from Fig. 8.11 that the damage efficiency determined in this way also decreases with increasing the laser radiation intensity in the focal spot. More detailed studies showed that the depth of the hole increased nonlinearly with increasing the number of pulses (Fig. 8.12). When the depth of the hole is small, the hole deepening velocity dh/dn is also small. This velocity increases with increasing the number of pulses, reaches a maximum, and then drops to zero when the depth h starts to saturate. The drop in $[dh/dEn]_{max}$ with increasing q is explained by plasma screening. This question was analyzed in more detail in [5, 14]. The damage efficiency in different gases was studied under experimental conditions similar to those in [13]. Figure 8.11 shows the dependence of the average damage efficiency on the laser pulse energy for stainless steel or $\tau = 70\,\mu$s and $d = 450\,\mu$m (the plate

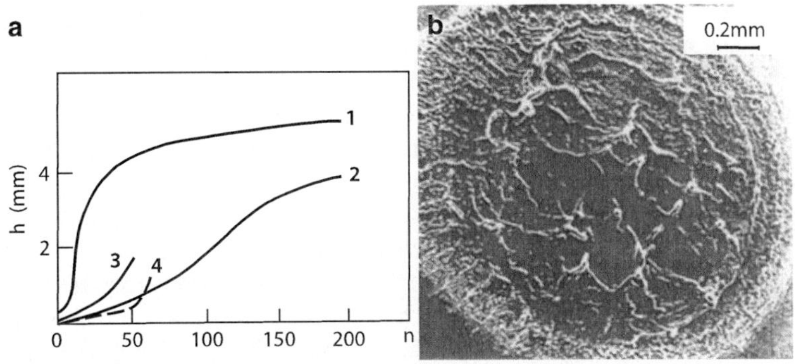

Fig. 8.12 (**a**) Dependences of the hole depth on the number of CO_2 laser pulses: (**1, 2**) $\tau = 70\,\mu s$, $d = 450\,\mu m$ for $E = 4\,J$ (**1**) and $E = 9\,J$ (**2**); (**3, 4**) $\tau = 40\,\mu s$, $d = 300\,\mu m$ for $E = 0.52\,J$ (**3**) and $E = 0.35\,J$ (**4**); (**b**) Structure of the metal surface obtained after irradiation by a $\lambda = 1.06\,\mu m$, $\tau_{0.5} = 300$ ns pulse in vacuum; the laser spot diameter is 2 mm

thickness was 1 mm). One can see that when the pulse energy (and, therefore, intensity) exceeds the threshold, the damage efficiency decreases for all gases.

For large q, the damage efficiency depends on the gas type, being maximal for helium and minimal for argon). The same was observed in the case of short pulses (see Fig. 7.26b). The value of Y decreases by a factor of 50 with increasing q. Such a drop cannot be explained by the change in the damage mechanism because the steel damage efficiency can change only by a factor of 5 for this reason. As the laser radiation power increases, the transparency coefficient also strongly deceases (see curve 4 in Fig. 7.26a). This drop is so large that the radiation power incident on the target even deceases with increasing he laser radiation power (Fig. 7.30). Under these conditions, the run of the curves becomes clear. The weakest attenuation of the light intensity occurs in helium and the strongest in argon.

Along with the study of integrated dependences, the damage of metals by laser pulses was investigated in more detail in a number of papers. In particular, the fraction of the removed material in the condensed phase was investigated. It was found by weighing directly solidified drops that upon drilling steel and aluminum by 0.6 ms pulses from a neodymium laser, up to 70% of the material was removed in the liquid state [15]. Similar measurements were also performed for different materials (steel, tungsten, zirconium) irradiated by a CO_2 laser [16]. To avoid the influence of the optical breakdown of air, experiments were performed in vacuum. The power density was $\sim$20 MW/cm^2, $\tau = 5\,\mu s$, and $r_f \sim 0.8$ mm. For such laser pulse durations, the target material should be removed due to evaporation, the liquid phase being absent (Fig. 8.2). However, the direct measurement of the mass of drops shows that in this case $\sim$70% of the target material is removed in the liquid phase. It seems that here the mechanism of nonuniform removal of the melt can be involved [17]. This mechanism is caused by the fact that the laser radiation intensity in the focal spot is usually nonuniform (multimode regime) and, in addition, inhomogeneities can appear because of the instability of the evaporation front. This can lead to the nonuniform removal of the melt with the characteristic size of inhomogeneities far smaller than the focal spot. The replacement of r_f in (8.22) by the inhomogeneity size will considerably reduce the value of τ_1, making possible the fountain or ejection regime. The estimates of the velocity of drops by expressions (8.3) and (8.44) for the conditions under study ($\alpha \approx 0.1$) give the value ~30 m/s. This value agrees with experimental data within an order of magnitude [16]. The decrease in the pulse duration by an order of magnitude down to $\tau = 0.5\,\mu\text{s}$ upon exposure of an aluminum target to UV radiation reduced the amount of the liquid fraction by several times [16]. This is consistent with the concepts developed above.

Other mechanisms of the melt removal are also possible at such high radiation intensities. The author of paper [24] explains this effect by a drastic increase in pressure on a melt film due to the intense plasma formation in material vapor. Upon irradiation of an aluminum target in vacuum by $\tau = 0.3\,\mu\text{s}$, $q \geq 350\,\text{MW/cm}^2$ pulses from a neodymium laser (the focal spot radius was $r_f = 1$ mm), pressure on the target surface, measured with a piezoelectric sensor, achieved 600 atm. This pressure causes a strong acceleration of the melt film, which is sufficient for exciting the Rayleigh–Taylor instability. This instability leads to jet ejections normally to the target surface, which is demonstrated by the photographs of solidified inhomogeneities on the target surface – stalagmites (Fig. 8.12b). The removal of the target material was also observed upon irradiation of carbon steels in the argon atmosphere by a 600-W argon laser. For $q \geq 1.8\,\text{MW/cm}^2$, an optical breakdown was observed in metal vapor, while for $q > 2.5\,\text{MW/cm}^2$, the melt was spattered in the form of ~0.1 mm drops [25]. Estimates by the cinegrams of the integral flow of drops showed that the energy removal by them was ~1% of absorbed laser energy. Thus, the role of spattering in the thermal balance is insignificant. The author of paper [24] explained this effect by excitation of capillary waves during steel evaporation in the focal spot. Note as a whole that due to the complexity of these phenomena, neither a quantitative theory nor a complete numerical model have been developed so far.

8.3.2 Material Processing by Ultrashort Pulses

In the previous chapters, we considered the ablation physics of surfaces irradiated by shorts and ultrashort pulses. The gas dynamics of vapors initiated by a single pulse was analyzed when a very thin material layer was removed per "shot", the surface remaining flat. In practice, the problems of precision and deep processing of surfaces without the formation of the liquid phase are often encountered. In this case, nanosecond and picosecond laser pulses are also often used. Recently, femtosecond lasers are also applied for this purpose. Figure 8.13 demonstrates that the probability of liquid phase formation is decreased with decreasing the pulse duration [26]. Consider briefly some properties of the damage (drilling) materials under these conditions. The radiation intensity in the focal spot of short laser pulses is many orders of magnitude higher than that for cw lasers, exceeding $\geq 10^{13}\,\text{W/cm}^2$.

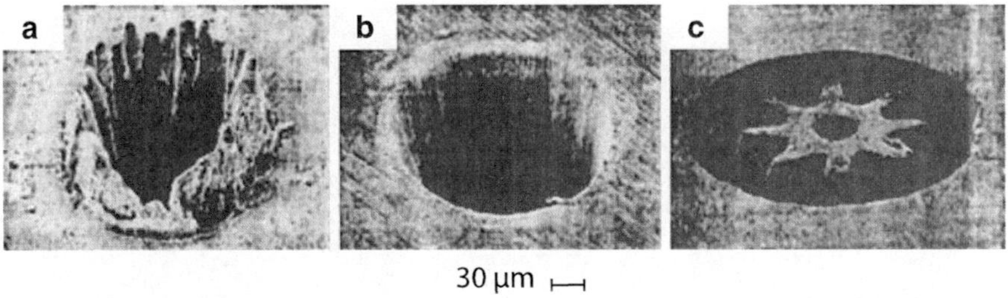

Fig. 8.13 SEM photographs of holes drilled through a 100 μm thick steel foil by (**a**) $\tau = 3.3$ ns, $\varepsilon = 4.2\,\text{J/cm}^2$ and (**b**) $\tau = 200$ fs, $\varepsilon = 0.5\,\text{J/cm}^2$ Ti:sapphire laser pulses at 780 nm and (**c**) of a gearwheel produced in a 1-mm thick polished copper target. Objects are imaged with an achromatic lens with a focal distance of $f = 150$ mm (demagnification factor is ∼30) on target ($\tau = 140$ fs, 700 μJ, 10,000 pulses)

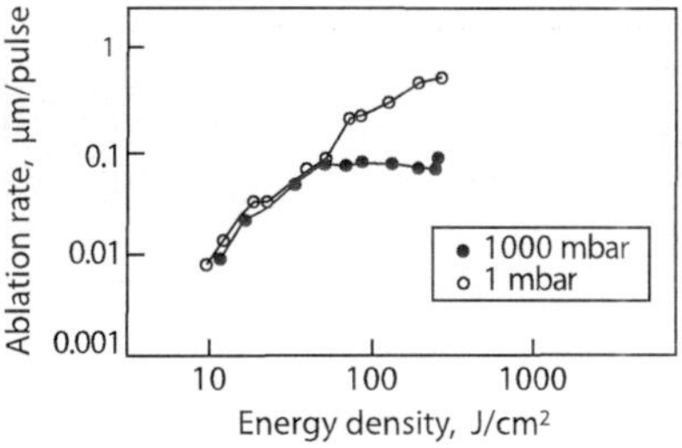

Fig. 8.14 Linear ablation rates achieved in 500-μm thick steel samples drilled by 120 fs pulses in air and vacuum (1 mbar)

The energy density is $\varepsilon \sim 1$–$100\,\text{J/cm}^2$ and $\tau \approx 100$ fs, the wavelength is ≤ 1 μm, and the pulse repetition rate is up to 10 KHz. It follows from (8.2) that the boiling of a surface occurs at such energy densities when the pulse duration is smaller than one femtosecond. Let us point out some features of the interaction of such pulses with matter. In the case of such short pulses, evaporation can be initiated by reducing the energy density by four orders of magnitude, i.e. according to the thermal model presented in Fig. 8.2, the operating point is located deeply in the evaporation regime. It would be expected that the energy loss in this case is small and the energy spent to remove the material should be close to the evaporation enthalpy. Recall that this enthalpy for carbon steels is $80\,\text{kJ/cm}^3$. However, as follows from Fig. 8.14, $X \approx 500$–$1{,}000\,\text{kJ/cm}^3$ even in vacuum (with respect to the absorbed power, $\alpha \approx 0.1$) [18, 19]. The saturation of the ablation rate observed in air is related to the nonlinear conversion of laser radiation to radiation with different spectral and angular distributions (which is conditionally called cone radiation [18, 19]). The influence of cone radiation on the ablation rate is demonstrated by Figs. 8.15 and 8.16. One can see that, when the conversion coefficient increases with pressure up to ∼1, the ablation rate at this pressure begins to decrease abruptly. However, the latter quantity decreases much faster. Consider now the specific energy consumption for ablation in vacuum. Because X exceeds L_b by an order of magnitude, this suggests probably that the usual thermal model is not valid for such short and intense pulses. As mentioned in Chap. 6, laser radiation heats the electron gas, while the rate

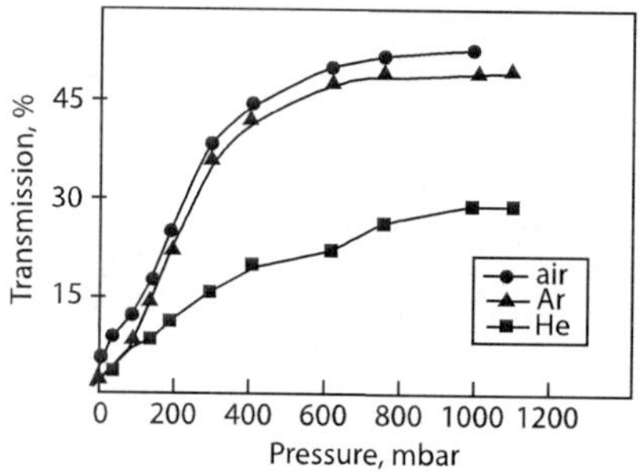

Fig. 8.15 Pressure dependences of the transmission of $\tau = 130$ fs, $\varepsilon = 260$ J/cm^2 pulses in different gases

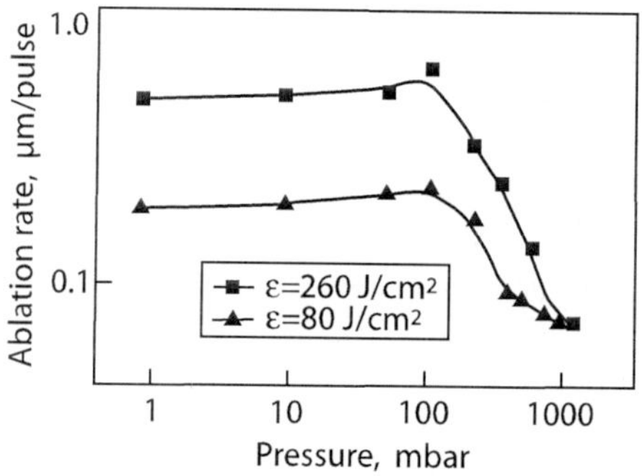

Fig. 8.16 Linear ablation rates obtained in a 500-μm thick steel sample irradiated by 120 fs pulses at different pressures

of energy transfer from electrons to a crystal lattice is finite. Therefore, energy can be accumulated in the electron subsystem, resulting in its overheating. The crystal lattice can be heated later, the situation being similar to that in nonequilibrium plasma described in Chap. 7. The penetration depth of laser radiation into metals is tens of nanometers. It determines the damage depth per pulse. This depth should increase logarithmically with the absorbed energy, which is confirmed by laser drilling of a copper plate (Fig. 8.14 [26]). Finally, despite a short pulse, energy losses due to heating and electron heat conduction can be significant. At present the interaction of ultrashort pulses with materials is being actively investigated.

The existence of some processes restricting the efficiency of action of ultrashort pulses is demonstrated by several facts. For example, during the perforation of very narrow (500 nm) holes in micron films by 248 nm UV laser pulses, the appearance of the melt could not be avoided even in the case of 160 fs pulses. The melt was produced both upon ablation of metal films and semiconductors (Si, Ge) [27] (Fig. 8.17). Another example is the restriction of the drilling depth of deep holes in metals and other solid materials (Fig. 8.18) and the appearance of the liquid phase on the walls of the hole (Fig. 8.19 [20]). The restriction of the depth of a hole was explained in [21] by the action of a plasma cloud on the material removal during repetitively pulsed laser drilling. In the absence of a target in regimes corresponding to Fig. 8.18, the optical breakdown of air by femtosecond pulses occurs rarely. In the repetitively pulsed regime, microscopic drops of the target material remain after

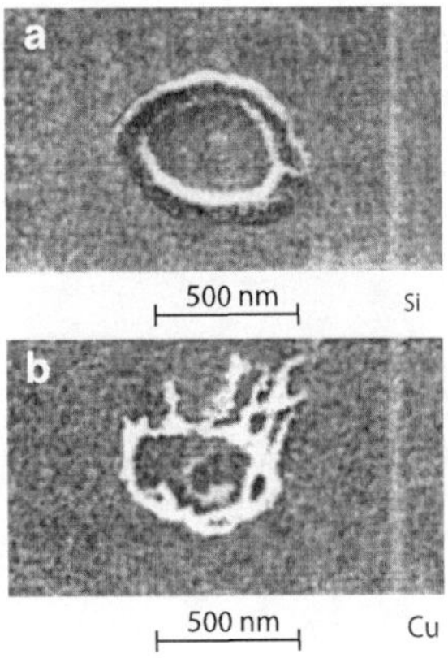

Fig. 8.17 Holes ablated in silicon (**a**) and copper (**b**) by 1.2 J/cm^2, 500 fs (**a**) and 160 fs (**b**) pulses

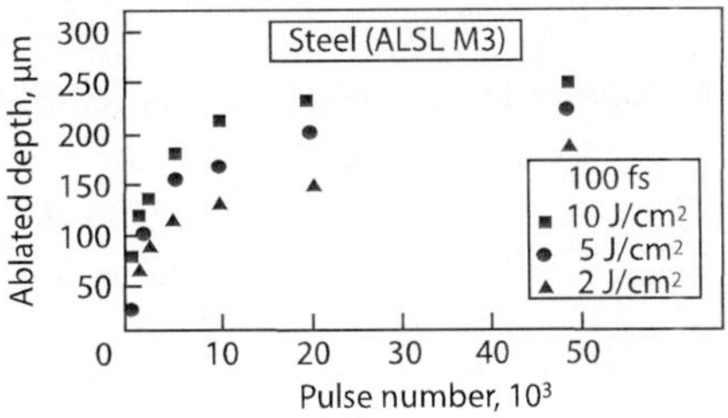

Fig. 8.18 Ablation curve for steel

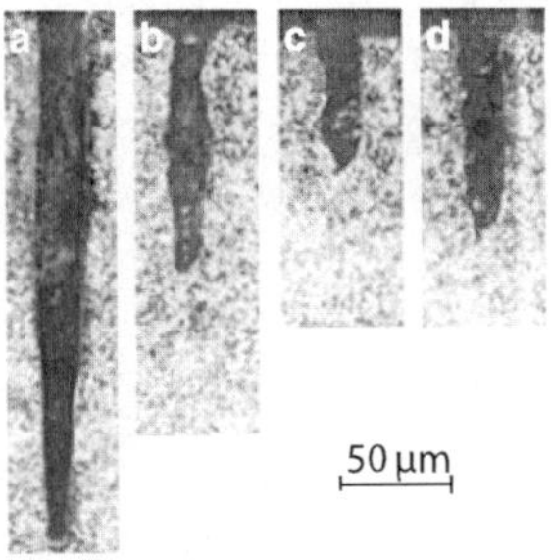

Fig. 8.19 Metallographic analysis of holes produced by femtosecond laser pulses in 52,100 steel: (**a**) 2 J/cm^2, 5,000 pulses (*bottom*); (**b**) 5 J/cm^2, 1,000 pulses; (**c**) 10 J/cm^2, 500 pulses; (**d**) 10 J/cm^2, 1,000 pulses (*etched*)

the previous pulse, on which a breakdown can easily occur. Thus, the shortening of a radiation pulse down to femtoseconds with the aim to avoiding the production of melt layers proves to be not always efficient [22, 23]. In this sense, it is better to increase the pulse duration up to 5–10 ps. In this case, thermal losses remain small and the nonlinear scattering of laser radiation decreases (Fig. 8.20).

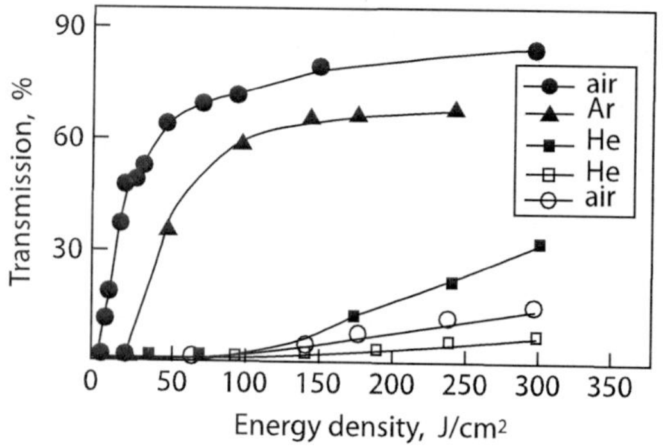

Fig. 8.20 Efficiency of conversion of incident 130 fs pulses (*black symbols*) and 1.5 ps pulses (*light symbols*) into the CE beam

8.3.3 Theoretical Models of Formation of Deep Keyholes in Metals by CO_2 Laser Radiation

As follows from experiments on the damage of metal targets by radiation from a pulsed CO_2 laser, the main feature of this process is the increase in the keyhole deepening speed with increasing the keyhole depth. We explained this effect by the increase in the radiation intensity at the keyhole bottom because of the reflection of radiation from walls and the decrease in the keyhole bottom area.

It was assumed for simplicity that the damage energy was independent of the radiation intensity and coordinates and was equal to the boiling enthalpy ρL_b and that the damaged material was removed from the keyhole without obstacles, not precipitating on the keyhole walls. Plasma phenomena were also neglected. In this case, the keyhole shape is described by the expression

$$\frac{dz}{dt} = \left[1 + \left(\frac{dz}{dr}\right)^2\right]^{-1/2} \frac{q_a}{\rho L_b} \tag{8.49}$$

where z is the coordinate of the keyhole wall; r is the keyhole radius; and q_a is the radiation intensity absorbed by the surface (Fig. 1.5b). We will calculate the distribution of the absorbed radiation in the simplest model of geometrical optics approximation (see Chap. 1). The calculation of radiation propagation in cavities of complex shapes is a complicated problem in itself [17]. Equation (8.49) does not describe the shape of a keyhole with concave regions [i.e. the dependence $z(r)$ should be unique]. The focal point was chosen over the surface at height h_0 so that $h_0\varphi_0 = r_f$, where φ_0 is the focusing angle and r_f is the focal spot radius. Figure 8.21 shows the time dependences of the hole depth for $h_0 = 5\,\text{mm}$, $\varphi_0 = 0.1$ and two different polarizations of radiation [29]. One can see that, when the keyhole is shallow, the deepening speed is the same in both cases and is equal to $v = \alpha P/\pi r_f^2 \rho L_b$. The difference for the two polarizations becomes noticeable when reflected beams are incident for the first time on the keyhole walls, for $h \geq d$. It is reasonable that beams with normal polarization destroy a metal stronger because

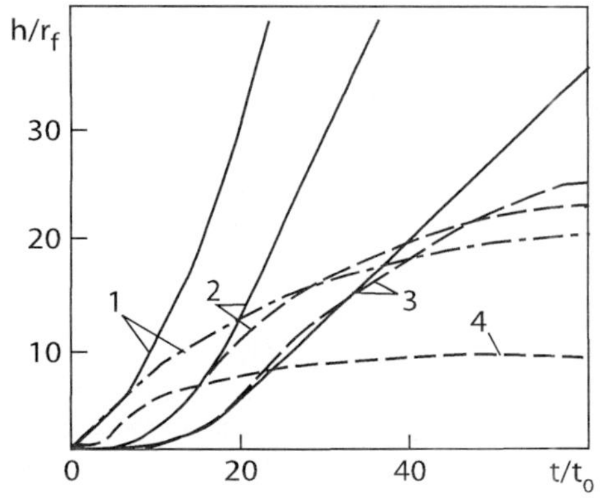

Fig. 8.21 Dependences of the drilling depth on the interaction time for different values of α: (**1**) $\alpha = 20\%$; (**2**) $\alpha = 10\%$; (**3**) $\alpha = 5\%$, $q_b = 0$; (**4**) $\alpha = 5\%$, $q_b/q = 0.3$; the *solid and dashed curves* correspond to the transverse and parallel polarizations

they lose smaller energy after reflection from keyhole walls (Fig. 8.21). As long as the keyhole is not too deep, the attenuation of the light beam is weak. Because of this, the deepening speed increases with increasing the keyhole depth. As shown in Sect. 1.2, the penetration depth of the E-mode beam with parallel polarization is of the order of d/α. When the keyhole depth exceeds his value, the beam is strongly attenuated due to absorption in the keyhole walls. As a result, the keyhole deepening speed begins to decrease. Figure 8.21 shows that this tendency is manifested for $h/d \geq 10$. Then, the dependence $h(t)$ take the asymptotic form $h \approx \sqrt{t}$. Note that the maximum deepening speed for $\alpha = 0, 1$ exceeds the initial (minimal) speed approximately by a factor of 7. Thus, the laser beam focusing by keyhole walls proves to be important for increasing the efficiency.

Taking into account the processes limiting the material damage, the keyhole depth ceases to grow with time. In the case of a cw laser, such a restriction should be determined by the heat conduction loss. Because the specific damage energy in the given model was assumed constant, the model is most convenient for the description of multipulse processing of samples by a CO_2 laser. It follows from (8.49) that in this case the depth h depends only on the total radiation energy. The case of a high pulse repetition rate, when the average heating of a sample is important, will be considered in Sect. 9.2. The damage of a target irradiated by a single pulse or many pulses will cease if none of the sites of the keyhole has time to be heated to the boiling temperature. Because the heating depth during the pulse time τ is considerably smaller than the keyhole radius (for $\tau = 10$–$100\ \mu$s), we can use the condition which is similar to the condition for boiling of a flat surface (8.21)

$$q_a(z, r) > q_b = c\rho T_b \sqrt{\chi/\tau}. \tag{8.50}$$

As mentioned above, we neglect the dependence of the specific damage energy on various parameters. Therefore, to take into account restriction (8.50) in (8.49), we should replace q_a by $q_a \theta(q_a - q_b)$, where $\theta(x)$ is the theta function

$$\theta(x) = 0, \quad x < 0 \text{ and } \theta = 1, \quad x > 0$$

In the case of drilling by a repetitively pulsed laser, the maximum value of the keyhole depth $h_{\max}$ will be determined only by the pulse power. It is clear that q_a increases with pulse energy; therefore, the maximum depth $h_{\max}$ will also increase. The values of q_a and q_b increase with decreasing the pulse duration, but q_a increases faster and, therefore, $h_{\max}$ increases with decreasing the pulse duration. If ε and τ are reduced simultaneously by maintaining the pulse power constant, the value of q_b increases and, therefore, $h_{\max}$ decreases. Note that, as q_b increases, the beam is "captured" in the keyhole after a shorter irradiation time (see curves 3 and 4 in Fig. 8.21). This is explained by the fact that the keyhole diameter d in this case proves to be smaller and, therefore, the "capture" condition $h \geq d$ begins to be fulfilled at a smaller keyhole depth.

Figure 8.21 presents the dependences $h(t)$ calculated for different values of the parameter α. At short irradiation times, h increases with α according to (1.31). However, at longer irradiation times the situation is reverse: the limiting keyhole depth increases with decreasing the absorption coefficient. This can be explained based on waveguide concepts (1.22), according to which the penetration depth of radiation and, therefore, the keyhole depth (at approximately constant diameter) increase with decreasing α. This also can explain a greater keyhole depth produced upon irradiation by light polarized perpendicular to the plane of incidence (Fig. 8.21). This polarization corresponds to the propagation of magnetic E-modes having the lowest decay [see (1.23)].

Note that the polarization dependence of the penetration depth of a laser beam into samples upon cutting metals was observed in [28]. This dependence was explained, as in [29], by the appearance of the waveguide multiple reflection regime in the keyhole.

8.3.4 *Waveguide Regime*

It is interesting to compare numerical calculations with simple analytic models of perforating holes, which take into account multiple reflections of light in the keyhole. It is shown in Sect. 1.2 that, along with the geometrical optics approach used in numerical calculations, the propagation of light in the keyhole can be described by using the waveguide approach. It is clear that the waveguide approach is valid if the keyhole is deep enough to neglect the influence of the keyhole bottom. In addition, the entrance region of the keyhole, where guided modes are produced, is also not considered. In this case, the light flux power P and the keyhole diameter $2a$ are described by the equations

$$\frac{\partial}{\partial z} P = \alpha_n P a^{-n} \tag{8.51}$$

$$\pi L_b \rho \frac{\partial}{\partial t} a^2 = \alpha_n P a^{-n} - W_T. \tag{8.52}$$

Here, the first equation describes decay in the waveguide due to absorption in the walls [see (1.24)]. For the E-waves (with parallel polarization), $n = 1$ and $\alpha_1 = \alpha/2$, while for the H_1-waves, $n = 3$, and $\alpha_3 = \lambda^2\alpha/8\pi^2$. We neglect the transformation of the H_1 mode to the waves of other types.

Equation (8.52) describes the energy balance: energy absorbed by walls is partially spent to destroy the material and is partially transferred to the sample due to heat conduction $W_T = 2\pi\kappa T_b$. The heat conduction loss of the laser beam energy restricts the keyhole growth. In the case of a repetitively pulsed laser, P is the average power, and the restricting condition of boiling should be introduced, as above. For this purpose, the first term in the right-hand side of (8.52) should be multiplied by

$$\theta\left(\frac{\alpha_n PQ}{2\pi\alpha^{n+1}} - c\rho T_b\sqrt{\chi/\tau}\right)$$

where θ is the theta function, Q is off-duty ratio. Equations (8.51) and (8.52) should be supplemented with initial and boundary conditions $d(t = 0) = d_0$ and $P(z = 0) = P_0$. The keyhole grows to the side $z < 0$. When the keyhole depth is far from the limiting depth determined by the mechanisms discussed above, these restrictions are not important. In this case, it is necessary to omit the last term in (8.52) or set $\theta = 1$. Then, the systems admits the self-similar solution [30]:

$$P(z,t) = P_0[1 + z/l(t)]^2 \tag{8.53}$$

$$a(z,t) = \frac{\alpha_1}{2} l(t)(1 + z/l) \tag{8.54}$$

where the depth l is defined by the expression

$$l(t) = \left(\frac{12}{\alpha_1^2}\frac{P_0 t}{\pi L_b \rho}\right)^{1/3}. \tag{8.55}$$

Thus, the keyhole deepens by the law $t^{1/3}$. In this case, the absorbed energy is incident at each instant inside the keyhole, so that $\partial P/\partial z = 0$ at the keyhole bottom. In this respect, such a model differs from the geometrical optics approach. In the latter case, the absorbed power at the keyhole bottom for not very deep keyholes is maximal (see Sect. 1.2). It seems that this explains the difference in the time dependences of the keyhole depth in these two cases. It follows from (8.54) that the keyhole wall is inclined with respect to the vertical at an angle of $\alpha_1/2$. When l approaches the limiting depth h, the dependence $l(t)$ changes. The limiting depth h for a cw laser can be found from the stationary solution (8.51) and (8.52):

$$P = P_0 + 2\pi\kappa T_b z \tag{8.56}$$

$$a = \alpha_1 P/2\pi\kappa T_b, \quad h = P_0/2\pi\kappa T_b.$$

One can see from (8.56) that the energy release over the keyhole depth is constant. In the case of a repetitively pulsed laser, we have

$$P = \left(\frac{\alpha_1 z}{2b} + \sqrt{P_0}\right)^2, \quad a(z) = b\sqrt{P(z)}, \quad h = 2b\sqrt{P_0}/\alpha_1,$$

$$b = [\alpha_1 Q\sqrt{\tau}/2\pi c\rho T_b\sqrt{\chi}]^{1/2}. \tag{8.57}$$

One can see from (8.57) that the energy release in the keyhole in the repetitively pulsed regime, as in the stationary case, decreases with the keyhole depth and is equal to zero at the keyhole bottom. It follows from (8.56) that the angle da/dz in the stationary regime is α_1, i.e. it is twice as small as that upon the keyhole formation. Thus, the keyhole first grows in depth and then in width.

In the case of repetitively pulsed processing, the angle is $da/dz = \alpha_1$, i.e. the final angle is equal to the initial one. Thus, the keyhole grows simultaneously both in width and depth.

The numerical solution of the system (8.51) and (8.52) confirms analytic relations. The stationary solution of (8.51) and (8.52) for the H-waves has the form

$$P = P_0 + 2\pi\kappa T_b z, \quad a = \left(\frac{\alpha_3 P}{2\pi\kappa T_b}\right)^{1/3}, \quad h = \frac{P_0}{2\pi\kappa T_b}. \tag{8.58}$$

While in the case of parallel polarization the value of h/a_0 is independent of power and is equal to $1/\alpha_1$, for normal polarization we have $h/a_0 = (P_0/2\pi\kappa T_b\alpha_3^2)^{1/3}$, i.e. h increases with power. Thus, the keyhole depth for both polarizations is the same, but the keyhole width for normal polarization is smaller.

A similar situation takes place in the case of repetitively pulsed processing. The solution of (8.51) and (8.52) for the H mode gives the expressions for the keyhole shape and depth:

$$a(z) = cP^{1/4}(z), \quad P(z) = \left[\frac{3}{4}\frac{\alpha_3 z}{c^3} + P_0^{3/4}\right]^{4/3}, \quad h = \frac{4}{3}\frac{P_0^{3/4}c^3}{\alpha_3},$$

$$c = [\alpha_3 Q\sqrt{\tau}/2\pi c\rho T_b\sqrt{\chi}]^{1/4}. \tag{8.59}$$

In this case, the keyhole bottom, as for a cw laser, is flat. The radiation intensity absorbed in the keyhole decreases from its entrance to bottom, where it vanishes. While the aspect ratio h/a_0 in the case of parallel polarization is independent of the laser power, in the given case is $h/a_0 = 4c^2\sqrt{P_0}/3\alpha_3$ and increases with power. By comparing depths from (8.57) to (8.59), we obtain

$$\frac{h_\perp}{h_\parallel} = \left(\frac{1}{3}\frac{2\pi Q P_0\alpha}{\lambda^2 c\rho T_b}\sqrt{\frac{\tau}{\chi}}\right)^{1/4} \tag{8.60}$$

where $h_{\perp}$ and $h_{\parallel}$ are keyhole depths for normal and parallel polarizations, respectively. While for a cw laser the depth ratio is $h_{\perp}/h_{\parallel} \approx 1$, in the given case this ratio weakly depends on the pulse power, and for $QP_0 = 10^5$ W and $\tau = 100\,\mu$s, this ratio for steel is $h_{\perp}/h_{\parallel} \approx 4$.

8.4 Physics of Deep Melting of Metals by Pulsed Radiation

In the previous section, we considered the drilling of holes in a plate by a laser pulse either due to evaporation or ejection of the boiling liquid by the vapor pressure. However, the material can melt without ejection. This phenomenon of deep pulsed melting is analogous to deep melting by a cw laser beam.

8.4.1 Pulsed Welding

Note that the passage from the drilling to melting regime (and vice versa) is determined by many reasons. It is clear from general considerations that the drilling regime should pass to the melting regime with decreasing the radiation intensity. In this case, only the central part of the laser beam penetrates into a target, while the rest of the beam only melts the target surface. The result of the beam action depends not only on the pulse intensity but also on its duration. If the pulse duration is long enough, the melting of keyhole walls becomes important and liquid is produced, which flows down to the keyhole bottom. As a result, the keyhole depth proves to be a nonmonotonic function of time. If the light pulse is long enough, the liquid completely fills the keyhole after switching off the beam. This was observed upon welding by a cw laser. Moreover, the liquid can also completely fill the keyhole when the laser beam is switched on, as was observed upon melting of paraffin by a low-intensity laser beam (Fig. 8.22).

The drilling (hole perforation) regime and deep melting regime differ from each other by the formation mechanism of the vapor channel (the necessity of existence of this channel during welding and deep melting is obvious). Radiation incident to the bottom of the channel in the case of deep melting evaporates the material, while radiation absorbed by the side walls melts it. If the volume of the melting zone by the pulse end considerably exceeds the channel volume, the deep melting regime

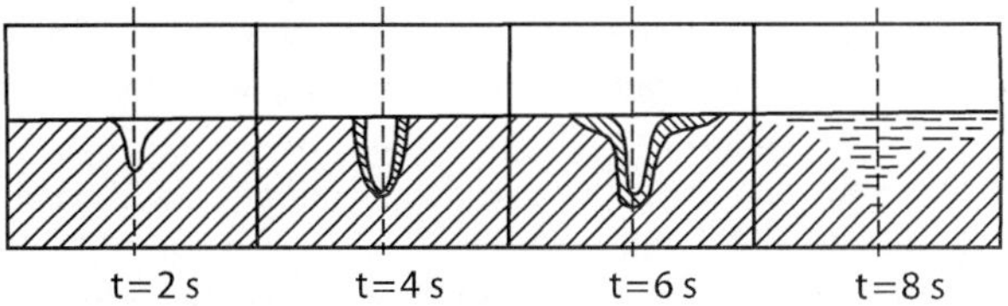

Fig. 8.22 Scheme of paraffin melting by the 25 W cw laser radiation for $d = 1$ mm

is realized. As shown in Sect. 8.3, melting deep inside a sample is restricted by the heat conduction loss. In this case, the channel depth approaches the stationary value; only a small fraction of the radiation power is spent for evaporation from the keyhole bottom and almost all power is spent for melting and heat conduction. Therefore, when the light pulse duration exceeds the establishment time τ_{st} of the stationary keyhole depth, the appearance of the deep melting regime can be expected. However, this regime can be also established in the case of a shorter pulse. This can occur if the width d_m of the melting zone exceeds the gas–vapor channel radius, i.e. the condition $\tau > \tau_m$ is fulfilled, where τ is the pulse duration; $\tau_m = d_m^2/4\chi$ is the time required for the melting zone width to grow to the value equal to the gas–vapor channel radius; and χ is the thermal diffusivity. The channel radius is usually close to the light beam radius r_f, and in this case $\tau_m = r_f^2/4\chi$.

Thus, if $\tau < \tau_m$, a narrow deep hole is formed in the sample, and if $\tau > \tau_m$, a narrow deep melting zone is formed. We will describe the properties of deep melting by using a simplified approach based on the thermal balance. As in the case of cw laser welding, we assume that the beam energy is spent completely due to heat conduction (the evaporation loss is small) and a keyhole deepens at a speed such that the temperature of the keyhole walls at each instant is equal to the boiling temperature. If a heat conduction wave gone off from the keyhole by a distance exceeding the keyhole radius, the keyhole depth for a sample at rest can be calculated from the expression

$$h = \frac{P}{2\pi\kappa T_b}\ln\frac{r_f + 2\sqrt{\chi t}}{r_f}, \quad 2\sqrt{\chi t} \geq r_f, \tag{8.61}$$

where r_f the focal spot radius. The depth growth speed $v = \mathrm{d}x/\mathrm{d}t$ can be found from (8.61) by differentiating with respect to time:

$$v = \frac{P}{4\pi\kappa T_b}\frac{1}{\sqrt{\chi t} + r_f/2}\sqrt{\frac{\chi}{t}}. \tag{8.62}$$

It follows from (8.62) that this speed decreases with time and vanishes when the keyhole depth ceases to grow, i.e. when the heat conduction wave propagates a distance approximately equal to h (a distance equal to χ/v in the case of a moving target). The average speed of keyhole deepening can be estimated by assuming in (8.62) that $t \approx r_f^2/4\chi$

$$v = Aq/c\rho T_b. \tag{8.63}$$

The numerical factor A can be found from comparison with experiments. It follows from (8.61) to (8.62) that at the beginning of the beam deepening process, when the heated zone is small, the keyhole depth grows with time even faster; however, this treatment is not valid at small times because not all radiation will be absorbed in the keyhole due to its insufficient depth.

The beam penetration speed close to (8.63) was obtained experimentally [31] during deep melting of various materials by 2 ms and 4 ms pulses from a neodymium

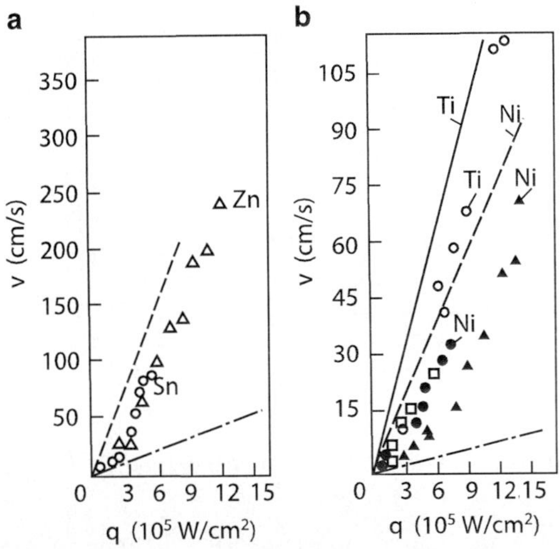

Fig. 8.23 Dependences of the deep melting speed on the neodymium laser radiation intensity for $d = 0.7$ mm: (**a**) Sn target, $\tau = 4$ ms (∘) and Zn target, $\tau = 2$ ms (*triangle*); (**b**) Ni target, $\tau = 4$ ms (•) and $\tau = 2$ ms (*black triangle*) and Ti target, $\tau = 4$ ms (*square*) and $\tau = 2$ ms (∘); the *dashed and solid straight lines* are calculated by (8.63) for $\alpha = 0.7$; the *dot-and-dash straight line* is calculated by (1.31) for $\alpha = 0.7$

laser (Fig. 8.23). The focal beam diameter on the surface was 0.7 mm. The dot-and-dash straight line shows the dependences of the beam penetration speed on the radiation intensity by assuming that the absorbed laser beam energy is completely spent for evaporation of the material. The dotted straight line shows dependence (8.63) for $\Lambda = 1$. One can see from this figure that the experimental value of the melting speed can be almost an order of magnitude higher than the evaporation wave velocity. After the end of the laser pulse, the keyhole is filled with a melt. The melting speed in this deep melting mechanism should depend on the pulse duration because the time τ_b when boiling starts is comparable with the pulse duration. According to (8.2), for $q = 0.3\,\mathrm{MW/cm^2}$ and $\alpha = 0.3$, we have $\tau_b \approx 1$ ms. One can see from Fig. 8.23b that the average melting speed for a longer 4 ms pulse (nickel sample) can be twice as small as that for a shorter 2 ms pulse. Experimental melting speeds for samples made of other materials prove to be close (which is probably explained by the approximate method of measuring the melting speed).

For CO_2 laser pulses in the same range of intensities and durations, similar properties were observed. Figure 8.24 shows that for large off-duty ratio $Q = T/\tau$ (the ratio of the repetition period T to the pulse duration τ), when pulses do not interact with each other, the melting depth is proportional to the pulse duration. Experimental data are consistent with the theory and are described by expression (8.63) with $\Lambda = 1/2$.

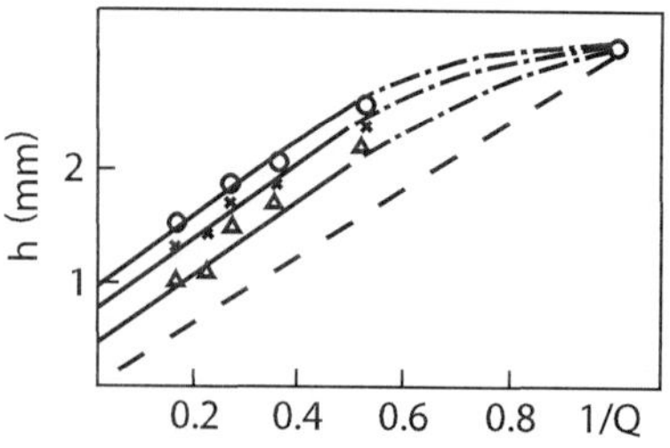

Fig. 8.24 Melting depth of stainless steel irradiated by 2.4 kW CO_2 laser pulses as a function of the off-duty ratio; $d = 1$ mm, $v = 1$ cm/s; laser pulses durations are $\tau = 4$ ms (Δ), $\tau = 7$ ms (*cross*), and $\tau = 10$ ms (*ring*) Calculations are preformed by (8.61)

8.4.2 Control of the Deep Penetration Melting Process

We considered above qualitative models of deep melting by a pulse neodymium laser beam, which were based on the total energy balance. Actually, the interaction of such pulses with metal depends on many factors such as the aperture angle of a laser beam, the focal spot size, the pulse energy, duration, and shape. Conventional commercial Nd lasers can melt a metal by a depth of ~2 mm per pulse. The attempts to increase the melting depth by increasing the laser energy lead to the ejection of the melt, resulting in the formation of a hole instead of welding. Studies have shown that the type of pulsed welding depends considerably on the pulse shape [32]. The problem is that, on the one hand, the recoil vapor pressure should deepen the keyhole bottom, and on the other hand, this pressure should not eject metal from the melt pool. Figure 8.25a,b demonstrate the dependence of the weld shape on the laser pulse shape. In all cases the duration, energy, and diameter of focusing were kept constant: $\tau = 12$ ms, $E = 20$ J, and $d \approx 0.33$ mm. One can see that, if the radiation intensity at the pulse onset was maximal, this resulted in the ejection of the melt from the keyhole. The symmetric pulse shape improves the weld quality. Good quality is obtained when the laser pulse is slightly symmetric. The displacement of the intensity maximum to the pulse end provides the maximum beam penetration depth, but also leads to the ejection of material. Similar dependences were obtained for many materials: steel, aluminum alloys, etc. It seems that this is related to the value of the buffer volume, i.e. of the melt pool produced before the laser beam penetration into a target [32]. This volume plays a stabilizing role in the formation of a gas–vapor channel. Note that model calculations based on other considerations also give a similar optimal shape of the laser pulse [33] (see below).

Apart from the pulse shape selection, point welding can be changed by varying spatial parameters such as the focal spot size and aperture angle β. Table 8.1 presents the results of the deep melting of instrumental chrome steel obtained upon irradiation by pulses of constant energy (12–13 J). The pulse shape is close to curve 9 in Fig. 8.25. The melting depth strongly depends on these external parameters. For minimal β but large focal spots, as in the case of the minimal spot size but large β,

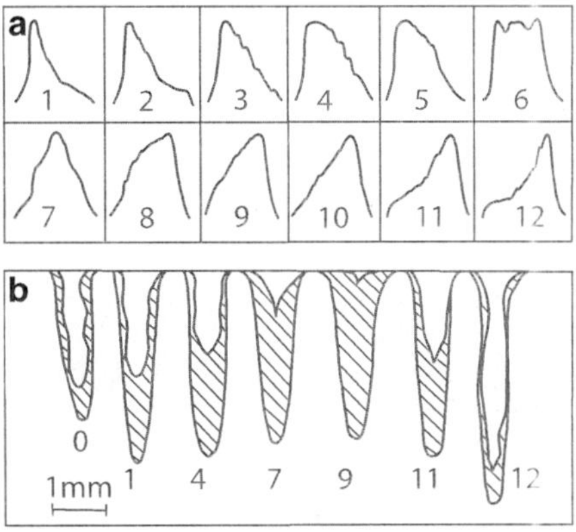

Fig. 8.25 (**a**) Shapes of investigated laser pulses; (**b**) longitudinal sections of melted zone in bearing steel for different pulse shapes

Table 8.1 Dependence of the melting depth on the focal spot diameter d and aperture β

	1	2	3	4	5	6	7
d, mm	0.20	0.40	0.80	0.14	0.18	0.24	0.32
2β, rad	0.24	0.12	0.06	0.09	0.07	0.05	0.04
h, mm	2.00	2.40	0.90	1.50	5.50	3.50	2.90

the melting depths are small. The minimal values of both β and d are contradictory (see Chap. 4). Because of this, the maximum depth is achieved for the intermediate values of these quantities.

By selecting these three parameters, the large weld shape parameter $\sim$15 was achieved. Such values could be achieved only upon welding by a CO_2 laser or electron-beam welding. As in the case of a CO_2 laser, a high aspect ratio of the weld is caused by the reflection of radiation from walls and the increase in the radiation intensity at the keyhole bottom (see previous sections and Figs. 8.12a and 8.21). This is confirmed by the direct measurements of the speed of beam penetration into a target, which is equal to $\sim$3.5 m/s when the melting depth is small ($\sim$1 mm). As the melting depth further increases, the beam penetration speed increases by a factor of 2–3, being $v \sim 9$ m/s for $h = 5$ mm [34]. These values are close to the data presented in Fig. 8.23, which were obtained for shorter pulses of duration $\sim$1 ms.

The influence of the laser pulse shape on the melting type and depth was considered in the numerical pulsed welding model [33]. The model is based on the equations of conservation of mass, momentum, and energy taking into account the melt viscosity, thermocapillary force, and recoil vapor pressure at the vapor-liquid interface. The calculations were performed for melting stainless steel and aluminum by pulses of different shapes, different durations (2.5, 5, and 7.5 ms) and different intensities (4.5, 7.5, and 10.5 GW/m^2). The focal spot diameter was constant and equal to 0.48 mm. The absorption of radiation was considered in the simplest case, with $\alpha = 0.3$ and $\alpha = 0.08$ for steel and aluminum, respectively. Figure 8.26 presents the dependences of the melting depth on the radiation power density for $\tau = 2.5$ ms and different laser pulse shapes. One can see that the melting depth

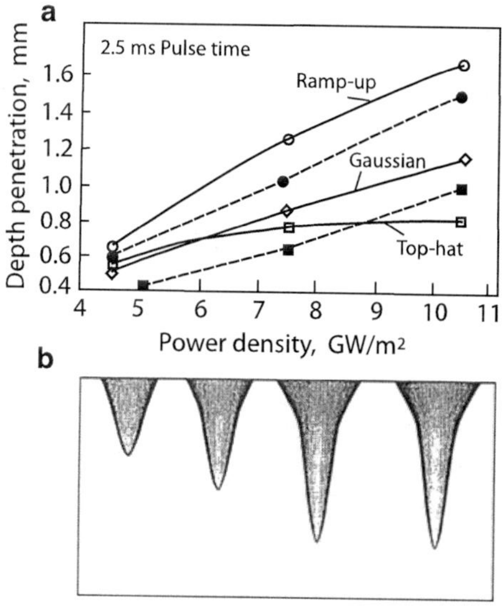

Fig. 8.26 Penetration depth for different pulses with the same parameters; (**a**) dependences of the penetration depth on the power density; (**b**) shapes of the melt pool for $\tau = 2.5\,\text{ms}$ and $q = 10.5\,\text{GW m}^{-2}$ (top-hat, Gaussian, ramp-down, ramp-up). The hatched symbols are experimental data

strongly depends on the pulse shape. Experimental data are presented by dashed symbols [35].

References

1. S.I. Anisimov, Ya.A. Imas, G.S. Romanov, et al., *Deistvie izlucheniya bol'shoi moshchnosti na metally (Interaction of High-power Radiation with Metals)* (Nauka, Moscow, 1970)
2. A.A. Vedenov, G.G. Gladush, *Fizicheskie protsessy pri lazrnoi obrabotke materialov (Physical Processes in Laser Material Processing)* (Energoatomizdat, Moscow, 1985)
3. B.A. Batanov, V.B. Fedorov, Pis'ma Zh. Eksp. Teor. Fiz. **17**, 348 (1973)
4. Ven H. Shui, Phys. Fluids **21**, 2174 (1978)
5. A.A. Vedenov, G.G. Gladush, S.V. Drobyazko, et al., Kvantovaya Elektron. **8**, 2154 (1981)
6. I. Smurov, L. Aksenov, G. Flomant, *ICALEO* (1993), p. 242
7. B.M. Zhiryakov, A.I. Korotchenko, N.I. Popov, et al., Kvantovaya Elektron. **2**, 1717 (1975)
8. S.I. Valyanskii, B.M. Zhiryakov, N.I. Lipatov, et al., Fiz. Khim. Obr. Mater. **5**, 12 (1978)
9. A.A. Samokhin, A.I. Korotchenko, N.I. Popov, Bull. Lebedev Phys. Ins. **6**, 393 (1986)
10. A.A. Samokhin, Quan. Elect. **10**, 2022 (1983)
11. N.N. Rykalin, I.V. Zuev, A.A. Uglov, *Osnovy elektronno-luchevoi obrabotki materialov (Fundamentals of Electron-beam Material Processing)* (Mashinostroenie, Moscow, 1978)
12. E. Sturmer, M. Von Allmen, J. Appl. Phys. **49**, 5648 (1978)
13. D.C. Hamilton, I.R. Pashby, Opt. Laser Techn. **11**, 183 (1979)
14. A.G. Borkin, G.G. Gladush, S.V. Dobyazko, et al., in *Vzaimodeistvie izlucheniya, plzmennykh i elektronnykh potokov s veshchestvom (Interaction of Radiation, Plasma and Electron Fluxes with Matter)* (Izd. TSNIIatominform, Moscow, 1984), pp. 97–98
15. M.G. Jones, G. Georgalas, A. Brutus, *Lasers in Material Processing*, ed. by M. Bass (North Holland Publishing, Amsterdam, 1983)

16. V.N. Anisimov, R.V. Arutyunyan, V.Yu. Baranov, et al., *Tezisy dokladov Vsesoyuznoi konferentsii po primenenniyu lazerov v narodnom khozyaistve (Abstracts of Papers, All-Union Conference on Laser Applications in National Economy)*, (Zvenigorod, 17–20 May, 1985), p. 133
17. R.V. Arutyunyan, V.Yu. Baranov, L.A. Bol'shov's, D.D. Malyuta, A.Yu. Sebrant, *Laser Materials Interaction*, (Nauka, Moscow, 1989)
18. S.M. Klimentov, T.V. Kononenko, P.A. Pivovarov, et al., Conference on Laser Processing of Advanced Materials and Laser Microtechnologies (Moscow, 22–27 June, 2002)
19. S.M. Klimentov, T.V. Kononenko, P.A. Pivovarov, et al., Proc. SPIE **5121**, 77 (2002)
20. V. Romano, H.P. Weber, G. Dumitru, et al., Proc. SPIE **5121**, 199 (2002).
21. T.V. Kononenko, S.M. Klimentov, V.I. Konov, Rossiisko-Frantsuzskii seminar (Russian-French Seminar Mathematic models and modelling of laser-plasma processes. 31.01-04.04.06, Moscow, Russia)
22. C. Fohl, D. Breitling, F. Dausinger, Conference on Laser Processing of Advanced Materials and Laser Microtechnologies (Moscow, 22-27 June, 2002)
23. C. Fohl, D. Breitling, F. Dausinger, Proc. SPIE **5121**, 271 (2002)
24. Yu.A. Chivel, Kvantiovaya Elektron. **18**, 1218 (1991)
25. A.I. Zaikin, A.V. Levin, SA.L. Petrov, S.A. Stranin, Kvantovaya Elektron. **18**, 706 (1991)
26. C. Momma, S. Nolte, B.N. Chichkov, et al., Appl. Surf. Sci. **109/110**, 15 (1997)
27. P. Simon, J. Ihlemann, Appl. Surf. Sci. **109/110**, 25 (1997)
28. E.D. Olsen, in Optoelectron. Techn. Vort. 5 Intern. Kong. Laser-81, Munchen, 1981, Berlin, 1982, p. 227
29. G.G. Gladush, E.B. Levchenko, A.A. Ezhov, Prikl. Mekh. Tekh. Fiz. **4**, 17 (1985)
30. A.A. Vedenov, O.P. Ivanov, A.L. Chernyakov, in *Vzaimodeistvie izlucheniya, plzmennykh i elektronnykh potokov s veshchestvom (Interaction of Radiation, Plasma and Electron Fluxes with Matter)* (Izd. TSNIIatominform, Moscow, 1984), pp. 79–80
31. M.S. Baranov, B.A. Vershok, I.N. Geinrikhs, Teplofiz. Vys. Temp. **3**, 566 (1975)
32. S.V. Kayukov, A.A. Gusev, Proc. 6th Europe Conference on Laser Treatment of Materials (Stuttgard, Germany, Sept. 16–18 (1994)
33. P.S. Mohanty, A. Kar, J. Mazumder, J. Laser Appl. **8**, 291 (1996)
34. S.V. Kayukov, A.A. Gusev, Quan. Elect. **22**, 811 (1995)
35. H.N. Bransch, D.C. Weekman, H.W. Kerr, Weld. Res. Suppl. 141s–151s, (1994)
36. A.Uglov, I. Smurov, L. Aksenov, Quantum Electronics, 18, 882 (1991).
37. L.I. Antonova, G.G. Gladush, A.F. Glova, S.V. Drobyazko, A.G. Krasyukov, V.S. Mainashev, V.K. Rerikh, M.D. Taran, Quantum Electronics **41**(5), 453 (2011)

Chapter 9
Interaction of Repetitively Pulsed Laser Radiation with Materials

Abstract The characteristics of the basic types of repetitively pulsed laser processing of materials – hardening of metals, welding and cutting of metals by tightly focused radiation with different off-duty ratios, and dividing remote cutting of metals by CO_2 lasers, are considered. The theoretical and numerical models of these processes are presented which qualitatively describe the basic properties and peculiarities of repetitively pulsed laser material processing. The instability of the deep melting of a metal by repetitively pulsed CO_2 laser radiation caused by the radiation channeling is described.

The interaction of a train of laser pulses with materials was already partially considered in previous chapters. In this chapter, we consider the features of this interaction in more details. Obviously, apart from a simple summation of the results of action of each pulse on a target, these results can affect each other. This concerns first of all their thermal influence. This influence is manifested in all the types of laser material processing: welding, cutting, drilling, surface hardening, etc. Repetitively pulsed radiation is characterized by the greater number of parameters than cw radiation. The additional parameters are the pulse repetition rate, the pulse duration, shape, and energy. This allows a more flexible control of laser material processing. Repetitively pulsed laser irradiation can be considered as the most general case of laser irradiation. Indeed, in the case of high pulse repetition rates, the repetitively pulsed regime passes to the quasi-continuous regime, while in the case of low pulse repetition rates, when the mutual influence of laser pulses in insignificant, this regime passes to the pulsed regime.

A great number of factors affecting the final result of repetitively pulsed laser irradiation and a great variety of relevant physical processes complicate a detailed analysis. Indeed, even in the case of a simplified approach taking into account only conductive heat transfer, the action of repetitively pulsed radiation from a movable heating source should be analyzed by using a three-dimensional nonstationary model.

G.G. Gladush and I. Smurov, *Physics of Laser Materials Processing*,
Springer Series in Materials Science 146, DOI 10.1007/978-3-642-19831-1_9,
© Springer-Verlag Berlin Heidelberg 2011

Repetitively pulsed lasers available at present allow the variation of the pulse duration from 0.1 to 10 ms (Nd:YAG laser) and from 0.01 to 500 ms (CO_2 laser). The pulse repetition rate can be varied from a few hertz to tens kilohertz, and the average power P can achieve tens of kilowatts. Note that conventional cw lasers can operate, as a rule, in the repetitively pulsed regime as well, but only at relatively long pulse durations (longer than 50 ms for Nd:YAG lasers and longer than 10 ms for CO_2 lasers). We begin to consider the action of periodically pulsed radiation with the simplest case, when the geometry of a sample being processed does not change (for example, during metal hardening).

9.1 Modeling of Thermal Processes During Repetitively Pulsed Irradiation of a Sample Surface

In most papers devoted to the investigation of thermal processes initiated by concentrated energy fluxes, either pulsed or cw irradiation regime was considered [1–3]. The properties of the repetitively pulsed regime of irradiation have been analyzed only in a few papers [4–7]. This is explained to a great extent by difficulties encountered in experimental (locality and short duration of effects) and theoretical (modeling of not only heating and melting, but also cooling and hardening) studies of these phenomena. Because of this, only heat conduction processes were taken into account in calculations, as a rule, whereas the change in the phase state and the related motion of phase boundaries were not considered in detail.

In the case of irradiation by intense energy fluxes, the evaporation process should be taken into account along with melting. Thus, it is necessary to analyze in the general form the motion of two phase boundaries of melting and evaporation. Note that the melting front velocity can take both positive and negative values, the latter corresponding to hardening.

The mathematical model analyzed below [8] describes the processes of heating, melting, evaporation, cooling and hardening caused by irradiation of a metal plate by an energy flow changing periodically in time. This model is similar to the mathematical model already discussed in Chap. 1. The model takes into account radiative and convective thermal losses on both sides of the plate. The melting process is simulated by the classical Stefan condition, while the evaporation process is simulated by using the Hertz–Knudsen law (see (8.33)).

Figure 9.1 [8] presents the results of calculations performed for a 1-mm thick steel plate, which is a massive body from the thermal point of view for irradiation regimes considered. The model neglects the temperature dependences of the thermal properties and absorption in the metal, the possible deformation of the melt surface by the vapor recoil pressure, and the interaction of the energy flux with the destruction products. The aim of this section is to analyze the dynamics of thermal processes depending on the energy parameters of repetitively pulsed irradiation.

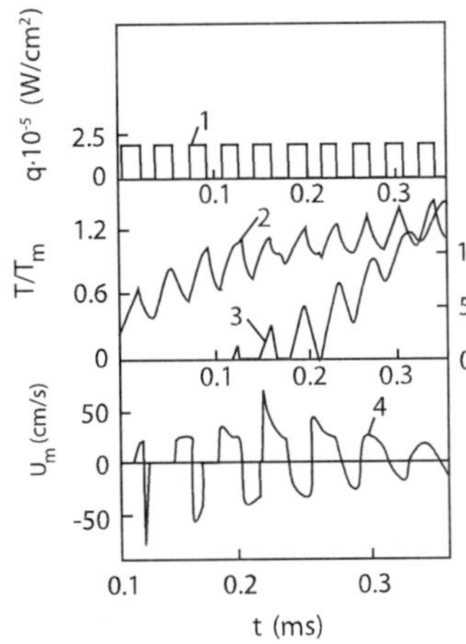

Fig. 9.1 Time dependences of the energy flux density (**1**), surface temperature (**2**), melt layer thickness (**3**), and melting (solidification) (**4**) ($q_0 = 2.5 \cdot 10^5$ W/cm^2)

9.1.1 Features of Thermal Processes and Phase Transitions During Repetitively Pulsed Laser Irradiation

Figure 9.1 presents the time dependences of the temperature of the heated surface, the melt layer thickness, and the melting (hardening) speed upon irradiation by ten rectangular energy pulses (with the constant flux density $q_0 = 2.5 \times 10^5$ W/cm^2) separated by time intervals equal to the pulse duration (the off-duty ratio is 2). The oscillations of the temperature of the heated surface correspond qualitatively to the energy flux structure. The maximum temperature is achieved at the moments of the end of energy pulses. Regular temperature oscillations are violated by hardening processes, which are manifested in the form of characteristic "ledges" – the horizontal parts of the temperature curve with values virtually equal to the melting temperature.

A specific feature of hardening processes is that first the overheating of the melt above the melting temperature almost disappears and then a drastic decrease in its thickness begins. In the general case, the amplitude of surface temperature oscillations decreases with increasing the energy flux density and the irradiation time Fig. 9.2. This is caused, on the one hand, by the increase in the minimal values of temperature achieved during the oscillation period due to the accumulation of heat near the heated surface and, on the other hand, by the ceasing of the increase in the maximum values of temperature due to the development of evaporation.

An important factor in the simulation of thermal processes induced in material by repetitively pulsed energy fluxes is the existence of the regime of established oscillations (quasi-stationary solution). The one-dimensional mathematical models of heating and melting predict the infinite growth of the material surface temperature, and only the consideration of surface evaporation gives the quasi-stationary solution [9].

A characteristic feature of the established oscillation regime induced in materials by a repetitively pulsed energy flux is the oscillation of all physical quantities at the

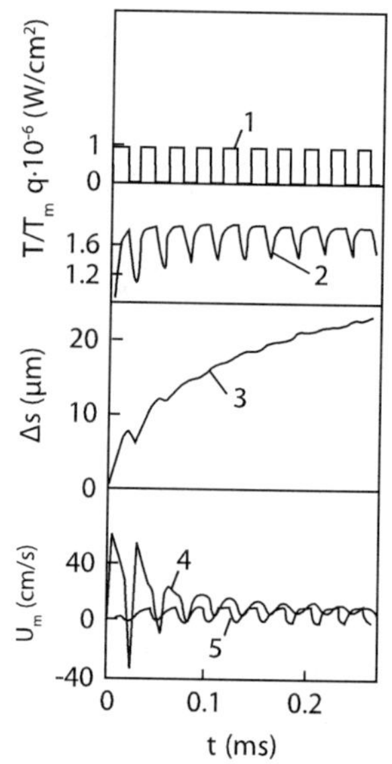

Fig. 9.2 Time dependences of the energy flux density (**1**), surface temperature (**2**), melt layer thickness (**3**), and melting (solidification) (**4**) and vaporization (**5**) front velocities for more laser pulse intensities

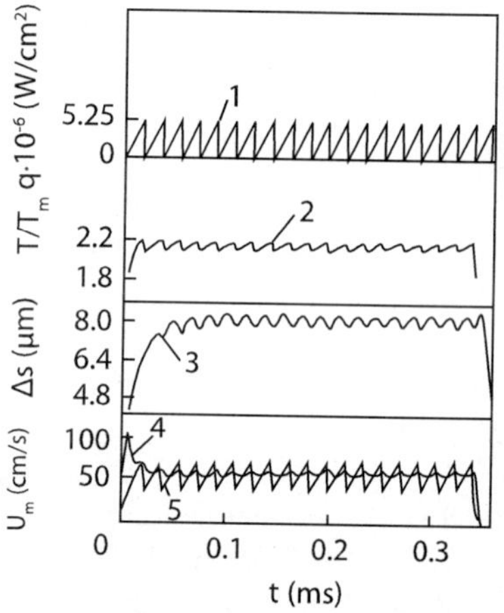

Fig. 9.3 Time dependences of the energy flux density (**1**), surface temperature (**2**), melt layer thickness (**3**), and melting (solidification) (**4**) and vaporization (**5**) front velocities for more laser pulse intensities and irradiation by triangle pulses

same frequency corresponding to the oscillation frequency of the energy flux, the average (quasi-stationary) quantities being invariable Fig. 9.3. The equality of the average evaporation and melting speeds provides a constant (on average) melt layer thickness.

After consideration of the general properties of the melting zone dynamics, we will analyze the dependence of this dynamics on the parameters of repetitively pulsed radiation, in particular, on the laser pulse shape. Let us compare the action of a train of rectangular pulses in Figs. 9.1, 9.2 with the action of triangle pulses in Fig. 9.4 [10]. One can see from Figs. 9.5 and 9.6 [8] that this action at the initial stage of a pulse train is quite considerable. This is confirmed experimentally (Fig. 9.4 [10]). The amount of energy in the surface area at the initial stage noticeably changes during one pulse.

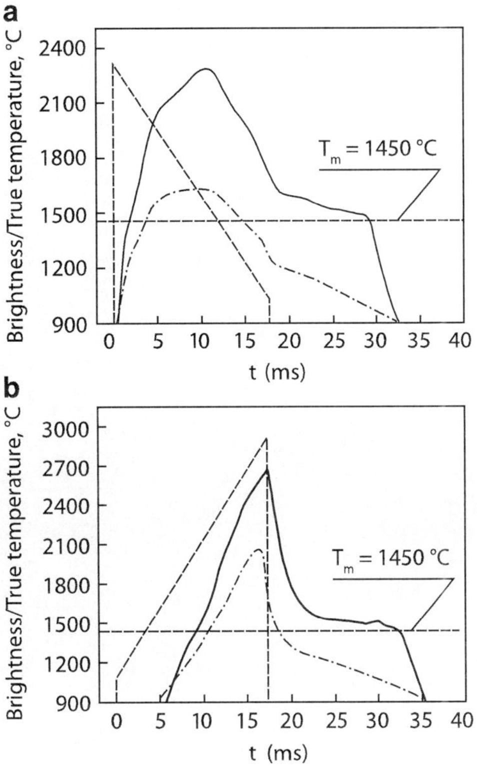

Fig. 9.4 Temperature evolution in the case of triangle $\tau = 18\,\text{ms}$, $E = 30\,\text{J}$ laser pulses. (*dotted curve*) pulse shape with decreasing (**a**) and increasing (**b**) energy flux densities, (*dot-and-dash curve*) brightness temperature, (*solid curve*) real temperature

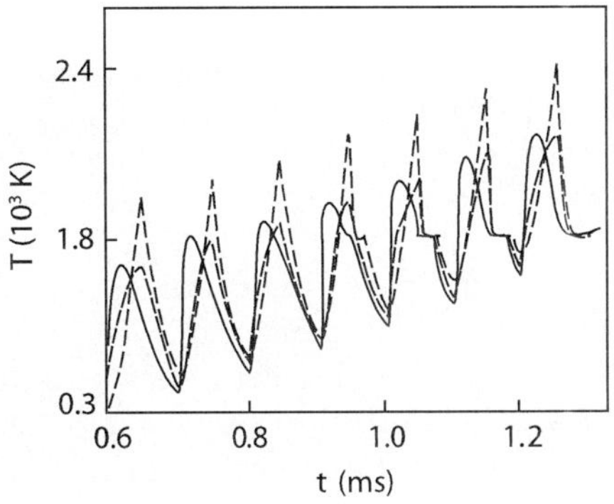

Fig. 9.5 Time dependences of the heated surface temperature. The *solid and dot-and-dash curves* correspond to irradiation by triangle pulses with decreasing and increasing energy flux densities, respectively ($q_{\max} = 2 \cdot 10^5\,\text{W/cm}^2$); the *dotted curve* corresponds to irradiation by rectangular pulses ($q_{\max} = 10^5\,\text{W/cm}^2$)

The oscillations of the melt thickness become close to each other with time (Fig. 9.7 [8]). At this stage, due to the increase in the melt thickness the heat content of the melt will be comparable with the laser pulse energy, and the time of heat transfer to the melting front also increases. All this leads to the averaging of temperature oscillations near the melting front. Therefore, the shape of laser pulses does not affect noticeably the melting front dynamics. As time further increases, the heat content of the melt becomes much higher than the pulse energy. Clearly, in this

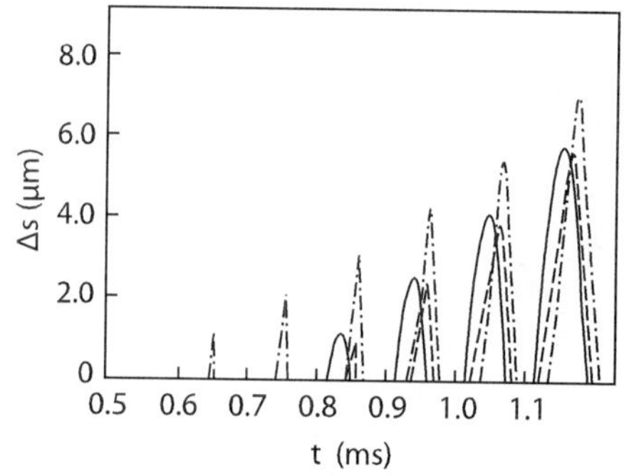

Fig. 9.6 The begining of time dependences of the melt layer thickness. The *solid and dot-and-dash curves* correspond to irradiation by triangle pulses with decreasing and increasing energy flux densities, respectively ($q_{max} = 2 \cdot 10^5$ W/cm^2); the *dotted curve* corresponds to irradiation by rectangular pulses ($q_{max} = 10^5$ W/cm^2)

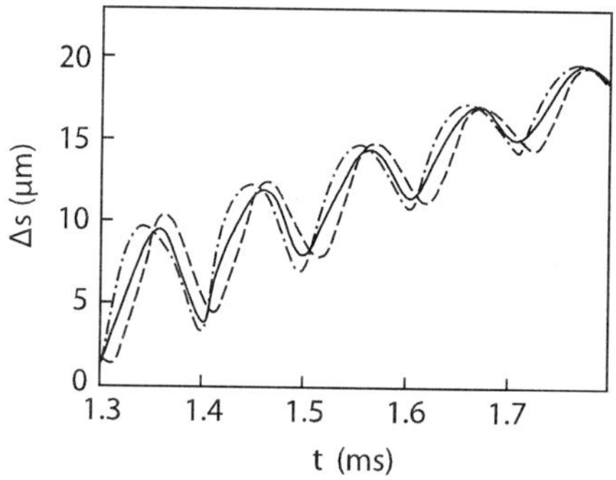

Fig. 9.7 Time dependences of the melt layer thickness. The *solid and dot-and-dash curves* correspond to irradiation by triangle pulses with decreasing and increasing energy flux densities, respectively ($q_{max} = 10^5$ W/cm^2); the *dotted curve* corresponds to irradiation by rectangular pulses ($q_{max} = 2.5 \cdot 10^4$ W/cm^2)

case the melting dynamics will be determined by the average pulse intensity rather than by the pulse shape and its energy. Nevertheless, some influence of the pulse shape is preserved. Indeed, because a triangle pulse with the increasing intensity provides more efficient evaporation, the melting zone thickness will be the smallest in this case due to these energy losses (Fig. 9.8 [8]). For the same reason, the melting zone will be maximal during cw irradiation with power equal to the average power of repetitively pulsed radiation (Fig. 9.9 [8]).

The picture of the dynamics of melting and evaporation phase transitions considered above can be useful for analysis of experiments on melting metal surfaces during repetitively pulsed laser evaporation. Such melting is required for the welding of cracks, the removal of surface tensions, and alloying. Thermal cycling can considerably change the dynamics of structural phase transitions, while the short-time evaporation of a surface can be used for its purification.

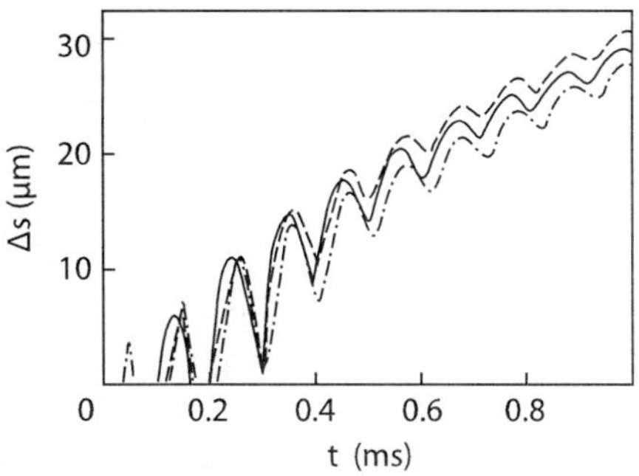

Fig. 9.8 Time dependences of the melt layer thickness. The *solid and dot-and-dash curves* correspond to irradiation by triangle pulses with decreasing and increasing energy flux densities, respectively ($q_{\max} = 5 \cdot 10^5$ W/cm^2); the *dotted curve* corresponds to irradiation by rectangular pulses ($q_{\max} = 2.5 \cdot 10^5$ W/cm^2)

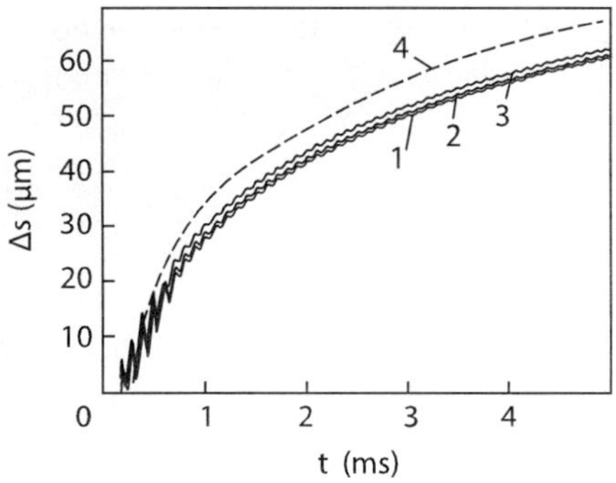

Fig. 9.9 Time dependences of the melt layer thickness upon irradiation by triangle pulses with increasing (**1**) and decreasing (**2**) energy flux densities ($q_{\max} = 5 \cdot 10^5$ W/cm^2), by rectangular pulses ($q_{\max} = 2.5 \cdot 10^5$ W/cm^2) (**3**) and by cw energy flux density ($q_{\max} = 1.25 \cdot 10^5$ W/cm^2) (**4**)

As an example important for practical applications of repetitively pulsed laser radiation, we consider the hardening of steels. We restrict our consideration to a simple approach taking into account only structural phase transitions, as in the study of the action of cw laser radiation (Chap. 2).

9.1.2 Thermal Model of Metal-Surface Hardening by Repetitively Pulsed Laser Radiation

Consider hardening under conditions that most strongly differ from the case of irradiation by a cw laser by the example of a repetitively pulsed 10.6-m CO_2 laser emitting pulses with the high off-duty ratio [11].

Physical processes accompanying repetitively pulsed laser processing of metals under such conditions considerably differ from processes proceeding during cw laser processing. First, this is explained by the fact that even for comparatively low average radiation power each of the pulses produces an optical breakdown near the surface and leads to the formation of a plasma plume. The latter can considerably affect the energetics of interaction of the laser beam with a target (see Chap. 7). Second, an important feature of repetitively pulsed laser processing is oscillations of the surface layer temperature with the amplitude achieving 1,500–2,000 C at the oscillation frequency of $\sim$100 Hz. Such a thermal regime can considerably change the dynamics of structural phase transitions in the processing zone. Similar phenomena were pointed out earlier in the case of usual thermal cycling of metals [12]. By using repetitively pulsed laser radiation, it is possible to control more flexibly structural transformations in the surface layer, by changing mechanical or chemical properties of materials in the desired direction. Integration in (2.3) was performed in three time intervals of target irradiation (see Chap. 2). As in the case of a cw laser, we will pass in integral (2.3) to the coordinate system moving with a sample. In this case, not one temperature oscillation will be observed at the observation point, as in Fig. 2.6, but $(d/v)f$ oscillations. We perform calculations for conditions close to experiments with steel 45. The thermal constants of this steel change weakly, which allowed us to use integral (2.3) for calculating temperature fields. We assume that $\kappa = 0.33$ W/cmK and $\chi = 0.06\,\text{cm}^2/\text{s}$ [13]. The absorption coefficient α is taken from experiment. The measured values of α are presented in Fig. 9.10. One can see that the absorption coefficient under our conditions considerably exceeds tabulated values $\alpha = 0.1$ (see Chap. 1). As mentioned above, this can be explained by plasma processes because experiments were always accompanied by the optical breakdown. These phenomena are discussed in more detail in Chap. 7.

Our calculations showed that the temperature oscillation amplitude drastically decreases with depth. The characteristic scale at which these oscillations are noticeable is $\delta \infty 2\sqrt{\chi\tau}$. Figure 9.11 presents temperature profiles over the hardening depth at the laser spot center at different instants of time ($v = 2$ mm/s), beginning from the moment of switching on of a successive pulse. As expected, the maximum of temperature on the surface is achieved by the pulse end. But

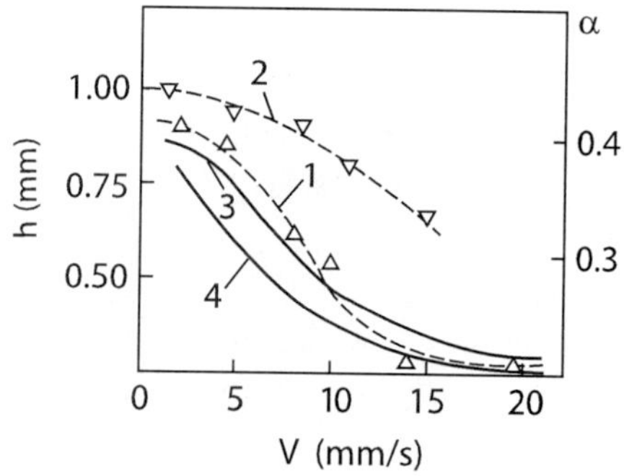

Fig. 9.10 Dependences of the hardening depth (**1**) and absorption coefficient (**2**) on the target travel speed [(**1**) experiment, (**3, 4**) calculation]; $E = 4.2$ J, $f = 200$ Hz. Calculations for repetitively pulsed radiation (**3**) and cw laser radiation of power $\overline{P} = Ef = 840$ W; $\alpha = 0.35$

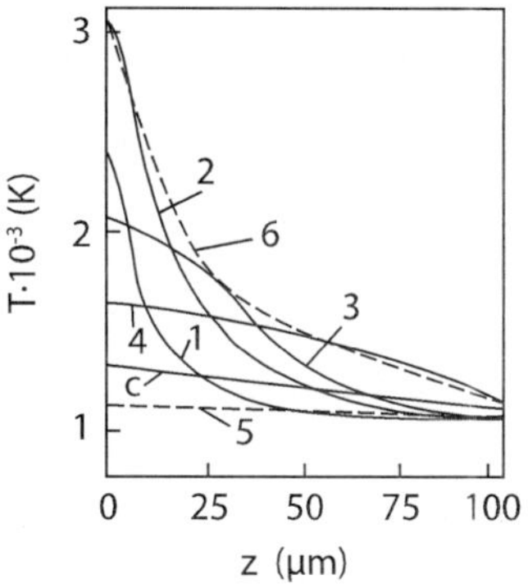

Fig. 9.11 Temperature distribution over depth at different instants of time ($x = y = 0$); $\alpha \overline{P} = 300$ W; $f = 200$ Hz, $v = 2$ mm/s): (**1**) $t_0 + 0.5\tau$; (**2**) $t_0 + \tau$; (**3**) $t_0 + 1.5\tau$; (**4**) $t_0 + 3\tau$. The instant t_0 corresponds to the onset of the next pulse. (**5**) $T_{\min}$; (**6**) $T_{\max}$; (C) calculation for cw laser radiation

the maximum temperature in the depth is achieved later, at different instants for different points. These curves are generalized by envelope curve 6 in Fig. 9.11 describing the distribution of the maximal temperature over the hardening depth. Curve 5 describing the dependence of the minimal temperature on the hardening depth is constructed similarly. The similar temperature profile obtained for cw laser radiation of power equal to the average power of a repetitively pulsed laser lies somewhat higher. The calculation of the hardened layer thickness h_h consists in the search for the maximum depth at which temperature T_γ is achieved. As shown in Chap. 2, the austenization temperature increases with increasing the heating rate and can considerably exceed the equilibrium value equal to 745°C. Because the relevant data are absent for the given experimental conditions, the choice of T_γ is arbitrary. We assumed in calculations that $T_\gamma = 800$°C.

For conditions presented in Fig. 9.11, $T_\gamma < \overline{T}_s$ ($\overline{T}_s$ is everage surface temperature), the type of hardening by repetitively pulse radiation will be close to cw radiation processing. We can use expression (2.5), where the intensity q should be replaced by the average value. As temperatures $T_{\max}$ is decreased and $T_\gamma > \overline{T}_s$, it should be taken into account that temperature oscillates only in a layer of depth $\sim\delta$. In this case, temperature decreases approximately linearly with the hardening depth:

$$T(z) \approx \overline{T}_s + \frac{2q\sqrt{\chi\tau}}{\kappa}\left(1 - \frac{z}{2\sqrt{\chi\tau}}\right) \tag{9.1}$$

Taking into account that $\overline{T}_s = 8\overline{P}\sqrt{\chi d/v}(\kappa\pi d^2)^{-1}$ and equating temperature (9.1) to T_γ, we can obtain the hardening depth

$$\frac{h_h}{2\sqrt{\chi\tau}} = 1 - \frac{\kappa T_\gamma}{2q\sqrt{\chi\tau}} + \sqrt{\frac{\tau f^2 d}{v}} \tag{9.2}$$

Expression (9.2) gives the lower bound of the hardening depth. To calculate the hardening depth more accurately, it is necessary to use the envelope in Fig. 9.11,

Fig. 9.12 Dependence of the hardening depth (**1**) on the pulse repetition rate for a target at rest for the number of pulses $N = 100$; (**2**) calculation, $\alpha = 0.35$, $E = 4.2\,\mathrm{J}$

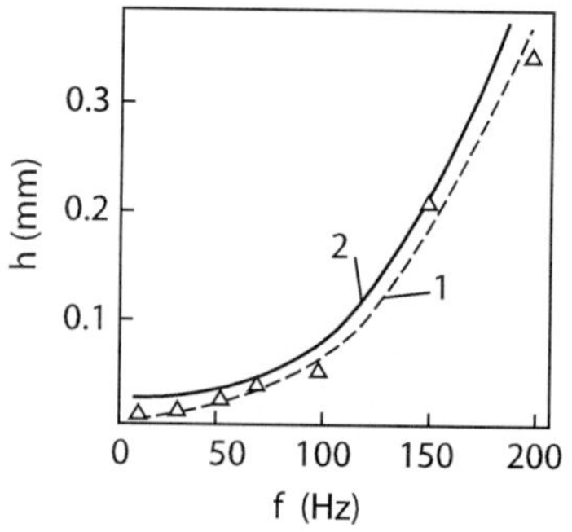

which can be got numerically. Numerical calculations are also necessary in the case of small Peclet's numbers because all analytic methods are valid only for $Pe \gg 1$.

Figure 9.10 presents numerical calculations of the hardening depth as a function of the sample travel speed. One can see that for small velocities the difference between repetitively pulsed and cw radiation is small. In this case, temperatures $\overline{T}_s$ and $T_{s\,\min}$ are high, exceeding T_γ. This explains a small difference between hardening depths. At high velocities, when $\overline{T}_s$ and $T_{s\,\min}$ are small, the difference between hardening depths increases. As a whole, the results of calculations are consistent with experimental data. Figure 9.12 demonstrates hardening depths of an immobile target irradiated by the specified number of pulses as a function of the pulse repetition rate and compares with experimental data. The increase in the hardening depth with the pulse repetition rate is explained by the accumulation of heat in the processed zone at the high pulse repetition rate. Thus, we see that the thermal model, as in the case of cw lasers, well describes the hardening of metals taking into account the efficient absorption coefficient.

9.2 Thermal Model of Deep Melting of Metals by Repetitively Pulsed Laser Radiation with Low Off-Duty Ratio

9.2.1 Thermal Model of Deep Melting of Moving Samples by Repetitively Pulsed Radiation

We considered above the problems of heating of a plane surface by a broad beam from a repetitively pulsed laser. A one-dimensional nonstationary problem was solved taking into account two phase transitions: melting-crystallization and evaporation (see Sect. 9.1.1). In the case of repetitively pulsed welding and cutting, the thermal problem is different. It is necessary to calculate the three-dimensional propagation of heat from a source penetrating inside a sample. As in the case of cw radiation (Chap. 4), we will use the model approach by replacing a laser beam by a linear heat source intersecting a finite-thickness plate normally to its surface.

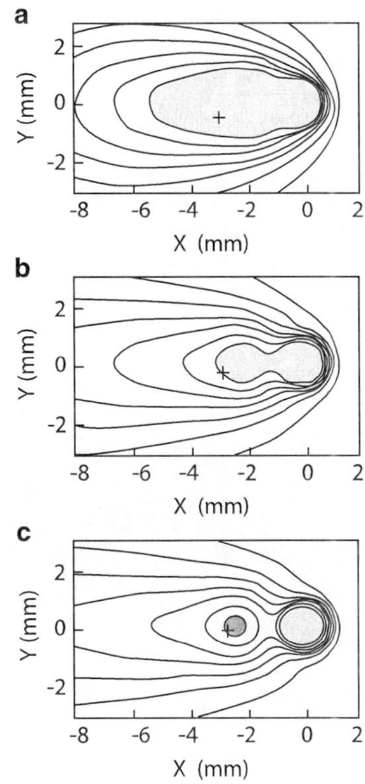

Fig. 9.13 Distribution of temperature oscillations established within 10 ms after the laser pulse end. The isotherms are drawn through 200 K. The melting zone is hatched. The plate travel speed is 3 (**a**), 4 (**b**), and 5 cm/s (**c**)

Table 9.1 Parameters of numerical calculations in Fig. 9.13

	Speed, cm/s	Power, kW	Pulse duration, ms	Energy, J	Period, ms
A	3	3.5	5	17.5	40
B	4	3.0	5	15.0	40
C	5	3.0	5	15.0	40

For simplicity, the phase transition heat was neglected. In this case, the problem becomes linear, and periodic solutions can be found by the superposition method [14]. Figure 9.13 presents the results of such calculations for repetitively pulsed radiation with the low off-duty ratio $\sim$8. The propagation of heat was considered in a 0.5 mm thick steel plate. The irradiation conditions are presented in Table 9.1.

One can see that the melting zone decreases with increasing the plate travel speed. In this respect the picture is similar to the case of cw radiation (see Chap. 4). When the travel speed is minimal, the shape of the melting zone is similar to that in the case of cw radiation. In this case, the target is displaced during the time between pulses by a distance that is approximately three times smaller than the melt pool length. As the travel speed is increased and the pulse power is decreased, the pulsed nature of the thermal source begins to be manifested. For $v = 4$ cm/s, the melting zone has a dumbbell-like shape. As the travel speed is further increased

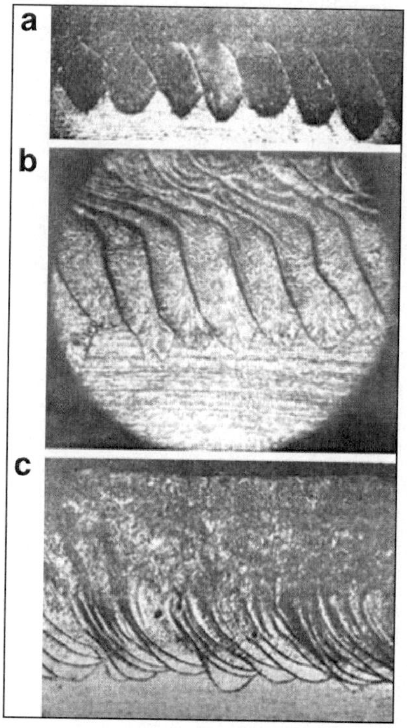

Fig. 9.14 Longitudinal section of a weld produced upon repetitively pulsed welding: $v = 1$ cm/s, $\tau = 10$ ms, $P_p = 2.5$ kW, 18-10 steel, the focal distance of a lens is 17 cm, the beam aperture is $D = 2$ cm, $Q = 7$ (**a**), 4 (**b**), 2 (**c**) (photograph by V.A. Fromm)

up to 5 cm/s, as expected, the melt pool consists of individual puddles. During a pause, the target travels a distance exceeding the melt pool length. In this case, a weld will be nonuniform over its depth or even not continuous. The experiment with insufficiently overlapped melting zones is presented in Fig. 9.14 ($Q = 7$).

Let us compare thermal models for repetitively pulsed and cw sources. By substituting data from Fig. 9.13a in the plot of the Swift–Hook–Jick model (Fig. 4.19), we see that the width of the melt zone for the repetitively pulsed source is close to the weld width in the cw regime.

9.2.2 *Thermal Model of Metal Welding with a Pulsed Laser with Low Off-Duty Ratio*

However, as we saw by considering laser welding, the deep melting mechanism includes not only the heat propagation process but should also provide the maintenance of a gas–vapor channel. At low frequencies, when the next pulse does not enter the gas–vapor channel produced by the previous pulse, the laser beam should produce such a channel anew. But the beam penetrates inside a heated or even melted metal. At a distance from the focal spot center exceeding the travel length of a thermal wave during the time between two pulses, the nearly stationary temperature distribution will be established. This distribution coincides with the temperature distribution produced by a thermal source in the form of a heated filament (similarly to the case of cw welding (4.21):

$$T_f = \frac{\overline{P}}{2\pi\kappa h} \ln \frac{2\chi/v}{r_f + \sqrt{4\chi/f}} \tag{9.3}$$

In (8.61), T_b should be replaced by $T_b - T_f$ because a keyhole should be heated up to the boiling temperature from T_f rather than from room temperature. By using equality (9.3), we obtain from (8.61)

$$h = \frac{1}{2}\frac{q\tau}{c\rho T_b} + \frac{P_i}{2\pi\kappa T_b Q} \ln \frac{2\chi/v}{r_f + \sqrt{4\chi/f}} \tag{9.4}$$

where P_i and $\overline{P} = P_i/Q$ are the pulsed and average powers, respectively. By comparing (9.4) and (8.61), we see that the keyhole depth increases linearly with increasing pulse repetition rate or average power. It follows from (9.4) that repetitively pulsed laser welding is more efficient than cw laser welding at the same average power. Indeed, expression (9.4) contains the additional term compared to expression (4.21) corresponding to cw welding regime. According to estimates, the first and second terms in (9.4) can be of the same order of magnitude (Fig. 9.15). In reality, at low cw welding powers the welding depth is considerably smaller than its theoretical value (see curves 4 and 5 in Fig. 9.15). Therefore, at low welding powers the advantages of repetitively pulsed laser welding compared to cw welding are more substantial in experiments than predicted by theory. One can see from Fig. 9.15 that the experimental dependence of the welding depth on the average power is also

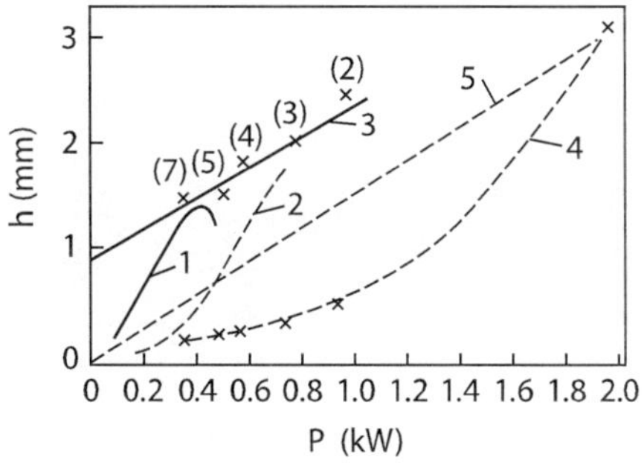

Fig. 9.15 Dependence of the weld depth on the average power of a CO_2 laser. The *dotted and solid curves* correspond to cw and repetitively pulsed laser welding, respectively; (**1, 2**) $v = 1.2$ cm/s, 304 steel [15]; (**3, 4**) $v = 1$ cm/s, 18-10 steel [15]; (**5**) calculation by (4.21); (**1**) $\tau = 4$ ms; (**3**) $\tau = 10$ ms; in parentheses is given the value of Q

linear. The slope of the straight lines to the x axis in experiments is 3.4 mm/kW, while according to (9.4), the slope for steel for $P_i = 24$ kW is 4.3 mm/kW.

Figure 9.15 demonstrates another advantage of repetitively pulsed laser welding compared to cw welding. The average power increases due to the increase in the pulse power at a constant off-duty ration $Q = 2.5$ [15]. One can see from (9.4) that the dependence of the weld depth on the pulsed power is linear. According to (9.4), the theoretical melting ability h/P for parameters from [15] is 3.4 mm/kW (the spot area was determined from the melting area [15]). The maximum value of h/P, found from experiments, is 3.8 mm/kW. It is interesting to study the dependence of h/P on Q. As the off-duty ratio is increased from $Q = 1$ (cw regime), the melting ability rapidly increases and should be determined in the regime under study, according to (9.4), by the expression

$$\frac{h}{\overline{P}} = \frac{T}{c\rho T_b S} + \frac{1}{2\pi\kappa T_b} \ln \frac{2\chi/v}{r_f + \sqrt{4\chi/f}}. \tag{9.5}$$

Expression (9.5) shows that, if Q is changed by changing period T, then $h/\overline{P} \approx Q$, while if Q is increased by decreasing the pulse duration, the ratio $h/\overline{P}$ should tend to a constant value with increasing Q. Indeed, Fig. 9.16 shows that $h/\overline{P}$ is proportional to the laser pulse repletion period [for $T \to 0$, according to Fig. 9.16, $h/\overline{P} = 1.8$ mm/kW, which coincides with theoretical value (9.5)]. Figure 9.16 also shows that the melting ability of repetitively pulsed laser radiation is independent of the pulse duration (the experimental data were obtained by maintaining constant either pulsed power or average power). This is also consistent with (9.5). However, curves 1, 2, and 3 show that $h/\overline{P}$ increases with decreasing τ. This can be explained by the low average and peak powers of laser radiation, when expressions (8.61) and (9.4) are invalid. Curve 3 was obtained for a low welding speed equal to 1 mm/s. It is possible that the welding mechanism changes at such low welding speeds and a channel is no longer collapses in the interval between pulses. In this case,

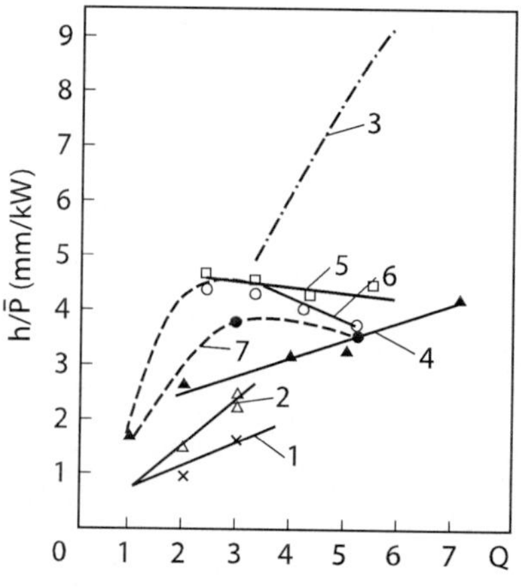

Fig. 9.16 Dependence of the specific melting on the radiation off-duty ratio: (**1**) $T = 10$ ms; (**2**) $T = 40$ ms, $\overline{P} = 0.7$ kW; $v = 1$ cm/s; (**3**) $T = 5.5$ ms; $v = 1$ mm/s, $P_i = 9$ kW [15]; (**4**) $\tau = 10$ ms, $v = 10$ mm/s; $P_i = 2.4$ kW; (**5**) $T = 30$ ms, $v = 2$ mm/s; $\overline{P} = 1.6$ kW; (**6**) $T = 30$ ms; $v = 4$ mm/s; $\overline{P} = 1.6$ kW; (**7**) $T = 5.5$ ms; $v = 4$ mm/s; $P_i = 9$ kW

to maintain the channel, the keyhole surface temperature should achieve only the boiling temperature during each pulse. Then the channel depth will increase with decreasing pulse duration, according to (9.13). Such a mechanism is typical for repetitively pulsed radiation with large off-duty ratios Q =10–100 (see Sect. 6.2). Generally speaking, low weld speeds $v \leq 1$ mm/s are not typical for cw laser welding, a noticeable part of a metal being removed from the weld at such speeds. We see that upon welding by single pulses at low weld speeds, the weld depth is considerably greater than upon welding at high speeds (cf. curves 7 and 3 in Fig. 9.16). The specific power proves to be small and no material ejection occurs. Figure 9.14 shows how with decreasing Q during welding by single pulses, the regime, in which the action of pulses on a target can be considered independently, passes to the regime in which a strong influence of a previous pulse is manifested by the arrival of the next pulse, i.e. the liquid has not time to harden during the interval between pulses. One can see that this process begins from the upper layers of the target where the average temperature of the metal higher.

The formation and collapse dynamics of a gas–vapor keyhole during the repetitively pulsed laser welding of stainless steel was studied in detail in special experiments [16]. We have already mentioned these experiments in Chap. 3 by considering plasma phenomena during metal welding. The scheme of experiments is presented in Fig. 9.17. The shape of a laser pulse was changed by shifting the auxiliary repetitively pulsed radiation from a Nd laser, as shown in Figs. 9.18 and 9.19. Figure 9.20 shows that the appearance of a gas–vapor keyhole is delayed with respect to the laser pulse. This delay in these figures is 0.7 ms. The keyhole growth rate estimated from these figures is 3–5 mm/s. The filling time of the keyhole after the laser pulse end can be estimated from the same figures. In experimental conditions under study, this time is ~0.4 ms. The estimate from expression (4.66) in Chap. 4 for the uniform collapse of a liquid cylinder gives the time ~0.5 ms, which is close to the experimental value. Figure 3.21 also shows that the keyhole is in the quasi-stationary regime and its depth corresponds to the shape pulse. This is clearly seen in Figs. 9.19 and 9.20. The maximum depth corresponds to the maximum instant power at thus time. Experiments also show that, if the laser power maximum corresponds to the pulse onset, this leads to strong oscillations of the keyhole. On the contrary, if the laser power maximum corresponds to the pulse end, the keyhole collapses rapidly and nonuniformly (see Fig. 9.21). This figure well illustrates the formation dynamics of large pores during the steel welding by repetitively pulsed radiation with a low off-duty ratio.

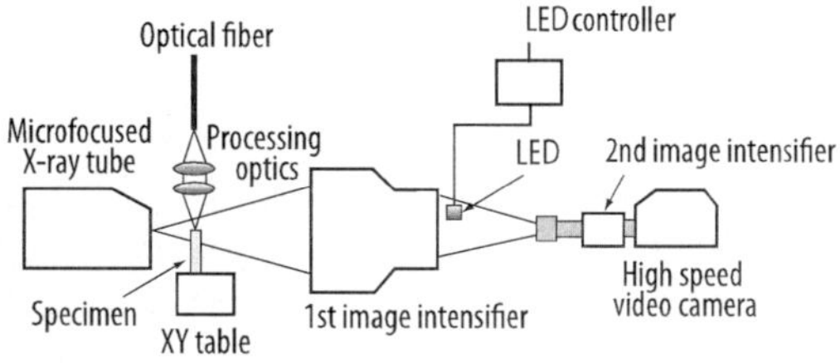

Fig. 9.17 Scheme of the experiment

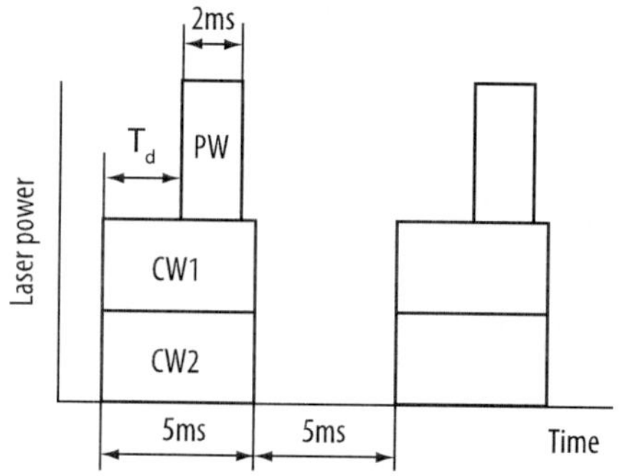

Fig. 9.18 Laser pulse shape obtained by the superposition of two rectangular laser pulses separated by the delay time T_d

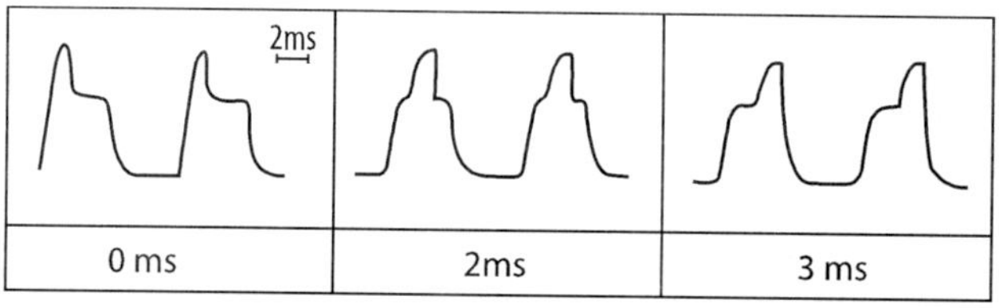

Fig. 9.19 Laser beam shapes measured for different time delays T_d

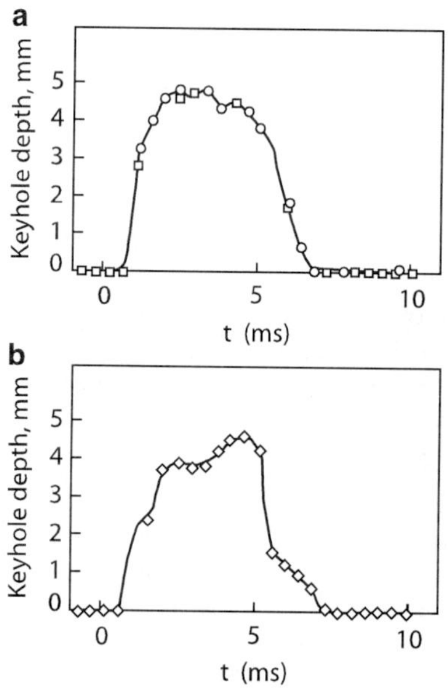

Fig. 9.20 Changes in the gas–vapor keyhole depth during irradiation by laser pulses of different shapes: $T_d = 0.2$ ms (**a**) and $T_d = 3$ ms (**b**)

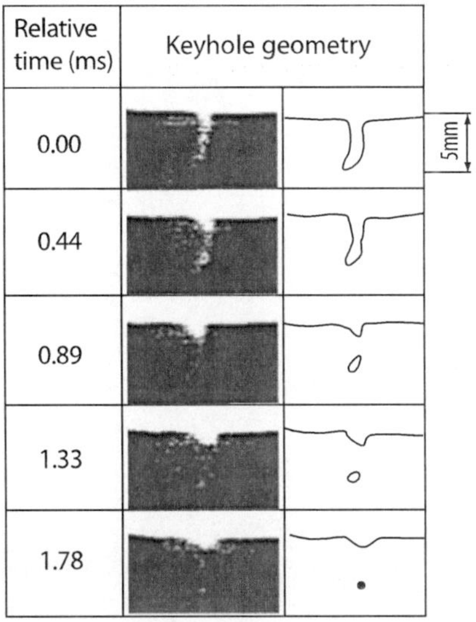

Fig. 9.21 Change in the gas-vapor keyhole shape during the formation of voids at the time delay $T_d = 3$ ms

9.3 Physical Processes During Welding of Metals by Repetitively Pulsed Laser Radiation with High Off-Duty Ratio

We considered in Sect. 9.2 the welding of metals by repetitively pulsed radiation under conditions when each light pulse produces deep melting in the metal. Consider now radiation with a higher pulse repetition rate $f \sim 300$ Hz and shorter pulses of duration $\tau \sim 100\,\mu$s. Such parameters are typical for $\sim$1 kW CO_2 lasers [17]. Under the action of such radiation on relatively slowly moving sample, the welding regime is produced. However, if the sample is moved rapidly enough, the welding regime passes to the cutting regime [18]. During welding by an electron-beam-controlled CO_2 laser with a high off-duty ratio $Q \approx 20$ the cutting regime was produced [19]. It seems that the welding mechanism at such radiation parameters differs from the mechanism of welding by single pulses considered in Sect. 9.1 (when each of the pulses incident on a sample produces a channel anew).

9.3.1 Theoretical Model

In the case that we will consider, the mechanism of the deep penetration of radiation into a sample is different: the channel, as in the case of continuous welding, does not collapse during the interval between pulses, existing till the arrival of the next pulse [20,21]. Let us assume that the gas–vapor channel in a metal has the form of a cone with the rear generatrix parallel to the incident light beam and the front generatrix tilted with respect to this light beam Fig. 9.22. The light beam incident on the front

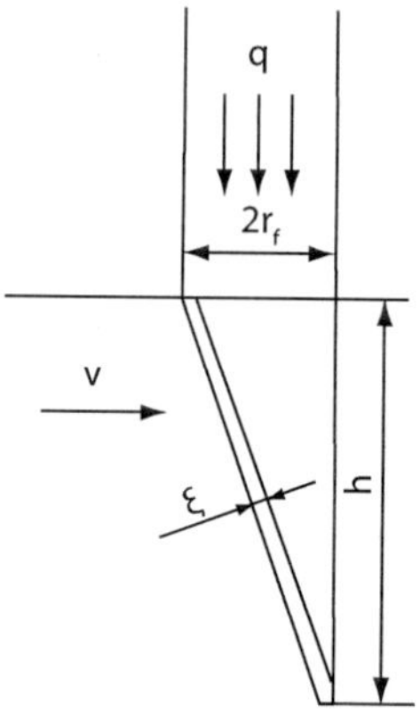

Fig. 9.22 Schematic view of a keyhole in the case of repetitively pulsed welding

wall of the keyhole heats and melts the metal during the pulse time by a certain depth. When the surface temperature achieves the boiling point, the melted metal layer from the front wall is driven over by vapor pressure to the rear wall, where it solidifies. Thus, the keyhole moves during a light pulse by a distance equal to the thickness of a metal layer transferred from the front to rear wall. This process is repeated during the next pulse. To estimate the amount of metal transferred from the front wall, we assume that the destruction process of the front wall is similar to the destruction of the plane surface of a metal considered in Sect. 8.1. The characteristic parameter determining the destruction rate is the specific destruction energy $X(\varepsilon)$ (see Sect. 8.1). Knowing $X(\varepsilon)$, we can determine the thickness ξ of the metal layer removed from the front to rear wall:

$$\xi = \varepsilon / X(\varepsilon)$$

where ε is the energy density absorbed by the front wall. The welding speed, i.e. the beam travel speed v on the sample surface is equal to the product of ξ on the pulse repetition rate f. Recall that $X(\varepsilon)$ is a nonmonotonic function of the energy density (see Sect. 8.1). For $\varepsilon \leq \varepsilon_b$, we have $X \to \infty$, and for $\varepsilon \geq \varepsilon_b$, we have

$$X(\varepsilon) \cong c\rho T_b. \tag{9.6}$$

Then, X increases again and for $\varepsilon \to q_1\tau$, we have $X \to L_b\rho$ [q_1 is determined from (8.10)]. In this case, material is removed from the wall in the form of vapor.

The energy density ε entering these expressions can be found by assuming that the pulse energy is uniformly distributed over the keyhole surface (and is uniformly absorbed in its walls). Such a distribution is caused by the multiple reflection of the laser beam from metal walls weakly absorbing CO_2 laser radiation. Recall that the assumption about the uniform distribution of absorbed energy on keyhole walls was used in the thermal model of continuous laser welding, which gives good results. It is clear at the same time that mass transfer from the front to rear wall can occur only if the energy incident on the front wall is greater than that incident on the rear wall. This difference is determined by many factors such as the beam focusing, the keyhole shape, etc. As follows from numerical calculations [22], under typical laser welding conditions, the elative difference k of radiation powers incident on the front and rear walls is $k \simeq \alpha \simeq 0.1$. The welding speed for $\alpha \ll 1$ should be proportional to α. Taking this into account, the expression for the welding speed can be written in the form

$$v = \alpha f \frac{E}{\pi \mathrm{d}h} X^{-1}(E/\pi \mathrm{d}h) \tag{9.7}$$

where E is the radiation pulse energy. The maximum depth $h_{\max}$ is achieved when the sample travel speed tends to zero or for $X(E/\pi \mathrm{d}h_{\max}) \to \infty$. Thus, $E/\pi \mathrm{d}h_{\max} = \varepsilon_b$, or, by using (8.21), we obtain

$$h_{\max} = \frac{E}{\pi d c \rho T_b \sqrt{\chi \tau}}. \tag{9.8}$$

As the depth h decreases $h < h_{\max}$, the speed rapidly increases and achieves the critical value v_κ, which can be found from (9.6) and (9.7) for $E/\pi \mathrm{d}h = \varepsilon_b$:

$$v_\kappa = \alpha f \sqrt{\chi \tau}. \tag{9.9}$$

For $v > v_\kappa$, the depth h drastically decreases with increasing v. This decrease in the welding depth is explained by the fact that the specific destruction energy increases with increasing the energy density, i.e. with decreasing h. When the specific destruction energy achieves the value equal to the heat of evaporation, the transfer of material will occur due to its evaporation on the front wall. The keyhole depth in this regime can be determined from the condition $\varepsilon = q_1 \tau$, which gives $h_1 = E/\pi d q_1 \tau$. The speed v_i can be found from (9.7), where we should set $E/\pi d h = q_1 \tau$ ans $X = L_b \rho$:

$$v_i = \alpha f \tau [c_s \chi^2 / r_f]^{1/3}. \tag{9.10}$$

For $v > v_i$, we have

$$h = \frac{\alpha f E}{\pi d v L_b \rho}. \tag{9.11}$$

One can see from (9.8) and (9.9) that the maximum welding depth and characteristic welding speed depend not only on the beam power but also on the pulse duration and energy.

As the pulse repetition rate f increases, the mutual influence of pulses appears well before the equality $\tau f = 1$ is achieved. Indeed, in the case of a high pulse repetition rate, the keyhole surface has no time to cool during the interval between pulses and this affects heating during the next pulse. At a distance from the keyhole equal (or greater than) $\sqrt{4\chi/f}$ the stationary temperature T_f is established. Because the speed of repetitively pulsed laser welding is usually small, we can assume that $\chi/v > h$. In this case, temperature at a distance from the beam axis equal approximately $r_f + \sqrt{4\chi/f}$ (where it is not changed with time) is determined by the beam average power $\overline{P} = \pi r_f^2 \varepsilon f$ and is equal to

$$T_f = \frac{\overline{P}}{2\pi\kappa h} \ln \frac{h}{r_f + \sqrt{4\chi/f}}, \quad (9.12)$$

[see (9.3)]. Because now the front wall of the keyhole is heated by each pulse from T_f rather than from room temperature, it is necessary to replace T_b in all expressions by $T_b - T_f$. Therefore, the maximum depth at high f will increase with increasing f. Indeed, it follows from (9.8) to (9.12) that

$$h_{\max} = \frac{E}{\pi d\sqrt{\chi\tau} c\rho T_b} + \frac{\overline{P}}{2\pi\kappa T_b} \ln \frac{h}{r_f + \sqrt{4\chi/f}} \quad (9.13)$$

The second term in (9.13) coincides with the expression for the depth of melting produced by a cw laser beam (4.21). This suggests that the melting ability of a repetitively pulsed laser at low welding speeds in higher than that of a cw laser. This is explained by the fact that during repetitively pulsed laser welding, the amount of heat delivered to an unmelted metal is smaller than during cw laser welding, and the weld proves to be narrower in the former case. Note that expression (9.13) is analogous to expression (9.4) for the repetitively pulsed laser welding depth produced by individual pulses. The first term in (9.13) is, however, greater by a factor of $r_f/\sqrt{\chi\tau}$ than the first term in (9.4), the value of $r_f/\sqrt{\chi\tau}$ achieving ~ 10 under typical welding conditions. This increase in the melting ability is explained by the fact that in the case of repetitively pulses laser welding, there is no need to produce a new hole during each pulse. At a high pulse repetition rate, the time-averaged temperature of keyhole walls can achieve the melting temperature. In this case, a metal overflowing to the rear wall will not solidify by the arrival of the next pulse. It can flow down, by filling the keyhole. Therefore, as in the case of a cw laser, the problem of removing the melt from the keyhole bottom appears. The critical pulse repetition rate f_κ can be found from the condition that keyhole depth (9.13) exceeds the depth determined from (9.12) for $T_f = T_m$:

$$f_\kappa = \frac{2}{r_f}\sqrt{\frac{\chi}{\tau}} \left(\frac{T_b}{T_m} - 1\right)^{-1} \ln^{-1} \frac{h}{r_f + \sqrt{4\chi/f}} \quad (9.14)$$

Thus, for low welding speeds $v < v_\kappa$ and $f < f_\kappa$, the melt zone width is close to the laser beam diameter. However, for speeds exceeding v_κ, when the vapor chanel depth considerably decreases, according to (9.12), temperature near the channel increases and can exceed the melting temperature. Thus, upon repetitively pulsed laser welding, the melt zone width increases with welding speed. Even at high welding speeds $v > v_i$, when h decreases with increasing welding speed not so strong, this properties remain valid. The dependence of the melt depth of the welding speed can be obtained from (9.11) to (9.12):

$$\frac{d_m}{2r_f} = \frac{\alpha f E}{2\pi r_f^2 v\rho L_b} \exp\left(-\frac{\kappa\alpha T_m}{r_f v\rho L_b}\right). \tag{9.15}$$

One can see from (9.15) that in the speed interval when $h/d_m > 1$, d_m increases with welding speed. Repetitively pulsed welding differs considerably in this respect from cw welding, where the melt zone width decreases with increasing welding speed [23], so that the aspect ratio h/d_m of welding increases (see Sect. 4.2).

The theoretical model of repetitively pulsed laser welding of metals considered above only qualitatively describes experimental data due to the complexity of the welding process.

In the case of a high off-duty ratio of laser radiation, the pulsed power incident on a keyhole can be so high that the metal melt will be ejected outside. To avoid this, it is necessary to impose certain restrictions on the welding regime. The equilibrium in the melt during the pulse period should take place:

$$v_0 T - \frac{aT^2}{2} = 0 \tag{9.16}$$

where v_0 is the melt velocity acquired under the action of vapor during a pulse, and a is the melt acceleration caused by the gravitational force and surface tension:

$$v_0 = IS/m, \; a = g + \sigma\frac{2\pi d}{m}$$

Here, I is the momentum per unit area, m is the melt mass, and σ is the surface tension coefficient. By substituting these expressions into (9.16), we obtain the condition of mechanical equilibrium on average

$$I = \frac{T\rho l}{2}\left[\frac{\sigma}{h\rho l} + g\right] \tag{9.17}$$

where l is the melt layer thickness, h is the keyhole depth, and ρ is the sample material density. For typical examples of repetitively pulsed welding with a large off-duty ratio, the action of the gravitational force can be neglected because $\sigma/h\rho l \gg g$. The calculation of the momentum depends on the laser pulse shape and can be performed numerically [24]. The qualitative study can be conducted by using approximate methods, as in Sect. 8.1. The vapor pressure is $p = qc_s/L_b$ and a metal starts boiling at the time $\tau_b = [c\rho(T_b - T_m)]^2\chi/q^2$ (see (8.2)). This gives the momentum

$$I = \frac{qc_s}{L_b}\{\tau - [c\rho(T_b - T_m)]^2\chi/q^2\}. \tag{9.18}$$

It is assumed here that the liquid metal does not crystallize between pulses. This requires a certain average power, which we will determine for simplicity from the linear moving source laser welding model (4.29). For low welding speed $vd_m/\chi < 1$, we have

$$Ef = 2\pi\kappa h T_m \ln^{-1}(4\chi/vd_m). \tag{9.19}$$

The conditions providing the melting of the specified depth and width depending on the welding speed can be determined from (9.17), (9.18), (9.19):

$$E = \pi c\rho(T_b - T_m)d_f h\sqrt{\chi\tau}\left[1 - \frac{\pi\sigma d_f L_b}{2c_s Ef}\right]^{-1/2}. \tag{9.20}$$

One can see from (9.19) to (9.20) that, as the welding speed is increased, the average power should be increased and the pulse energy decreased. At the fixed welding speed, the pulse energy should decrease with decreasing pulse duration τ. These dependences are also predicted by more accurate calculations (Fig. 9.23a,b,c) [24]. If these conditions are not fulfilled, welding can pass to cutting or to some intermediate regime characterized by frequent melt ejections. We will show below

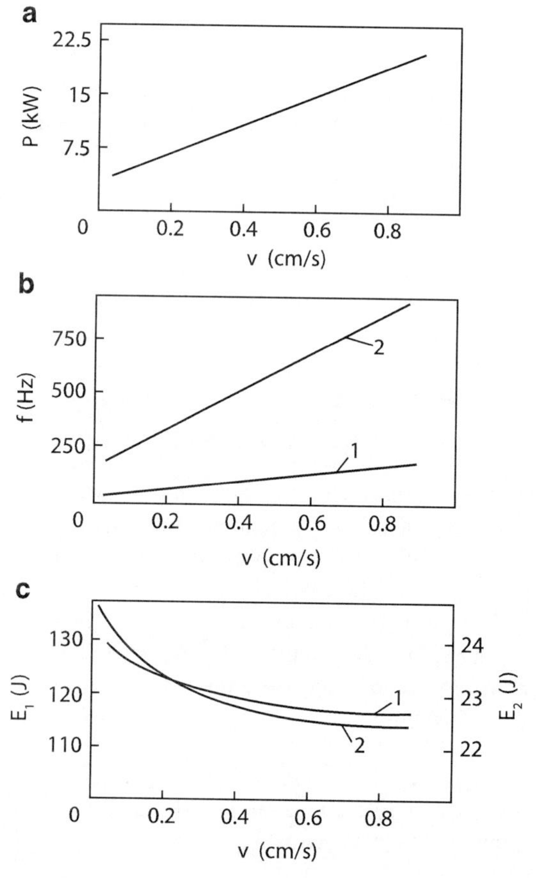

Fig. 9.23 Theoretical dependences of the parameters of repetitively pulsed radiation on the beam travel speed for a 5-cm thick stainless plate for the melt zone diameter $d = 0.25$ cm. (E) energy absorbed per pulse (c); (f) pulse repetition rate (b); ($P = Ef$) average absorbed power (a). Curves 1 and 2 correspond to rectangular pulses with $\tau = 1$ and $\tau = 0.1$ ms, respectively

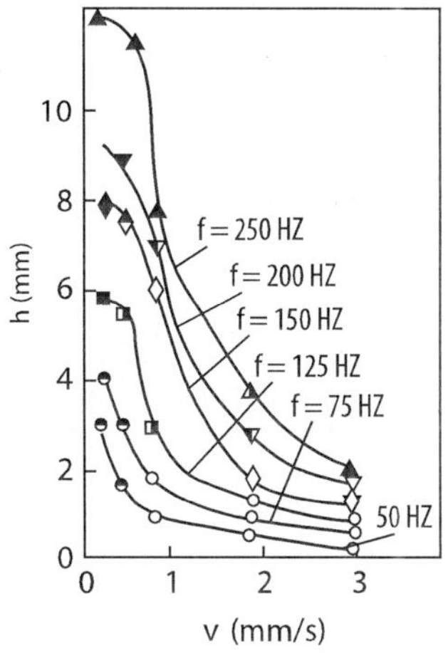

Fig. 9.24 Dependences of the melting depth on the target travel speed at different pulse repetition rates ($E = 3.2\,\text{J}$, $\tau = 60\,\mu\text{s}$). The *open circles* correspond to cutting, *filled circles* correspond to welding; *half filled circles* correspond to the intermediate case

that, if for example, the pulse energy is not reduced with increasing welding speed, welding indeed passes to cutting (see Fig. 9.24).

9.3.2 Experimental Studies

In experiments [18, 25], a fast-flow transverse-discharge repetitively pulsed CO_2 laser was used. By varying the composition and total pressure of the gas mixture, the laser pulse duration was varied from 4 to 900 μs. The pulse energy could be varied from 0 to 5 J, and the pulse repetition rate – up to 300 Hz.

Laser radiation was focused in a spot of diameter $d_f \sim 0.4$ mm on a horizontal target by a KCl lens with a focal distance of 16 cm. The welding depth was increased by blowing a small amount of helium into the welding zone. The sample travel speed was varied from 0.04 to 3 mm/s. The main experiments were performed with stainless steel samples.

Figure 9.24 presents the dependences of the welding depth on the speed at different pulse repetition rates [21]. The laser pulse energy and its FWHM were 3.2 J and 60 μs. respectively. One can see from this figure that the welding depth h decreases with increasing speed approximately as $1/v$. The theoretical dependence of the maximum welding depth on the pulse repetition rate is determined by expression (9.13). Figure 9.25 presents theoretical and experimental maximum welding depths obtained by the extrapolation $v \to 0$. As the pulse repetition rate f is increased, the experimental maximum welding depth increases approximately proportionally to f.

At a high welding speed $v \approx 3$ mm/s and low welding speed $v \approx 0.16$ mm/s and low pulse repetition rates $f \simeq 50$–70 Hz, the passage from welding to cutting was observed (see Fig. 9.24). The reason for this effect was explained in the previous section. Figure 9.26 shows that the maximum welding depth at a constant pulse repetition rate increases linearly with increasing pulse energy, in accordance with theoretical dependence (9.13).

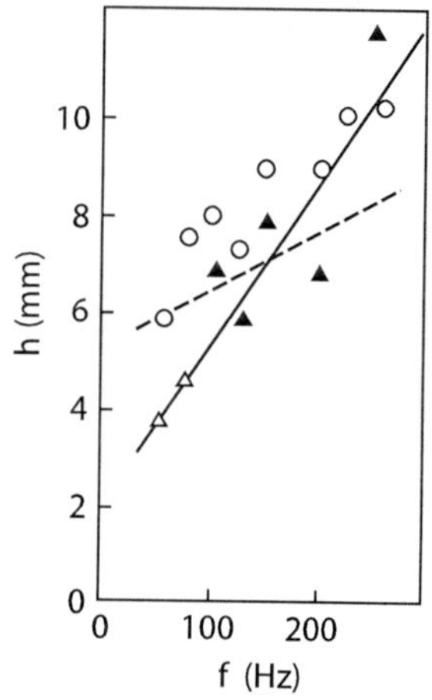

Fig. 9.25 Dependence of the melting depth on the pulse repetition rate ($E = 3.2$ J, $\tau = 60\,\mu$s, $v = 0.016$ mm/s). Calculation by (9.13) (*dotted curve*). Experiments with a laser beam focused on a target surface (*ring*) and in a depth of 1 mm, (*black triangle*); (*light triangles*) cutting regime; (*solid curve*) dependence averaged over points

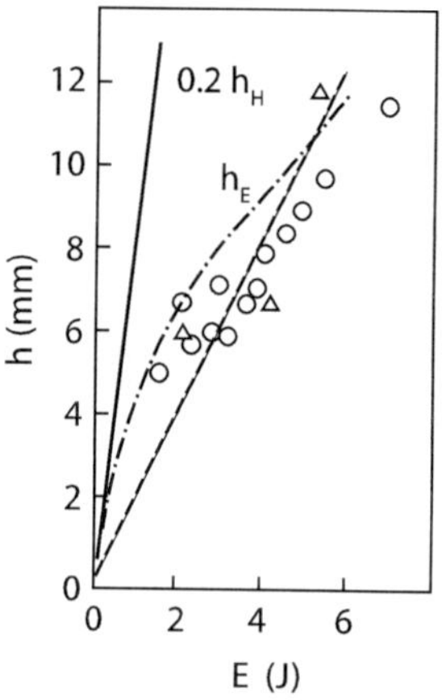

Fig. 9.26 Dependence of the melting depth on the pulse energy [21]; $f = 150$ Hz, $v = 0.16$ mm/s, $\tau = 60\,\mu$s): (*dotted curve*) calculation by (9.13) for $d = 0.8$ mm; (*dot and dash curve*) calculation by (1.22) and (9.13); (*solid curve*) calculation by (9.21); experiments with a laser beam focused on a target surface (Δ) and in a depth of 2 mm (o)

A large melting ability of repetitively pulsed laser welding is related to a small width of the weld, i.e. to a high aspect ratio h/d_m. As shown in Sect. 1.3, due to absorption in keyhole walls, laser radiation cannot penetrate deeply into metals even at high radiation energies. If the value of h exceeds the characteristic absorption length of light in the keyhole, this means that the depth h is no longer increases with increasing power [see (1.22)]. For the E wave (the electric field in a wave is directed along the keyhole radius), $l_E = 1/\alpha_e = d/\alpha$, where α is the absorption coefficient at the normal incidence. For $h > l_E$, the keyhole growth with increasing power slows down and occurs logarithmically. To estimate the maximum welding depth, we will assume that it is equal to the absorption length of light in the keyhole $h = d/\alpha$. By determining diameter d from this relation and substituting it into (9.13), we find the maximum depth of the keyhole in which the E wave propagates. One can see from Fig. 9.26 that experimental keyhole depths are close to limiting depths if the incident light excites the E wave in the keyhole. If the weakly decaying H_{01} mode is excited in the keyhole, the limiting depth will be considerably larger. In this

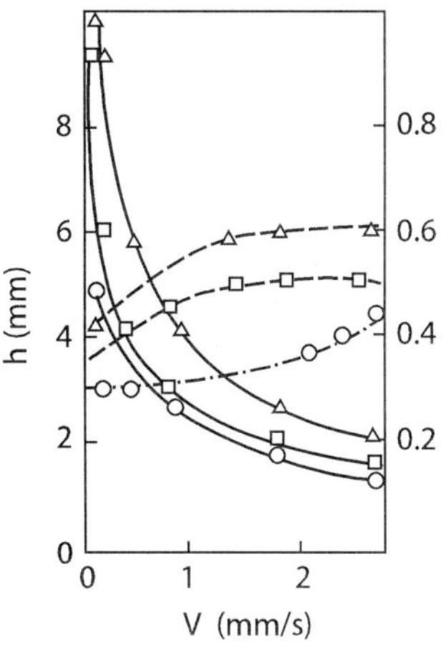

Fig. 9.27 Dependences of the weld depth (*solid curve*) and width (*dotted curve, dot and dash curve*) on the welding speed for $f = 150\,\text{Hz}$ (*triangle*), $f = 200\,\text{Hz}$ (*square*), and $f = 300\,\text{Hz}$ (*ring*); $P = 400\,\text{W}$; $\tau = 60\,\mu\text{s}$

case, the decay length is $l_H = d^3/\alpha_H$, where $\alpha_H = \lambda_0^2\alpha/2\pi^2 \approx 0.5 \times 10^{-8}\ \text{cm}^2$, λ_0 is the wavelength of light [see (1.23)]. By substituting the value of d into (9.13), we obtain from the condition $h = l_H$ the approximate expression

$$h_H = \left(\frac{E}{\pi c\rho T_b\sqrt{\chi\tau}}\right)^{3/4} \alpha_H^{-1/4}. \tag{9.21}$$

Figure 9.26 shows that, when the H wave propagates in the keyhole, the penetration depth can be considerably increased compared to the case of the E wave. To obtain such a regime, it is necessary to improve focusing and to use azimuthally polarized laser radiation (the electric field vector of the wave should touch keyhole walls) (see Fig. 1.6).

It follows from (9.13) that the welding depth at a constant power should increase with increasing pulse energy and decreasing f. One can see from Fig. 9.27 that at a low welding speed and constant average power, the welding depth increases with increasing pulse energy. This figure also shows that the melt zone width increases with increasing speed according to theoretical dependence (9.15).

The aspect ratio of repetitively pulsed welding is rather high over the entire range of welding speeds, achieving especially large values at low speeds (Fig. 9.28a,b) (the maximum experimental value is $h/d_m = 25$).

9.3.3 Dynamics of a Weld Pool Upon Repetitively Pulsed Irradiation

To study the melt zone shape during repetitively pulsed welding, copper pins of diameter 1 mm were pressed into a steel sample, which were oriented perpendicular to the welding direction. Figure 9.28b presents the longitudinal section of a weld for this sample. One can see from the copper alloy position that the melt zone

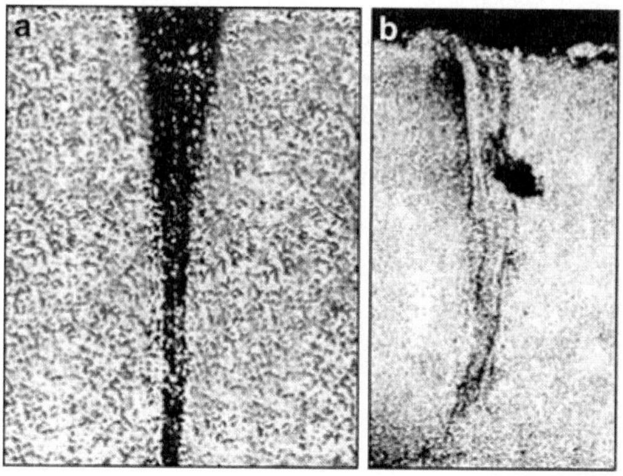

Fig. 9.28 Transverse (**a**) and longitudinal (**b**) sections of a weld for $f = 200$ Hz. $\tau = 60\,\mu$s: (**a**) $P = 400$ W, $h = 10$ mm; (**b**) $P = 600$ W, $h = 12$ mm; dark spot on the photograph corresponds to copper pin

of length of a few millimeters is produced behind the laser beam. The copper distribution pattern shows that vortex motions of the liquid metal appear in the melt zone, which cause the metal transfer both in the horizontal and vertical (along the beam) directions. However, these flows do not flood a gas–vapor channel after switching off the laser beam, as occurs during cw laser welding of metals. No traces of copper were noticeable in front of a pin. This circumstance and also the absence of the filling of the vapor cannel by liquid (after switching off the laser) suggest that the liquid is mixed only during the action of the laser pulse, when the vapor pressure and the pressure of a liquid metal jet are large. The action of the gravitational force during the interval between pulses proves to be insufficient for filling the keyhole with the melt. The surface tension force of the melt (which is located only on the rear wall of the keyhole) can probably prevent the filling of the vapor cannel with liquid. Note that the longitudinal size of the vapor channel (along the welding speed direction) exceeds by several times the transverse size, which is close to the weld width.

The microstructure and phase composition in the zone exposed to laser radiation were investigated by the methods of X-ray phase analysis, optical, and scanning electron microscopy. The photographs of longitudinal stainless steel sections [26] presents the bands separating metal portions carried by each pulse are distinctly observed. The distance between boundaries $\sim$3–6 μm is close to the distance over which the sample is displaced during the time between pulses. Such a structure of the solidified metal confirms the assumption that the metal is carried from the front wall by each pulse and solidifies on the opposite wall of the keyhole. It seems that these bands represent crystallization fronts caused by periodic laser action.

The properties of mass transfer during repetitively pulsed laser welding were investigated in [27] by the method used for filming a gas–vapor keyhole during cw laser welding (Chap. 4).

The 4-J, 100-μs pulses with a pulse repetition rate of 200 Hz were focused by a lens on the edge of a sample consisting of titanium and transparent dielectric plates put together and mounted vertically. A laser beam was directed on a metal plate so that the metal melting occurred near the boundary with the dielectric. The welding

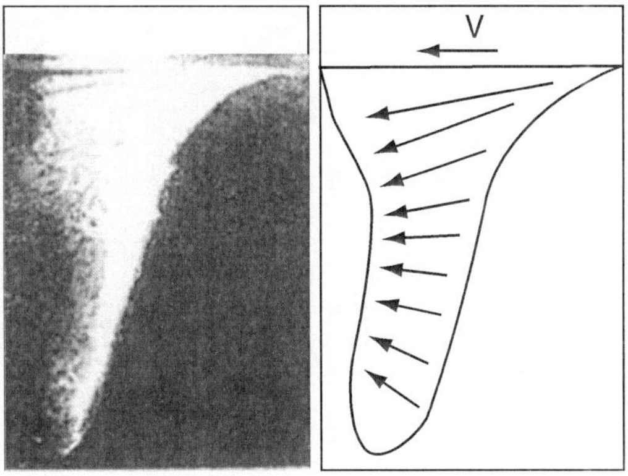

Fig. 9.29 Keyhole shape for small (<1 mm/s) welding speeds (**left**) and the direction of the liquid metal flow along the side wall of the keyhole (**right**)

process was observed with the help of a movie camera through the transparent dielectric plate.

The filming rate was 100 and 3,000 frames/s. Neutral optical filters were used during filming and helium was used as a protective gas which was blown into the metal melting zone.

The behavior of the melt during the pulse action and in the interval between pulses was different. In addition, it is interesting to study the motion of the melt under the action of averaged forces during pulses and intervals between them. Figure 9.29 presents the photograph of the weld zone and the scheme of the melt flow over the side wall of the keyhole at a low welding speed $v_w = 1$ mm/s, when the keyhole shape and the direction of melt flows almost do not change during a long time $t > (d/v_w)$.

The scheme of the melt flow in Fig. 9.29b was obtained from records of the melting process (the filming rate was 100 frames/s). The measured averaged melt flow rate on the side wall of the keyhole was ~1.5 mm/s.

One can see from Fig. 9.29 that the front wall of the keyhole is tilted with respect to incident beam, the tilt angle in the rhizosphere being smaller. These records demonstrate the flow of liquid in different parts of the welding pool Fig. 9.29b.

While at low welding speeds $V_w < 1$ mm/s the keyhole shape in a photographic record is relatively constant during a long time $t > (d/v_w)$, as the welding speed is increased above 1 mm/s, welding passes to "quasi-cutting". In this case, a step projection is periodically formed on the front wall of the keyhole, which rapidly descends to the rhizosphere at the velocity ~1–2 cm/s (Fig. 9.30), and sharp projections are formed in the weld root. The character of their formation is similar to that observed during cw laser welding (see Fig. 4.26).

The behavior of the liquid metal between pulses was observed by using high-speed filming (3,000 frames/s). The analysis of the photographic records obtained in these experiments showed that within ~1 ms after the pulse end, the melt begins to move in the backward direction from the rear to front wall of the keyhole. The velocity of this motion is ~3–5 cm/s. During cooling of the welding pool, this motion decays. The backward flow of the melt can be caused by thermocapillary

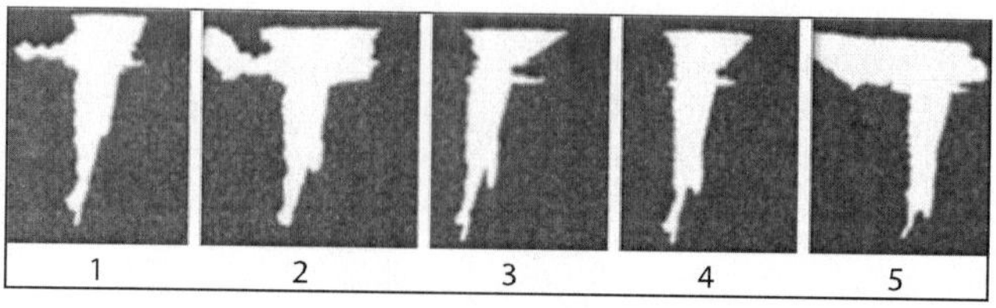

Fig. 9.30 Cinegram of variations in the shape of the front wall of a keyhole during welding in the quasi-drilling regime ($v \gg 1$ mm/s). The filming rate is 50 frames/s

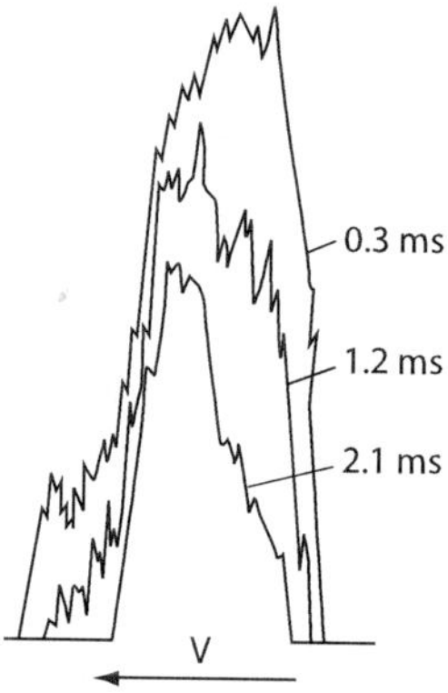

Fig. 9.31 Displacement of the temperature maximum along the side wall of a keyhole during cooling in the interval between pulses

convection, which was considered in Chap. 2. This flow can appear if the rear wall became hotter than the front wall. The photometry of the photographic records across the keyhole at the half-width showed that after the laser pulse end the front wall cools off faster than the rear wall (see Fig. 9.31). This is explained by the fact that behind the rear wall the heated part of the target is located, while in front of the front wall the cold metal is located. The backward motion of the melt begins when the temperature maximum is displaced to the rear wall.

It seems that the lower efficiency of repetitively pulsed laser welding with a high off-duty ratio (see below) compared to cw welding can be explained by the backward motion of the melt during the interval between pulses, which requires to transfer repeatedly the same portion of the liquid metal from the front to rear wall.

Figure 9.32 presents the values of the specific power per metal unit depth melting P/h, which is reciprocal to the melting ability. At low welding speeds, the minimal specific energy consumption for repetitively pulsed laser welding proves to be lower than the minimal specific consumption for cw welding [28] and electron-beam welding in vacuum [28]. Because the melting ability of repetitively pulsed welding is high only at sufficiently low welding speeds, the efficiency hv/P of repetitively pulsed welding proves to be low (Fig. 9.32). The value of hv/P is equal to the

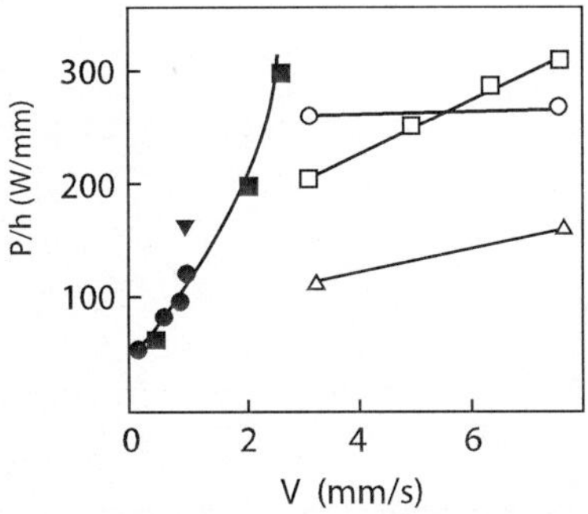

Fig. 9.32 Comparison of the energies spent for melting a 1-mm thick steel plate at different welding regimes: (*ring, square*) cw laser welding for $P = 1$ and $P = 1.2$ kW [28]; (*triangle*) electron-beam welding in vacuum [28]; (*black ring*) repetitively pulsed laser welding ($P = 400$ W, $f = 170$ Hz); (*black square*) ($P = 500$ W, $f = 200$ Hz) [18]; (*black triangle*) [25]

tangent of the angle between a straight line drawn from the coordinate origin to a point of curve P/h and the ordinate.

Due to the importance of these questions, we will consider them in more detail. One can see from Fig. 9.32 that the specific power consumption is minimal in the limit of small welding speeds. This allows us to estimate P/h from (9.13):

$$\overline{P}/h = \frac{\pi d c \rho T_b f \sqrt{\chi\tau}}{1 + \frac{df}{2}\sqrt{\tau/\chi}\ln\frac{h}{d/2+2\sqrt{\chi/f}}}. \tag{9.22}$$

It follows from (9.22) that the specific power consumption is a nondecreasing function of d. For $d < d^*$, P/h decreases linearly with decreasing d, where

$$d^* = \sqrt{\chi/(\tau f^2)}\ln^{-1}\left(\frac{h}{d/2 + 2\sqrt{\chi/f}}\right). \tag{9.23}$$

It follows from (9.22) that, as d and the pulse repetition rate f are decreased, a decrease in the consumption for melting can be expected. This was verified in experiments on welding of stainless steel and aluminum by using a low-power repetitively pulsed Nd:YAG laser [29]. The shorter-wavelength radiation from this laser compared to a CO_2 laser can be focused to a smaller spot. For the same purpose, experiments were performed at low average powers 4.2, 8, and 13.4 W, $\tau =$ 140 μs, and $f = 25, 33$, and 50 Hz. The focal spot diameter was 0.2 mm for $f = 25$ and 33 Hz and 0.4 mm for $f = 50$ Hz. One can see that $d < d^*$ for parameters used in experiments [see (9.23)]. The transverse sections of welds on AD16 aluminum alloy and stainless steel are shown in Fig. 9.33. The processing of these data gives the dependences P/h under different welding conditions (Fig. 9.34). One can see that, as the focal spot and pulse repetition rate are increased, the power spent for melting the unit depth increases in qualitative accordance with (9.22). It follows

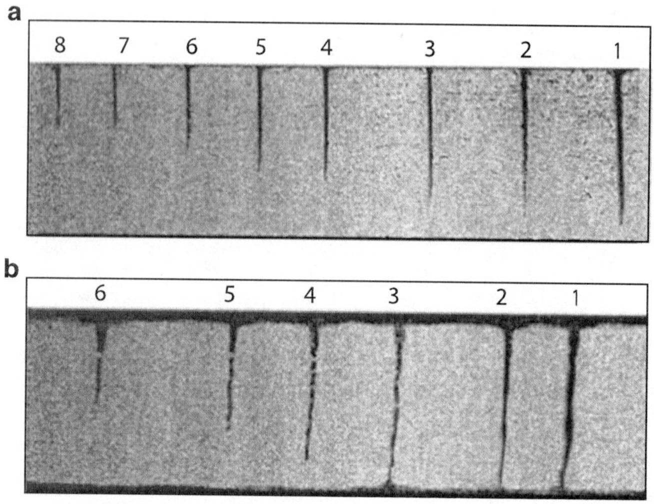

Fig. 9.33 Photographs of the cross sections of a weld in a 4.6-mm thick AD16 aluminum alloy sample (**a**) and in a 3.2-mm thick stainless steel sample (**b**) for $W = 8$ W, $f = 33$ Hz, and $v = 0.1$ (1), 0.2 (2), 0.3 (3), 0.4 (4), 0.5 (5), 0.6 (6), 0.8 (7), and 1 mm/s (8)

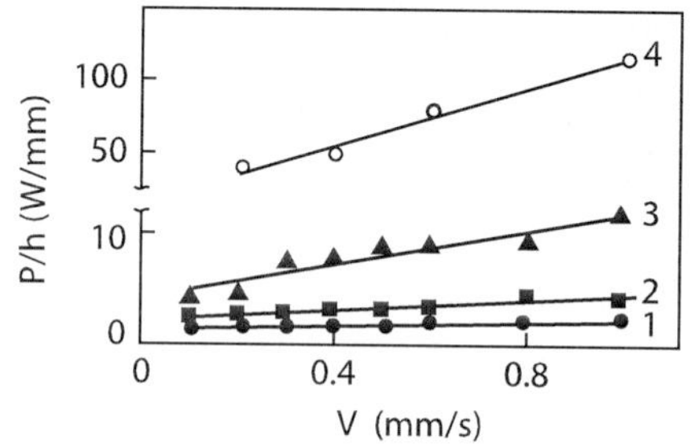

Fig. 9.34 Dependences of the specific power expenditure P/h on the travel speed v of a steel sample for $P = 4.2$ W and $f = 25$ Hz (**1**), $P = 8$ W and $f = 33$ Hz (**2**) $P = 13.4$ W and $f = 50$ Hz (**3**), and for a repetitively pulsed CO_2 laser (**4**) in Fig. 9.32

from this figure that the specific consumption slowly increases with the welding speed. This leads to the increase in the welding efficiency with increasing welding speed. Note that in the conditions under study (Fig. 9.34) it is possible to connect 0.4 mm^2/J, whereas under conditions in Fig. 9.32 this value is considerable lower (0.01–0.03 mm^2/J). Note finally that, as in the case of cw welding, the thermal efficiency of repetitively pulsed welding increases with the welding speed and achieves the same values (cf. Figs. 9.35 and 4.20).

As for the welding of freakish materials such as aluminum alloys, the specific power consumption for melting under the given conditions is close to that for the melting of steel. By the way, the estimate by (9.22) gives close values of $\overline{P}/h \sim 5$ W/mm for both materials, which is close to the experimental value. As for the

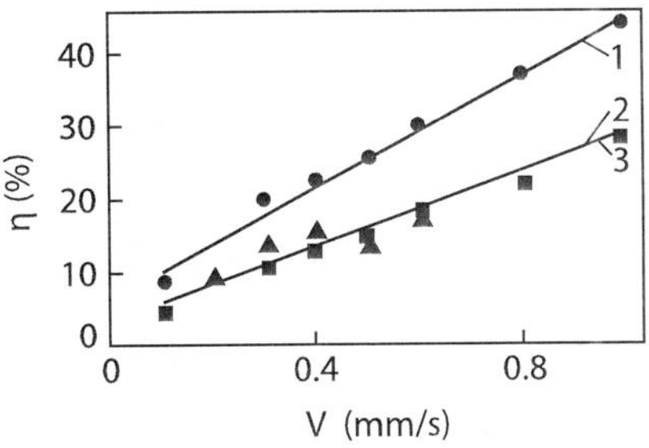

Fig. 9.35 Dependences of the welding efficiency on the travel speed V of a steel sample for $P = 4.2$ W and $f = 25$ Hz (**1**), $P = 8$ W and $f = 33$ Hz (**2**), $P = 13.4$ W and $f = 50$ Hz (**3**)

thermal efficiency, it is for Al is three times lower than that for steel welding. This is explained by the fact that the heat conduction of Al considerably exceeds that of steel (see table in Chap. 1). Thus, the welding of metals by repetitively pulsed lasers with properly selected parameters can be considerably more efficient than cw laser welding. It seems, however, that the weld quality in the case of repetitively pulsed laser welding is inferior, but this problem is beyond the scope of our book.

9.4 Drilling and Cutting of Metals by Repetitively Pulsed Radiation

9.4.1 Properties and Mechanism of Metal Cutting by Repetitively Pulsed CO_2 Laser Radiation

It was shown in Sect. 9.3 that irradiation of a metal by a long enough, low-intensity light pulse leads to deep melting. If a pulse is short and its intensity is high enough, a hole is formed in a material. This type of interaction was considered in Chap. 8. The drilling efficiency can be increased in practice by using enough high pulse repetition rates. In this case, if the average radiation power supplied to a sample is high enough, the mutual influence of pulses will be considerable. First of all, this will result in an increase in the sample temperature near the produced hole. Clearly, for any destruction (drilling) mechanism, the increase in the temperature should lead to the increase in the hole depth with increasing pulse repetition rate at the same total energy. This effect was really observed for refractory metals (Fig. 9.36 [30]). Experiments were performed under conditions close to those for welding (see Sect. 9.3). The volume of a removed material increases slower with the pulse repetition rate than the hole depth (see curve V in Fig. 9.36). Therefore, the material removed from the keyhole bottom partially remains on the side walls of the hole. This can be caused by the increase in the radiation intensity on the keyhole bottom. Indeed, the mutual influence of pulses should be manifested first of all near the keyhole bottom, where the heat release is maximal. This influence on the walls proves to be weaker, which leads to the partial deposition of the materials on them.

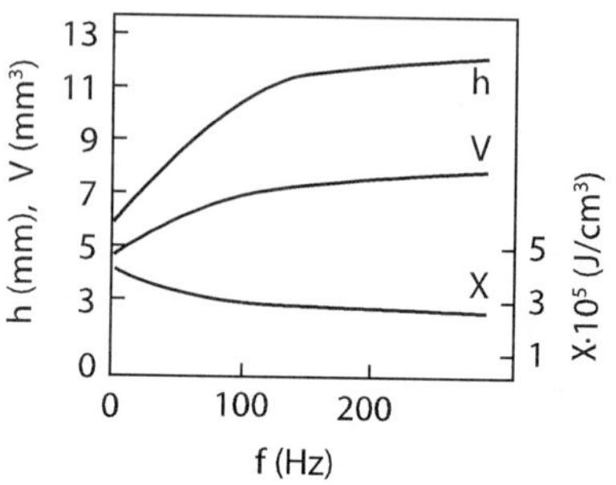

Fig. 9.36 Dependences of the keyhole depth h and volume V and the specific energy expenditure X on the pulse repetition rate f for $N = 500$, $E = 4\,\mathrm{J}$, and $\tau = 80\,\mu\mathrm{s}$

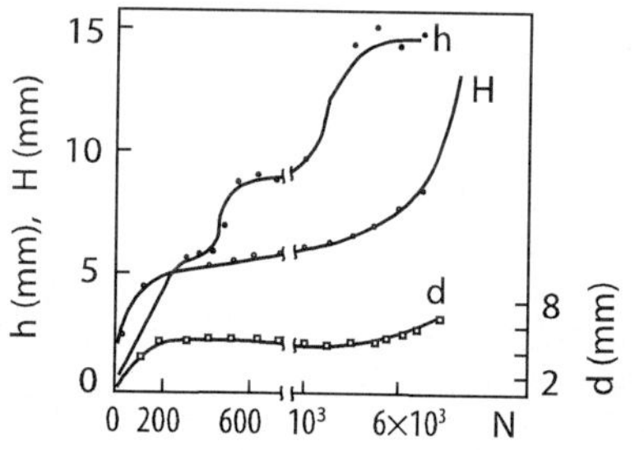

Fig. 9.37 Dependence of the keyhole depth h in stainless steel on the number N of pulses. Dependence of the keyhole depth H and diameter d in quartz on the number of pulses N, (·) $f = 300\,\mathrm{Hz}$, $E = 2\,\mathrm{J}$, $\tau = 80\,\mu\mathrm{s}$; (o, x) $\tau = 40\,\mu\mathrm{s}$, $f = 200\,\mathrm{Hz}$, $E = 5\,\mathrm{J}$

It is interesting to compare the dependence of the hole depth on the number of pulses at different pulse repetition rates. We can compare the curve for h in Fig. 9.37 ($f = 300\,\mathrm{Hz}$) and curve 2 in Fig. 8.12a ($f = 0.1\,\mathrm{Hz}$). While for $f = 0.1\,\mathrm{Hz}$, the hole depth ceases to grow for $N \geq 100$, for $f = 300\,\mathrm{Hz}$ the saturation appears for $N \geq 10{,}000$. In this case, the limiting drilling depth exceeds the depth of drilling by single pulses by 2–3 times. Figure 9.37 shows that for $N = 500$–1,000, the laser beam deepening slows down (curve h goes on an intermediate plateau). The further increase in the depth for $N > 1{,}000$ is accompanied by a considerable increase in the input diameter of the hole and a change in the keyhole shape. Below, we use the hole depth in the intermediate plateau region ($N = 500$–1,000) as the limiting hole depth. Figure 9.38 shows the dependences of this depth on the pulse repetition rate for different pulse energies. The dependences of the limiting hole depth on the average radiation power for different pulse repetition rates are presented in Fig. 9.39. One can see that the drilling process is determined not only by the average radiation power but also by the pulse energy E. For example, for $E = 4\,\mathrm{J}$ ($f = 100\,\mathrm{Hz}$), the depth is almost three times larger than the depth for $E = 1\,\mathrm{J}$ ($f = 400\,\mathrm{Hz}$), i.e. the same tendency is observed as for laser welding (see Fig. 9.27). It seems that the mechanisms of the limiting laser beam deepening into a metal for repetitively pulsed welding and cutting are close. The mass transfer direction in these cases is

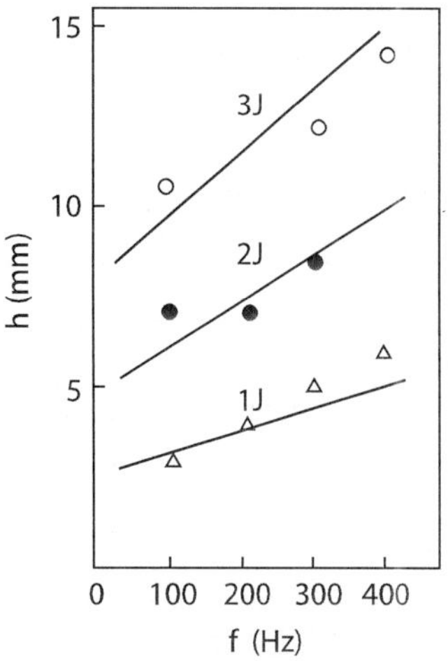

Fig. 9.38 Dependences of the limiting drilling depth on the pulse repetition rate for different pulse energies E. The curves are calculated by (9.13) for $\tau = 80\,\mu s$, $d = 0.5$ mm; points are experimental measurements

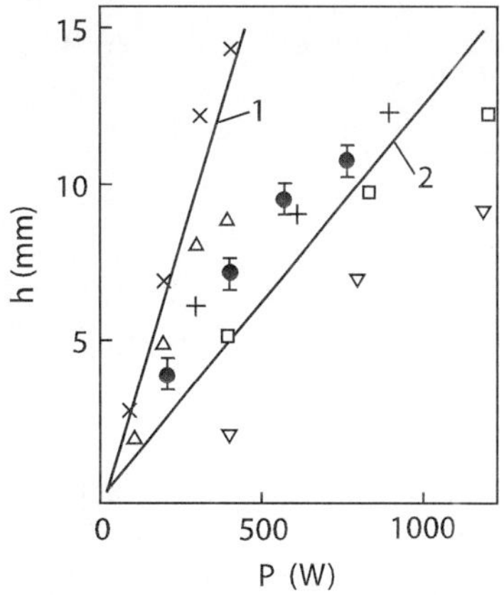

Fig. 9.39 Increase in the limiting drilling depth with increasing the output power of a repetitively pulsed CO_2 laser at different pulse repetition rates: $f = 100$ Hz (He) (*cross*), 100 Hz (air) (*triangle*), 200 Hz (He) (*black ring*), 300 Hz (He) (*plus*), 400 Hz (He) (*square*), 400 Hz (air) (*inverse triangle*). The curves are calculated by (9.13) for $f = 100$ Hz (**1**) and 300 Hz (**2**) for helium

probably unimportant. In this connection we will describe the drilling process by expression (9.13) for the limiting depth of penetration of a laser beam into a metal. Figures 9.38 and 9.39 present curves calculated by (9.13) for $d = 0.5$ mm. One can see that these curves are in qualitative and partially in quantitative agreement with experiments. This is also confirmed by the dependences of the depth of a weld and laser drilling on the number of pulses for the same repetition rates and the same pulse energies Fig. 9.40 (in the case of welding, $N = fd/v$, where d is the focal beam diameter, and v is the sample travel speed).

It follows from Fig. 9.39 that the drilling of steel in air is less efficient than in helium. The difference is greater for high pulse energies, which is related to the plasma screening of laser radiation (Fig. 8.11). The influence of a plasma plume is more clearly seen in the case of drilling of transparent quartz. The change in the keyhole shape was recorded by movie-taking through the side surface of the sample. Figure 9.37 presents the dependences of the keyhole depth and diameter on

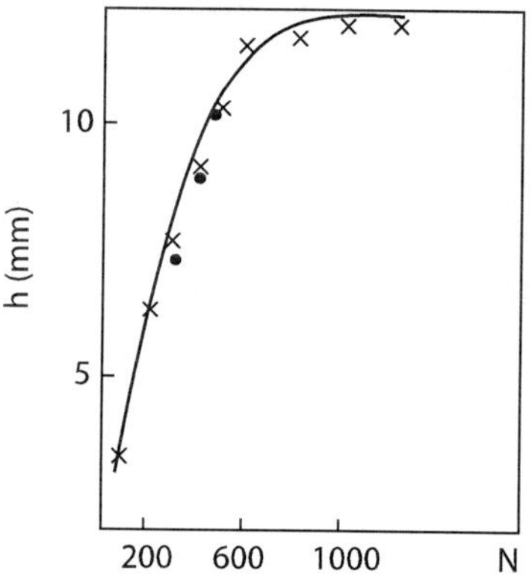

Fig. 9.40 Dependence of h on the total number N of pulses during drilling (*black ring*) and welding (*cross*) for $N = f \cdot d/v$, $E = 3.2$ J, $\tau = 80\,\mu$s, $f = 250$ Hz

the number of pulses. For $N > 500$, the keyhole growth rate decreases. However, for $N > 2{,}400$, the deepening of the beam considerably accelerates. The plume size and brightness begin to decrease for $N > 2{,}000$, and for $N > 2{,}700$ the plume disappears. This probably leads to the increase in the growth rate of h and in the average keyhole diameter.

9.4.2 Gas Assisted Laser Cutting of Metals by Repetitively Pulsed Radiation

It was shown in the previous section that the blind repetitively pulsed cutting of metals can change repetitively pulsed welding under certain conditions. This occurs when either the sample travel speed is increased or the pulse energy is increased. In practice, intense gas jets are used during through repetitively pulsed cutting, as in the case of cw lasers. The cutting is performed by using repetitively pulsed CO_2 and neodymium lasers. The cutting by radiation from Nd laser at low pulse repetition rates (tens Hz) gives large depths at speeds that are considerably lower than in the case of cw radiation cutting. Here, the analogy with repetitively pulsed and cw radiation welding is seen (Fig. 9.32). The cut depth increases with increasing the pulse energy and decreasing its duration (Fig. 9.41 [31]). Here, the analogy with repetitively pulsed welding is also seen (Figs. 9.26 and 9.27).

Because, unlike welding, during cutting the rear wall of a gas-vapor keyhole is absent, the propagation of radiation in the keyhole also differs from the cases of welding and drilling. During welding and drilling, multiple reflections of beams from the front and rear walls are very important. As a result, the structure of incident radiation during cutting acquires a more important role. This concerns, first of all, the caustic length, which is proportional to the focal distance F of a lens (see Sect. 4.5.3). Figure 9.42 shows that the cutting depth grows with increasing focal distance [31]. This growth ceases when the focal spot size, which also increases with increasing F, becomes large. The latter reduces the laser radiation intensity.

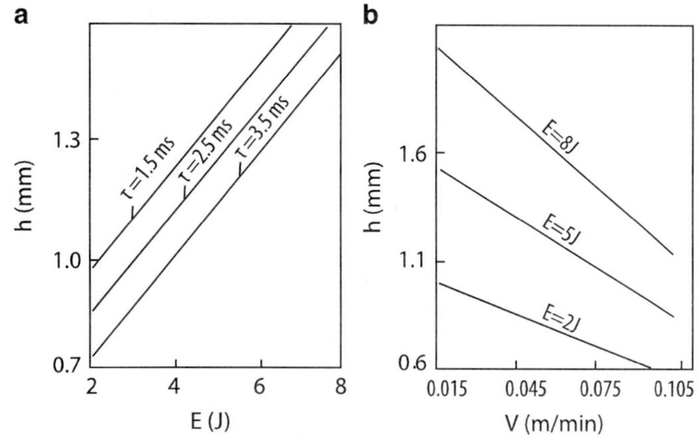

Fig. 9.41 Dependencies of the cut depth on the laser energy for different pulse durations (**a**) and on the cutting speed for different pulse energies (**b**)

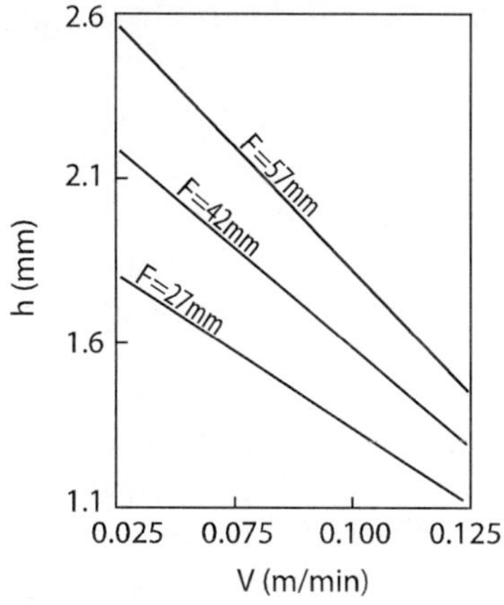

Fig. 9.42 Dependencies of the cut depth on the cutting speed in the repetitively pulsed regime for different focal distances of a lens

Thus, the deepest cuts and grooves are produced when long-focus lenses are used at low cutting speeds and high intensity (short and high-energy pulses).

A comparison of the parameters of cuts produced by repetitively pulsed and cw lasers showed that repetitively pulsed laser cutting produces narrower cuts, thereby providing a higher accuracy. The roughness of a cut and the deviation of its edges from perpendicularity during repetitively pulsed cutting are also smaller [31]. However, the efficiency of repetitively pulsed cutting is considerably lower than that of cw laser cutting due to the low cutting speed in the former case (Fig. 9.41b). Here, we also see analogy by comparing repetitively pulsed and cw welding regimes (see Fig. 9.32).

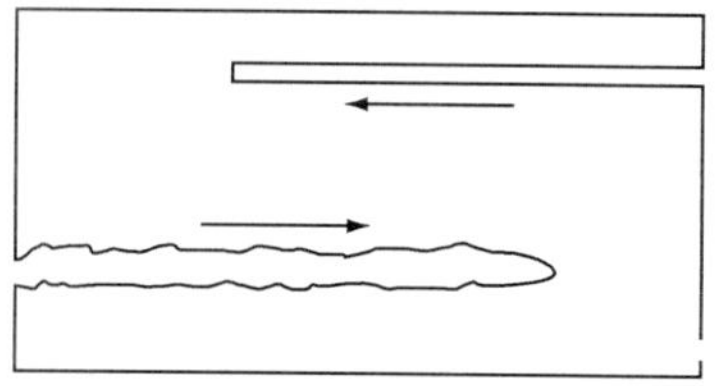

Fig. 9.43 Dependence of the cut shape on the target travel direction for $P = 300\,\text{W}$, $f = 300\,\text{Hz}$, and $V = 3\,\text{mm/s}$, $\tau = 40\,\mu\text{s}$ (3-mm-thick St3 plate)

It is known that cw laser cutting in an oxygen flow at low speeds ($v \leq$ 1–10 cm/s) often passes into the so-called uncontrollable regime of metal burning (see Chap. 5). The regime of high-quality repetitively pulsed CO_2 laser cutting also can be unstable. It considerable depends on the process organization, in particular, on the combining of the gas jet and light beam. Figure 9.43 shows the dependence of the steel plate cutting regime on the cutting speed direction [32].

9.4.3 *Modelling of the Instability of Deep Laser-Beam Penetration into a Moving Target*

It was pointed out above that through and blind repetitively pulsed material cutting is characterized by the roughness of the cut walls or oscillations of the groove depth. Defects in the weld root are also often observed during repetitively pulsed welding. Repetitively pulsed welding and cutting, unlike these processes during cw irradiation, are characterized by comparatively small amounts of the melted metal. As mentioned above, during repetitively pulsed welding the channel is not filled with liquid metal after switching off laser radiation. Figure 9.30 shows the sharp boundary of the leading edge of a gas–vapor channel, on which an indentation descending to the weld root is formed. All these demonstrate a small thickness of the melt on the front wall compared to the characteristic dimensions of the channel and inhomogeneities of its shape. This suggests that an instability mechanism can exists which is not related to the presence of the liquid phase. This is also confirmed by the groove shape in organic glass (Fig. 9.44), where the destruction occurs without the formation of the liquid phase at all. The exclusion of the liquid phase from consideration significantly simplifies the problem. This approximation was used to study the influence of the laser radiation polarization on the shape and speed of deep drilling of holes by a stationary laser beam (Chap. 8). The dynamics of the laser beam motion during deep penetration into a sample should be considered in the three-dimensional space.

This process was studied by the three-dimensional numerical simulation of ablation of a material cut by radiation from a repetitively pulsed laser [33]. The total energy flux was simulated by a few thousands (2,000–3,000) infinitely thin "elementary rays". The laser pulse energy in each of the elementary rays was the same. The beam of elementary rays is emitted from a point located at a distance of z_f from a target surface (Fig. 9.45). The focal point moves parallel to the target surface

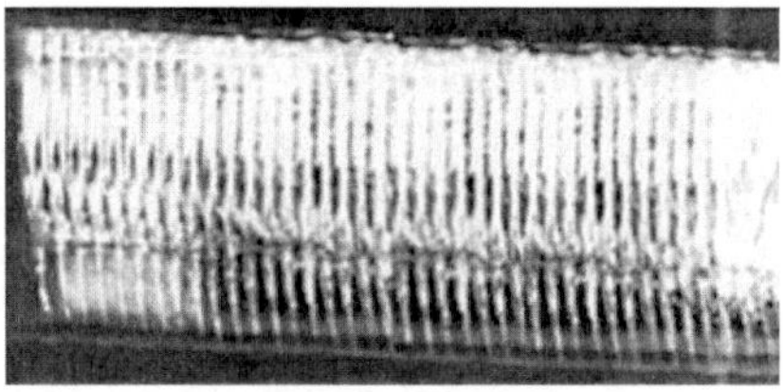

Fig. 9.44 Organic glass samples cut by a repetitively pulsed CO_2 laser beam

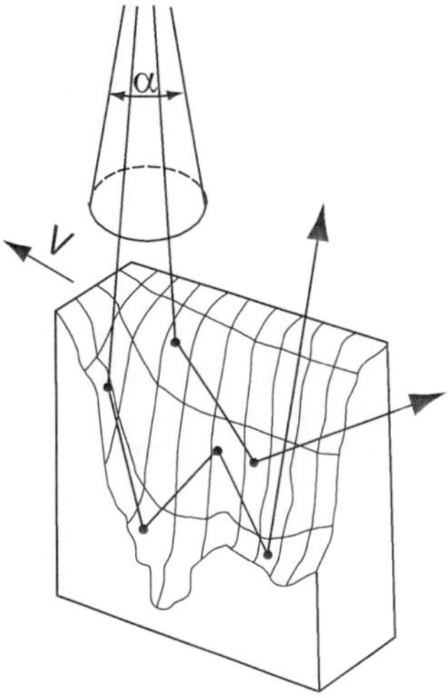

Fig. 9.45 Scheme of the interaction of a laser beam with a target and paths of elementary rays inside a cut

at the speed v. The density of elementary rays in the target plane was assumed Gaussian ($\sim \exp(-r^2/r_f^2)$), where r_f is the focal spot radius on the target surface. The beam divergence angle α was assumed small enough ($tg\alpha = 2r_f/z_f \approx 0.2$) (Fig. 9.45).

The path of each elementary ray was calculated inside a cut at each time step by using the laws of geometrical optics. During each reflection of an elementary ray, a part of its energy equal to $\alpha(\Theta) \cos \Theta$ is absorbed (where Θ is the angle between the ray direction and the normal **N** to the cut surface at the reflection point. The value of $\alpha(\Theta = 0)$ is ≈ 0.1, which corresponds to the level of absorption of normally incident CO_2 laser radiation by metals. Because a keyhole is three-dimensional, the correct description of the dependence of the absorption coefficient on the angle of incidence is a complicated problem. An elementary ray experiences several reflections. It can have normal polarization with respect to one face and parallel or intermediate polarization with respect to another. It is assumed for simplicity that radiation has circular polarization and the deviation of the electric field direction from parallelism to the keyhole surface is neglected.

In calculations the three-dimensional surface of the cut cavity is simulated by in calculations 3–5 thousands triangular faces. The beam travels speed was chosen so that it was displaced by the distance $\sim r_f/50$ during the time between pulses.

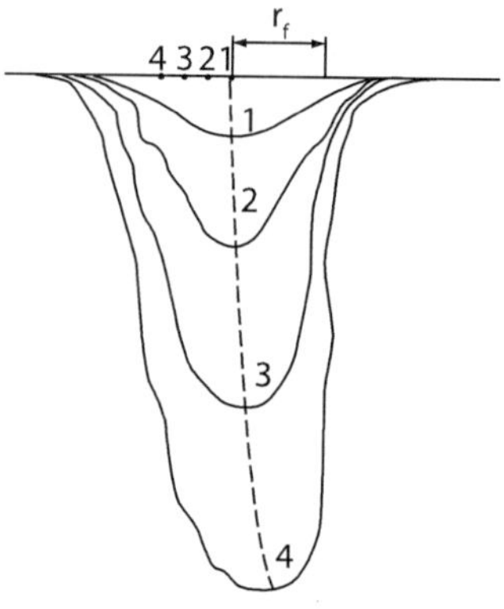

Fig. 9.46 Longitudinal sections of a keyhole in a target at the beginning of cutting. (**1–4**) Positions of the beam center (*black ring*) and the corresponding keyhole profiles at the successive instants of time. The dotted curve is the trajectory of the groove depth maximum

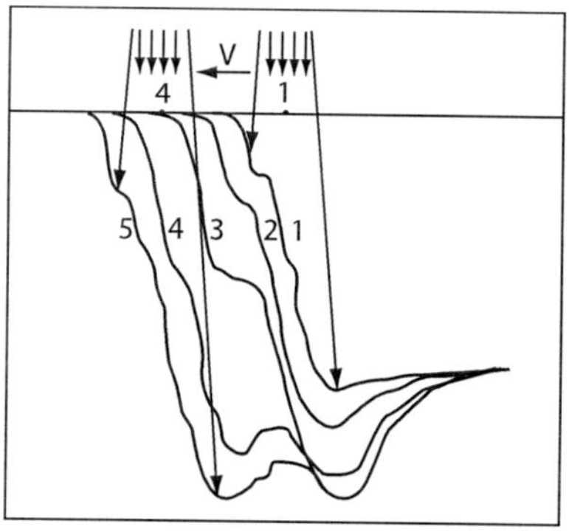

Fig. 9.47 Formation of the next tooth at the cut bottom. (**1–5**) Groove section profiles at successive instants of time. The beam positions for sections 1 and 4 is shown (the arrows are drawn at the e^{-1} intensity level)

The total laser pulse energy E_0 was chosen within (1–10) $\rho L_b r_f^3$, where ρL_b is the sample unit volume vaporization energy.

Figure 9.46 presents cross sections along the symmetry plane of a channel formed by a laser beam penetrating into a sample. The beam size and positions of its center at instants corresponding to cross sections 1–4 are shown. $E_0 \approx 4\rho L_b r_f^3$. One can see that the depth (dotted line) moves oppositely to the beam propagation direction. The laser beam is "captured" in a crater formed at the initial moment of cutting. Due to the angular dependence of the absorption coefficient and rereflection, the laser energy is mainly absorbed near the cut bottom and on the rear wall. This wall becomes steeper with time. The position of the depth maximum first lags behind the beam center, and then, due to the deflection of the rear wall of the channel, the depth maximum begins to move oppositely to the beam propagation direction.

Figure 9.47 presents the longitudinal cross sections of the front part of the cut at successive instants of time. The beam positions for curves 1 and 4 are also shown. The laser pulse energy in this calculation was set equal to $\sim\rho L_b r_f^3$. One can see that a shelf is formed on the front wall at the initial moment. As the beam center approaches it, the shelf broadens and descends by forming the next tooth at the cut bottom after the beam "departure". The light is manly absorbed in the shelf and the tooth nearest to the laser beam. Due to rereflection, the radiation is "captured" in the growing tooth, similarly to the case shown in Fig. 9.46. The oscillation amplitude of the cut depth was set equal to $\sim 0.1h$ in calculations, where h is the depth averaged over several periods. The oscillation frequency is determined by the presence of small inhomogeneities in the upper front wall of the cut. The mechanism of oscillations is not discussed within the framework of this model, their amplitude was set equal to $\sim 0.1r_f$.

Thus, "teeth" can be formed on the front wall of the cut, and the groove bottom after the propagation of the beam proves to be wavy. This phenomenon is of the general type, being determined only by the optical properties of the beam-absorbing surface system and the beam motion with respect to the target. In practice, the oscillations in the beam penetration depth exist in a very broad range of conditions. They are manifested in the processing of metals and dielectrics at power levels differing by tens and hundreds times.

9.5 Damage and Remote Cutting of Metals by a Repetitively Pulsed Laser

The remote metal cutting by cw laser radiation was described in Chap. 5. It was pointed out that the "bottleneck" of this process is the removal of a material from the cut. In the case of cw radiation, this removal "requires a large width of the melt zone, which reduces the process efficiency. It is obvious that the modulation of even low-average intensity laser radiation can provide the short-term boiling up of the metal surface and the removal of the melt due to the vapor recoil pressure. It remains unclear under which conditions this additional mass transfer compensates energy losses for vaporization and what is the efficiency of this process.

9.5.1 Formulation of the Problem

In Chap. 8 we considered the mass transfer produced a single laser pulse, when the melted region depth was significantly smaller than its width and also smaller than the plate thickness. In the case of remote cutting, a single pulse cannot damage the surface of a steel plate. Therefore, it is very important to heat the interaction zone up to a rather high temperature close to the melting temperature, after which the damage of the surface begins. Thus, the cutting process in this case cannot be represented as the mechanical summation of the material removal by single pulses. Because mass transfer occurs at a high average temperature of the heated interaction zone, apart from the melt splash induced by vapor, which was mentioned above, other mechanisms are also possible. When a plate is melted through, the vapor recoil pressure can extrude the melt forward along the beam direction. And, finally, the melt can simply flow out under the action of the gravity force.

We considered in the previous section metal drilling and welding by repetitively pulsed laser radiation without the use of gas jets. Instead of welding, blind cutting regimes were sometimes observed (see Figs. 9.24 and 9.25). However, processing was performed in these cases by tightly focused radiation, when a single pulse could damage the metal surface. In this case, the radiation intensity was increased by more than an order of magnitude due to focusing. The weld or blind weld had a high aspect ratio, i.e. the physical situation differed qualitatively as a whole from the case of remote cutting. A repetitively pulsed CO_2 laser used for studying remote cutting emitted 20–500 μs, 0.1–8 J pulses at pulse repetition rates of 0.1–100 Hz. The average laser radiation power was 1 kW and the output beam diameter was 45 mm. The laser radiation energy distribution in the focal spot has the form of a trapeze with sharply falling edges and the nonuniformity at the top ~20% [34]. The absorption coefficient of the target averaged over a pulse train was determined by measuring the target temperature during a laser pulse with a chromel-alumel thermocouple.

9.5.2 Experimental Results

The remote repetitively pulsed cutting of steel plates in these experiments was accompanied by the appearance of a plasma plume. The radiation intensities at which an optical breakdown is induced near the target weakly decrease with increasing the focal spot area and approach thresholds measured in [43] for very large irradiated regions ($1–2\times10^5$ W/cm^2) (see [35]).

Figure 9.48 presents the external view of a steel plate with cuts produced in different irradiation regimes. The width d_c of the cuts is always 2–3 times smaller than the focal spot size. On the cut edges the solidified melt rollers are observed. The upper and lower rollers (when plates are oriented vertically) are identical. The cutting during which a sample is moved upward or downward is similar to the cutting during the horizontal motion of a sample. As the speed is increased, the value of d_c monotonically decreases (see Table 9.1). When the critical speed is achieved, the cut width drastically drops down to zero and cutting passes to welding. As a rule, a plate is melted through in this case. An indentation was observed on the face of the plate, while the rear side remained weakly perturbed. This passage was often accompanied by a series of holes Fig. 9.48.

The width of a cut is usually not important for practical applications, the important parameter being the critical cutting speed v_{cr}. Figure 9.49 presents the dependence of this speed on the laser radiation power. As expected, the critical speed increases with increasing the average radiation power. The average power was varied either by changing the pulse energy at a fixed pulse repetition rate or by varying the pulse repetition rate for a fixed pulse energy. One can see that the critical speed is mainly determined by the average laser power. This is additionally confirmed by the independence of v_{cr} on the focal spot size d_f at a constant average laser power ($f = 100$ Hz, $E = 3.2$ J, $\tau = 70$ μs). The focal spot diameter was changed from 3.8 mm to 1.7 mm; in this case, the average width of the cut also deceased by half.

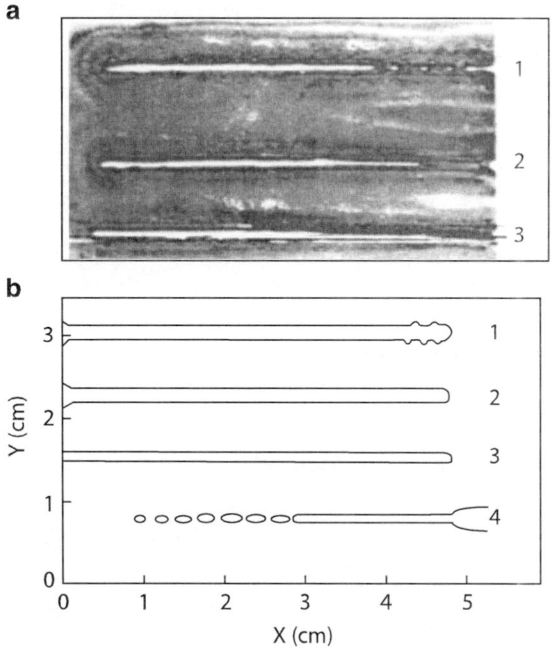

Fig. 9.48 View of experimental and calculated cuts of 0.5-mm-thick steel plates. (**a**) (1) $P = 370$ W, $d_f = 3$ mm, $f = 100$ Hz, $\tau = 70$ μs; (2) $P = 280$ W; (3) $P = 235$ W. (**b**) Calculations: $P = 380$ W, (1) $V = 0.4$ mm/s, (2) $V = 1$ mm/s, (3) $V = 2$ mm/s, (4) $P = 310$ W, $V = 6$ mm/s

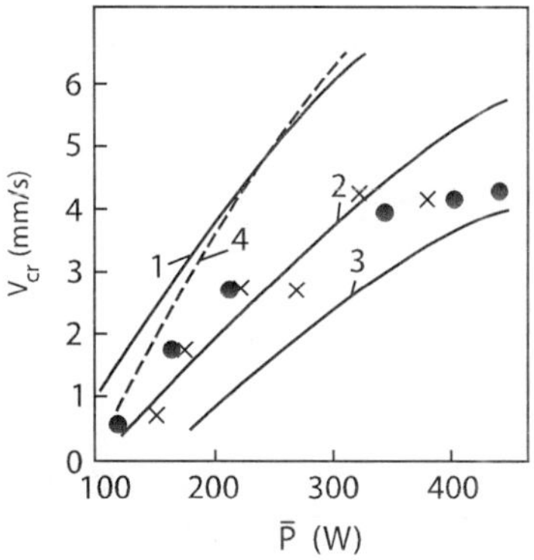

Fig. 9.49 Dependences of the maximum cutting speed on the average laser power: (*cross*) experiment, $f = 100$ Hz, $\tau = 70$ μs, $d_f = 3$ mm; (*black ring*) experiment, $E = 2.2$ J, $\tau = 70$ μs, $d_f = 3$ mm. The solid curves are calculated for $f = 100$ Hz, $\alpha \sim 30\%$ (1), $\alpha \sim 20\%$ (2), and $\alpha = 15\%$ (3). The dashed curve is calculated for $E = 2.2$ J and $\alpha = 30\%$

Table 9.2 presents the dependence of the specific cutting energy on the plate travel speed. One can see that this energy, taking the absorption coefficient into account, is considerably smaller than the specific vaporization energy for steel $\rho L_b = 80$ kJ/cm^3. The latter fact suggests that the removal of a material during remote cutting is caused not only by the evaporation of the material. Before proceeding to the discussion of experimental results, it is reasonable to consider the mechanism of the process, its model and numerical calculations based on this model.

Table 9.2 Conditions and parameters of remote repetitively pulse cutting

$\overline{P}$ (W)	τ (μc)	V (mm/c)	X (kJ/cm^3)	d_c (mm) Experiment	d_c (mm) Simulation
		0.5	250	2.2	2.3
		1.0	100	2.0	2.3
380	70	2.5	80	1.5	1.5
		0.6	240	1.8	1.5
		1.0	170	1.4	1.5
280	70	2.5	150	0.9	1.5
		1.0	400	1.3	2.2
		2.0	200	1.1	1.5
280	150	4.0	150	0.6	1.0

9.5.3 *Numerical Model*

In the repetitively pulse irradiation regime with a large off-duty ratio, the temperature distribution and shape of the cut in a moving metal plate are three-dimensional and nonstationary. The problem can be simplified by using a number of considerations [34]. Due to a short duration of the laser pulse, the temperature distribution can be represented by the sum of two functions: the slow function $\overline{T}$ and the fast δT (see, for example, Figs. 9.1, 9.2, 9.3). Temperature δT strongly fluctuates in a thin surface layer. Therefore, this heating can be considered one-dimensional and it is described by the known analytic expression (see Chap. 1). The average temperature $\overline{T}$ changes under the action of many pulses and is determined from the balance of the average power and heat removal due to the heat conduction and sample displacement. Due to a slow variation of temperature $\overline{T}$, its smooth distribution along the surface, and a small thickness of the plate, the latter can be considered thermally thin. Thus, temperature $\overline{T}$ will be determined by solving the two-dimensional heat conduction equation, as in the case of remote cutting by cw radiation (see Chap. 5):

$$c\frac{\partial \overline{T}}{\partial t} - \kappa\left(\frac{\partial^2 \overline{T}}{\partial x^2} + \frac{\partial^2 \overline{T}}{\partial y^2}\right)$$

$$= \alpha\frac{\overline{q}}{h}\exp\left(\frac{(x-vt)^2+y^2}{r_f^2}\right) - \frac{\varepsilon\sigma_B}{h}\overline{T}^4 - \frac{q_v}{h}\Theta(T_b-\overline{T}) - \frac{q_c}{h}, \quad (9.24)$$

where c and κ are the heat capacity and heat conductivity; q and α are the intensity and absorption coefficient for laser radiation; v and r_f are the travel speed and radius of the light spot; σ_B is the Stefan–Boltzmann constant; ε is the thermal radiation coefficient, and Θ is the Heaviside function. The last three terms in the right-hand side of (9.24) describe thermal radiation from the surface, and thermal losses due to evaporation and natural convection of air, respectively. The last two processes can be described in the simplest way because in ours case, as in the case of cw radiation cutting, the qualitative description is sufficient [24]:

$$q_v = 0.2 n_s \overline{V} L_b = \frac{2.56 \times 10^3}{\sqrt{\overline{T}}}(p_s - p_A),\ p_A = 760.$$

The saturated vapor pressure P_s is described by the expression [24]

$$p_s = \frac{4 \times 10^{\left(13.27 - \frac{19710}{\overline{T}}\right)}}{\overline{T}^{1.27}}.$$

Pressure is expressed in mm Hg, temperature in Kelvin degrees, and numerical expressions are presented for iron. The convective cooling is described by the expression [36]

$$q_c = \kappa \frac{\overline{T} - \overline{T}_G}{H^{1/4}} \left(\frac{g}{\nu^2}\right)^{1/4}$$

where κ and ν are the heat conductivity and kinematic viscosity of air; H is the characteristic vertical size of a sample; $\overline{T}_G$ is the temperature of surrounding air; and g the acceleration of gravity.

Because the boundaries at the plate and cut edges are free, we have $\partial\overline{T}/\partial n = 0$, where n is the normal to boundaries. The initial condition is $\overline{T} = \overline{T}_G$. For simplicity, the phase transition energy in (9.24) during material melting is neglected, as is often accepted in laser technology problems (this energy for iron does not exceed 30% of the heat content at the melting temperature). For the same reason, thermal constants were assumed invariable and close to the parameters of stainless steel ($\kappa = 0.3$ W/cm K, $\kappa/c = \chi = 0.13\,\text{cm}^2/\text{s}$). The absorption coefficient was measured to be ~20%. The thermal radiation coefficient was 0.3. The fluctuating temperature of the surface layer is described by the known expression (8.1)

$$\delta T = \frac{2\alpha q_0}{\kappa}\sqrt{\frac{\chi}{\pi}t},\ t < \tau \tag{9.25}$$

where χ is the thermal diffusivity. As a result, the surface temperature T_s is determined by the sum $T_S = \overline{T} + \delta T$. By equating T_S to the boiling temperature, we find the change in the boiling boundaries with time:

$$\overline{T}(t, x, y) + \delta T(t) = T_b \tag{9.26}$$

Because the beam energy in the boiling zone is spent to evaporate the material, this reduced the energy heating the sample. Now the power $\tilde{P}$ spent for heating is determined by the expression

$$\tilde{P} = \overline{P}\left(1 - \frac{1}{\tau\pi r_f^2}\int_{t_0}^{\tau} S_b(t)\mathrm{d}t\right),\ \tau > t > t_0 \tag{9.27}$$

where the boiling area S_b is calculated from (9.26); t_0 is the boiling onset time; $S_b(t_0) = 0$. The value of $\tilde{P}$ determines the average intensity entering (9.24):

$$\bar{q} = \tilde{P}/\pi r_f^2. \tag{9.28}$$

Thus, the system of equations (9.24)–(9.28) describes the behavior of the surface temperature after each of the pulses. Knowing the sizes of the melting and boiling zones, we can estimate approximately the melt mass removed per pulse. Excitation of the melt flow by a low-intensity laser beam with a large focal spot has a number of specific features compared to the regimes considered above (Chap. 8). In this case, the subsonic flowing out of vapor to the atmosphere can occur. The approximate condition of passing from the subsonic to supersonic flow is the requirement $q > q^*$ (1.43)

$$q^* = c_s L_b \frac{p_A}{kT_b} m \tag{9.29}$$

where m is the mass of atoms of the material vapor; c_s is the sound speed in the metal vapor; p_A is the atmospheric pressure; and L_b is the vaporization enthalpy.

For the iron vapor parameters, we have $q^* \approx 10^5$ W/cm^2. If $q > q^*$, the vapor recoil pressure is proportional to the radiation intensity (see Sect. 1.3.4):

$$p = \alpha q c_s / L_b. \tag{9.30}$$

If $q < q^*$, the recoil vapor pressure is smaller:

$$p = \left(\frac{\alpha q}{L_b}\right)^2 \frac{kT_b}{p_A m}. \tag{9.31}$$

For $q = q^*$, expressions (9.30) and (9.31) are equivalent.

Both these regimes can be realized during repetitively pulsed remote cutting. If melting is blind, the melt will move along the plate surface under the action of the vapor recoil pressure, and the material will be carried out toward the laser beam. In the model of thermally thin plate, such a region is small, and therefore we will consider only the extrusion of the melt along the laser beam. This can occur in the model for $r_m > h$, where r_m is the characteristic size of the melt zone. When $S_b > 0$, after the end of the pulse action, the boiling region will receive the momentum $p(\tau - t_0)S_b$. This will result in the increment of the melt motion velocity

$$\Delta v = \frac{p(\tau - t_0)}{\rho h}$$

where ρ is the target material density. During the interval T_p between pulses, a liquid film will shift by the distance $\Delta v T_p = p(\tau - t_0)T_p/\rho h$.

According to (9.30), we obtain the displacement of the film during the interval between pulses:

$$\Delta h = \frac{\alpha q c_s(\tau - t_0)}{L_b \rho h} T_p.$$

We assume that the film is ruptured when its total displacement produced by pulses is h. Finally, if the pulse intensity is low and the pulse repetition rate is high, the melt can be removed without the involvement of vapor due to flowing out under the action of the gravity force, as was assumed during remote cw laser cutting (Chap. 5). Let us use the melt removal model describing remote cw cutting: if the maximal vertical size of the melt zone achieves the critical value d_{cr}, the total melt region is removed from the plate. The value of d_{cr} is described by expressions (5.14).

9.5.4 Comparison of Numerical Calculations with Experiment

Figure 9.48 presents the calculated shapes of a cut of a steel plate of thickness $h = 0.5$ mm and length 100 mm for different cutting speeds. One can see that the cut width decreases with increasing speed at a constant power. Because the width of the thermal front is $\chi/v \sim 1$ cm, as the laser beam approaches the edge of the plate approximately by the same distance, the temperature in the spot increases. As a result, the size of the melting zone and the cut width increase. Such a boundary effect is also always observed in experiments (cut 1, 2 in Fig. 9.48a). The initiation of cutting at high speeds is accompanied by periodic piercing of separate holes and then the cut is stabilized after the heating of the plate (cut 4 in Fig. 9.48b). This result also agrees with experiments. Obviously, this instability has the thermal nature because it is also observed during remote steel cutting by cw laser radiation (Chap. 5).

The calculated cut widths are compared with experimental widths in Table 9.1. Note that the cut width for the same speed increases with increasing the average output laser power both in experiments and calculations. The calculations of cuts for different focal spot sizes, close to experimental values, showed that cutting parameters weakly depend on the focal spot size. This also agrees with experiments. Figure 9.49 presents the dependences of the maximal cutting speed on the average laser power in two cases: for the constant pulse repetition rate $f = 100$ Hz or for the constant pulse energy $E = 2.2$ J. One can see that these dependences are close, which confirms the key role of the average laser power in remote repetitively pulsed laser cutting. A comparison of calculations with experiments is presented for different values of the absorption coefficient. One can see that the best agreement is achieved for $\alpha = 20\%$, which corresponds to the approximate measurements of the absorption coefficient. Note that mass transfer under these conditions caused by the gravity force would produce a cut of width 5.2 mm, which does not agree with experiments and calculations.

It is interesting to compare the remote cutting of steel plates by repetitively pulsed and cw radiation (see Chap. 5). Because these experiments were performed

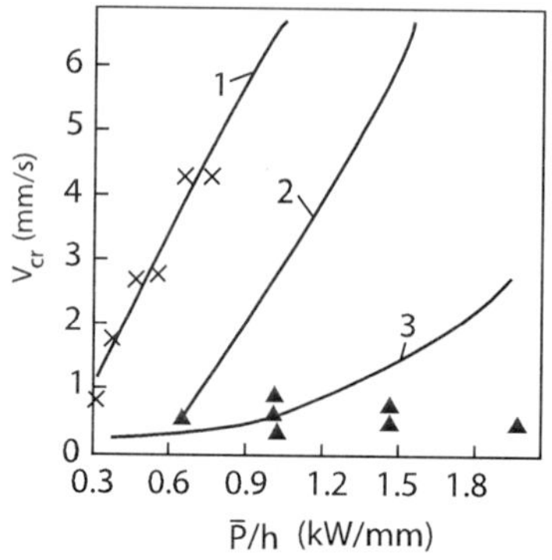

Fig. 9.50 Comparison of cw and repetitively pulsed cutting efficiencies: (**1**) experiment and calculation of the maximal cutting speed for $f = 100$ Hz, $\alpha = 20\%$, $\tau = 70\,\mu$s, and $h = 0.5$ mm; (**2**) calculation of the maximal cw radiation cutting speed under the same conditions (see Chap. 5); (**3**) calculation of the maximal cw radiation cutting speed for $h = 2$ mm. Experimental points are located in the calculation cutting region (see Chap. 5)

under different conditions, it is necessary to compare limiting speeds at the same specific powers (see Fig. 9.50). The slope $\tan\alpha = V_{cr}/(P/h)$ of the curve drawn from the coordinate origin to any point of curves is equal to the cut area per unit energy spent for cutting, i.e. to the cutting efficiency. One can see that the efficiency of the repetitively pulsed process exceeds the cw radiation cutting efficiency by several times. Thus, for $P/h = 1$ kW/mm, the repetitively pulsed cutting efficiency is 6 mm^2/kJ, while the cw cutting efficiency is 2 mm^2/kJ. For thicker plates ($h =$ 2 mm), the cw radiation cutting efficiency decreases own to 0.5 mm^2/kJ.

Thus, we have shown that the regimes are possible when the additional energy spent for the vaporization of a material during repetitively pulsed laser cutting are compensated by the intensification of the process of melt extrusion from the interaction zone. As a result, repetitively pulsed laser cutting with a high off-duty ratio proves to be considerably more efficient that cw radiation cutting.

9.5.4.1 Remote Material Processing by Two Lasers

Although repetitively pulsed remote metal cutting is more efficient than cw laser cutting, it requires the average power and the spot size providing the boiling of the sample surface at least at the pulse end. However, the focusing of high-power laser radiation at large distances is rather complicated. In this case, it is convenient to use two lasers. The first laser should have a high average power, while the second one can have low output power, but its radiation should be tightly focused. We will describe remote cutting under these conditions by using the same model as for one laser. The numerical code was modified to take into account the action of CO_2 and Nd lasers. The pulse durations of both lasers were 100 μs, while the pulse repetition rate was varied from 50 to 100 Hz. The CO_2 laser radiation was focused into a spot

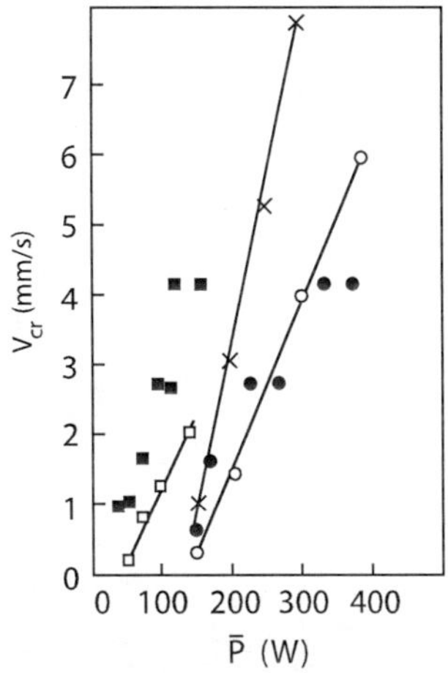

Fig. 9.51 Dependences of the critical cutting speed for a 0.5-mm thick stainless steel plate on the average laser energy. The focal spot size for a CO_2 and Nd lasers is 2 mm and 0.5 mm, respectively. The laser pulses are separated by the delay time ~120 μs. (*square*) Numerical results for $\alpha = 0.35$, $d = 3$ mm for the CO_2 laser; $\alpha = 0.7$, $d = 1$ mm for the Nd laser; $f = 100$ Hz, the average power of the Nd laser is 18 W. (*Black square*) Experimental data; (*black ring*) experimental data for cutting performed only by the CO_2 laser; (*ring*) numerical calculations for the CO_2 laser, $\alpha = 0.3$; (*cross*) numerical calculations for the CO_2 laser ($\alpha = 0.3$) and Nd laser ($\alpha = 0.6$), $f = 100$ Hz. The average power of the Nd laser is 10% of the average power of the CO_2 laser

of diameter 2 mm, while the Nd laser radiation was focused into a 0.5-mm spot. The time delay between laser pulses was 120 μs. Figure 9.51 demonstrates the calculated dependences of the maximum cutting speed on the average power for one CO_2 laser and for two lasers, the Nd laser power being 10% of the CO_2 laser power [37]. One can see that the cutting speed doubles.

A similar situation takes place for remote metal drilling and cutting by cw and repetitively pulsed lasers [38]. The parameters of lasers were selected so that each laser separately could not perforate holes or cut metals. In experiments [38], a multitube quasi-stationary CO_2 laser was used. The laser power was varied from 1 to 3 kW. The pulse duration achieved 4 s. A repetitively pulsed Nd:YAG laser emitted 0.35 J, 100 μs pulses with a repetition rate of 50 Hz. The CO_2 and Nd:YAG laser beams were focused into a spot of diameter 1 and 0.2 mm, respectively. Both beams were combined into one spot. Figure 9.52 shows the dependence of the melting depth of X18H9T stainless steel plates on the time of irradiation by the CO_2 laser alone. One can see that its intensity is insufficient to make a hole in the plate. Figure 9.53 (left) shows the photograph of the cross section of a solidified melt pool. Because the laser radiation intensity is low, the material did not vaporized. The melt pool should be semi-circular because its width considerably exceeds the focal spot diameter. However, the transverse size of the pool is larger than its depth, which probably suggests that thermocapillary vortices are excited (see Fig. 2.26 in Chap. 2). The picture changes strongly when the Nd:YAG laser is simultaneously

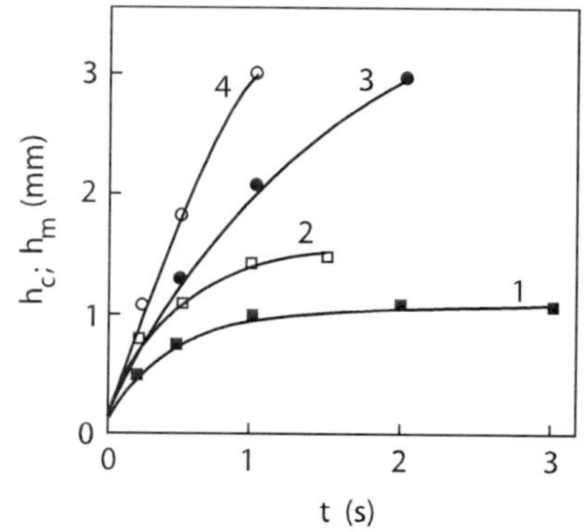

Fig. 9.52 Time dependences of the melt depth h_m (1, 2) and keyhole depth h_c (3, 4) for $P = 1.75$ (1, 3) and 2.4 kW (2, 4), and $h = 3$ mm

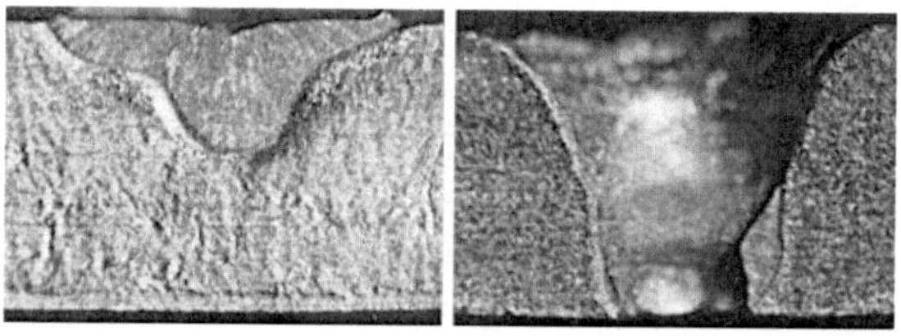

Fig. 9.53 Photographs of a solidified melt and a hole; $t = 1$ s, $P = 2.4$ kW, $h = 3$ mm

used (Figs. 9.52 and 9.53 (right)). Although its average power is less than one percent of the CO_2 laser power, the penetration depth of the beam increases by three times and melting passes to drilling. The pulse intensity in the focal spot of the Nd:YAG laser is 6 MW/cm^2, which is sufficient to initiate vaporization. As for the melting efficiency (the melting depth per unit power), this combined drilling is inferior by an order of magnitude to repetitively pulsed drilling by one CO_2 laser (see Figs. 9.38 and 9.39). This is explained by the fact the drilling by two lasers is remote when the focal spot of the main laser is large and its emission is stationary. The increase in the melting efficiency of the stationary beam when weak repetitively pulses radiation is added emphasizes the efficiency of mass transfer due to the vapor recoil pressure even in the case when the melt pool is considerably larger than the focal spot of a tightly focused beam.

9.5.5 Remote Damage of Metals by Radiation from High-Average-Power Lasers

As pointed out in Chap. 5, to perform remote laser cutting of metal constructions in practice, laser radiation powers of tens kilowatts are required. The approximation of thermally thin plates in this case is invalid. To find efficient cutting regimes and optimal conditions, a model is required that would allow us to select the proper values of numerous parameters involved in the remote cutting of plates of an arbitrary thickness by high-power repetitively pulsed radiation. Before considering cutting in these cases, we will analyze the types of a melt flow under different conditions.

The excitation of a melt flow by a low-intensity laser beam with a large focal spot has a number of specific features compared to the regimes considered above. Because of the large focal spot and short pulse duration, the nonviscous melt flow produced in this case will be unsteady [3]. We will show below that because of the large average power of a laser, the melt has no time to solidify between radiation pulses, and therefore the liquid can flow out by inertia during the interval between pulses if the flowing out time $r_m/v < 1/f$ (r_m is the melting zone radius, and v is the melt flow rate).

Consider the cutting threshold, i.e. the minimal laser power at which cutting with a specified speed begins. In this case, metal vapor escapes from the surface with a subsonic speed, and vapor recoil pressure p rapidly decreases with decreasing the absorbed power [see expression (9.31) and its application threshold (9.29)]. Because of a low liquid flow rate, a pulsed regime will be established in the melt pool with the average flow rate determined by the averaged Bernoulli equation:

$$\bar{v} = \left[2 \left(\frac{q_v}{L_b} \right)^2 \frac{k_B T_b}{p_A} \frac{(\tau - t_b) f}{m\rho} \right]^{1/2} \tag{9.32}$$

where t_b is the time required to heat the surface up to the boiling temperature and ρ is the plate material density. Because we are interested in cutting of samples of thickness above 5 mm at speeds (equal to the melt flow rate) $v \geq 0.5$ mm/s, the melt flow will not be viscous, which was taken into account in deriving expression (9.32). All the above said concerns the region of a plane which is not melted through. Thus, we will use these results to consider the remote cutting of thick plates. To analyze the cutting of comparatively thin plates by high-power repetitively pulsed radiation, we will use the model of a thermally thin plate developed above for low-power lasers.

9.5.6 *Remote Cutting Model for Thick Plates*

It is clear beforehand that the situation with the cutting of plates of thickness comparable with the typical cut width is rather complicated due to a great number of physical processes and the difficulty of their quantitative three-dimensional description. However, this situation is of interest for practical applications, because it is important to know limiting cut depths that can be achieved for the specified parameters of laser radiation and cutting speed.

Obviously, to initiate cutting, a plate should be melted through by the irradiation time end $t_{\text{end}} \sim d_f/v$ and the material should be removed from the melt pool. Both these processes can occur simultaneously, resulting in a complicated three-dimensional shape of the melt pool. Such problems require time-consuming calculations. Therefore, it is reasonable to simplify the consideration of these processes with the aim of elucidating their features only qualitatively. Thus, we assume below that thermal and mass transfer processes develop independently [39]. Clearly, this will give the power required for cutting a plate of the given thickness.

To find the through melting condition, we will us the expression for the temperature distribution in a plate approaching a laser beam at the speed v (cutting speed) along the coordinate x [24]

$$T(x,y,z) = \frac{1}{\kappa}\left(\frac{\chi}{\pi^3}\right)^{1/2}$$
$$\times \overline{P}\int_0^\infty \mathrm{d}\tau \sum_{n=0}^{\infty}\left\{\exp\left[-\frac{(z+2hn)^2}{4\chi\tau}\right]+\exp\left[-\frac{[z-2h(n+1)]^2}{4\chi\tau}\right]\right\}$$
$$\times[\tau^{1/2}(r_f^2+4\chi\tau)]^{-1}\exp\left[-\frac{(x+v\tau)^2+y^2}{r_f^2+4\chi\tau}\right], \tag{9.33}$$

where z is the coordinate directed inside the plate; χ is the thermal diffusivity; κ is the heat conductivity; $T(x,y,z)$ is the plate temperature; $\overline{P}$ is the average absorbed laser radiation power; and h is the plate thickness.

It is necessary to find the coordinate x_0 at which the melting depth is maximal. The average power should be chosen to provide the maximum melting depth equal to the plate thickness, thereby obtaining the through melting conditions depending on the speed, the plate thickness h, and the focal beam radius r_f. Because expression (9.33) is too cumbersome, it is impossible to obtain expressions for x_0 and through melting power even for the minimal number of the terms of the series. For this reason, these quantities were found numerically. The absorbed powers $\overline{P}_1$ required for melting the plate bottom are presented in Fig. 9.54.

For a liquid metal to have time to flow out during the irradiation time d_f/v, the liquid flow rate should be no less than the plate motion speed, i.e. the cutting speed.

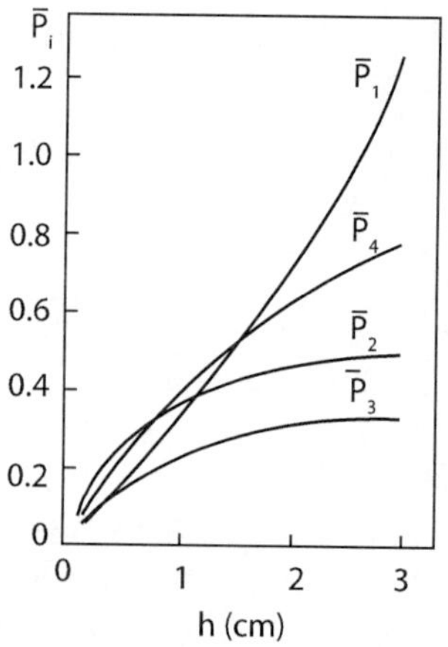

Fig. 9.54 Radiation power $\overline{P}$ required for remote cutting of steel plates of different thickness at a speed of 1 mm/s^{-1}, $f = 100$ Hz, $\tau = 170\,\mu$s, $r_f = 1$ cm; $\overline{P}_i = \overline{P}_i/2\pi^{3/2}\kappa T_m r_f$ is the normalized absorbed laser radiation power (T_m is the melting temperature of the material)

For low radiation powers, we will use expression (9.32), where the boiling time t_b can be found from expression (9.25) determining the surface temperature upon pulsed heating:

$$t_b = \frac{\pi^2 \kappa^2 \Delta T^2 (f\tau)^2 r_f^4}{\overline{P}^2 \chi}. \tag{9.34}$$

In our case, the fluctuation of the surface temperature is $\Delta T = T_b - \overline{T}$, where $\overline{T}$ is the average temperature of the plate surface in the irradiated zone. By assuming that $\overline{P}_2$ radiated surface is heated up to the boiling temperature by the end of each pulse, and assuming that $t_b = \tau$, we find from (9.34) the average power at which the irradiated surface is heated up to the boiling temperature by the end of each pulse:

$$\overline{P}_2 = \pi\kappa(T_b - \overline{T}) r_f^2 f \left(\frac{\tau}{\chi}\right)^{1/2}. \tag{9.35}$$

The average surface temperature $\overline{T}$ depends on the plate thickness and other parameters of the problem. When the absorbed laser power exceeds $\overline{P}_2$, all the excess power is spent for material vaporization:

$$\overline{P}_V = \overline{P} - \overline{P}_2. \tag{9.36}$$

Intensity q_V in expression (9.32) is determined by power $\overline{P}_V$:

$$q_V = \frac{\overline{P} - \overline{P}_2}{\pi r_f^2 (\tau - t_b) f}. \tag{9.37}$$

By substituting (9.37) into (9.32), we obtain the condition for the flowing out of a liquid melt during the irradiation time. The critical power $\overline{P}_{\rm cr}$ at which the melt is removed is determined from the equation

$$\frac{v^2 p_A m \rho L_b^2/2}{kT_b} = \frac{(\overline{P}_{\rm cr} - \overline{P}_2)^2}{(\pi r_f^2)^2 f(\tau - t_b)}. \tag{9.38}$$

Taking (9.34) and (9.35) into account, we obtain from (9.38) for $h \to 0$ and $\overline{P}_2 \to 0$ the limiting critical average power

$$P_{\rm cr} = \frac{v L_b}{2} \pi r_f^2 (f\tau)^{1/2} \left(\frac{p_A m \rho}{kT_b}\right)^{1/2}. \tag{9.39}$$

In the opposite case, when $P_{\rm cr} = \overline{P}_2 + \Delta P$ and $\Delta P / P_2 < 1$, we determine the power $\overline{P}_{\rm cr}$ from (9.38) approximately:

$$P_{\rm cr} = \overline{P}_2 + \frac{2v^2 (\pi r_f^2)^2 f\tau}{\overline{P}_2} \left(\frac{p_A m \rho}{kT_b}\right) \frac{L_b^2}{2}. \tag{9.40}$$

Let us determine the radiation power $\overline{P}_2$ at which the surface "boils up" by the laser pulse end. For this purpose, we find the average temperature of the focal spot surface by using integral (9.33) for $x = y = z = 0$. For convenience, we will express $\overline{T}$ in terms of the power $\overline{P}_3$ at which the surface temperature achieves the melting temperature T_m:

$$T(0) = \frac{\overline{P}}{\overline{P}_3} T_m. \tag{9.41}$$

The power $\overline{P}_3$ is calculated numerically from (9.33). Its dependence on h is presented in Fig. 9.54. At little h surface temperature must be clouse to bottom temperature, it is seen in Fig. 9.54. Thus, in this case, the plate can be considered optically thin up to $h \leq 7$ mm. This value is close to the limiting thickness obtained by simple estimates. Note that powers $\overline{P}_1$ and $\overline{P}_3$ for a thermally thin plate are described by the expression [40]

$$\overline{P}_1 = \overline{P}_3 = 2\pi h \kappa T_m \left(\ln \frac{4\chi}{v r_f} - 0.16 \right)^{-1}.$$

By substituting (9.41) into (9.35), we obtain

$$\frac{\overline{P}_2}{\overline{P}_3} = \frac{\pi \kappa r_f^2 (\tau f)(\chi \tau)^{-1/2} T_b}{\pi \kappa r_f^2 (\tau f)(\chi \tau)^{-1/2} T_m + \overline{P}_3}. \tag{9.42}$$

One can see from Fig. 9.54 that the curve described by expression (9.42) lies approximately 1.5 times higher than the curve described the power $\overline{P}_3$ at which the average temperature in the focal spot is equal to the melting temperature. Estimates by expression (9.40) show that $P_{\mathrm{cr}} \approx \overline{P}_2$ in the case under study, i.e. to remove the melt due to the vapor recoil pressure, it is necessary only slight to exceed the power at which the focal spot surface boils up at the pulse end.

If a plate is oriented vertically and a laser beam horizontally, the gravity force can be also involved in the melt transfer. Recall that the role of this force is decisive in the remote cw laser cutting of steels (see Chap. 5). We consider the action of the gravity force by using the model that gave good agreement with experiments on the cw laser cutting of thermally thin plates.

To calculate the threshold power at which the melt zone width achieves d_{cr} (along the vertical), it is necessary to find x_0 for (9.33) at which the melt zone width y_0 is maximal and then to calculated this width. By equating it to d_{cr}, we can determine the laser power $\overline{P}_4$ at which the melt will flow out from the interaction zone. The value of d_{cr} is calculated from expression in Chap. 5 5.14. The curve for $\overline{P}_4$ is presented in Fig. 9.54. One can see from this figure that up to the plate thickness ~1.7 cm, the melt removal leads to a thorough cut. If the laser power is close to $\overline{P}_4$, then for $h > 1.7$ cm, the cut should be blind because melting will be blind by the instant of melt removal. If the laser power is close to $\overline{P}_1$, the cut should be through.

Thus, we see from Fig. 9.54 that up to $h \approx 1.2$ cm in the case under study, the threshold melt removal powers due to the gravity force and vapor recoil pressure are close. Because the plate is melted over the entire thickness, a through cut is produced after the melt removal. At large plate thicknesses for $\overline{P} \approx \overline{P}_2$, the surface "boils up", but the plate is not melted through, and therefore the cut will be blind. By increasing the power up to $\overline{P}_1$, we can expect to obtain a through cut. Finally, for $h > 1.7$ cm and $\overline{P} \approx \overline{P}_1$, both mechanisms of the melt removal are involved, and the cut should be through. Because curve $\overline{P}_1$ increases with h considerably faster than $\overline{P}_2$ and $\overline{P}_4$, to obtain a through cut per pass, it is necessary to increases strongly (exponentially) the laser radiation power.

It is clear from general considerations that all the threshold powers should increase with increasing the cutting speed. The power $\overline{P}_1$ at which the plate bottom is melted increases most strongly with increasing the cutting speed. Calculations show that, for example, the values of $\overline{P}_2$, $\overline{P}_3$, and $\overline{P}_4$ increase by 20%–30% when the cutting speed is doubled, whereas the power $\overline{P}_1$ doubles. At high cutting speeds, $\overline{P}_4 > \overline{P}_2$, i.e. the cutting threshold under these conditions is determined by the melt removal due to vapor recoil pressure. One can see from (9.42) that it is easier to perform remote cutting when the off-duty ratio is high, while in the case of a constant off-duty ratio, it is better to use longer pulses.

We considered above the initiation of cutting, i.e. the dependence of power thresholds on the plate thickness. Below, we consider the dependences of the cutting speed threshold on the focal spot size by the example of a thermally thin plate ($h = 7.5$ mm) and some properties of this cutting process.

9.5.7 *Thin Plates*

Because the focal spot of a MLTK-50 laser beam is large [41] ($d_V > d_{cr}$), we will use the model of thin-plate cutting (9.24). Figure 9.55 presents cut widths calculated for different plate thicknesses and different cutting speeds. The cut width decreases with increasing the cutting speed and the plate thickness. Calculations of the cutting of a 5 mm thick plate at a speed of 1.3 mm/s showed that 2/3 of radiation pass through the cut. This radiation can be sufficient for the simultaneous cutting of two other identical plates.

It is interesting to compare cuttings by cw and repetitively pulsed CO_2 lasers. It was shown above that for $P < 0.5$ kW the repetitively pulsed cutting is more efficient than cw cutting. Figure 9.56 the calculated dependences of the maximum cutting speed on the focal spot size. These dependences can be of current interest because the focal spot diameter of the MLTK-50 laser beam under real remote cutting conditions is estimated as 1–3 cm [41]. Let us assume, for example that the plate thickness is $h = 7.5$ mm, the absorbed power is $\sim$4.5 kW, $\tau = 170\,\mu$s, and $f = 100$ Hz. Recall that the critical width d_{cr} at which the melt flows out under the action of the gravity force is 15 mm in this case (see Chap. 5). The data in Fig. 9.56 are obtained for repetitively pulsed and cw lasers of the same power.

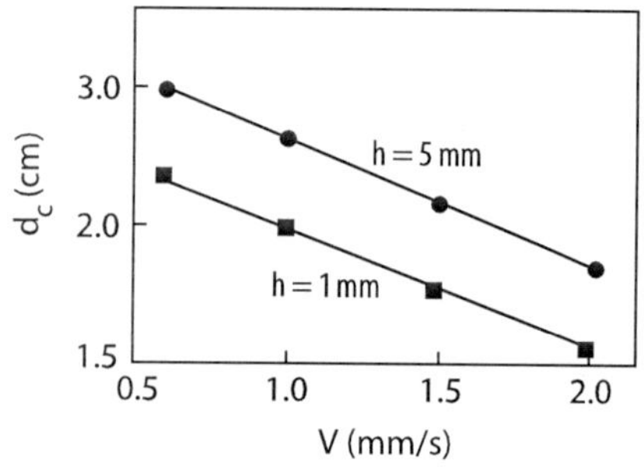

Fig. 9.55 Calculated dependences of the cut width d_c on the cutting speed for plates of different thicknesses h for the absorbed laser radiation power $\sim$4.5 kW, $f = 100$ Hz, $\tau = 170\,\mu$s, and $r_f = 1$ cm

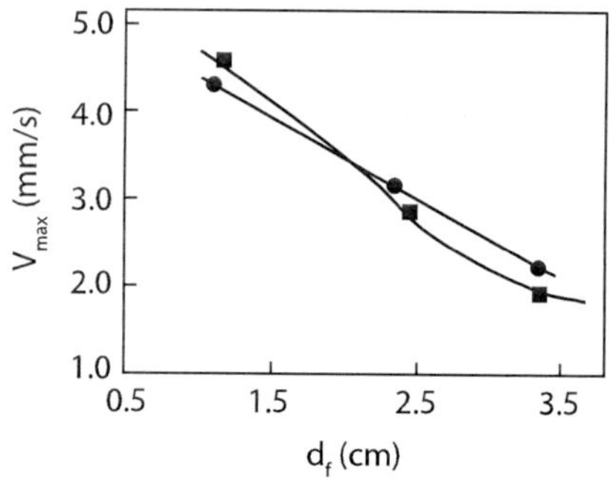

Fig. 9.56 Maximal steel cutting speeds V_{max} for cw (*black square*) and repetitively pulsed radiation (*black ring*) calculated for different focal spot sizes. $h = 7.5$ mm, $f = 100$ Hz, $\tau = 170\,\mu$s, the absorbed radiation power is $\sim$4.5 kW

Calculations showed that the cutting efficiencies of these lasers are close. It seems that the energy spent for material vaporization under these conditions does not lead to a considerable additional melt-mass transfer.

Consider cut shapes produced during cw and repetitively pulsed cutting at different focal spot sizes. One can see from Fig. 9.57 that in the case of cw irradiation of the plate, the cut shape is typical for this cutting regime (Chap. 5). The material flows out from the melt pool when the latter achieves the maximum size d_{cr}. In this case, the melt temperature on the symmetry axis at the instant of melt removal is approximately 1.5 times higher than the melting temperature.

In the case of repetitively pulsed cutting, the cut width is considerably smaller than that for cw cutting. The melt is removed faster when the zone size did not achieve the maximum value d_{cr} (Fig. 9.57b). Due to the vaporization loss, the maximum average temperature of the melt is lower than that in the cw regime, being close to the melting temperature. As the focal spot diameter is increased, the radiation intensity and the influence of vapor recoil pressures decrease. As a result, repetitively pulsed cutting is first similar to cw radiation cutting. Then, the boiling region size increases due to plate heating, and the role of the vapor recoil pressure increases. The mass removal due to the gravitational force is changed to mass transfer under the action of vapor recoil pressure. Thus, the cutting parameters under study lie in the region of changing regimes, and therefore the limiting speeds of cw and repetitively pulsed cutting differ insignificantly within the framework of our model.

Experimental data on the remote cutting of metals by the MLTK-50 laser are not numerous and are mainly of the technological type related to practical applications

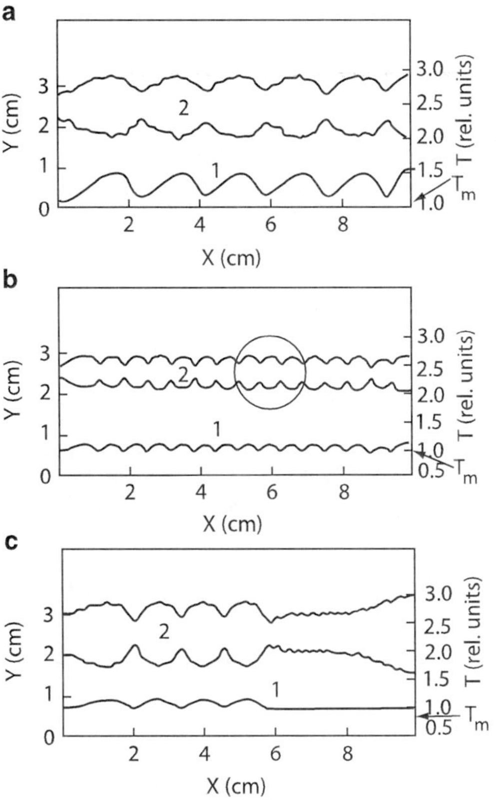

Fig. 9.57 Temperature profile (**1**) and cut shape (**2**) calculated for cw (**a**) and repetitively pulsed (**b, c**) radiations. The average absorbed power is ~4 kW, $h = 7.5\,\text{mm}$, $d_f = 1.5$ (**a, b**) and 2.5 mm (**c**). The rest of the conditions are as in Fig. 9.56. The circle indicates the focal spot shape

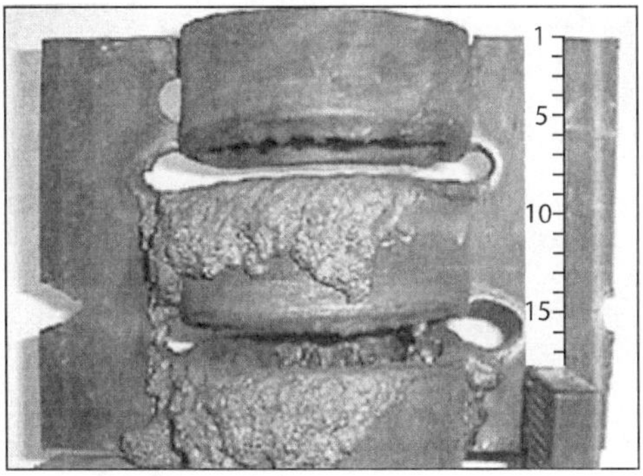

Fig. 9.58 Result of irradiation of a steel tube of diameter 150 mm and wall thickness 12 mm by an MLTK-50 laser. $E_p = 250\,\text{J}$, $\tau = 180\,\mu\text{s}$, $f = 100\,\text{Hz}$. The distance from the laser to the tube is 30 m, $\tilde{P} = 25\,\text{kW}$

[41]. As for the scientific aspect of the problem, it was pointed out in [41] that a 2 cm thick steel plate can be cut at a speed of up to 2 mm/s. According to our model, 7 kW of the absorbed radiation power is sufficient for cutting such a plate at a speed of 2 mm/s (see Fig. 9.54). This value is lower than the experimental absorbed power [41], which suggests that this experimental point lies in the calculated region of cutting. Figure 9.58 demonstrates the cut shape produced upon cutting of steel tubes by ~30 kW repetitively pulsed CO_2 laser [42].

References

1. N.N. Rykalin, *Raschety teplovykh protsessov pri svarke (Calculations of Thermal Processes in Welding)* (Mashgiz, Moscow, 1951)
2. A.A. Uglov (ed.), *Vozdeistvie kontsentrirovannykh potokov energii na manerialy (Interaction of Concentrated Energy Fluxes with Materials)* (Nauka, Moscow, 1985)
3. A.A. Uglov (ed.), *Fiziko-khimicheskie protsessy obrabotki materialov kontsentrirovannymi potokami energii (Physicochemical Processing of Materials by Concentrated Energy Fluxes)* (Nauka, Moscow, 1989)
4. A.A. Vedenov, G.G. Gladush, *Fizicheskie protsessy pri lazrnoi obrabotke materialov (Physical Processes in Laser Material Processing)* (Energoatomizdat, Moscow, 1985)
5. A.A. Glytenko, B.Yu. Lyubov, Inzh.-Fiz. Zh. **53**, 642 (1987)
6. M.M. Yakunin, Teplofiz. Vys. Temp. **26**, 759 (1988)
7. V.V. Bashenko, A.E. Lavrov, V.A. Lopota, Fiz. Khim. Obrab. Mater. **4**, 56 (1988)
8. A.A. Uglov, I.Yu. Smurov, A.M. Lashin, A.G. Gus'kov, *Modelirovanie teplofizicheskikh protsessov impul'snogo lazernogo vozdeistviya na metally (Simulation of Thermal Processes of Pulsed Laser Action on Metals)* (Nauka, Moscow, 1991)
9. S.I. Anisimov, Ya.I. Imas, G.S. Romanov, Yu.V. Khodyko, *Deistvie izlucheniya bol'shoi moshchnosti na metally (Interaction of High-Power Radiation with Metals)* (Nauka, Moscow, 1970)
10. M. Doubensskaia, I. Smurov, Appl. Surf. Sci. **252**, 4472 (2006)
11. G.G. Gladush, S.V. Drobyazko, V.I. Myanko, Yu.V. Pavlovich, A.N. Yavokhin, Poverkhnost. Fiz., Khim., Mekh. **7**, 115 (1988)
12. F.K. Fedyunin, *Termotsiklicheskaya obrabotka stali i chugunov (Thermal Cyclic Processing of Steel and Cast Irons)* (Nauka, Moscow, 1977)
13. I.K. Kikoin (ed.), *Tablitsy fizicheskikh velichin. Spravochnik (Handbook of Physical Quantities)* (Atomizdat, Moscow, 1976)
14. N. Groc, I. Smurov, M. Amouroux, Private communication (1992)
15. V.M. Smirnov, Avtom. Svarka **2**, 12 (1983)
16. S. Fuijinaga, H. Takenaga, N. Narikiyo, et al., J. Phys. D Appl. Phys. **33**, 492 (2000)
17. A.A. Vedenov, *Fizika elektrorazryadnykh lazerov (Physics of Electric-Discharge Lasers)* (Energoizdat, Moscow, 1982)
18. A.A. Vedenov, G.G. Gladush, S.V. Drobyazko, et al., Izv. Akad Nauk SSSR Ser. Fiz. **47**, 1473 (1983)
19. A.P. Averin, N.G. Basov, E.P. Glotov, et al., Dkl. Akad. Nauk SSSR **266**, 844 (1982)
20. A.A. Vedenov, G.G. Gladush, S.V. Drobyazko, *Svesoyuznoe soveshchanie po primeneniyu lazerov v tekhnologii mashinostroeniya (All-Union Conference on Laser Applications in the Machine Building Technologies)* (Nauka, Moscow, 1982), pp. 63–64
21. A.A. Vedenov, G.G. Gladush, S,V. Drobyazko, et al., Kvantovaya Elektron. **12**, 60 (1985)
22. G.G. Gladush, E.B. Levchenko, A.A. Ezhov, et al., *Svesoyuznoe soveshchanie po primeneniyu lazerov v tekhnologii mashinostroeniya (All-Union Conference on Laser Applications in the Machine Building Technologies)* (Nauka, Moscow, 1982), pp. 69–71
23. H. Mario, I. Miyamoto, E. Kawabata, et al., J. Jap. Weld. Soc. **50**, 404 (1981)
24. R.V. Arutyunyan, L.A. Bol'shov, V.V. Vityukov, V.P. Kiselev, Preprint IAE 40023/16, 1984
25. G.I. Levin, Avtom. Svarka **9**, 29 (1980)
26. A.A. Vedenov, G.G. Gladush, S.V. Drobyazko, et al., Preprint of Kurchatov Inst. 4123/9. M. 1985
27. S.V. Drobyazko, Yu.V. Pavlovich, Yu.M. Senatorov, R.D. Seidgazov, Avtomat. Svarka **12**, 417 (1987)
28. Y. Arata, H. Mario, I. Miyamoto, Tool Eng. **25**, 24 (1981)
29. G.G. Gladush, A.F. Glova, S.V. Drobyazko, Kvantovaya Elektron. **36** (2006)
30. A.A. Vedenov, G.G. Gladush, S.V. Drobyazko, Yu.V. Pavlovich, Yu.M. Senatorov, Kvantovaya Elektron. **13**, 1473 (1986)

31. A.G. Grigor'yants, A.A. Sokolov, Lazernaya rezka metallov. "Lazernaya tekhnika i tekhnologiya" (Laser Metal Cutting. "Laser Equipment and Technology"), Manual for Educational Institutions.
32. A.G. Borkin, A.A. Vedenov, G.G. Gladush, et al., *Vzaimodeistvie izlucheniya, plazmennykh I elektronnykh potokov s veshchestvom (Interaction of Radiation, Plasma and Electron Fluxes with Matter)* (Izd. TsNIIatominform, Moscow, 1984), pp. 95–97
33. G.G. Gladush, A.N. Yavokhin, Proceedings III International Conference on Trends in Quantum Electronics, Bucharest, 1988, pp. 290–291
34. G.G. Gladush, S.V. Drobyazko, N.B. Rodionov, et al., Kvantovaya Elektron. **30**, 1072 (2000)
35. S. Markus, J.E. Lowder, P.L. Mooneg. J. Appl. Phys. **47**, 2966 (1976)
36. B. Gebkhart, I. Dzhamuriya, R. Makhadzhan, et al., *Svobodno-kovektivnye techeniya, teplo I massoobmen (Free Convective Flows, Heat and Mass Exchange)* (Mir, Moscow, 1991)
37. G.G. Gladush, S.V. Drobyazko, N.B. Rodionov, Yu.M. Senatorov, Abstracts, International Conference of Quantum Electronics, Bulgaria, 2003
38. A.F. Glova, S.V. Drobyazko, N.B. Vavilin, E.M. Shvom, Kvantovaya Elektron. **32**, 169 (2002)
39. G.G. Gladush, M.V. Rodionov, Kvantovaya Elektron. **32**, 14 (2002)
40. Yu.I. Dudarev, A.V. Kazakov, M.Z. Maksimov, V.P Nikonenko, Fiz. Khim. Obr. Mater. **2**, 24 (1998)
41. A.G. Krasyukov, V.G. Naumov, V.M. Shashkov, et al., Khim. Neftegaz. Mashin. **5**, 52 (2001)
42. V.G. Vostrikov, A.G. Krasyukov, V.G. Naumov, V.M. Shashkov, et al., Gaz. Promyshlennost (2001)
43. V.B. Ageev, S.G. Burdin, I.N. Goncharov, et al., Kvant. Electron. **10**, 774 (1983)

Index

G.G. Gladush and I. Smurov, *Physics of Laser Materials Processing*,
Springer Series in Materials Science 146, DOI 10.1007/978-3-642-19831-1,
© Springer-Verlag Berlin Heidelberg 2011